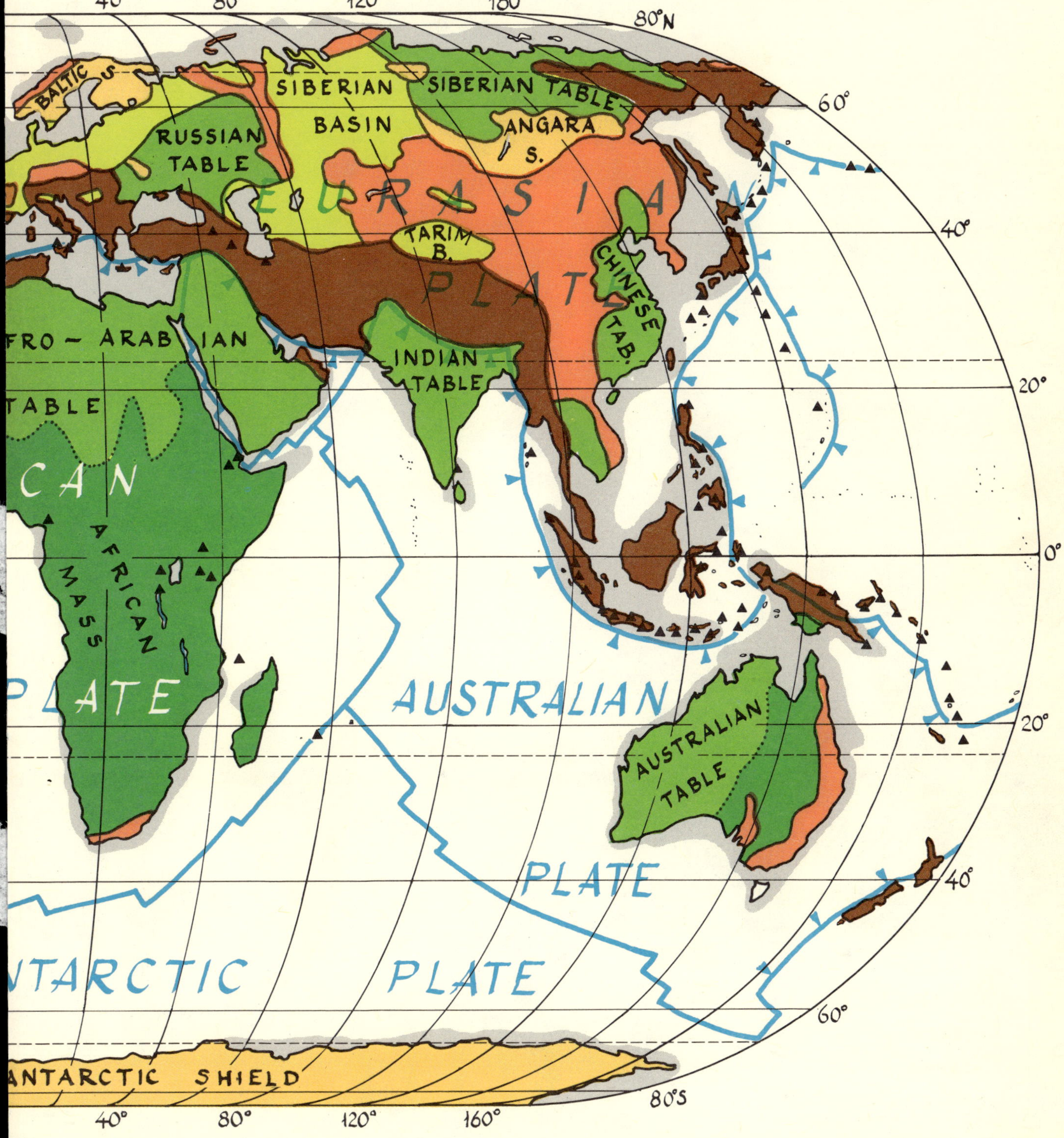

40°
80°
120°
160°
80°N
BALTIC S.
SIBERIAN BASIN
SIBERIAN TABLE
ANGARA S.
RUSSIAN TABLE
EURASIAN PLATE
TARIM B.
CHINESE TAB.
60°
40°
20°
0°
20°
40°
60°
FRO-ARABIAN TABLE
INDIAN TABLE
CAN
AFRICAN MASS
PLATE
AUSTRALIAN PLATE
AUSTRALIAN TABLE
NTARCTIC PLATE
ANTARCTIC SHIELD
80°S
40°
80°
120°
160°

Physical Geography

SECOND EDITION

Clyde P. Patton
University of Oregon

Charles S. Alexander
University of Illinois

Fritz L. Kramer
Portland State University

Duxbury Press North Scituate, Massachusetts
A Division of Wadsworth Publishing Company, Inc.
Belmont, California

1974 by Wadsworth Publishing Company, Inc., Belmont, California 94002. All rights reserved. No part of this book may be reproduced, stored in a retrieval system, or transcribed, in any form or by any means electronic, mechanical, photocopying, recording, or otherwise, without the prior written permission of the publisher, Duxbury Press, a division of Wadsworth Publishing Company, Inc., Belmont, California.

L.C. Catalog Card No.: 73-89886
ISBN: 0-87872-022-7

Physical Geography, Second Edition, was edited by Margaret Kearney. The cover was designed by Oliver Kline.

Printed in the United States of America

1 2 3 4 5 6 7 8 9 10 — 78-77-76-75-74

Preface

The aims of this second edition of *Physical Geography* remain the same: to describe the major physical processes that shape the surface of our earth and to present the distribution of the resulting surface features. Some topics have been given more than the usual shrift. Such subjects as continental drift, the role of mass wasting, the significance of the ocean basins, the heat budget of the earth's atmosphere, adiabatic processes, the mechanism of the monsoon, and the impact of man on vegetation all seemed to call for a more complete treatment than is normally found in introductory texts. Nevertheless, we have tried to be as brief and clear as we were able without damaging continuity or interrelatedness. In short, the text and the illustrations, which are an integral part of the narrative, are meant to be a reasonably complete and up-to-date introduction to the field of physical geography.

We have tried to improve this second edition of *Physical Geography* by deleting some of the less useful material, by improving some of the existing features and by adding some entirely new sections. Many of the illustrations are new and the world maps have been enlarged and reworked to make them more readable and more useful. Metric equivalents have been added to English units in an effort to familiarize the reader with the metric system. The various components of the physical environment have been placed in the context of systems analysis wherever appropriate. New sections on tornadoes, or urban climates, on forest climates, and on soil classification have been added. Finally, there is an entirely new chapter dealing with environmental quality and the relevance of physical geography to that all-important topic.

Thanks are due to all those who have helped imporive this text by pointing out errors and suggesting changes. More specifically, Sarah and Denise Patton made the early stages of revision into an easy and happy occasion for their father; Margaret Kearney spent long and devoted hours in making the manuscript much clearer and more graceful than it would otherwise have been; and our families and colleagues have been a consistent and necessary source of moral and intellectual support. We thank them all.

Contents

1. Introduction   1

Processes that Shape the Natural Environment   2
Natural Systems   3
Summary   6

2. The Heat Balance of the Earth   7

Introduction   7
Incoming Solar Radiation   8
Terrestrial and Atmospheric Radiation   15
The Heat Balance Over the World   19
Summary   21

3. Temperature   22

Introduction   22
Surface Temperature   23
The Temperature Aloft   37
Summary   39

4. Pressure and Winds   41

Introduction   41
Causes of Pressure Differences in the Atmosphere   43
The Existing Planetary Pattern of Pressure   50
The Planetary Wind System   59
Models of the General Circulation of the Atmosphere   64
Summary   65

5. Evaporation and Condensation   67

Introduction   67
Evaporation   70
Condensation   72
The Adiabatic Process   76
The Importance of Lapse Rates to Climate   84
Condensation Forms   84
Summary   85

6. Air Masses, Fronts, and Storms   88

Introduction   88
Classification of Air Masses   89
Description of Air Masses   91
Atmospheric Disturbances and Associated Weather   93
Summary   104

7. Precipitation   106

Introduction   106
World Distribution of Precipitation   102
Summary   120

8. The Classification of Climates   121

Introduction   121
The Köppen Classification   125
Summary   132

9. World Distribution of Climates   134

Introduction   134
Tundra Climates   134
Frost Climates   138
Steppe and Desert Climates   141
Tropical Rainforest Climates   147
Tropical Monsoon Climates   150
Tropical Savanna Climates   154
Humid Mesothermal Climates   156
Mediterranean Climates   158
Mid-Latitude Marine Climates   160
Humid Continental Climates   163
Subarctic Climates   166
Special Climates of a Non-Zonal Nature   167
Summary   171

10. Introduction to Landforms   173

Introduction   173
Earth Materials   176
Tectonic Processes   184
Summary   190

11. Weathering and Mass Wastage   191

Introduction   191
Weathering   191
Mass Wastage   195
Summary   200

12. Gradation by Streams   202

Introduction   202
Landforms Resulting from the Work of Streams   205
Summary   215

13. Gradation by Glaciers   217

Introduction   217
Past Glaciations   219
Glacial Gradation   224
Landforms Resulting from the Work of Glaciers   225
Summary   232

14. Gradation by Wind   233

Introduction   233
Wind Erosion   234
Landforms of Wind Gradation   235
Summary   238

15. The Classification of Landforms   239

Introduction   239
Landforms as Part of a Dynamic System   240
Landform Properties   242
Landform Categories   243
Summary   244

16. Plains   246

Introduction   246
Plains of Slight Relief   246
Plains of Moderate Relief   253
Plains with High Relief Features   258
Summary   261

17. Hills and Mountains   263

Introduction   263
Hills and Low Mountains   263
High Mountains   269
Summary   274

18. The Continental Margins and the Ocean Basins   276

Introduction   276
Factors of Shore Formation   277
Classification of Shore Lines   281
Types of Shores   283
The Continental Shelf   287
The Ocean Basins   289
Summary   290

19. The Oceans 291

Introduction 291
Characteristics of Ocean Waters 291
Water Masses 300
Summary 304

20. The Environment and Distribution of Vegetation 306

Introduction 306
Plant Communities and Factors Affecting Their Distribution 307
The Distribution of Vegetation 314
Summary 325

21. The Nature and Formation of Soils 326

Introduction 326
The Nature of Soils 326
Soil-Forming Processes 332
The Soil System 336
Summary 338

22. Soil Types and Their Distribution 339

Introduction 339
Zonal Soils 340
Nonzonal Soils 348
A New Soil Classification 349
Summary 351

23. Environmental Quality and Physical Geography 353

Introduction 353
Water Pollution 354
Atmospheric Pollution 357
Pollution and the Environment 359
Summary 367

Appendix: Location and Maps 368

Introduction 368
Location 369
Maps 375
Summary 388

Glossary 389

Index 398

Physical Geography

1 Introduction

One important thread of Western thought has always been a curiosity about our environment — the earth and its physical features. Homer, Hippocrates, Herodotus, Aristotle, Strabo, Pliny, and Ptolemy, writers who span the period between 800 B.C. and A.D. 200, all had something to say about the form and order of the land, air, and waters of the earth. Their works were both speculative and practical. They built theories about the shape and size of the earth and about the underlying causes of climatic differences. At the same time, they described the parts of the world known to them or to other travelers. This attention to the natural landscape certainly predates the Hellenistic Age (323-30 B.C.) for the more ancient civilizations of the Near East were much occupied with the literal worship of Mother Earth. The myths of Babylonia and the theology of ancient Egypt both involve descriptions of the natural order in the cosmos, with a strong focus on the land itself. The obvious interest of these floodplain dwellers in the rhythm of their rivers led them to evolve cosmological theories of the creation and divine regulation of river, wind, and sea.

Nor has attention to the physical earth been limited to the West. Confucius (551-478 B.C.), to name but a single example from the Orient, dealt with the land at considerable length in his teachings; and the Moslem civilization produced a number of travelers and writers, of whom Ibn Batuta (1304-1377) is an early, well-known example.

This very old concern with nature has often taken the form of attempts to answer three recurrent questions about the relationship between man and the habitable earth:

Is the earth, which is obviously a fit environment for man and other organic life, a purposefully made creation? Have its climates, its relief, the configuration of its continents influenced the moral and social nature of individuals, and have they had an influence in molding the character and nature of human culture? In his long tenure of the earth, in what manner has man changed it from its hypothetical pristine condition?[1]

These are important metaphysical and practical questions, but we cannot begin to answer them until we know how the natural environment has developed and continues to evolve. The concern of this book is therefore with the processes that shape the natural environment. We will not dwell on naïve and exaggerated notions of the influence of climate and landforms on human thought and action which have gradually given way to a much greater attention to man's role in changing the face of the earth. This is not to say that the physical part of geography is easily separable from the whole. It is not as simple as it might seem to distinguish between the natural and the artificial, the physical and

---

[1] Clarence J. Glacken, *Traces on the Rhodian Shore*, University of California Press, Berkeley, 1967, p. vii.

the cultural, the primeval and the man-made, when we look at the world around us. The physical landscape appears to be easily definable — it consists of the totality of landforms, climate, soils, and vegetation. The cultural landscape appears equally clear-cut — it consists of all of the works of man that give character to a particular place. However, a closer look at the problem or a closer look at the landscape itself shows that any particular form cannot so easily be put into one or the other of these two broad categories.

Consider the Southern Pine Forest of the Atlantic and Gulf Coastal Plain, for example. It appears to be a natural part of the physical landscape in the sense that it is presently closely associated with a particular climate and a particular sandy soil. Human activity in this forest area, however, has been geared to grazing animals and extracting naval stores from the trees. These activities are greatly helped by deliberate, periodic burning, which has contributed to the creation of a distinctive cultural landscape — the Pine Forest. The pines are quickly replaced by broadleaf deciduous trees where these human activities are no longer practiced.

Despite the fact that the natural and cultural imprints on a particular landscape are not always easily separated, the *processes* that shape the earth's surface are readily distinguished into natural and cultural categories. It is this distinction that allows us to treat the study of the physical environment as a separate topic within the field of geography.

There are two ways of studying the physical processes that create different landscape patterns on the surface of the earth: one is to start with the pattern and work back to processes; the other is to start with processes and work toward the pattern. In the first case, one is dealing essentially with interpretation; in the second, with explanation. We have chosen to emphasize process, not because we think the interpretation of patterns is of secondary importance in geography, but simply because we feel that an understanding of processes is a prerequisite to interpretation. Explanatory description of patterns must, in fact, be one of the major aims of any introductory text in geography, and we have devoted considerable attention to the patterns as well as the processes.

The purpose of this book is, therefore, to describe the natural environment from which mankind derives its sustenance and a good deal of its esthetic stimulation, by explaining the natural processes and interpreting the resulting natural patterns on the land. The physical environment, in contrast to the man-made landscape, appears to be relatively permanent and unchanging, but the distinction is largely a matter of time scales: the natural patterns change more slowly than cultural ones, although there are striking exceptions such as the sudden appearance of a volcanic island in mid-ocean or the over-night change from Indian summer to arctic blizzards and deep snows.

Processes that Shape the Natural Environment

The physical phenomena in the landscape and their relatively slow change are the result of the interaction of a number of processes over time. These processes are set in motion by three kinds of underlying forces. First are forces arising from the unequal distribution of solar energy on the earth's surface. **Solar radiation**[2] received at the surface varies with place and time, thereby setting up inequalities of energy that in turn set other processes in motion. Second are forces that derive their energy from within the earth, called **tectonic forces.** Their nature is not completely understood, but the impact of tensional and compressional forces (that is, forces that respectively pull apart and squeeze together portions of the earth's crust) is readily observable in the world's landforms. Third is the force of **gravity** that is related to the earth's mass.

Energy gradients[3], resulting from the unequal distribution of heat from the sun, result in areal variations[4] in temperature. Temperature gradients in turn are primarily responsible for differences in atmospheric pressure. In addition, pressure gradients explain air movements. The pattern of winds underlies the distribution of precipitation on the surface of the earth. In other words, each of the climatic elements has its own pattern of distribution, but that pattern is very strongly influenced by the basic areal distribution — that of solar radiation.

The tectonic forces have created continental platforms and ocean basins, and initial relief from which evolve hills, mountains and plateaus, as well as depressions which become collecting basins for debris from the uplands. Operating with unequal impact on

---

[2]Bold-face items are defined in the glossary

[3]*A gradient* is simply the rate of change with respect to distance of any measurable quantity. Thus a steep energy gradient (or temperature gradient, or pressure gradient) is one in which energy (or temperature, or pressure) changes rapidly in a short distance. A weak gradient implies a small change over a relatively long distance. When we talk of elevation, we use the word *slope* to express the notion of gradient. Thus a steep slope and a steep gradient of elevation are synonymous expressions.

[4]Areal variation is a common phrase on the tongue of geographers. It refers to the unequal distribution of any phenomenon over some area. The phrase *spatial variation* refers to the same kind of distribution, but emphasizes a three-dimensional distribution in space.

different parts of the world and at different times, these tectonic forces not only create the framework of the major landforms but also determine differences in rock types and provide the basic structural variations in the earth's crust as well as initial relief.

On sloping land, gravity provides the force necessary to move weathered rock downhill and to cause streams to flow, thus eroding the land surface. It is also basic to the creation of atmospheric pressure and to air movement in the atmosphere.

All of these forces affect every physical process with which a geographer deals and hence all contribute in some degree to the observable areal variation of physical phenomena on the earth's surface; but many variations result mainly from the operation of a single one of these forces. Thus some differences arise primarily from the action of the sun on the atmosphere, whereas others stem largely from forces internal to the earth, the tectonic forces and gravity. This distinction, even though it holds only in a general way, is the basis for one of the fundamental divisions of physical geography into climatology (the study of climates) and geomorphology (the study of the form of the land), which appears in this book.

Gradients of solar energy, the tectonic forces and gravity supply the primary forces necessary for the development of landform and climatic patterns. Other forces also affect the distribution of climate and landforms; among these, the most important are those caused by the earth's rotation and those created by friction between the earth's surface and wind. All of the forces discussed are also involved in the development of patterns of vegetation and soils.

The distinctions between the various forces just described are easily made and readily understood, but the many processes they set in motion operate interdependently and with varying degrees of intensity. Furthermore, every element of the physical environment is conditioned by many diverse processes which derive their energy from some or all of the forces operating in the earth's crust or in the atmosphere directly above it. As a result, we often need to keep in mind every aspect of the natural environment in order to describe and understand any single element. Consider, for example, the movement of air over the Great Plains of the United States. The basic force that sets the air in motion is that derived from differences in air pressure which result from complex atmospheric processes, including the effect of different kinds of ground surfaces on the nature of the air above them. The direction of this surface wind depends not only on the direction of the pressure gradient, but also on the deflection resulting from the rotation of the earth underneath the wind. Friction within the air, and particularly between air and ground, affects wind direction as well as wind speed, and the strength of the frictional forces is determined by the character of the air flow and by the nature of the surface over which it passes. Rugged terrain and forested land will slow down horizontal air movement; in addition, the wind may be funneled in a particular direction by the form of the land. Finally, the content of the moving air will depend on its past history and on the nature of the surface over which it has recently blown. Evaporation from a hot and wet surface may have supplied the air with lots of water vapor; or the air may have picked up a load of dust from a bare surface of small dry particles. In short, as simple a phenomenon as a local wind cannot be understood unless one considers not only the atmospheric processes at work, but also the landforms, the vegetation, and even the soils that have affected its movement.

The processes at work in creating a particular kind of wind are evidently many and quite varied, but the form of the phenomenon created is simpler to deal with. For this purpose we may think of the *descriptive elements* of the physical environment which include: *climate* (of which the most significant items are temperature, wind and precipitation); *topography* (in the narrow meaning of slope and relief[5]); *rock type* (the nature of the rock and its physical and chemical character); *rock structure* (the way the rocks are arranged at the surface of the earth); *biota* (the vegetation cover and the animal population in a particular place); and *soils*.

These descriptive elements, when taken together, define precisely the form of the natural landscape, but they are also factors in the formation of the natural landscape since they condition the rate at which the basic processes operate and also the degree of interdependence between various processes. In this context an additional variable, *time*, must also be considered. Thus we speak of *climate*, *topography*, *rock type*, *rock structure*, *biota*, *soils* and *time* as the **geographic factors.**

Natural Systems

The relationship between processes, geographic factors and the descriptive elements of the landscape is shown in Figure 1-1. In the figure, the major forces are seen to activate certain processes (only the most important are listed) which in turn work changes in the elements of the physical environment. The elements, because they have been changed, thereupon affect the rate at which the processes work. This sort of inter-

---

[5]Relief refers primarily to differences in elevation within a given area.

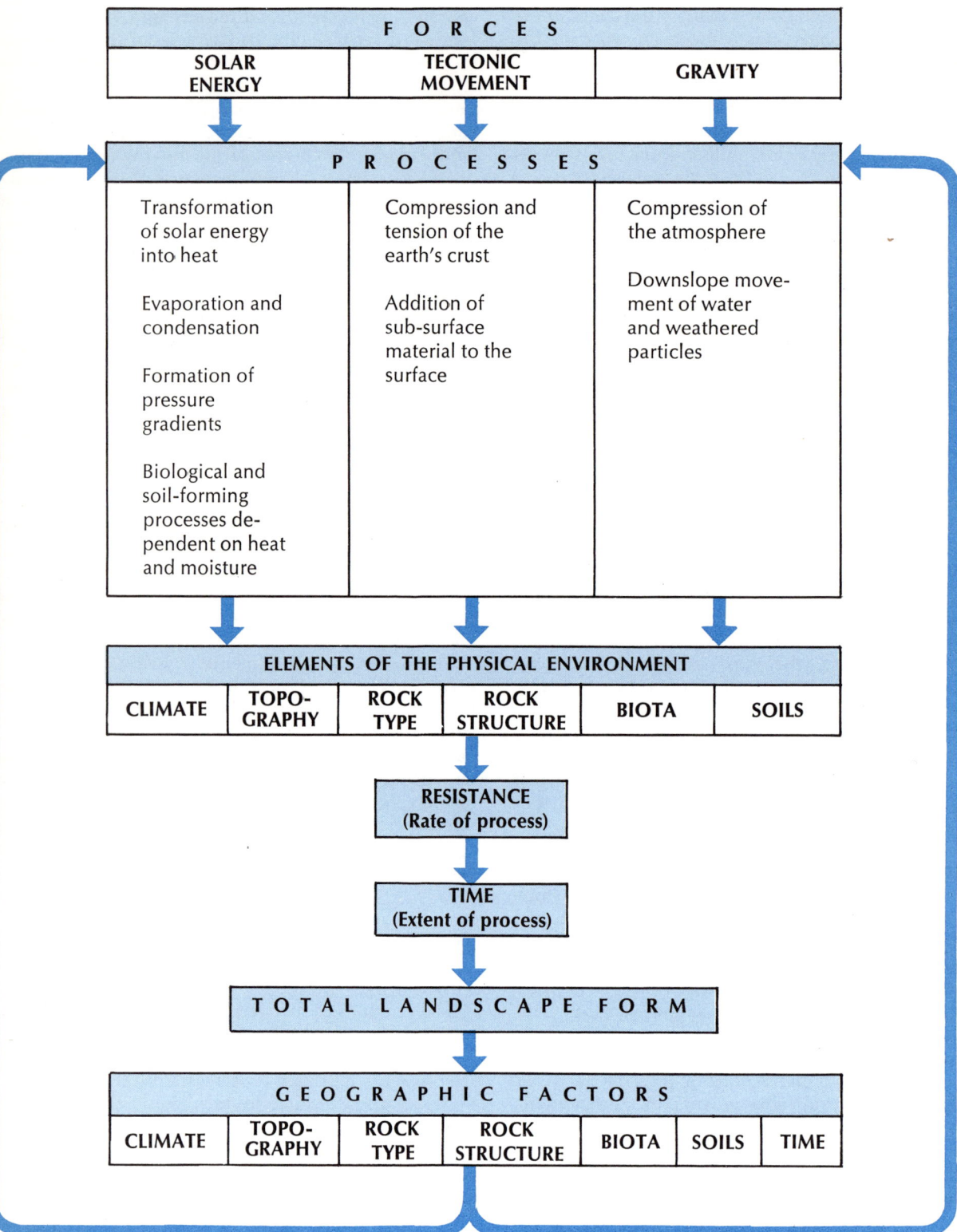

Fig. 1–1. A model of the natural landscape system. The forces set in motion processes which consequently change the elements of the system. The changed elements, in turn, affect the way the processes operate.

connection is often described as a *system*. A **closed system** is one in which every element of the system is affected only by other elements within the system; there is no input of any kind from outside the system, nor any output that is not incorporated within the system. Many laboratory experiments try to create essentially closed systems, with as few elements as possible, in order to analyze the relationships between the included elements, while excluding all or most of the extraneous elements. In nature, however, such conditions rarely exist; **open systems** are much more common than closed ones.

Open systems share several basic characteristics. First there is an *input* of energy and mass from outside the system and an *output* of energy and mass from within the system. Second, the input from outside, even when it remains constant over time, sets up *changes* within the system. Third, there are *resistances* to change within the system. Fourth, any change in resistance or in the character of any element of the system will result in *feedback*; a **positive feedback** is one that accelerates the change; a **negative feedback** slows down the rate at which a system is changing. Fifth and last, most open systems tend toward a *state of equilibrium*, in which output equals input and all parts of the system are in a mutual state of adjustment.

Bear in mind that every system we can construct is only a convenient way of organizing data and our thoughts about the data. Thus we often divide a system (such as the heat budget of the earth and atmosphere) into convenient sub-systems (such as the heat budget involving short-wave solar radiation; the heat budget involving long-wave terrestrial and atmospheric radiation; and the non-radiative heat budget). But just as we can partition systems into smaller sub-systems, so also can we think of any system as being a sub-system of a larger system, at least until we consider the whole universe as the ultimate system.

The value of such an approach is that it focuses attention on the interconnection of all the features (be they forces, processes or elements) of a natural system. Let us consider three examples of natural systems to illustrate a systems approach to physical phenomena.

The first example is one of the most fundamental of all climatic systems, the transfer of energy which takes place at the surface of the earth. This system, called the heat budget or heat balance of the surface, includes income of energy or heat from the sun, from the atmosphere, and from below (from the subsoil on land, and from water at depth in the oceans). The heat budget also includes outgo of heat from the surface to the atmosphere and to the subsurface. Any change in the total income of heat will change the output as well; an increase of heat at the surface will raise the temperature and consequently increase radiation from the surface to the air above it and conduction into the soil or convective transfer into the water below the surface. There are, however, resistances within the system. Reflection from clouds or from the surface itself prevents a good deal of energy from being used by the surface. The atmosphere absorbs nearly a fifth of the incoming solar radiation. The ground takes up heat very slowly and with great difficulty but water bodies store enormous quantities of heat with relative ease. All of these resistances change as other parts of the system change. Thus, a fresh clean snow cover may reflect over 80 percent of the radiation that strikes it. But in middle latitudes the snow will eventually melt away in spring because of the heat applied to it by warm air masses from outside the region. Once the snow is gone the reflection from the ground may drop to less than 15 percent. Two kinds of feedback are involved. The reflectivity from a snow or ice-covered surface is a positive feedback operating locally. The cold conditions which allow the snow to exist are intensified by the increased reflectivity from that very snow cover. The relationship between heat accumulation at the surface and heat flow away from the surface is a negative feedback operating on a much wider scale. Here the increased outflow of heat from the surface tends to destroy the heat accumulation which caused it in the first place. Other kinds of feedback take place as well. Higher temperature at the surface means more evaporation into the air; it also tends to accentuate air movement in the atmosphere. The net result will be more cloudiness which in turn affects the temperature at the surface both by reflecting incoming solar energy and by trapping outgoing radiation from the earth. As we shall see in the next chapter, all of these elements tend to be in balance if we consider the system as a whole. There are obvious differences in income and outgo of heat between day and night; equally obvious are the variations between summer and winter or between polar and equatorial latitudes. But if these daily and seasonal cycles and the differences between latitudes are averaged out, we find that the surface exports just as much heat as it imports, and that the amount of heat present results in an average surface temperature of about 67°F (19°C) that scarcely changes from one year to the next. The heat budget, seen as a system, thus tends to an equilibrium of which the nearly constant surface temperature is only the most striking example.

The heat budget of the earth's surface has illustrated each of the characteristics of an open natural system, inputs and outputs of energy, resistances and changes within the system, positive and negative feedbacks, and a tendency to reach a state of equilibrium. Before we leave this example we should emphasize the fact that any given state of equilibrium may be short-lived or even unattainable because of changes of input from outside the system. For instance a change in solar output, or even a change in the nature of the solar radiation (more ultraviolet radiation than normal, for example) will obviously change every item in the heat budget of the surface. If the change levels off at some new but constant value, however, the surface system will probably reach a new state of equilibrium (with higher or lower temperatures, more or less cloudiness, greater or lesser absorption by the atmosphere, and so on).

Our second example consists of a comparison of two very small and highly localized systems. The setting is that of a series of low hills in southern Illinois. These hills have an east-west alignment so that one slope faces north and the other faces south. Let us think of the several north-facing slopes as one system and the south-facing slopes as another. If we examine the descriptive components of these two systems we find them to be remarkably similar: climatic differences are slight from one hill to another; the topography (Greek *topos* place + *graphein* to describe)[6] is roughly the same (the slopes have about the same steepness and the differences in elevation are everywhere less than 200 feet, or about 60 meters, between the hill tops and the valley bottoms); rock type and rock structure are roughly the same on north and south-facing slopes; and the soils, while not exactly the same, are similar. The vegetation cover, on the other hand is markedly different. The system of south-facing slopes consists of a variety of species of oak, whereas the north-facing slopes support a forest rich in white oak, sugar maples and beech, species generally found two or three hundred miles (about 400 km) farther north. Evidently there must be a greater difference between the two systems than our generalized description led us to believe. The lower angle of the sun's rays on the north-facing slopes results in considerably less heat accumulation than in the south-facing system, as explained in the next chapter. The difference in heat is reflected in the temperature and also in the moisture content of the ground. Both slopes receive nearly the same amount of precipitation, but the warmer south-facing slopes have more energy available to evaporate water from the ground. Thus the local climate is one of contrast between cool and wet north-facing slopes and warmer but drier south-facing slopes, a contrast which on flat ground is found only over distances of several hundred miles, as shown by the difference between the vegetation on the two slopes. The oaks that tolerate relatively dry conditions exist in a system with one sort of equilibrium; the sugar maples and beeches that tolerate colder climates exist in another system with a different equilibrium.

Our last example illustrates a system that oscillates between two entirely different states of equilibrium. It involves one of the theories proposed to explain the advance and retreat of ice caps during the last million years. It is discussed in more detail in Chapter 13, but its main points can be used here to illustrate the notion of feedback. According to the theory, the locking up of water on the land in the form of continental ice sheets during a period of glaciation, lowered sea level enough to isolate the Arctic Ocean, at least from the warm surface currents of the North Atlantic. This isolation probably cooled the Arctic Ocean to the point where so little evaporation took place that not much snow fell on surrounding lands. The reduced snowfall caused the glaciers to wane, thus raising sea level. The Arctic Ocean was once again invaded by warmer water and the increased evaporation nourished the ice cap until sea level dropped and the whole cycle repeated itself. There are objections that can be raised to this hypothesis, but it does illustrate negative feedback. The loss of water by the oceans which is a result of glaciation on the land has an eventually negative effect on the growth of glaciers. By the same token, the disappearance of the glaciers, by raising sea level, eventually leads to their rejuvenation.

Summary

These examples, which have been chosen from very different kinds of systems, have emphasized the fact that patterns in the physical landscape are complex because they are the result of many processes operating interdependently and with varying degrees of intensity. In the succeeding chapters we will describe these physical processes and explain the major features of the resultant patterns on the surface of the earth.

---

[6] The etymologies of words of Greek and Latin origin are given to aid the reader in fixing the meaning of these words in his mind. Some of the words such as *antipode* are words that were part of classical Greek; others, like the word thermometer, were made up much later from classical Greek or Latin roots and had no counterparts in the original languages. Such words are identified by the letters I.S.V. (International Scientific Vocabulary), in dictionaries.

2 The Heat Balance of the Earth

Introduction

In our discussion of climate we will be concerned much more with external solar forces operating in and through the atmosphere than with the internal **tectonic forces** of the earth's crust. To be sure, we will have to talk about climates in discussing landforms; and landforms have to be considered in discussing world climates. For the most part, however, the impact of internal forces on the surface of the earth is tectonic, whereas the impact of external forces is climatic.

Climate, then, is conditioned by atmospheric forces; and the simplest way to define climate is to say that it consists of the total impact upon our senses (and measuring devices) of all that is going on in the layer of air directly above and around us. In some cases, we want to consider what is happening in a very thin layer of the earth's crust as well. Temperature, precipitation, and the like, which appear to be individual phenomena, are the *climatic elements*. If we focus on the phenomena as they appear at a particular instant, we then speak of the *weather*. Further narrowing of our point of view leads to a classification of the *elements of weather*.

The Importance of Radiation to Climate

The main concern of physical geography is with climate and the climatic elements (rather than with weather and its elements) and particularly with the processes whereby differences in climate are created within the atmosphere. The principal source of energy for all these atmospheric processes is the **radiation** received from the sun. The rate at which this incoming radiation strikes a horizontal surface is called **insolation** (Latin *sol* sun). The importance of insolation becomes apparent when we examine all climatic phenomena in a general way. The various climatic elements, such as temperature, pressure, evaporation, and rainfall, have two main characteristics. Each has a particular magnitude or *intensity* at a given place; and each has *gradients*, or rates of change, from one place to another. All these climatic intensities and gradients are determined by the intensity of insolation and by its gradient from place to place. Gradients of insolation are responsible for the temperature gradients found on the earth's surface. Temperature gradients determine, ultimately, pressure gradients. Pressure gradients and the winds associated with them determine where air masses are formed and where they move. And so it goes, down to the smallest item of weather and of climate. The word "climate" itself comes from a Greek root, *klima*, meaning "inclination" or "slope," hence also "grade" or "gradient." The importance of insolation is therefore paramount, particularly when we consider that it is not only the ultimate cause of climatic differences, but that it also affects all phenomena, such as natural vegetation and soils, which are conditioned by climate in one way or another.

The Definition and Measurement of Radiation

Every object in the universe (except those at a temperature of absolute zero) radiates energy in the form of waves. The length of the wave determines the character of the radiation. Thus radiation with very short wavelengths, such as X-rays or ultraviolet radiation are not visible to the human eye and differ in other respects from radiation of moderate wavelength which are visible. Within the narrow spectrum of visible radiation, the human eye distinguishes colors according to the wavelength — violet, indigo, blue, green, yellow, orange, red, in order of increasing wave length. Extremely longwave radiation, such as the infrared is not visible but is perceived as heat. Most objects radiate at various wavelengths but have a concentration at some particular length, called the *wavelength of maximum energy*, which determines the characteristics of the energy, such as its intensity, its ability to penetrate different substances or its impact on living organisms. The higher the temperature of the radiating object, the shorter the characteristic wavelength and the more intense the radiation.

Heat can be transferred by means other than radiation. **Conduction** refers to the relatively slow transfer of energy that takes place when one molecule bangs into another. For example, a spoon placed in a hot liquid warms up by conduction. **Convection** is a much more rapid process. It occurs in liquids and gases, when the heat of one part of the medium is transferred to another by turbulence.

Thus, as we have seen, the temperature of a radiating body determines both the amount of radiation and the wavelength of maximum energy. The higher the temperature, the greater the radiative output and the shorter the wavelength. As a result, the radiation from the sun, although it includes both short waves of the ultra-violet type and longer waves of the infrared variety, has its maximum intensity, because of its high temperature, in that part of the electromagnetic spectrum with relatively short wavelengths. Emission of radiation from the earth and from the atmosphere is not only much less than that from the sun, as we might expect from the much lower terrestrial and atmospheric temperatures, but it is also in the form of much longer waves. This distinction between solar short-wave and terrestrial long-wave radiation has important consequences for the heat balance of the earth.

Incoming Solar Radiation

The unit most commonly used to define radiation, as well as other forms of energy or heat, is the **calorie** (Latin *calor* heat). One calorie is the amount of heat or energy required to raise the temperature of 1 gram of water by 1° on the centigrade scale. A calorie of heat spread over a square centimeter of surface (1 cal/cm^2) is the unit we will deal with most. For the sake of simple notation, 1 cal/cm^2 is called a **langley,** after S. P. Langley, a 19th-century physicist and astronomer who did pioneering work in radiation. The langley (ly) measures magnitude of radiation; for rates of radiation, we speak of so many langleys per minute (ly/min) or langleys per year (ly/yr), depending on the time period we are considering. The relationships among the various rates are simple. One ly/min is the same as 60 ly/hr; 1 ly/hr equals 24 ly/day, and so on.[1]

Insolation is measured by a *pyrheliometer*, (Greek *pyros* fire + *helios* sun + *metron* measure) an instrument which measures radiation indirectly, by recording the temperature effects of the incoming radiation on an absorbing surface, such as a silver disk. Note the analogy between the operation of this instrument and the definition of a calorie; both get at radiation or heat indirectly through temperature.

Factors Affecting the Amount of Solar Radiation that Reaches the Atmosphere

The average temperature of the surface of the sun is about 11,000° F (6,000° C);[2] at that temperature the sun radiates at a rate of about 56 x 10^{26} gram calories per minute in all directions equally. This radiation is spread over spheres of increasing radii and hence of increasing areas, the further it gets from the sun. At the average distance of the earth from the sun, which is 93 million miles, the sphere over which the solar radiation is spread has an area of nearly 28 x 10^{26} square centimeters. Therefore the intensity of the solar radiation that strikes any point on that sphere is 56 x 10^{26} gram calories divided by 28 x 10^{26} square centimeters, or 2

---

[1] For the sake of comparison with more familiar English units, the following identities are given: 1 horsepower equals approximately 10,000 calories per minute; 1 langley per minute equals approximately 65 watts per square foot.

[2] See Fig. 3-1 and p. 23 for the relations between various temperature scales.

gram calories per minute per square centimeter. The average amount of radiation reaching the outer edge of the atmosphere is thus about 2 langleys per minute on a surface at right angle to the incoming rays of the sun. This value is called the **solar constant,** even though it varies a little because of slight changes in the sun's output of energy. The amount of energy reaching the atmosphere varies even more as a result of the eccentric orbit of the earth around the sun. The eccentricity of the earth's orbit is such that the earth is closest to the sun around January 1 and most distant from the sun on about July 1. Consequently, solar radiation is about 3 percent higher than average in the winter season of the Northern Hemisphere (which is the summer season in the Southern Hemisphere) and about 3 percent below average in the Northern Hemisphere summer (the Southern Hemisphere winter). The fact that the amount of solar energy reaching the earth is lowest during July indicates that the small variation due to orbital eccentricity is relatively unimportant to the heat balance of the earth since July is the warmest part of the year in the Northern Hemisphere.

Factors Affecting the Proportion of the Solar Constant that Strikes a Horizontal Plane at the Surface of the Earth

Seasonal and latitudinal differences in the amount of insolation received at the surface of the earth are determined mainly by factors other than variations in solar distance and solar output. The three most significant factors are the angle at which the sun's rays strike the surface, the duration of sunlight during any 24-hour period, and the amount of *depletion* of the insolation that gets through the atmosphere.

Altitude of the Sun. The angle of the sun's rays with respect to a horizontal surface (or to the horizon), called the **altitude of the sun,** varies from place to place and from one time of year to another because of the spherical shape of the earth and because of the fixed tilt of its axis as it orbits the sun. The simplest effect of the spherical shape is illustrated by Fig. 2-1. The solar constant of 2 langleys per minute refers to the amount of radiation received upon a surface at right angles to the rays of the sun. Thus, on the surface of the circle in the diagram, there will be 2 calories on every square centimeter during each minute of time. Only the radiation passing through this circle will strike the spherical earth, and this energy will disperse over the entire surface of the globe. Since the surface of the globe is four times as large as that of the perpendicular circle, the average insolation on the globe, or the rate at which radiation is received, is only one-fourth of the solar constant.

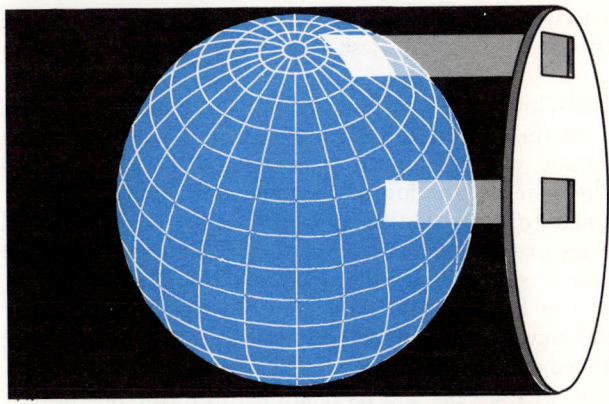

Fig. 2-1. The relation between radiation incident at right angles, the spherical shape of the earth and differences in receipt of insolation at different latitudes. All of the radiation that will strike the earth and its atmosphere must pass through the circle at the right. Since the area of the circle is one-fourth the area of the sphere, only one-fourth of the solar constant is available per unit area of the spherical surface. The two windows illustrate another aspect of this relation. They allow equal amounts of radiation to pass through; but the radiation from the vertical rays is spread over a small area on the globe, and hence is more intense than the radiation from the oblique rays which is spread over a larger area.

This reduction is only an average for the earth as a whole, however. The same effect is shown at a particular place at a particular time of the year and of the day in Fig. 2-1. On March 21, the sun's rays at noon strike the surface of the earth at the equator at an angle of 90°. At higher latitudes, the angle of the sun is less; and the radiation, incident on a surface vertical to the sun's rays, is spread over a horizontal area on the earth considerably larger than the area radiated at the equator; hence, the insolation per unit area is much less.

The more nearly vertical the sun, the smaller the surface over which a given amount of radiation is spread; hence the greater is the intensity. The altitude of the sun obviously changes from sunrise to sunset, but we disregard this daily variation and consider only the seasonal variation of the position of the sun at noon. The noon altitude of the sun is one index of the total radiation received during the day. At any given place, the higher the altitude of the noon sun, the more radiation if all other things are equal.

We need know only two things to determine the altitude of the noon sun at some given place, which we will call A. We must know the latitude of A and the time of year. From the date, we can determine the latitude at which the noon sun is directly overhead. At any point on that latitude the altitude of the sun is obviously 90°. If place A is 1° away from this latitude, the altitude of the sun at A is 1° less than 90°. If A is 10°

distant from the latitude with the overhead sun, the altitude at A is 10° less than 90°, and so on and so forth.

Obviously, in a problem of this kind, we expect to be given the latitude of the place A, but what about the latitude where the sun is overhead? This latitude is called the **declination of the sun,** and it may be determined for any day of the year, either from nautical tables or from an *analemma* (Greek *analemma* sundial) such as is often printed on globes. In Fig. 2-2, for example, one looks for the date along the figure-eight of the analemma and reads off the appropriate latitude along the edge of the diagram. The declination at certain times of the year and its relationship to the position of the earth with respect to the sun are also shown in Table 2-1. Note that the names for the various positions of the sun at particular dates were obviously coined by Europeans who were oblivious to the

Table 2-1. Declination of the Noon Sun at the Solstices and Equinoxes

Position of the earth with respect to the sun	Date	Declination of the sun
Winter solstice	Dec. 22	23½°S. (Tropic of Capricorn)
Vernal equinox	Mar. 21	0° (equator)
Summer solstice	June 22	23½°N. (Tropic of Cancer)
Autumnal equinox	Sept. 23	0° (equator)

Southern Hemisphere. The shortest day of the year at the **winter solstice** occurs on December 22, only in the Northern Hemisphere. The shortest day in the Southern Hemisphere occurs on June 22, for which the term **summer solstice** is obviously a misnomer. Similarly, the seasonal designation of the **equinoxes** (when day and night are of equal length) are inappropriate for the Southern Hemisphere.

The following example will show how we find the altitude of the noon sun in a given latitude on a given day of the year.

What is the altitude of the noon sun in 12° N latitude on December 22?

The declination of the sun is 23½° S.

The place in question, 12° N latitude, is 35½° away from 23½° S.

The altitude of the sun is, therefore, 35½° less than 90° or 54½°

Length of Daylight. The second factor affecting the amount of radiation that reaches the earth's surface is the obvious one of duration of insolation, or length of daylight. The length of daylight varies, like altitude, with season and with latitude. This variation is explained by the relationship of the **circle of illumination** to the parallels of latitude on various days of the year. The circle of illumination is the great circle separating the dark from the sunlit side of the earth. Figure 2-3 shows the position of the circle of illumination at the equinoxes and solstices and indicates what portion of each parallel lies on the sunlit side of the circle on the specified dates. During the course of the year there is a large variation in length of daylight in high latitudes and a much smaller range as one approaches the equator. The equator itself is exactly bisected by the circle of illumination at all times; therefore the length of daylight is always twelve hours.

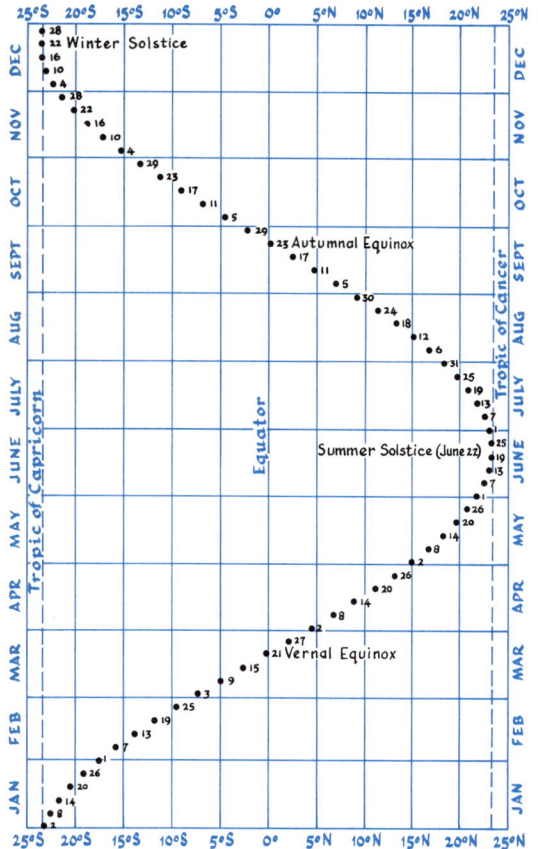

Fig. 2-2. This simplified version of an analemma shows the declination of the sun (the latitude in which the sun is directly overhead at noon) for every sixth day of the year, starting with January 2. Note the slow change in declination near the Solstices and the more rapid change near the Equinoxes.

Effect of Atmosphere on Insolation. The third factor affecting the insolation received at the surface of the earth is the depletion of incoming radiation by the atmosphere. The atmosphere is made up of a number of gases. The invariable, or permanent, gases are so well mixed that their proportions are practically the same

The Heat Balance of the Earth

mosphere. Water vapor is more important to climate than any of the other gases. It is not only the source of rain and snow, but it also absorbs six times as much solar radiation as all the other gases combined, despite its relative rarity. Furthermore, it accounts for nearly all the absorption of radiation from the earth.

The disposition of incoming solar radiation is shown in Fig. 2-4. The lower left-hand portion, shown in gray, relates to *income* of solar energy; the upper right-hand portion, also shown in gray, indicates *outgo* of terrestrial or atmospheric radiation. In other words the diagram shows the heat budget of the earth atmosphere system. Each number refers to an expenditure of energy (negative numbers) or to an income of energy (positive numbers). As we shall see, the whole set-up must be in balance, with expenditures equalling income, both for the atmosphere and the surface of the earth taken separately. An imbalance would imply that the earth or atmosphere (or both) were getting constantly richer or poorer in energy. Such a situation would result in constantly increasing (or decreasing) temperatures at the surface of the earth or in the atmosphere, which we know not to be the case. Temperatures have changed, but only very little even during glacial and interglacial periods, so that we may assume that the heat balance represents an equilibrium of income and outgo of energy.

What happens to insolation as it strikes the atmosphere? Let us take 100 units of incoming solar radiation and see what happens to them under average conditions (Fig. 2-4). Averaging necessarily masks both diurnal and seasonal variations, as well as differences from latitude to latitude; but it will give us an idea of the processes at work. If 100 units of short-wave radiation — short-wave because of the high temperatures of the sun — strike the atmosphere, the following approximate proportions are disposed of as indicated:

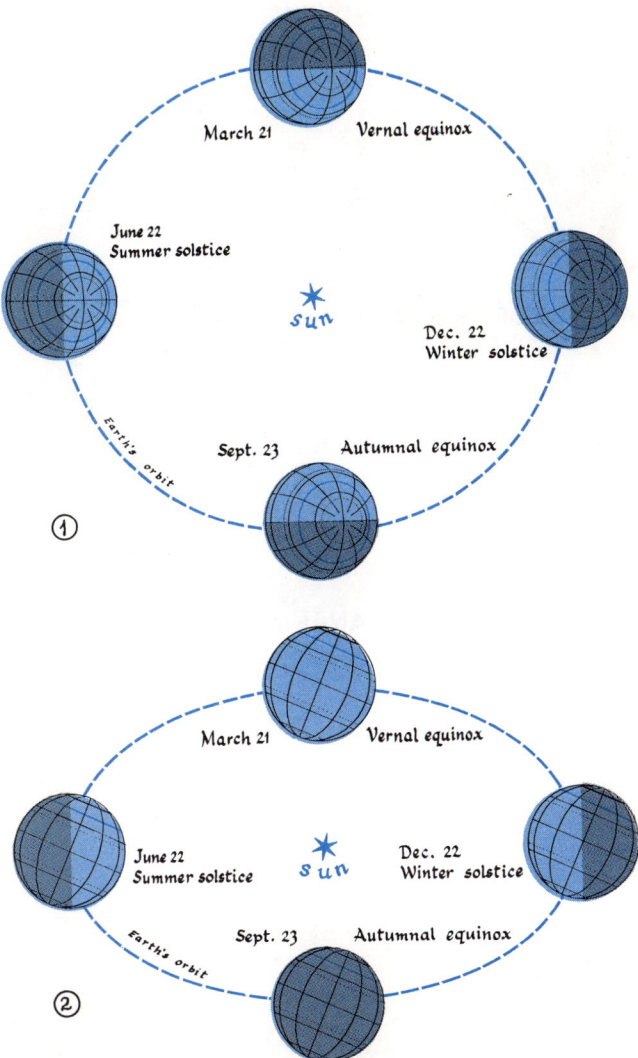

Fig. 2-3. Changing aspect of the circle of illumination during the year. Note that at the two equinoxes, the circle passes through the poles, hence every place on earth has twelve hours of darkness and twelve hours of light. At the solstices, the circle leaves one of the polar caps completely in the dark and the other one completely in the light, even as the earth rotates about its axis.

throughout the depth of the atmosphere. The amounts of other, variable, gases decrease rapidly with height and their proportions vary from time to time and from place to place. The major invariable gases are nitrogen and oxygen, which respectively make up 78 percent and 21 percent of the volume of the atmosphere. All the other permanent gases, such as carbon dioxide, helium, and hydrogen, account for only 1 percent of the volume of air. The most important variable gas is water vapor, which is present in amounts ranging from practically nothing to 3 percent of the volume of the at-

1. *Twenty units are absorbed by the atmosphere, half directly from the sun, the other half after being reflected from cloud tops. (These two items are drawn as one quantity on the left-hand side of Fig. 2-4.)*

2. *Twenty-five units are reflected from massive objects, mainly from cloud tops, though some are also reflected from the surface of the earth itself. (These two kinds of reflection are also drawn as one quantity on the left-hand side of Fig. 2-4.)*

3. *Thirty units are reflected from small particles in the air, mainly air molecules and the tiniest dust particles. This radiation is scattered in the form of diffuse radiation. Of the 30 units, 10 are scattered up and 20 are scattered downward. This downward component of the diffuse radiation is often called* **sky radiation.** *It gives the sky its characteristically blue color and is the source of heat and light for all objects in the shade. The light in a room with exposure away from the sun is*

12 Chapter 2

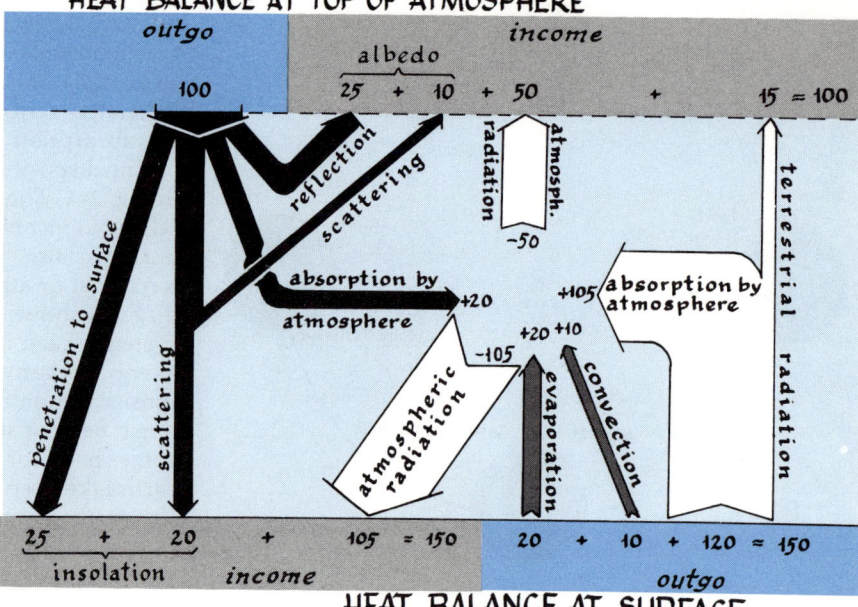

Fig. 2-4. The heat balance of the atmosphere. Short-wave solar radiation is shown by black arrows; long-wave atmospheric or terrestrial radiation by white arrows; non-radiative heat transfer by dark gray arrows. Income of energy at the top of the atmosphere or at the surface is shown in the gray boxes; outgo in the blue boxes. The numbers are percentages of the incoming solar energy at the top of the atmosphere. Positive numbers indicate income of energy; negative numbers, outgo.

an expression of diffuse (or scattered, or sky) radiation. The amount of scattering depends upon the amount of dust, dirt, water vapor, and the like present in the atmosphere and is called the **turbidity** of the atmosphere.

4. The remaining 25 units penetrate directly to the earth's surface and, together with the 20 units scattered downward, make up the insolation of the surface.

From the foregoing it can be seen that 35 units, or 35 percent, of the incoming radiation are returned to space. This part of the short-wave radiation is called the **albedo** (Latin *albedo* whiteness) of the earth. Another 20 percent goes to heat the atmosphere; and 45 percent reaches the surface.

Remember that, to this point, the discussion of heat balance applies to average conditions; that is, we assume a cloud cover of about 50 percent of the sky, and we further assume water-vapor and dust content of the air that is average for the earth as a whole and for the year. Any change in cloudiness or in the moisture content of the air will change all the percentages mentioned above. Moreover, we assume an average altitude of the sun, and an average length of daylight.

The average heat balance does not take into account other very important variations. So far, we have referred to 100 units of short-wave radiation that traverse an average distance through the atmosphere. But Fig. 2-5 shows that oblique rays must pass through the atmosphere for a longer distance than vertical rays and are thus much more depleted by reflection and scattering. The altitude of the sun has a double importance to incoming radiation. It determines not only the extent of the surface over which the solar intensity is spread but also the amount of depletion by the atmosphere, since the depletion is proportional to the length of traverse through the atmosphere. In the following discussion we shall refer, positively, to the

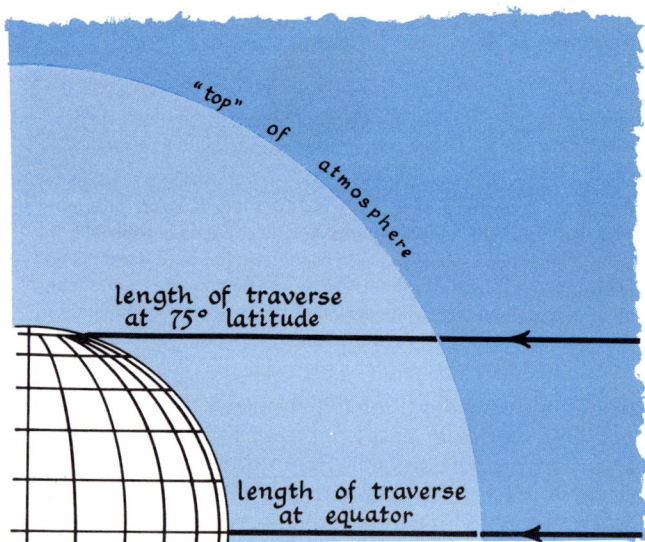

Fig. 2-5. The effect of the spherical shape of the earth and its atmosphere on the length of traverse of light rays. Note that oblique rays must pass through the atmosphere for a longer distance than vertical rays, and are thus much more depleted by reflection and scattering, by the time they reach the surface of the earth.

transmission by the atmosphere or its *transmissivity* rather than negatively to the depletion by scattering and reflection.

Annual and Latitudinal Cycles of Incoming Radiation

All the factors that affect the magnitude of insolation at the surface vary with latitude and with season. Figure 2-6 shows this variation graphically. Note that the ordinates (vertical axes) represent percentages of the maximum possible value, whether this value be altitude of the sun, length of daylight, transmission of energy by the atmosphere, or insolation reaching the ground. Thus, the highest value in each graph is labeled 100 and

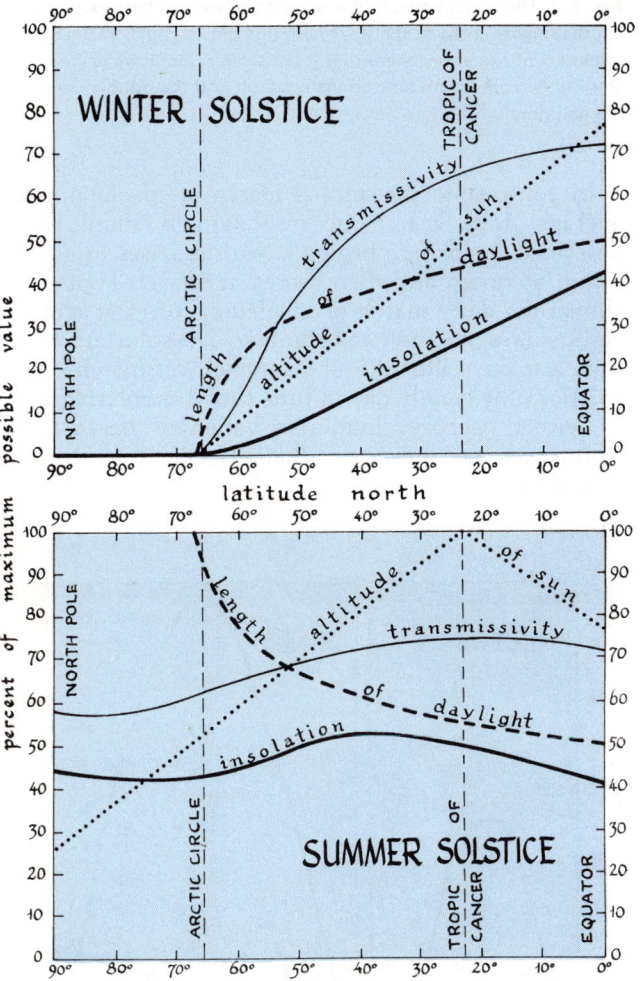

Fig. 2-6. The seasonal variation of factors affecting insolation received at the surface. Note that in winter all the factors increase in effectiveness from the Arctic Circle to the equator, whereas in summer, the length of daylight and altitude of the noon sun have opposite gradients so that the differences in insolation are much smaller between high and low latitudes.

may represent 24 hours of daylight, a noon-sun altitude of 90°, a transmission of 100 percent by the atmosphere, or 1440 langleys, which would be the highest possible insolation during a 12-hour day.[3] Figure 2-6 shows the rapid decrease, in winter, from the equator to the Arctic Circle, of altitude, of length of day, and of percentage transmission by the atmosphere. Bear in mind that the decrease in transmissivity is due largely to the lower angle of the sun and to the consequently greater length of traverse by the sun's rays (Fig. 2-5) and not to increases in cloudiness and turbidity, since in computing the curves of transmissivity, average cloudiness and average turbidity were assumed. In any case, the result is a steep gradient of insolation, from a theoretical maximum at the equator of about 42 percent of 1440 langleys (or 615 langleys) to a minimum of zero from the Arctic Circle northward.

In summer, on the other hand, the gradient of the altitude of the sun is equally steep, but the values are everywhere higher, with a maximum at the Tropic of Cancer. The gradient of transmissivity is very weak; there is little difference between pole and equator. The gradient of length of day is reversed; the higher latitudes enjoy longer days. The net result of these summer gradients is that the gradient of insolation is much weaker than in winter. In 40° N latitude, about 50 percent of the maximum possible insolation reaches the surface; at 75° N latitude, about 44 percent — a much smaller difference than in winter. In fact, during the summer, there is more decrease of insolation from the mid-latitudes to the equator than toward the poles.

The relatively large difference in insolation between equator and pole in winter (about 616 langleys compared with none) and the relatively small difference between mid-latitudes and Arctic regions in summer (about 730 langleys compared with 624 langleys) is one of the basic climatic facts. All climatic phenomena have, on the average, much weaker gradients in summer than in winter. This weakening is most obvious in the case of temperature but is also reflected in all phenomena whose intensity depends ultimately on the strength of the energy gradient from equator to pole.

Figure 2-7 shows the annual march of insolation, again assuming cloudless skies and average turbidity, at various latitudes.[4] Note the double cycle of the

---

[3] 1440 langleys equals 2 langleys per minute multiplied by 720 minutes, the number of minutes in a 12-hour day.

[4] The notation "$q = 0.8$," in Fig. 2-7 and Table 2-2, means that we assume a turbidity such that eight-tenths of the solar radiation gets through one unit thickness of the atmosphere; eight-tenths of the remainder gets through a second unit thickness; and so on. A unit thickness is the length of traverse by vertical rays. Oblique rays may have to go through several unit thicknesses (see Fig. 2-5).

equator curve: the two maxima coincide with the two equinoxes when the sun is overhead, and the two minima coincide with the solstices when the overhead sun is respectively 23½° north and 23½° south of the equator. All places between the equator and the two tropics have two maxima and two minima. From the Tropic of Cancer northward, there is but a single maximum and a single minimum of insolation with progressive steepness of increases of insolation in the spring and of decreases in the fall. The result is that the curve representing the annual march of insolation at the North Pole is much more peaked than the rounder curve of the Tropic of Cancer. This merely reflects the greater extremes in high latitudes and is summarized in Table 2-2.

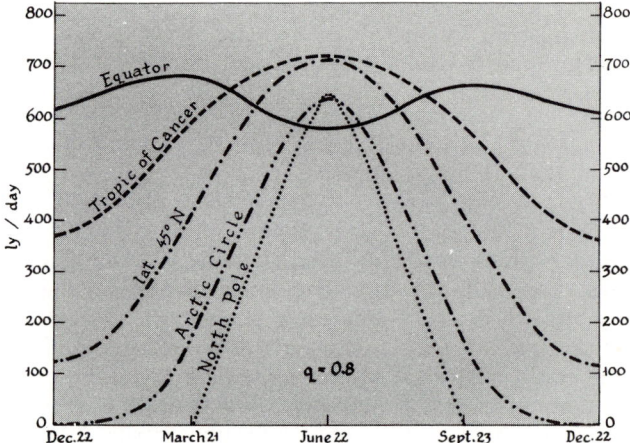

Fig. 2-7. The annual march of insolation in various latitudes, assuming that eight-tenths of the solar radiation gets through each unit thickness of the atmosphere that it penetrates. Note the decreasing amplitude from North Pole to the equator, and the double curve at the equator.

Table 2-2. Annual Range of Insolation of Selected Latitudes

Place	Insolation (ly/day)*		
	Maximum	Minimum	Range**
Equator	672	577	95
Tropic of Cancer	719	365	354
45° N latitude	718	118	600
Arctic Circle	639	0	639
North Pole	643	0	643

*Assuming q = 0.8.
**Difference between maximum and minimum.

Notice in Fig. 2-7 that places as far north as 45° have more insolation than those at the equator during May, June, and July. The reason for this phenomenon is plainly seen in Fig. 2-6. The altitude of the sun and the transmissivity of the atmosphere are about the same for the equator and 45° N latitude on June 22, but the length of daylight is about 10 percent greater at 45°; hence the insolation is also larger.

One further point should be made about solar radiation. Remember that in talking about insolation, we not only have assumed cloudless skies and average turbidity but have also assumed that the sun makes the same angle with the ground at all places in the same latitude. This assumption is valid only if the ground is horizontal. If the sun shines on a slope, the angle it makes with the ground may be much greater or much less than the altitude above a horizontal plane, depending on whether the slope faces south or north. In extreme cases, as in steep, narrow valleys, this factor may be considerable. In French-speaking Switzerland, the valley dwellers call the contrasting slopes *ubac* (for the shady slope) and *adret* (for the sunny slope); similarly, the German language uses the terms *Sonnenseite* and *Schattenseite* to distinguish between them.

In contrast to the annual march of insolation, the average daily march of insolation is more easily described. Insolation begins at sunrise, rises to a maximum at noon, and then ceases at sunset. Figure 2-8 shows the daily march of insolation for clear and for cloudy days in winter and summer. The solid curves for the clear days illustrate the symmetrical march of insolation; the cloudy day in June shows more irregularity, either because cloudiness varied or because the clouds were not uniform in their effect on the incoming radiation.

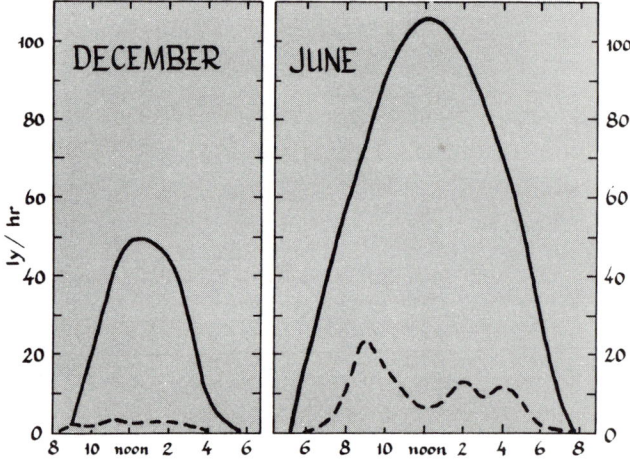

Fig. 2-8. The daily march of   on clear (solid line) and cloudy (dashed line) days near Donner Pass, California.

Distribution of Solar Radiation at the Earth's Surface

The average annual receipt of solar radiation at the surface of the earth is shown in Fig. 2-9 by **isolines** whose units are kilolangleys per year (kilo is the Greek-derived prefix for 1000 used in the metric system; thus a kilolangley or kly is equal to a 1000 ly, just as a kilometer or km is equal to 1000 m). The map shows several outstanding patterns. The controlling effect of solar altitude and of length of day is reflected in the generally latitudinal arrangements of the isolines, with the lowest values of insolation in the highest latitudes. The importance of the atmosphere in determining the amount of solar radiation that reaches the surface is illustrated by the fact that the latitudinal pattern is broken up in two important respects: the highest amounts of insolation are not found near the equator, and oceanic values are lower than those over the land.

The continents receive less insolation near the equator than do places as far poleward as 40° N or 40° S latitude. The Congo Basin, in Africa, gets about 150 kilolangleys per year, which is approximately the amount received in San Francisco; parts of the Amazon Basin receive less than 120 kilolangleys per year, an average exceeded in New York City. This relatively low equatorial insolation results from the cloudy character of the atmosphere. The dry air of the subtropical deserts, on the other hand, allows maximum penetration of solar radiation. The maxima are therefore over the continents, in latitudes of 20° to 30°. The highest value shown on the map is slightly above 220 kilolangleys per year, in the eastern Sahara.

In middle and high latitudes, the effect of the greater cloudiness over the ocean causes the isolines to bend poleward over the land and equatorward over the sea. This pattern indicates higher radiation in any given latitude on land than over the oceans. Some of the lowest values, below 70 kilolangleys per year, are found over the cloudiest parts of the oceans. Parts of the North Pacific, North Atlantic, and Southern Oceans, in latitude 60°, receive as little insolation as the poles.

Aside from these major regional distortions, Fig. 2-9 does not exhibit some of the intricate patterns of the other types of climatic maps of the world; the form and elevation of the land do not affect incoming radiation as significantly as they do other climatic elements; nor is the radiation received at any particular spot affected by the amount received elsewhere, whereas temperature and other climatic elements are so affected. Also, the map is highly generalized because it is based on the records of only a small number of stations. Yet it is more than adequate when it comes to portraying the major effects of latitude and atmospheric blockage on the distribution of incoming solar radiation.

Terrestrial and Atmospheric Radiation

So far, we have discussed only the incoming short-wave radiation and have described only part of the heat balance of the earth and the atmosphere. As we have seen, 35 of 100 units of incoming radiation are reflected or scattered back to space. The other 65 units are used to heat the earth and the atmosphere. Since the earth is not warming up appreciably, it follows that 65 units of heat must in some way be returned to space. The second part of the heat balance deals with the processes whereby this return is accomplished. We refer once more to Fig. 2-4.

The Remaining Components of the Heat Balance

The return of 65 units of heat to space is accomplished as follows:

1. *One hundred twenty units are radiated from the heated earth. The earth is able to radiate 120 units even though it receives only 45 units from the sun because it also receives radiation from the atmosphere (see 2, below). The magnitude of this long-wave radiation — long-wave because of the relatively low terrestrial temperatures — depends on the temperature of the earth, just as the amount of any other radiation depends on the temperature of its radiating body. In fact, we determine the value of the terrestrial radiation from the average temperature we assign to the earth. Since this temperature is about 65° F, we estimate the radiation to be about 120 units.*

2. *Of these 120 long-wave units, only 15 escape through the atmosphere; the other 105 are directly absorbed. The crucial importance of water vapor is illustrated by this inability of long-wave radiation to escape to space. Of the incoming short-wave radiation, 45 percent gets through the atmosphere, but only 12.5 percent (15 out of 120 units) of the outgoing long-wave radiation is lost to space, largely because of the effectiveness of water vapor in absorbing long-wave energy. This differential behavior of the atmosphere toward radiation of different wavelengths is called the* **greenhouse effect,** *by analogy with the panes of glass in a greenhouse, which allow insolation to reach the plants but retain the heat by intercepting the long-wave radiation from inside the structure.*

3. *The atmosphere radiates 50 units of long-wave energy out to space and 105 units down to the surface.*

We have now dealt with all the radiative elements of the heat balance; and if we attempt to total up our heat

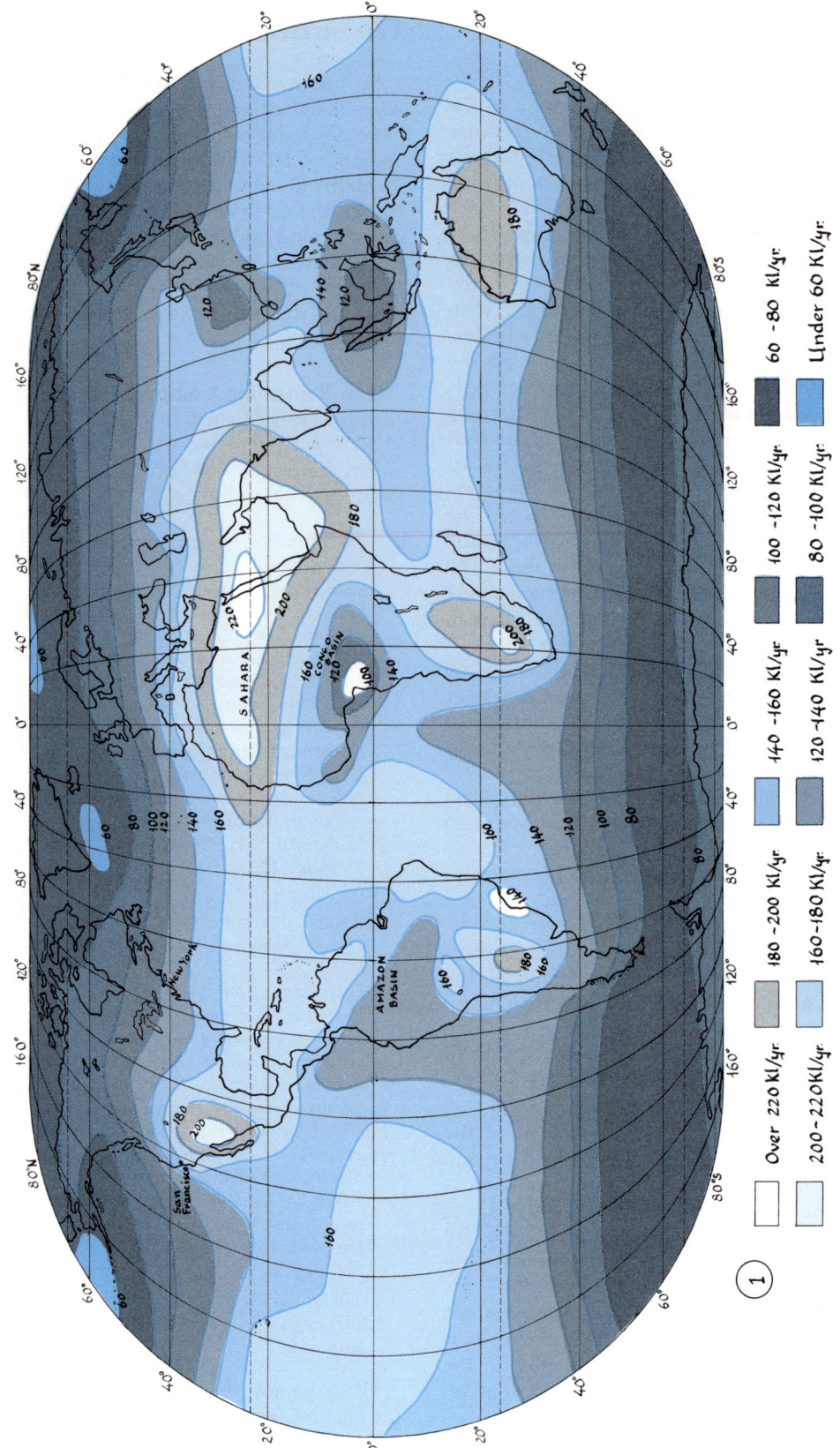

Fig. 2-9. The average annual income of solar radiation at the surface of the earth.

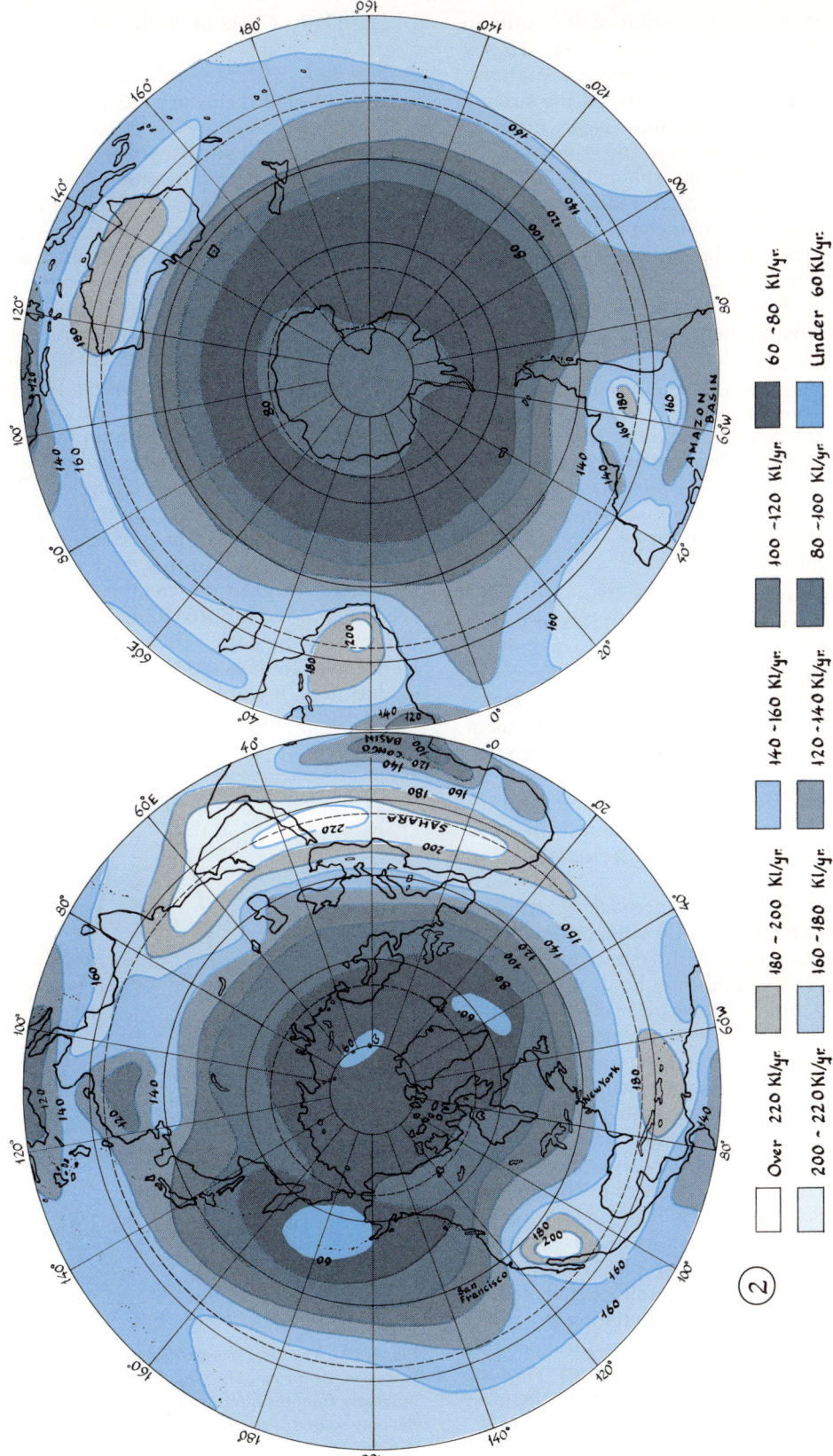

Fig. 2-9. The average annual income of solar radiation at the surface of the earth. The units attached to the isolines are kilolangleys per year. For some comparisons, it is more convenient to consider the average daily radiation, which is usually expressed in langleys per day. To convert the values on the map to langleys per day, multiply them by 1,000/365 or about 2.7.

budget, we find that there is an output of 100 units to space (25 by reflection, 10 by scattering, 15 by terrestrial radiation, and 50 by atmospheric radiation), which balances the income of 100 units. At the surface of the earth, the income of radiation is 45 units of short-wave radiation from the sun and 105 units of long-wave radiation from the atmosphere. Against this total income of 150 units, the earth radiates only 120 units. The difference between income and outgo of radiation at the surface of the earth is called the **net radiation** and amounts to 30 units, on the average. The heat represented by this surplus at the surface is disposed of in the following manner:

1. *Twenty units of heat are used to evaporate water at the surface. When this moisture condenses in the atmosphere, the same amount of heat is given off by the condensation process.*
2. *Ten units of heat are conducted from the ground to the lowest layers of air and are transferred farther up by the turbulence of the air.*

Now all three sectors are in balance. The earth gains 150 units, by various means, and loses 150 units. The atmosphere gains 155 units and disposes of 155 units. One hundred units are returned to space to balance the income of 100 units. But bear in mind that as far as radiation alone is concerned, the earth has a surplus of 30 units, the net radiation; the atmosphere, however, has an average deficit of 30 units, which it makes up by receiving 20 units through condensation and 10 units through turbulent transfer from the ground. In other words, there is a heat balance, but a *radiative imbalance*.

This discussion of the heat balance clearly implies a system through which energy is flowing, and which is assumed to be in equilibrium with respect to all forms of energy transfer. This equilibrium holds not only for the earth as a whole, but for every place on the surface, though the form of the steady state reached varies from place to place, and indeed from season to season at any single place as the inputs of the system change with time and place. Thus every place has its particular heat budget, and every change in income or atmospheric conditions creates a negative feedback loop which changes some or all of the other items of the budget until it is brought back into balance. If income decreases, because of lower angles of the sun and shorter days in winter, the surface at a particular place must either borrow energy from elsewhere, or failing that, reduce its own expenditures (which means radiating at a lower temperature). Fortunately there are few natural positive feedbacks, such as the processes involved in releasing atomic energy; otherwise we would have very unstable climates, indeed.

The Daily Cycle of Radiation

The daily march of solar and atmospheric radiation for the period February 23-25, 1939, at Riverside, California is shown in Fig. 2-10. Just as in Fig. 2-8, insolation begins at sunrise, rises to a maximum at noon, and ceases at sunset. The amount of atmospheric radiation is lowest at sunrise and highest at noon, when the temperature of the atmosphere is highest; but it remains practically constant throughout the day and the night, since the atmosphere exhibits little diurnal temperature variation. The total insolation is the sum of solar and atmospheric radiation; and the curve representing its daily march, is flat during the night, when atmospheric radiation alone provides heat to the surface, and peaked during the day. The total heat received directly from the sun on the particular days in question averaged 350 langleys; heat received from atmospheric radiation was 500 langleys. The supplementary 500 langleys of heat are particularly important at night, when there can be no receipt of energy directly from the sun. Without this moderating effect of the atmosphere, night-time temperatures would be exceedingly low, approaching but not reaching the minima on the dark side of the moon, whose "night" lasts about 28 days. At the Poles which have 6 months of darkness, it would be as cold as in outer space, were it not for atmospheric radiation.

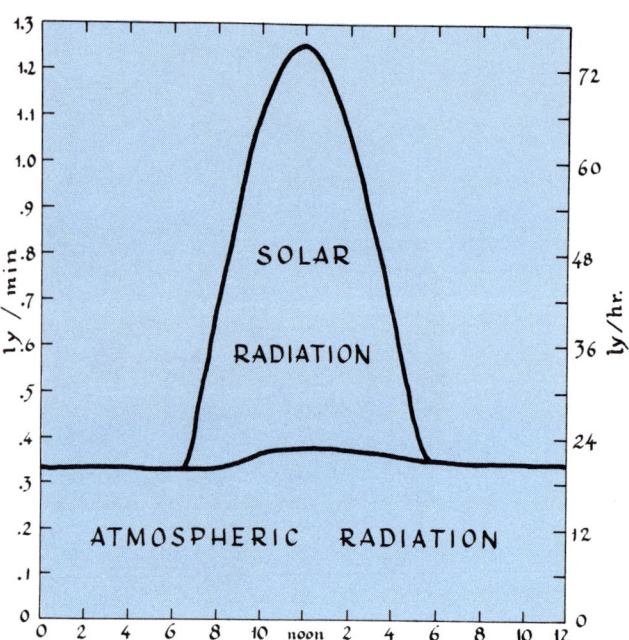

Fig. 2-10. The daily march of solar and atmospheric radiation at Riverside, California. The large share of energy reaching the surface indirectly from the atmosphere is evident. At night it is obviously the only source of energy.

Summary of the Average Heat Balance

This discussion of the heat balance points up several important considerations:

1. *The atmosphere is heated primarily by terrestrial radiation and not by direct insolation. Compare the 20 units received from the sun with the 135 units received from the earth's surface (105 by radiation, 20 by latent heat, 10 by turbulent transfer).*
2. *The earth radiates 120 units even though its solar income is but 45 units, thanks to the greenhouse effect, which accounts for 105 units of atmospheric radiation to supplement the 45 received from sun and sky. This atmospheric radiation is particularly important at night, when it becomes the major source of heat at the surface of the earth.*
3. *The heat used in evaporating water is an important component of the heat balance since it transfers 20 units of energy from the surface to the atmosphere.*

The Heat Balance Over the World

All the elements of the heat balance can be summarized into: (1) the net radiation, (2) the heat used in evaporation or given off in condensation, and (3) the heat transferred by conduction and turbulence. We have seen that, on the average, the net radiation at the surface comes to about 30 units and that evaporation and turbulent transfer take up 20 and 10 units respectively, so that there is a balance of heat. For individual places, however, the magnitudes of these three items may be very different; in fact, in high latitudes in the winter, there is normally a negative value of net radiation (that is, there is more radiation outgo than income). To counterbalance this deficit, the gradient of turbulent transfer of heat is also reversed, so that it is the atmosphere that gives up some of its heat to the surface, and not the other way around.

In addition, the three items may not balance at all during a particular period, although they tend to balance for the year as a whole. The reason for this seasonal imbalance is that some of the surplus radiation of summer can be disposed of in the form of heat storage in the soil. By the same token, the stored heat is released in winter to make up for the deficit in net radiation.

These aspects of the heat balance are shown in the examples presented in Fig. 2-11. They also show the geographical variation in the balance. The first example shows the diurnal march of the heat-balance components at Leningrad (60° N latitude) in July. Notice that more heat is used for evaporation ($S \rightarrow E$) than is transferred to the atmosphere by turbulence ($S \rightarrow A$). This differential is characteristic of moist climates everywhere. Notice also that since the observations are for July, there is a surplus of net radiation, except for very small amounts during the night; and there is, therefore, considerable transfer of heat to the soil from 6 A.M. to 4 P.M. ($S \rightarrow R$). In the second example, a desert station, where no evaporation takes place, the curve of turbulent transfer ($S \rightarrow A$) shows much higher values, as is indicated in the diagram for Pakhta-Aral, in Russian Turkestan.

The other four examples in Fig. 2-11 show the annual march of heat balance for a few selected places and also include the temperature curves for reference. Usually, temperature is measured about 5 feet above the surface; therefore, such measurements relate to the lowest layer of the atmosphere and not to the surface itself. For this reason, the temperature curve should closely approximate the curve of turbulent transfer, since the lowest part of the atmosphere is warmed primarily by turbulent transfer.

Manaos, on the Amazon (3° S latitude), shows an equatorial double maximum and double minimum of net radiation. Because of high temperature and much rain at all seasons, **latent heat** of evaporation is almost equal to the net radiation throughout the year; the amount of turbulent transfer ($S \rightarrow A$) is therefore relatively small, rising to a slight maximum in October, when the evaporation slackens during a period of relative dryness. Because of the very small annual variation of insolation, heat transfer to and from the soil is negligible in equatorial regions and is not shown on the graph. Note the fair accordance between temperature march and turbulent transfer.

Aswan, in Egypt (24° N latitude), represents desert conditions in the subtropical latitudes. Evaporation is negligible at all times of the year; heat transfer to and from the soil is slight. The net radiation is almost all converted into turbulent transfer, hence the high values of temperature. Much of the heat of desert regions can be attributed to the lack of evaporation. Aswan is sufficiently far from the equator that it does not have two maxima of net radiation, and the single maximum of radiation is reproduced in the single high maximum of temperature.

Lisbon, Portugal (39° N latitude), represents Mediterranean conditions. Its annual range of net radiation is much higher than those of the two low-latitude stations and about equal to that at Turukhansk. There is very little transfer of heat to and from the soil. The major characteristic of the heat balance is the high value of the heat used for evaporation in the wet spring and early summer. Evaporation is much reduced in late summer and fall, after the ground

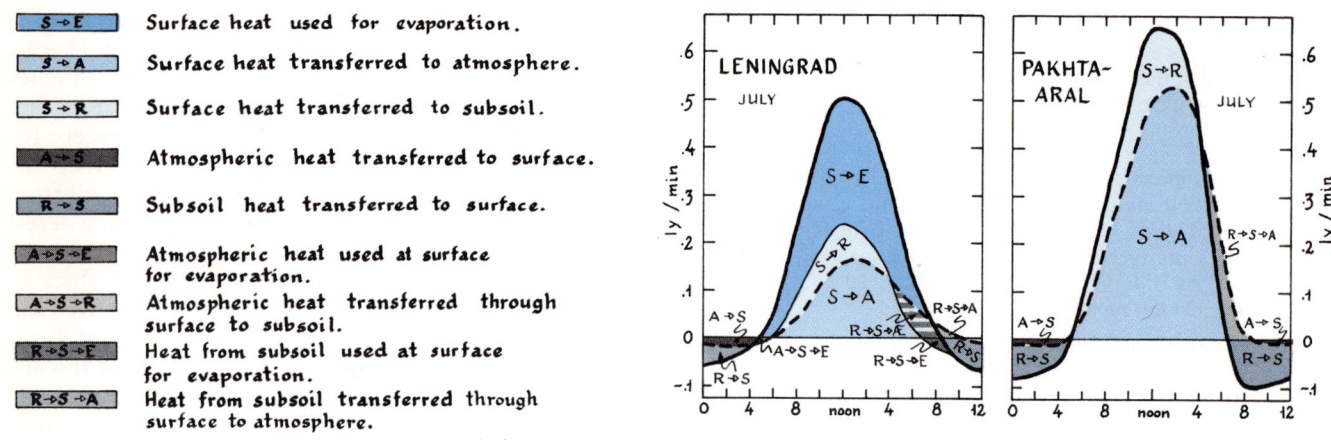
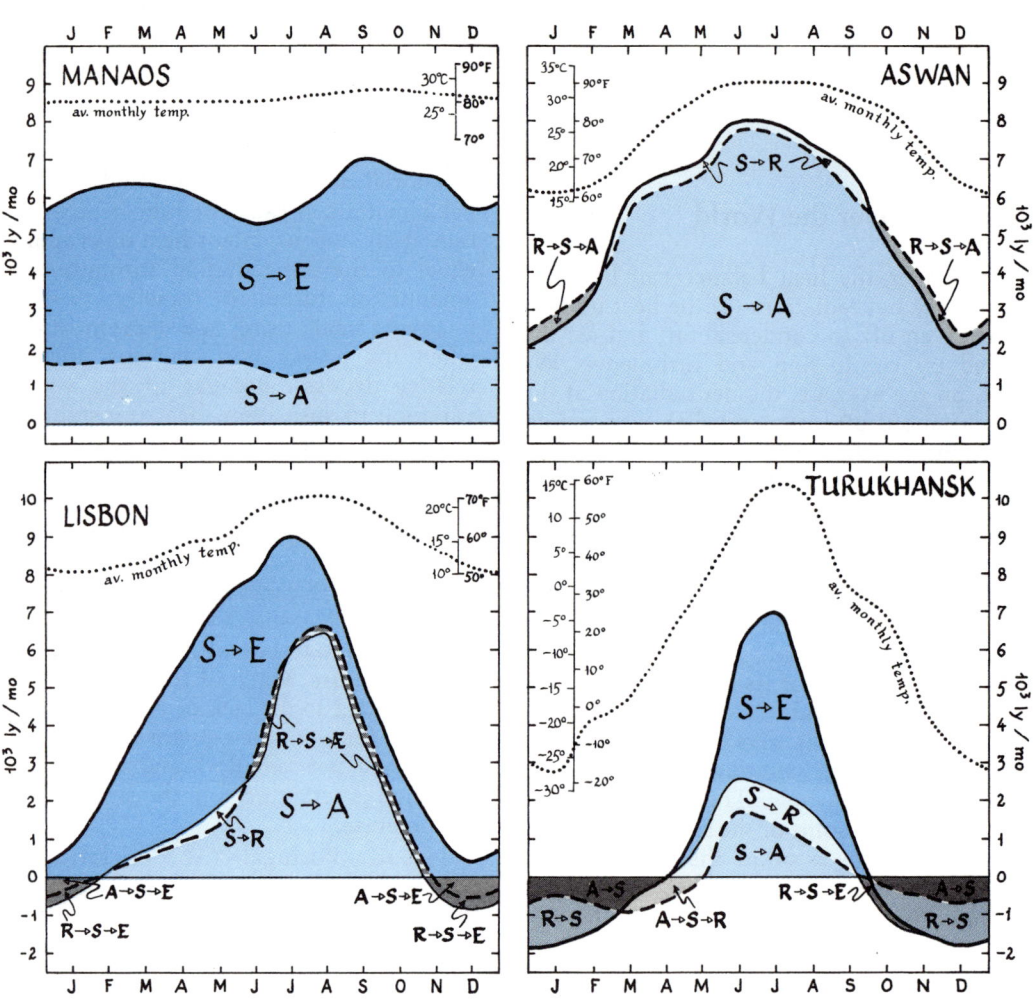

Fig. 2-11. The daily march of the components of the heat balance at Leningrad (60°N latitude) and Pakhta-Aral (40°N latitude), and the annual march of the components of the heat balance at Manaos (3°S latitude), Aswan (24°N latitude), Lisbon (39°N latitude), and Turukhansk (66°N latitude).

dries out; consequently, turbulent transfer ($S{\to}A$) rises to a pronounced maximum in August, as does the average monthly temperature.

Turukhansk, on the Yenisey River in Siberia (66° N latitude), represents a continental station in high latitudes. The net radiation falls considerably below zero in winter but has a summer maximum slightly greater than that for Manaos and almost as large as that of Aswan. The annual march of the heat used for evaporation is identical to that of the net radiation. The rain comes in summer, as does the heat; hence, the evaporation is highest in July. There is considerable transfer of heat into the soil in summer ($S{\to}R$) and an equal withdrawal from the soil in winter ($R{\to}S$), reflecting, of course, the surplus radiation in summer and the deficit in winter. The turbulent transfer also shows a deficit in winter, indicating a downward transfer of heat from the atmosphere to the ground ($A{\to}S$ and $A{\to}S{\to}R$). In other words, heat is being imported from elsewhere by way of the atmosphere. In summer the reverse is true. The temperature curve does not accord very closely with the curve of turbulent transfer because of the disturbing effect of import and export of heat by air masses.

These examples do not give a complete picture of the world but are presented merely to show possible variations in the heat balance from place to place. They indicate the importance of considering the amount of heat used in evaporation and the amount of transfer to and from the soil in trying to estimate how much heat is available to warm the lower part of the atmosphere. Any discussion of the temperature regime of a specific place must be prefaced by at least some general ideas on the annual marches of the components of the heat balance.

Summary

In this chapter on the heat balance of the earth and atmosphere, words such as "budget," "income," "outgo," and "balance" were used with about the same meanings they would have if we were talking about a family budget, with its income and expenditures.

First we talked about the primary income — the short-wave radiation from the sun, excluding all the deductions (reflections and scattering). Then we examined the expenditures — long-wave radiation from the earth and from the atmosphere, including the internal payments from earth to sky and back.

We found a balanced budget as far as radiation was concerned, but only if we considered earth and atmosphere as one family unit. Between the two, there is normally a surplus at the surface and a deficit in the atmosphere. This imbalance is taken care of by payments of nonradiative character — latent heat of evaporation and condensation, conduction and turbulent flow — from the earth to the sky.

Turbulent flow was stressed to aid comprehension of the temperature relations to be discussed in the next chapter, for it is the temperature of the lower atmosphere that is normally meant when we use the term "temperature" without qualification. And this lowest layer is heated by conduction and turbulent transfer from the surface.

These notions can also be thought of in terms of an open system with variable inputs of energy, changing resistances and negative feedback of various sorts. The resulting states of equilibrium represent the heat balance under conditions of insolation, albedo, and atmospheric turbidity appropriate to a particular place at a particular time of the year.

Suggestions for Further Reading

Budyko, M. I., *The Heat Balance of the Earth's Surface*, trans. N. A. Stepanova, Office of Technical Services, U.S. Dept. of Commerce, Washington, D.C., 1958.

Byers, Horace R., *General Meteorology*, 3rd ed., McGraw-Hill Book Company, Inc., New York, 1959, pp. 11-44.

Riehl, Herbert, *Introduction to the Atmosphere*, McGraw-Hill Book Company, Inc., New York, 1965, pp. 31-48.

Seller, William D., *Physical Climatology*, The University of Chicago Press, Chicago, 1965, pp. 11-81, 100-26.

3 Temperature

Introduction

Definitions

Temperature (Latin *temperatura* due measure, temper, temperament) is such a commonly experienced phenomenon that it would seem unnecessary to define it beyond the layman's intuitive notion. But even the dictionary definition — degree of hotness or coldness — is not very enlightening for it introduces the concept of "heat" without calling attention to the very important difference between heat and temperature.

Heat refers to the amount of energy in a system, and **temperature** to the availability of such energy. If two bodies have different temperatures, there will be a flow of heat from one to the other until the temperatures are equalized. A given flow of heat, however, will not increase the temperature of all substances by the same amount. The number of calories of heat required to raise the temperature of 1 gram of a substance by 1° C, which differs widely among common substances, is called the **specific heat** of that substance. The specific heat of water is 1, since, as we mentioned earlier, a calorie is defined as the heat required to increase the temperature of 1 gram of water by 1° C. Almost all other common substances have a specific heat lower than 1 (that is, as compared with water, each needs fewer calories to raise its temperature by a given amount). A few examples are given in Table 3-1. Among the common elements, only hydrogen has a specific heat higher than that of water. The climatic significance of the high heat capacity of water lies in the ability of water to store and transfer large amounts of energy without excessive changes in temperature.

Table 3-1. Specific Heat of Selected Substances

Substance	Specific heat
Water	1.00
Ice	0.50
Wood	0.42
Air	0.24
Basalt	0.20
Land surfaces (average)	0.20
Granite	0.19
Iron	0.11

Instruments

The thermometer (Greek *therme* heat + *metron* measure) dating from Galileo's construction of a simplified model in 1593, measures the effect of temperature on the volume of liquids or metals. The most common thermometers utilize mercury or alcohol. While various specialized thermometers are used to measure the effect of temperature on electrical resistance or on the amount of distortion of two metal strips with different coefficients of expansion, we will be concerned primarily with the familiar liquid-in-glass thermometers which are used to measure climatological temperatures. Unless otherwise specified, air temperature is measured by a thermometer in a ventilated, shady place, usually 5 feet (about 1.5 m) above the ground. The readings do not, therefore, refer to ground temperatures but to the temperature of the lower part of the troposphere.

Units of Measurement

There are several units of measurement for calibration of thermometers. The oldest system is that of G.D. Fahrenheit, invented in 1714. The temperature of melting snow and salt, the lowest temperature that he could achieve in his laboratory, was given a value of 0; normal body temperature, as Fahrenheit estimated it, was given a value of 100; as every other temperature was fixed on this scale, the temperature of boiling water thus had a value of 212. In the centigrade scale (invented by the Swedish astronomer Celsius in 1742), the values of 0 and 100 were assigned to the freezing and boiling temperatures of water. The Celsius scale is part of the metric system of measurement; but it has no inherent advantages over the Fahrenheit scale other than its more common use by scientists and by most of the world. The correspondence between the two scales is shown in Fig. 3-1.

To convert a temperature from centigrade to Fahrenheit, we multiply by 9/5, then add 32°. To convert a temperature from Fahrenheit to centigrade, we first subtract 32°, then multiply by 5/9. Note, however, that this method of conversion applies only to a single temperature reading; that is, to a point on the thermometric scale. For temperature intervals, such as temperature ranges and the like, it is only necessary to multiply by 9/5 to get Fahrenheit from centigrade; and 5/9 for the reverse. For example, to convert a range of 20° C (say between 10° C and 30° C), we could convert each end of the range to its Fahrenheit value. Using the formula, we get 10° x 9/5 =18°, 18° + 32° = 50°F; and also 30° x 9/5 = 54°, 54°+32°= 86° F. Subtracting 50° from 86° F gives us a range of 36° F as equivalent to a range of 20° C. But we can get the same equivalence directly by multiplying 20° by 9/5, which also equals 36°. The reason we need not add 32° is that 32° must be added to each temperature, at both ends of the range, so that the net effect on the range is nil.

A thermometric scale that is commonly used in physics, chemistry, and meteorology moves the centigrade scale downward by about 273°, establishing its zero at the lowest possible temperature (absolute zero). To convert from this scale, called *absolute* or *Kelvin* (after the 19th-century physicist William Thomson Kelvin) to centigrade simply subtract 273° from the absolute temperature. Absolute zero thus corresponds to —273° C and —460° F (Fig. 3-1).[1]

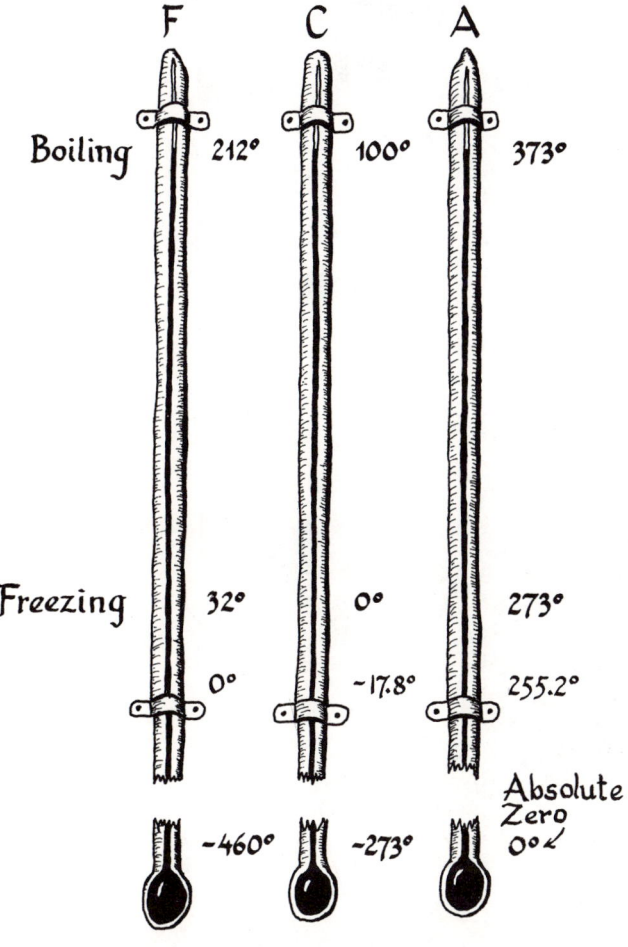

Fig. 3-1. The Fahrenheit, Celsius, and Kelvin temperature scales. The zero point on each scale (freezing point of salted water, freezing point of pure water, and absolute zero) is shown along with its equivalents on the other two scales. Also shown are the three temperatures corresponding to the boiling point of water on each scale.

Surface Temperature

The term "surface temperature" does not actually refer to the temperature of the earth's surface but rather to the temperature of the air approximately 5 feet above the surface. We measure and discuss the

---
[1]*Actually —273.2° C and —459.7° F, to the closest tenth of a degree.*

24 Chapter 3

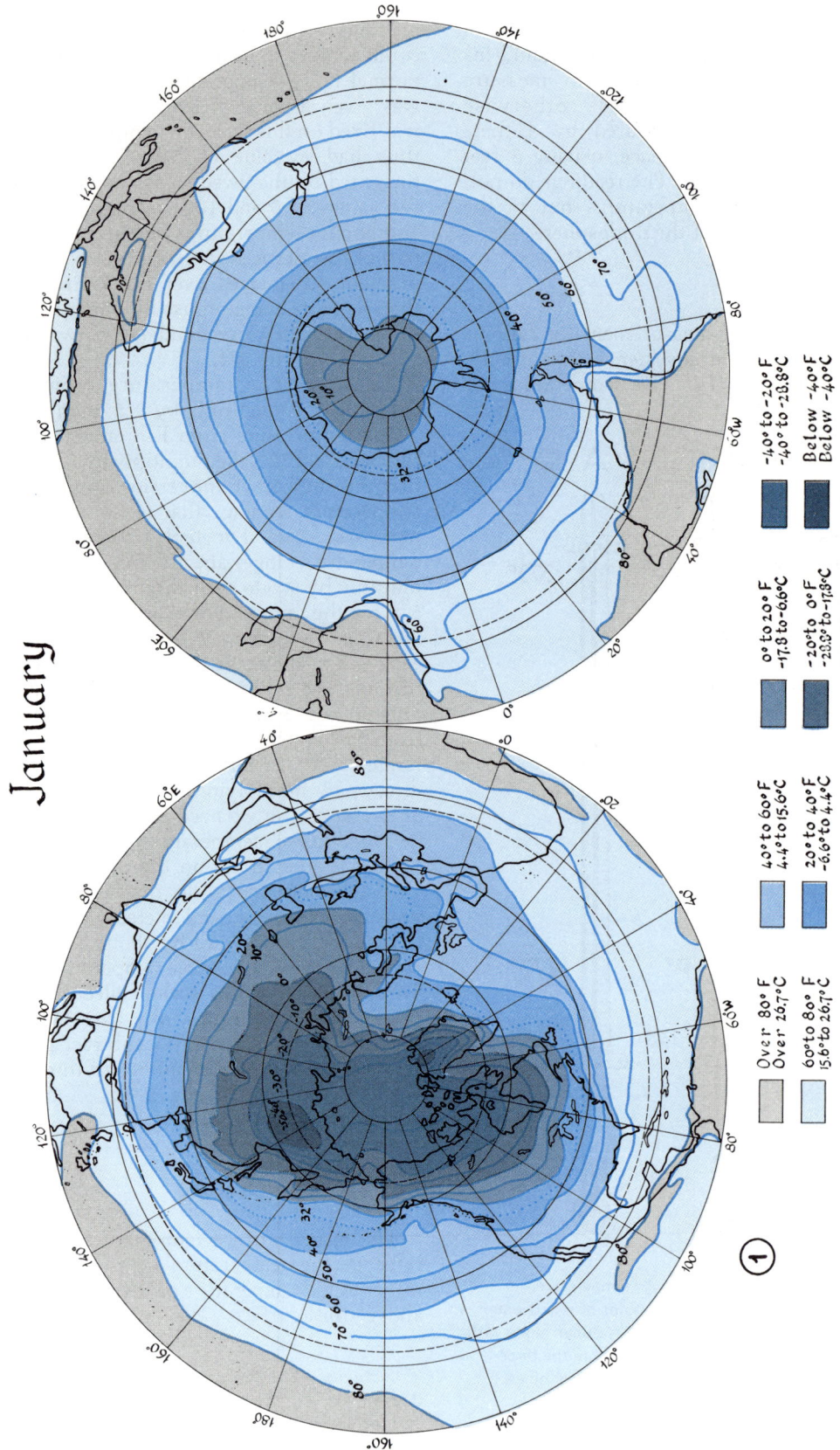

Fig. 3-2. January (1) isotherms (°F).

Temperature 25

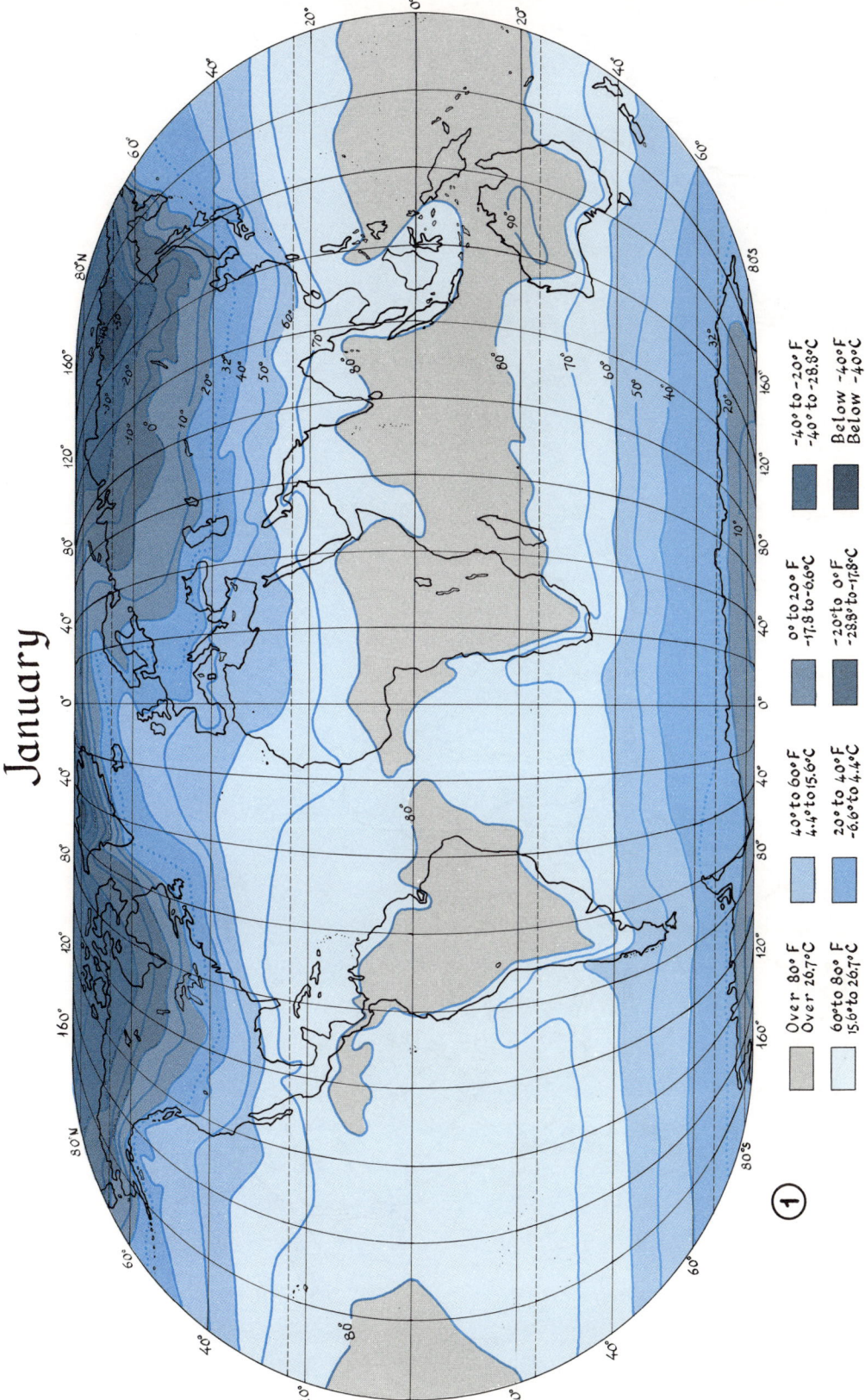

Fig. 3-2. January (1) pp. 24 and 25, and July (2) pp. 26 and 27, isotherms (°F). Temperatures have been reduced to equivalent sea-level values, and are shown at intervals of 10°F. The isotherm of 32°F (0°C) is also shown and is represented by a dotted rather than a solid line.

26 Chapter 3

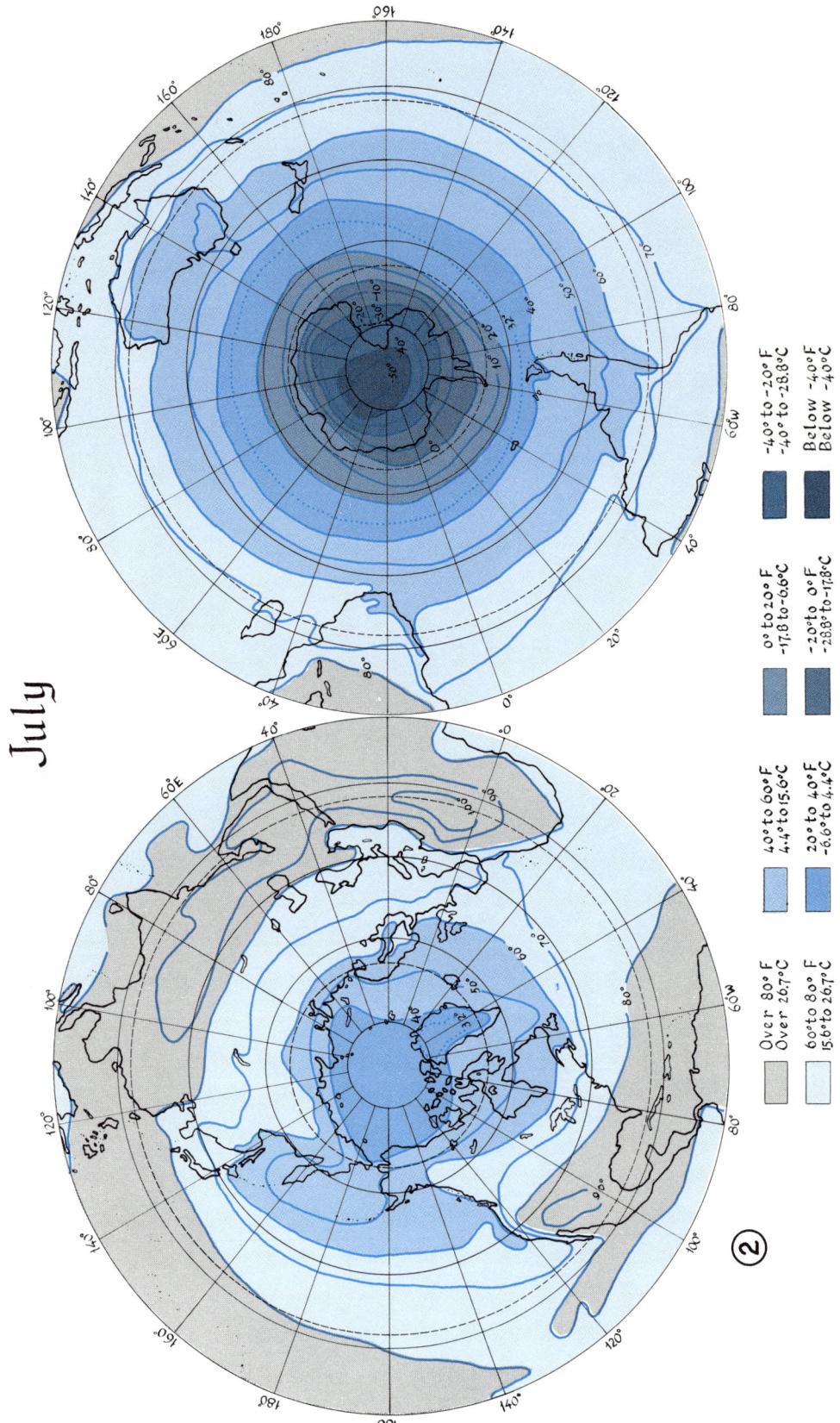

Fig. 3–2. July (2) isotherms (°F).

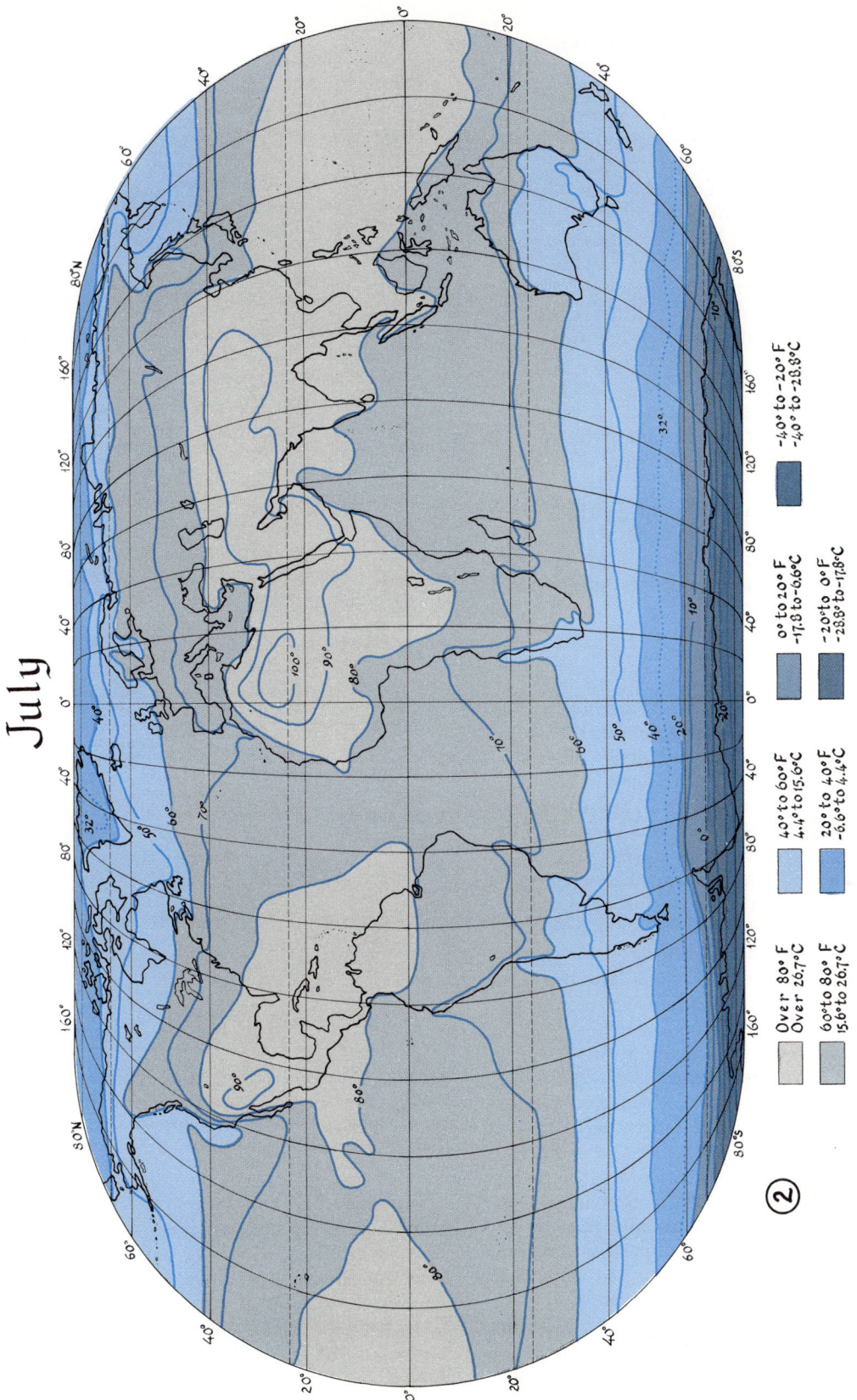

Fig. 3-2. July (2) isotherms (°F).

temperature at this distance from the earth simply as a matter of convenience. The temperature pattern of the surface itself is exceedingly complicated by differences created by various kinds of surfaces. A grass plot has measurably different temperatures from the concrete driveway next to it, but measurements taken 5 feet above the ground eliminate much of this complication. With this measurement convention, general patterns of temperature are more evident; but obviously, many important and interesting variations remain hidden.

What is the general pattern, and how does it change with the seasons? The maps of temperature for summer and winter (Fig. 3-2) illustrate the major features of the world distribution by means of **isotherms** (Greek *isos* equal + *therme* heat), or lines connecting points of equal temperature, for the months of January and July. The factors that affect the world distribution of surface temperature are best understood by keeping these maps in mind and in view.

The six principal factors affecting the distribution of surface temperature are (1) insolation, (2) elevation above sea level, (3) differential behavior of land and sea, (4) movement of air masses, (5) ocean currents, and (6) cloudiness.

Insolation

The most important factor affecting temperature is insolation, or the amount of solar radiation received at the surface. Insolation is almost directly related to latitude. Only differences in cloudiness and in turbidity from place to place (in the same latitude) disturb the exact relationship between insolation and latitude. This similarity is clearly demonstrated by the correspondence of the isotherms, in a general way, with the parallels of latitude. The seasonal shift of insolation from one latitude to another is also reflected in the corresponding shift of isotherms. The curves depicting the annual march of temperature in various latitudes (Fig. 3-3) clearly show the relationship between the annual marches of insolation and of temperature. Insolation is shown by bar graphs because the values are those of discontinuous totals (averaged over two-week periods); temperature is shown by a smooth curve because the values are those of continuous mean temperatures.

For Fairbanks, Alaska, near the Arctic Circle, the bar graph of insolation shows the great contrast between nearly zero values during the period of extremely short days and very low sun, and fairly sizable values of insolation during the long days of summer. As a result, the range of insolation is large during the year and has a very rapid rate of change in spring and fall. The temperature curve, though it lags — both its maximum and minimum come a little later than the respective solstices — has the same general shape as the bar graph.

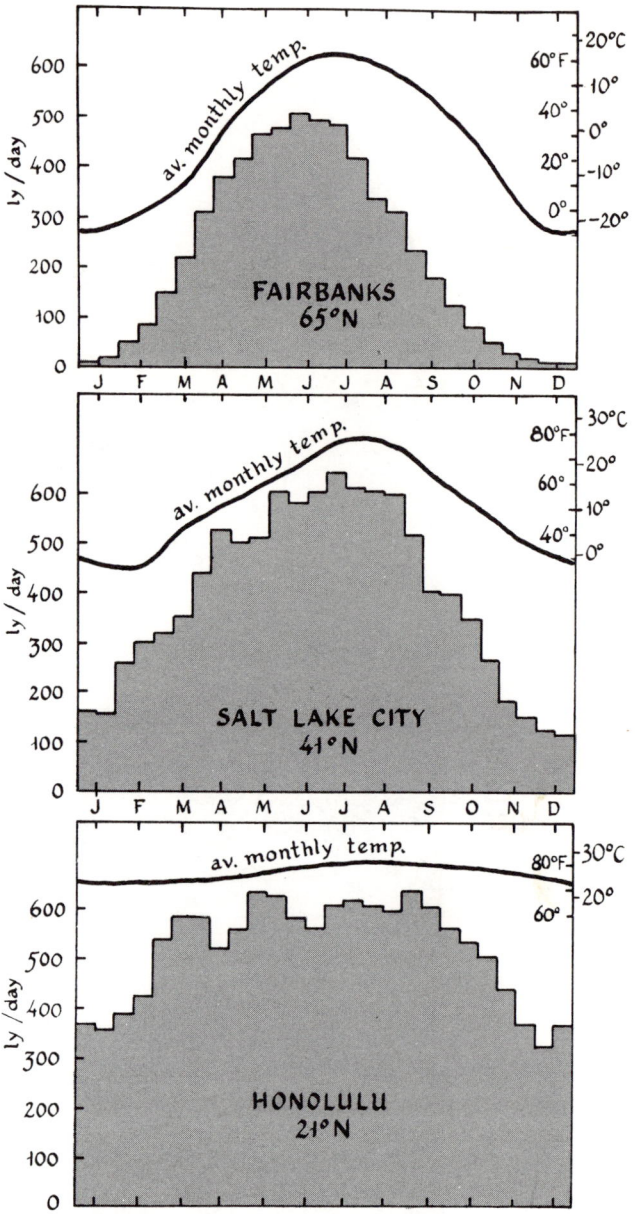

Fig. 3-3. The annual march of temperature and insolation at Fairbanks, at Salt Lake City, and at Honolulu. Note the decreasing range of insolation from Fairbanks to Honolulu, and the corresponding flattening of the temperature curve.

For Salt Lake City, Utah, in the mid-latitudes, the bar graph indicates almost as large a range as for Fairbanks. This range of insolation is due to the cloudy winters and clear summers of Utah. The year-round higher insolation is reflected in the greater temperatures. Once again, there is a lag, more pronounced than for Fairbanks, of the maximum and minimum temperatures behind the summer and winter solstices, respectively.

For Honolulu, Hawaii, a little south of the Tropic of Cancer, the bar graph of insolation reflects the small range of insolation in this latitude. Any tendency toward a double maximum is masked by the irregularity of the biweekly averages. There seem to be cloudy periods in April and in June. The temperature curve shows very little amplitude; and the delay of the extremes of the annual temperature march with respect to the solstices is quite marked.

The three graphs show the general correspondence between the annual march of insolation and that of temperature. With decreasing latitude, the winter insolation increases rapidly; the summer insolation, not nearly so much. In the same way, winter temperatures rise from −10° F (−23° C) at Fairbanks to 72° F (22° C) at Honolulu — a difference of 82° F (45° C). However, in summer, the difference in temperature between the two places is only 18° F (10° C). In other words, from high to low latitudes, the temperature curves take on a flatter appearance, reflecting the smaller annual range.

Elevation Above Sea Level

Elevation above sea level is so important in determining the temperature of a place that a map of actual temperature in mountainous regions would reflect elevation so closely as to resemble a contour map and thus obscure any general pattern of temperature distribution. For this reason, most world maps, including those in this text, do not show actual temperatures but rather temperatures reduced to sea level; for every 1,000 feet of elevation, about 3.6° F are added to the actual temperatures (about 0.65° C for every 100 m). The effect of elevation on temperature will be discussed in the section of this chapter dealing with the vertical distribution of temperature. For the moment, it is sufficient to know that the reduction of temperatures to sea level on the world maps nullifies the effect of elevation on the pattern of temperature.

Differential Behavior of Earth Surfaces with Respect to Incoming Radiation

As we have already seen in the discussion of the heat balance for a number of land stations (p.20), there are notable differences in the disposition of the net radiation from one place to another. These are differences in the amount of heat transfer from and to the soil, and differences in the amount and the annual march of heat used in evaporation. The net radiation itself is affected by differences in **albedo**. Great as these differences between various land surfaces are, they are not nearly as significant as the difference between the two basic kinds of surfaces that absorb radiation — land surfaces on the one hand and water surfaces on the other. Let us examine these differences between continent and ocean without, however, overlooking the variation within each category.

Differences in Albedo. One distinction between land and sea is the amount of reflection. The estimated average albedo used in the heat-balance diagram is 35 percent. But fresh snow may have an albedo as high as 90 percent, whereas the crown of a pine forest can have an albedo of only 1 percent. Similarly, over the oceans, the range of albedos is marked. A water surface that is roughened by waves may have an albedo as low as 4 percent even when the sun is only 47° above the horizon. On the other hand, smooth water and low sun result in albedos comparable to those over fresh snow. These figures, quoted to give some idea of the great range in albedo over both land and sea, refer only to the reflection from the surface itself. Since there are also differences in cloudiness associated with the different climates of continent and ocean, the pattern is further complicated. The effect of albedo on temperatures is, therefore, difficult to assess. The large albedos over snow certainly reduce the amount of energy available for warming the atmosphere above the snow pack, but whether the land as a whole has a significantly different albedo from that of the sea is a moot point.

Differences in Evaporation. Although evaporation is sometimes limited by insufficient energy, the most common limiting factor is the amount of moisture available at the surface. It is obvious, therefore, that there is more evaporation from ocean surfaces than from continents. As we have seen, use of energy in evaporation reduces the amount of heat available for turbulent transfer from the surface to the atmosphere; hence the air temperatures over oceans are lowered, particularly in winter when evaporation is greater than in summer. Higher evaporation results from the fact that the temperature of the sea is higher than that of the air above it, whereas in summer the air is warmer than the water.

Differences in Transparency. One obvious and striking difference between land and sea with respect to the disposition of the incoming energy lies in the fact that land surfaces are practically opaque to solar radiation, so that all of the incoming heat is concentrated in the top fraction of an inch of the ground. In the sea, one-third of the radiation penetrates 10 feet (3 m) below the surface of the water; one-tenth of the radiation penetrates 30 feet (9 m). Obviously, there will be a smaller rise of temperature in water per unit input of heat, since the heat is distributed through a larger volume of water than of land.

Differences in Specific Heat. The specific heat of water is five times as large as that of land surfaces (1.0 compared with 0.2). This difference means that the same amount of heat gain applied to the same mass of water and of land results in a temperature increase that is approximately five times greater for the land than for the water. If the heat is withdrawn from land or water, the temperature decrease is five times greater in the land than in the water. Once again, there is a smaller change in temperature in the water body per unit change in heat than in the land mass.

Differences in Heat Transfer. Transferability of heat by turbulent **convection** in water is by far the greatest cause of differential behavior between land and sea. In the soil, the only means of transferring heat downward is by radiation or by the slow process of molecule-to-molecule **conduction**. As a result, diurnal and annual temperature changes are very large, because they are confined to a very shallow layer of ground. For instance, the diurnal cycle of temperature is hardly felt 18 inches (about 45 cm) below the surface. In the sea, on the other hand, heat is transferred downward not only by radiation and conduction, but also by the mixing of water from various depths. As a result, the diurnal and annual temperature ranges are small, because the temperature changes are spread over a large volume of water. In the water, the diurnal temperature changes are felt to a depth of about 100 feet (30 m) as compared to a few inches on land. Parenthetically, we might point out the superior position of the atmosphere to *both* land and sea in this regard. The diurnal changes are felt as high as 5,000 feet (1.5 km) above the ground, and annual temperature changes are felt to the top of the **troposphere**. On the other hand, the specific heat of air is much less than that of water.

One should distinguish between summer and winter convection of water in the oceans. In summer, the upper layers of water are made lighter because of warming. Mixing takes place only if some outside force is present. Winds, tides, currents, and differences in salinity cause some mixing, so that high temperatures are found fairly uniformly distributed throughout a layer whose depth depends on the strength of the mixing agents (Fig. 3-4). Below this isothermal layer lies another layer of uniform temperatures extending to some depth. Between the two isothermal layers, there is a very rapid drop in temperature with increasing depth, the **thermocline** (Greek *therme* heat + *klima* slope or gradient). The thermocline is very evident to swimmers in shallow bodies of water, where it lies close to the surface.

In winter the thermocline is absent. The cooled surface layers, being denser than those below, sink to take their place; warmer water rises to the surface, and it in

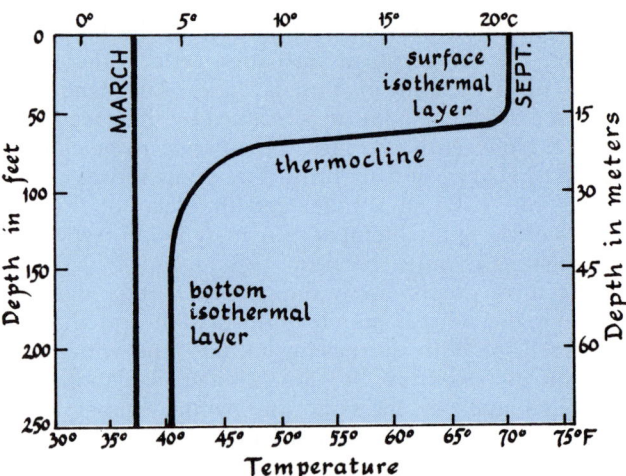

Fig. 3-4. The variation of temperature with depth in Lake Michigan (after Phil E. Church). The straight line on the left shows the isothermal condition in March (the temperature is the same at all depths). In September, there are two isothermal layers separated by a rapid decrease in temperature within a shallow layer—the thermocline.

turn is cooled. The result is a deep isothermal layer and, of course, temperatures that remain relatively high because of this large reservoir of heat. Ice can form eventually, but only after the whole body of water has been cooled nearly to the freezing point. At about 39°F (4°C) water attains its maximum density. Further cooling creates a lighter layer that stays on top of the water body and freezes if the temperature dips below 32° (0°C) in fresh water and slightly lower values in salt water (depending on its saltiness). This is why water freezes from the top down rather than from the bottom up. Otherwise oceans would freeze from the bottom, summer warmth would reach the ice with difficulty and the oceans would gradually become masses of ice.

Effect of Land and Sea Surfaces on Temperature. The effect of differences in specific heat, transparency and turbulent transfer, and, to a lesser degree, albedo and evaporation makes the water more conservative than the land with respect to temperature. The summers are cooler than over continents in the same latitude; the winters are warmer. This conservatism also applies to the diurnal march of temperature. From the climatic point of view, the effect of the oceans is that of a gigantic cooling system in summer and a heating mechanism in winter. The effect on temperatures is so profound that we distinguish maritime from continental climates as two basic types.

The annual march of temperature at St. Louis, Missouri (38° N latitude), and those for portions of the Atlantic and Pacific Oceans at about the same latitude

(Fig. 3-5) clearly reveal the important differences in temperature between ocean and land. The amplitude of the annual march (the annual range of temperature) is fully three times as great at St. Louis as at either of the oceanic stations. In addition, other characteristics of the temperature march distinguish the curves. The maximum and the minimum of temperature come much later over the oceans. This delay of the extremes is another facet of the conservative nature of temperature relations over water. The more heat stored, the greater the delay of the seasonal maximum. In a very general way, continental stations experience the extremes of their annual march of temperature about one month, or slightly less, after the extremes of insolation (normally the solstices); oceanic stations, three months after the solstices; and coastal stations, at some intermediate date.

The effect of land and sea surfaces is broadly shown on the world isothermal map. The conservative nature of sea surfaces with respect to temperature means that the oceans are warmer in winter than the continents in the same latitude, whereas the continents are relatively warmer in summer. Hence, the isotherms bend poleward over the oceans in winter and over the continents in summer; they bend equatorward over the oceans in summer and over continents in winter. As shown in Fig. 3-2, the major deviation of the isotherms from the parallels of latitude is caused by the poleward and equatorward displacement due to land and sea differences.

Movement of Air Masses

Climates are not necessarily manufactured at the place where they are experienced. If they were, the solar insolation and the kind of surface would be the most important considerations. In fact, the term *"solar climate"* is used to designate a climate that is largely the result of local influences. At most places however, much of the climate is imported and thus reflects conditions at some distance. A station that is continental with respect to its physical location may have a maritime climate if the air masses over it come largely from the ocean. By the same token, a coastal station so located that air masses come only from the interior of its continent will have a continental climate, despite its position close to the ocean.

Temperatures, therefore, are affected by air-mass movement. The most important consistent movement of air at the surface on our planet is the eastward (west to east) flow of air in the mid-latitudes, from 30° to 60° in both hemispheres. To be sure, this movement is occasionally reversed, and local exceptions abound, but the generalization is true. Bear in mind that winds flowing from west to east are called *west* winds.

It follows from the above that places in the mid-latitudes import their climate from the west. The western parts of the continents are overrun by maritime air masses; the eastern parts by air masses that have acquired their characteristics over the interiors of the continents. The west coasts, being maritime, have warmer winters and cooler summers than the continental east coasts; and, of course, they have a smaller temperature range than the east coasts. This effect is shown by the temperature graphs for San Francisco and for Washington, D.C. (Fig. 3-6). St. Louis also serves as a standard of comparison. Note how close Washington's temperatures are to those of St. Louis and how dissimilar to those of San Francisco. West winds bring oceanic temperatures to San Francisco, but the same winds carry continental conditions to Washington.

Air masses do not, however, penetrate all continents with equal ease. The western **cordilleras** of the Americas (Coast Ranges, Cascades, Sierra Nevada, Rocky Mountains, Sierra Madre Occidental, Andes) are oriented along a north-south axis, which is more or less at right angles to the most common flow of the air masses. As a result, the importation of maritime temperature regimes is limited to a narrow coastal strip, and continental influences are felt on the eastern sides of the mountains, even though they are very close to the Pacific Ocean.

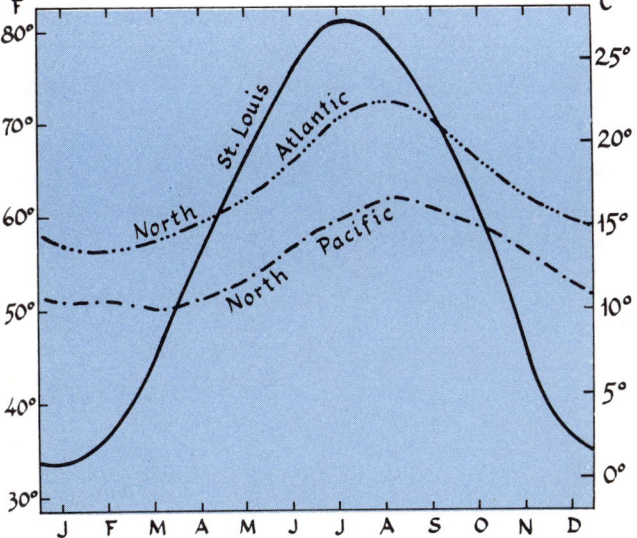

Fig. 3–5. The annual march of temperature at St. Louis and over the North Pacific and North Atlantic Oceans. The curve depicting North Atlantic temperatures has the same shape as the one for the North Pacific, even though it is warmer at all times of the year. The curve for St. Louis has an entirely different shape; it has a much greater amplitude, and its maximum and minimum both come earlier in the year than over the oceans.

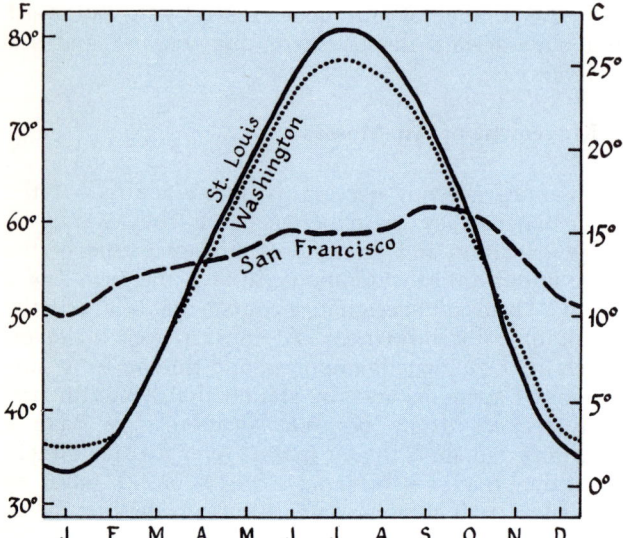

Fig. 3-6. The annual march of temperature at St. Louis, at San Francisco, and at Washington, D.C. Washington is very similar to St. Louis despite its nearness to the Atlantic Ocean, but San Francisco is clearly oceanic since it is dominated by North Pacific air masses.

Note the sizable increase of continentality from San Francisco to Modesto, cities which are less than 80 miles (129 km) apart, but separated by the Central Coast Ranges of California (Fig. 3-7). Summers are much warmer and winters colder at Modesto. In addition, the maximum annual temperature comes earlier than at San Francisco.

Along the west coasts of mid-latitude in the Old World, the situation is different. In Europe (Africa hardly extends into the mid-latitudes), the major mountain ranges are arranged in an east-west direction; consequently, air masses can penetrate into the continent relatively easily. Marine influences, therefore, are felt for some distance inland.

This difference, created by the dissimilar arrangement of mountain masses in the Old and New Worlds, is illustrated by the three sets of annual marches of temperature shown in Fig. 3-8. All the stations shown are within 40 miles (64 km) of the 48th parallel of latitude, except Guernsey, which is 100 miles (161 km) north. In each pair, the stations are at comparable altitudes above the sea and are about the same distance from the ocean. Their locations with respect to mountain ranges are shown in the accompanying sketch. There is slight difference between Tatoosh Island, just off the northwestern tip of the state of Washington, and Guernsey Island in the English Channel, just off the French coast. Guernsey is warmer than Tatoosh Island in summer, but the annual range is of the same order of magnitude at both places. The two inland stations — Wenatchee, Washington, on the eastern flanks of the Cascades, and Orléans in central France — have very dissimilar temperature curves. Orléans has much more moderate winters and summers, reflected in the much smaller annual range at the French station. Farther inland, the difference becomes even more marked. Glasgow, Montana, on the High Plains, is separated from the ocean not merely by 700 miles (over 1,100 km), but also by the Cascades and the Rocky Mountains. Vienna, equally far from the ocean, is relatively open to incoming air. The annual range of temperature at Glasgow is characteristic of continental stations because it is separated from the ocean by several mountain ranges. That at Vienna, although much less oceanic than the range at Guernsey, nevertheless reflects the ability of Atlantic air to temper the annual march, particularly during the winter months.

These notions about the relative effectiveness of air movement are also illustrated in Fig. 3-2 which shows that the isotherms in western North America and northwestern Europe are more closely parallel to the coast than to the parallels of latitude, indicating that the temperature changes much more quickly from the coast toward the interior than it does from north to south. If these isotherms are closely spaced, as they are in western North America, they indicate that the inland penetration of marine air is slight and is confined to a

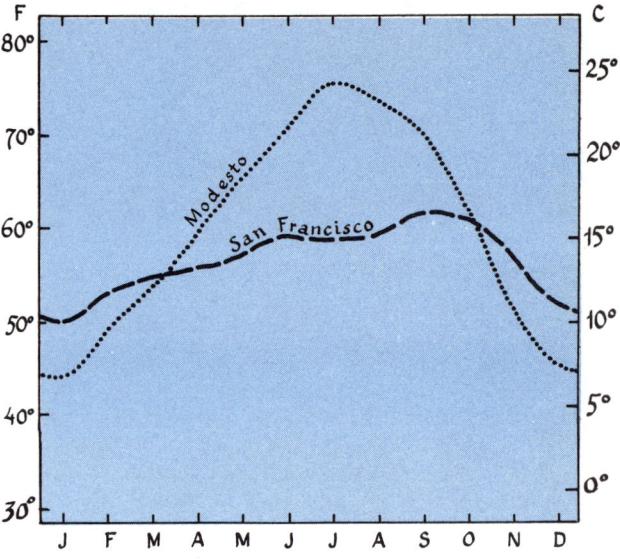

Fig. 3-7. The annual march of temperature at San Francisco and at Modesto, California. The rapid increase of continentality inland is expressed by the more peaked temperature curve and earlier occurrence of the maximum at Modesto, which is only 80 miles (129 km) from San Francisco.

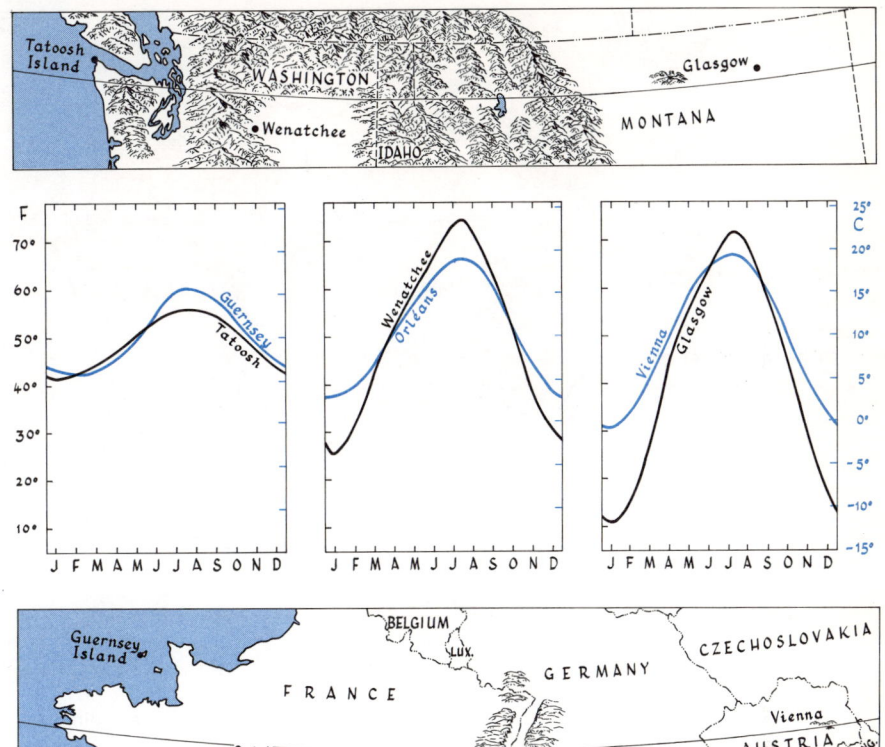

Fig. 3–8. The annual march of temperature along the 48th parallel in North America and in Europe. The top and bottom diagrams show the relative locations of three North American and three European stations. The middle part of the figure shows the increasing difference between the North American and European stations as one goes inland.

narrow coastal strip. Over northwestern Europe, however, the isotherms are much more widely spaced, which indicates that the inland penetration of marine air is quite effective in bringing marine temperatures to the interior of the continent.

Ocean Currents

The major circulation of the oceans will be sketched in Chapter 19. In this discussion, we shall be concerned largely with those currents that flow along the margins of the continents. The effect of such currents is not limited to temperature; it is also felt in the precipitation pattern. However, we shall be concerned here only with their effects on temperature. Surface currents which seem relatively fast compared to other movements in the ocean, are nevertheless fairly slow. The name **drift**, often applied to them indicates that they move slowly, usually at rates less than 3 miles per hour (about 1.3 m per second). They are primarily set in motion by wind. Their direction is further affected by the earth's rotation and by the trend of the coastlines on which they impinge. These forces combine to cause huge eddies that move clockwise in the Northern Hemisphere and counterclockwise in the Southern Hemisphere. Fig. 3-9 shows the average circulation of these eddies over a period of several years. The similarity of this pattern to the average pattern of circulation of the wind is striking (compare Fig. 4-14, p. 61). It suggests a causal relation between wind and water movement.

In simplest terms, the wind first moves water by frictional drag as it strikes the water surface. Water is also moved by wind pressure on waves, especially on small ripples. Air pressure is greater on the windward slope (the slope facing the wind) of the ripple than it is on the lee slope (the slope away from the wind). This unequal pressure over rippled surfaces forces the water to move downwind (away from the wind). Energy is also imparted to the water surface by the pounding of spray once the surface of the sea becomes sufficiently agitated by the wind. It should be kept in mind that the discussion here deals with the *movement of surface water by wind*, not with the generation of waves. Waves are generated by the wind, but they move across the water surface much more rapidly than the water itself moves.

Once the water is set in motion, how does it become part of the great oceanic circulation? A look at the maps of average seasonal pressure (Fig. 4-11) and of prevailing winds (Fig. 4-14) reveals part of the answer. There is an extensive **high-pressure** cell over each of the major oceans. (The reasons for the presence of oceanic high-pressure cells and the relationship between these cells

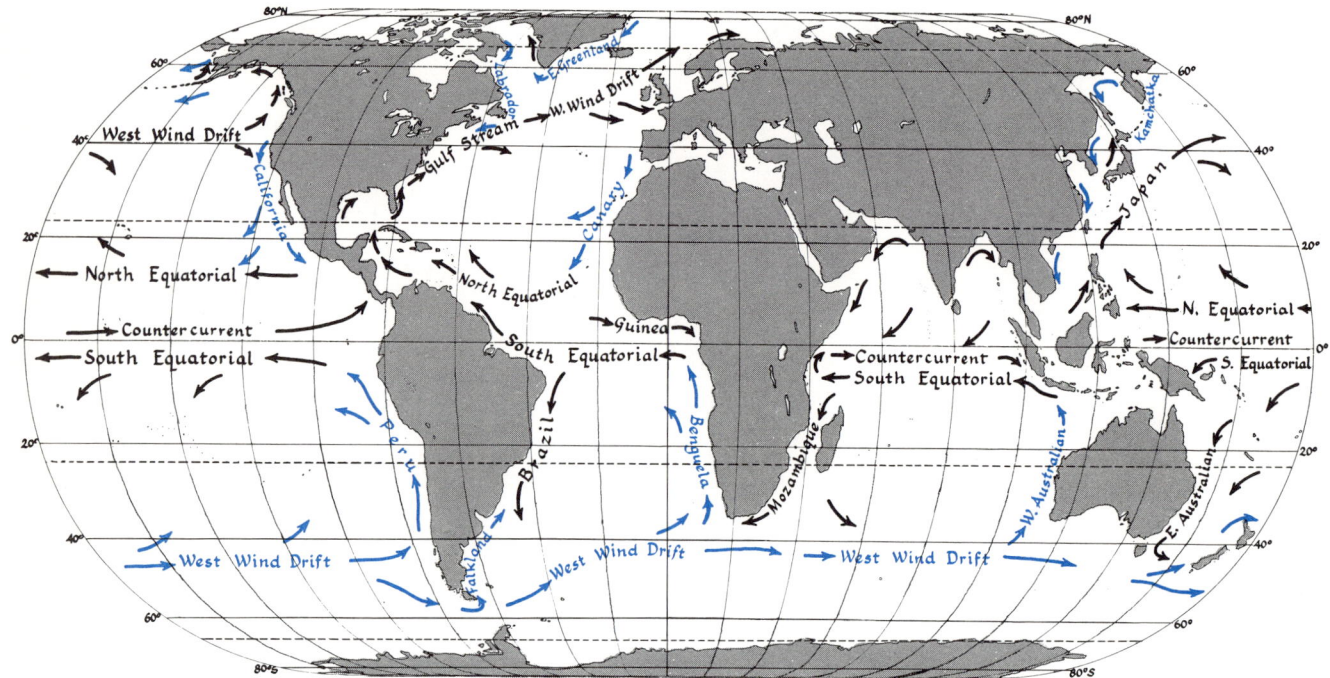

Fig. 3-9. The surface ocean currents. The blue arrows show the location and direction of cold currents; the black arrows show the location and direction of warm currents.

and the surface wind pattern are discussed in Chapter 4.) The prevailing winds that develop around these nearly permanent high-pressure cells approximate a gigantic whirl of air. This generalized and large-scale motion, which is clockwise in the Northern Hemisphere and counterclockwise in the Southern Hemisphere, is imparted to the waters beneath it. Viewed very roughly, the result is westward movement in the equatorial part of the whirl, poleward-moving water in the western part, eastward movement in the polar part, and equatorward movement in the eastern part of the whirl. Currents moving poleward are warm currents since they carry warm water poleward from areas of its normal occurrence. For analogous reasons, currents that flow equatorward are termed cold currents. Figure 3-9 clearly indicates the presence of warm currents in the western parts of the oceans along mid-latitude east coasts, and cold currents along mid-latitude west coasts. The designation of the equatorial currents as warm is conventional. These waters are neither warmer nor colder than one might expect in the latitudes of their occurrence; they simply partake of the general warmth of equatorial waters. Between the equatorial currents of the two hemispheres, there exists a countercurrent that flows eastward, mainly as a result of the slight, but measurable, pileup of water along equatorial east coasts. This countercurrent is generated by the strength and the persistence of the westward-flowing equatorial currents.[2]

These currents, which are shown schematically in Fig. 3-10, all have their real counterparts in Fig. 3-9. The east-coast warm currents and the west-coast cold currents are generally named after the coastal areas they bathe. Their names are shown on both maps. The equatorial currents and the West Wind Drifts (so named because the drifts or currents are propelled by west winds blowing from west to east) are identified by adding the name of the appropriate ocean basin and hemisphere.

The polar branches of the oceanic whirls differ one from another, even though they all are referred to as West Wind Drift. The differences are most marked between the two hemispheres. The southern oceans are not divided into separate oceans by land masses, as are the North Atlantic and North Pacific; consequently, the West Wind Drift is a circumplanetary current. In addition, it is a zone of decidedly cold water because of its proximity to Antarctica. The North Pacific, on the other

---

[2]Convention requires that winds be designated by the compass point *from* which they blow; currents, by the compass point *to* which they blow. Thus an east wind and a westerly current both represent motion from east to west.

hand, has only narrow connections with the Arctic Ocean. The West Wind Drift in the North Pacific is hardly affected by cold polar waters and retains the relative warmth of the Japan Current. The North Atlantic has a slightly different configuration: The arrangement of the land masses and islands allows the West Wind Drift to split into a southern branch, which becomes the Canary Current, and a northern arm, which transports the warmth of the Gulf Stream beyond the northern tip of Scandinavia before its characteristics are destroyed by loss of heat and by **subsidence.**

The subsidence of warm currents in polar latitudes may be the cause of the several cold currents that flow out of polar waters along east coasts. It is conceivable that the Labrador, Greenland, and Kamchatka Currents represent an outflow of water caused by the intrusion of subsiding warm water. It is more difficult to explain the cold currents off the east coasts of South America and Africa by the same mechanism, but their relative position and temperature are similar.

The similarity between the distribution of observed currents and the simplified model is striking, given the differences in configuration among the various ocean basins. The northern part of the Indian Ocean is an exception. In this part of the oceanic world, a seasonal reversal of currents takes place, reflecting the seasonal, or **monsoonal,** shift in wind direction. During the winter monsoon, the currents in the northern Indian Ocean move in a southwesterly direction, while in the summer monsoon, the currents move in a northeasterly direction into the Arabian Sea and into the Bay of Bengal. Seasonal changes related to seasonal wind shifts are found in other oceans, but the Indian monsoon is the only one that has an appreciable effect on the general pattern of ocean currents.

As we have seen, the middle latitude west coasts of all the continents are bathed by cold currents; the east coasts, by warm currents. The cold currents affect temperatures mainly in summer; the warm currents, in winter. In fact, many of the cold currents along the western coasts of continents are hardly discernible in winter. In the mid-latitudes, therefore, air imported from the oceanic west onto the west coast of a continent has its summer coolness intensified by passage over one of the cold currents, such as the California, Peru, Canary, or Benguela Current. Note, in Fig. 3-2, the equatorward bending of isotherms in regions of cold currents, particularly during the summer, and the poleward bending of the isotherms where there are pronounced warm currents, such as the West Wind Drift in the North Atlantic in winter. Because of the prevailing westerly winds in mid-latitudes, the cold currents affect air that moves east from ocean to continent; the warm currents influence mainly air masses that do not move over the land. The West Wind Drift is an exception. The air that is warmed by passage over this current significantly modifies the climate of northwestern Europe. The West Wind Drift in the North Pacific affects the coast of southern Alaska in the same way.

The annual march of temperature in the North Atlantic and North Pacific Oceans illustrates the effect of currents (Fig. 3-5). Although the annual marches apply to nearly identical latitudes, the Atlantic station is about 10° warmer than the Pacific station. If the quadrangles to which the temperatures apply are located on the map,[3] it will be seen that although the latitudes are similar, the Pacific quadrangle lies in the zone of the California Current and the Atlantic quadrangle is in the area occupied by the Gulf Stream. It is no wonder that the Atlantic example is about 10° warmer throughout the year.

A further effect of ocean currents is revealed by the

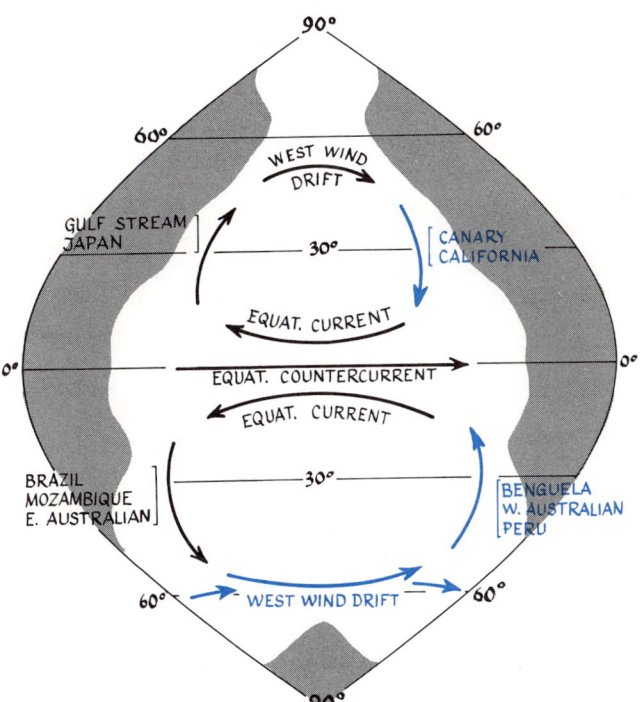

Fig. 3-10. An idealized model of surface currents in the two hemispheres. Once again, blue arrows show cold currents and black ones, warm currents.

---

[3] The North Atlantic values are from the quadrangle 40°–45° N, 40°–45° W; those for the North Pacific are from the quadrangle 40°–45° N, 140°–145° W.

36   Chapter 3

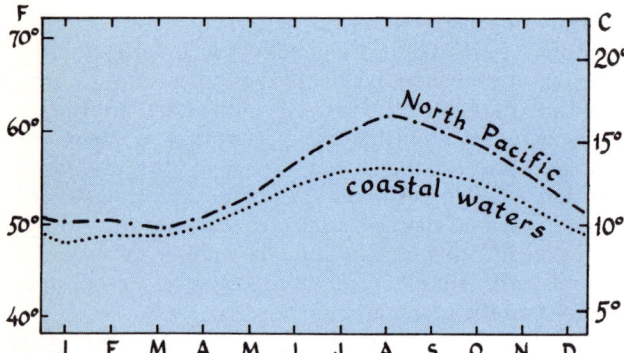

Fig. 3-11. The annual march of temperature over the North Pacific Ocean and over coastal waters. Note that temperature differences are most marked in summer when the coastal waters are influenced by the cold California Current and by upwelling.

comparison of the annual march of temperature over the North Pacific and over water closer to the Oregon shore (Fig. 3-11). The coastal area is a little colder in winter than the North Pacific, a condition to be expected due to a continental influence on maritime air near the coast. However, the inshore zone is also colder than the North Pacific in the same latitude in summer, a phenomenon that cannot be explained by continentality, since we expect warmer summers over land than over oceans. Summer temperatures near the coast are depressed because between the shore and the oceanic quadrangle lies a zone of even colder temperatures than those found over the California Current. The coldness of the water (and of the air overlying it) is due to **upwelling** of cold bottom water in a narrow coastal strip. This upwelling comes about because of the deflection of all freely moving bodies on the earth's surface to the right in the Northern Hemisphere (see Coriolis force, p. 44 ff). The southward-flowing California Current is thus deflected to the west in a series of smallish eddy movements. This pull of the water away from the coast would lower the sea surface, except that bottom water comes to the top to replace the deflected surface layers. The water immediately off the coast of California and Oregon is very cold as a result. The cooling effect of this narrow strip of water is felt not only along the west coast of North America but also along most west coasts bathed by cold currents; and it is evidenced not only by reduced summer temperatures but also by increased cloudiness.

The major ocean currents show up quite clearly in Fig. 3-2 as pronounced bending of the isotherms over restricted parts of the oceans. The effectiveness of the cold Peru Current or of the warm West Wind Drift in the North Atlantic is shown by the strong equatorward bending of the former and the equally strong poleward bending of the latter. These two currents maintain their relative coldness or warmth all year long; hence the bending is discernable on both the isothermal maps. Other currents, such as the Japan Current, show their warmth only in winter; or as in the case of the California Current, they appear as cold currents only in summer.

Cloudiness

In a sense, the effect of cloudiness is part of the effect of different amounts of insolation. However, cloudiness affects not only the income of radiation but also the amount of terrestrial radiation that escapes to space and the percentage that is trapped by the atmosphere. The common recognition that clear nights are cold nights reflects the important effect of clouds on outward radiation. This effect is illustrated by the daily curves of temperature at Karachi, Pakistan, shown in Fig. 3-12. In July, the principal portion of the radiation available for heating the ground comes from solar rather than from atmospheric radiation. Anything that cuts down the solar radiation, such as a cloud deck, reduces the temperature. In January, atmospheric radiation becomes more important than solar radiation. A cloud deck in January intercepts much of the outgoing terrestrial radiation. This means that the atmosphere has more energy available for reradiation to the ground and that therefore surface temperatures are higher on cloudy days than on clear days.

Obviously, the effect of clouds during the day is different from that at night. Daytime conditions are roughly similar to those in July; nighttime is somewhat

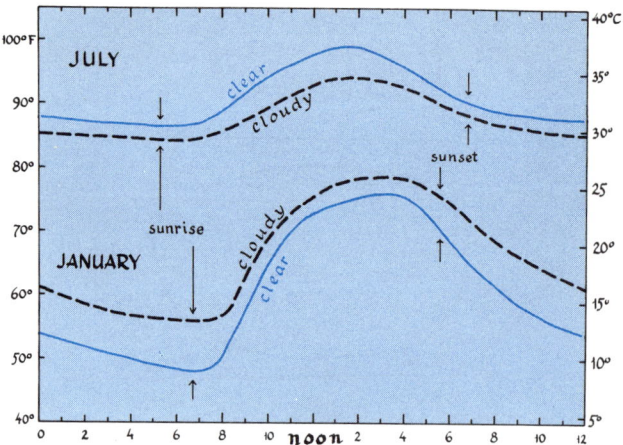

Fig. 3-12. The diurnal march of temperature at Karachi, Pakistan, on clear and cloudy days in summer and winter. Clear days are warmer than cloudy ones in summer, particularly so during the day. Cloudy days are warmer than clear ones in winter, and especially at night.

analogous to January. The temperature differences between clear and cloudy days in July are, therefore, much greater during the day than at night; whereas in January the greatest difference is at night.

The Daily March of Temperature

The curves for Karachi also serve to illustrate a typical daily march of temperature. Note, first of all, the times of sunrise and sunset marked on the graph. The lowest temperatures of the day occur just after sunrise; the highest temperatures, somewhere after two o'clock in the afternoon. There is a lag between temperature and radiation in the daily cycle as well as in the annual cycle, but there is less symmetry in the daily march than in the annual march. Temperatures decrease gradually from the afternoon maximum until shortly after sunrise, usually 16 hours later. Thereafter, the temperature increases very rapidly once the sun is above the horizon. The period of increasing temperature lasts only about 8 hours, from sunrise minimum to early afternoon maximum. This lack of symmetry is clearly shown in the skewed January curves but is also present in those for July. One thing to bear in mind about the annual march of temperature is that it is, in a sense, made up of 365 sets of daily marches, of which the two examples shown here are merely representative of summer and winter conditions.

The Temperature Aloft

The atmosphere is commonly divided into several rather distinct layers. Since these subdivisions are based largely on temperature differences, we may conveniently employ them to illustrate the distribution of temperature with height. Figure 3-13 describes a *standard atmosphere*, that is, an atmosphere for which certain average conditions are assumed. These average, or standard, conditions include a surface temperature of 60° F (15° C), a normal **lapse rate** of temperature, and a troposphere 7 miles (11 km) thick. Descriptions of the principal characteristics of the subdivisions shown in Fig. 3-13 follow.

Principal Subdivisions of the Atmosphere

The Troposphere. In the troposphere (Greek *tropos* a turn or change + *sphaira* sphere or ball), the layer of the atmosphere lying closest to the earth, temperatures decrease as height above sea level increases, to about 5 miles (8 km) above the poles and about 10 miles (16 km) over the equator. Even though other parts of the atmosphere may affect climate, the troposphere is climatically the most important layer, as it is here that most of the significant weather disturbances take place. The troposphere has two major subdivisions: (1) the lower troposphere and (2) the middle and upper troposphere.

The lower troposphere is a layer extending from the earth's surface to a somewhat indeterminate level, 1,500 to 6,000 feet (about 450 to 1,800 m) above the ground. This zone is affected by friction with the surface and reflects the diurnal temperature range of the surface. Temporary increases of temperature with elevation are common near the surface, although they are not shown as part of the average conditions depicted in Fig. 3-13.

The middle and upper troposphere are practically uninfluenced by friction or by diurnal temperature changes, although annual temperature differences are still felt. The upper limit of the troposphere, and

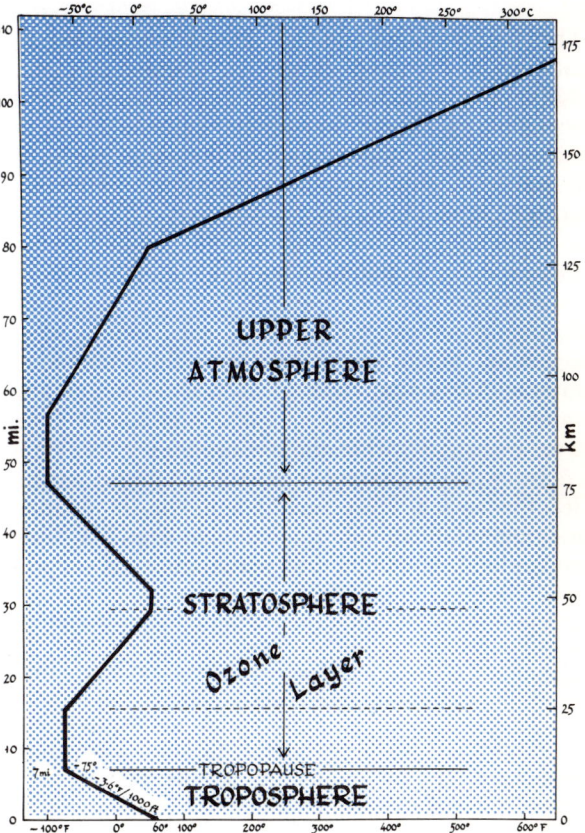

Fig. 3-13. The standard atmosphere, showing variation of temperature with elevation. Note the consistent decrease of temperature with elevation in the troposphere (local and temporary exceptions have been averaged out). The isothermality of the stratosphere is limited to its lowest part; there is a rapid increase in temperature in the ozone layer between 20 and 30 miles (about 30 to 50 km) aloft.

therefore the level at which temperatures no longer decrease with height, is called the **tropopause** (Greek *tropos + pausis* a ceasing). Broadly speaking, it also marks the upper limit of the influence of yearly or seasonal temperature changes at the surface.

The Stratosphere. The **stratosphere** (Latin *stratum* to be spread out + Greek *sphaira*) differs from the troposphere in not being subject to convectional overturn of air. Hence the contrasting names, troposphere (the sphere of overturning air) and stratosphere (the layered, or spread-out sphere). The stratosphere is characterized by nearly uniform temperatures of around 75° F below zero (-60° C), extending from the tropopause to about 15 miles (24 km) above the earth. There are certain portions, above 15 miles, that are much warmer due to the absorption of solar radiation by layers of ozone.[4] Above the zone of ozone production, the stratosphere is again isothermal with temperatures similar to those below the ozone layer.

The Upper Atmosphere. In this region there is great variation of temperature, with some layers as hot as 2,000° F (over 1,000° C). The heat content, however, is negligible because of the very few gas molecules present in this most rarefied part of the atmosphere. We will not be much concerned with the temperature stratification in the upper atmosphere.

Temperature in the Troposphere

Having described the principal subdivisions of the atmosphere, let us return briefly to the troposphere. Figure 3-13 shows an average decrease of temperature of about 3.6° F for every 1,000 feet (0.65° C per 100 m) of height within the troposphere. This average decrease with increasing distance from the earth is called the *normal lapse rate of temperature*. Maps of world temperature, such as those in Fig. 3-2, incorporate reductions of observed temperatures to supposed

---

[4]Ozone is a gas which differs from oxygen by having three atoms of the element oxygen in each molecule (O_3) instead of two (O_2). Ordinary oxygen molecules (O_2) break up into two oxygen atoms (O) when subject to ultraviolet radiation from the sun. If such an atom (O) collides with an oxygen molecule (O_2) under certain conditions, ozone (O_3) is formed. But ozone is very unstable in the presence of sunlight, and it breaks up into O_2 and O once again, as it absorbs solar radiation. And so the process of formation and destruction of ozone proceeds. In the stratosphere the two processes are in balance, but there is little ozone in the troposphere, except where it is created by the artificial chemistry of the air above certain industrial establishments.

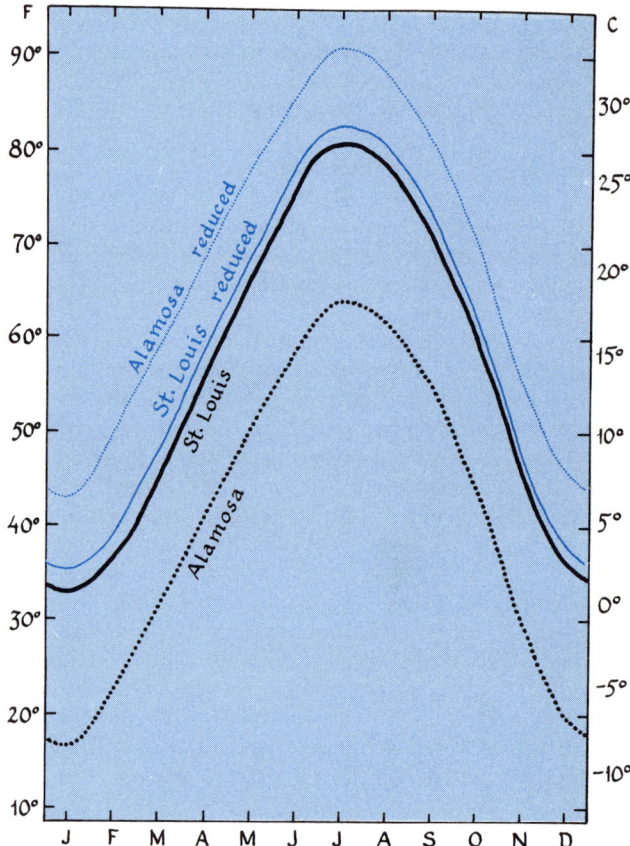

Fig. 3-14. The annual march of temperature, actual and reduced to sea level, at St. Louis and at Alamosa, Colorado. The actual temperature at St. Louis is consistently higher than at Alamosa; but when the standard correction for elevation is applied, Alamosa appears warmer than St. Louis.

sea-level values; the reductions are based on the assumption that a normal lapse rate exists. Figure 3-14 illustrates both the effect of elevation on temperature and the method employed for reduction of temperatures to sea level. The two heavier curves depict the annual march of observed temperature at St. Louis, Missouri (elevation 568 feet or 173 m), and at Alamosa, Colorado (elevation 7,536 feet or 2,297 m), both in approximately the same latitude. The difference between these two curves clearly shows the effect of elevation on temperature. The lighter lines on the diagram represent the temperatures reduced to sea level (by the addition of 3.6° F for every 1,000 feet of elevation). Our standard formula for reduction of temperatures to sea level shows that Alamosa is warmer than St. Louis when due consideration is given to its higher elevation. Another way of putting it is to say that we expect two stations to differ in temperature by 3.6° F for every 1000 feet (or

0.65° C for every 100 m) of difference in elevation, but Alamosa is only 13° F (7° C) colder than St. Louis in March and no more than 17° F (9° C) colder in June. Since Alamosa is about 7,000 feet (2100 m) higher than St. Louis, the lapse rate between the two places is not 3.6° F/1,000 feet (0.65° C/100m) as assumed, but varies between 1.9° F/1,000 feet (0.35° C/100 m) in March and 2.5° F/1,000 feet (0.45° C/100 m) in June.

One of the principal factors that disrupt normal conditions of the temperature lapse rate is strong heating and cooling of the surface. When the earth's surface is very cold, the difference between the surface temperature and that at the top of the troposphere is reduced; hence the lapse rate is smaller. With strong surface heating, the lapse rate is larger. Figure 3-15 shows the lapse rate on a summer day at Oakland, California. The tropopause dividing stratosphere from troposphere was about 10 miles (16 km) up. In its upper layers, the troposphere had a lapse rate very close to normal. Near the surface, however, two things had occurred to disturb the normal situation (see inset, Fig. 3-15). The air through which this vertical sounding of temperature had been taken had come from the northwest, passing over the cold upwelling water off the coast before reaching Oakland. In the process, cooling from below had taken place, and the lapse rate had been reduced to the point where there was no longer a decrease of temperature with elevation. Instead, temperatures increased with elevation in the zone between 1,000 and 3,000 feet (300 and 900 m). Such an increase in temperature with elevation is called an **inversion,** or inverted lapse rate, since it is not normal. In the levels below 1,000 feet (300 m), something else had occurred. Here, despite the very short stay of the air over land, there had been significant heating — so much heating, in fact, that the lapse rate was substantially greater than 3.6° F/1,000 feet (0.65° C/100 m) in the first 1,000 feet (300 m) above the surface. One word of caution about terminology: It is customary to speak of a very large lapse rate as a "steep" lapse rate and of smaller lapse rates as "gentle." But a rapid decrease of temperature with elevation, when plotted on a height-temperature diagram, such as Figs. 3-13 and 3-15, shows up as a line of gentle slope in the common sense of the word. The importance of the lapse rate to other climatic items will be made clear in chapter 4.

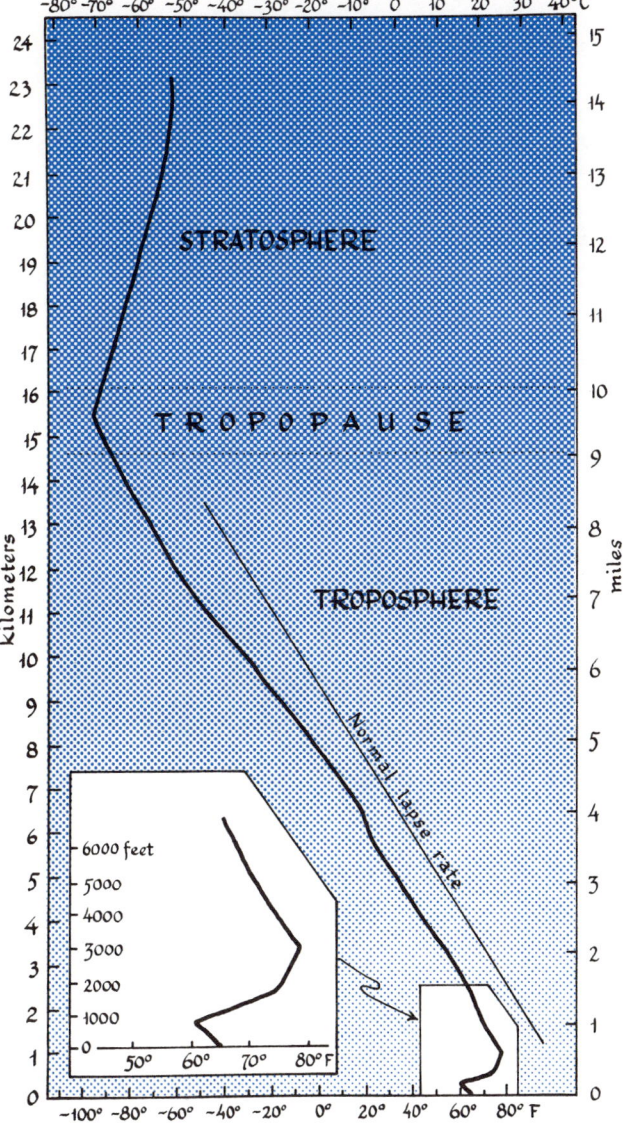

Fig. 3-15. The lapse rate of temperature at Oakland, California, on July 14, 1953. The normal lapse rate is found in the upper troposphere, but there is a marked inversion near the ground, which is shown in more detail in the inset.

Summary

The factors that affect the horizontal distribution of temperature on the surface of the earth have been discussed in this chapter. The six major factors are insolation, elevation, land and sea contrasts, air movements, ocean currents, and cloudiness. They can be summarized by considering the series of temperature graphs illustrated in the chapter. A comparison of St. Louis with Honolulu or with Fairbanks (Figs. 3-3 and 3-5) shows the effect of insolation on determining temperature differences. A comparison of St. Louis with Alamosa, in about the same latitude but 7,536 feet (2,297 m) above sea level (Fig. 3-14), shows the effect of elevation. Comparison of St. Louis with a North Atlantic position and with a North Pacific position (Fig. 3-5) shows the effect of land and sea contrasts. Comparison of St. Louis with San Francisco and with Washington, D. C. (Fig. 3-6) shows the importance of air movement in determining temperatures. A comparison of the North Atlantic curve with the North Pacific curve (Fig. 3-5) illustrates the effect of ocean currents. A further contrast between the Pacific position and coastal waters (Fig. 3-11) reveals the importance of upwelling. The upwelling is associated with fog formation, so that the very depressed summer temperatures near the coast reflect not only the reduction due to import of cold air masses but also the blocking of radiation by summer fogs. Figure 3-12 illustrates the effect of cloudiness on temperature.

Even though the factors that affect the horizontal distribution of temperature have been described separately, they all operate together. Major emphasis was given to the annual march, made up of the monthly means of temperature, but the diurnal regime is no less a part of the climatic character of a place.

The vertical arrangement of temperature has considerable climatic significance, particularly with regard to precipitation and clouds, as we shall see in a later chapter. With respect to temperature, we can distinguish the troposphere, in which temperatures normally decrease with elevation, from the stratosphere in which temperatures are roughly the same at all levels. Within the troposphere itself we can separate a lowermost layer into which the surface temperature differences between day and night are transmitted, and an upper layer in which only seasonal temperature differences can be detected.

The horizontal and vertical gradients of temperature are basic to the creation and maintenance of atmospheric pressure differences from place to place. The following chapter will be concerned with a most important climatic element — atmospheric pressure — and its associated pattern of winds.

Suggestions for Further Reading

Byers, Horace R., *General Meteorology*, 3rd ed., McGraw-Hill Book Company, Inc., New York, 1959, pp. 45-60.

Landsberg, Helmut, *Physical Climatology*, Pennsylvania State University, University Park, Pa., 1941, pp. 108-122.

Petterssen, Sverre, *Introduction to Meteorology*, 2nd ed., McGraw-Hill Book Company, Inc., New York, 1958, pp. 251-264.

Trewartha, G.T., *An Introduction to Climate*, 3rd ed., McGraw-Hill Book Company, Inc., New York, 1954, pp. 26-51.

4 Pressure and Winds

Introduction

Definitions

The force exerted on a surface by the overlying weight of the atmosphere is called **atmospheric pressure.** The gases that make up the atmosphere rapidly decrease in density with increasing altitude, and the pressure also decreases rapidly. Three miles (4.8 km) above the surface, only 50 percent of the atmosphere remains; at 6 miles (9.7 km), only 25 percent; and at 12 miles (19.3 km), only 5 percent. The atmosphere has no clearly defined upper limit; it simply thins out gradually. Even 100 miles (160 km) from the surface there are some traces of atmospheric gases and therefore, some atmospheric pressure.

Winds are the more-or-less horizontal movements of air with respect to the earth's surface. The direction and the speed of the movement are relative to that surface. To an extraterrestrial observer, air moving at the same speed and in the same direction as an observer on earth (still air) would appear to be undergoing complicated gyrations. By the same token, air that seems to an observer in space to move in a straight line would appear to an observer on earth to be twisting in a spiral. The terrestrial observer, carried about in circular motion by the rotation of the earth, attributes this spiral flow not to himself but to the wind. Later we will consider the relationship between pressure and winds.

Instruments

Atmospheric pressure is measured by a barometer (Greek *baros* weight + *metron* to measure), an elegantly simple instrument invented in 1643 by Evangelista Torricelli, a student of Galileo. Torricelli's barometer consisted of a pan of mercury (the most convenient liquid of great density) into which he upended a tube full of mercury, making sure that none of the liquid escaped from the tube in the process. If we repeat Torricelli's experiment with a shorter tube, one less than two feet long, nothing unexpected happens; the upended tube remains full of mercury. But if we use a tube longer than 30 inches, the mercury drops in the tube until it stands about 30 inches (or 760 millimeters) above the mercury in the pan (Fig. 4-1). A vacuum has been created in the top of the tube because the pressure of the atmosphere on the mercury in the pan is just enough to keep the liquid in the tube at a maximum height of 30 inches (760 mm). In short, the weight of the column of air in the atmosphere is counterbalanced by the weight of a column of mercury 30 inches (760 mm) long. If the atmospheric pressure should rise a little (that is, if there were more air than normal overlying the pan), the mercury in the tube would also rise. If the pressure should decrease, and it will decrease significantly if you move upward in the atmosphere, the level of the mercury in the tube would also drop. In 1648, Blaise Pascal, the French philosopher and mathematician, proved this fact

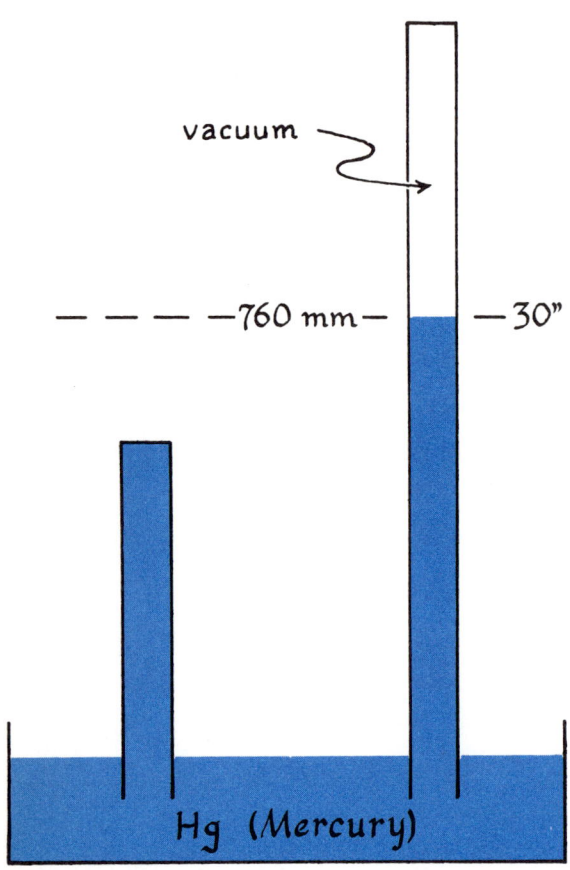

Fig. 4-1. Torricelli's experiment. The atmospheric pressure at sea level supports a column of mercury 30 inches (760 mm) high. If the tube is shorter than 30 inches, no vacuum is created, but in a longer tube, a vacuum appears and the height of the mercury can be used to measure pressure, since it will rise or fall as the pressure on the mercury increases or decreases.

to his satisfaction by sending his brother-in-law with just such a barometer to the top of Puy-de-Dôme, a 4,800-foot (1,465 m) mountain in central France. Modern barometers are no more than accurate elaborations of these early instruments.

Wind direction was first observed by means of vanes in Italy in 1650 and the wind speed was measured by noting the velocity of any visible particles carried by the wind. The modern *cup* anemometer (Greek *anemos* wind + *metron*) measures wind speed by noting the rotation speed of a shaft to which three cups are attached.

Two aspects of the definition of wind are important: (1) The wind is named according to the direction from which it blows — an east wind, for example, blows from east to west. (2) The term "wind velocity" implies a direction as well as a speed — do not use the terms "speed" and "velocity" synonymously if you wish to be accurate.

Units of Measurement

Although it is inappropriate to measure force in linear units, until recently measurements of atmospheric pressures were expressed in inches of mercury (or millimeters of mercury), just as observed on a barometer. Today, meteorologists use the **millibar** (mb) as the unit of force to describe pressure. A millibar is the equivalent of a force of 1,000 dynes exerted on 1 square centimeter of surface.[1] Average sea-level pressure is about 29.92 inches (760 mm) of mercury or 1,013 millibars.

The daily weather map of the U.S. Weather Bureau shows pressure by means of **isobars** (Greek *isos* equal + *baros*), or lines connecting points of equal atmospheric pressure. The reproduced weather map in Fig. 4-2 shows that the units used are, in fact, millibars. Rather than show actual pressures, the isobars show pressures reduced to sea level. This reduction is even more important than the analogous reduction of temperature because pressure variations related to elevation are much greater than pressure variations in the horizontal plane. On the world weather map for a single day, the observed sea-level pressures are seldom above 1,055 millibars or below 955 millibars, a variation of 100 millibars.[2] On the other hand, the average decrease of pressure with elevation is about 34 millibars for every 1,000 feet (11 mb per 100 m), at least in the lowest 5,000 feet (1,500 m) of the atmosphere. A difference in elevation of 3,000 feet (900 m), therefore, produces a difference in pressure of about 100 millibars. A typical daily range in sea-level pressure over the whole world is thus no more than the range observed in a 3,000-foot (900 m) interval of elevation. The relatively small horizontal variations in pressure are important — in fact, they are all-important in determining wind velocity — but they are negligible when compared with vertical pressure differences and would be completely unidentifiable if the vertical changes were not eliminated by the reduction of pressures to sea level.

Wind directions are indicated either by angular units (0° representing a north wind, 90° an east wind, and so on) or by the familiar points of the compass (N, NNE, NE, and so on). Wind speeds are expressed in a variety of units, including *knots* (a knot is a speed of 1 nautical mile per hour, about 1.15 statute miles per hour or 0.5 meters per second), miles per hour, feet per second, or

---

[1]A dyne is the force necessary to produce an acceleration of 1 centimeter per second in every second, in an object whose mass is 1 gram.

[2]Pressures as low as 890 millibars have been observed near the centers of tropical hurricanes, but such marked depressions are relatively short-lived features.

Pressure and Winds 43

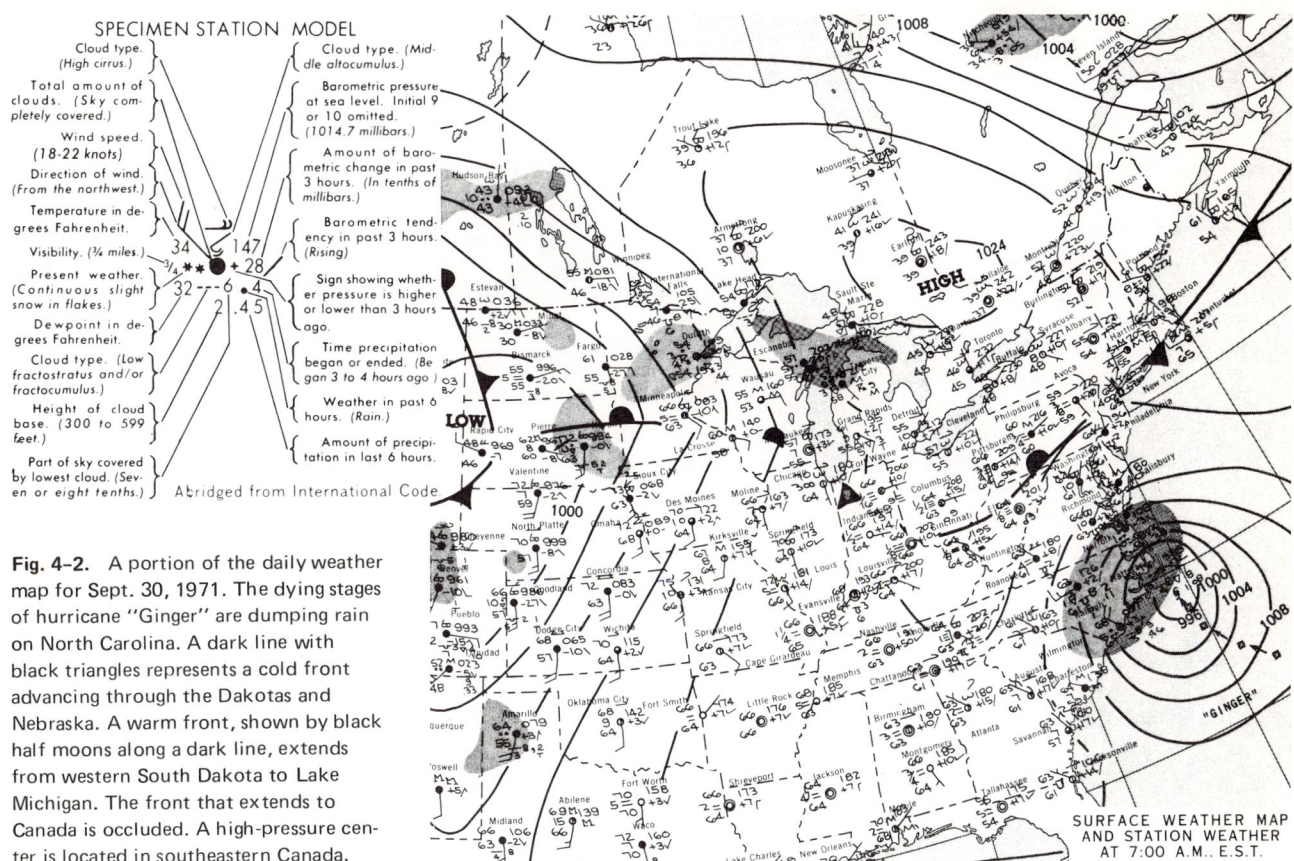

Fig. 4-2. A portion of the daily weather map for Sept. 30, 1971. The dying stages of hurricane "Ginger" are dumping rain on North Carolina. A dark line with black triangles represents a cold front advancing through the Dakotas and Nebraska. A warm front, shown by black half moons along a dark line, extends from western South Dakota to Lake Michigan. The front that extends to Canada is occluded. A high-pressure center is located in southeastern Canada.

meters per second. Lines of equal wind speed on a map are called **isotachs** (Greek *isos* equal + *tachys* swift).

Causes of Pressure Differences in the Atmosphere

Thermally Produced Pressure Patterns

Pressure differences on the surface of the earth are basically the result of temperature gradients, which, as we have seen, reflect the gradients of insolation. How do temperature differences from one place to another affect the pressure? This effect and other phenomena are explained by certain equations that physicists refer to as the gas laws. *Boyle's law*, for instance, states that the pressure and the volume of a gas vary inversely; that is, any increase in pressure results in a decrease in volume if the temperature remains constant. *Charles's law* states that volume and temperature vary directly if the pressure remains the same. If we keep the pressure constant, any increase in the temperature produces an increase in the volume and thus reduces the density[3] of the gas; and, of course, a decrease in the temperature reduces the volume and increases the density. More commonly, we say that hot air rises and cold air sinks. *Poisson's equation* states that if the volume of a gas is constant, the pressure and the temperature vary directly. These three physical laws describe the relationships among the pressure, temperature, and volume of gases. Since density is the inverse of volume for a unit mass, the laws also describe density changes when pressure or temperature changes, though, of course, the relationships are inverse. In this chapter, we are concerned largely with Boyle's law and Charles's law; in the next chapter, we will refer to Poisson's equation.

To examine the relationship between a temperature gradient and a pressure gradient, let us imagine, first of all, a portion of the earth's surface with uniform temperature. Because the temperature is uniform at the surface, it will also be uniform at 1,000 feet (300 m) above the surface. Of course, it will be about 3.6° F

---

[3] density = $\dfrac{\text{mass} (\cong \text{weight})}{\text{volume}}$, and inversely, volume = $\dfrac{\text{mass}}{\text{density}}$.

(2° C) colder if normal conditions prevail. At 2,000 feet (600 m), the temperature will be uniformly 7.2° F (4° C) colder, and so on. In other words, although there is a predictable decrease of temperature upward, there are no temperature differences in a horizontal direction at any level. In a cross section, this temperature pattern would be represented by a series of horizontal isotherms. The pressure pattern reflects this temperature pattern and is represented by the series of horizontal isobars in Fig. 4-3 (1).

Now let us suppose that one part of the surface is heated or the other cooled (it makes no difference which, as long as there is a temperature difference). Over the warmer part of the surface (labeled W on the rest of the diagrams in Fig. 4-3), the warmer air increases in volume and expands, resulting in a pressure pattern over W like the one shown in Fig. 4-3 (2). Over the unwarmed part of the surface (labeled C on the diagram), the pressure remains unchanged for a while. Aloft, however, the pressure is no longer uniform in a horizontal direction; therefore, air moves from warm to cold regions at this upper level. This situation is represented by Fig. 4-3 (3), in which the isobars are no longer horizontal at some distance from the surface. The result of the movement from warm to cold regions is more air over C and less over W; therefore, the pressure at C increases and that at W drops because of the change in weight of the overlying atmosphere. Now there is a pressure gradient at the surface as well as aloft; and along the surface, air moves from colder to warmer regions. This situation is shown in Fig. 4-3 (4). The cold air moving from C to W at the surface forces the warm air at W to rise, and there is corresponding sinking of air over C. This is a *convectional* circulation of air, from hot to cold aloft and from cold to hot at the surface, with rising air in heated regions and subsiding air in cooled regions. This movement also implies a **high-pressure** center at the surface over the cold region, and a **low-pressure** center over the warm region. Aloft there is higher pressure over the warm region than above the cold one; we thus speak of upper-air highs or lows. Bear in mind that all pressure designations are *relative*; a low-pressure center at the surface, even though it has lower pressure than other places at the

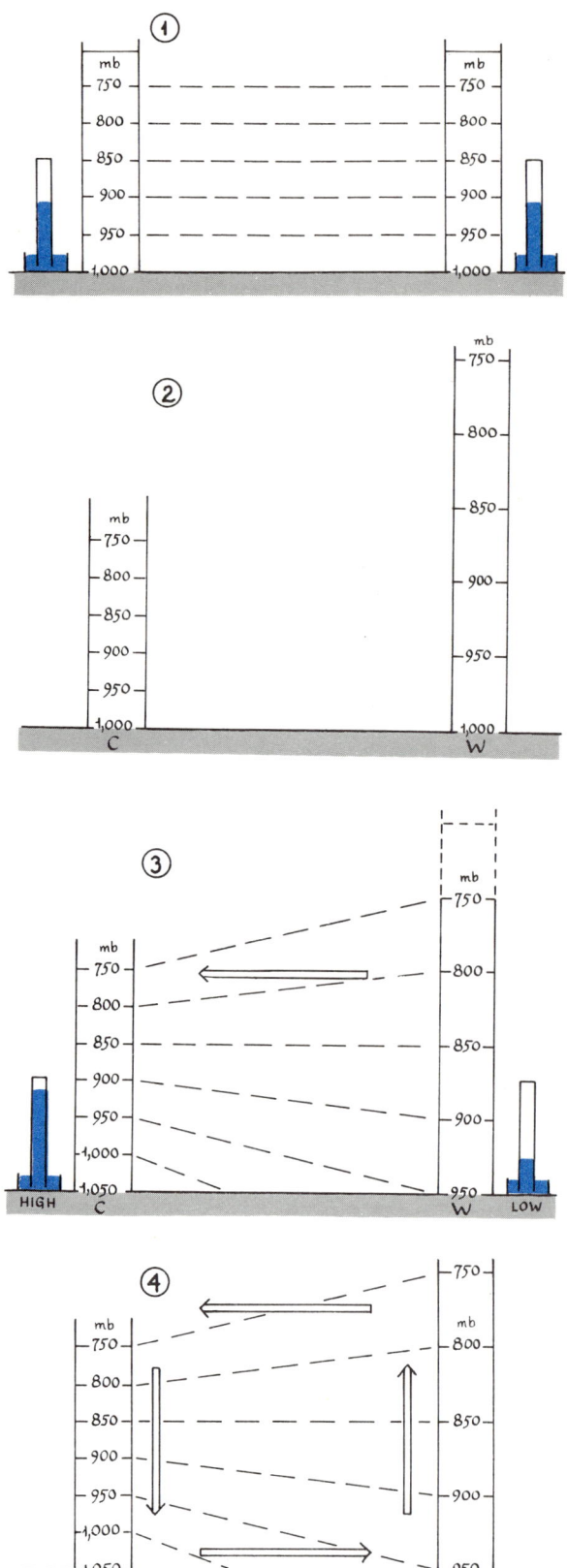

Fig. 4-3. The development of a thermally produced pressure gradient. The dashed lines are isobars, and the solid arrows represent air movement. In (1) the isobars are parallel to the surface because there are no temperature differences at the surface, and therefore, no pressure differences (as shown by the two barometers). In (2), the heating at W has caused the rising air to expand, thus changing the pressure distribution above W. In (3), air begins to move because the pressures are no longer the same at any particular horizontal level. As a result air piles up over C and its pressure increases, while that over W decreases. The resultant set of isobars and air movement is shown in (4).

ground, has higher pressure than any place aloft. Likewise an upper-air high has lower pressure than any place near the ground; it is its pressure relative to other places aloft that makes it a high.

We might imagine the equatorial regions as the warmed regions and the polar regions as the cooled regions of Fig. 4-3. We could even construct a theoretical picture of the circulation of the air and pressure patterns as shown in Fig. 4-3 (4), although we must keep in mind that in reality the observed pressure pattern differs from a simple, thermally produced convection, except in certain situations. For instance, islands tend to have relatively lower pressure at the surface than the oceans during the day and higher pressure at night, because they are relatively warmer than the surrounding ocean in the daytime and colder at night. The resulting onshore winds during the day and offshore winds at night are referred to as the *land and sea breeze*.

To a lesser degree, the heating and cooling of large continental masses differ from temperature changes in their surrounding oceanic expanses, and this differential explains the reversal of pressure from summer to winter. In summer, pressures over the relatively warm continents tend to be lower than those over the oceans. In winter, continents are colder than the oceans around them and consequently tend to have higher pressures than the oceans. This seasonal reversal of pressure and therefore also of winds is called a **monsoon** (Arabic *mausim* a time, a season). Figure 4-4 shows the shift in wind direction associated with the Indian monsoon. In January (Fig. 4-4 (1)) the prevailing direction is from the north and northeast as indicated by the direction of the arrows and by their lengths (which are proportioned to the frequency of the wind from any particular direction). In July (Fig. 4-4 (2)) the winds are mainly from the southwest; they are also somewhat stronger (as indicated by the thicker arrows), a local contradiction of our generalization about weaker climatic gradients in summer. In any case, the considerable pressure changes in the Asiatic monsoon are limited to a very shallow layer (perhaps no deeper than 5,000 feet or 1,500 m) and cannot explain all the wind patterns, let alone the other weather phenomena that are associated with a typical monsoon (see Chapters 6 and 7).

Dynamically Produced Pressure Patterns

Although horizontal pressure gradients, that is the presence of high- and low-pressure centers on the surface of the earth, are basically caused by the presence of temperature gradients, there are secondary factors as well. Winds blowing into a certain area may result in a pileup of air sufficient to create a local high-pressure center and outflow may produce a low-pressure center; such a pressure pattern is dynamically produced by

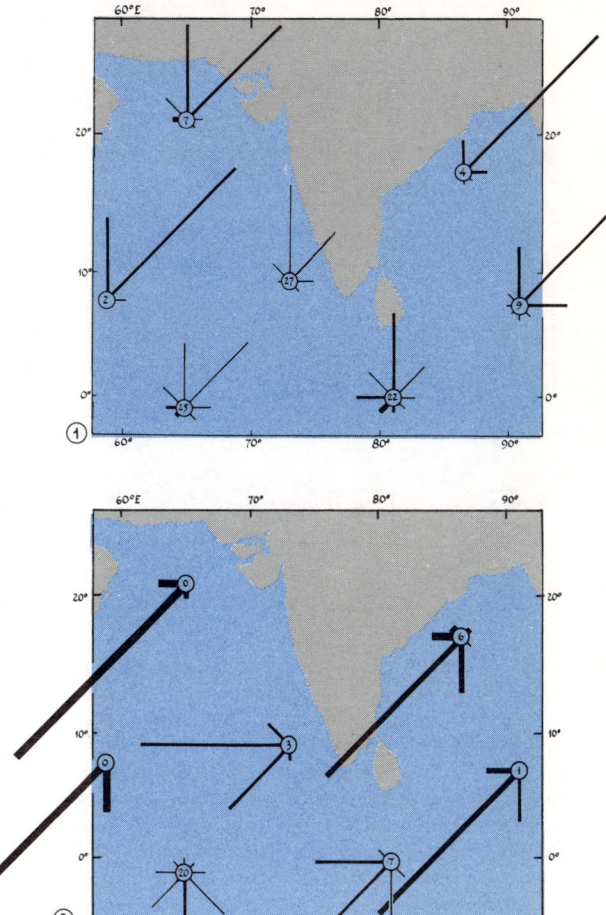

Fig. 4-4. The wind pattern of the Indian monsoon in January (1) and in July (2). The length of the wind arrows is proportional to frequency of occurrence, and their thickness varies with the strength of the winds. A long thick arrow from the southwest to the circle representing a particular station, for instance, stands for a wind that blows frequently *and* with high average speed from the southwest. A long thin arrow means frequent but light winds; and so forth. The percentage of calms is shown in the center of the wind roses.

movement of air. In order to understand the action of winds in producing pressure, we need to consider more closely the relation between pressure gradients and wind movements.

Pressure Gradients and Wind Movements: The Coriolis Force. So far, we have been speaking of pressure gradients and winds as if they were exactly analogous to slope gradients and water flowing down the side of a hill. The situations are different, however, largely because the very flow of air from a high-pressure center toward a low-pressure center tends to break up the pressure gradient itself, apparently creating an impermanent situation in which pressure

46 Chapter 4

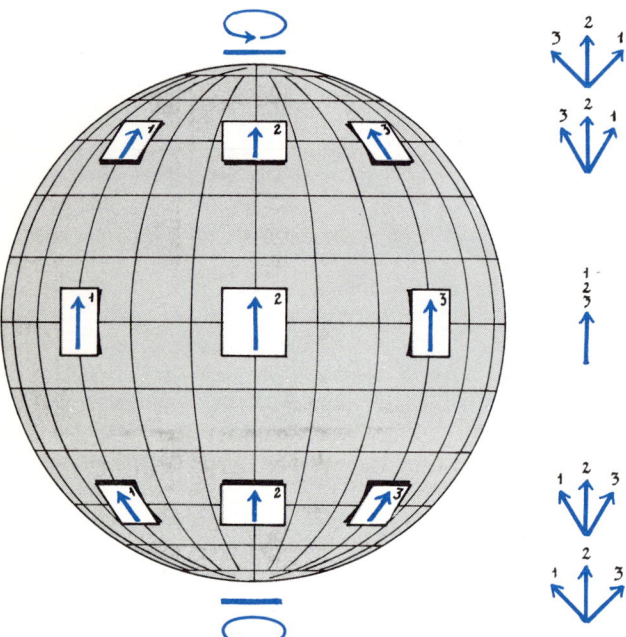

Fig. 4–5. The Coriolis effect as seen from outer space. Since the earth rotates from west to east, the horizon planes labelled *1, 2* and *3* represent three successive views, in that order. If we combine the north arrows of each plane at a common center, as shown at the right, we can see that in the Northern Hemisphere the north line appears to move counterclockwise from *1* to *3*; at the equator there is no shift; and in the Southern Hemisphere there is an apparent rotation clockwise from *1* to *3*. The two additional sets of arrows on the right show the situation at the poles (for which the horizon planes are just straight lines in this view of the earth. Note that the rotation of the arrows is zero at the equator, reaches a maximum at the poles, and is in opposite directions in the two hemispheres. This apparent rotation of horizon planes is referred to as the Coriolis force or Coriolis effect.

differences are destroyed about as fast as they are created. Yet, any map of seasonal pressure (for example, Fig. 4-11) clearly shows permanent centers of high and low pressure. What prevents destruction of the highs by the outflow of air and of the lows by inflow? Something must interfere with the free flow of air. The interference is caused by the rotation of the earth's surface.

Figure 4-5 indicates how the rotation of the earth interferes with the movement of any freely moving particles (including, of course, air particles). An orthographic projection of a hemisphere of the earth shows the parallels and meridians of the geographic grid as visualized at a considerable distance from the earth. A series of small square planes, drawn tangent at various points, are shown as they would appear to an observer in outer space who is looking directly at the center of the hemisphere. Note that the tangent plane at that point is the only one that appears square. A north-south line, aligned exactly with the meridians, is drawn on each of these planes.

Notice what happens as the earth rotates from west to east. Consider first the three tangent planes in the middle latitudes of the Northern Hemisphere. From left to right, they represent the successive views of the same point on the globe to the fixed observer in outer space, as the earth rotates beneath him. As you can see from the arrows at the right of the drawing, the north-south line appears to the observer to rotate counterclockwise as the point moves from west to east. Now imagine an air particle moving freely and in a straight line across the tangent plane. To the observer in space, the air particle appears to move in a straight line, while the tangent plane appears to rotate counterclockwise under the moving air. If, however, the point of vantage is shifted from outer space down to the earth, things seem quite different. As we move from west to east, to us the tangent plane appears to be motionless, but the direction of any freely moving air seems to rotate *clockwise* with respect to the apparently non-moving earth. The contrast between this view and that of the observer in outer space illustrates, among other things, the relativity of direction of motion. The same principle is illustrated by the belief of the ancients that the sun revolved around the earth from east to west. Copernicus found it simpler to explain the mechanism of the solar system by having the earth rotate from west to east with respect to a fixed sun.

Because the rotation of the earth causes the grid lines of the Northern Hemisphere (by means of which we judge directions) to rotate counterclockwise, freely moving bodies in the Northern Hemisphere appear to be deflected in a clockwise direction. For the Southern Hemisphere, the north-south line in the similar group of three tangent planes seems to the outer-space observer to be rotating in a clockwise direction, and for the earth-bound observer, wind direction seems to be deflected counterclockwise with respect to a fixed earth grid. At the equator, the north-south line does not rotate; hence the earthbound observer does not imagine any deflection of the wind, either clockwise (as in the Northern Hemisphere) or counterclockwise (as in the Southern Hemisphere).

In the diagram, the planes tangent at the poles show up only as lines, since they are drawn as viewed by an observer positioned over the equator. Nevertheless, it should be easy to imagine the north-south lines drawn on these planes. As the earth rotates, so do the lines; and at the poles, they rotate more than at any other latitude. Therefore, winds appear to be deflected more at the poles than elsewhere.

In the consideration of the apparent deflection of winds resulting from the rotation of the earth, the following points are important:

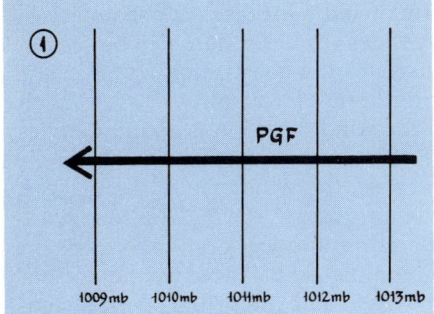
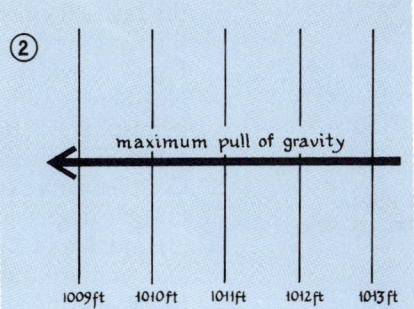
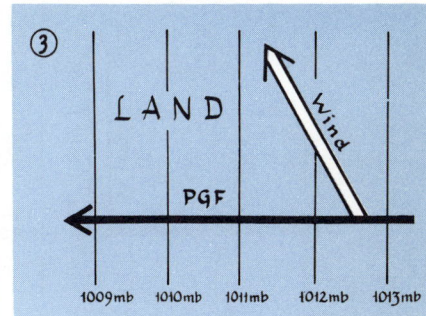
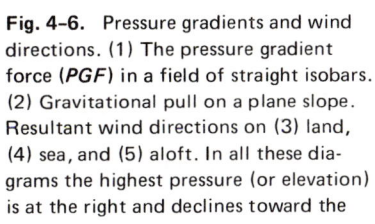
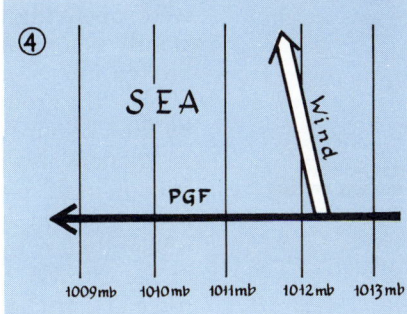
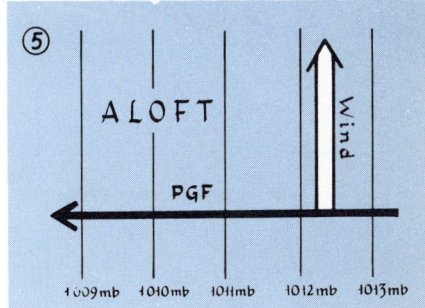

Fig. 4-6. Pressure gradients and wind directions. (1) The pressure gradient force (*PGF*) in a field of straight isobars. (2) Gravitational pull on a plane slope. Resultant wind directions on (3) land, (4) sea, and (5) aloft. In all these diagrams the highest pressure (or elevation) is at the right and declines toward the left.

1. For convenience of discussion, the apparent horizontal deflection due to the earth's rotation is ascribed to the **Coriolis force**,[4] even though no real force is operating; the deflection is only apparent and is apparent only to an earthbound observer.
2. Because of the Coriolis force, moving objects seem to be deflected in a clockwise direction north of the equator and counterclockwise in the Southern Hemisphere.
3. The Coriolis force is greatest at the poles, and is not evident at the equator.
4. The Coriolis force is greater for rapidly moving objects than for slowly moving objects.

Winds Aloft and at the Surface. How is the pattern of winds affected by the Coriolis force? Figure 4-6 (1) shows a particular pressure distribution at a particular level. In this situation the isobars happen to be straight and evenly spaced. The direction of the greatest pressure change in a given horizontal distance and the force exerted by this pressure change, or gradient, are represented by the arrow labeled *PGF*, for **pressure gradient force**. By analogy with slope gradients and the flow of water downhill (Fig. 4-6 (2)), the isobars compare with contour lines, and the *PGF* is comparable to the force of gravity acting down the steepest slope. In both instances, the steepest gradients are indicated by lines drawn at right angles to the isobars or to the isohypses. Water set in motion by the force of gravity flows downhill approximately in the direction shown by the arrow in Fig. 4-6 (2), but not exactly, because the Coriolis force deflects the flow clockwise (counterclockwise in the Southern Hemisphere). The speed of the moving particles in water flowing downslope is very slow compared with that of most wind movement, and since this speed is further reduced by friction between the water and the ground, the effect of the Coriolis force is negligible.[5] The speeds of the winds blowing "down" the pressure gradient are much greater, and friction is much less; consequently, the Coriolis force exerts enough clockwise deflection that the winds cross the isobars at angles considerably less than 90°, the degree depending on the roughness of the terrain. Over land, the angle between wind and isobars is about 30° (Fig. 4-6 (3)); over the oceans it may be only 15° or even less (Fig. 4-6 (4)); and above certain elevations (ranging from 3,000 to 5,000 feet or 900 to

---

[4] Named after the French mathematician G. G. de Coriolis (1792-1843), who first demonstrated quantitatively the existence of this force in 1835.

[5] There is some evidence, though quite inconclusive, that streams in the Northern Hemisphere tend to erode their right banks more than their left banks, since the clockwise deflection would force the water toward the right bank. In the Southern Hemisphere, it may be that the left banks are subject to slightly more erosion.

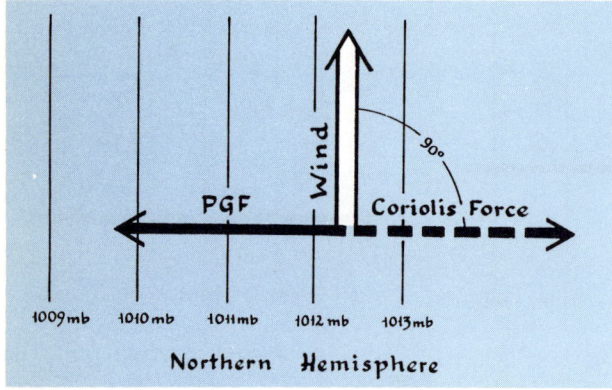

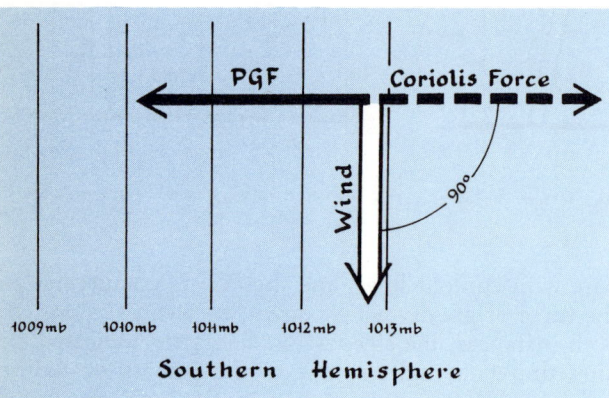

Fig. 4–7. The geostrophic wind in the Northern and Southern Hemispheres. The pressure gradient force (*PGF*) is at right angles to the isobars and the Coriolis force must be equal to the *PGF* and opposite in direction. Since the wind direction must be at right angles to the Coriolis force, it must therefore be parallel to the isobars.

the faster the resulting wind. One other consequence of the Coriolis force (besides the deflection) can be seen in Fig. 4-7, which illustrates the geostrophic wind in the Northern and Southern Hemispheres. In both hemispheres, since the wind direction is at right angles to the *PGF*, the *PGF* must be completely balanced by the Coriolis force; otherwise, the winds would be inclined to some degree in the direction of the *PGF* (as in Fig. 4-6 (3) and 4-6 (4)). Therefore, the accelerative force of the *PGF* on the winds is completely balanced by the Coriolis force, and the air moves at an equilibrium speed dependent upon the strength of the *PGF*. If the *PGF* alone were operating, wind velocities would increase constantly as long as the existing pressure pattern were not broken up.

Near the ground, the *PGF* is not completely balanced by the Coriolis force because the reduction in speed due to friction also reduces the Coriolis effect; but since friction itself serves as a brake on the speed, equilibrium is also reached. Notice in Fig. 4-8 that the Coriolis force is slightly less than the *PGF* (as indicated by the length of the arrows), but the combined effect of the Coriolis force and friction just balances the *PGF*. Bear in mind that the Coriolis force must be shown as acting at 90° from the wind direction and that friction is opposite in direction to the wind. In this situation, equilibrium is reached only if the winds blow at an angle to the isobars. Figure 4-8 represents the same situation as Fig. 4-6 (3) and 4-6 (4).

So far, we have considered only the movement of air resulting from a pressure distribution that can be shown by straight isobars. Much more commonly, the isobars

1,500 m), the winds no longer cross the isobars but blow parallel to them (Fig. 4-6 (5)). Note that from land to sea and from sea to upper air, the amount of friction decreases, with a resultant increase in wind speed and, consequently, an increase in the effect of the Coriolis force. Notice also that the maximum deflection from the *PGF* is 90°. A deflection greater than 90° would cause the winds to blow against the pressure gradient (upslope), which is impossible, since the Coriolis force is only an apparent force and can only result in apparent changes in direction.

The winds aloft closely approximate the **geostrophic wind**, a frictionless wind that blows parallel to straight isobars. The speed of such a wind is proportional to the intensity of the pressure gradient, which, in turn, is reflected by the closeness of the isobaric spacing. The closer the isobars, the steeper the pressure gradient and

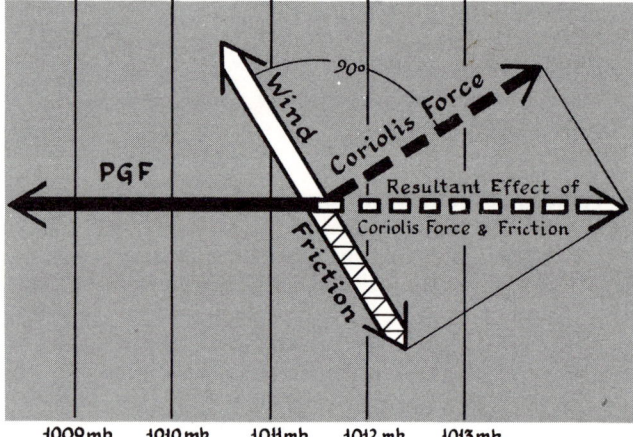

Fig. 4–8. The balance between pressure gradient, Coriolis, and frictional forces at the surface. Friction has reduced the wind speed and thus also the Coriolis force. But the resultant of these two forces is sufficient to counter-balance the *PGF* if the wind direction is shifted as shown. The wind now blows across the isobars.

Pressure and Winds 49

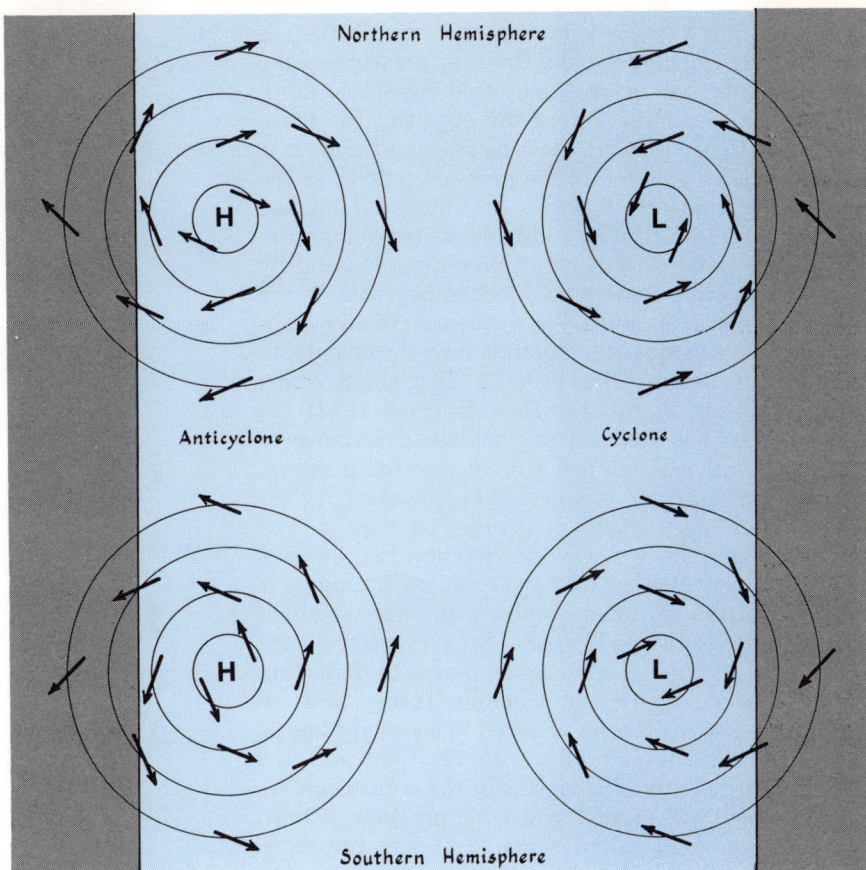

Fig. 4–9. Wind flow in the two hemispheres. The gray represents land masses. There is a clockwise flow around Northern Hemisphere high pressure centers (anticyclones) and Southern Hemisphere low pressure centers (cyclones). Conversely, there is a counterclockwise flow about Northern Hemisphere lows and Southern Hemisphere highs.

are curved, and the wind blowing along curved isobars is called the **gradient wind.** The velocity of the gradient wind depends not only on the *PGF*, the Coriolis force, and the forces resulting from friction, but also on centripetal and centrifugal forces, depending on whether the air is moving around a center of high pressure or low pressure.

Since gradient wind flow is so much more common than geostrophic wind flow, we need to remember only the patterns of gradient wind flow. The simplest way to remember the way in which the wind blows in a given situation is to visualize the direction of the pressure gradient and imagine a deflection of about 60° clockwise for land, a little more over sea, and a full deflection of 90° in the upper air. Of course, the clockwise deflection applies only in the Northern Hemisphere. For Southern Hemisphere wind patterns, the deflection is counter-clockwise from the direction of the *PGF*. We can always check the general accuracy of wind direction by the rule of thumb known as *Buys Ballot's law*[6]: in the Northern Hemisphere, low pressure will be to the left of an observer facing downwind; in the Southern Hemisphere, low pressure will be to the right.

The terminology of wind flow distinguishes between **cyclones** (Greek *kyklos* circle) and **anti-cyclones.** "Cyclone" is the term properly applied to any center of low pressure. The wind flow about such a center is termed cyclonic whether it is counterclockwise, as in the Northern Hemisphere, or clockwise, as in the Southern Hemisphere (Fig. 4-9). Similarly, centers of high pressure are termed anticyclones, and the wind flow about them is said to be anticyclonic, again whether it is clockwise, as in the Northern Hemisphere, or counterclockwise, as in the Southern Hemisphere. The various resultant winds in Fig. 4-9 represent the approximate wind flow over oceans and over land. Notice the greater angle at which the arrows cross the isobars over land; aloft, the arrows would be nearly parallel to the isobars.

---

[6]C. H. D. Buys Ballot (rhymes with "rice pallet") was a Dutch meteorologist who first published his "law" in 1857, sometime after it had been discovered by the American meteorologists William Ferrel and James Coffin.

50 Chapter 4

Pressure is related to wind much as the chicken is related to the egg. We usually think of pressure as the cause of winds, just as we usually find it easier to think of the chicken coming before the egg. But for those pressure patterns that are dynamically produced, it is obviously easier to think of the pressure pattern as the result of the winds. (In this case, the chicken is hatched from the egg.) In fact, we could have started our discussion of the relationship between pressure and wind by assuming a wind pattern and deducing from it the resulting pressure patterns. For the benefit of any skeptics, a simple example will illustrate how the discussion might have proceeded. In Fig. 4-10 (1), a single arrow depicts some sort of initial air flow. In Fig. 4-10 (2), the wind direction has been shifted clockwise in response to the Coriolis force. The open arrows are used to represent the tendency for this deflection to affect the air everywhere along the line of general movement. The constant swing in a clockwise direction results in a deficit of air on the left and a surplus on the right. If enough air piles up on the right, a pressure gradient from right to left is established. This pressure gradient (Fig. 4-10 (3)) counteracts any tendency to deflection toward the right by the Coriolis force; and an equilibrium flow will be established in a straight line as before, but with high pressure on the right and low pressure on the left. This is exactly the conclusion we reached when we started with the pressure pattern (compare Fig. 4-10 (3) with Fig. 4-7). We can summarize this reversal of cause and effect by comparing the atmosphere to a machine with a tight and sensitive feedback mechanism. Every change in wind affects the pressure distribution; every change in pressure influences the velocity of the wind, and so it goes, in an endless cycle.

The Existing Planetary Pattern of Pressure

Now that we have introduced the major factors that affect the creation of pressure differences and gradients and the associated wind velocities, we can turn to a discussion of the existing patterns in our atmosphere and the reasons for them.

Interpretation of Seasonal Sea-level Pressure Maps

Figure 4-11 shows the isobars of average pressure for January and July, reduced to sea level. Two characteristics of such maps of average sea-level pressure must be well understood before the description of their major features has any meaning: (1) The isobars are based on monthly averages, and (2) pressures are reduced to sea-level values.

All sorts of pressure observations have been lumped together to yield a single monthly value for the isobars. This derivation of values is analogous to that for the

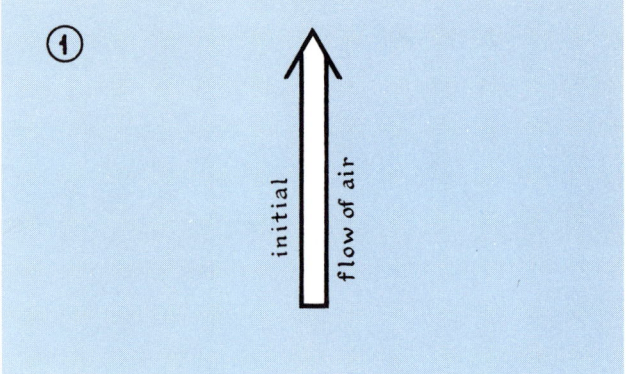
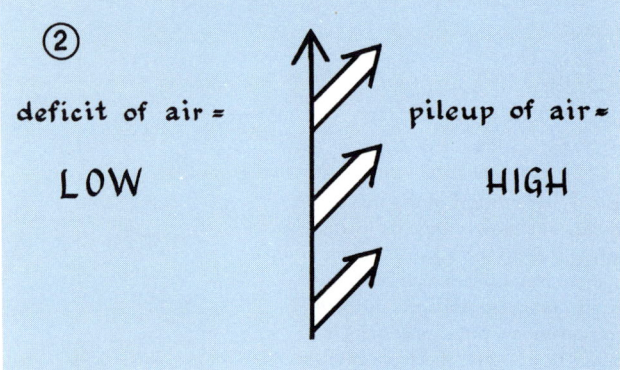
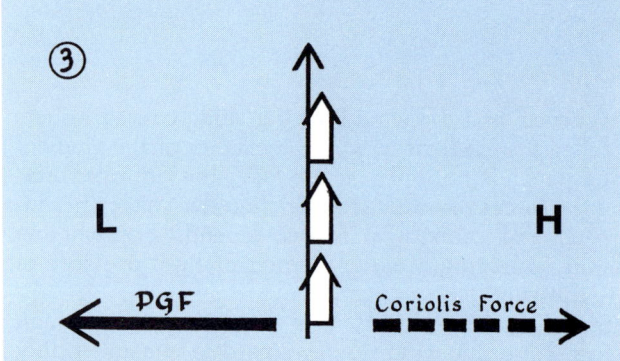

Fig. 4-10. Pressure patterns resulting from straight wind flow in the Northern Hemisphere. An initial flow of air (1) is deflected clockwise, thus creating low pressure to the left of the flow and high pressure to the right (2). This situation can also be represented by showing Coriolis force exerted to the right and balanced by an equal *PGF* toward the left (3).

temperature maps. As you recall, the temperatures during the day, at night, on hot days, and on cold days were combined. The resulting average could be interpreted only by keeping in mind the diurnal range of temperature and by making some allowance for possible hot and cold spells. However, although it occurs in many places, diurnal variation is relatively unimportant for pressure averages. The significant variations occur over irregular lengths of time, and what is more difficult (from the point of view of interpretation of the patterns), slight changes in pressure, which scarcely affect the average, may create significantly different weather situations. In other words, the average gives no indication of day-to-day variation.

We can illustrate this problem using two examples from the July map (Fig. 4-11 (2)). Over the North Pacific, the map shows the well-developed North Pacific high, centered about 35° N latitude and 155° W longitude. North of the high is an east-west zone of lower pressure between the North Pacific high and a cell of high pressure overlying the north polar regions. This trough of low pressure is, in fact, occupied by high pressures most of the time but is crossed at intervals by well-developed low pressure centers moving from west to east. The weak low portrayed on the map is based on pressure values obtained by averaging rather frequent but fairly weak high pressure with less frequent but strong low pressure. It does not tell us much about the actual day-to-day situation near the Aleutian Islands. On the other hand, the North Pacific high is consistently present on most daily weather maps; the average high pressure describes the daily situation very well.

The second problem met in the interpretation of the maps of seasonal pressure is the reduction of pressure to sea-level values. Without such a reduction, the pressure map is nearly the same as an elevation map. After all, one of the common ways of determining elevation above sea level is by means of an altimeter; and the altimeter is nothing more than a barometer with an elevation scale substituted for the pressure scale. To show pressure differences that do not result from differences in elevation from place to place, we must refer all pressure to some common elevation. Sea level is the most obvious plane of reference, since the seas occupy three-fourths of the earth's surface. Therefore, while pressures over the oceans are shown just as they are, pressures measured over the continents must be increased by amounts that depend primarily on the elevations of the recording stations. At the beginning of the chapter, we noted that pressures decrease by 34 millibars for every 1,000 feet (11 mb for every 100 m) of elevation above sea level. But this change is only approximate and is valid only for an atmosphere of average density. Any change in the density of the air produces changes in the pressure-elevation relationships. The density of the air depends principally on its temperature, and adjustment for the density can be made by including the temperature in the calculations of reduced pressure values, a method which works well enough if we know the pressure at sea level and want to estimate the pressure at some point in the atmosphere. The temperatures of the air overlying a given place can be measured or at least estimated, and the pressure aloft can thus be calculated. On the other hand, if we are given the pressure of some station at considerable elevation, reducing such pressure to the sea-level value is difficult. For instance, consider the situation at the South Pole, whose elevation is about 9,000 feet (2,750 m) above sea level, and where during the International Geophysical Year, many meteorological observations were taken. If we wish to reduce the observations of pressure at the pole to sea-level values, it is necessary to imagine the hypothetical density of the air column between the station and sea level! Even if we can conceive of such a column of air, there are no logical temperatures that can be assigned to it; but some estimate of temperature is needed to compute the density of the fictitious air, which, in turn, is needed to estimate pressure. As we can see, the procedure whereby pressures recorded at considerable elevations are reduced to sea-level values is by no means simple and is certainly not very reliable for elevated continents — particularly when the continent has abnormally low temperatures. It would be much easier and perhaps more logical to reduce pressures (and temperatures) to some specified high elevation such as 30,000 feet or 10,000 m rather than to sea level. This would have the advantage of eliminating the need to imagine lapse rates of temperature in fictitious columns of air since there would be some atmosphere between every station, no matter what its elevation, and the imaginary surface whose elevation is 30,000 feet above sea level.

The maps of average pressure reduced to sea-level values must, therefore, be interpreted with some care, since the averages may mask some of the most significant changes of pressure from day to day and since the reduction to sea-level values is often unreliable. Nevertheless, Fig. 4-11 does indicate something about the prevalent pressure conditions; and it is much easier to interpret than the mass of data — thousands of daily weather maps — from which it was derived.

Some General Notions About the World Pattern of Pressure

The pressure maps can be best understood and remembered if we bear in mind three generalizations about them. First, there is a rough zonality (east-west

52 Chapter 4

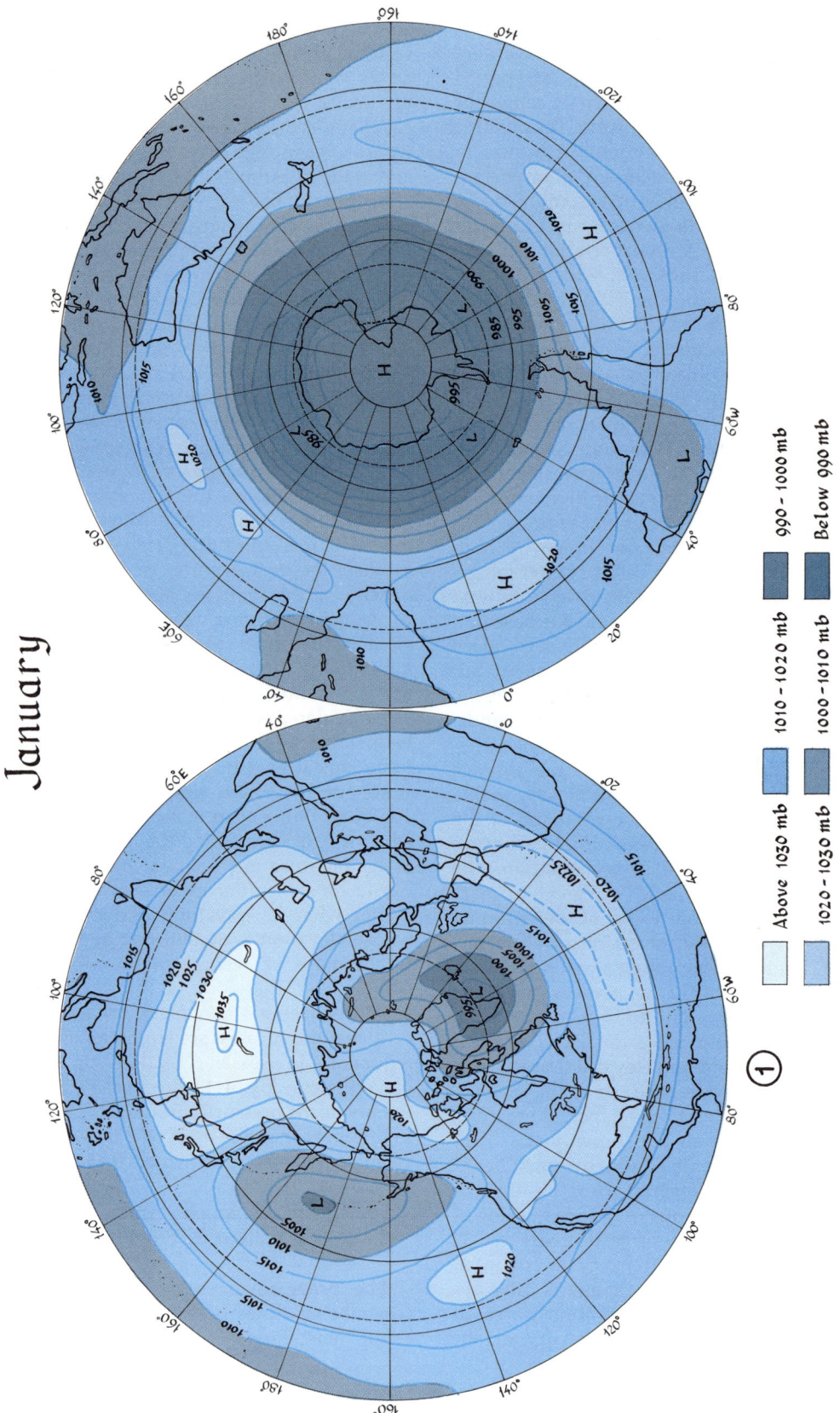

Fig. 4-11. January (1) isobars for the world.

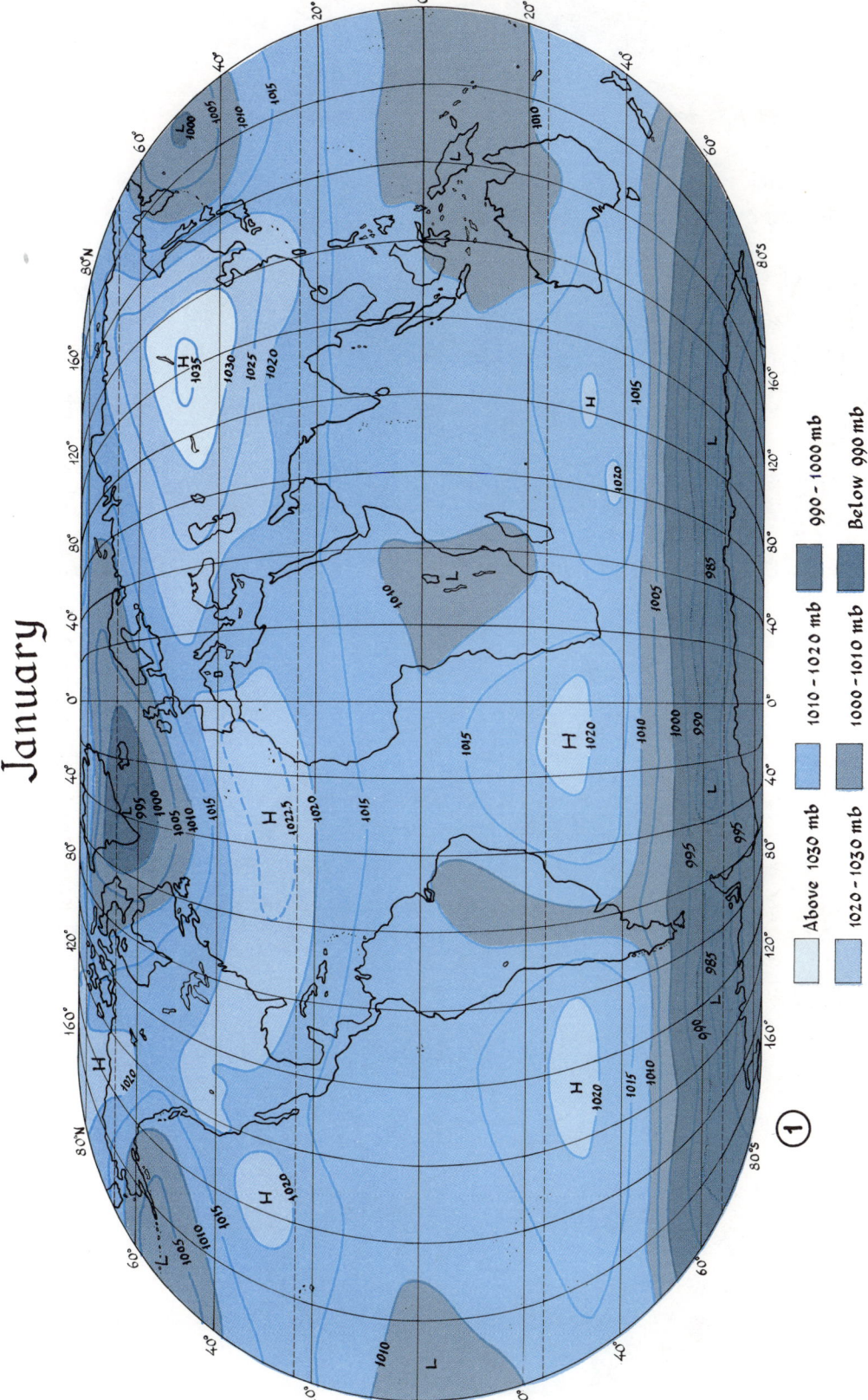

Fig. 4–11. January (1) pp. 52 and 53, and July (2) pp. 54 and 55, isobars for the world. The pressure values have been reduced to equivalent pressures at sea level, and are given in millibars. The solid lines represent isobars at intervals of 5 mb; the dashed lines are intermediate values at 2.5 mb intervals to emphasize certain pressure patterns.

54 Chapter 4

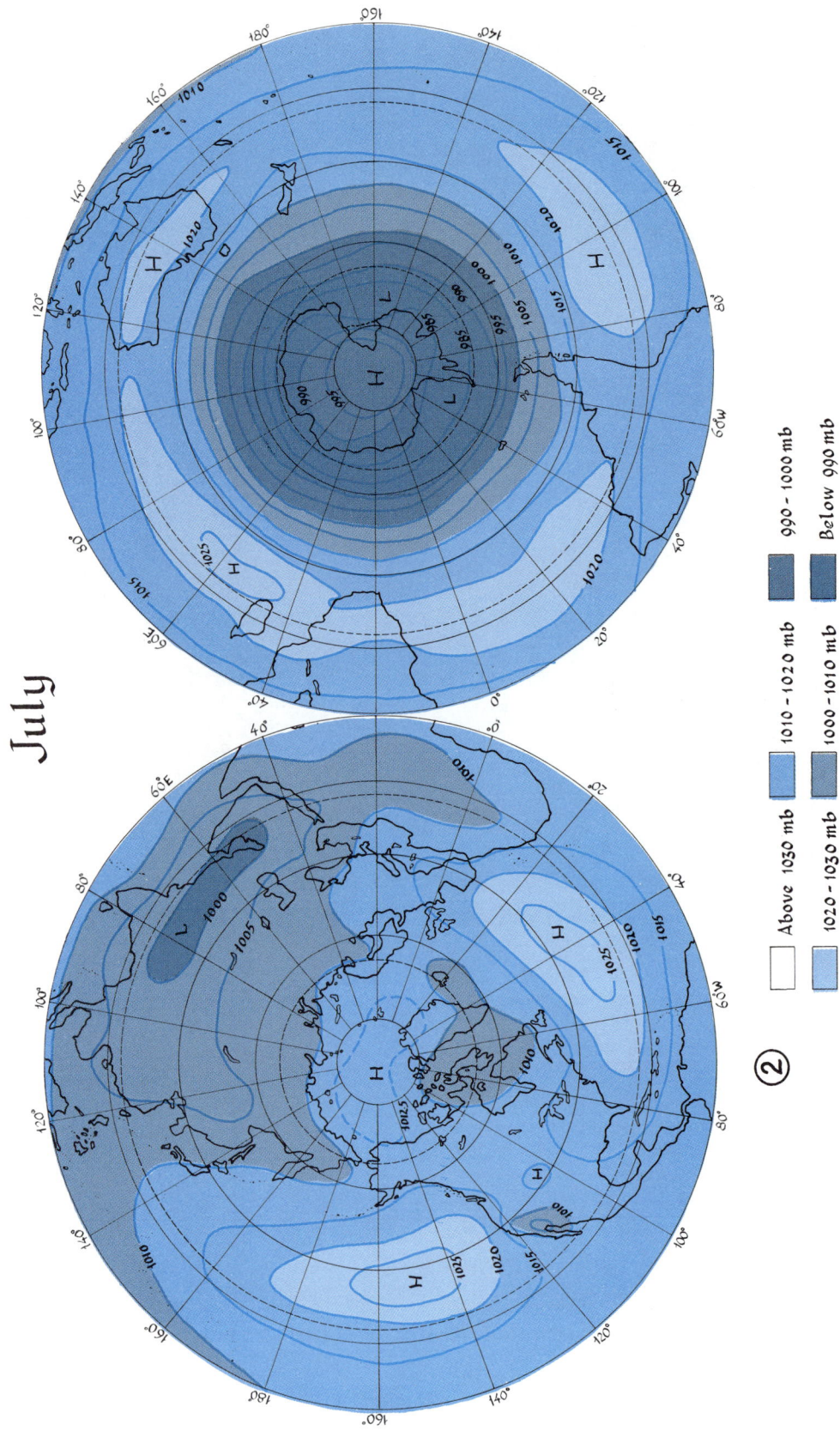

Fig. 4-11. July (2) isobars for the world.

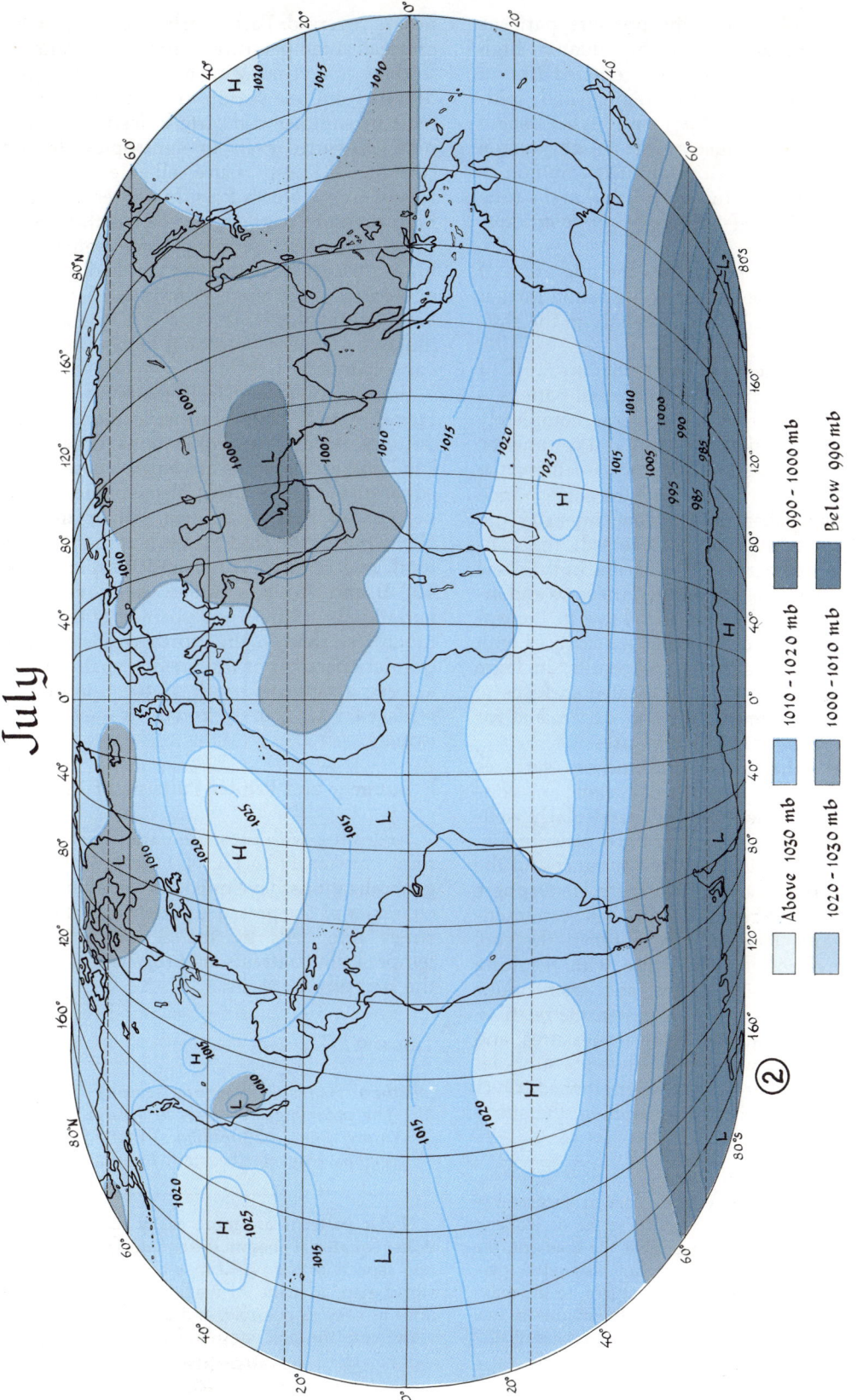

Fig. 4-11. July (2) isobars for the world.

orientation) of the isobars and the pressure patterns they describe: low pressure over the equator, high pressure over the subtropical latitudes (centered around 30°-35°), low pressure over the higher and midlatitudes (centered around 60°-70°), and high pressure over the polar regions. Because of the roughly zonal character of the pressure distribution, it is common to speak of the *equatorial low*, the *subtropical highs* (one in each hemisphere), the *mid-latitude* or *subpolar lows*, and the *polar highs*.

The second generalization involves the migration of these pressure zones. In January they are found farther north than in July. This shift is due mostly to the shift of temperature zones.

The third generalization concerns the breakup of pressure zones into pressure cells. The zonal pattern is very much distorted, and in some places completely destroyed, by the distribution of oceans and continents. In summer, the relatively hot continents intensify the lows and weaken or destroy the highs; the correspondingly cooler oceans weaken the lows and intensify the highs. In winter, the situation is reversed: the continents are relatively colder than the oceans in the same latitude; the colder land masses weaken the lows and intensify the highs, while the oceans behave in the opposite way. This behavior of land and sea surfaces with respect to the overlying atmosphere results, in large part, from the general tendency for heated surfaces to produce low pressures immediately above them and for cooled surfaces to produce high pressures.

The differential behavior of land and sea surfaces modifies the zonal pattern into one of oceanic and continental pressure cells. These cells are particularly well-developed and most easily seen over the oceans. In Fig. 4-11, observe the subtropical highs of summer and the subpolar lows of winter over the Northern Hemisphere oceans. In both cases, the zonal pressure has been intensified by the oceanic location[7]. In summer, the cool oceans tend to produce high pressure with resulting strength of the subtropical highs and tremendous weakening of the subpolar lows. In winter, the relatively warm oceans are associated with lower pressures; and the subtropical highs are weakened, while the subpolar lows are enormously enlarged and strengthened. Over the continents of the Northern Hemisphere, the situation is reversed. Particularly over Asia — the continent of greatest temperature contrast from summer to winter — there is a striking reversal of surface pressure conditions from one season to the other. The heated continent in summer completely destroys any evidence of high pressure, even in subtropical latitudes, and a continental low-pressure cell extends from the equator almost to the North Pole. In winter, on the other hand, high pressure reigns from the North Pole practically to the equator because low surface temperatures produce high pressure near the surface, and low pressures disappear from the map, even in the latitude of the subpolar lows. North America is a smaller continent and illustrates these seasonal pressure changes less conspicuously.

So far, we have dealt mainly with the Northern Hemisphere. The Southern Hemisphere pressure patterns fit our generalizations only approximately. Land and sea contrasts are much less strong south of the equator. In the Northern Hemisphere, land occupies 40 percent of the total area, while in the Southern Hemisphere, the land area is less than 20 percent of the total. In fact, there is no significant land mass between 40° S and 70° S latitudes. As a result, the seasonal variation in strength and position of the pressure cells is much less than in the Northern Hemisphere; and the cellular character of the pressure pattern which is weak at best, disappears south of 40° S latitude, where it is replaced by a purely zonal, or east-west, pattern of isobars.

Summer and Winter Patterns of Pressure

In dealing with pressure patterns, it will be helpful to refer constantly to Fig. 4-11 and to bear in mind the generalizations developed above; that is, (1) the zonal character of the pressure pattern, (2) the breakup of the zones into cells by the action of land and water temperature contrasts, and (3) the seasonal migration of the pressure patterns.

January. (Fig. 4-11 (1).)

Northern Hemisphere (winter season)
1. The polar high[8] consists of a ridge of high pressure (more than 1,020 millibars) that crosses the polar regions from northeastern Eurasia to the Yukon

---

[7] The intensity of a pressure system refers to its pressure gradients (which are read on the map in terms of isobaric spacing — the closer the isobars, the steeper the gradient, the more intense the pressure cell); but it can also refer, particularly if we are describing very large features, to the maximum pressure at the center of a high or to the minimum in a low. The two notions are closely related, even if not identical.

[8] Polar patterns are more easily seen on the polar projections; equatorial ones on the equatorial view of the world. The *two* aspects of the world map are shown for these and other world maps to make the reading and interpretation of the isolines as easy as possible, and also to show how different the view from the pole may be from that over the equator.

area of North America; it joins the two coldest parts of the world in January.

2. The subpolar low-pressure belt is broken by the continents into the Aleutian low (with pressures less than 1,000 millibars centered in about 50° N latitude), and the Icelandic low (with pressures less than 995 millibars centered in about 60° to 65° N latitude). These well-developed low-pressure cells are centered over the parts of the oceans that are abnormally warm for their latitudes in this season.

3. The subtropical high-pressure belt is strongest over the cold continents, where it extends all the way to the polar high. The pressure is much higher over Asia (1,035 millibars) than over North America (1,023 millibars). The high-pressure cells over the oceans are fairly weak and are located relatively far to the south, but they do occupy the area between 15° N and 40° N latitudes, and reach maxima of more than 1,020 millibars over the Pacific Ocean and more than 1,022.5 millibars over the Atlantic Ocean.

Southern Hemisphere (summer season)

4. The equatorial low extends into the Northern Hemisphere to about 15° N latitude but is mainly found south of the equator, extending southward for considerable distances over the warm continents of the Southern Hemisphere.

5. The subtropical high is broken into three cells, one over each of the southern oceans. The centers of the highs are in about 30° to 35° S latitude and attain pressures in excess of 1,020 millibars.

6. The subpolar low is practically continuous around the world in 60° to 65° S latitude, with pressures less than 985 millibars.

7. There may be some evidence of a polar high, though the indicated pressures are quite modest (995 millibars). The sea-level stations operating in Antarctica during the International Geophysical Year seldom observed pressures greater than 1,000 millibars.

July. (Fig. 4-11 (2).)

Northern Hemisphere (summer season)

1. The polar high has practically disappeared from the map with the warming of the Arctic Ocean in summer. A weak ridge of high pressure separates the Asiatic low, but pressures are less than 1,020 millibars.

2. The subpolar low-pressure belt has almost disappeared from the oceans. A weak remnant of the Icelandic low is centered just north of 60° N latitude, but no trace remains of the Aleutian low over the North Pacific. By contrast, a tremendous low-pressure system covers all of Asia and extends from high latitudes all the way to the equator.

3. The subtropical high-pressure belt in summer is well developed over the oceans, in contrast to the winter situation. The North Pacific high and the Azores high are both about 5 millibars stronger in July than in January and are both centered about 5° north of their weaker January counterparts.

4. The equatorial low lies almost entirely north of the equator in July. The center of low pressure lies over the oceans in about 10° to 15° N latitude and is much farther north of the low pressure over Asia and North America.

Southern Hemisphere (winter season)

5. The subtropical highs seem to be more strongly developed over the oceans than over the cold continents! This behavior does not fit the generalized pattern we have seen elsewhere. Probably the smallness of the continents and their predominantly low latitudes do not allow enough cooling of air masses to affect the pressure patterns in the same way as in the Northern Hemisphere.

6. The subpolar low is a continuous belt in about 65° to 70° S latitude.

7. The polar high is again represented by very modest pressures, 995 millibars and above.

The Pressure Patterns Aloft

At some distance from the ground, the pressure relationships change, as we saw in Fig. 4-3. The major changes involve the gradual replacement of the equatorial low by high pressure aloft and the transformation of the polar high into a zone of low pressure. At elevations of about 10,000 feet (3 km), the cellular character of the sea-level isobars is gone; the influence of land and sea contrasts, which loomed so large in the sea-level pressure patterns, is practically extinguished at this distance from the earth. Figure 4-12 shows the isobaric patterns at an elevation of 20,000 feet (6 km) above sea level in the Northern Hemisphere. At this elevation, the cellular character of the sea-level circulation has been completely obliterated in January (Fig. 4-12 (1)). The subpolar low has merged into a single low-pressure cell that completely covers the polar regions and extends far to the south. The subtropical high is part of the belt of relatively high pressures that completely circumscribes the earth in equatorial latitudes. The July map (Fig. 4-12(2)) shows some remnants of the subtropical high and a generally weaker gradient of pressure than in January.

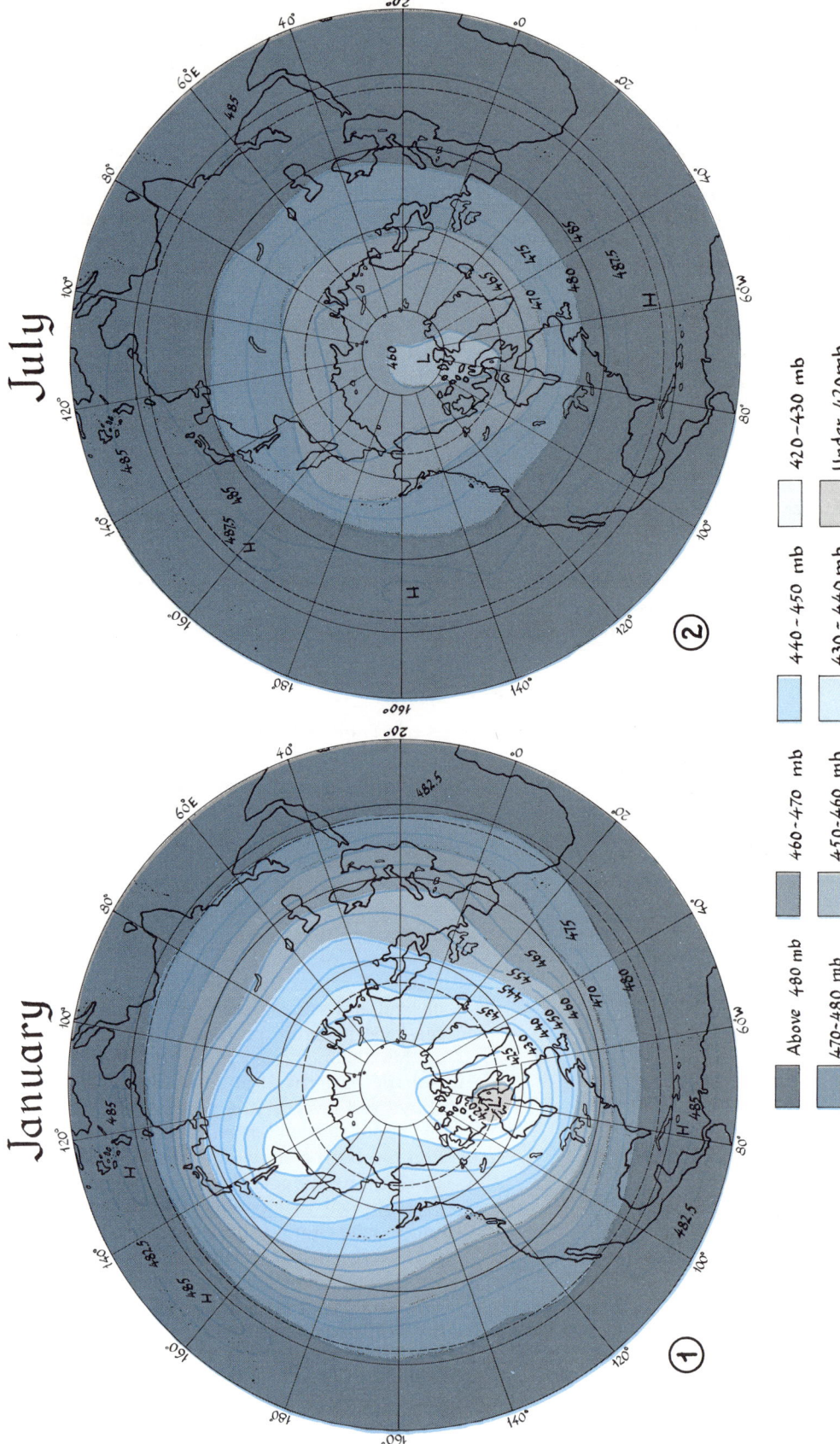

Fig. 4-12. January (1) and July (2) isobars at 20,000 feet (6 km) in the Northern Hemisphere. Note the nearly circular pattern of isobars in each season. Land and sea contrasts are not reflected at this distance from the surface. Note especially the much more intense pressure gradient in winter as shown by the more closely spaced isobars in (1).

The Planetary Wind System

Aloft, where friction is negligible, the Coriolis force has maximum effectiveness; and the wind movement is parallel to the isobars. The huge low-pressure cell that extends over almost all the Northern Hemisphere is thus associated with westerly winds; and the circulation of air is almost entirely zonal rather than cellular. The strength of this zonal circulation varies seasonally and from place to place in the atmosphere. The highest mean wind speeds occur in winter, at about 40,000 feet (12 km) above sea level and in 30°N latitude (also in 30° S latitude). The westerly winds in this part of the atmosphere, which are unique in their strength and persistence, are known as the **jet stream** and have received considerable attention from meteorologists since their discovery during World War II. Near the core of the jet, speeds of 150 miles per hour (67 m per second) are common; speeds greater than 400 miles per hour (179 m per second) have been recorded; and even the mean velocity is more than 90 miles per hour (40 m per second).

The mean velocity of the jet stream is about twice as great in winter as in summer. It is generally stronger in the Southern Hemisphere than in the Northern Hemisphere; it is usually located farther poleward in summer; and it varies in strength along its course around the earth. In addition, it has a meandering path, so that part of the stream may be found at considerable distances north or south of its normal position. This meandering is reflected in the sinuous character of the isobars at high elevations.

The mean position and the mean strength of these winds hide the fact that there is, at a given moment, more than one zone of very strong westerly winds. In fact, there are at least two jets in each hemisphere. One maximum of wind flow, located directly above the surface subtropical highs in 30° latitude, is called the subtropical jet. The other maximum is located directly above the polar front. The significance of these jets to surface weather and their connection with the polar front will be touched on in Chapter 6.

The winds at the surface are more difficult to describe than those aloft. In the first place, the cellular pressure pattern of the surface is more complicated than the zonal pattern aloft. In the second place, frictional effects vary considerably, particularly from land to sea, so that the amount of deflection by the Coriolis force is variable. Finally, surface winds are affected by the major features of the earth's land surface. In a narrow canyon, for instance, the surface wind can blow only upstream or downstream, regardless of the pressure pattern. For all these reasons, surface winds are not only more difficult to describe but also much less consistent in velocity.

Certain major patterns are quite apparent, however, particularly over the oceans. The zonal nature of pressure patterns, its modification into a cellular pattern by land and sea contrasts, and its seasonal migration are all reflected in the wind patterns. But before we describe the several maps depicting the surface pattern of winds, it is necessary to say something about the ways in which a great number of measurements of wind velocity may be averaged into a single value. On first thought, it might seem that averaging speeds would be a straightforward business once an array of reliable wind speeds had been accumulated. One could simply add all the observed speeds and divide by the number of observations. This computation would give the average speed of the wind and serve admirably to convey a picture, although limited, of wind strength. Quite often, however, it is not the wind speeds that are important but rather how much air is being transported past the place in question. For instance, suppose that at noon the wind velocity was recorded as 5 miles per hour from the north; whereas at one o'clock the recorded velocity was also 5 miles per hour, but from the south. The mean wind speed for the two observations is obviously 5 miles per hour, but the mean transport of air is nil.

Wind velocity is usually depicted by a **wind rose** such as those shown in Figs. 4-4, 4-13 (1), and 4-15, in which the directions of wind flow are shown by lines whose lengths are proportional to the frequency of occurrence of the wind from the particular direction. The average speed of wind from a particular direction is also shown on the wind rose, usually by means of feathers on the end of the bars (Fig. 4-13(1)) or by some other symbol such as the thickness of the line, as shown in Figs. 4-4 and 4-15. Two kinds of averages may be derived from

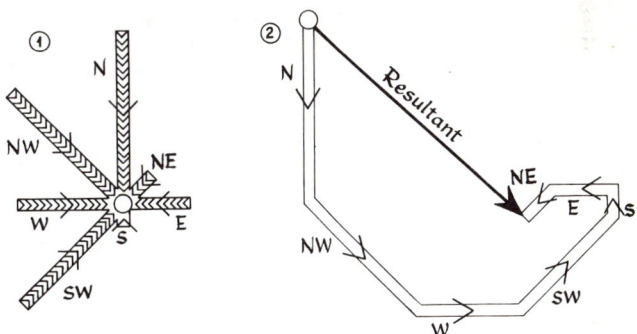

Fig. 4-13. Wind rose (1) and wind resultant (2). In (1) the length of the bars is proportional to the mean transport of air (frequency times speed). In (2), the resultant wind is developed geometrically, by taking the transport from each direction in turn and tacking the beginning of one line onto the end of the other. In this example, we show *N* winds first, then *NW*, and so on.

the wind rose. One may simply pick the direction from which the wind blows most often (the *prevailing wind*), and may average all of the speeds regardless of direction (the *average wind speed*). Another way to reduce the data shown by a wind rose to a single value is to compute the direction and strength of the **resultant wind.** To obtain the resultant wind, one considers both frequency of occurrence and strength of wind by computing the air transport from each of the directions. For example, a wind blowing for 30 hours at an average rate of 6 miles per hour results in a transport of air of 180 miles (30 × 6). The net transport from the north is obtained by subtracting values for south winds from those for north winds. Values for northwest winds are distributed between north and west winds, and so on. The result is a speed and a direction that represent the net average flow of air rather than the prevailing direction and the average speed. Figure 4-13(2) shows how such a resultant can be obtained graphically. Note that, on this diagram, all the arrows, including the one that represents the resultant wind, indicate air transport (in miles or some other unit of length) and not speed. To obtain the speed of the resultant wind, one must divide the number of miles of transport by the number of hours of observation.

This discussion of the problems of representing data that often vary considerably from day to day, even from hour to hour, was necessary to aid understanding of maps of seasonal wind velocities. It should also alert you to problems inherent in the representation of any variable quantity by a single abstracted value, such as an arithmetic mean or a resultant velocity.

Despite all these qualifications, however, the seasonal maps of wind velocity provide much useful information, particularly portions of the maps that deal with winds over the oceans. The major patterns of wind movement over the oceans are shown in Fig. 4-14. Continental winds have been omitted, largely because of the complicating effect of channeling of air movement by landforms and also because of complications brought about by variable effects of friction. The amount of frictional drag largely determines the deviation of wind direction from the pressure gradient, and localized differences in friction are reflected in localized variations in wind direction, which may or may not have much to do with the general circulation of the atmosphere.

On the January map, two zones of westerly winds are clearly apparent. One, in the Southern Hemisphere, is circumpolar; the other is broken up by the Northern Hemisphere land masses into North Pacific and North Atlantic segments. These strong but relatively inconsistent winds, called the **westerlies,** are part of the flow around the equatorward portions of the subpolar lows.

Easterly winds are evident in tropical and subtropical latitudes. These are the **trade winds.**[9] They are weaker but much more consistent than the westerlies and represent the equatorward flow out of the subtropical highs. The relationship between pressure and wind is evident in the counterclockwise patterns (reflecting anticyclonic movement about a high-pressure cell) over the three Southern Hemisphere oceans. It is less clear-cut in the North Pacific and North Atlantic Oceans, although there is a suggestion of clockwise movement of air in the eastern part of the North Pacific. The centers of high pressure, called the **horse latitudes,** are mostly represented in the Northern Hemisphere by weak easterly winds that rapidly give way, northward, to the westerlies. Between the southeast and the northeast trades is a zone of weak, variable easterly winds, known as the **doldrums** because here sailing ships were often becalmed.

The wind pattern in July shows several striking changes from that in January, particularly in the Northern Hemisphere. The weakening of the subpolar lows and the corresponding intensification and northward movement of the subtropical highs are reflected in the disappearance of cyclonic (counterclockwise) flow and the dominance of anticyclonic (clockwise) flow over both the North Atlantic and the North Pacific. The westerlies in summer are weaker, are found farther north, and tend to be more southwesterly than westerly. The northeast trades are also found somewhat farther to the north. Although they retain the consistency of the winter season, they are somewhat weaker in summer.

Superimposed upon these shifts in position, in consistency, and in type of flow is a general weakening of the air movement in summer, which corresponds to the weakening of the pressure gradient between pole and equator from winter to summer. This weakening, in turn, is related to the weakening of the temperature gradient.

Another striking change in the Northern Hemisphere circulation is the difference between flow in winter and flow in summer over the Indian Ocean. You will recall that earlier we spoke of this reversal of the winds from north or northeast in winter to southwest in summer as the monsoon. Superficially, it would appear that the monsoon circulation could be easily explained in terms of the pressure changes over the Asiatic continent. The north winds of winter in

---

[9]The word "trade" in this sense, comes from the identical medieval Dutch word meaning a course, a track, or a path. The courses, or tracks, of sailing vessels depended on the direction and constancy of the trade winds.

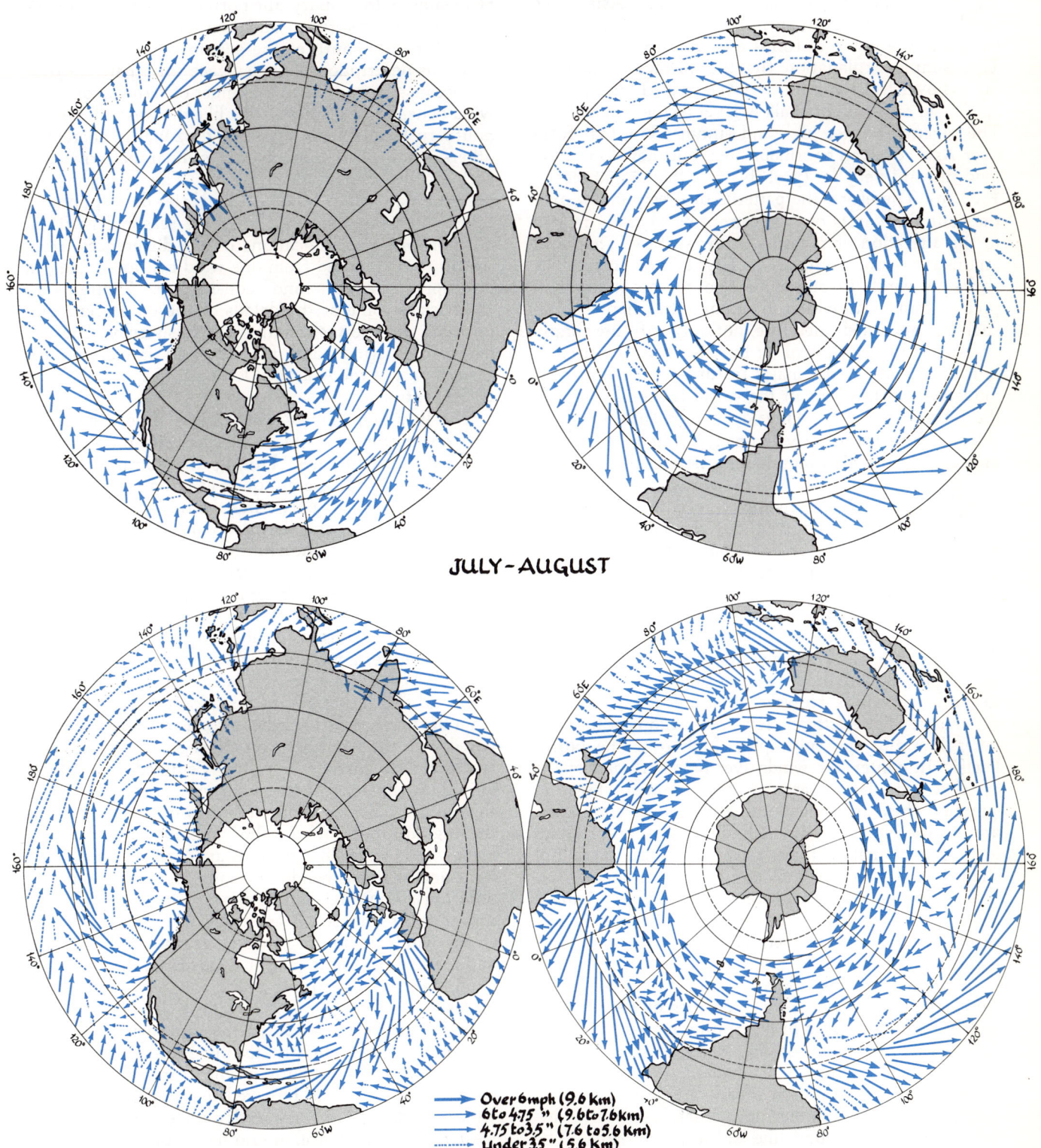

Fig. 4-14. January–February (1) and July–August (2) wind patterns over the oceans. Adapted from Sir Napier Shaw, Manual of Meteorology, Vol. 2, 2nd ed. (London and New York: Cambridge University Press, 1936), pp. 244–247, and used by permission.

southern Asia are often described as the direct result of the formation of the huge high over Asia. The southwest winds of summer are similarly related to the flow into the low that forms over northwestern India. This simple explanation fits the surface wind conditions over India and Burma-Malaya fairly well, but it applies to only a very shallow layer of the atmosphere. The gigantic winter high and summer low that form over Asia cover an enormous area, but they do not extend upward very far (probably less than 5,000 feet or 1500 m). Therefore, the patterns of precipitation over the southern coasts of Asia cannot be so easily related to the alteration of pressure at the surface. The word "monsoon," unfortunately, is applied not only to the shifts in wind direction but also to the changes in precipitation from winter to summer. In Chapter 7, we shall be concerned, therefore, not only with the simply described surface flow of air but also with the circulation aloft. One of the major features of the upper-air movement in this part of the world is the blocking of the jet stream by the Himalayas. The subtropical jet shifts northward in early summer, but its movement is impeded by the high mountains. As a result, there is a rather sudden change from winter conditions to summer conditions when the jet eventually "jumps" the Himalayan barrier. Without the Himalayas, one would expect a more gradual change from the winter pattern of north winds to the summer pattern of southwest winds.

Other regions experience seasonal wind shifts similar to the monsoon conditions of southern Asia. If we compare the winter and summer conditions of wind flow over Japan, we will indeed see a change from northwest to southwest winds. However, this shift is explained by the change in strength and relative positions of the subpolar low and the subtropical high over the North Pacific. In winter, the winds over Japan are part of the cyclonic circulation around the Aleutian low; in summer, they are part of the anticyclonic circulation around the North Pacific high. The monsoon of Japan is thus explained without reference to the large but shallow highs and lows of the Asiatic continent.

The Southern Hemisphere exhibits fewer changes from January to July (from summer to winter in this case) than does the Northern Hemisphere. This pattern is, of course, analogous to the small seasonal changes observed on the pressure maps. It is also related to the similar summer and winter intensities of temperature and pressure gradients in the Southern Hemisphere as compared with those of the Northern Hemisphere. Some slight changes are evident, however. The westerlies of the southern oceans are shifted slightly northward from summer to winter, as is the anticyclonic flow around the subtropical highs. The southeast trades in July extend to the equator and northward, so the doldrums are mainly in the Northern Hemisphere. In January, the northeast trades invade the Southern Hemisphere, and the doldrums are thus located mainly south of the equator.

In this description of the surface wind patterns over the oceans, we have used a number of terms to describe wind belts. Some of these belts, like the westerlies of the Southern Hemisphere, are strongly zonal in character; others are broken into distinct cellular patterns over each of the major ocean basins. The analogy with pressure patterns is obvious. The brief descriptions of these winds involved not only the seasonal shifts in position but also indications of the differences, from season to season and from one wind belt to another, in direction, strength, and consistency. Figure 4-15 emphasizes some of the details of these differences in position, direction, strength, and consistency. The two maps show wind roses for July and January in the eastern part of the Atlantic Ocean, from about 50° N to 15° S latitudes.

The July map (Fig. 4-15(1)) shows five wind belts. The northernmost wind rose represents the southern part of the westerlies. During the summer in this latitude, the winds are not quite as strong or consistent as they are farther north, but the fairly strong winds, the lack of calms, and the predominance of flow from the west are clearly shown. Proceeding southward, the next two locations are typical of the horse latitudes, near the center of the subtropical high (the Azores high, in the North Atlantic). There is little consistency of direction, and calms occur about one-fifth of the time. At the southern margin of the high, between 30° N and 20° N latitudes, are the northeast trades. At least one wind rose illustrates the remarkably consistent flow from the northeast, the rather strong winds (when compared with the wind belts on either side of the trades), and the near absence of calms. Centered near 10° N latitude is the zone of the doldrums. The wind rose in this latitude displays some features of the trades, but the high proportion of calms and the occurrence of some winds from all directions are characteristic of the light and variable winds normally associated with the doldrums. The three southernmost wind roses all lie in the zone of the southeast trades. Few calms, fairly strong winds (increasing in average speed southward), and marked predominance of winds from the southeast — all these are descriptive of the Southern Hemisphere trades.

In January (Fig. 4-15(2)), the same wind belts are shown displaced slightly to the south. The greater strength of the pressure gradient in the Northern Hemisphere winter is reflected in the greater average wind speeds and the reduced number of calms. The westerlies, for instance, are no more consistent than in July, but they are much stronger. The winds of the

Pressure and Winds 63

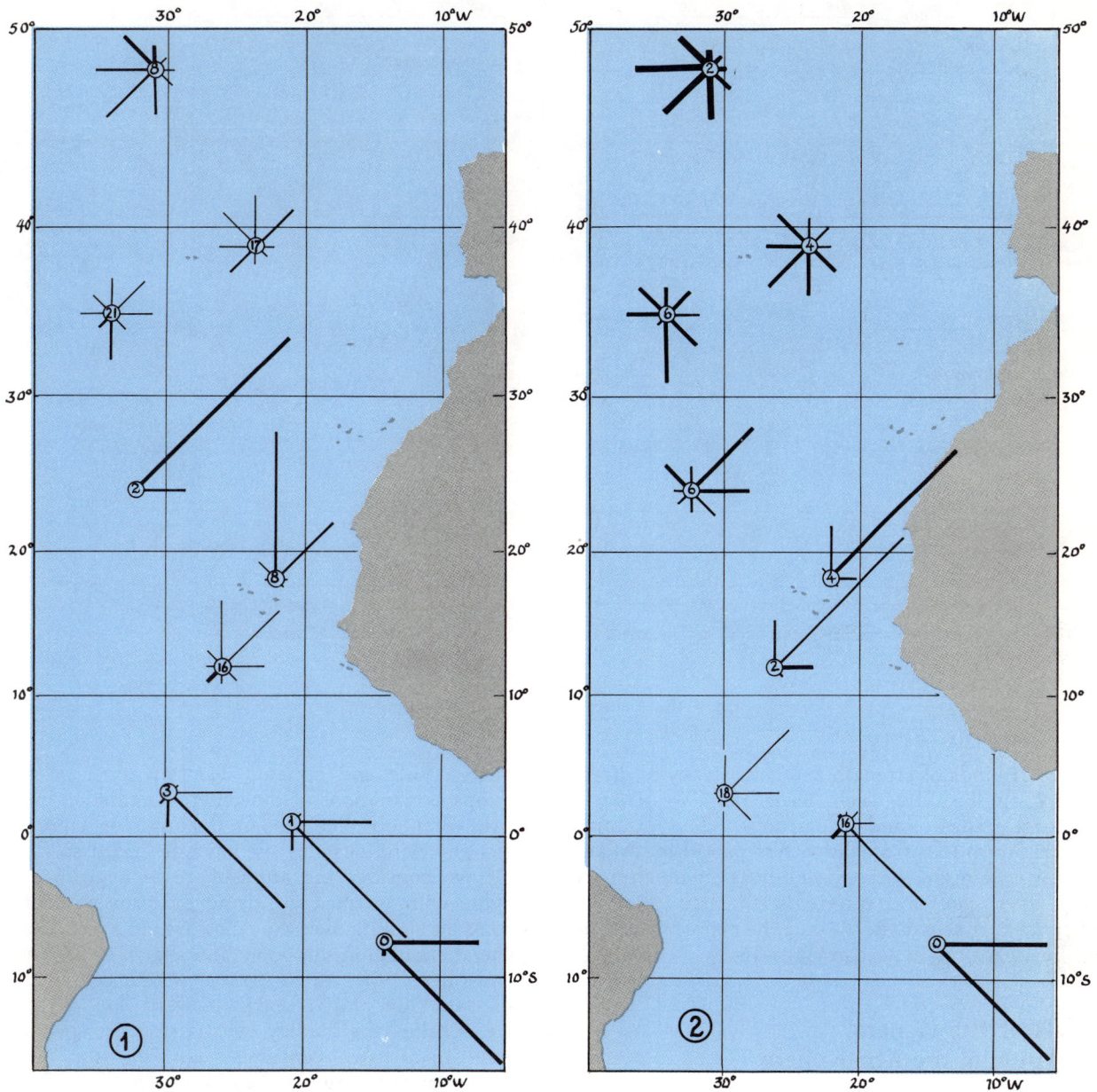

Fig. 4-15. Wind roses over part of the Atlantic Ocean in July (1) and January (2). As in Fig. 4-4, the length of the wind arrows is proportional to frequency of occurrence, and their thickness varies with the strength of the winds. The percentage of calms is shown in the center of the wind roses.

horse latitudes are also stronger, with fewer calms but with the same diversity of direction as in July. The northeast trades show the same features as in summer, but they are shifted southward to 10° to 20° N latitudes. The doldrums, indicated by the two roses with very high values of calms, have also migrated southward but are still found north of the equator.

Finally, the southeast trades are illustrated by the strong and consistent winds of the southernmost wind rose on the map.

It is evident from the map, and from the short discussion of the wind roses on it, that the various belts of winds do exist, though there is considerable gradation from one type into another. It is important to under-

64　Chapter 4

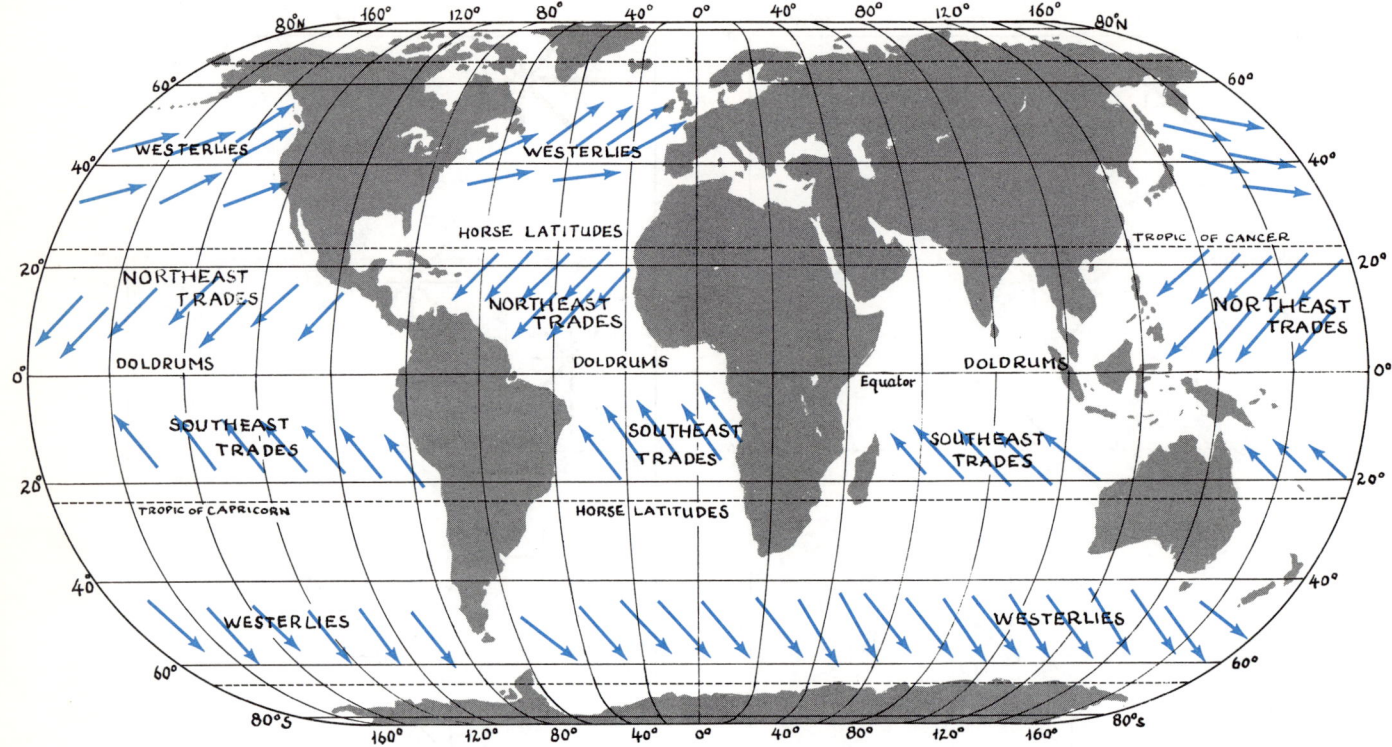

Fig. 4-16. Major wind belts over the oceans.

stand what kinds of strength and consistency of direction are implied by such terms as "westerlies," "northeast trades," and "doldrums." The example shown in Fig. 4-15 is from the eastern part of the Atlantic Ocean, where the various wind belts are much more clearly developed and differentiated than they are farther west. A schematic view of the major wind belts over the oceans can be seen in Fig. 4-16.

Models of the General Circulation of the Atmosphere

Thus far in our discussion of pressure and winds, we have talked about temperature contrasts on the surface of the earth (and to a lesser extent, temperature contrasts in the atmosphere) as being the ultimate causes of pressure differences. We have also spent considerable time describing the relationship between pressure gradients and wind velocity. However, we have not said much about the processes involved in the formation of the subpolar lows, of the subtropical high-pressure belt at the surface, and of the jet streams aloft. Since these are three of the most important phenomena in the general circulation of the atmosphere and since none of them can be explained in terms of temperature gradients between pole and equator, we now turn briefly to various explanations of the general circulation in order to see what light they shed on the origin of the pressure belts and wind systems we have described so far.

If we consider the atmosphere as a gigantic heat engine with warmed air rising over the equator and cooled air sinking over the poles, we get a picture of the general circulation that looks like Fig. 4-17 (1). In each hemisphere, the single convectional cell with its poleward flow aloft and equatorward flow at the surface is called the Hadley cell. It was first postulated more than two centuries ago by the English meteorologist George Hadley in an attempt to describe the trade winds at the surface and the antitrades aloft. The Hadley cell does not, however, describe the pressure and wind patterns in the mid-latitudes. The cell must be modified by taking into account the roles played by the Coriolis force and by friction. The Coriolis force makes the poleward-moving air appear as west winds and the equator-bound air as east winds, which has the effect of breaking up the Hadley cell into a number of smaller ones, since the surface wind will not reach its equatorial destination nor the wind aloft its polar destination before they are deflected 90° off course by the Coriolis force. In addition, friction slows the movement so that several cells are required if air is to be transported from pole to equator and vice-versa. This transport is necessary to explain the temperature differences between pole and equator which, though large, are much smaller than what might exist if the two zones did not exchange heat in some way.

In the middle of the last century, William Ferrel suggested the existence of a mid-latitude cell, in addition to Hadley's tropical cell; and this middle cell is sometimes called by his name. The Swedish meteorologist Carl-Gustav Rossby made the most significant recent contribution to the understanding of the general circulation by introducing the notion of three cells (Fig. 4-17(2)) to explain the various zonal pressure and wind belts. In the figure, the directions of the winds are shown. The typical cell, characterized by rising air in the equatorial regions and subsiding air near 30° latitude, is thus seen as the major cause of equatorial low pressure and subtropical high pressure as well as the trade winds at the surface (northeasterly in the Northern Hemisphere; southeasterly in the Southern Hemisphere) and the antitrades (westerly winds) aloft. Low pressure in the higher mid-latitudes results from the rising air in the polar branch of the Ferrel cell and the equatorial branch of the Polar cell. Finally the Polar High is explained by subsidence over the polar region. Note, however, the serious conflict between theory and fact in the higher parts of the atmosphere in mid-latitudes. According to Rossby's hypothesis, which was developed before much was known of air movement aloft, there is equatorward flow of air in the upper part of the Ferrel cell. The Coriolis force would deflect this flow into easterly winds; however, this is precisely the location of the jet stream, a zone of maximum westerly flow! Obviously, Rossby's explanation must be changed, at least in this respect.

More recent notions, which are exemplified by the work of the Finnish meteorologist E. Palmén, are shown graphically in Fig. 4-17 (3). The polar cell has been abandoned in favor of a zone of turbulent circulation in which the cells are horizontal rather than vertical; the polar front has been moved equatorward to fit its actual position more closely; and the various jet streams have been added to the picture. Although there is still considerable uncertainty about the details of the general circulation, the main features of Fig. 4-17(3) accord closely with the observed facts.

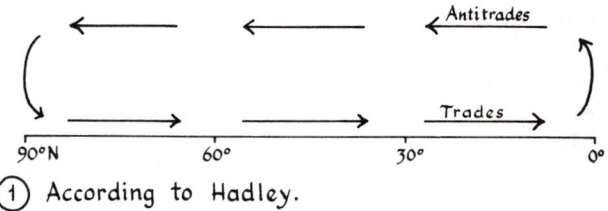

① According to Hadley.

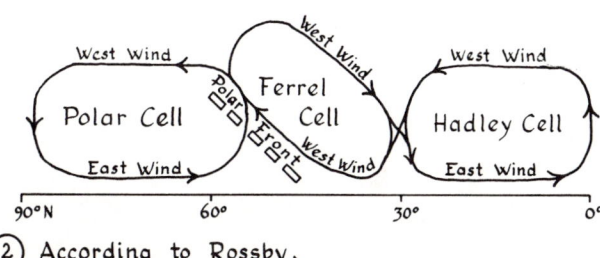

② According to Rossby.

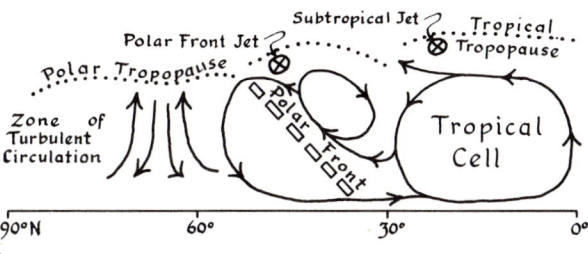

③ According to Palmén.

Fig. 4-17. Models of the general circulation of the atmosphere. (1) Shows the simple Hadley cell extending from equator to pole. (2) Represents the three interacting cells visualized by Rossby. (3) Is a slightly different pattern, in which the cross-sectional cell of previous models is replaced by vertical turbulence in the polar regions.

Summary

In this chapter, we began by defining atmospheric pressure as the force exerted by the weight of the overlying air. Differences in pressure from place to place occur principally because of differential heating and cooling of neighboring areas, but also because of the movement of air. These thermally or dynamically produced pressure differences set off air movements. The air movements do not, however, take place directly from high pressure toward low pressure because of the apparent deflection created by the rotation of the earth. This apparent horizontal deflection, the Coriolis force, results in a clockwise deflection of winds in the Northern Hemisphere and a counterclockwise deflection in the Southern Hemisphere. The deflection is greatest with rapidly moving bodies near the poles, and it decreases toward the equator and as the speed of the wind drops. As a result, winds aloft tend to be parallel to the isobars, while winds near the surface cross the isobars at a small angle, which is smaller over oceans than over land because of the greater friction encountered over the continents.

The result of the differential heating and cooling of the earth's surface and of the apparent horizontal deflection due to the earth's rotation is a zonal pattern of isobars and winds. The major features of this zonal arrangement are the subpolar lows and the subtropical highs of the Northern and Southern Hemispheres and the equatorial low. These pressure belts migrate with the sun and are broken into continental and particularly oceanic cells by the contrasting temperatures of land and water. The pattern of winds over the oceans reflects the pressure situation rather well: cyclonic whirls around the low-pressure cells, anticyclonic movement about the high-pressure cells. Over the land, the correspondence between pressure and winds is reduced by the effects of friction and of local relief features.

The general circulation of the atmosphere is best described, in terms of present knowledge, by using a cross-sectional model that consists of two vertical cells — the Hadley cell and the Ferrel cell — and a horizontal cell in high latitudes. In the equatorial and mid-latitude cells, air transport is largely north and south; in the polar cell, the turbulent movement of air results mainly in up-and-down transport. Two other important components of the general circulation are the polar front and the various jet streams.

So far, in our climatic discussion, we have seen how solar energy is distributed over the world, in terms of the heat budget of particular places; how insolation affects temperature; and how temperature differences create pressure gradients and thus air movements. We shall turn next to the content and behavior of water in the atmosphere, which will complete our discussion of the elements of climate and prepare us to turn to the classification and distribution of climates.

Suggestions for Further Reading

Byers, Horace R., *General Meteorology*, 3rd ed., McGraw-Hill Book Company, Inc., New York, 1959, pp. 193-231.

McDonald, James E., "The Coriolis Effect," *Scientific American*, Vol. 186, No. 5 (May 1952), pp. 72-78.

Rossby, C. G., "The Scientific Basis of Modern Meteorology." *Climate and Man*, Yearbook for 1941, U. S. Dept. of Agriculture, 1941, pp. 559-655.

Wexler, Harry, "The Circulation of the Atmosphere," *Scientific American*, Vol. 193, No. 3 (September 1955), pp. 114-125.

5 Evaporation and Condensation

Introduction

Definitions

Evaporation is the change of water from the liquid to the gaseous state; **condensation** is the reverse process. Although the terms are easily defined, the two phenomena are part of complex processes whereby water changes its state and moves between the surface of the earth and the atmosphere. We must, therefore, define and explain many terms that are used in discussing these processes. We shall begin with terms that refer to measures of moisture content in the air.

The amount of water vapor (water in the gaseous state) in the air can be indicated in an absolute sense by a number of measures, of which the simplest is **vapor pressure.** Like all other gases, water vapor exerts pressure. The vapor pressure is merely that part of the total atmospheric pressure due to the water vapor present in the air. Another measure, the **absolute humidity,** indicates the density of the water vapor. It refers to the weight of water vapor per volume of air and is normally expressed as so many grams of water vapor per cubic meter of air. The numerical value of the absolute humidity changes with each change in atmospheric pressure, even though there is no addition or removal of water vapor, since a change in pressure brings about a proportional variation of volume. The **specific humidity** refers to the amount of water vapor present per unit mass[1] of air and is normally expressed as so many grams of water vapor per kilogram of air. This measure is more useful than the absolute humidity because its numerical value changes only when the water vapor in the air increases or decreases. A fourth way to express the absolute vapor content of the air is by the **mixing ratio.** The mixing ratio is similar to the specific humidity, differing only in that the weight of water vapor is expressed per unit weight of dry air (that is, the weight of the air less the weight of the water vapor in that air). A mixing ratio of 4 grams per kilogram, therefore, indicates the presence of 4 grams of water vapor for each kilogram of dry air (or for each

---

[1] Mass and weight are the same as long as the force of gravity is constant. But an astronaut weighs less on the moon than on earth because the moon's gravitational pull is lower (since the mass of the moon is smaller than that of the earth) even though the mass of the astronaut is practically the same wherever he goes.

1.004 kg of air, including the 4 g of water vapor). Although the mixing ratio differs very little in either definition or numerical value from the specific humidity, it is preferred by meteorologists because it can be expressed exactly in terms of the vapor pressure and the total atmospheric pressure in a much simpler way than specific humidity. We shall use the mixing ratio as our measure of the absolute content of water vapor in the air.

The moisture content of the air can also be expressed in a relative way by the **relative humidity** (R.H.), which refers to the amount of water vapor present compared with the capacity of the air to accommodate water in the gaseous form. This capacity of the air can also be expressed in a number of ways, of which the simplest is the **saturation vapor pressure.** The saturation vapor pressure is the maximum vapor pressure that can exist in the atmosphere. Beyond this maximum, any increase in the vapor pressure oversaturates the air with water vapor, and the excess vapor condenses if there are nuclei of condensation present in the air (see p. 72). The saturation vapor pressure, or maximum capacity of the air varies with the temperature of the air.[2] Figure 5-1 shows the saturation vapor pressure, in millibars, for various temperatures. Note that with each temperature increase of 10° C, the capacity of the air almost doubles.

The relative humidity of a mass of air is the vapor pressure of that air mass expressed as a percentage of the saturation vapor pressure. As a formula it is written

$$R.H. = \frac{\text{vapor pressure}}{\text{saturation vapor pressure}} \times 100.$$

Figure 5-1 shows that it is possible to compute the relative humidity of the air if the vapor pressure and the temperature of the air are given. For example, if a certain mass of air has a temperature of 10° C and a vapor pressure of 6 millibars, we can determine that its relative humidity is 50 percent. At that temperature, as the graph shows, the saturation water vapor pressure is about 12 millibars. The vapor pressure was given as 6 millibars; therefore,

$$R.H. = \frac{6 \text{ millibars}}{12 \text{ millibars}} \times 100 = 50 \text{ percent}.$$

Since we shall be using the mixing ratio rather than the vapor pressure as our measure of moisture content of the air, we shall consider the relative humidity in terms of the mixing ratio. The formula then becomes

---

[2]Total atmospheric pressure also affects the saturation vapor pressure, but this effect can be disregarded because of the relative smallness of atmospheric pressure changes at any given elevation

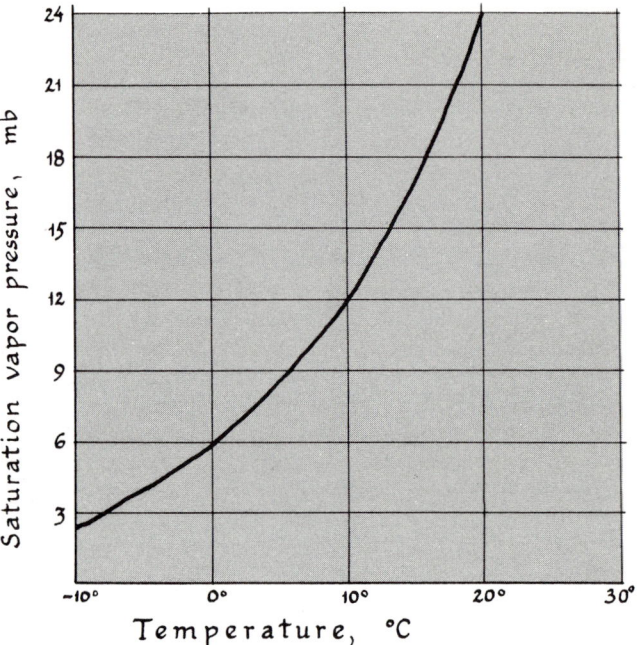

Fig. 5-1. The relationship of saturation vapor pressure to temperature. Note that with each temperature increase of 10°C, the capacity of the air to hold moisture almost doubles.

$$R.H. = \frac{\text{mixing ratio}}{\text{saturation mixing ratio}} \times 100.$$

Strictly speaking, this formula gives a value for R.H. that is slightly below its value defined in terms of vapor pressure, but this difference is negligible. Figure 5-2 is analogous to Fig. 5-1, except that it shows the relationship of the **saturation mixing ratio** to temperature. The inset table shows the same relationship by means of approximate numerical equivalents. By using this table, we should be able to calculate the relative humidity when the mixing ratio is given or the mixing ratio when we know the relative humidity. Two examples illustrate this simple procedure.

On a given day the temperature is 90° F (32.2° C) and the mixing ratio of the air is measured as 18 grams per kilogram. What is the relative humidity? According to the formula,

$$R.H. = \frac{\text{mixing ratio}}{\text{saturation mixing ratio}} \times 100.$$

Substituting the known values, we get

$$R.H. = \frac{18 \text{ g/kg}}{30 \text{ g/kg}} \times 100 = 60 \text{ percent}.$$

The value for the saturation mixing ratio was obtained from the table, which tells us that, at a temperature of 90° F (32.2° C), the saturation mixing ratio is 30 grams per kilogram. By solving the equation we obtain a value of 60 percent for the relative humidity.

On a given day the temperature is 80° F (26.7° C) and the relative humidity is 50 percent. What is the mixing ratio on that day?

$$R.H. = \frac{\text{mixing ratio}}{\text{saturation mixing ratio}} \times 100.$$

Substituting the known values, we get

$$R.H. = \frac{\text{mixing ratio}}{22 \text{ g/kg}} \times 100.$$

Multiplying both sides of the equation by 22 grams per kilogram and dividing both sides by 100, we get

$$\frac{50 \times 22 \text{ g/kg}}{100} = \text{mixing ratio}.$$

The mixing ratio in question is therefore 11 grams per kilogram.

Another measure of moisture content of the atmosphere is the dew point. The word "dew" refers, of course, to condensation; and the word "point," to a point on the thermometer. The dew point, in other words, is the temperature at which condensation is expected to occur. It is, therefore, the temperature at which the saturation mixing ratio is equal to the actual mixing ratio; and, of course, if the saturation mixing ratio and the actual mixing ratio are equal, the relative humidity is 100 percent. The dew point depends entirely on the water vapor content of the air, and not at all on its temperature, as the following examples should make clear.

What is the dew point of a mass of air whose mixing ratio is 3.4 grams per kilogram? The table in Fig. 5-2 shows that the saturation mixing ratio is 3.4 grams per kilogram when the temperature is 30° F (-1.1° C). The dew point is therefore 30° F (-1.1° C). If the temperature of this mass of air is 50° F (10° C), it will need to be cooled by 20° F (11.1° C) before saturation takes place and condensation occurs. If the temperature of this mass of air is 40° F (4.4° C), a cooling of only 10° F (5.5° C) will produce condensation. In either case, the dew point remains at 30° F (-1.1° C). The dew point will change only if the moisture content changes.

How much will the dew point change if the mixing ratio of an air mass increases from 2.2 to 5.1 grams per kilogram? The dew point of a mass of air whose mixing ratio is 2.2 grams per kilogram is given in the table as 20° F (-6.7° C). The dew point corresponding to a mixing ratio of 5.1 grams per kilogram is 40° F (4.4° C). The dew point will, therefore, rise by 20° F (11.1° C).

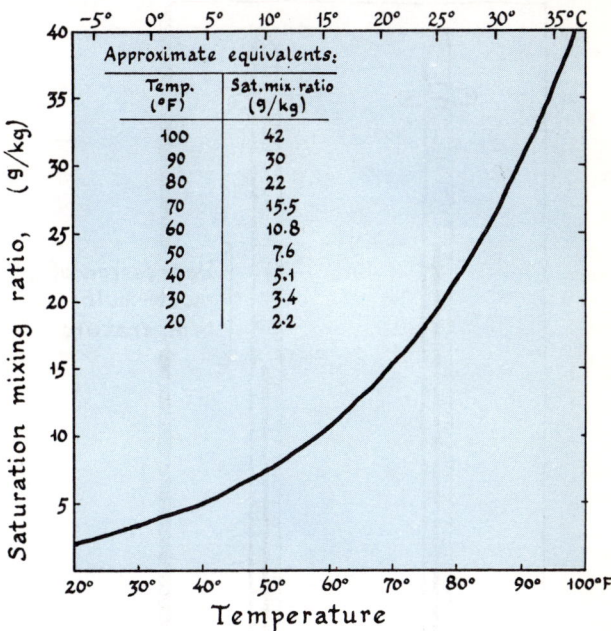

Fig. 5-2. The relationship of saturation mixing ratio to temperature. The inset table shows the relationship by means of approximate numerical equivalents.

Instruments

Various instruments are used to measure the amount of water vapor present in the air; none measures the rates of evaporation and of condensation directly. The hygrometer (Greek *hygros* wet + *metron*) consists essentially of a human hair whose sensitivity to changes in the moisture content of the air results in twists and turns that are transmitted to an appropriately calibrated dial. There are a number of variants of this instrument, many of which use indicators other than human hair, but hygrometers are not as accurate as barometers or thermometers. A more commonly used instrument is the psychrometer (Greek *psychros* cold + *metron*). The principle involved in the psychrometer is illustrated in Fig. 5-3. One of the two thermometers is a normal (dry-bulb) instrument; the other, the wet-bulb thermometer, has its bulb packed in some absorbent material that is kept wet and well ventilated. As we shall see, evaporation reduces the temperature of the material around the wet bulb by an amount that depends on the moisture content of the air. The numerical value of this moisture content can be obtained from appropriate tables, which generally give the relative humidity that corresponds to a given dry-bulb temperature and a given depression of the wet-bulb temperature below that of the dry-bulb thermometer. Once we know the temperature of the air (dry-bulb temperature) and its relative humidity, it is

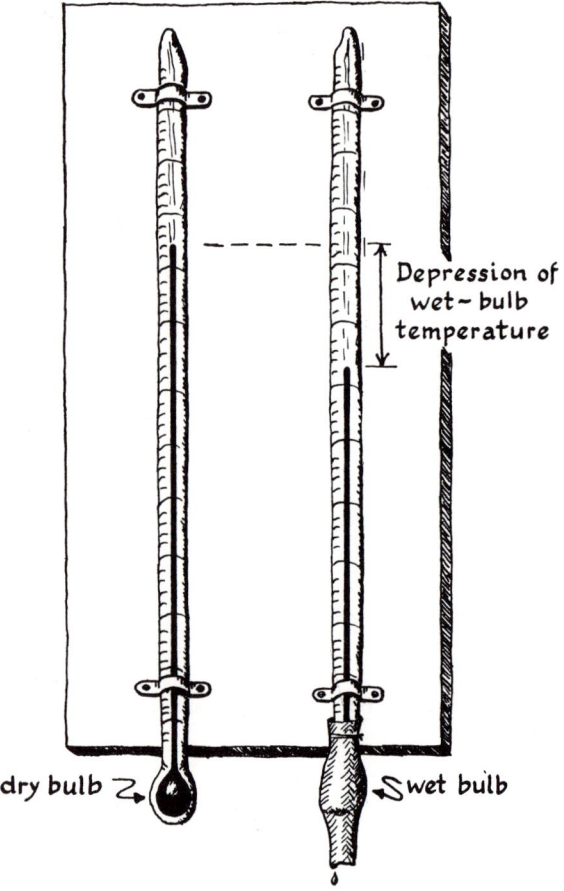

Fig. 5-3. Wet-bulb and dry-bulb thermometers. The temperature in the wet-bulb thermometer is lower because of the heat lost in evaporating the water around the bulb.

easy to calculate its mixing ratio, vapor pressure, or any other measure of humidity.

Measuring evaporation is a difficult process. The U. S. Weather Bureau uses an evaporation pan of standard width and depth. The loss of water from the pan is balanced against any additions of water by rain or snowfall, and the net loss hopefully represents the evaporation. Differences in exposure, in wind conditions, and in the nature of the surrounding ground make for widely varying measurements in reasonably similar air masses. In addition, the evaporation pans measure the rate of evaporation from water surfaces only and cannot be used to describe the evaporation from moist soil. More complicated instruments, often referred to as evaporimeters, attempt to measure the evaporation from the soil and the transpiration from plants by isolating a portion of the topsoil and measuring all the possible water inflow; any loss in the moisture content of the system must then be due to evaporation. The few such evaporimeters in existence unfortunately do not constitute a world-wide network. The problem in measuring evaporation is that simple instruments are inaccurate and measure only evaporation from water bodies, whereas the accurate instruments are complicated and not widespread or uniform.

The measurement of condensation is relatively simple mainly because we measure its end products, snow and rain. All rain gauges are variants of a simple bucket with graduations marked on its side to indicate the depth of the water collected. Snowfall has generally been measured directly by means of stakes, but recently developed measuring systems utilize radioactivity. A Geiger counter and a microphone are hung above a piece of radioactive cobalt that has been buried near the surface. As snow begins to fall, the counter records less radioactivity because of the intercepting effect of the snow. Since the amount of interception depends directly on the water content of the snow, the strength of the signals sent out by the counter can be translated into a numerical value of the water content of the fallen snow. This simple, but accurate device has revolutionized the forecasting of runoff resulting from the melting of snow in high elevation river basins. Radioactive substances may also prove useful in the measurement of evaporation. Water has an isotope, sometimes called heavy water because of its slightly greater atomic weight, that has already been used in measuring evaporation from lakes and other inland waters.

Evaporation

The kinetic energy of a liquid — the motion of the molecules of the liquid — is reflected in its temperature. The higher the temperature, the greater the kinetic energy possessed by the liquid. In this view, evaporation results from the escape of the faster molecules from the liquid. The energy necessary to counteract the restraining molecular forces at the surface of the liquid allows only the faster-moving molecules to leave the liquid. The kinetic energy of molecular motion in the liquid is decreased, since only the slower molecules are left behind, and this loss in energy is reflected in a decrease of temperature. If some of the faster molecules that have evaporated return to the liquid state, the kinetic energy of the liquid is increased, and the temperature rises.

The cooling effect of evaporation is, of course, well known. It is, among other things, the mechanism whereby body temperature is kept down through the evaporation of perspiration. The heat lost in evaporation is called the **latent heat of evaporation.** The return of heat to a liquid through condensation is referred to as the **latent heat of condensation.**

Factors Affecting Evaporation

From the foregoing, we can see that temperature is the principal factor affecting evaporation from a water surface. As the water temperature is increased, more and more of the molecules will have sufficient energy to break the surface molecular barrier and to enter the atmosphere as water vapor. Some of the water vapor, however, will condense and return to the liquid state. Thus, when we speak of evaporation, we normally mean the net evaporation; that is, the difference in number between the molecules that evaporate and those that condense. After a time, however, there will be just as much condensation going on as evaporation. A thin layer of air has become saturated, and there will be no more net evaporation.

If the temperature of the water increases indefinitely, the temperature of the layer of air above it likewise increases. This increase in air temperature is accompanied by an increase in the saturation vapor pressure, and the greater capacity of the air to accommodate water vapor results in continued evaporation from the water. Despite the increasing concentration of water vapor in the air, evaporation continues as long as the temperature of the water keeps the saturation vapor pressure higher than the actual vapor pressure in the air. Eventually, all the water molecules will evaporate, and we say that the boiling point has been reached.

The role of temperature in the evaporating process is obvious. As long as water is heated by some outside source of energy, evaporation continues to the boiling point. Other factors that affect the rate of evaporation are revealed when we examine the evaporating process under conditions in which the water temperature remains constant. Evaporation is resisted to some degree by total atmospheric pressure, against which the kinetic energy of the evaporating molecules must be balanced. At a given place, however, momentary changes and even yearly changes in atmospheric pressure are relatively slight. Practically speaking, the relatively small changes in pressure have little effect on evaporation; changes in pressure, at least at a given elevation, hardly affect the rate of evaporation.

Evaporation is also affected by the molecular forces at the surface of the water. These forces are practically constant. They decrease slightly when the surface of the water becomes curved rather than flat as in individual water droplets; but this decrease is noticeable only in very small water particles, such as tiny droplets, in which the radius of curvature is extremely small.

The concentration of water vapor in the air above an evaporating surface is another variable that affects the net rate of evaporation. This concentration varies considerably. It depends not only on the amount of evaporation into the layer but also on the wind conditions. A strong wind sweeps away a nearly saturated layer and replaces it with air that has a much lower vapor pressure. Under such conditions, evaporation continues for some time. If there is no wind, the saturated air loses some of its water vapor by diffusion upward; but this is a slow process, and the air will remain nearly saturated. Only a rise in the temperature of the water will increase evaporation in such a case.

All these factors indicate that maximum evaporation occurs under conditions of strong daytime insolation, which normally results in high temperatures of the ground surface and the development of a layer of turbulent air. The high temperatures increase the rate of evaporation, as we have seen; and the turbulence transports the water vapor upward, so that saturation does not put an end to net evaporation. During the night, when cold surface conditions are often accompanied by still air, the rate of evaporation is much reduced. The amount of water vapor in the air may be higher near the ground at night, however, because of the lack of turbulent transport of water vapor away from the moist surface of the ground. Relative humidity, by the same token, is almost invariably greatest during the late night hours, when the temperature is lowest. While low temperatures reduce the rate of evaporation, they have a much more pronounced effect on the saturation vapor pressure, so that only a small amount of water vapor will saturate the cold air.

Potential Evaporation. In addition to the factors so far considered, the rate of evaporation also depends upon the supply of water available. The limitation placed upon evaporation by the availability of evaporable water is so obvious that it requires little amplification. As far as climate is concerned, it is sufficient to distinguish between *actual evaporation* and *potential evaporation*. Potential evaporation refers to the amount of water vapor that would be evaporated if there were an optimum amount of water available at all times. Plants also transfer a considerable amount of water into the atmosphere by transpiration, and this phenomenon is included in the more general term **potential evapotranspiration.**

Both actual and potential evapotranspiration play important climatic roles. In Chapter 2, in our discussion of heat balance, we saw that an important part of the surplus of radiative energy at the surface of the ground is disposed of by actual evaporation. In the average heat balance of the earth's atmosphere (Fig. 2-4 p. 12), considerable transfer of energy occurs due to loss through latent heat of evaporation at the surface of the earth and gain by latent heat of condensation in the atmosphere. This transfer, which is given a numerical value of 20 units in Fig. 2-4, amounts to about 13 percent of all the heat losses from the surface (and about the same percentage of all the heat gains by the atmosphere).

The importance of actual evaporation to the heat balance at a particular place and at a particular time may be much greater than is implied in the average heat balance. In Fig. 2-11 p. 20, the differing impact of evaporation is portrayed graphically. At Manaos, on the Amazon, for example, evaporation disposes of a large percentage of the radiative surplus at all times of the year. At Lisbon, Portugal, the role of evaporation is still great, but there is considerable difference between evaporation in April and in September. At Aswan, Egypt, the actual evaporation is negligible and plays no role in disposing of the radiative surplus.

Potential evaporation is more important in the water balance than in the heat balance. In determining the water balance of a portion of the earth's surface, we need to consider not only the supply (in the form of rain, snow, and water stored in the soil) but also the demand, (represented by potential evaporation, or rather evapotranspiration.) A place with very little precipitation may have considerable water surplus if the potential evapotranspiration is low. Conversely, a place with abundant rainfall may have a considerable deficit of water if the potential evapotranspiration is greater than the supply of water.

World Distribution of Evaporation

Several problems arise if we try to portray the differences in evaporation from place to place on the surface of the earth. Do we wish to map actual or potential evaporation? If it is actual evaporation we are after, we are faced with the fact that there are very few reliable measurements. In the absence of such measurements, we could estimate evaporation if we knew the amount of energy available for evaporation, the amount of water available for evaporation, and the amount of turbulence in the lowest layers of the atmosphere. Unfortunately, the magnitude of these variables over much of the earth is not well known. The isolines of actual evaporation shown in Fig. 5-4 represent the best estimates available and are based on indirect measurements of energy and moisture availability.

Potential evaporation presents a simpler problem, perhaps, since the availability of moisture need not be considered in determining the water demands. Even so, there are also formidable barriers in the way of constructing a map of potential evaporation. One approach to the problem of depicting the world pattern of evaporation is to estimate the evaporation from climatic elements that are measured and recorded in some detail. Average monthly temperature and precipitation are the two most commonly observed climatic elements. An American geographer, C. W. Thornthwaite, developed a scheme whereby the potential evapotranspiration can be estimated from monthly temperatures. The insert of Africa in Fig. 5-4 shows isolines of potential evapotranspiration based on his system, and is included to show the difference between actual and potential values, at least for that continent.

Condensation

We have already defined condensation as the process whereby water vapor changes into liquid water. The term also refers to the end product of the condensation process, the condensation forms themselves. Condensation may occur on surface objects that are cooled below the dew point of the air above; dew and frost are examples of this type of condensation. Some condensation particles are so small that they float freely in the air where they are formed; clouds, fog, and smog are composed of such particles. Other condensation forms, such as rain, snow, sleet, and hail, fall to the ground as precipitation.

Factors Affecting Condensation

Condensation is not simply the reverse of evaporation. Evaporation goes on constantly from any moist surface, even when the net evaporation is nil. Condensation, on the other hand, does not take place automatically when saturation is reached in a particular volume of air. The water vapor needs some kind of surface to condense on. Condensation will not take place in air from which all impurities have been removed; some impurities must be present to act as *condensation nuclei*. For practical purposes, only two kinds of nuclei are important. Most condensation nuclei are small salt particles, the by-product of evaporation from ocean spray. Generally, the lower atmosphere contains enough such salt particles that condensation normally occurs when saturation is reached. The other major source of nuclei of condensation is industrial pollution of the air. Oxides of sulfur and of phosphorus are common by-products of industrial processes. These acid condensation nuclei often produce a condition halfway between smoke and fog; the oppressive and corrosive condensation form that results is called smog. Some nuclei of condensation have such a strong chemical affinity for water that condensation will take place around them before saturation is reached; these are called **hygroscopic nuclei.** Salt is only mildly hygroscopic; but silver iodide, for instance, and some industrial pollutants, such as the anhydride of sulfuric acid, are extremely hygroscopic.

Condensation is also a more complicated

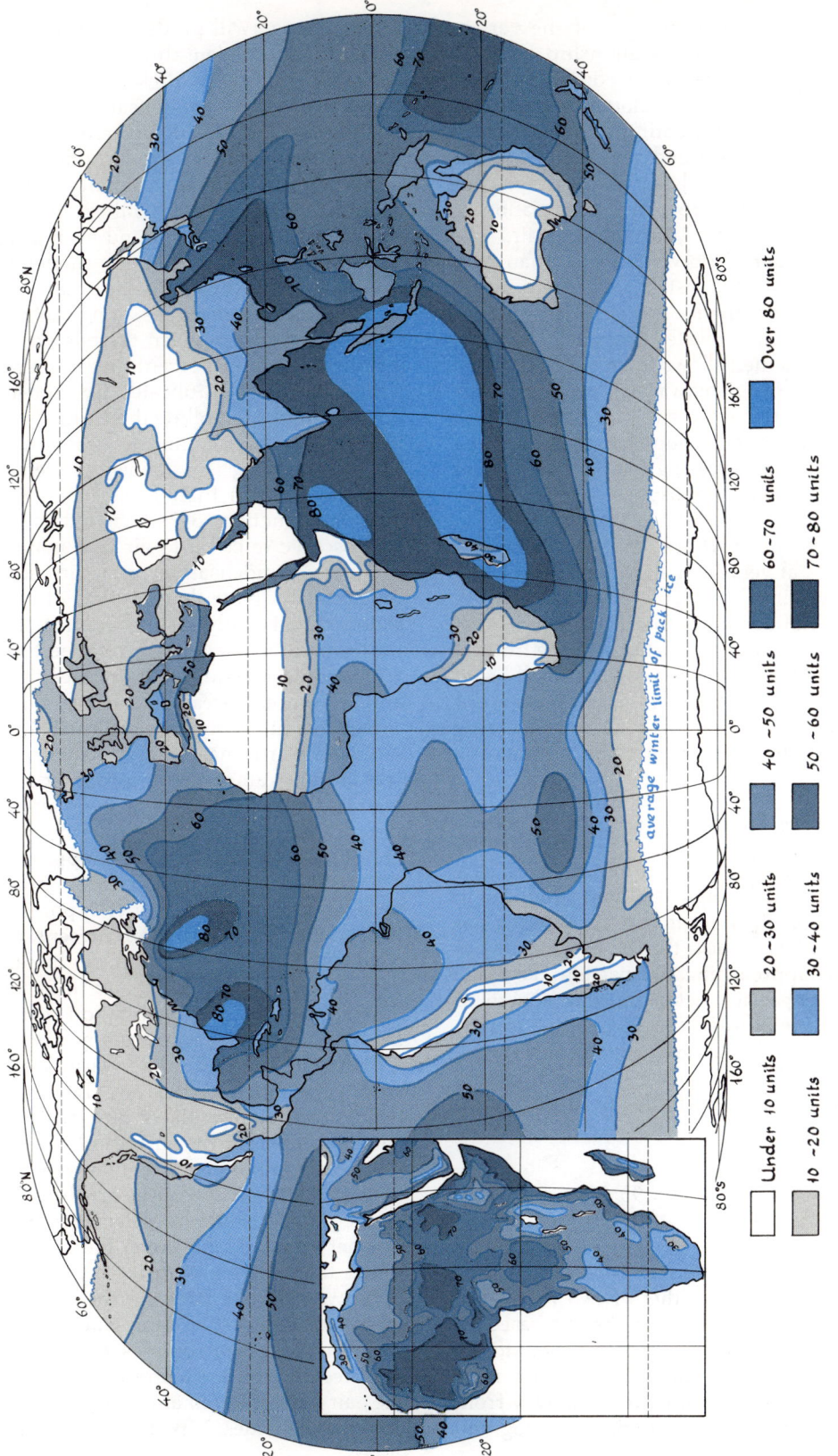

Fig. 5–4. The average annual evaporation over the world. The insert shows the average potential evapotranspiration over Africa. All values in units of inches.

phenomenon than evaporation because of the variety of condensation forms and the relationship between them. It makes a great difference, of course, whether the condensation is in the form of clouds or precipitation; but the transition from floating condensation to falling precipitation is complex and little understood. It is clear, however, that ice crystals, or frozen droplets of water, play a very important role in the formation of particles large enough to fall through the air. The saturation vapor pressure is lower over ice than over water. As a result, when ice crystals are present, water vapor condenses around the ice in fairly large quantities, at the expense of the water droplets. The large size of the ice particles then allows them to fall through the air. As they reach lower and warmer levels of the atmosphere, they melt and continue to precipitate as rain. This process is thought to account for most types of heavy rain or snow when freezing temperatures occur somewhere in the lower troposphere. This theory works well enough in middle or high latitudes, but in lower latitudes, where there are no freezing temperatures except at very high elevations some other mechanism is required for precipitation to fall. Meteorologists do not agree whether it is the presence of unusually large hygroscopic particles in the tropical atmosphere, or whether it is the growth of small water droplets into larger ones by a random process of coalescence, that is the major factor in changing floating particles into tropical and equatorial precipitation.

Condensation in the Atmosphere

With this general background in mind, let us now turn to the ways in which condensation is produced in the atmosphere. We shall begin by considering air masses that are relatively moist; that is, air masses with relative humidities approaching 100 percent. To make things as concrete as possible, let us imagine a parcel of air with a temperature of 90° F (32.2° C) and a mixing ratio of 28.5 grams per kilogram. The table in Fig. 5-2 indicates that at 90° F (32.2° C) the saturation mixing ratio is 30 grams per kilogram. In the air of our example, therefore,

$$\text{R.H.} = \frac{28.5 \text{ g/kg}}{30 \text{ g/kg}} \times 100 = 95 \text{ percent.}$$

The relative humidity of the air in our example may be increased to 100 percent in either of two ways. The mixing ratio (moisture content) can be increased in some fashion from 28.5 grams per kilogram to 30 grams per kilogram, or the saturation mixing ratio (capacity of the air) can be lowered in some way from 30 grams per kilogram to 28.5 grams per kilogram.

Either change will produce a relative humidity of 100 percent, and condensation will normally occur if condensation nuclei are present in the air. We could, of course, produce condensation without raising the relative humidity to 100 percent by introducing enough hygroscopic nuclei to allow condensation to take place at 95 percent relative humidity. Under natural conditions, condensation takes place with relative humidities of 95 percent and even lower, if there is an unusual concentration of the moderately hygroscopic salt particles in the air. This condition is often fulfilled by sea fogs. Summer fogs along the Pacific coast of the United States form when the relative humidity reaches a value of 92 to 95 percent. In certain extreme cases this situation can be duplicated by allowing so much industrial smoke to pollute the air that smog appears even when the relative humidity is as low as 80 percent.

Cloud seeding is a deliberate attempt to create condensation at less than 100 percent relative humidity. The introduction of silver iodide or other highly hygroscopic nuclei into air that is not quite saturated produces condensation and, in some cases, even precipitation. Note that this way of producing condensation usually only results in either fog or smog, and nearly saturated air (with relative humidities close to 100 percent) is required to produce significant amounts and widespread areas of condensation.

Relative humidity may be brought to 100 percent by mixing a warm, nearly saturated air mass, such as the one in our example, with a cold air mass. Table 5-1 illustrates what happens if two air masses of equal weight and volume are thoroughly mixed. The mixing of the two air masses would appear to produce considerable condensation. In fact, there are 4.2 grams per kilogram of excess water vapor in the mixed mass. If condensation nuclei were present, this excess would condense, and precipitation might fall in fairly large amounts. There are too many unlikely circumstances, however, for such precipitation actually to occur in nature. First, a very warm and nearly saturated air mass is essential. Second, that warm air must be mixed with very much colder air. Two air masses with a temperature differential of 60° are unlikely to be found cheek-by-jowl on the surface of the earth; and even if that nearly impossible situation were to come about, the two air masses, because of their widely differing temperatures, would have densities so different that the warmer air would glide over the colder and denser mass rather than mixing with it.

As we shall see later on, the coming together of air masses of contrasting temperatures, is, indeed, one of the prime causes of condensation and precipitation, but the precipitation thus produced is due to the fact that the air masses remain separate because of their different densities. Mixing of air, as illustrated in Table

Table 5–1. Hypothetical Mixing of Two Air Masses

Characteristic	Air mass A	Air mass B	A and B mixed
Temperature (° F)	90.0 (32.2° C)	30.0 (−1.1°C)	60.0 (average of 90 and 30) (15.6° C)
Mixing Ratio (g/kg)	28.5	1.5	15.0 (average of 28.5 and 1.5)
Saturation Mixing Ratio (g/kg)	30.0	3.4	10.8 (at 60° F or 15.6° C)
Relative Humidity (%)	95.0	44.1	>100.0 (15.0/10.8 × 100)

5-1 practically never occurs. The main reason for introducing this insignificant condensation-producing process is to furnish further examples of the relations among temperatures, saturation mixing ratios, mixing ratios, and relative humidities.

A more obvious way of raising the relative humidity of an air mass to 100 percent is by addition of water vapor. In our example, the relative humidity was 28.5 grams per kilogram/30 grams per kilogram x 100, or 95 percent. If the mixing ratio could be increased from 28.5 grams per kilogram to 30 grams per kilogram, the desired relative humidity would result. The mixing ratio of the air can be increased only by increased evaporation into the air. To evaporate an additional 1.5 grams of vapor into each kilogram of air requires about 840 calories of energy.[3] Even if this amount of energy were available, the condensation resulting from the additional evaporation would be limited to a very shallow layer of air. Continuous application of tremendous amounts of energy would be necessary for extensive condensation, not to mention precipitation. Adding moisture to the atmosphere is not a very effective way of producing condensation. Condensation forms produced by this means are even rarer and more localized than those resulting from the abundant presence of hygroscopic nuclei. Occasionally, air that is very cold and has a correspondingly low saturation mixing ratio blows over a very warm sea whose relatively high heat content supplies the necessary latent heat of evaporation. The combination of cold air and warm seas is found during the winter over parts of the Arctic Ocean. Sometimes there is enough condensation to form a low fog called *arctic sea smoke*.

Reducing the capacity of the air to hold moisture is a much more effective way of producing condensation than raising its moisture content. Reducing the saturation mixing ratio of an air mass is accomplished simply by lowering its temperature. Suppose we have an air mass whose temperature is 70° F (21.1° C) and whose mixing ratio is 10.8 grams per kilogram. The table in Fig. 5-2 indicates that the saturation mixing ratio of air at 70°F (21.1°C) is 15 grams per kilogram. The relative humidity of this air is, therefore, 10.8 g/kg/15g/kg x 100, or about 70 percent.

To increase the relative humidity from 70 percent to 100 percent, we would have to add 4.7 grams of water vapor to each kilogram of air. The same result can be accomplished by lowering the temperature of the air from 70° to 60° F (from 21.1° to 15.6° C), since at 60°F (15.6° C) the saturation mixing ratio is only 10.8 grams per kilogram.[4] It is much simpler, in nature, for the temperature to drop 10° F (5.6° C) than for 4.7 grams of water to evaporate for every kilogram of air.

Thus, the most efficient way of producing condensation is by lowering the temperature of an air mass to its dew point. There are three important cooling processes that produce condensation under the proper conditions; **radiational cooling, cooling by advection** and **adiabatic cooling.**

Radiational cooling takes place when incoming radiation at the surface is very slight and outgoing radiation has exhausted the heat content of the ground. Such conditions are best satisfied by a cold winter night with clear skies and little wind. There is no solar radiation, of course; and the clear skies have absorbed very little of the outgoing terrestrial radiation, so that atmospheric radiation back to the ground is very low. The temperature at the surface of the ground drops rapidly under such circumstances. The lower layers of the atmosphere are cooled by contact; and, in the absence of wind, this cooling may be considerable in a shallow layer of air next to the ground. Moderate or strong winds tend to spread the coldness through a large volume of air and may possibly bring warm air from aloft down to the surface. The absence of such winds is, therefore, necessary for intensive radiational cooling in the lowest layers of the atmosphere. If the cooling is pronounced, the temperature of the air near the ground may drop below its dew point; and condensation will take place, nearly always as a low fog of some sort, which is a common nighttime or early morning phenomenon in winter and is often called

---

[3]The latent heat of evaporation is approximately 600 calories per gram at 32° F, decreasing to about 540 calories per gram at 212° F. At 90° F it is about 560 calories per gram.

[4]In a more general way, we say that we have lowered the temperature of the air to its dew point, which is the temperature at which the saturation mixing ratio is exactly equal to the actual mixing ratio of the air mass in question.

ground fog. It is most frequently found in low-lying areas into which cold air may drain from the immediate surroundings, thus accentuating the effects of radiational cooling.

Cooling by advection, which also produces fog rather than precipitation, takes place when a warm air mass moves over a cold surface. In winter, *advection fogs* are generated by warm, moist oceanic air blowing over ice- or snow-covered land surfaces. The oceanic air is cooled below its dew point by contact with the colder land surface. In summer, advection fogs are formed over cold water bodies by the passage of warmer air. Sometimes the warm air is continental in origin, as in the case of the advection fogs that occur over the Great Lakes in May and June. In other cases, the warm air is oceanic, as in the case of the *sea fogs* that are the result of the eastward passage of warm, moist air from the central Pacific Ocean over the cold, upwelling coastal waters of California and Oregon. The same kind of fog results when air that has been warmed over the Gulf Stream moves to the colder waters in the vicinity of the Atlantic coast from northeastern Maine to Greenland.

So far, in our discussion of condensation-producing mechanisms, we have only been able to generate a little fog and hardly any precipitation. And even the densest ground or sea fogs eventually lift as soon as there is some source of energy available for reversing the cooling process by heating the surface. Most of the condensation in the atmosphere and almost all the precipitation is due to a third cooling mechanism, adiabatic cooling.

The Adiabatic Process

The Relation Between Pressure and Temperature in a Gas

In Chapter 4, we had occasion to refer to the gas laws, with particular reference to Charles's law which states that volume and temperature vary directly if the pressure remains the same. Here we are concerned with Poisson's equation, developed by the French mathematician, Siméon Poisson (1781-1840), which states that if the volume of a gas is kept constant, the absolute temperature will vary as the 0.288th power of the pressure[5]; that is,

$$\frac{T_2}{T_1} = \left(\frac{P_2}{P_1}\right)^{0.288}$$

where P_1 and T_1 refer to some original values of pressure and temperature (in degrees of the absolute temperature scale), and P_2 and T_2 to subsequent values of the pressure and temperature in the same gas.

The important thing to remember about Poisson's equation is that the temperature of any gas increases if the pressure is increased, and decreases if the pressure is reduced. We have already mentioned a number of times that changes in atmospheric pressure at particular places are relatively insignificant, even if we consider the extremes of pressure that may occur over a long period of time. Therefore, such small pressure changes hardly affect the temperature. Pressure does change very rapidly with elevation, however. A mass of air displaced upward experiences a decrease in pressure that amounts to about 34 millibars for every 1,000 feet (11 mb per 100 m) of upward movement. The rate of decrease of pressure is much less at very high elevations, but the values indicated are approximately true for the lowest 10,000 feet (3 km) of the troposphere. If we use Poisson's equation to calculate how much the temperature drops if we decrease the pressure by 34 millibars, we find the answer to be about 5.5° F. This decrease in temperature of 5.5° F for every 34 millibars of pressure drop can also be expressed as a decrease of 5.5° F/1,000 feet of elevation, since a rise of 1,000 feet corresponds approximately to a drop in pressure of 34 millibars. This figure (5.5° F/1,000 feet or 1° C/100 m), which is a useful approximation for much of the troposphere, is called the **adiabatic rate** *of change of temperature with elevation*. The word "adiabatic" (Greek *adiabatos* not passable; that is, not passable to heat changes) refers to the fact that the temperature changes involved are independent of the temperature of the atmosphere through which a rising mass of air is passing. Upward movement of air is sufficiently rapid that we can consider the rising air to be thermally insulated from its surroundings. Correspondingly, a pressure drop of 11 mb (resulting from a rise of 100 m) will reduce the temperature by about 1° C.

There are two important points to keep clearly in mind when thinking of the adiabatic rate. The first is that the temperature change described takes place entirely within a rising (or sinking) air mass, and has nothing whatever to do with the fact that air aloft normally has lower temperatures than air layers near the ground. The differences in temperature between layers

---

[5] The 0.288th power of a number is very roughly the 1/3 power of that number (since 0.288 is approximately 1/3). The 1/3 power is the cube root. Therefore, in a very approximate fashion, Poisson's equation tells us that the absolute temperature of a gas varies as the cube root of the pressure. In other words, a given increase in pressure will result in an increase in absolute temperature which is proportional to the cube root of the pressure increase. For example, if the pressure doubles, the absolute temperature will increase by about $\sqrt[3]{2}$ or approximately 1.2 times.

of air at different distances from the surface of the earth are due to the fact the atmosphere is heated principally by terrestrial radiation. The greater the distance from the stove of the earth, the colder the air, and we are referring to this phenomenon when we speak of a **lapse rate** of temperature (chapter 3 p. 37). The adiabatic rate, on the other hand, refers to temperature changes that take place within a rapidly rising or sinking air mass as that air mass experiences a decrease or an increase in pressure.

The second point to bear in mind about the adiabatic rate is that, although it is expressed as 5.5° F/1,000 feet (1° C/1,000 m) and is referred to as the adiabatic rate of change of temperature with elevation, it is pressure, not elevation (except indirectly), that governs the temperature change.

The Saturation Adiabatic Rate

While it is true that the rapid rise or fall of air masses insulates them from the temperatures of the air through which they move, they cannot escape one influence of the environment — the effect of condensation within a rising mass. As we have just seen, a rising mass experiences a decrease in pressure, which lowers the temperature of the rising air. A rise of 1,000 feet (300 m) lowers the temperature 5.5° F (3° C); a rise of 10,000 feet (3 km) lowers it 55° F (about 30° C). In all but the driest air masses, a drop in temperature of 20° or 30° F (11.1° to 16.7° C) brings the air below its dew point, and condensation takes place. Latent heat of condensation is then released within the air mass, and the temperature of the air increases. As a rough average, the temperature goes up about 2.5° F as a result of the condensation that takes place in each 1,000 feet of rise after saturation (or about 0.45° C per 100 m). The latent heat of condensation thus increases the temperature by about 2.5° F/1,000 feet (0.45° C/100 m), while adiabatic cooling lowers it about 5.5° F/1,000 feet (1° C/100 m). The result of these opposing processes is a drop in temperature of about 3° F/1,000 feet (0.55° C/100 m). This rate of temperature change with elevation, which applies only when an air mass is saturated and condensation takes place, is called the **saturated adiabatic rate,** or wet adiabatic rate. The adiabatic rate under unsaturated conditions is referred to as the **dry adiabatic rate,** to distinguish it from the other; but the word "dry" is really a misnomer, since the dry adiabatic rate applies to any unsaturated air mass, even one with a very high moisture content.

The Adiabatic Process in Nature

Adiabatic cooling is the most effective method of producing condensation and accounts for almost all precipitation in nature, because adiabatic cooling is a continuous process that lowers the temperature of an air mass as long as there is a pressure decrease. The decrease in pressure, furthermore, is easily achieved through the simple mechanism of lifting. The three situations in which air masses are forced to rise are called *orographic*, *frontal*, and *convectional*. Since each of the three produces its own characteristic kind of precipitation, we shall describe them briefly in turn.

Orographic Precipitation. The most obvious kind of lifting is that which occurs when an air mass encounters a mountain range and is forced to rise over it. This kind of lifting is called *orographic*, (Greek *oros* a mountain + *graphein* to describe), and the accompanying condensation is referred to as **orographic precipitation.** Figure 5-5 illustrates the principal adiabatic relationships involved in orographic lifting. The diagram depicts a mountain range almost 6,000 feet (about 1,800 m) high, close to the ocean and lying at right angles to prevailing westerly winds. This situation is typical of many west coast areas of the United States and Canada. The four small boxes describe the temperature and moisture conditions of oceanic air moving across the mountains. The initial conditions are pictured at point A. The air has a temperature of 61° F (16.1° C) and a mixing ratio of 7.6 grams per kilogram. This air, although fairly moist, is by no means saturated, as the saturation mixing ratio corresponding to a temperature of 61° F (16.1° C) is 11.2 grams per kilogram. Its relative humidity is, therefore, only about 68 percent.

From point A, the air mass is forced to ascend the slope of the mountain, and adiabatic cooling lowers its temperature at the rate of 5.5° F/1,000 feet (1° C/100 m). By the time the air mass has risen 2,000 feet (610 m), its temperature has dropped 11° F (6.1° C), to 50° F (10° C). The table in the diagram indicates that the saturation mixing ratio corresponding to 50° F (10° C) is 7.6 grams per kilogram. The air mass has become saturated. This situation is shown at point B, which is called the condensation level for obvious reasons. From point B, the air continues to rise as a saturated mass, and it cools according to the saturated adiabatic rate. In the 4,000-foot (1219 m) ascent that takes the air from the condensation level to the summit, the temperature drops at the rate of 3° F/1,000 feet (0.55° C/100 m); the saturation mixing ratio decreases correspondingly; and the excess moisture condenses and falls as precipitation. By the time the air reaches the summit, the temperature has dropped to 38° F (3.3° C), and the saturation mixing ratio is, therefore, only 4.7 grams per kilogram. Since the excess moisture has been disposed of as precipitation, the mixing ratio cannot be higher than the saturation mixing ratio and hence is

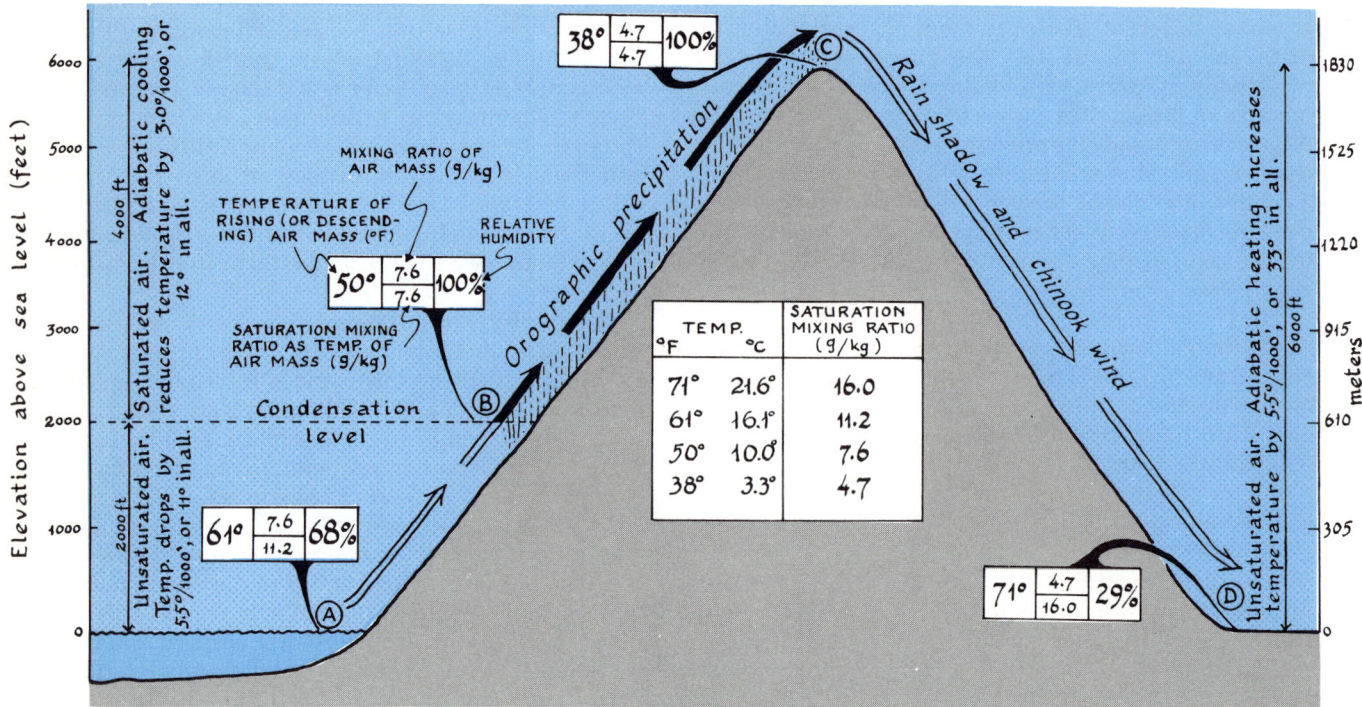

Fig. 5-5. Orographic precipitation, rain shadow, and chinook winds. The particle of air starts at A, with a temperature of 61°F and a relative humidity of 68%. By the time it has risen 2000 feet to B, its temperature is 50° and its relative humidity has reached 100%. At the summit, C, its temperature has dropped to 38°, but its relative humidity is still 100%, despite the decrease in mixing ratio from 7.6 g/kg at B to 4.7 g/kg at C. As the particle descends the leeward slope, it warms up to 71° and its relative humidity drops to 29%, at D.

also 4.7 grams per kilogram. This situation is depicted at point C. The air mass is now 23° F (12.8° C) colder than it was at point A, and every kilogram of air involved in the rising mass has dropped 2.9 grams (7.6 − 4.7 = 2.9) of water vapor in the form of orographic precipitation on the windward slope of the range.

Adiabatic cooling and condensation are phenomena associated with rising air. The right-hand part of Fig. 5-5 illustrates the effects of **adiabatic heating.** On reaching the summit of the mountain, the air mass will not necessarily flow down the slope as shown. Whether it flows downhill or continues to flow eastward at the elevation it attained at the summit depends on the character of the air masses found in the lee of the mountain and on the kind of eddies that develop in the air mass as it crosses the summit. The air masses in the lee of a mountain, such as the one depicted, are normally cold in winter and warm in summer. The descending air could not readily displace the cold, dense, winter air and would normally flow above it rather than move downslope. With the right kind of eddies, however, frictional drag pulls the air down the lee slope as indicated in the diagram. In such a case, the air is warmed adiabatically; and since the warming raises the satura-tion mixing ratio, the mass is no longer saturated, and adiabatic heating proceeds at the rate of 5.5° F/1,000 feet (1° C/100 m). If the air moves downward 6,000 feet (1,829 m), the total heating will be 33° F (18.3° C); and the air will have a temperature of 71° F (21.7° C) by the time it reaches the bottom of the slope. This situation is indicated at point D. The air is warm (10° F or 5.6° C warmer than at A) and dry (the loss of moisture by orographic precipitation on the windward slope has reduced the relative humidity to 29 percent). Such a warm dry wind is called a **chinook**[6] in the western parts of the United States. In southern California, it is often referred to as a *Santa Ana*, after the mountain range down which it flows. It is also common in the Alps, where it is known as the *Föhn*. The warmth and dryness of the Föhn are so marked that the name is applied in German to any hot-air drier, particularly the kind that women sit under in beauty parlors. The chinook, or Föhn, is not a condensation form, of course; it is mentioned here only as a by-product of

---

[6] "Chinook" is probably an Indian word meaning "snow eater."

orographic rain or snow. It should be pointed out that the impression of warmth given by the chinook is not entirely the result of latent heat of condensation. Repeated fluctuations of 40° F (22.2° C) or more within a space of a few hours often characterize Föhn conditions. These temperature contrasts are due to the displacement of cold continental air by the Föhn wind and not to the difference between air that has been adiabatically heated and air that has not.

Lifting by orographic means is thus an important condensation-producing mechanism. Its importance is easily deduced from any precipitation map, which will show that all mountainous areas have more rain than the surrounding lowlands. It is an especially effective generator of snow, since the precipitation falls at high elevations where the air and ground temperature is often below freezing.

There is one additional characteristic of the lee side of a mountain such as the one depicted in Fig. 5-5. The reduced moisture content of the air means that precipitation is unlikely east of the summit, and the area is said to be in the **rain shadow** of the mountain. Rain-shadow areas have no necessary connection with chinook winds. The lack of rain on the leeward slope is simply due to the loss of moisture on the windward slope and is present whether the air is dragged downslope or not.

Frontal or Cyclonic Precipitation. A second situation in which air masses are forced to rise — thereby being adiabatically cooled below their dew point — involves the meeting of two masses of air with contrasting temperatures. The warmer air either rises above the cold or is displaced aloft by underrunning cold air. Usually, both active overrunning and passive displacement take place in different parts of the zone along which the contrasting air masses meet. We shall discuss this phenomenon at greater length in Chapter 6. It is necessary, however, to anticipate that discussion by pointing out that the boundary between two air masses is called a **front** and that low-pressure centers, or cyclones, develop along certain portions of a front. Hence, the rain or snow that condenses as a result of the upward movement or displacement of a warm air mass when it encounters a colder mass is called **frontal, or cyclonic precipitation.** Frontal action is wholly or partly responsible for about 90 percent of the precipitation in nonmountainous areas outside the tropics; and even in mountainous areas, frontal activity is often the major phenomenon involved in starting precipitation. In other words, both frontal and orographic effects are necessary for the very high rainfall and snowfall that often occur on windward slopes of sizable mountain barriers.

Frontal precipitation and the weather conditions that characterize the passage of fronts are such familiar and important parts of the climate of mid-latitude areas that we shall devote considerable time to them. For the moment, however, let us complete our discussion of precipitation types.

Convectional Precipitation. A third method of producing adiabatic cooling by lifting an air mass is to heat the ground. We have already seen, in the discussion of thermally produced pressure patterns, that we can expect large-scale upward movement of air in regions where the surface is heated. The air rises because of its reduced density, the pressure decreases, the temperature drops correspondingly, and eventually, the saturation mixing ratio of the rising air mass becomes lower than the prevailing mixing ratio. Condensation ensues, often in the form of towering cumulonimbus clouds; and precipitation will occur if the air mass is adequately supplied with moisture and if there is an adequate mechanism for the formation of water droplets large enough to fall through the rising air. The sequence of events is similar to what occurs in orographic and frontal precipitation, but the rain tends to be more localized, and the vertical dimensions are more striking. **Convectional precipitation** is characteristic of equatorial and tropical regions, but it is also familiar as the summer thunderstorms of the mid-latitudes. The word "convectional" refers, of course, to the thermal convection that causes the air to rise. Whether air rises or not, under given conditions, is a slightly more complicated problem than may appear at first, and we shall discuss this important point next.

The Relation Between Prevailing Lapse Rates and the Adiabatic Process

We must, once again, differentiate between the wet and dry adiabatic rates, on the one hand, and the lapse rate of temperature, on the other. The wet and dry adiabatic rates describe a change in temperature within a rising (or sinking) air mass. The temperature change is a fixed function of the pressure changes and occurs inside the air mass, regardless of outside conditions. The lapse rate of temperature describes the differences in temperature between air masses of differing elevations above the surface of the earth. It is best, at first, to think of the air masses as stationary. What happens, however, when some part of the atmosphere rises and moves through the stationary layers of air? We shall deal with this question by considering the upward motion of a very small volume of air, simply to emphasize the difference between the little bit of air that is rising and the rest of the atmosphere, which is not involved in this uplift.

80 Chapter 5

Let us first consider three different lapse rates, each of which is typical of summer conditions over North America in about 45° N latitude. These lapse rates are shown as dotted lines in Fig. 5-6 (and are indicated by columns 2 and 3, 6 and 7, and 10 and 11 of Table 5-2).

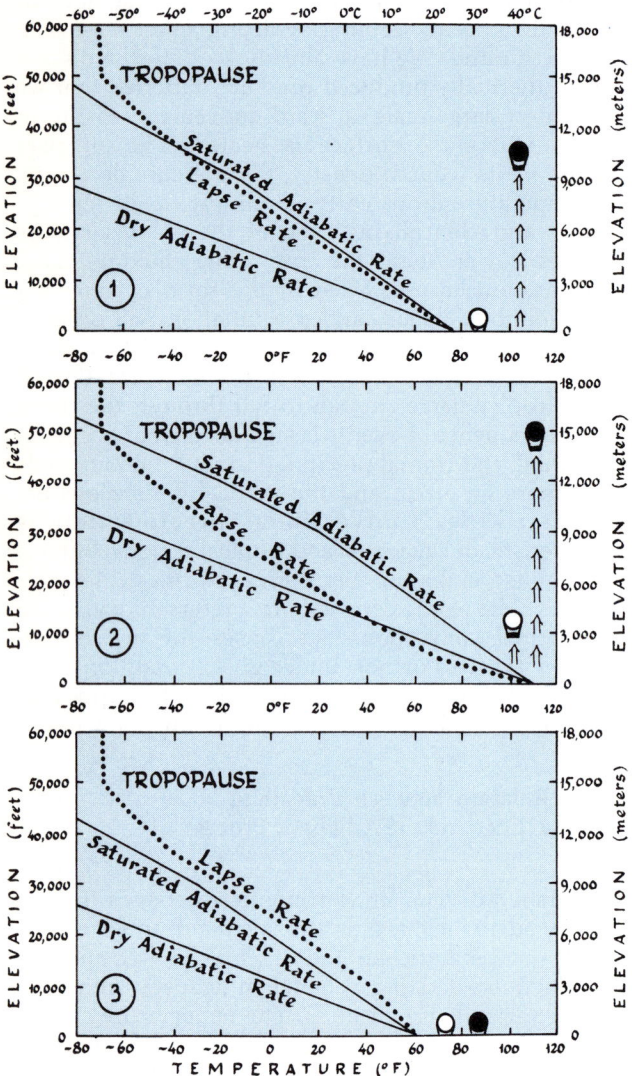

Fig. 5-6. Conditionally unstable (1), unstable (2), and stable (3) lapse rates. Under conditionally unstable lapse rates (1), the unsaturated balloon, shown in white, never leaves the ground, since the dry adiabatic rate produces temperatures everywhere lower than those of the ambient temperature given by the lapse rate; the saturated balloon, shown in black, rises so long as the saturated adiabatic rate gives temperatures greater than those of the ambient air given by the lapse rate. With unstable lapse rates (2), both balloons rise, although only the saturated one reaches the tropopause. With stable lapse rates (3), neither balloon rises, since both the dry and saturated adiabatic rates yield temperatures lower than those of the ambient air, as given by the lapse rate.

The dots serve as a reminder that the temperature sounding on the graph represents the temperatures at each level of the atmosphere. Each dot can be considered as representing the temperature of a particular level and not the continuous temperature of a single rising air mass. Figure 5-6 (1) shows normal summer conditions. The surface temperature is about average; and the lapse rate, at least near the ground, is 3.5° F/1,000 feet or 0.64° C/100 m (the normal lapse rate is 3.6° F/1,000 feet or 0.65° C/100 m). Above 10,000 feet the lapse rate decreases to 3.0° F/1,000 feet (0.55° C/100 m) and above 40,000 feet to 2.0° F/1,000 feet (0.36° C/100 m). Above 50,000 feet (15 km), the temperatures no longer decrease; beyond that level lies the stratosphere, with nearly uniform temperatures.

Figure 5-6 (2) shows what happens when there is strong heating at the surface. The surface temperature has been increased to 110° F (43.3° C), and the lapse rate is, therefore, quite steep. Near the ground, it is very steep — 6° F/1,000 feet (1.09° C/100 m); from 10,000 feet to 30,000 feet (3 to 9 km), it is much less steep but still steeper than under normal conditions. Beyond 30,000 feet (9 km), the temperature and its lapse rates are the same as under normal conditions. In our example, the effect of surface heating does not extend beyond this elevation. This is probably a generous estimate of the effectiveness of transmission of temperature changes at the ground to air aloft.

Figure 5-6 (3) depicts the temperature situation under conditions of surface cooling. The temperature at the surface is only 60 °F (15.6° C), and the lapse rate in the lower part of the troposphere is consequently reduced to 2° F/1,000 feet (0.36° C/100 m). Surface cooling is effective only up to about 10,000 feet (3 km); above that level, conditions are normal. Surface cooling does not affect upper air temperatures as much as surface heating. We shall see why in a moment.

Now that we have described the lapse rates, let us return to the problem of a rising small volume of air. To make things as easy to visualize as possible, let us imagine a balloon especially constructed for our purpose. It has several unusual features: the envelope is a perfect insulator; instead of helium, it contains ordinary air; and the passengers riding inside the balloon are weightless. In short we have constructed a closed system with regard to temperature changes (outside temperatures do not affect those inside the balloon, and vice-versa) but open with respect to pressure, since changes in pressure outside the balloon will affect the air pressure inside. The trip in the balloon begins not on a day with normal temperature conditions, but rather on a day with very high surface temperatures. You can imagine yourself to be a passenger inside the balloon; or, if you are more cautious and comfort-minded, you can follow the described action by

Evaporation and Condensation 81

Table 5-2. Lapse Rates Aloft Under Selected Conditions of Surface Heating. (The lapse rates in metric units are direct conversions of those in English units. Because of rounding errors, they do not correspond exactly to the lapse rates calculated by dividing temperature differences in °C by elevation differences in km.)

(1) Temperature Conditions with Average Surface

Elevation (1)		Temp. of air (2)		Lapse rate (3)		Temp. inside unsaturated balloon (4)		Temp. inside saturated balloon (5)	
(feet)	(km)	(°F)	(°C)	(°F/1,000 ft.)	(°C/100 m)	(°F)	(°C)	(°F)	(°C)
0	0	75	23.9			75	23.9	75	23.9
1,000	0.3	71.5	21.9	3.5	0.64	69.5	20.8	72	22.2
10,000	3	40	4.4	3.5	0.64			45	7.2
20,000	6	10	−12.2	3.0	0.55			15	−9.4
30,000	9	−20	−28.9	3.0	0.55			−15	−26.1
40,000	12	−50	−45.6	3.0	0.55			−55	−48.3
50,000*	15*	−70	−56.7	2.0	0.36				
60,000	18	−70	−56.7	0.0	0.00				

(2) Temperature Conditions with Surface Heating

Elevation (1)		Temp. of air (6)		Lapse rate (7)		Temp. inside unsaturated balloon (8)		Temp. inside saturated balloon (9)	
(feet)	(km)	(°F)	(°C)	(°F/1,000 ft.)	(°C/100 m)	(°F)	(°C)	(°F)	(°C)
0	0	110	43.3			110	43.3	110	43.3
1,000	0.3	104	40.0	6.0	1.09	104.5	40.3	107	41.7
10,000	3	50	10.0	6.0	1.09	55	12.8	80	26.7
20,000	6	15	−9.4	3.5	0.64	0	−17.8	50	10.0
30,000	9	−20	−28.9	3.5	0.64			20	−6.7
40,000	12	−50	−45.6	3.0	0.55			−20	−28.9
50,000*	15*	−70	−56.7	2.0	0.36			−70	−56.7
60,000	18	−70	−56.7	0.0	0.00				

(3) Temperature Conditions with Surface Cooling

Elevation (1)		Temp. of air (10)		Lapse rate (11)		Temp. inside unsaturated balloon (12)		Temp. inside saturated balloon (13)	
(feet)	(km)	(°F)	(°C)	(°F/1,000 ft.)	(°C/100 m)	(°F)	(°C)	(°F)	(°C)
0	0	60	15.6			60	15.6	60	15.6
1,000	0.3	58	14.4	2.0	0.36	54.5	12.5	57	13.9
10,000	3	40	4.4	2.0	0.36				
20,000	6	10	−12.2	3.0	0.55				
30,000	9	−20	−28.9	3.0	0.55				
40,000	12	−50	−45.6	3.0	0.55				
50,000*	15*	−70	−56.7	2.0	0.36				
60,000	18	−70	−56.7	0.0	0.00				

*Tropopause

reference to Fig. 5-6 (2) or to columns 6, 8, and 9 of Table 5-2.

The first important point to be made is that absolutely nothing happens when the passengers first get into the balloon. The air at the ground (and inside the balloon) is considerably warmer than the air aloft, so why does this hot air not rise? Hot air is less dense than cold air, but only when pressures are equal. The difference in density between the surface air and that at 1,000 feet (305 m) is counterbalanced by a pressure difference that keeps the two air layers in their respective places. It may occur to you that the pressure difference is fairly constant (about 34 millibars/1,000 feet), but temperature differences may be quite great. If this is so, it should be possible to increase the surface temperature (and/or decrease the temperature aloft) to such an extent that the pressure gradient can no longer counterbalance the large temperature gradient you have created. In that case, the air at the surface should rise spontaneously. If the temperature contrast between the surface and the air at 1,000 feet becomes greater than 18.8° F (or greater than 3.43° C/100 m), the pressure gradient will, indeed, no longer compensate for the enormous differences in density. Putting things in slightly more formal language, we say that if the lapse rate is greater than 18.8° F/1,000 feet (3.4° C/100 m), the air is **autoconvectively unstable.** The tremendous lapse rates that are required for autoconvective instability are seldom found in the atmosphere. They may exist in a very shallow layer of air, perhaps no deeper than an inch, which lies directly above the ground. For our purposes we can neglect their existence.

By this time, the balloonists are understandably impatient about their lack of progress on a day when the lapse rate is very large (6° F/1,000 feet or 1.09° C/100 m), although not nearly large enough for autoconvection. So, they hail a passerby to give them a push and up they go. What happens inside the balloon from then on depends on whether the air in the balloon is saturated or unsaturated. Let's take the case of the dry air first. If the balloon rises to 1,000 feet (304 m), the inside temperature drops by 5.5° F (3.1° C) because of the decrease in pressure between sea level and the higher elevation. Another 1,000-foot (305 m) rise depresses the temperature inside the balloon by an additional 5.5° F (3.1° C). And so on, as long as the balloon continues to rise and the air inside remains unsaturated. This situation is shown by the two columns marked (8) in Table 5-2 and by the line marked "dry adiabatic rate," in Fig. 5-6 (2). A look at either the table or the diagram will tell you that as the balloon goes up, the air inside becomes warmer than the air outside. The higher the balloon goes, the greater this contrast becomes. This is easily seen in the table by comparing columns 6 and 8. On the graph, you should compare the dry adiabatic rate with the dotted line representing the lapse rate of the air temperature.

If the air inside the balloon is warmer than that outside and the pressure is the same for both, it follows that the inside air is lighter than that outside. Consequently, the balloon rises of its own accord once an initial upward push has been applied. Except for the necessary presence of some sort of triggering mechanism, the situation is the same as that described for autoconvective instability. The balloon rises to 12,500 feet (3.8 km); above that level, the outside temperature becomes warmer than that inside the balloon.

If the balloon is full of saturated air, the ascent is much the same. For every 1,000 feet or rise, the inside temperature drops 5.5° F (1° C/100 m). However, the drop in temperature lowers the saturation mixing ratio, the excess moisture condenses, and the released latent heat of condensation increases the temperature by about 2.5° F. The net decrease in temperature is thus only 3° F for every 1,000 feet of rise (0.65° C/100 m). The temperatures inside the saturated-air balloon are shown in column 9 of Table 5-2 and by the saturated adiabatic line on the diagram. Except for greater contrasts in temperature between inside and outside air, the situation is the same as with the unsaturated-air balloon. The relatively warm temperatures inside the saturated-air balloon cause it to rise through the relatively cold outside air, but in this case the balloon rises all the way to the tropopause before the outside temperature becomes as warm as that inside.

When we have a steep lapse rate, such as the one under consideration, which allows both dry and saturated air to rise of their own accord once they are started upward, we say that the air is *absolutely unstable.* Usually we omit the word "absolutely" and simply speak of **instability,** *unstable air*, and, conveniently but somewhat illogically, of *unstable lapse rates.* Any lapse rate greater than 5.5° F/1,000 feet (but less than 18.8° F/1,000 feet) fits the bill (between 1° C and 3.4° C/100 m). Unstable air, since it promotes upward movement, is an important factor in producing precipitation. To understand the tremendous implications of instability for climate, we need to consider two other kinds of lapse rates.

However, one last word should be said about the ascending balloonists. Their ascent does not continue forever. It takes place only as long as the lapse rates are unstable (that is, only in that part of the troposphere where the lapse rates are greater than 5.5° F/1,000 feet or 1° C/100 m). Above 10,000 feet (3 km), the heating at the ground is felt only slightly, and much smaller lapse rates prevail (see column 7 of Table 5-2). By 20,000 feet (6 km), the air inside the unsaturated-air balloon would be considerably colder than the air outside, but ascent would have stopped long before this level was reached, because the air inside the balloon would no longer be lighter than the air outside, which would occur at about 12,500 feet (3.8 km) above the surface (Fig. 5-6 (2)). In the saturated-air balloon, the crew would be carried all the way to the tropopause — not only frozen but thoroughly drenched in the bargain. Notice that the saturation adiabatic rate increases with elevation. In Fig. 5-6 this is reflected by a slight leveling off of the line. Between 30,000 and 40,000 feet (9 and 12 km), it is already 4° F/1,000 feet (0.73° C/100 m); above 40,000 feet (12 km), it increases to 5° F/1,000 feet (0.91° C/100 m). The reason is quite simple. At these levels almost all the moisture has already been condensed out of the air, and the release of latent heat of condensation becomes very small. In fact, at high elevations, the saturated adiabatic rate comes very close to being equal to the dry adiabatic rate.

So much for unstable lapse rates. Figure 5-6 (3) and columns 10 to 13 in Table 5-2 indicate what happens to a rising mass of air on a day with surface cooling and consequently with very small lapse rates. Perhaps we had better leave the balloonists stranded where they are and proceed simply from the graph and table. The dry and saturated adiabatic lines in Fig. 5-6 (3) and the appropriate columns in Table 5-2 show clearly enough what happens when the lapse rate is very low. At any level above the ground, the temperature inside the balloon (either a dry-air or a saturated-air balloon) is less than that of the air outside; the air in the balloon is denser than its surroundings, and so the balloon sinks back to the surface. External forces can push the balloon up; but the higher the balloon goes, the denser its air becomes relative to its surroundings. As soon as there is no outside push, it sinks back to the ground.

When we have a small lapse rate that prevents both dry and saturated air from rising, we say that the air is *absolutely stable*. Once again, we normally omit the word "absolutely," and speak simply of **stability**, *stable air*, or *stable lapse rates*. Any lapse rate smaller than 3° F/1,000 feet (0.55° C/100 m) creates absolute stability. Occasionally, the cooling at the ground is sufficiently pronounced to create an inversion of normal conditions; that is, the air at the ground may actually be colder than that above it. Such an increase in temperature with elevation is called an **inversion** for obvious reasons. Bear in mind that we have described decreases in temperature aloft as positive lapse rates. For instance, a lapse rate of 2° F/1,000 feet (0.36° C/100 m) means that the temperature is 2° F less at 1,000 feet than at the surface (or 0.36° C colder at 100 m than at the ground). Obviously then, increases in temperature with elevation are described as negative lapse rates. An inversion, therefore, is marked by a lapse rate that is less than zero — or, in other words, a negative lapse rate. The important thing to remember about inversions is that they are simply the most stable of all lapse rates. Lapse rates less than 3° F/1,000 feet (0.55° C/100 m) but greater than 0° F/1,000 feet (0° C/100 m) are also stable, of course, but they are not inversions.

Figure 5-6 (1) and columns 2 to 5 of Table 5-2 describe the situation in the atmosphere when the lapse rate of temperature is nearly normal. A look at either the table or the diagram will show that the situation of the unsaturated-air balloon is the same as it was with stable lapse rates, whereas the saturated-air balloon moves upward just as it did in the situation with unstable lapse rates. The saturated-air balloon does not rise quite to the tropopause, but otherwise the situation is similar.

Lapse rates of intermediate value — between 3° F/1,000 feet and 5.5° F/1,000 feet (between 0.55° C and 1° C/100 m) — which allow saturated air to rise but prevent the upward movement of unsaturated air are said to be *conditional*. Whether such lapse rates create instability or stability is conditional on their being saturated or unsaturated. We speak, in such cases, of **conditional instability**, *conditionally unstable air*, or *conditionally unstable lapse rates*.[7]

We can summarize what we have just learned about lapse rates as follows:

1. *Lapse rate less than 3° F/1,000 feet (0.55° C/100 m) — stability.*

    A lapse rate less than 0° F/1,000 feet (0° C/100 m) is given a special name, "inversion," but it represents merely an extreme form of stability.

2. *Lapse rate between 3° F/1,000 feet and 5.5° F/1,000 feet (between 0.55° C and 1° C/100 m) — conditional instability.*

    The air is stable when unsaturated but unstable if saturated.

3. *Lapse rate greater than 5.5° F/1,000 feet (1° C/100 m) — instability.*

    A lapse rate greater than 18.8° F/1,000 feet (3.4° C/100 m) creates autoconvective instability in the air — a rare situation.

These relationships are important enough to be memorized. They have far-reaching consequences for the climate of the world. We shall be talking about stability and instability throughout the rest of the chapters dealing with climate. At this point, it might be well to elaborate some of the more significant effects of lapse rates.

---

[7]It is immaterial whether we call an air mass conditionally stable or conditionally unstable. The choice is one of point of view, as in the expressions "half full" and "half empty."

The Importance of Lapse Rates to Climate

The most important climatic effect of lapse rates is their influence on precipitation. We have just seen that cooling below the dew point is the only important way of producing condensation. Of the cooling mechanisms, adiabatic cooling is the only one that results in sizable precipitation. But adiabatic cooling depends on pressure decreases brought about through the ascent of air masses. Stability of the air prevents any such ascent and thus makes precipitation extremely unlikely. Instability of the air, on the other hand, abets upward movement and thus makes precipitation quite likely if there is a reasonable amount of water vapor in the air. The normal lapse rate is neither stable nor unstable; it is conditional upon the moisture content of the air. Saturated air masses are unstable in this situation; unsaturated ones are stable. Precipitation is likely in the former case; it is unlikely in the second case (both on the grounds of stability and of lower moisture content).

There are two prerequisites for precipitation: water vapor in the air and instability or conditional instability which allows the air to ascend and become adiabatically cooled. Lack of moisture is a limiting factor for interior portions of the continents, far removed from oceanic sources of moisture; but the limiting factor in most cases and most places is stability. If we confined our discussion of precipitation to a single causal factor, we could make much more sense of the global pattern of rainfall and snowfall by considering only the stability or instability of the air in various parts of the world than we could by considering only the moisture content of the air. One example should help to make this point clear. Table 5-3 gives certain relevant monthly averages at San Francisco.

As you can see, July is the wet month by any standard, except for the amount of rain. Its relative humidity averages 6 percent higher than that of January, and the mixing ratio is appreciably larger. But July, although foggy, is practically rainless. The stable lapse rates, which result partly from cooling of the air by passage over cold upwelling offshore waters, allow condensation in the cold air but prevent any vertical movement and thus preclude precipitation. In January, the conditionally unstable lapse rates, which are associated with cyclonic storms, allow vertical movement of saturated air where warm and cold air masses meet. The result is a moderate amount of rain. The factor that limits precipitation at San Francisco is obviously not moisture content but rather the degree of stability of the atmosphere. San Francisco is something of a pat example, but other places exhibit the same phenomenon to a considerable degree.

Table 5-3. Average Moisture Conditions at San Francisco in Winter and in Summer

	January	July
Relative Humidity*	85%	91%
Mixing Ratio*	5.4 g/kg	8.1 g/kg
Lapse Rate*	Conditional	Stable**
Condensation Forms	4.7 in of rain (119 mm)	0.0 in of rain (< 2 mm); frequent fog

*At 5:00 A.M.
**Inversion

While the effect of lapse rates on precipitation is most important, there are a number of other effects that are only relatively less significant. Stability prevents the upward movement not only of the water vapor in the air but also of everything else that is normally transported upward by convective turbulence. Under average conditions or with unstable lapse rates, dust and other pollutants are carried upward and their concentration is diluted to the point where they are neither irritating to the nose and eyes nor hazardous to health. Under conditions of stability, there is no upward movement of air and, therefore, no upward dispersal of these irritants, which become concentrated near the surface in a form called smog. A heavy, dark combination of pollutants and fog, smog is a serious health hazard, even known to result in some deaths.

Still another consequence of the lapse rate of temperature is the speed with which an air mass acquires the characteristics of the surface over which it has recently moved. A cold dry air mass overlying a relatively warm ocean surface, for instance, rapidly becomes saturated and is warmed to considerable levels if the prevailing lapse rate is unstable. A stable lapse rate, on the other hand, allows only very slight upward penetration of moisture and of oceanic warmth. Since heating creates instability, the effects of surface heating are felt much farther aloft than those of surface cooling, which is associated with stability. That is the reason for showing normal conditions as low as 10,000 feet (3 km) above the cooled surface in Fig. 5-6 (3), whereas the effects of surface heating are shown to extend beyond 20,000 feet (6 km) in Fig. 5-6 (2).

Condensation Forms

The prevailing lapse rate plays a crucial climatic role. In the case of condensation, the influence of lapse rates results from adiabatic cooling being the most effective way to achieve saturation in an air mass. One final way of underlining the relationship between the degree of stability of the atmosphere and the kind of condensation that is likely to take place is by classifying condensation forms, which are sometimes called **hydrometeors** (Greek *hydros* water + *meteoros* things in the air,) according to the following scheme.[8]

A. *Hydrometeors of the Stratiform Type*

These condensation forms occur when lapse rates are stable and therefore without convection or extended vertical motion. The stratus clouds, as the name implies, are uniform, unbroken sheets. They do not necessarily cover the whole sky; but where they occur, the cloud sheet is continuous, without any tendency to break up into small cloudlets.

1. *Fog forms*
   a. *Radiation ground fog, smog*
   b. *Advective fog*

2. *Cloud forms* — *Elevation of cloud base*
   a. *Low stratus* (**St**) — *Below 3,000 feet (0.9 km)*
   b. *Nimbostratus* (**Ns**) — *3,000-8,000 feet (0.9-2.4 km)*
   c. *Altostratus* (**As**) — *8,000-18,000 feet (2.4-5.5 km)*
   d. *Cirrostratus* (**Cs**) — *Above 18,000 feet (5.5 km)*

3. *Precipation*
   a. *Fine drizzle rain*
   b. *Light, powdery snow (fine flakes)*

B. *Hydrometeors of the Limited Convective Type*

These condensation forms occur when lapse rates are either unstable or conditionally unstable in a layer of limited depth, which is topped by a markedly stable layer of air, frequently even by an inversion. They represent the break-up of the stratiform clouds into cellular patterns. Each cloudlet marks a cell of rising air; the breaks between cloud cells indicate where the air is subsiding.

1. *Practically no fog forms*
2. *Cloud forms* — *Elevation of cloud base*
   a. *Stratocumulus* (**Sc**) — *1,000-8,000 feet (0.3-2.4 km)*
   b. *Altocumulus* (**Ac**) — *8,000-18,000 feet (2.4-5.5 km)*
   c. *Cirrocumulus* (**Cc**) — *Above 18,000 feet (5.5 km)*
3. *Precipitation*
   a. *Light drizzle showers*
   b. *Light snow flurries*

C. *Hydrometeors of the Unlimited Convective Type*

These condensation forms occur when the atmosphere is sufficiently unstable for vertical convection to penetrate to a high elevation.

---

[8]This classification is very slightly modified from that in H.C. Willett and F. Sanders, *Descriptive Meteorology*, 2nd ed., Academic Press, Inc., New York, 1959, and is used by permission.

1. *Practically no fog forms*
2. *Cloud forms* — *Elevation of cloud base*
   a. *Cumulus* (**Cu**) — *3,000-10,000 feet (0.9-3. km)*
   b. *Cumulonimbus* (**Cb**) — *3,000-10,000 feet (0.9-3. km)*
   c. *Cirrus* (**Ci**) — *20,000-40,000 feet (6-12 km)*
3. *Precipitation*
   a. *Moderate or heavy frontal and orographic rain & snow*
   b. *Moderate or heavy convective showers and snow flurries*

As you can see from the classification, the condensation forms depend principally on the lapse rates. Stable lapse rates are associated with clouds that are sheet-like in appearance and have considerable horizontal development but do not extend vertically into the atmosphere. These are the clouds of the stratus family. There may also be fog, but very little precipitation is likely. With unstable lapse rates, the clouds take on considerable vertical development. These are the clouds of the cumulus family. Fog is rare, but precipitation is commonly moderate to heavy when it occurs. With intermediate lapse rates, the likelihood of precipitation is small, and the amounts are likely to be negligible. The various cloud forms are illustrated in Fig. 5-7.

Summary

In this chapter we have discussed the principal processes involved in condensation; evaporation was more easily explained, but as we saw, it is difficult to measure. We used three different kinds of measures of the moisture content of the atmosphere: the mixing ratio, an indication of the absolute humidity of the air; the relative humidity, which relates the absolute moisture content to the capacity; and the dew point.

The processes whereby condensation occurs can best be summarized by the following outline:

A. *Condensation occurring at less than 100 percent relative humidity due to presence (or addition of) hygroscopic nuclei*

B. *Condensation produced by mixing air masses of different temperatures*

C. *Condensation produced by adding moisture to the air*

D. *Condensation produced by reducing the capacity of the air to hold moisture — that is, by reducing the temperature to or below the dew point*
   1. *Cooling by radiation (radiation fog)*
   2. *Cooling by advection (advection fog)*
   3. *Adiabatic cooling; normally produced by lifting an air mass and thereby decreasing pressure*
      a. *Orographic lifting*
      b. *Cyclonic or frontal lifting*
      c. *Convectional uplift*

Fig. 5-7. Selected cloud forms. At the left, clouds of stratiform and limited convective type which yield little or no precipitation. Top to bottom, altocumulus, stratocumulus, and low stratus. At the right, clouds of unlimited convective type which yield moderate to heavy precipitation. Top to bottom, dense cirrus flowing out of cumulonimbus, massive cumulonimbus, and fair-weather cumulus (with cirrus clouds aloft). *Photographs by R. L. Day.*

Of these processes, adiabatic cooling is by far the most important condensation-producing mechanism. The likelihood of vertical movement of air — and therefore of adiabatic cooling, condensation, and ultimately precipitation — depends on the relationship between the existing lapse rate at a particular place and at a particular time and the adiabatic rate of change of temperature with elevation.

The critical values of the existing lapse rates are listed below, together with the degree of stability and the condensation forms associated with them:

1. *Lapse rate less than 3° F/1,000 feet (0.55° C/100 m) — stability: fog, stratiform clouds, very light rain or snow.*
2. *Lapse rate between 3° F/1,000 feet and 5.5° F/1,000 feet (0.55° C and 1° C/100 m) — conditional instability: variety of condensation forms, including the possibility of considerable precipitation, particularly in connection with orographic uplift.*
3. *Lapse rate greater than 5.5° F/1,000 feet (1° C.100 m) — instability: no fog, cumuliform clouds; possibility of very heavy snowfall or rainfall.*

The importance of the lapse rate is not limited to the principal role it plays in determining the likelihood of precipitation; it is also felt in such matters as air pollution, transformation of air masses from marine to continental character, and the like.

Suggestions for Further Reading

Byers, Horace R., *General Meteorology*, 3rd ed., McGraw-Hill Book Company, Inc., New York, 1959, pp. 428-457.

Riehl, Herbert, *Introduction to the Atmosphere*, McGraw-Hill Book Company, Inc., New York, 1965, pp. 67-89.

Willett, Hurd C., and Frederick Sanders, *Descriptive Meteorology*, 2nd ed., Academic Press, Inc., New York, 1959, pp. 82-117.

6 Air Masses, Fronts, and Storms

Introduction

In Chapter 4 we described, in a broad way, the major movements of air over the surface of the earth. That discussion of the general circulation of the atmosphere stressed only the movement of air and did not deal with the other climatic elements affected by these movements. In this chapter we shall be concerned with more detailed description of the climatic results of the transport of heat, of moisture, and of the degree of stability of moving masses of air.

It should immediately become obvious that we need a finer instrument than the rather blunt weapon of the general circulation. Some meteorologists, in fact, distinguish among several approaches to the problem, referring to such concepts as the zonal and cellular pressure patterns and the associated wind belts as the statistical approach. By this they mean that the basic tools of description — the Azores high or the trades, for instance — represent statistical averages of some sort. As we have seen, these averages may not be adequately descriptive of day-to-day weather or even of the long-run climate.

In another approach, which involves the study of air masses, the focus is on the original characteristics of some mass of air, its modification while moving from its source region, and the type of weather it brings along its path of travel. This second approach was developed largely in Scandinavia during World War I under the leadership of the Norwegian meteorologist Tor Bergeron.

Other students of climate, particularly tropical meteorologists, doubt that either the statistical or the air-mass approach can produce meaningful descriptions of the pattern and the chronology of weather and climate in the world. They stress the perturbations of the pressure field, claiming that slight-to-moderate changes in the prevailing pattern of isobars are the key to climate and weather, particularly in low latitudes. The importance of the tropics in World War II and the accumulation of observations at the ground and aloft gave impetus to this school of thought and at the same time created considerable controversy over the best approach.

We shall be using here parts of all three approaches, since they all shed light on the geographic distribution of climates. Basically, however, we used the statistical aproach in the previous chapter; in this chapter we shall concentrate on Bergeron's air-mass approach while we are studying middle-latitude climates — and we shall hope for the best when we come to the tropics.

Definitions

Much of the power of the concept of air-mass analysis depends on the definition we use and on the implications that follow from the definition. You may feel that the whole thing is quite simple indeed; but bear in mind that this simple idea was not effectively formulated until the time of World War I; and that simple ideas, in any case, are often more powerful and revolutionary than sophisticated complexities. An air mass is a fairly large body of air whose physical properties (notably temperature and moisture content) are relatively uniform in the horizontal plane. The words "fairly large" indicate the continental scale of the air masses; a body of air with a diameter smaller than 1,000 miles (1,600 km) does not qualify as an important air mass. The words "relatively uniform" indicate that the gradients of temperature and moisture within the air mass (say, within a few hundred miles) are much smaller than those between two adjacent air masses.

Two more definitions are necessary before we can go on to the classification and description of typical air masses. The boundary between two air masses of contrasting temperature or of moisture content or of both is called a front. Strictly speaking, a front is the boundary along the surface of the earth; the boundary aloft is called a frontal surface. Finally, the original home of the air mass, the area that gave the air mass its initial characteristics, is called the **source region.**

Implications of the Definitions

The definitions of air masses, of fronts, and of source regions carry with them certain important implications. The notion of relative uniformity over fairly large areas implies a set of fairly large and relatively uniform source regions. This means, of course, that source regions are either oceans or continents — never a mixture of both. It also means that either high latitudes or low latitudes can serve as source regions; but the mid-latitudes are almost completely eliminated, since they are regions of almost constant air movement that are partly covered by snow and partly snow-free and that are crossed, over the oceans, by both warm and cold currents. All these contrasts of the mid-latitudes are detrimental to the formation of air masses with horizontal homogeneity, so that only polar or tropical areas function as source regions. One other feature affects the formation of air masses. In regions of low pressure, the inflow piles up air of contrasting characteristics; around high-pressure areas, outflowing air spreads the acquired properties outward. The zones of high pressure, therefore, serve as the principal source regions. Furthermore, an air mass must remain over its source region long enough to acquire the characteristics of the ground surface, not only in its lowest layers, but to a considerable distance aloft as well. In addition, if the air mass is to maintain its identity when it leaves the source region, it must move fast enough that its physical properties, though modified to some extent by passage over new terrain, retain much of their initial character. The notion that a source region imparts certain native characteristics to an air mass is not very useful if every body of air takes on, like a chameleon, the character of every piece of ground it crosses.

From these implications, we gather that air masses overlie their source regions for considerable periods before moving rather rapidly to new locations. When viewed in the large, the atmosphere is not in constant motion; rather, there are periods of days or even weeks when there is no large-scale transport of air. After such a quiescent interval, outbursts of polar air occur, and tropical air flows poleward to compensate for the movement of polar air toward the equator.

Instruments

One of the innovations of air-mass analysis, when it was first applied by the Scandinavian meteorologists, was the great attention paid to the three-dimensional nature of weather processes. This approach necessitates collection of upper-air data in addition to those of traditional surface observations. The instruments used are familiar: a barometer, a thermometer, and a hygrometer.[1] The three instruments, together with a small radio transmitter, are built into a container that protects them from rain and sun but allows free access of air. The container is attached to a balloon that drifts with the wind as it rises. Such a rig, called a *radiosonde*, was first constructed by the Russian meteorologist Moltchanoff in 1928. Radiosonde observations are called *raobs*, and when the path of the radiosonde balloon is tracked from the ground to get wind speed and direction, the combined technique is referred to as *rawinsonde*.

Classification of Air Masses

One of the easiest ways to become familiar with the characteristics of and differences among air masses is to learn the simple classification described in the next few paragraphs. This classification utilizes two characteristics to differentiate air masses: (1) the ver-

---

[1]The barometer is not, however, a mercury tube but an aneroid barometer, in which air pressure works a delicate spring. The thermometer is bimetallic rather than the more common mercury or alcohol type.

tical distribution of temperature and (2) the moisture content. Both of these may be estimated, to some degree of accuracy, from the character of the source region and of the terrain crossed by the air mass.

Classification of Source Region

The first part of the classification identifies the source region as being either maritime or continental and, in addition, either polar or tropical. The letters *m*, *c*, *P*, and *T* are used to designate the various possibilities. Thus, an *mP* air mass is one whose source region is maritime and polar, a *cP* air mass originated in a continental polar environment, and so on. The first two letters of the classification tell us quite a bit about the temperature near the ground and the moisture content, particularly if we also know whether we are dealing with a winter or a summer air mass. The other important trait is the degree of stability of the air as determined by its lapse rate. The lapse rate is conditioned not only by the thermodynamic situation over the source region, but also by thermal and dynamic changes the air mass undergoes after it leaves the source region.

Classification According to Stability

Two major factors affect the degree of stability of an air mass: thermal processes and air movement. The stability near the ground is governed by thermal processes. If the air is heated from below, it becomes unstable; cooling by the ground creates stability, and if the cooling is strong enough, an inversion may develop. These are essentially the conditions shown in Fig. 5-6 (2) and 5-6 (3). Generally, a cold air mass (that is, an air mass that is colder than the underlying surface) is warmed and thus made unstable. The letter *K* is applied to such cold air masses ("kalt" is the German word for cold). Warm air masses, on the other hand, are usually cooled by the underlying ground and thus made stable. Such air masses are designated by the letter *W*. The designations *K* and *W* should be considered as indicating instability and stability, respectively. It is true that they also describe the temperature of the air mass relative to the ground, but the degree of stability created by these temperature relationships near the surface is the more important characteristic. Lapse rates near the ground, then, are indicated by the capital letters *K* and *W*.

Lapse rates aloft depend more on the movement of the air than on the thermal situation and are said to be dynamically conditioned. This term means, in a very general way, that the rising air characteristic of low-pressure areas is conducive to unstable lapse rates, whereas the subsiding air of high-pressure areas (particularly of the eastern portions of such anticyclones) produces markedly stable lapse rates aloft. A subsidence inversion may form at some distance from the ground under conditions of strongly subsiding air. In the classification, the degree of stability aloft is indicated by the lower-case letters *s* and *u* (for stable and unstable, respectively). Occasionally, an air mass is neutral in the sense that it is neither particularly stable nor unstable. In such a case, the third or fourth letter of the classification is simply omitted. Thus we speak of an *mPKs* air mass if it is markedly stable aloft and near the ground; or *mPK* if it is markedly stable only near the ground; or *mPs* if it is stable aloft only, or *mP* if it is neither very stable nor unstable at any level.

We may classify any air mass by some combination from the four sets of letters defined above. Let us consider, for example, an air mass formed over northern Canada in winter. How should we classify it? The first two letters are easy: this is obviously continental polar air and should be designated *cP*. What about the relative temperatures of the air and of the ground? As we have seen, continental areas in high latitudes experience the coldest winter temperatures in the world. It follows that the air mass overlying this coldest of all possible surfaces must be warmer than the ground. The air should, therefore, be classified as *W*, which indicates its stable lapse rate near the ground. What about lapse rates aloft? In winter, continental interiors are zones of high pressure and of generally subsiding air; hence, the air mass is stable aloft as well as near the ground, and the letter *s* is called for. The complete classification is *cPWs*. What does this tell us about the air mass? The most obvious characteristic of a *cPWs* air mass in winter is the low temperature near the cold ground. Slightly higher up, the temperature is probably higher. The next most obvious characteristic is the dryness of the air. The air not only is continental in origin, but comes from a cold continent as well. This means that the rate of evaporation is exceptionally low because little moisture is available in the ground and little heat is available to evaporate it. Furthermore, the moisture on and in the ground is probably locked up in the form of snow or ice and cannot enter into the **hydrologic cycle.** The low temperature of the air, moreover, is associated with a low moisture-holding capacity, so that even saturated air would not contain much water vapor. For all these reasons the air will be very dry. If we look at the second part of the classification, we see that the air is stable near the ground and aloft. Hence, upward movement of air is impeded, and cooling of the air below the dew point by adiabatic means is out of the question. The likelihood of precipitation is slight, primarily because the air is stable throughout and secondarily because the air is dry. Clear, cold days and nights are the characteristic weather conditions associated with a *cPWs* air mass.

Another example will conclude our discussion of the classification of air masses. Let us classify, in this instance, an air mass formed in summer over the middle of the Sahara. How should we classify this air mass? It is obviously a *cT* air mass because of the continental and tropical nature of the Saharan source region. Furthermore since the subtropical deserts are the warmest parts of the earth's surface in summer, by definition, the air must be colder than the underlying surface. The third letter of the classification is thus *K*. The Sahara is an area of low pressure in summer, so the lapse rate aloft is unstable. The complete classification is therefore *cTKu*. What does this tell us about the air mass? The air mass is probably dry, although the high surface temperatures are conducive to high rates of evaporation, and the high temperature of the air indicates a high capacity to accommodate water vapor. The air is unstable throughout, and we can expect considerable vertical movement of air, which would lead to considerable precipitation, except for the low moisture content and particularly the low relative humidity of the air. Under certain conditions, the strong surface heating produces vertical convective currents, known as *dry thermals*, that stop short of the condensation level. These currents may reach heights of 10,000 feet (3,000 m) above the ground. If the moisture content of the air is high enough and if the vertical ascent of the air is pronounced enough, the condensation level will be reached; and cumulus clouds will form, without, however, producing precipitation. This condition is known as *cumulus convection*.

Description of Air Masses

The preceding paragraphs have given examples of air masses, illustrating the usefulness of the classification. The major emphasis was on the source region. Detailed discussion of air masses necessitates discussion of all the possible kinds of source regions, of the most important trajectories of air masses, and of the modifications they undergo during their displacements. In addition to this focus on individual air masses, we should examine the climatic implications of weather patterns that result from meetings of markedly different air masses. For our purpose, we must limit the discussion to a general description of the Northern Hemisphere source regions in winter and in summer and to the phenomena associated with the meeting of air masses, with only incidental attention being paid to other trajectories. Figure 6-1 shows air-mass source regions in January and July. Two kinds of air masses, equatorial and arctic, that were not included in the scheme of classification are shown on these maps. Equatorial air masses, designated by the letter *E*, are similar to tropical air masses. The major distinction, and the reason for showing separate source regions for *E* and *T* air masses, is that equatorial air is associated with **trade winds** as they approach the equator (the *ending trades*), with low pressure, and with the **Intertropical Convergence Zone** (the **ITC**), about which there will be more later. The tropical air masses are more diverse in occurrence and characteristics. Arctic air (*A*) is similar to *cP* air, particularly in winter. In summer, however, the arctic air masses preserve the winter characteristics of the *cP* masses, whereas *cP* air undergoes considerable modification of both temperature and stability aloft.

Winter Air Masses

With this introduction in mind, we can make certain descriptive statements about the source regions in the two seasons. Bear in mind that Fig. 6-1 shows air mass *source regions* and not the extension of any particular air mass. Thus *cP* air masses *originate* in the *cP* source region, but they may move far beyond the limit of the source region itself. That is why the fronts, which represent a meeting place of contrasting air masses, are not shown along source region boundaries, but rather at the place where the contrasting air masses meet most frequently.

Let us begin with the January map (Fig 6-1 (1)). In January, the source regions of *cP* air consist of the Arctic Ocean and the adjacent continental areas. The Arctic Ocean is a source region for *cP* air because it is covered by ice or snow and thus does not behave like a maritime surface. Both the ocean and the adjoining land are therefore very cold surfaces overlain by high pressure at this time of year. The southern limit of the source region is set approximately along the line where the snow cover lasts for three months. The pressure and temperature characteristics of the source region result in very stable air masses — often with inversions. Because of the coldness and dryness of the ground surfaces, the air masses surrender both heat and moisture to the ground. Dry stable air masses of the kind developed here yield little precipitation. Stratiform clouds are the only common condensation form.

The high-latitude oceans are the source regions for *mP* air, but as you can see from Fig. 6-1 (1), the source regions are limited to the eastern parts of the North Atlantic and North Pacific. The western parts of the oceans are too much influenced by the eastward and southeastward movement of *cP* air to be considered source regions. They are, therefore, labeled "transitional" on the map and will be considered separately. The *mP* source regions, aside from their very small area, have other common characteristics. The surface

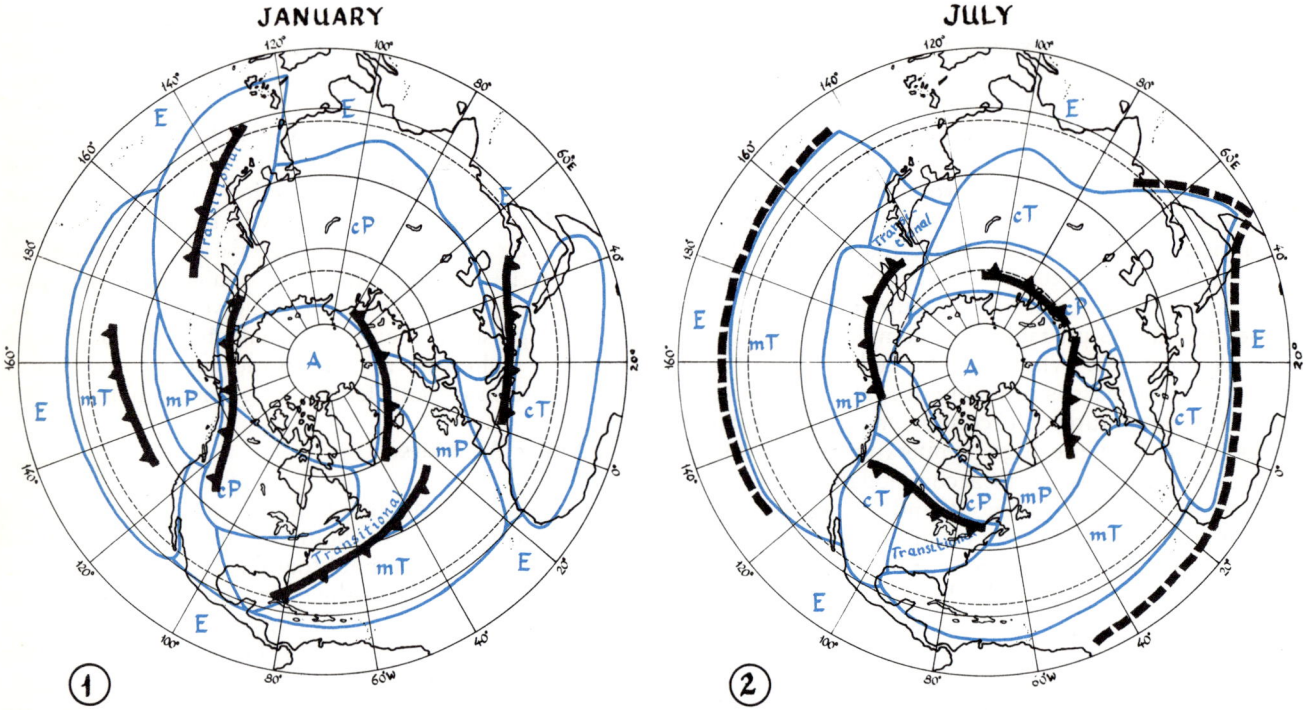

Fig. 6-1. Northern Hemisphere air-mass source regions in January (1) and July (2). Fronts are shown by the heavy black symbols; the Intertropical Convergence Zone (ITC) is shown by a heavy black dashed line in (2). Unlabeled areas are not source regions.

temperature is abnormally high for the latitude, mostly because of the warmth of the West Wind Drift in both the Atlantic and the Pacific. The warm waters of these currents impart mild temperature and high humidity to the air. Because prevailing low pressure results in unstable air masses, considerable precipitation can be expected from these moist air masses.

Northern Africa is the only Northern Hemisphere source region of cT air. North America south of 30° N latitude is too narrow to serve as an important source region; southern Asia is too much influenced by polar air masses. The North African cT air is warm and dry; the circulation is moderately anticyclonic, and subsidence inversions aloft make the air stable.

The subtropical high-pressure areas over the Atlantic and Pacific Oceans are the source regions for mT air masses. The ocean surface has uniformly high temperature, but aloft there is a marked contrast between the eastern and western portions of these two oceanic regions. In the east, the high-pressure cells are characterized by subsiding air and by inversion from 1,500 to 3,000 feet (about 450 to 900 m) above the sea; in the west, ascending air carries moisture upward and creates instability. This distinction is reflected in the fact that mTs air brings practically no rain to the west coasts of the continents (on the eastern edges of the subtropical highs). La Paz, in Baja California (24° N latitude), receives only 0.1 inches (2.5 mm) of rain in January; Villa Cisneros, in Spanish Sahara (also 24° N latitude), has about the same amount. At the western edges of the highs, however, mTu air brings considerable rain to the east coasts of the continents. Miami, Florida (26° N latitude), receives 2.5 inches (64 mm) in January; Taipei, Formosa (25° N latitude), receives 3.5 inches (89 mm). An important point to bear in mind regarding these air masses is that the mT air which invades North America and western Europe in mid-latitudes comes from the western part of the source region and not from the eastern sector where stable air masses prevail; these air masses bring relatively warm and rainy weather. This continental invasion by mT air is one way in which heat and water vapor are transported poleward in winter.

The areas marked "transitional" on the map consist of regions where cP air flows eastward over adjacent oceans or southward onto tropical lands and seas. In either case, the cP air is rapidly transformed into warm, humid air masses with greatly steepened lapse rates. The role of these regions in the formation of fronts and cyclones will be touched on later.

The last source region shown on the map is that of equatorial air masses. We have already indicated the similarity of such air masses to tropical air and the reason for showing the source region separately, despite the similarity. The major features of the equatorial air masses are the instability associated with the ascending air of the equatorial low-pressure belt and the high moisture content. This combination of characteristics results in a strong vertical development of clouds (the unlimited convective clouds) and heavy rain.

Summer Air Masses

In summer, there is a general shift poleward of all the source regions, with an enlargement of the maritime source regions at the expense of transitional areas and with an increase in size of cT source regions at the expense of cP regions.

In July, the source region for cP air masses is limited to the Arctic Ocean and to the continents above 55° N latitude. Typical cP air is associated with high pressure. Over the Arctic Ocean, cPW air forms under the influence of the mildly anticyclonic circulation at the surface. Over the North American and Eurasian land masses, cPK air is the rule, and when passing high-pressure centers invade the continents, the continental polar air masses may spread southward well beyond the boundary shown in Fig. 6-1 (2).

The mP source regions in summer extend much farther to the west and are much larger than in winter. They occupy the areas of weakly cyclonic circulation in the Aleutian and Icelandic portions of the Pacific and Atlantic Oceans. Relatively cool and moist air, with surface stability but with occasional moderate instability aloft, may bring rain to the west coasts of North America and Europe. This is particularly true of the air that comes from the northern part of the North Atlantic source region.

Source regions for cT air are much more extensive in summer than in winter. Even the relatively narrow North American continent supports a fairly extensive source region, extending across the mountainous western half of the continent. In the Old World, cT air flourishes from the Sahara to Mongolia. The high temperatures of the surface tend to create instability in the lower layers; but because of low moisture content in relation to the extremely high moisture-holding capacity and the presence of high pressure overlying the surface lows, this is an area of little precipitation.

The mT source regions in summer are much larger and extend farther northward than in winter. Except for higher summer temperatures, however, the air masses are very similar in the two seasons. They are dominated, in summer as in winter, by the North Atlantic and North Pacific centers of high pressure. As a result, extreme stability is the rule in the eastern portions of the source regions, where there is strong subsidence. The western margins of the source regions are unstable aloft.

The transitional areas of the summer map differ somewhat in location and considerably in character from the transitional areas of winter. In January, transitional areas are found over oceans where strong flow of cP air negates the maritime character of the surface. The summer areas, on the other hand, are continental zones modified by inflow of mT air from adjoining oceans. The two major areas of transition in summer are found in the eastern part of the United States and along the east coast of Asia between 40° N and 50° N latitudes.

As do most of the other source regions, equatorial air-mass source regions extend farther north in summer than in winter. This northward migration is associated with the poleward movement of the equatorial belt of low pressure. The major characteristics of air masses in this part of the world are the instability of the ascending air and the high moisture content up to considerable elevations. Heavy cloud cover and much convective rainfall are the natural results of this combination. These equatorial air masses penetrate far inland in the southern and eastern parts of Asia, where they are referred to as the summer monsoon.

Atmospheric Disturbances and Associated Weather

The weather, particularly the precipitation regimes of the various parts of the world, is most closely conditioned by the kinds of atmospheric disturbances peculiar to each place. The term "atmospheric disturbance" is used instead of the simpler word "storm" not out of pedantry, but because the longer term can be used to include both the "storms" of mid- and high latitude and the much more gentle disturbances of the tropics. Furthermore, atmospheric disturbances include both frontal phenomena (that is, phenomena associated with the interaction of different air masses) and air-mass phenomena (that is, phenomena occurring within a single air mass). We shall begin with extratropical (Latin *extra* outside of + *tropical*) frontal disturbances. Most of these disturbances are associated with low pressure, and particularly with cyclones; but bear in mind that the term "cyclone," whether applied to middle latitude or equatorial storms, covers a wide range of phenomena, of which the intense tropical cyclone (or hurricane) and the middle latitude tornado (sometimes called a cyclone, and often outwaited in a cyclone cellar) are only extreme examples.

Frontal Phenomena in the Middle and High Latitudes

As we have seen, the boundary between air masses of distinctly different characteristics is referred to as a front. Since the boundary is seldom very sharp and often extends over a distance of 100 miles or more, the term "frontal zone" is more descriptive, but the shorter term is commonly employed to describe such zones of relatively rapid transition of temperature and of moisture content.

Weather phenomena of great importance to the climate of the globe develop along some of these frontal zones, particularly where there is a strong temperature contrast between adjoining masses of air. The strongest temperature contrasts are found along the boundary between polar and tropical air. Frontal activity also takes place elsewhere, but we shall start our discussion by considering the most clear-cut case. Let us begin by considering the origin and the nature of the disturbances that develop along this boundary between warm tropical air and cold polar air.

The Structure of Extratropical Cyclones.

Figure 6-2 shows the sequence of events along the boundary between a cold and a warm air mass. The development of an **extratropical cyclone** is called **cyclogenesis** (Greek *kyklos + genesis*.) At first the boundary, or front, is fairly straight, but if the winds in the warm sector are moving eastward more rapidly than those in the cold air mass, a cyclonic circulation (counterclockwise in the Northern Hemisphere) is set up by the motion of the warm winds relative to the cold (Fig. 6-2 (1)).[2] The cyclonic circulation can be seen both in the plan view and in the block diagram. The latter shows the structure of the front not only at the surface but also at upper levels. The cold air mass is seen to consist of a wedge of air that underlies part of the warm air. This relationship is the natural result of the greater density of cold air. In fact, it is the difference in density rather than the difference in temperature that results in well-developed frontal structures of the kind we are about to describe.

In Chapter 4, we described how pressure gradients result in particular kinds of wind velocities and also how a particular speed and direction of wind creates its appropriate pressure gradient. Here the cyclonic circulation creates a local low-pressure center and further deforms the front into a wave-like structure (Fig. 6-2 (2)). The wave is carried eastward, as part of the general west-wind circulation of the mid-latitudes, at a rate that averages about 25 miles (40 km) per hour. All the weather phenomena associated with such an extratropical cyclone (sometimes called a wave cyclone or mid-latitude cyclone) therefore move from west to east at a fairly rapid pace. Slow-moving extratropical cyclones, which are most common in summer, when the whole atmospheric circulation is slowed down, cross the United States in about 6 days; faster ones may take as little as 3 days.

The wave consists, at first, of two distinguishable parts (Fig. 6-2 (2)). The western member of the wave represents the leading edge of the cold air mass. It is, therefore, called a *cold front* and is indicated on weather maps either by a blue line or by the triangular symbols shown in the figure. The eastern member of the wave marks the leading edge of advancing warm air and is called a *warm front*. It is designated either by a red line or by semicircular symbols.[3] The two fronts and the associated wind movement are shown not only in plan view and by means of a block diagram, but also in a cross section along the lines labelled AB in 6-2 (1), CD in 6-2 (2), and EF in 6-2 (3). Notice the peculiar shape of the isobars in the plan view. They are roughly circular in the cold sector but straight in the warm sector, with kinks at the fronts. The isobaric kinks correspond, of course, to the sudden shift in wind direction as either of the two fronts is crossed. The block diagram shows the increased complexity of the frontal surface aloft, including the overriding of the cold air by warm air above the warm front. The cross sections are most useful since they show the upper-air features of the warm and the cold fronts and, in addition, allow for the representation of the attendant weather phenomena. Let us look at the cross sections in some detail.

The weather, most particularly the condensation forms resulting from frontal disturbances, is greatly influenced by the degree of stability of the air involved. Since we are dealing with two different air masses and since there are three major categories of stability (instability, conditional instability, and stability), at least nine different situations are possible. Fortunately, we can simplify things by lumping instability and conditional instability into one category. Furthermore, the cold air is most likely stable to begin with, and the flow of warm air above the cold intensifies this stability. In any case, the degree of stability of the cold air is of minor importance to precipitation, since the cold air does not contain nearly so much moisture as the warm air. Therefore, we need consider only two cases: instability and stability in the warm air mass.

---

[2] In the figure, the cold air is moving southwestward. It could as well have been shown as blowing toward the east, as long as its speed was less than that of the westward flow of the warm air. The same result would be achieved by a southward outburst of polar air.

[3] You may also refer to Fig. 4-2 which shows an actual extratropical cyclone and fronts as portrayed on a weather map.

Air Masses, Fronts, and Storms 95

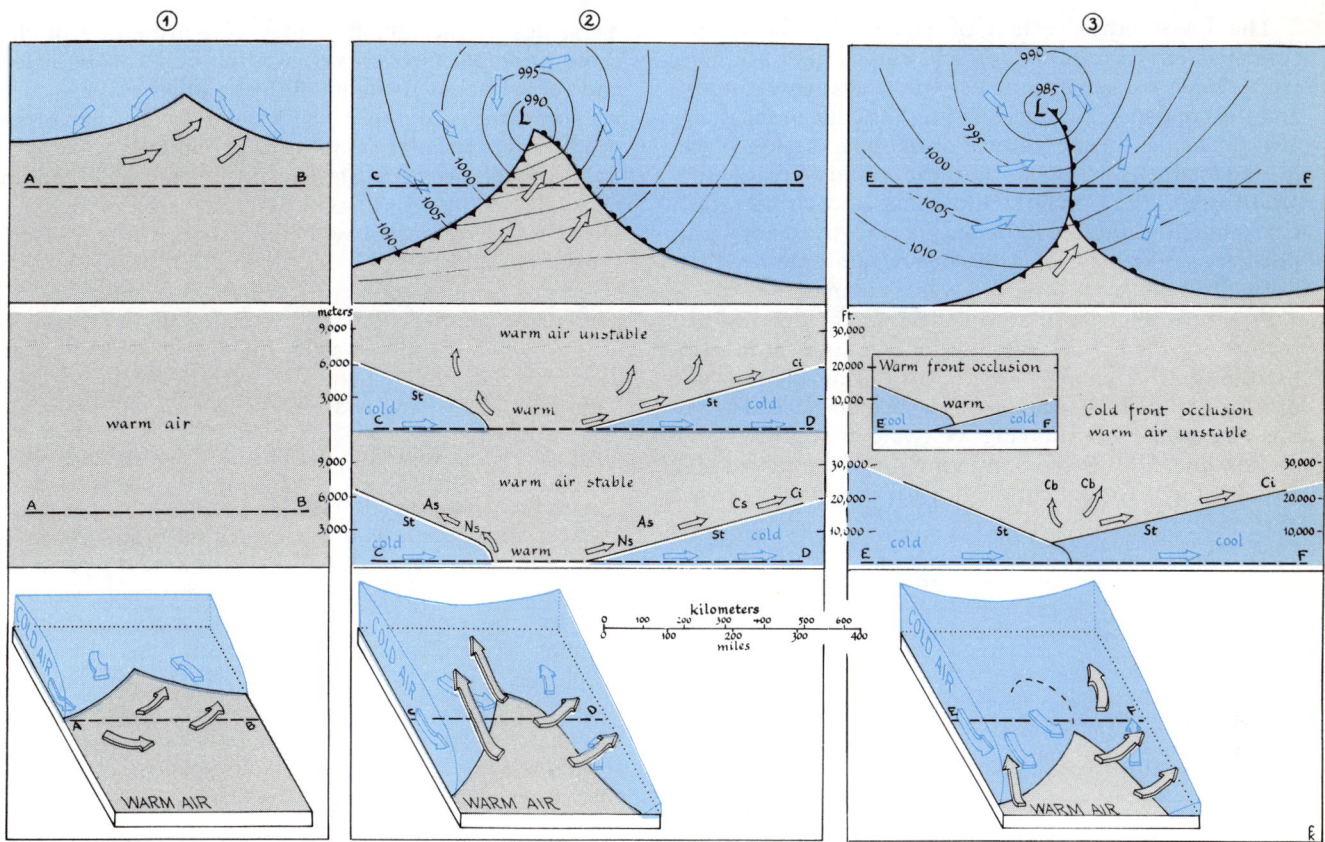

Fig. 6–2. The development of an extratropical cyclone along the polar front. The beginning of cyclogenesis (1); a fully developed cyclone (2); and occluded fronts (3). The top view, in each case, is a bird's eye view of the storm, showing isobars as light black lines, with appropriate pressure values; cold and warm fronts by heavy lines with black triangles and half-circles, respectively; air movement by open arrows (blue for cold air, black for warm); and the areas dominated by the cold and warm air masses in blue and gray, respectively. The middle views are vertical cross-sections along the dashed line indicated in the plan view above them, with added designations of cloud forms and air mass characteristics. The bottom views are attempts at showing these storms in three dimensions.

The upper cross section (Fig. 6-2 (2)) shows the condensation forms that develop in an unstable air mass. The vertical scale has been exaggerated about 25 times to show things clearly. The slopes of the two fronts, though they appear gentle enough on the cross section, are actually 25 times less steep than shown. Warm fronts normally have slopes that vary from 1:100 to 1:400; cold fronts, from 1:25 to 1:100. Along the warm front, the aggressive warm air rises smoothly, without much frictional drag, above the underlying cold wedge. Since the warm air is unstable, unlimited convective condensation forms develop. Near the surface front, we find cumulonimbus clouds and gentle, widespread rain (or snow if the cold air is below freezing); farther to the east and at higher elevations, we find cirrus clouds. To an observer on the ground, the cirrus is often the first indicator of the approach of an extra-tropical cyclone. Below the warm front, in the stable cold air, we find stratus clouds, which, incidentally, completely obscure the warm-sector cloud forms. The rain in the cold sector comes from the clouds above, not from the stratus within the cold air mass.

The cold front has not only a steeper slope but also a nose-like protuberance where it is in contact with the ground, which is the result of frictional drag by the ground on the cold air as it underruns the warm sector. This underrunning displaces warm air upward directly above the front. Again, cumulonimbus or other unlimited convective cloud forms develop; but in contrast to the situation over the warm front, the convective activity is limited in areal extent. The rain (or snow) of the cold front is much more localized and tends to be more intense than that which is associated with the warm front.

The lower cross section of Fig. 6-2(2) shows the condensation forms developing in a stable air mass. The fronts have the same form, as with unstable warm air, and the condensation forms within the cold air mass are also the same. But no vertical development takes place in the stable warm air; and the clouds are of the stratiform variety, mainly altostratus and cirrostratus. Consequently, there will be only a fine drizzle or light powdery snow instead of the heavy rain or snow that normally falls from unstable warm air masses.

A simple exercise that will help to fix the various features of the extratropical cyclone in your mind is to list the weather changes that occur during the passage of a frontal disturbance over a particular place. This is most easily done by moving westward along the cross-section line AB (instead of moving the storm eastward). A convenient format for such a list is shown below.

Position relative to weather storm	East of warm front	Near the warm front	Within the warm sector	Near the cold front	West of cold front
Pressure					
Wind direction					
Temperature					
Cloud forms					
Precipitation					

The life cycle of an extratropical cyclone does not end with the kind of structure just described. The final stage, shown in Fig. 6-2(3), comes about from the fact that the cold front usually overtakes the warm front, at least near the center of the extratropical cyclone. In such a case, the warm air is completely displaced aloft; temperature contrasts along the ground are limited to the difference between the temperature of the cold air in the forward part of the storm and that of the cold air in the rear; and the clouds and rain of the two fronts are concentrated in one place, as shown in the cross section. The displacement of the warm air aloft is called an **occlusion** (Latin *ob* toward+*claudere* to shut), and the new frontal situation is referred to as an *occluded front*. The cross section shows a cold-front occlusion with unstable warm air aloft. The small inset represents a warm-front occlusion. The difference between the two types of occlusions lies in the density relation between the cold air in the rear and that in the forward portion of the disturbance. In the cold-front occlusion, the colder air to the west undercuts the somewhat warmer air to the east. In the warm-front occlusion, the slightly warmer air to the west rises over a wedge of colder air to the east.

Climatic Importance of Extratropical Cyclones. The principal climatic importance of frontal phenomena lies in the fact that they constitute one of the three principal mechanisms whereby air can be lifted and cooled below its dew point (the other two are convection and orography). Orographic lifting often produces spectacular amounts of precipitation; convective activity seems all-important in summer and in tropical regions, but some sort of frontal activity seems to be connected with the majority of rain- and snow-producing storms. In the mid-latitudes, the role of fronts is all-important. Probably as much as 80 percent of all mid-latitude precipitation is frontal in origin, which is the reason for this rather detailed description of the structure of extratropical cyclones. Let's now see where and when this frontal activity is concentrated.

The Polar and Arctic Fronts. The word "front" is used in three slightly different connections: (1) in a very general sense, to describe any boundary between air masses; (2) in a much more restricted way, to describe the small-scale features of the extratropical cyclone, as in the terms "warm front," "cold front," "occluded front"; and finally, (3) in an intermediate sense, to define the major zones of frontal activity where extratropical cyclones are most likely to develop, as in the terms "polar front" and "arctic front." We shall use the term in this last sense.

We have seen that the prerequisite for full-fledged extratropical cyclonic activity is a marked contrast in density between air masses. The necessary differences in density can be provided only by air masses with distinctly different temperatures. The most obvious and important zone along which marked temperature contrasts are found is the boundary between polar and tropical air-mass source regions. In winter, the greatest contrast is between cP (including arctic) air and mT air, because the polar-tropical contrast is intensified by the coldness of the continents and the relative warmth of the oceans. In fact, the North Pacific and North Atlantic Oceans are so warm in winter, relative to the neighboring continents, that some of the major fronts develop along the boundary between continental polar and maritime polar source regions. Most of the winter fronts, however, are found in transitional source regions, since these are precisely the areas that lie between the breeding grounds of polar and tropical air masses. In summer, latitudinal temperature contrasts are much reduced, as we have seen; and the fronts are found only in very high latitudes, where arctic source regions retain part of their winter coldness and thus become markedly different from neighboring areas, whether these be mP areas of mild warmth or cP source regions that are considerably warmed because of their continental character.

These general remarks can be clarified and made

more specific by looking back at Fig. 6-1, which shows the location of the major fronts in winter and in summer, all of which are in high or middle latitudes. Figure 6-1(1) shows the relationship between air-mass boundaries and the six major fronts in winter. Where the front lies along the southern boundary of arctic air, it is called an Arctic front; elsewhere, it is referred to as a Polar front. This difference in nomenclature is of no great significance to our discussion. Notice the position of these fronts, however. The Pacific Arctic front lies along the northern edge of the Pacific Ocean, at the southern limit of arctic and cP source regions. The Atlantic Arctic front occupies a similar position at the northern edges of the Atlantic Ocean. The Pacific Polar front, the Atlantic Polar front, and the Mediterranean front are found in transitional areas lying between polar and tropical air-mass source regions. The Pacific Polar front has, in addition, a secondary position, located in the mid-Pacific, far from polar or arctic air masses.

Figure 6-1(2) shows the high-latitude location of the fronts in summer. The Pacific Polar front has migrated northward to about the winter position of the arctic front. The Atlantic Polar front has also moved northward to about the winter position of the Arctic front. Two new fronts have appeared, the Canadian Polar front, at the southern edge of the cP source region in North America, and the Siberian Arctic front, which can be thought of as a continental extension of the Atlantic Polar front.

The cyclonic precipitation associated with the fronts is found over a wide area on either side of the fronts themselves, and, of course, the seasonal movement of the arctic and polar fronts results in a similar shift in latitude of the area of cyclonic precipitation. The band of frontal rain and snow extends about 10° on either side of the mean positions shown in Fig. 6-1. There is a similar belt of cyclonic storms in the Southern Hemisphere. It is more continuous and more uniform in latitude because of the great preponderance of ocean surfaces in the Southern Hemisphere; but the seasonal migration follows the same pattern. In summer (January), the fronts are far to the south, and cyclonic rainfall affects only the southern tip of South America and South Island of New Zealand. In winter (July), the fronts are closer to the equator, and cyclonic precipitation is experienced as far north as the southern tip of Africa and the southern parts of Australia. We shall have more to say on the subject of precipitation distribution in Chapter 7.

Tornadoes. Although tornadoes are not frontal phenomena, they are such well-known atmospheric disturbances of the middle latitudes that a brief description is not out of place. A **tornado** is a severe local storm, so concentrated in its intensity as to cause un-

Fig. 6–3. A tornado developing from the cloud system of a severe thunderstorm near Freeman, South Dakota, June 1, 1965. *Photograph courtesy of the National Oceanic and Atmospheric Administration.*

believable damage along its narrow and erratic path. It consists of violently rotating columns of air that show up as funnels of whirling dust and debris that open up into unlimited convective cloud forms of the cumulonimbus type (Fig. 6-3). The vortex, or active center of the storm, is small, normally a few hundred yards (or meters) in diameter, and consists of winds estimated to approach 300 miles (nearly 500 km) per hour. The cause of these fantastically intense but small cyclones is not well understood, nor is their world dis-

Fig. 6–4. The path of a tornado is narrow, often less than 500 feet wide, but the damage as shown in this picture is nearly total, as is the evident grief of one of its victims. *Courtesy of the National Oceanic and Atmospheric Administration.*

tribution known, although they occur on all continents, and also over the oceans in the form of water spouts. In the United States they are most common in the Great Plains and in the Southeast, and are practically unknown west of the Rocky Mountains. They occur mainly in late spring and early summer, usually in late afternoon. They are almost always associated with very steep lapse rates resulting from intense heating of the ground. Between 1915 and 1950, tornadoes killed over 200 persons and caused over 13 million dollars of damage per year in the United States alone (Fig. 6-4).

Atmospheric Disturbances Within the Tropics

Tropical areas experience changes in air masses, some of them quite as striking as those found in the mid-latitudes. Polar air masses seldom penetrate as far equatorward as the tropics, but parts of Africa and Australia are on the border of *cT* and *mT* source regions, so that there are changes of air mass of some importance. Even here, however, the change from one air mass to another is slow and seasonal and does not resemble the abrupt, week-to-week or even day-to-day changes found in higher latitudes. There are, therefore, no fronts and no frontal precipitation of the kind we have described above.

To be sure, in much of the tropics there is an alternation of air masses, which is almost as rapid as that brought about by the passage of an extratropical cyclone; but only two air masses are involved, and they differ only slightly in their meteorological characteristics. These are the *mT* and the equatorial air masses shown in Fig. 6-1. The *mT* air, also called trade air, consists of a warm and moist layer several thousand feet thick and an upper layer of much drier and quite stable air. This upper layer consists of air that has subsided in the subtropical high-pressure belt — hence its dryness and stability. Equatorial air, also called doldrum or monsoon air, is warm and moist to very high levels and is quite unstable compared to trade air.

As you can see, the surface characteristics of the two air masses are much the same so that weather changes from one air mass to the other are not likely to be noticeable. Nor is the boundary between them to be thought of as a front, since there is little temperature difference and hence no density contrasts between them. In Fig. 6-1(2) a zone of frontal action is shown between them, but it is called the **Intertropical Convergence Zone (ITC)** rather than a front. The ITC, which lies between the trade winds of the two hemispheres, generates disturbances because of the opposed directions of the two wind systems and not because of temperature contrasts between the air masses involved. The ITC is not shown in Fig. 6-1(1) since it is found south of the equator in winter; in the summer, it is found from 5° to 20° north of the equator, as shown in Fig. 6-1(2). The major role of the ITC is the creation of a zone of **convergence** and, therefore, of ascending air. The upward movement of unstable, moisture-laden air results in convective rainfall of considerable magnitude.

As a matter of fact, this heavy charge of moisture is one of the distinguishing characteristics of tropical and equatorial air masses. The contrast between tropical and extratropical weather phenomena is, therefore, a contrast between weather that is developed along the boundary of two air masses with decidedly different temperatures and weather that is developed without the benefit of air-mass contrasts but in air masses so heavily laden with water vapor that any slight convergence will produce rain and squalls.

The ITC is the principal feature that produces convergence, but day-to-day weather is conditioned by more localized zones of converging and rising air. The localization of weather is determined by a number of atmospheric disturbances that occur either within the ITC or on its immediate borders. For our purposes, we shall divide these disturbances into five types: (1) convergence resulting from the surge of the trades, (2) convergence resulting from transverse (north-south) waves in the belt of easterly winds, (3) convergence produced in weak tropical cyclones, (4) convergence in tropical hurricanes, and (5) the monsoon circulation.

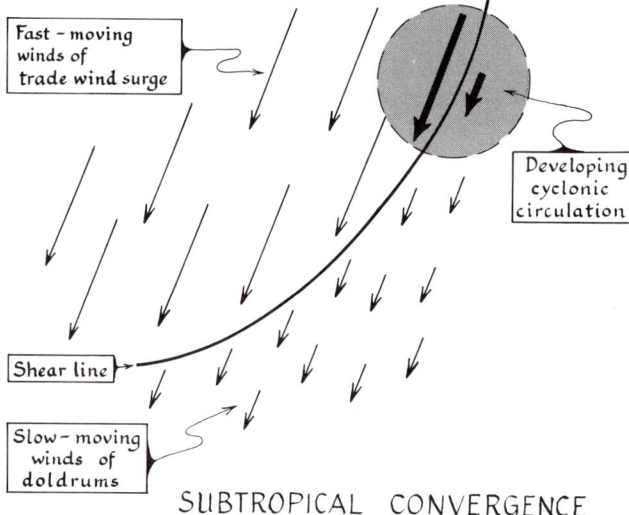

Fig. 6-5. A surge of the trades. The faster moving air northwest of the shear line creates a cyclonic (or counter-clockwise, in the Northern Hemisphere) movement when it meets the slower moving air southeast of the line.

Surge of the Trades. "Surge of the trades" refers to occasional strengthening of the trade winds as a result of high-latitude disturbances whose effect extends into the tropics without, however, any invasion of the tropics by extratropical air masses. Figure 6-5 shows the wind situation that develops. Winds from the subtropical high-pressure area increase in speed over a local area for a few days; equatorward of the surge, the air movement continues to be much slower; the surging air begins to overtake the slower mass; and the inevitable result is a pileup of air, or convergence. Considerable convective rainfall may accompany the forward edge of the surge. In addition, cyclonic movement develops along the boundary between surging and stationary air. This *cyclonic shear* aids the process of convergence and rising air.

Easterly Waves. Transverse **easterly waves** in the pressure pattern within the tradewind zone are reflected in a poleward bulge of the isobars (Fig. 6-6(1)), which indicates poleward extension of equatorial low pressure. The winds follow a sinuous course corresponding to that of the pressure pattern. East of the wave, the isobars converge toward the trough, hence there is convergence of air; west of it, there is **divergence**. These transverse waves move slowly westward (10 to 15 miles per hour or about 15 to 25 km per hour) along the boundary between mT and equatorial air masses. Thus, good weather precedes the passage of the wave because of the divergence of the air; and bad weather with convective rainfall follows. Figure 6-6(2) shows the structure of one of these waves aloft. The cross section of wind speeds and the air-mass boundaries indicate the upper-air structure. To the west, in the zone of divergence, we see subsiding air; and the boundary between the upper trade air and the lower doldrum air is quite low. There is not only divergence but dry air as well. In the center, we see the wave trough (the area of lowest pressure). The dry trade air is displaced aloft to considerable altitudes by the upward movement of air, which is itself the result of low-level convergence. Just to the east of the trough, therefore, we find moist air ascending. The inevitable result of such ascent is seen in the weather symbols at the bottom of the cross section. Notice that rain showers, thunderstorms, and cumulonimbus clouds lie to the east of the trough; fair weather and cumulus clouds lie to the west. Over the western parts of the tropical oceans, there is a more or less continuous passage of transverse waves and, therefore, an alternation between fair and rainy weather. These unspectacular disturbances in the normal pressure pattern of the atmosphere actually account for a large part of the weather changes and of the precipitation in the tropics.

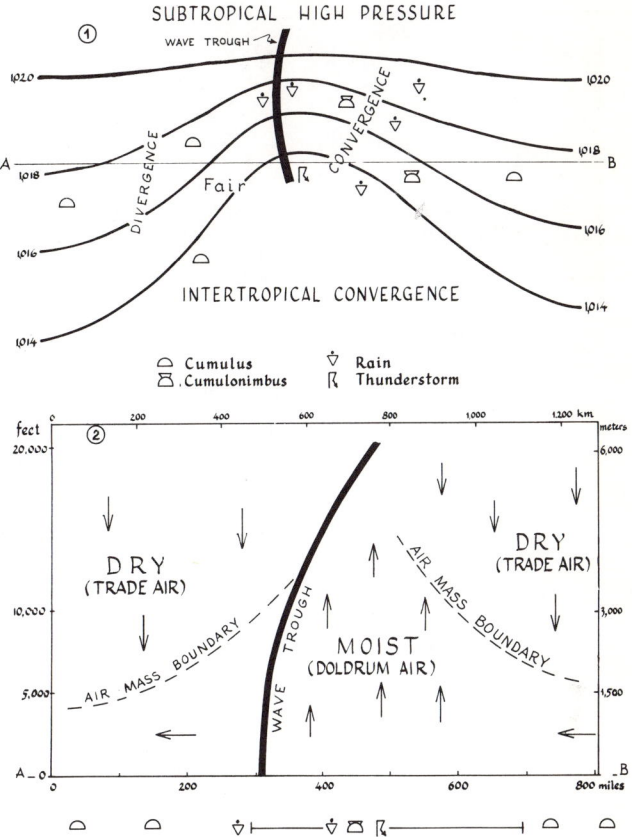

Fig. 6-6. A transverse wave. (2) is a vertical cross-section of (1). Note the bad weather (rain, thunderstorms, and cumulonimbus clouds) in the zone of convergence, behind or east of the wave trough.

Tropical Cyclones. Sometimes the transverse troughs of low pressure develop into low-pressure centers with closed isobars at the surface. These centers, however, have very low pressure gradients compared to those in extratropical cyclones. Nevertheless, convergence and rising air produce bad weather all out of proportion to the strength of the low. Such weak **tropical cyclones** are frequent in the immediate vicinity of the ITC but are rarely found in the trade-wind belt. Wind speeds are usually less than 25 miles per hour (11 m/sec or 40 km/hour).

Tropical Hurricanes. Once in a while, the low-pressure center intensifies and wind speeds pick up. Intermediate storms with wind velocities from 25 to 75 miles (40 to 120 km) per hour are common. Occasionally, the intensity of the pressure gradient increases until wind speed reaches hurricane velocity. An arbitrary limit of 75 miles (120 km) per hour has been set to dis-

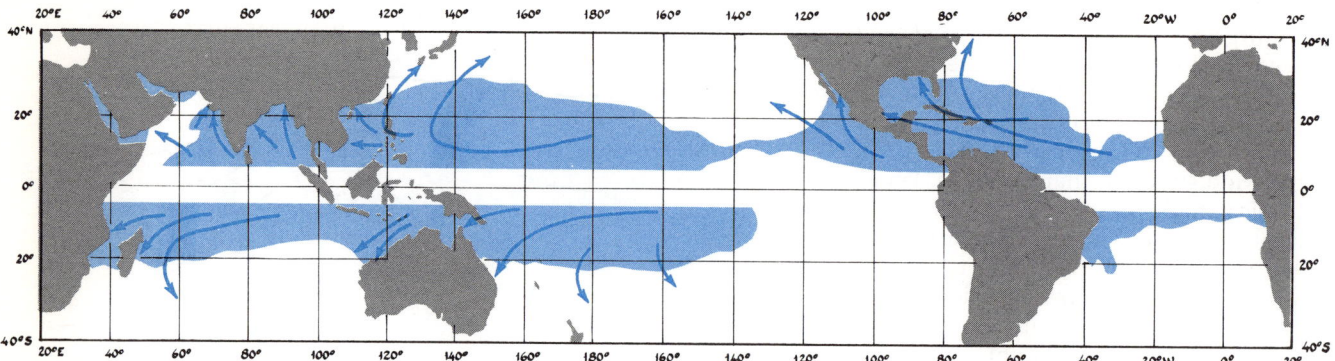

Fig. 6–7. Source regions of tropical hurricanes. Areas shown in blue are more than 5° of latitude away from the equator, and have average sea surface temperatures greater than 80°F (26.7°C) in their warmest month.

tinguish between mere tropical storms and the destructive disturbances we call **hurricanes**. We shall take a few paragraphs to describe the origin, structure, and movement of hurricanes, which represent one extreme of a class of tropical phenomena that range through weak tropical storms to the relatively unspectacular transverse waves at the other extreme.

Intense tropical cyclones are called by different names in various parts of the world. In the West Indies, the term "hurricane" (West Indian *hurakan* evil spirit) was borrowed from the Indians by the Spaniards and French and later became part of the English language. In southeastern Asia, tropical cyclones are called typhoons (Chinese *tai-fung* big wind). In the Bay of Bengal, the British meteorologists used the word *cyclone*. Off the coasts of Australia, they are called *willy-willies*. Whatever the name, the characteristic features are the same, and we shall describe them all under the single heading of hurricane.

It is relatively easy to describe the gross features of hurricanes and to draw a map of their general area of occurrence and movement; but it is very difficult to say precisely what causes them or to predict accurately the path of any individual storm. Nevertheless, certain general statements about the nature and the origin of hurricanes will be useful in understanding their distribution and climatic effect.

Hurricanes develop from preexisting tropical storms. A modest center of low pressure may become more intense, largely through the action of cyclonic whirling of air. The energy required to maintain the very high wind speeds about the low-pressure center of a hurricane comes from the latent heat of condensation, heat released by the condensation of water vapor in the hurricane itself. As long as air circling about the center of low pressure is prevented from filling in the low by the deflective action of the **Coriolis force**, and as long as there continues to be an abundant supply of energy by way of latent heat of condensation, the storm continues to intensify until the wind reaches hurricane force. One additional condition must be satisfied for the genesis of such a tropical storm: steep lapse rates are essential. The instability of the warm oceanic air permits considerable updraft. Coriolis deflection cannot keep all the air out of the low-pressure center, and, therefore, there must be some ascent of air over the low to maintain and intensify the low pressure and the associated cyclonic movement. Only under conditions of instability is it possible to have large-scale updraft and outflow aloft of sufficient magnitude to maintain the extremely low pressure of a hurricane and to supply the water vapor for condensation in the storm. The extreme character of the low pressure can be easily illustrated. The lowest pressures in an extratropical cyclone seldom drop below 980 millibars, or about 33 millibars below average sea-level pressure. In most hurricanes, on the other hand, the pressure drop is more than 100 millibars; a well-developed tropical cyclone can register pressures less than 900 millibars. This condition represents the removal of 2 million tons of air for every square mile occupied by the low pressure (or about 700,000 metric tons per km²).

The conditions described above can be satisfied only over very warm oceans and far enough from the equator that the Coriolis force can deflect the winds from the low-pressure center of the storm. Most hurricanes originate more than 5° from the equator in areas where the sea-surface temperature exceeds 80°F (26.7°C). Figure 6-7 shows the oceanic areas where both these conditions are met. Note the close correspondence between these areas and the zone of origin of hurricanes. Notice also in the South Atlantic the very small area with sea-surface temperatures above 80°F (26.7°C) and the corresponding absence of hurricanes. Hurricane development takes place during much of the year but is overwhelmingly concentrated in late summer and fall, when ocean temperatures are at their maximum. August and September are the favored months in the Northern Hemisphere; February and March in the Southern Hemisphere.

In its early stages, a hurricane remains almost stationary. Only after it has become a well-established center of low pressure can it move with the prevailing winds. Light variable winds without a well-organized movement in a particular direction are essential for hurricane formation. Strong prevailing winds disperse the water-vapor concentration, destroy the steep lapse rates, and mix the warm surface water downward. All these actions, of course, strongly inhibit the initial formation of the storm. This first stage is usually completed close to the point of origin within 2 to 10 days.

In the second stage, which consists of rapid intensification of the hurricane, the vortex, or active center of the storm, reaches its maximum intensity and also its smallest size. In the center of the vortex, there now develops a small zone of calm and clear conditions, called the eye of the hurricane, around which the clouds are arranged in a spiral form. Many storms never complete this stage and disintegrate after a week or so, but others grow to a maturity of fearsome destructive power (Fig. 6-8).

In its mature stage, a hurricane no longer intensifies by contraction. The storm area expands to a radius of about 200 miles (about 320 km), vast masses of air are absorbed into the hurricane, and the pronounced circular symmetry of the earlier stages is distorted by the fact that the winds become much stronger on the side of the storm where circling wind and moving storm have the same direction. The mature hurricane has three characteristic paths (Fig. 6-7). A few move continuously westward, dying out over tropical coasts in low latitudes. Most hurricanes, however, turn away from the equator rather sharply at about 25° of latitude; of these, most tend to recurve to the east, although a few continue directly northward or southward (Fig. 6-9).

The final, or decaying stage of a hurricane is brought about either by passage over land, where it loses the supply of moisture necessary to maintain its energy supply, or by a change of direction away from the equator and especially to the east. Although this recurvature may result in faster movement of the storm as a whole (up to 400 miles or 640 km per day), it also acts as a brake on the cyclonic whirling that is necessary for the very high wind speeds inside a hurricane. A decaying hurricane often gets involved with extratropical storms as it leaves the tropics and may produce fantastic amounts of precipitation.

The greatest destruction of the works of man by winds, waves, or rain associated with hurricanes obviously takes place over land, usually along coasts. The New England hurricane of September 1938, for example, caused more than 400 million dollars of damage and killed more than 500 persons. Ironically, such

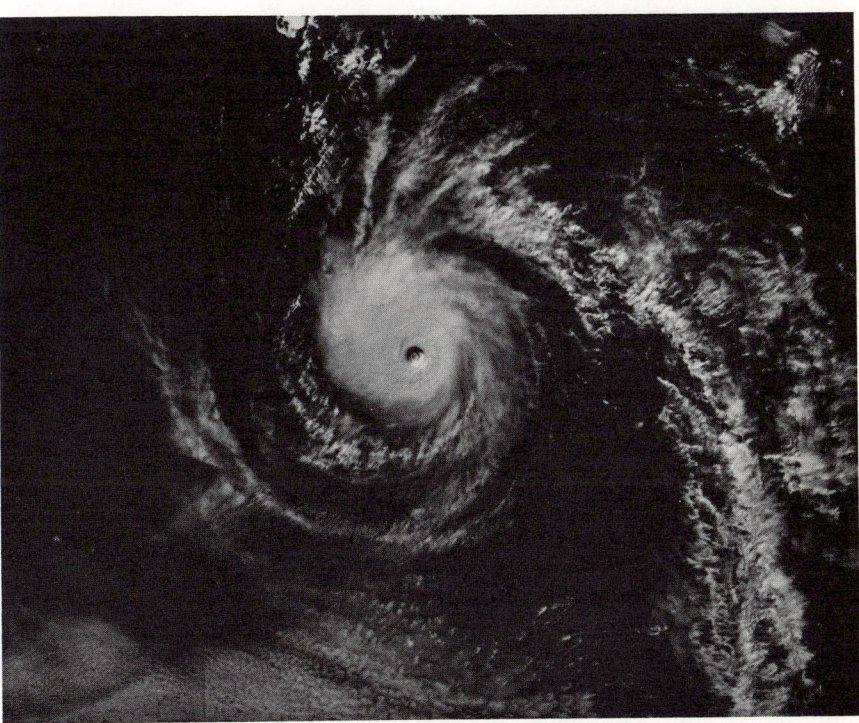

Fig. 6-8. A tropical hurricane, Hurricane Ava on June 7, 1973 as photographed by NOAA-2 weather satellite. The spiral pattern of clouds reflects the counter-clockwise movement of wind in a Northern Hemisphere low-pressure center (central pressure, 914 mb). The eye of the storm shows up as a tiny circular hole in the cloud cover. *Courtesy of the National Oceanic and Atmospheric Administration.*

102 Chapter 6

Fig. 6-9. Hurricane Agnes. On June 17th the hurricane is south of Florida and has just begun its re-curvature northeastward. By June 18th the hurricane center is approaching the west coast of Florida, and by the 19th it is well inland. The enormous amount of precipitation generated by such storms is hinted at by the density and size of the cloud structure. Note, incidentally, the layer of advective fog lying just off the Pacific coast from Canada to Mexico. *Photographs courtesy of the National Oceanic and Atmospheric Administration.*

destruction occurs when the hurricane is beginning to decay; nevertheless, the wind damage may be enormous (winds of more than 200 miles or 320 km per hour have often been observed). In final stages of hurricanes, waves and ocean swells can reach levels more than 20 feet (6 m) above normal; and rainfall can reach such incredible totals as 46 inches (1,168 mm) in a single day (Baguio, Philippines, in July 1911) or nearly 100 inches (2,500 mm) in a four-day period (Silver Hill, Jamaica, from November 5 through 8 in 1909).

Monsoon Circulation in the Tropics. One last but very important element of air-mass circulation in the tropics is that of the **monsoon**. The word "monsoon" itself has acquired a variety of meanings. It may refer to high-level winds in the stratosphere or to the winds near the surface; it may refer to a seasonal alternation of wind directions, wherever it is found, or only to the southwest winds of summer over the Indian peninsula; it may refer to a variety of similar phenomena lumped together, as in the term "Asiatic monsoon," or it may be highly localized, as in "Indian monsoon." Nor are the explanations of this important mechanism, which affects the lives of half the world's population, very much clearer. We shall define a monsoonal circulation as one in which there is a change of nearly 180° in the prevailing wind direction from one well-defined season of the year to another. To simplify somewhat, we shall be concerned primarily with alternations of wind direction that occur near the surface of the earth and especially with those affecting southern and eastern Asia, where the phenomenon is best developed.

The seasonal alternation of relatively dry conditions in winter and very much wetter conditions in summer is an obvious climatic characteristic of large parts of southern and eastern Asia. Such alternation is also found in lesser degree and over smaller, more fragmented patches elsewhere. Associated with this seasonal shift in wind direction is the seasonal change in sea-level pressure, particularly marked over the Asiatic mainland. We have already remarked on the intensity and extensiveness of the Siberian high in winter and the equally pronounced character of the low-pressure center over northwestern India and Pakistan in summer. It is easy enough to explain the formation of these semi-permanent cells by the action of continental cooling and heating with consequent subsidence and high pressure or ascent and low pressure. It is also easy enough to ascribe the monsoon to the outflow of air from the Siberian high in winter and to the indraft into the Indian low in summer. This is, in fact, the classical explanation of the monsoon: winter drought associated with outflow of air from the continent and summer rains corresponding to the inflow of air from the surrounding oceans. The maps of sea-level pressure and wind movement (Figs. 4-4, 4-11, and 4-14) do indeed show an anticyclonic flow out of the Siberian high in winter and a cyclonic flow into the Indian low in summer. Certain objections to this simple explanation, however, have already been mentioned in Chapter 4. One of the most important is that the two contrasting centers of action are very shallow — so shallow, in fact, that neither the winter high nor the summer low extends more than a few thousand feet (a kilometer at the most) above the surface. The crest line of the Himalayas, on the other hand, lies above 16,000 feet (nearly 5,000 m) for more than 1,000 miles (1,600 km). This formidable east-west barrier thus lies athwart the surface winds. It is, therefore, impossible to ascribe the direction of the winter winds over India to the action of the Siberian high, whose height is only a fraction of that of the Himalayas. Similarly, the Indian low can have practically no effect on wind direction in Siberia and northeastern Asia. Figure 4-12 corroborates these statements. At 20,000 feet (6 km), the pressure pattern shows no trace of continental high in winter or of continental low in summer; westerly winds blow uninterruptedly from west to east around a polar low-pressure center.

What, then, is the mechanism that creates the seasonal wind shift over India, Malaysia, and the Far East? Part of the explanation for the monsoon circulation is found in the local effect of the continental pressure cells. Although the Siberian high and the Indian low are shallow, they are extensive and strong and thus affect wind movement over large areas, despite the limits set by the Himalayan barrier.

More of the explanation can be gleaned from a consideration of the seasonal shift, on a planetary scale, of the general circulation. Figure 6-10 shows winter and summer positions of air-mass boundaries and frontal zones, as well as the generalized wind flow. It is a version of Fig. 6-1 on a somewhat larger scale.

The winter situation is quite simple (Fig. 6-10 (1)). It consists of anticyclonic outflow of cP air from the Siberian high, north of the Himalayas. South of the mountains and extending eastward over China and Japan is a weak branch of the polar front. Aloft, the jet stream also occupies a position south of the Himalayas, thus adding a wind barricade to the orographic barrier. Very little cP air from Siberia can penetrate into India. The ITC is to the south of the equator. There is little convergence along the polar front, though some extratropical cyclones of Mediterranean origin do occasionally move down the Ganges Valley, bringing winter rains to northern India. Between the polar front and the ITC, wind movement is from the north or

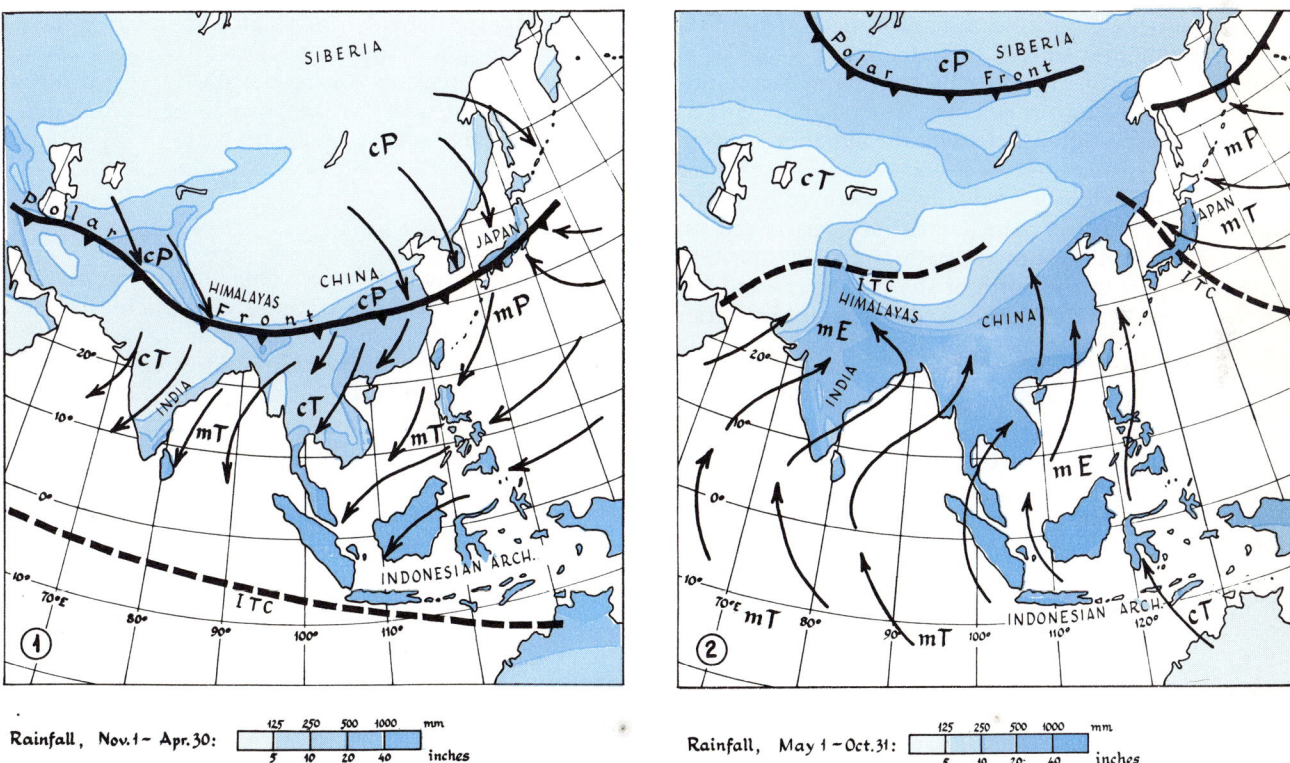

Fig. 6-10. Surface winds and precipitation patterns during the winter (1) and summer (2) monsoons of southern and eastern Asia. The intensity of precipitation in each season is shown by the intensity of the blue screen. The arrows indicate the prevailing surface wind direction. Also shown are the average positions of the Intertropical Convergence Zone (ITC) and Polar Front as well as the dominant air masses. These maps are drawn on the stereographic projection in order to show directions truly.

northeast; the air masses are varied; they include such different air masses as *cT*, *mP*, or *mT*, hence the label "transitional" in Fig. 6-1(1). The *cT* air masses consist of stable trade-wind air and bring about dry conditions over most of the Indian peninsula and Malaysia, except for the eastern coastal fringes, which are under the influence of *mT* air. The Philippines and the Indonesian archipelago are overlain by either *mP* or *mT* air and receive considerable rainfall during this season. Southern China, Japan, and southern Korea also receive appreciable amounts of precipitation, in part due to the presence of frontal activity along the polar front and in part because the *cP* air flowing out of the Siberian high becomes unstable and picks up moisture while crossing the Japan and China Seas.

It is obvious from the above that there is much precipitation in southern and eastern Asia, despite the preponderance of dry continental air masses and of air masses that have had a very short trajectory over water. Very wet conditions are limited to the island archipelagoes of the southeast, and very dry conditions are limited to the area north and west of an arc extending from southern India to northeastern Siberia.

In summer (Fig. 6-10 (2)), the most obvious features are the large-scale northward displacement of the ITC and a more modest shift of the polar front. The ITC is found very far northward partly because of heating of the continent during the summer. The position aloft of the jet stream has a lot to do with the situation as well. For a variety of reasons, the jet remains fixed south of the Himalayas while it gradually weakens during the summer. Beyond a certain point, however, it can no longer maintain its southerly path and shifts position rather abruptly to the north of the Himalayas and the Tibetan Plateau. In other words, the orographic barrier created by the Himalayas rather markedly modifies the general circulation of the atmosphere in Asia, since it forces the jet stream to occupy a position either south of the mountains or far to the north, beyond the Tibetan Plateau. The consequence of this modification of the normal circulation is that, in summer, the jet stream that separates polar and tropical air masses aloft shifts sufficiently far north to allow the ITC to penetrate the heated interior of Asia to a much higher latitude than it attains anywhere else. In addition, the ITC is broken into two branches.

The general inflow of air into the ITC from the south brings widespread rain to southern and eastern Asia and to the Indonesian archipelago. The onset of this rain is usually quite sudden — it is often referred to as the burst of the monsoon — and coincides with the sudden shift of the jet stream. The rains, moreover, are strongest where orographic barriers lie in the path of the southwest winds — along west or southwest coasts, in contrast to the concentration of rain along east coasts in winter. Nor is rainfall continuous during the summer; it depends on the occurrence of the convergence phenomena discussed earlier in this chapter. It is worth noting on Fig. 6-10 (2) that the summer rains in Japan are of different origin from Indian or Chinese precipitation, since Japan is largely north of the ITC and is overlain by *mT* air from the north and central Pacific.

Five different kinds of monsoonal regimes thus emerge in this part of the world. On the Indian peninsula, there is a marked contrast between winter drought and summer rain, despite the occurence of limited amounts of cyclonic rainfall in the Ganges Plain and some tropical rain along the east coast and on Ceylon. From Thailand eastward and northward to southern Korea, and southern Japan, a slightly different regime is experienced. Summer conditions are roughly similar to those over India; but the winters, though less wet than the summers, are by no means dry. In northern China and in Manchuria, the contrast between winter drought and summer rains is nearly as marked as in India, but both seasons are drier. Japan's regime resembles that of southern China, but the summer rains come from quite a different source. Finally, the Indonesian archipelago stands out as an area of very heavy rainfall throughout the year. The differences among these various regimes are obviously due to differing locations with respect to the position of frontal zones and surface air movement.

Summary

In this chapter we have examined air masses and their significance to the climates of particular places. The notion of air masses implies that the atmosphere remains quiescent long enough for large volumes of air to acquire some of the temperature and moisture characteristics of the underlying source region. Air masses are classified by the general character of their source regions and by the temperature lapse rates which develop as a result. The major categories of the classification employed here include the following:

1. *Continental* air masses and *maritime* air masses.
2. *Polar* air masses and *tropical* air masses.
3. Air masses with *stable* lapse rates *near the ground* and those with *unstable* lapse rates *near the ground*.
4. Air masses with *stable* lapse rates *aloft* and those with *unstable* lapse rates *aloft*.

In general, the major cause of instability near the ground is strong heating of the surface; stability near the ground is induced by surface cooling. Instability aloft is most often associated with low pressure systems at the surface, convergence, and rising air masses; stability aloft is produced by anticyclonic circulation at the surface, divergence, and subsidence of the air.

One of the major consequences of air-mass contrasts is the formation of frontal zones along which the atmosphere is sufficiently perturbed to produce rapid day-to-day changes in the weather. In the middle and high latitudes, particularly along the polar front, the contrast between cold polar air masses and the warmer tropical air masses creates a localized cyclonic circulation, the extratropical cyclone. The rise of warm air over an underlying layer of colder, denser air along the warm front and the upward displacement of warm air by underrunning cold air along the cold front produce cyclonic precipitation, the major form of extratropical precipitation.

In equatorial and tropical latitudes, the Intertropical Convergence Zone plays a role that is somewhat similar to that of the polar front. In these latitudes, air-mass contrasts are weak, but the air is often heavily laden with moisture and relatively unstable. All that is needed under these conditions to produce copious rain is some minor triggering action to induce convergence and its consequent convective rain showers. Such mechanisms as the surge of the trades, easterly waves, and weak tropical cyclones produce much of the rain and most of the day-to-day weather changes in the neighborhood of the ITC. Tropical hurricanes are an extreme form of atmospheric perturbation along the ITC which results in extremely low pressure, very strong winds, and abundant precipitation.

Monsoonal circulation in the tropics, particularly that over southern Asia, can only partly be explained in terms of seasonal alternations of continental pressure systems. More important is the seasonal shift of the general circulation. Much of the heavy summer precipitation in monsoonal areas can be explained by the poleward migration of the ITC. This combination of pressure alternation and seasonal shift of the general circulation results in at least five different types of monsoon in southern and eastern Asia and in a variety of similar types elsewhere.

Suggestions for Further Reading

Namias, Jerome, "The Jet Stream," *Scientific American*, Vol. 187, No. 4 (October 1952), pp. 27-31.

Pédelaborde, Pierre, *The Monsoon*, trans. M. J. Clegg, Methuen & Co. Ltd., London, 1963.

Petterssen, Sverre, *Introduction to Meteorology*, 2nd ed., McGraw-Hill Book Company, Inc., New York, 1958, pp. 189-250.

Riehl, Herbert, *Introduction to the Atmosphere*, McGraw-Hill Book Company, Inc., New York, 1965, pp. 143-195.

Willett, Hurd, C., and Frederick Sanders, *Descriptive Meteorology*, 2nd ed., Academic Press, Inc., New York, 1959, pp. 244-256.

7 Precipitation

Introduction

The interconnection of all atmospheric processes is reflected in the distribution of rain and snow over the surface of the earth. To understand why it rains at a particular place and at a particular time, one needs to know the moisture content of the air, its degree of stability, and the kind of atmospheric disturbance that is the immediate cause of the drizzle, downpour, or deluge. This knowledge depends, in turn, on a consideration of the factors affecting evaporation and condensation, both locally and at some distance; the thermal and dynamic conditions that affect stability, both near the ground and aloft; and the components of the general circulation. These items involve practically every topic discussed so far. Their relevance to any consideration of the distribution of precipitation should become clear in the sections of this chapter that deal with the factors affecting the spatial and temporal differences in distribution. We shall begin, however, with a few important definitions.

Definitions

Precipitation (Latin *praecipitatio* a headlong falling down) in its meteorologic sense is condensed or frozen water vapor that falls to the ground. It includes mainly rain and snow, but other **hydrometeors** such as hail, sleet, or fog drip, which may have considerable local significance, are also included.

Instruments

The instruments used to measure precipitation range from the extremely simple to the sophisticated and complex. At one end of the scale is the simple rain gauge consisting of a bucket into which the precipitation falls; the observer merely measures the depth of the water in the bucket after each fall. Equally simple is the snow course for measuring snowfall. Before and after each snowfall, an observer records the height of the snow at each of a series of stakes set in line.

At the other end of the spectrum are instruments such as recording rain gauges and an apparatus that utilizes radioactive cobalt for the measurement of the water content of snow. The recording rain gauge is so arranged that whenever the weight of the falling precipitation reaches a fixed amount, it trips a circuit and the fact is recorded on a chart. The total number of marks on the chart during a given period thus corresponds to the total precipitation. Cobalt gauges were described in Chapter 5, p. 70.

Many other kinds of instruments are used to measure precipitation — for example, tipping-bucket gauges that tip in the direction of the wind in order to measure more realistically the amount of falling water.

The examples given must suffice as illustrations of the variety of apparatus in use. It is not possible to discuss adequately in a few words the problems inherent in attempts to catch, at the ground, an amount of precipitation that is representative of the falling snow or rain. The most obvious problems connected with interpretation of precipitation records include the disturbing effect of wind, the blocking of gauges by snow, the distortion of the catch by nearby buildings or trees, and the effect of the gauge itself on the wind.

Units of Measurement

Precipitation is normally recorded as a depth, or thickness, of water. A rainfall of 1 inch indicates that enough water has fallen to cover the land uniformly to a depth of 1 inch. In most of the world outside the United States, the metric system is used, and the depth of precipitation is normally expressed in millimeters. Strictly speaking, precipitation amounts, such as 4.4 inches or 112 millimeters, are rates, since the period during which the amount fell must be specified. Thus, the 112 millimeters may have fallen in 1 month, or in 1 year, or in 1 century. Precipitation should therefore always be expressed as a certain depth of water per given length of time. Lines of equal precipitation on a map are called **isohyets** (Greek *isos* + *hyetos* rain).

Snowfall is measured in the same way. That is, the depth of the snowfall within a given period is indicated. When snow and rain are combined to give total precipitation, however, snow depth is converted to equivalent water depth. For the conversion, the snowfall is divided by 10, since on the average a layer of snow 10 inches (250 mm) deep when melted yields a layer of water only 1 inch (25 mm) in depth. "Wet" snow may have a much higher water equivalent (values up to 20 percent are common), whereas "dry" snow may have a water equivalent of less than 10 percent. There is no fixed relationship between snow and its water equivalent; the 10:1 ratio is merely an average.

Major Categories of Precipitation

The distribution of precipitation over the world becomes more meaningful if the major causes of condensation in the atmosphere are kept in mind. We may summarize Chapter 5: most of the condensation that yields precipitation is the result of cooling; **adiabatic cooling** is by far the most important cooling process in the production of condensation; uplift of air accounts for almost all the adiabatic cooling; and uplift is aided by **convergence** and **instability** of air but hindered by **divergence** and **stability**.

We can, therefore, categorize precipitation according to the form of the uplift responsible for the adiabatic cooling that produces the condensation. There are four major kinds of atmospheric situations in which uplift resulting in precipitation takes place. We have already discussed the resulting precipitation types but repeat them here for emphasis.

The first type, **orographic precipitation,** (see Fig. 5-5), occurs when air masses are forced upslope by the prevailing pressure pattern. The slope may be that of hills or mountains, or it may be the edge of a plateau. The amount of precipitation depends on the stability of the air, the moisture content of the air, and the height of the ascent. When the air is forced over a sizable mountain range, the maximum precipitation usually occurs below the summit, but still on the windward side. As we have already noted, the moisture-holding capacity of air decreases very rapidly at high temperatures and much less quickly when it is colder. For instance, a 10° drop in temperature from 100° F to 90° F (37.8° C to 32.2° C) results in a reduction of the water-vapor holding capacity of 12 g/kg, whereas a 10° drop from 30° F to 20° F (-1.1° C to -6.7° C) brings about a reduction of only 1.2 g/kg. Heavy precipitation is aided by a rapid drop in the moisture-holding capacity and consequently tends to occur long before the cold temperature of the summit is reached. Furthermore, if the air masses are unstable or even conditionally stable, the air rises not only upslope but also straight up, in which case there is more precipitation where the vertical movement takes place and correspondingly less farther up the mountain. On the other hand, since the general movement of air is upslope, this localized adiabatic precipitation also tends to be blown toward higher elevations. In any case, the location of maximum precipitation depends on the original water vapor content of the air, on its degree of stability, and on the slope of the mountain.

A second type is **frontal precipitation,** which occurs when air masses of contrasting temperature and, consequently, contrasting density meet. We have already discussed the various kinds of frontal precipitation and the major aspects of its seasonal and spatial distribution. We can add the fact that it is essentially a convergence phenomenon in that the precipitation is the result of the convergence of air masses along a frontal zone.

A third type is **convectional precipitation.** Surface heating produces unstable lapse rates; the air rises; and adiabatic cooling of the ascending air does the rest. It is essential, however, that the rising air contain at least a moderate amount of water vapor; otherwise, no precipitation occurs. This type of precipitation is also associated with convergence. The surface heating produces a local low-pressure center, and air converges to replace that which ascends.

The fourth type is **equatorial convergence precipitation.** This category includes the various phenomena discussed in Chapter 6 under the headings "surge of the trades," "easterly waves," "tropical cyclones," "tropical hurricanes," and "monsoon circulation." We may emphasize, once again, that the weakly convergent nature of some of these disturbances is compensated for by the strong charge of moisture and generally unstable condition of the atmosphere.

World Distribution of Precipitation

We shall try to make the map of annual precipitation (Fig. 7-1) as intelligible as possible by listing the factors responsible for the variations in precipitation from place to place. This, of course, is the same arrangement we have tried to follow for all the other climatic elements. In the case of precipitation, however, the seasonal regime is more complex and less easily understood than that of temperature, for example. For this reason, we shall look first at annual precipitation and then discuss seasonal regimes separately.

Areas of Heavy Precipitation

Let us consider first the areas of the world that receive more than 80 inches (about 2,000 mm) of precipitation per year. What combination of factors is needed to produce so much rain or snow? The major items, all of which should be very familiar by now, are (1) occurrence of unstable lapse rates throughout the year, (2) orographic accentuation of precipitation, (3) nearness to warm water with a high rate of evaporation, and (4) position within a zone of storm tracks. These four conditions, in one way or another, account for the presence of heavy precipitation. Not all four conditions need be fully satisfied to get rainfall in excess of 80 inches (2,000 mm) a year, but every place with heavy precipitation is characterized by at least two of the four items listed above.

The importance of unstable lapse rates does not need reemphasizing, but it is necessary to stress the importance of year-long instability. Many places are overlain by air that is unstable during warm summer days; but if winter air masses at those places are stable, significant precipitation will occur only in summer, and the annual totals will not reach the magnitude we are discussing here.

The importance of orographic effects becomes obvious when we consider the places in the world that have the highest average annual precipitation. Among the best known are Cherrapunji, on the southern foothills of the Himalayas (426 inches or 10,820 mm per year); Silver Hill, in the uplands of Jamaica (410 inches or 10,414 mm per year); and Waialeale, a 5,000-foot (1,500 m) mountain on the island of Kauai (probably the wettest spot on earth, with 476 inches or 12,090 mm per year).

Nearness to water is often considered the most important rain-producing factor; however, it is probably far less important than stability or orography. Many continental areas at relatively great distances from the sea have considerable precipitation, and many oceanic areas are deficient in rainfall.

The fourth item listed, position within a zone of storm tracks, needs some elaboration. In this context, storm tracks include the zone of extratropical cyclones, as well as the zone of weak tropical storms and of tropical hurricanes. These are shown for the Atlantic Ocean in Fig. 7-2. Bear in mind that this factor is closely bound up with the degree of instability of air masses. The major storm tracks, tropical or extratropical, are zones where convergence of some kind is taking place. This means, in a general way, that there is ascent of air and, consequently, low pressure at the surface. These conditions all favor instability. While it is possible to have stable lapse rates in extratropical cyclones, the air masses, particularly in the warm sector, are much more commonly unstable or conditionally unstable. In the tropics, the Intertropical Convergence Zone (ITC), within which most tropical storms develop, is by definition a zone of instability. It is true that a tropical hurricane, once it has reached maturity, can carry its own conditions of extremely low pressure, rapidly ascending air, and quite unstable lapse rates even into areas of marked stability. In general, however, hurricanes avoid the extreme stability of the subtropical high-pressure centers. They may penetrate regions of otherwise mediocre rain-producing potential and thus bring considerable amounts of rain to places that are not conspicuously marked by instability or orographic effects. This is true not only of hurricanes but also weak tropical cyclones.

The foregoing can be summarized by indicating that both instability and position within a zone of storm tracks are necessary but not sufficient conditions for very heavy precipitation. The other two factors, orography and closeness to sources of moisture, are neither necessary nor sufficient, but the presence of either one greatly increases the likelihood of large quantities of rainfall. With these notions in mind, we can now turn to Fig. 7-1, which reveals four characteristic locations of very high precipitation: along the equator, along west coasts in mid- and high latitudes, in monsoon Asia, and along subtropical east coasts. We shall examine each of these in turn, discussing in some detail the combination of factors responsible for the heavy rainfall.

The Zone of Equatorial Convergence Rainfall. The location of very heavy rainfall close to the equator corresponds rather closely to the average position of the ITC. This narrow zone is nearly continuous over the oceans but is broken up on the continents. Over the Pacific Ocean there is a relatively narrow belt centered about 5° N latitude; its extension over the American continent is limited to a very narrow western coastal fringe in Central America and to the Chocó coast and the Atrato Valley of Colombia. This zone, although north of the equator, is within the limits of the ITC all year round.[1] The high precipitation is easily explained in terms of nearness to storm tracks, year-round instability, and closeness to warm water. On the land, orographic effects add their contribution: for example, Buenaventura, on the Chocó coast, averages 281 inches (7,137 mm) a year.

Over continental South America, the heat of the Amazon Basin pulls the ITC to an equatorial location. The presence, during most of the year, of convergence phenomena in an unstable and saturated atmosphere results in very high precipitation without benefit of orographic effects, which are limited to the eastern slopes of the Andes. These conditions are not duplicated elsewhere; the Amazon Basin is the only major example of very high precipitation over a continental lowland.

Heavy rainfall also occurs along the northeastern coast of South America from the Guianas to northern Brazil. The separation of the two rainy regions by a drier corridor is perhaps explainable by the invasion of stable conditions associated with the subtropical high in the South Atlantic, which exerts considerable influence over the South American continent in winter (July).

The pattern of heavy precipitation over the Atlantic Ocean resembles the distribution over the Pacific. There is a band of heavy precipitation, similar in width and position, which impinges on the African coast from Sierra Leone to Liberia. This coastal strip at the foot of the Fouta Djallon Plateau has a distinctly monsoonal regime. In summer, it lies within the ITC, and unstable equatorial air flows from the southwest; in winter, the ITC is far to the south, and the dominant air mass consists of stable trade-wind air from the northeast. Summer air masses are thus unstable, of oceanic origin, and subject to convergence within the ITC; winter air masses are stable and of continental origin. Tower Hill, in Freetown, Sierra Leone, for example, receives a total of 71.2 inches (1,808 mm) in the combined months of July and August but only 0.6 inches (15 mm) in January and February.

Farther to the east, the rainfall pattern is similar to that over the west coast and the interior of South America. The high rainfall along the Nigerian and Cameroon coasts is roughly similar to that along the Colombian coast; the situation in the Congo Basin is analogous to that in the Amazon Basin. Orographic effects over the Cameroon Highlands are probably more necessary for the production of annual totals in excess of 80 inches (2,000 mm) than similar effects in the Colombian Andes. Rainfall is less over the Congo than over Amazonia for a variety of reasons. The Amazon Basin is more open than the Congo Basin to inflowing maritime air masses in all directions but west. Furthermore, Amazonia has a source of moisture, in the Caribbean Sea to the north, that the Congo lacks.

Precipitation over equatorial coasts of East Africa is remarkably different from that over similar regions in South America. The contrast between the very dry conditions of the Somali coast and the rainy climate of the Guianas and of adjoining coastal Brazil is partly due to the different orientation of the coastlines. The American coast is at right angles to the prevailing direction of the ending trades; the African coast is parallel to both the northeast winds of winter and the southwest winds of summer. Equally significant is the fact that the African coast is predominantly under the influence of subsiding air masses with stable lapse rates.

The Indian Ocean reproduces the Pacific and Atlantic patterns, except that the ITC and, therefore, the zones of heavy equatorial rainfall are located south of the equator. East of the Indonesian archipelago, both the ITC and the equatorial zone of high precipitation shift north of the equator.

Frontal Precipitation of Mid-Latitude West Coasts. A second characteristic location of very high precipitation is in middle and higher latitudes along west coasts, in the domain of extratropical cyclones. Frontal precipitation from these storms is not as heavy as the equatorial rainfall of the ITC; precipitation in excess of 80 inches (2,000 mm) is limited to places where mountains produce orographic precipitation by obstructing the prevailing westerlies, generally between 40° and 60° latitude.

There are four such places shown on the map. Along the west coast of North America, a fairly extensive area of heavy precipitation extends from southern Alaska to northern California, with a maximum development along the coast of British Columbia and adjacent waters. A similar zone appears in southern Chile, but its poleward limit is determined by the southern tip of South America, in about 55° S latitude. Northwestern Europe has a small and discontinuous zone of very high

---

[1] The location of the ITC north of the equator, even in January, is due to the relative coldness of Southern Hemisphere oceans and particularly to the effect of the very cold waters of the Humboldt Current. The same situation prevails in the Atlantic Ocean.

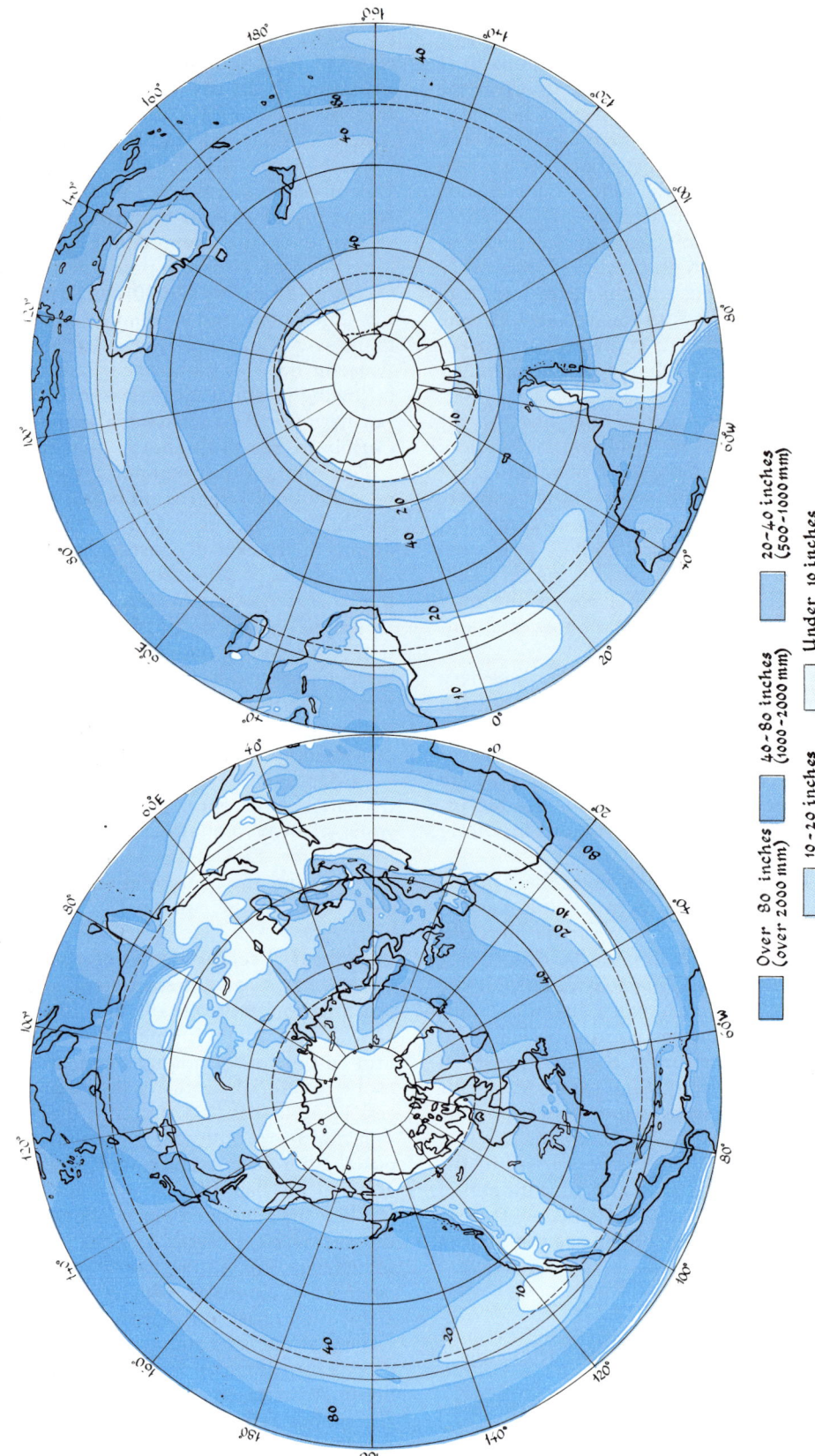

Fig. 7-1. Average annual precipitation.

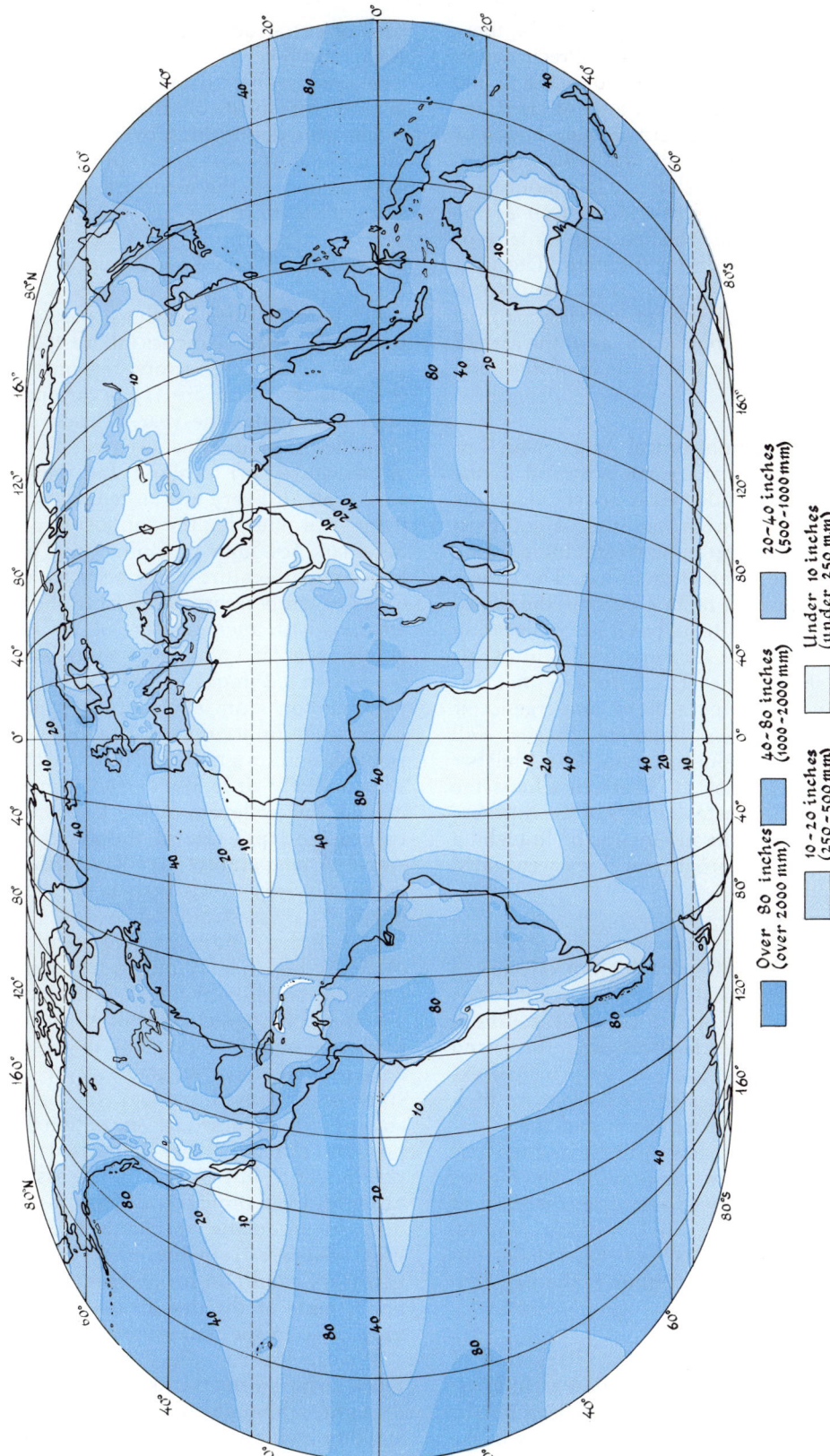

Fig. 7-1. Average annual precipitation. The darker the screen, the greater the intensity of precipitation. The darkest screen shows areas of more than 80 inches (2000 mm); the lightest shade, areas of less than 10 inches (250 mm). The three intermediate shades represent 10–20 inches (250–500 mm), 20–40 inches (500–1000 mm), and 40–80 inches (1000–2000 mm).

frontal precipitation. From about 40° N to 55° N, there are no major barriers to westerly winds. Precipitation along the coasts is, therefore, less than 80 inches (2,000 mm) a year, except for a very small area of the Cantabrian Mountains in northwestern Spain. North of 55° N, however, the influence of the polar front is stronger; and the effect of the relatively low mountains of Northern Ireland and of Scotland, as well as the volcanic plateau of Iceland and the higher mountains of southern Norway, can be seen on the map. The west coasts of Tasmania and New Zealand, from about 40° S to 45° S, represent the fourth area of very high frontal precipitation. The poleward boundary, as in the case of Chile, corresponds to the southern limit of the land areas.

Monsoonal Rainfall in Subtropical Asia. Much of the heavy precipitation on the earth is found in the equatorial and mid-latitude west-coast locations described above. One important addition is southern and eastern Asia, where orographic effects increase the rainfall potential of the summer monsoon. The causes and distribution of rainfall in this part of the world have been discussed in Chapter 6, but it bears repeating that the summer monsoon of Asia involves an unusually far northward shift of the ITC. It follows that unstable equatorial air masses in a zone of convergence, if forced to rise over sizable mountain ranges, will precipitate enormous quantities of rain. The Malabar Coast of India at the foot of the Western Ghats, the Himalayan foothills, the coastal borders of the Burmese and Malayan ranges, and the mountains of Indochina are obvious examples of orographically accentuated monsoonal precipitation.

Subtropical East-Coast Precipitation. Finally, there are a number of areas with very high precipitation along east coasts in subtropical latitudes. The east coast of Central America, parts of the coast of southern Brazil, the east coast of Madagascar, and the island chain off the east coast of Asia are the principal examples. We have already indicated the reasons for high precipitation on subtropical east coasts and their contrast with the drier west coasts, which, in the same latitudes, are dominated by very stable air masses (Chapter 6). Furthermore, these areas are often lashed by tropical hurricanes (Fig. 6-7). Weak tropical cyclones following the same tracks bring smaller but more frequent amounts of precipitation to these eastern coasts.

Summary. We can summarize this discussion of the distribution of very high precipitation as follows: high precipitation is found in four characteristic locations: (1) along the ITC, (2) on west coasts in the zone of extratropical cyclones, (3) in parts of monsoon Asia, and (4) on subtropical east coasts. All these locations involve unstable or conditionally unstable air masses for most of the year; orographic features localize the distribution over land; almost all the areas are oceanic or coastal, except for the Amazon and the Congo; and all the areas are associated with a major zone of storm tracks.

Regions of Very Low Precipitation

Let us now turn to those parts of the world that receive less than 10 inches (about 250 mm) of precipitation a year. The factors responsible for such low precipitation rates are (1) occurrence of stable lapse rates throughout the year, (2) rain-shadow conditions in the lee of mountains, (3) great distance from sources of moisture, (4) position far removed from storm tracks, and (5) very low air temperatures. The first four items are simply the opposite of the conditions necessary for high precipitation. The fifth condition, low air temperatures, has been added because of the effect of temperature on moisture-holding capacity. An air mass with a temperature of 20° F (-6.7° C), for example, even if completely saturated, can contain no more than 2.2 grams of water per kilogram of air. The precipitation from such an air mass cannot be very heavy even under the most favorable conditions of instability and orographic lifting. Furthermore, instability is unlikely since low surface temperatures, by cooling the lowest layers of the troposphere, significantly reduce the lapse rate of the air, thus making the air stable. A temperature inversion — the most stable of all lapse rates — is a very likely result of such cooling from below.

Not all five conditions need be fully satisfied in areas receiving less than 10 inches (250 mm) of precipitation per year. Stability of the air is the most important condition and is characteristic of all but a few of the really dry areas of the world. The dryness of some interior deserts that have unstable lapse rates during the summer is caused by lack of moisture rather than stability, but these are exceptional. A rain-shadow position can be a very important factor in reducing precipitation but is not a necessary condition for drought. Even orographic lifting is permissible if other drought-producing mechanisms are strong. Nor is great distance from the ocean a sufficient or even a necessary condition for low precipitation. On the other hand, none of the dry areas of the world are located very close to the major zones of storm tracks.

We can see from the above discussion that only distance from storm tracks is a necessary condition for drought, although it is not a sufficient condition; stability and low temperatures are sufficient conditions but not necessary; rain-shadow effects and distance from the ocean are neither necessary nor sufficient conditions but serve to accentuate dryness in certain parts of the world. Figure 7-1 clearly shows the importance of each of these factors. None of the dry areas are near storm tracks; their greatest concentration is in areas of stability or low temperatures, in the eastern parts of subtropical high-pressure centers and in polar lands; other areas of moisture deficiency are more irregularly distributed, either far from water or in the rain shadow of high mountains or both. Let us look at these patterns in detail.

Dry Areas of the Subtropical High-Pressure Belts. In Chapter 6, we emphasized the subsidence of air in the eastern parts of the oceanic high-pressure cells in subtropical latitudes. The stability of the air created by this subsidence is greatly accentuated by the effect of cold currents. The major cold currents of the world, in fact, coincide with the eastern portions of the high-pressure cells, since it is the anticyclonic circulation over the subtropical oceans that drives cold water from relatively high latitudes toward the equator along the west coasts of the continents. This coincidence of low surface temperature and subsiding anticyclonic circulation creates some of the greatest aridity on the surface of the earth. Five major areas exemplify these conditions. Their seaward extension depends on the strength and coldness of the current with which they are associated, their landward extension is determined by the presence or absence of mountain ranges parallel to the coast.

The desert of Baja California extends but a moderate distance over the subtropical Pacific Ocean; although the California Current has important effects on the climate of the two Californias, it is almost entirely a summer phenomenon and thus of only moderate overall effectiveness. The desert is limited on the east by the Sierra Madre of western Mexico; its northward prolongation lies beyond the sphere of the subtropical high pressure and has a different origin. Despite the small size of this dry area, annual precipitation drops below 3 inches (76 mm) in the central part of the peninsula of Baja California.

The Atacama Desert extends from northern Chile almost to the equator. This penetration into latitudes normally occupied by convergence rains of the ITC and the remarkable westward extension into equatorial latitudes bear testimony to the coldness and year-long intensity of the Humboldt Current. Few places are as dry over such long distances as the Atacama Desert. Arica, in northern Chile, has had less than 1 inch (25 mm) of precipitation during 29 years of observations. This average of 0.03 inches (less than 1 mm) per year is remarkable enough, but in certain localities for which we have no climatic records, the memory of the inhabitants includes no rain whatsoever!

The subtropical desert of northern Africa, associated with the Canary Current, has a considerable extension over the Atlantic Ocean; to the east, it merges with the Sahara and stretches, without significant interruption, from the Atlantic Ocean to the Indian Ocean. This enormous area of deficient precipitation 1,000 miles (over 1,600 km) broad and 6,000 miles (nearly 10,000 km) long, is the world's largest desert; Antarctica has a greater area with less than 10 inches (250 mm) of precipitation annually but is not classified as a desert for obvious reasons.

The oceanic portion of the desert of northern Africa is comparable in origin to the desert of Baja California and the Atacama Desert. Anticyclonic circulation is also found over the continental Sahara and over Arabia, despite the presence of a shallow cell of low pressure during the summer period of extremely high temperatures. The absence of mountain barriers along the west coast of North Africa allows the stability of the subtropical high-pressure cells to extend inland indefinitely, whereas the Andes of South America limit such conditions to a narrow coastal strip. Many places in the Sahara have less than 1 inch (25 mm) of rain per year, but probably none can match the more localized aridity of northern Chile.

Southern Africa and the South Atlantic Ocean have their own subtropical desert. The reasons for the extraordinarily large expanse of low precipitation over the ocean are not well known and can be only partly attributed to the Benguela Current. On land, the more typical coastal desert is known as the Namib Desert; its inland extension, slightly less arid, is the Kalahari Desert.

Australia, the fifth and last area of subtropical drought, differs from the four other regions in that the cold current off its west coast is neither very strong nor very cold. As a result, dry conditions are for the most part limited to the interior of the continent, which is dominated by stable anticyclonic air masses during most of the year.

Dry Areas of the Mid-Latitude Continental Interiors. The continental interiors of mid-latitudes represent a second important location of very dry climates. Winter air masses are cold and very stable, but summer air masses may become unstable because of heating from below. In both seasons, however, the air is very dry, since maritime air masses that reach the interior have lost their moisture either after moderate precipitation over a long continental trajectory or after heavy precipitation on the windward side of mountains, which thus cast a rain shadow on the land. The distribution of interior deserts, is therefore, strongly influenced by the size and position of the continents and of their landforms.

In North America, the interior desert covers a relatively small area in the lee of the Sierra Nevada and Cascade Mountains (the Sonoran and Mojave Deserts in the south and the Great Basin in the north). The potential source of relatively unstable and moist air is the Pacific Ocean in winter and the Gulf of Mexico in summer. Dryness in winter, despite the nearness of the Pacific Ocean, is the result of the rain-shadow position; summer drought must be ascribed to the interior location with respect to the Gulf of Mexico — an isolation somewhat accentuated by the various ranges of the Rocky Mountains.

The interior deserts of the Old World, Russian Turkestan and the upland basins or plateaus of Sinkiang, Tibet, and Mongolia, are far more extensive and much drier than their North American counterparts. Distance from sources of moisture is the principal drought-producing mechanism, even though the pattern is somewhat distorted by the alignment of mountains. Orographic rainfall over the Pamir Knot and the Tien Shan divides the interior dry lands into two distinct parts; the Himalayas limit the northern extension of monsoonal rain and allow the dry conditions to penetrate southward almost to the crest of the range.

The Southern Hemisphere has no large continental interiors in mid-latitudes. Patagonia is in the same latitude as the interior deserts of the Northern Hemisphere, but its drought is ascribable entirely to its position with regard to the Andes and the prevailing westerly winds.

Dry Areas of the Polar Regions. The most extensive, contiguous areas having less than 10 inches (250 mm) of precipitation annually are Antarctica and the Arctic Basin. Almost every factor tending to limit precipitation is in operation in the polar regions. Anticyclonic circulation, subsidence, and stable lapse rates are present throughout the year. The coldness of the surface accentuates the stability of the overlying air and limits the amount of evaporation from the small area of open water in summer. The coldness of the air masses further reduces the likelihood of precipitation by limiting their moisture-holding capacity. The West Wind Drift in the North Atlantic is responsible for the only major exception to these conditions. Its warm waters and the relatively warm air masses overlying it extend the zone of mid-latitude oceanic precipitation beyond 80° N latitude. The West Wind Drift in the North Pacific does not play a similar role, because its northward transport of warm water is arrested and deflected by the Aleutian Islands. The Southern Hemisphere Drift is a cold current and thus has little effect on the Antarctic region of low precipitation.

It may seem odd that the world's greatest accumulation of ice should be found in regions where precipitation is low. Remember that ice accumulates wherever snowfall exceeds snow melt. Annual snowfall over the polar regions may amount to less than 10 inches (250 mm) of water equivalent, but snow melt is even less.

Summary. We can describe the pattern of distribution of very low rainfall in terms of three characteristic locations: (1) in the eastern parts of the oceanic subtropical high-pressure cells with an eastward continental extension where the relief permits, (2) in the continental interiors or rain-shadow positions of mid-latitudes, and (3) over the polar regions. Each of these locations is distant from the ITC and from the zone of extratropical cyclones. Except for continental interiors, these are places characterized by year-long stability of air masses. Subtropical stability is accentuated by the presence of cold currents; polar stability by the low surface temperatures.

Seasonal Variation of Precipitation

The preceding discussion and Fig. 7-1 are meant to draw attention to places with extremes of precipitation. What about the rest of the world? Most places with moderate amounts of precipitation are characterized by a seasonal contrast in the factors affecting precipita-

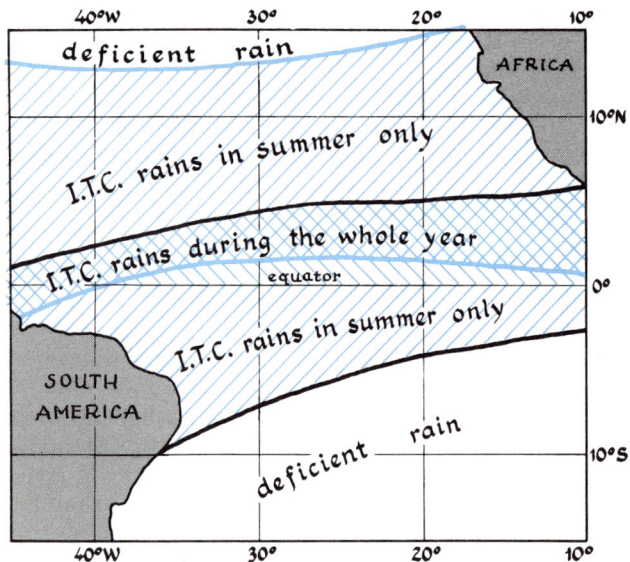

Fig. 7-2. The seasonal migration of the Intertropical Convergence Zone (ITC) in the Atlantic Ocean. The lines from lower left to upper right represent rain in summer (Apr.–Sept. in the Northern Hemisphere; Oct.–Mar. in the Southern Hemisphere); the lines from lower right to upper left represent winter rain.

tion. It is, therefore, appropriate to discuss them under the general heading "seasonal variation of precipitation." This convenient scheme has its drawbacks because some seasonal variation is found everywhere, even in regions of very high precipitation. The extremely high rainfall of monsoon Asia, to cite the most flagrant example, occurs largely in the six warmest months; the rest of the year is dry. In general, however, the regions of extremely high or low precipitation, outlined in Fig. 7-1, are also regions of relatively minor seasonal contrasts of precipitation. On the other hand, the areas of more moderate precipitation are areas of more marked seasonality. There are some exceptions, of course. The eastern seaboard of the United States and northwestern Europe, for instance, have moderate rainfall but lack seasonality. Seasonality of precipitation can be attributed to the fact that many of the places are dominated by conditions that are characteristic of the very wet parts of the world in one season of the year, whereas they share the atmospheric lot of very dry regions in the other. The seasonal shift in position of the ITC and of the polar front accounts for a large part of this seasonality.

The resulting pattern of rainfall is best illustrated over the oceans in low latitudes, where neither seasonal contrasts of surface temperature nor irregularities of relief distort the planetary pattern. The Atlantic Ocean will serve as an example (Fig. 7-2). The effective limits of convergence rains in the ITC at the time of its most southward position in January are about 2° S and 5° N. In July, the limits have migrated northward to 2° N and 14° N, respectively. It is obvious that the ocean between 2° N and 5° N is under the influence of the ITC during the whole year; this is where the highest precipitation and the least seasonality are found. From 2° S to 2° N, the ITC rains fall mainly during the Southern Hemisphere summer; from 5° N to 14° N, rainfall is concentrated in the Northern Hemisphere summer. Thus, the central belt of high rainfall is flanked by two zones of lesser precipitation and greater seasonality. Beyond 14° N and 2° S, even drier conditions prevail.

This notion is generalized in Fig. 7-3, which includes the effects of the seasonal migration of the ITC, the subtropical highs (STHP), the polar front rains (PF), and the stability of the Arctic and Antarctic regions (A). Bear in mind that although high rainfall is associated with the ITC and PF and drought with the STHP and A, they do not have exactly the same latitudinal limits everywhere in the world. For example, the ITC occupies a more northerly position in the Pacific Ocean than in the Indian Ocean. Nor do they all extend completely around the world. The STHP is particularly broken up, and its stability is limited to the eastern parts of the oceanic cells. Nevertheless, Fig. 7-3

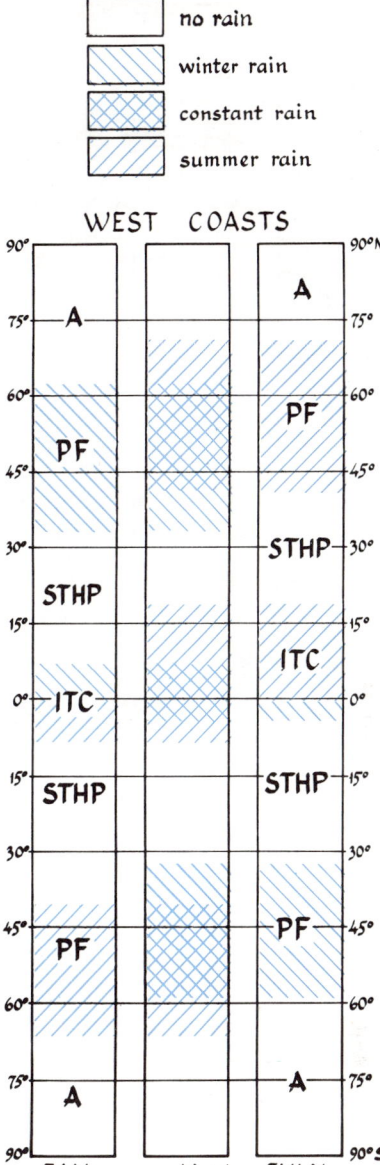

Fig. 7-3. Theoretic latitudinal arrangement of precipitation regimes along west coasts. The left-hand column shows the location of belts of rainfall associated with the January position of the Polar front and ITC, hence winter rain in the Northern Hemisphere and summer rain in the Southern. The arrangement of the analogous belts in July is shown in the right-hand column. The center column depicts the overlap of the two other columns and thus portrays the characteristic pattern of rainfall along west coasts. From north to south, we have a dry polar zone, then summer rains, then year-round precipitation (between about 45° and 60° N latitude), then winter rains (the Mediterranean regime), then subtropical droughts, followed by summer rains, and finally year-round rainfall near the equator. The same pattern (but slightly offset in latitude) can be seen in the Southern Hemisphere.

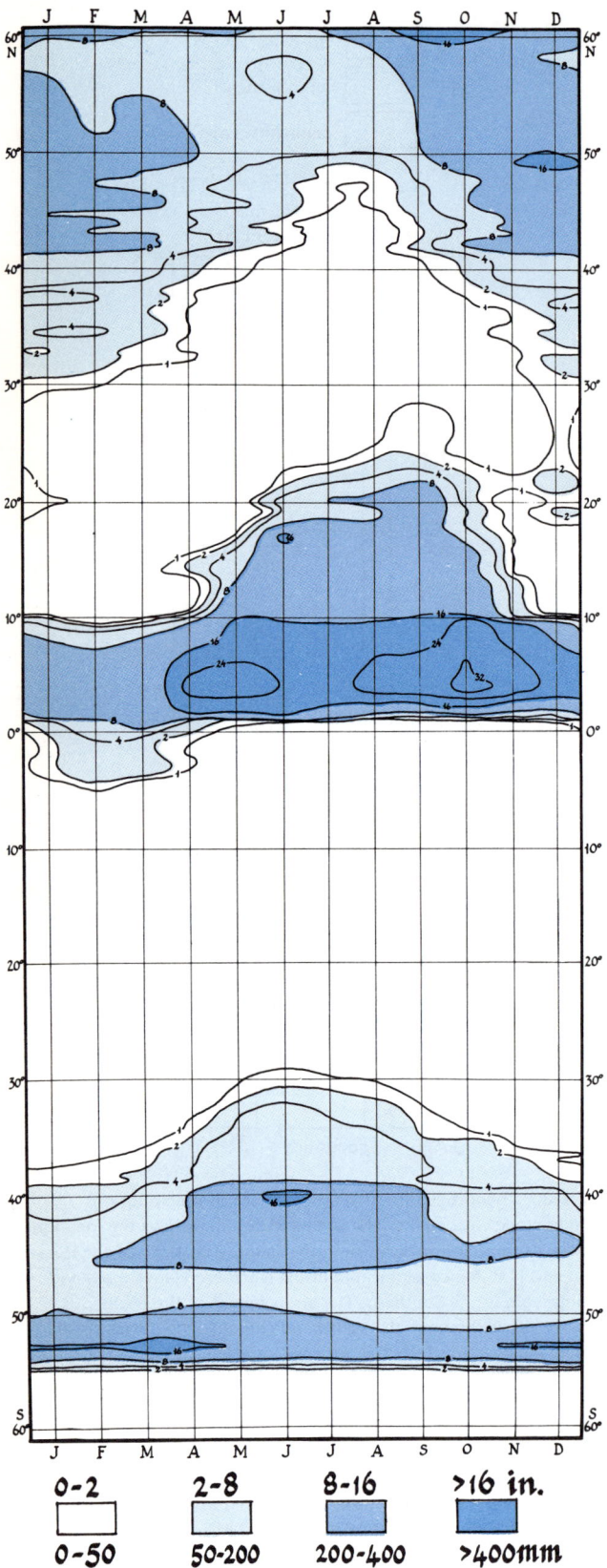

represents the precipitation regimes along west coasts quite well. The systematic succession of regimes and their symmetrical arrangement about the equator are the result of the systematic migration of the drought-producing or rain-producing centers of action in both hemispheres.

Along west coasts, the equatorial belt of year-round rainfall is flanked by two regions of winter drought and summer rain. These in turn give way to deserts without appreciable rain in either season. Farther poleward, the deserts merge into regions affected by the PF only in winter, so that they are dry in summer but have some precipitation during the winter. Beyond these are the regions of year-round cyclonic rains of the PF. In higher latitudes still, the PF is present in summer only, Arctic conditions predominating in winter, so that the precipitation is concentrated in the warmer months. Finally, in the most poleward locations, we find very little precipitation at any time.

The schematic and theoretical arrangements of Fig. 7-3 are more realistically portrayed in Fig. 7-4, which shows the rainfall regimes along the west coasts of the Americas. The long axis represents latitude extending from about 60° N to 60° S. The short axis represents the course of the year, with January 1 at the left and December 31 at the right. The seasonal course of precipitation for coastal stations in various latitudes can thus be read off the graph and compared with Fig. 7-3. For example, the area of both winter and summer rainfall in about 60° N latitude appears in Fig. 7-4 as a regime with high precipitation throughout the year, even though there is a pronounced maximum in September and October. Farther south, in 40° N latitude, Fig. 7-3 shows a zone of winter rain only. This regime is clearly seen in Fig. 7-4 — in 40° N latitude, the winter months have more than 5 inches (125 mm) of precipitation, but the summer is very dry. In 5° N latitude, Fig. 7-4 shows the equatorial regime of the ITC — heavy rainfall throughout the year, but with a double maximum in May and in October.

The advantage of Fig. 7-4 lies not only in the fact it

Fig. 7-4. Seasonal regimes of precipitation along the west coasts of the Americas. Isohyets in inches of precipitation per month. This graph is based on existing station records. By following a horizontal line, in any latitude, one can read off the month-to-month variation in precipitation in that latitude. Thus in 40°S latitude, we see that January and February both have between 2 and 4 inches of rain, March has more than 4 inches but less than 8, April and May have more than 8 but less than 16 inches, June has more than 16 inches, and so on. One can also read off the latitudinal variation in any given month by following the appropriate vertical line. Thus in February the places with less than one inch of precipitation are between 30°N and 10°N and between 5°S and 37°S; whereas in July the corresponding zones are between 45°N and 24°N and between 1°N and 30°S.

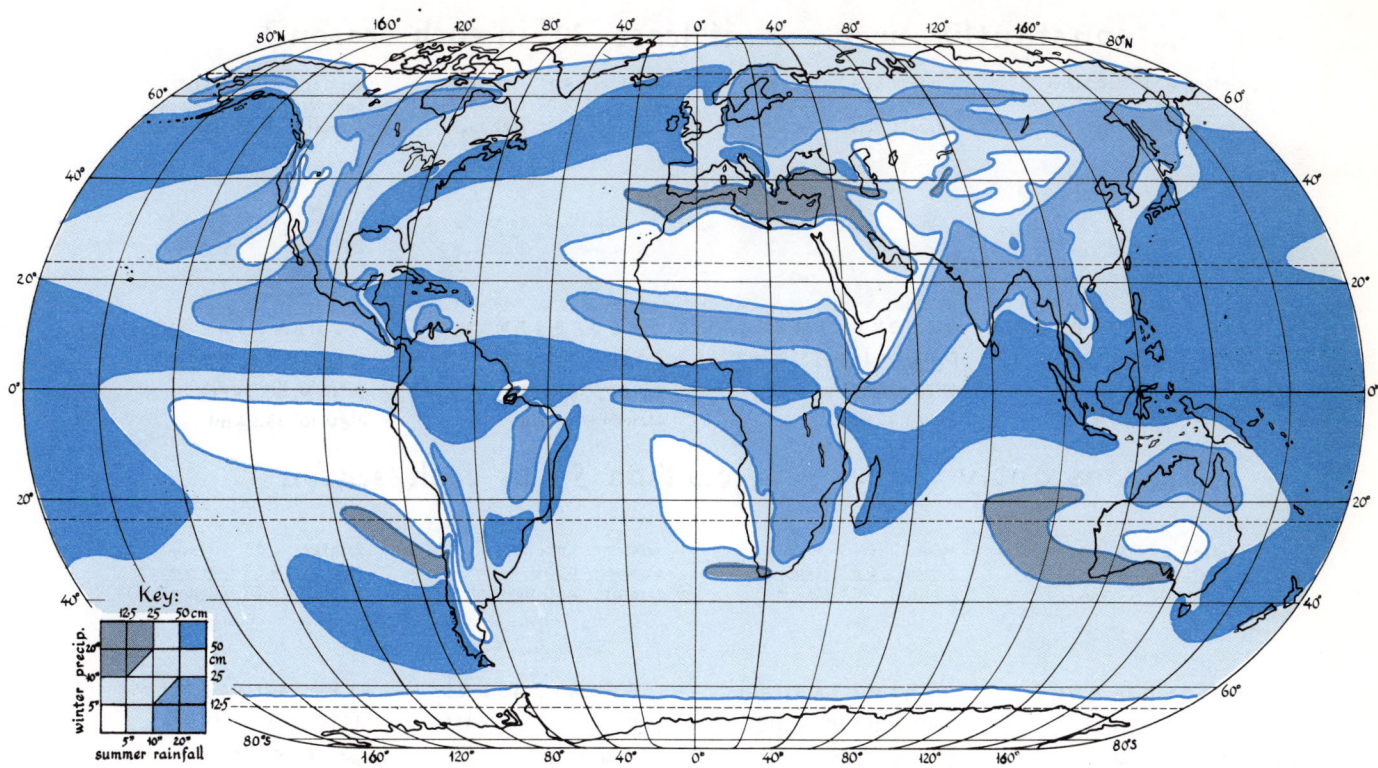

Fig. 7-5. The distribution of seasonality of precipitation over the world. Note the very extensive distribution of moderate precipitation regimes over the ocean. Places with heavy rainfall in all seasons are also largely oceanic, and are found in the location of the polar front, particularly in the Northern Hemisphere, and of the ITC. Summer rainfall regimes are largely continental; winter rainfall regimes are the least extensive of all.

represents a real situation along a real coast and shows details not included in the theoretical and schematic diagram, but also in that it demonstrates graphically the shift from one regime to another as one progresses from one latitude to another.

Let us now enlarge our view to include the whole world. Figure 7-5 is a schematic representation of various types of seasonal regimes of precipitation. It is schematic in that it attempts to reduce the complexity of rainfall regimes to five categories. To do this conveniently, we can consider only two seasons of the year — a winter season extending arbitrarily from November 1 to April 30 and a summer season from May 1 to October 31. This nomenclature is reversed for the Southern Hemisphere, of course.

The five categories used are as follows:

1. *Regions of constantly high precipitation (at least 20 inches or 500 mm of precipitation in each season).*

2. *Regions of constantly low precipitation (less than 5 inches or 125 mm of precipitation in each season).*

3. *Regions of pronounced summer rainfall (winter precipitation less than 10 inches or 250 mm, winter precipitation more than 10 inches or 250 mm and more than twice the summer precipitation).*

4. *Regions of pronounced winter rainfall (summer precipitation less than 10 inches or 250 mm, winter precipitation more than 10 inches or 250 mm and more than twice the summer precipitation).*

5. *Regions without extremes of precipitation or marked seasonality (precipitation not fitting any of the other four categories).*

Each of the five categories of precipitation regimes is illustrated by four examples in Fig. 7-6.

The map of seasonal regimes is schematic in yet another sense. Seasonal rainfall data for the oceans are sparse and unevenly distributed. The lines on the map are, therefore, somewhat tentative approximations of the actual distributions. Nevertheless, we may begin our discussion of Fig. 7-5 with an encouraging note — Notice that there is a marked correspondence between this map and the one of total annual precipitation (Fig. 7-1).

The first category, areas of constantly high precipita-

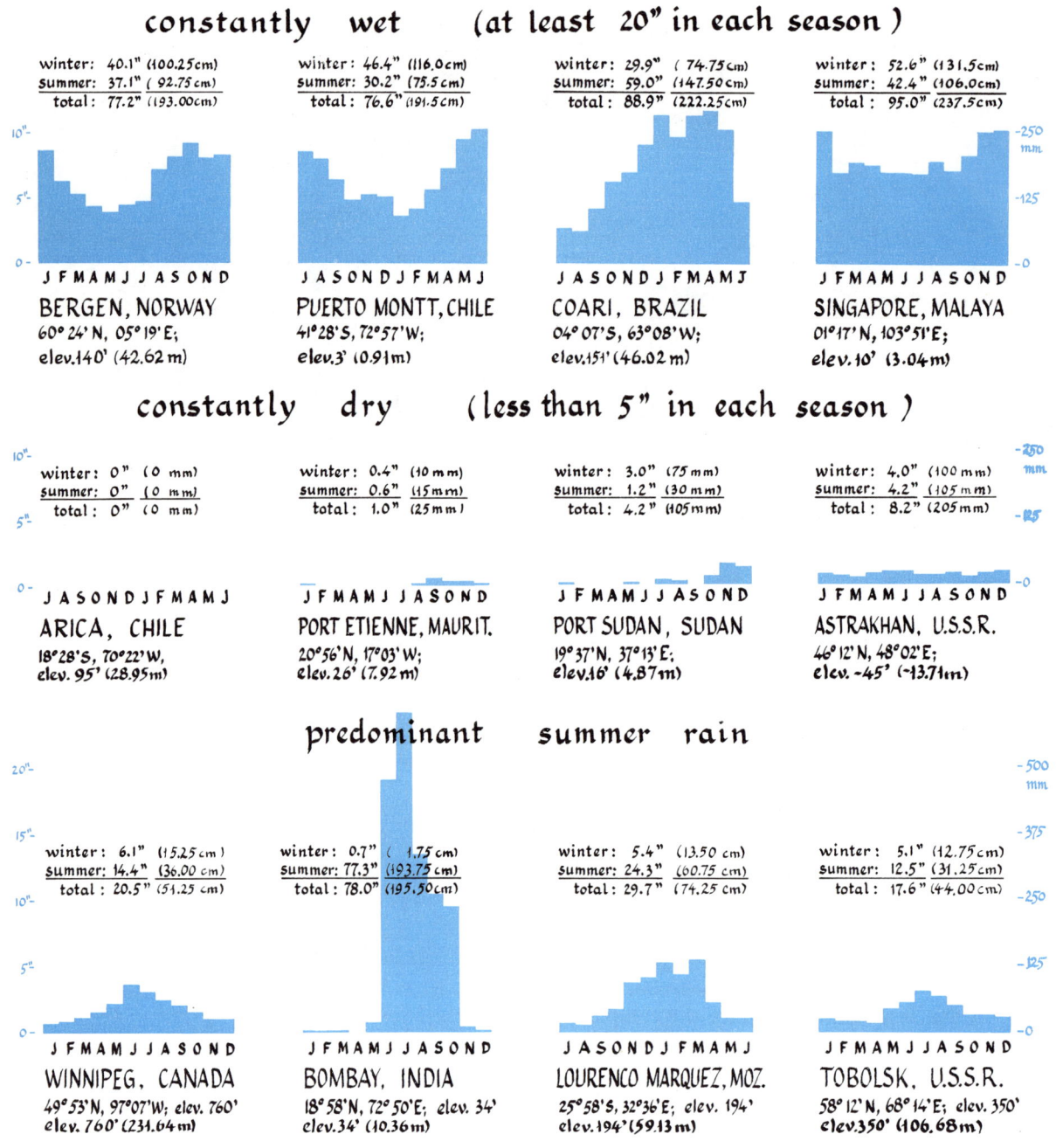

Fig. 7-6. Examples of the five types of precipitation regimes.

tion, fits the distribution of the 80-inch (2,000 mm) isohyet in Fig. 7-1 rather well, except that it excludes the regions of very high monsoonal rainfall in India and along the east coast of Middle America. It also includes much larger areas over the mid-latitude oceans. Cyclonic rainfall of the polar-front zone does not reach 80 inches (2,000 mm) a year except in favored orographic situations, but it is high enough in both seasons to qualify large areas as having constantly high precipitation.

The second category, areas of constantly low precipitation, fits the distribution of the 10-inch (250 mm) isohyet very closely and for obvious reasons. If a place has less than 5 inches (125 mm) in summer and less than 5 inches (125 mm) in winter, it cannot very well have more than 10 inches (250 mm) during the

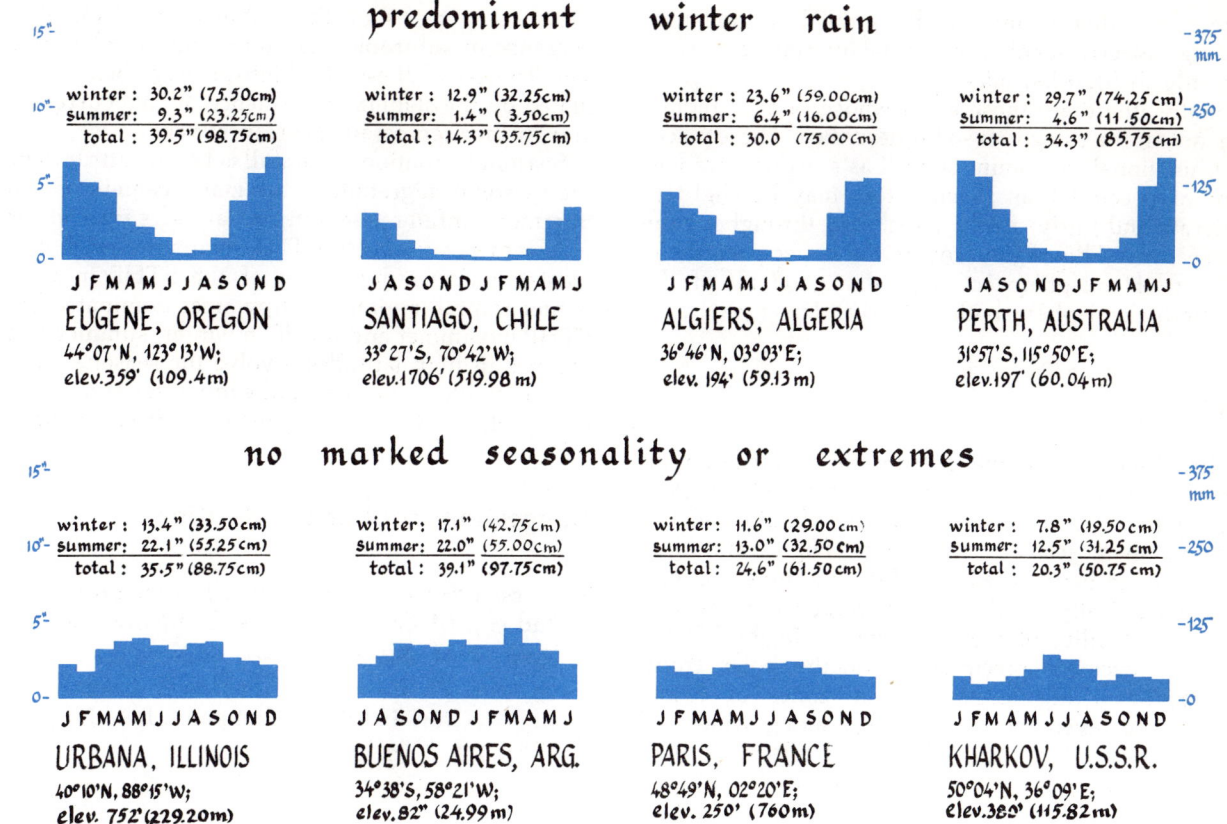

year. Remarkably, the area covered by Category 2 is not much smaller than the area outlined by the 10-inch (250 mm) isohyet. Although the area of constantly low precipitation, as we have defined it, cannot be larger than the area bounded by the 10-inch (250 mm) isohyet, it could be considerably smaller.

The other rainfall regimes have no counterpart on Fig. 7-1. They can be referred to Fig. 7-3, however. The latitudinal arrangement of varying seasonal regimes shown in the diagram is more or less faithfully reproduced along west coasts, with some minor variants. There is no zone of summer rainfall in high latitudes of the Southern Hemisphere. Summer rainfall, except in the zone of the ITC, is a continental phenomenon. The absence of continents in the Southern Hemisphere limits the development of summer rainfall regimes to low latitudes. Nor does there seem to be any great development of areas of constantly high precipitation over the southern oceans. More records of precipitation might reveal a different situation, however.

The theoretical pattern of Fig. 7-3 does not correspond to the distribution along east coasts for reasons already indicated and is further distorted by the differential behavior of the Northern and Southern Hemisphere oceans.

If we take the distribution of constantly wet and constantly dry areas as corresponding closely to the very wet and very dry areas of Fig. 7-1, the pattern of seasonality shown on Fig. 7-5 can be described in fairly simple terms. The areas of winter rain are poleward of the subtropical drylands along west coasts and adjacent oceans. This zone is not very large and is more oceanic than continental, except where the Mediterranean Sea allows considerable inland penetration of cyclonic rains in winter.

The areas of summer rain occupy two different locations. In low latitudes, summer rains are found on either side of the equatorial belt of high precipitation. Summer rainfall regimes separate the constantly wet climates near the equator from the perennially dry areas of the STHP on the land, but their development over the oceans is slight. The other major location of summer rainfall consists of the continental interiors in mid- and high latitudes of the Northern Hemisphere, generally north of the interior drylands. Regions of pronounced summer rainfall are thus much more continental in location than those of pronounced winter precipitation. The tendency for oceans to develop high pressure in summer, when the hotter continents are more likely to be covered by low pressure, and the reversal of this situation in winter obviously play a large role; other factors, such as the low moisture-holding capacity of cold continental interiors, are effective only in high latitudes.

Finally, there are those areas whose rainfall regime does not fit the categories so far described. These places are transitional in amount as well as seasonal arrangement of precipitation. Precipitation may be slight to moderate and fairly evenly distributed throughout the two seasons. This regime covers the larger part of the Southern Hemisphere oceans and a good bit of the Northern Hemisphere, both on land and over water.

Summary

This discussion of the distribution of precipitation over the surface of the earth has focused on the factors making for deficiency or abundance of rain and has outlined their seasonal migration. High rainfall is associated, in general, with surface low pressure and, more specifically, with the convergence rains and general instability of moist air masses in the ITC or with the cyclonic precipitation of the polar front. Orographic lifting localizes some zones of heavy precipitation even more specifically along windward coasts. Low rainfall is directly related to surface high pressure in terms of the stability and lack of convergence in subtropical latitudes and in polar regions. The dryness of these two high-pressure belts is accentuated by the stabilizing influence of cold currents and low surface temperatures, respectively.

Seasonal variation in rainfall is largely attributable to the seasonal migration of the major centers of action. Summer rainfall regimes in low latitudes mark the most poleward position of the ITC in summer (and also the most equatorward position of the STHP in winter); winter rainfall regimes occupy areas dominated by the STHP in summer and the PF in winter; summer rainfall regimes in high latitudes involve the seasonal alternation of stable arctic conditions in winter and the much less stable conditions along the polar front in summer.

Suggestions for Further Reading

Petterssen, Sverre, *Introduction to Meteorology*, 2nd ed., McGraw-Hill Book Company, Inc., New York, 1958, pp. 265–282.

Trewartha, G. T., *An Introduction to Climate*, 3rd ed., McGraw-Hill Book Company, Inc., New York, 1954, pp. 139–148.

8 The Classification of Climates

Introduction

There are probably as many possible classifications of climate as there are purposes for which they might be devised. The general purpose of a classification is, of course, to introduce some order into the existing diversity, to produce a scheme of organization which sheds some light on what otherwise is simply a catalogue of facts. To a geographer, a classification has the important added advantage of allowing him to map the distribution of the item classified, for it is only by ignoring unimportant and irrelevant differences that a readable and understandable map can be constructed at a world scale. It would be absurd to proceed by simply overlaying isohyets, isotherms, isobars, and so on. Not only would the map be too complex to be useful, but it would not show many important properties of climate, such as its annual variation. Therefore, to produce a single, readable map we must group together climates of reasonably similar characteristics, so as to distinguish between trivial and significant differences from place to place.

What significant climatic differences should a classification emphasize and what trivial ones should it ignore? There is no simple answer, not only because of our incomplete knowledge of climatic processes, but also because climatic classifications can have widely different purposes. A classification that emphasizes the surplus or deficit of water at the surface of the earth while neglecting moisture conditions in the atmosphere can be judged only in the context of its potential use. A farmer might find great use for a classification that tells him only about the water balance in his fields, while the aviator, on the other hand, would no doubt prefer a classification that focuses more on visibility. Even when only a single element is in question, various users may have quite different views of a particular classification. A tourist's concern with precipitation, for example, might involve only the average number of rainy days within a restricted period of time; a gardener's concern might be more with the duration of rainless periods; whereas an engineer building a bridge would have to consider extremes of precipitation in order to enable his bridge to withstand floods of considerably more than average intensity.

What kind of classification best suits our purpose? At our level of generalization we need a classification that enables us to distinguish the major climates of the world. It must be a classification that is based on climatic data that are available for most of the world. It must also divide the surface of the earth into few enough climatic regions to be shown distinctly on a fairly small-scale map whose major features of distribution can be appreciated by the map reader. At the same time, these regions must be numerous enough to bring out the major climatic distinctions. We are faced, therefore, with two constraints on our choice of classifications. We are limited by the availability of data, and we must compromise between detail and clarity.

The first of these limitations forces us to deal essentially with monthly averages of temperature and precipitation or at least values derived from them. Classifications based on other climatic elements or even on probabilities and extremes of temperature and precipitation (rather than averages) are not very helpful if one needs to show the pattern of climate over the whole world, since other data are available for only a relatively few places. The second condition precludes naively simple schemes such as the following. Let us consider the average monthly temperature of the coldest month as one of the criteria; the average temperature of the warmest month as another; and the average annual precipitation as the third. The simplest way to proceed might be to set up categories that are easy to remember, such as 10° intervals of temperature and 10-inch intervals of precipitation. If we look at the actual temperatures on the earth's surface, however, we would see that it takes 19 categories to accommodate the range of temperatures for the coldest month, another 14 to take care of the warmest month temperatures, and at least 20 categories of precipitation, even if we lumped all places with more than 200 inches (5,000 mm) of precipitation per year into one category. This simple scheme would therefore produce 19 times 14 or 266 separate categories of temperature alone. To be sure, 66 of these are illogical since they describe climates in which the coldest month is warmer than the warmest month. Of the remaining 200 categories, a good many would not include any real places, but even so we would be left with a very large number of real climatic categories.

Figure 8-1 illustrates this simplistic approach to climatic classification and also describes the distribution of the world's surface area among the various categories of temperature. The percentages shown in Fig. 8-1 reveal a number of things. First is the large number of categories of climate. Seventy-four categories may not seem like an excessive number until one tries to give each category a name. Even if we reduce the categories to 29 (by increasing the intervals from 10° to 20°), we still have a formidable problem of nomenclature.

The graph also shows the remarkable predominance of warm climates. The cumulative totals for the coldest month show that in more than half the world the temperature of the coldest month exceeds 55° F (13° C), and in nearly a third it exceeds 70° F (21° C).

A third item of interest is the relative isothermality of most of the world's climates, that is the small annual range of temperature. The stair-step line in the upper left-hand part of the graph marks off those climates in which there is no difference at all between the average temperature of the warmest and coldest months, and the closer a point lies to that line the more isothermal its climate. Nearly 60 percent of the world has annual ranges of temperature that are less than 15° F (8° C), and in the more oceanic Southern Hemisphere, 75 percent of the surface experiences annual ranges of less than 15° F (8° C). If these facts are surprising it is because we are apt to neglect the Southern Hemisphere when we visualize climates, and we often overlook the great extent of equatorial and tropical oceans. All of these places tend to have warmer winters, and hence lower annual ranges of temperature than we are used to on the continents of the Northern Hemisphere.

Now it is true that the graph we have constructed is of some use in describing the world's climates, but it certainly fails as a scheme of classification, since even by extending the temperature intervals to 20, we would still have 29 categories of climate, without considering any climatic elements other than the mean monthly temperature of the two extreme months. Obviously a climatic classification could be devised by extending the temperature intervals farther or by choosing intervals that are not equal over the whole range to be categorized. Such a classification is shown in Fig. 8-2; it is not a very good scheme since it is still limited to monthly temperatures of the extreme months. We could make up this deficiency by having three or so categories of humidity for each of the temperature groupings. For example, dry climates with less that 15 inches (381 mm) of precipitation, wet climates with more than 50 inches (1,270 mm) and moderate climates in between. But obviously 15 inches (381 mm) of rain has entirely different consequences for the water budget or for plant growth in the tropics than it would in higher latitudes; and 15 inches (381 mm) during the winter is a different story altogether from 15 inches (381 mm) that fall in summer.

What we are trying to prove is that schemes of classification such as we have been describing are never very satisfactory compromises between too much detail and too much generality. Yet how can we get a climatic classification that is useful in distinguishing the major climates of the world? What kinds are there? The most obvious is the *morphologic* variety that classifies according to the form of the climate, or according to the amount and distribution of precipitation during the year, or according to the temperature of various months, and so forth. An alternative is the *genetic* classification, in which the fundamental elements are those that indicate the origin of the climate rather than its form. For instance a genetic scheme might be based

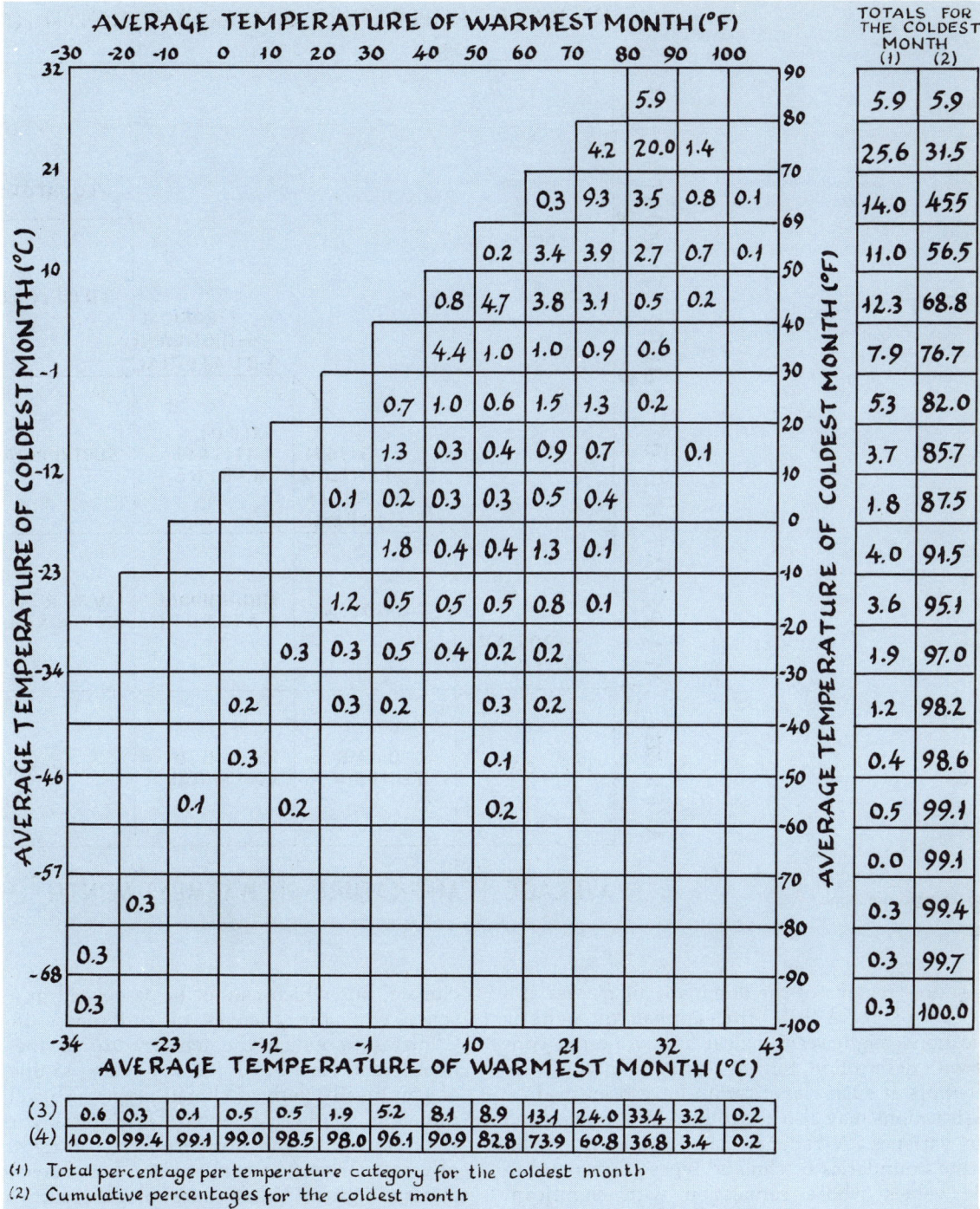

Fig. 8-1. The areal distribution of the earth's surface according to temperature of the extreme months. The values indicated in the various boxes represent the percentage of the earth's surface that experiences temperatures of the coldest month between the temperatures of the lines at the top and bottom of the box; and warmest month temperatures between the temperature of the lines at the left and right of the box. The figures at the right (and at the bottom) show overall percentages for each interval of winter (and summer) temperatures; they also give the cumulative percentages starting with the warmest temperatures.

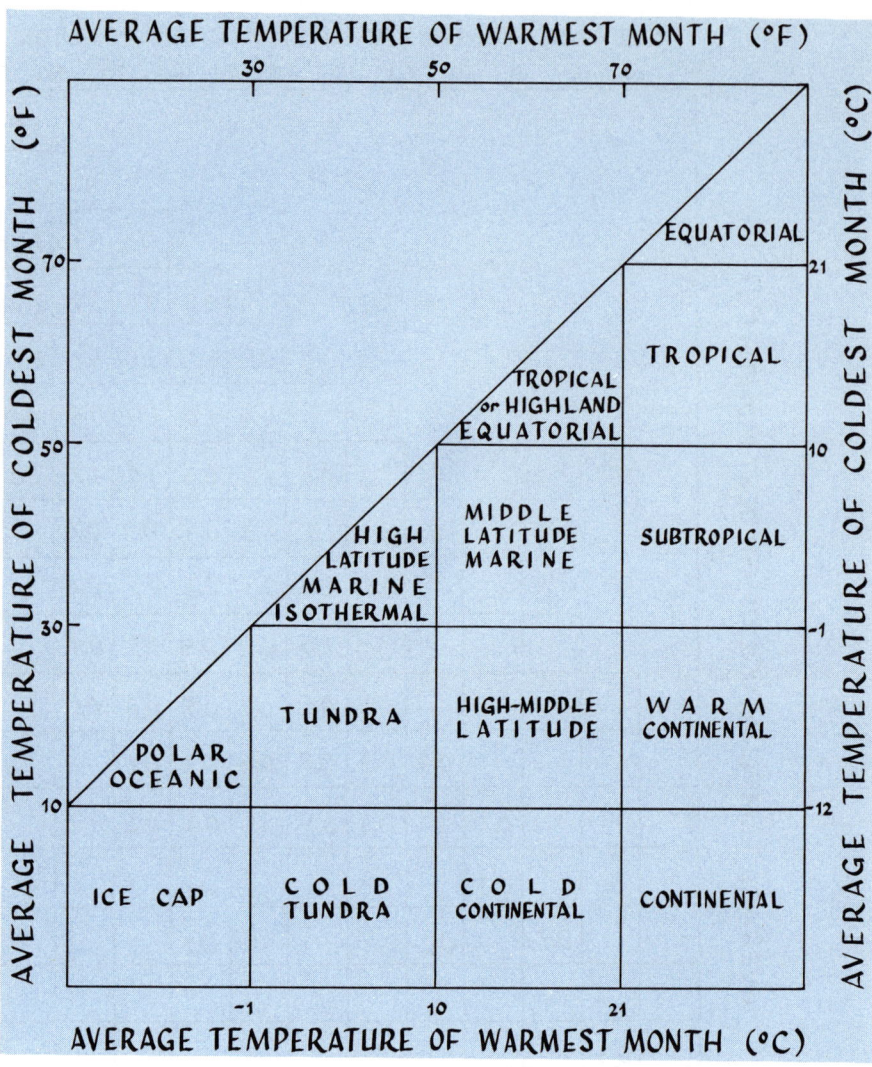

Fig. 8–2. A classification of climates based on temperature alone.

on the source regions of the dominant air masses at a particular place. While the appeal of genetic classifications might seem logical to anyone not concerned with description, for one reason or another very few attempts at such classification have been made.

Classifications may also be grouped into *rational* and *empiric* varieties. A rational classification is one in which the boundaries of climatic types correspond to climatic values whose connection with significant changes in climate can be demonstrated. Unfortunately there are not many such values. The complete absence of precipitation is an obvious one, as is the freezing point of water, or a temperature such as 39° F (4° C) which marks the approximate maximum density of water. Most climatic classifications use empiric delimitations; that is, they use values that appear to work satisfactorily in delimiting major differences in climate, but which cannot be associated in a physical sense with the changes or differences in climate. Climates in which the temperature of the warmest month is at least 50° F (10° C) belong to this category of empirically derived climatic types. This climate, or rather its southern boundary in Eurasia, corresponds very closely to the northern limit of trees, but the physical connection between 50° F and tree growth is dim and indirect. Most morphologic classifications are therefore largely empiric; genetic schemes may attempt to be more rational. But the terms morphologic or descriptive should not be thought of as synonymous with the word empiric.

Of the many systems of classification that have been developed, mostly in the period between 1890 and 1950, two seem to have stood the double test of usefulness and popularity. The oldest and most systematic, first described in an embryonic form in 1898, is that of the German botanist and climatologist Wladimir Köppen (1846-1940). His classification is used for the best wall maps of world climate, and maps based on his system are found in most American atlases and in almost every textbook that deals with climate. It is the morphologic classification that we shall use to describe the climates of the world. A full description of the other widely used classification, that of C.W. Thornthwaite, goes beyond the aims of this book. Its scheme of calculating the water balance, another example of an open system in the physical environment, is nevertheless utilized to describe the climate of two stations in the next chapter. Thornthwaite's method focuses primarily on determining and classifying the elements of the water balance, but it can also be used to classify climates at a world scale.

The Köppen Classification

The advantages of the Köppen classification are due to a number of factors. It uses only mean monthly or annual values of precipitation and temperature, and these criteria are simple and unambiguous. Furthermore, the use of these criteria results in only a dozen major categories of climate. Yet these few groups divide climatic conditions over the world in a satisfactory and realistic way; that is, the variations in climate within a single category are necessarily large (since there are only twelve groupings) but acceptable at the world scale. In addition, every important distinction based on criteria other than precipitation and temperature (for example, evaporation) or even on nonclimatic criteria (for example, natural vegetation or soils) is reasonably well reflected by the classification.

The concordance between climates according to Köppen and certain associations of natural vegetation should not be too surprising, since the classification was originally based on a scheme for subdividing the world into five large zones on the basis of their natural vegetation. Furthermore, Köppen deliberately used the limits of certain plant associations as the boundaries of some of his categories. The system is thus in many ways a classification of vegetation. It is climatic only in that Köppen used temperature and precipitation criteria to define the boundaries of the vegetation zones and hence of his climatic categories. This characteristic of the classification is emphasized by the use of names referring to vegetation for a number of the categories.

The botanical aspect of the classification appears in the following description of its categories and bounding criteria. The original basis of the classification was a division of the world in 1874 by a Swiss botanist, Alphonse de Candolle, which took the following form:

A *Zone of the megatherms (heat-loving plants of tropical regions).*

B *Zone of the xerophytes (drought-resistant plants of the subtropical interior deserts).*

C *Zone of the mesotherms (plants flourishing under moderate temperatures).*

Table 8-1. Major Categories of the Köppen Classification

Climatic symbol	Climatic category	Definition based on vegetation	Climatic definition	Quantitative limit
A	Tropical climates	Warm enough and moist enough for tropical plants	No winter season	No month colder than 64°F (18°C)
B	Dry climates	Too dry for tree growth	Deficient precipitation	Annual rainfall less than a certain amount (which depends on the temperature and the seasonal concentration of the rain.)
C	Mesothermal climates	Temperature and moisture suitable for mesothermal plants	Mild winter	Coldest month between 27°F (-3°C) and 64°F (18°C)
D	Microthermal climates	Temperature and moisture suitable for microthermal plants	Cold winter	Coldest month below 27°F (-3°C)
E	Polar climates	Too cold for tree growth	No summer season	No month above 50°F (10°C)

Table 8-2. The Polar Climates

Climatic symbol	Climatic category	Definition based on vegetation	Climatic definition	Quantitative limit
ET	Tundra	Too cold for trees but enough summer warmth for some vegetation	No summer season, but some monthly temperatures above freezing	No month above 50°F (10°C), but at least one month above 32°F (0°C)
EF	Frost	Too cold for any vegetation growth	No mean monthly temperatures above freezing	No month above 32°F (0°C)

D   Zone of the microtherms (plants flourishing under low temperatures).

E   Zone of the hekistotherms (cold-resistant plants of polar regions).

You will note that the alphabetical order corresponds to the rough latitudinal arrangement of the five zones from equator to poles. De Candolle also had in mind a symmetrical arrangement in the two hemispheres; so that from pole to pole, one should find the following order: E–D–C–B–A–B–C–D–E.

Köppen took over this arrangement, complete with letter symbols but with a substitution of climatic names for those referring to plants and a substitution of quantitative[1] as well as qualitative climatic definitions for the botanic ones. The major climatic categories of Köppen's classification thus appear as shown in Table 8-1.

We shall treat these subdivisions not in their alphabetical (or roughly latitudinal) order but in the order which we would use to classify a particular place, given its mean monthly temperatures and also its precipitation.

We first determine whether the station has a treeless climate, either polar or dry. Polar (E) climates have been defined as associated with places where the mean monthly temperature never rises above 50°F (10°C); they are further subdivided as in Table 8-2. The stations that follow will serve as examples.

Does Verkhoyansk in northeastern Siberia have a polar climate? How about Barrow, Alaska? In which of

---

[1]The various limits of the Köppen classification were, of course, expressed in metric units (centigrade degrees for temperature and centimeters for the depth of precipitation). In the discussion these values have been converted to English units, but with an attempt to keep the round numbers that Köppen used. Thus, 18°C is rendered as 64°F (not 64.4°F, which would be its exact equivalent).

Verkhoyansk, U.S.S.R.
(67° 33' N, 133° 24' E; elevation 402 feet or 123 m)

	Temp.		Precip.	
	°F	°C	in.	mm
Jan.	-58	-50	0.2	5
Feb.	-48	-44	0.1	3
Mar.	-22	-30	0.1	3
Apr.	8	-13	0.2	5
May	35	2	0.3	8
June	55	13	0.9	23
July	59	15	1.1	28
Aug.	51	11	1.0	25
Sept.	36	2	0.5	13
Oct.	6	-14	0.3	8
Nov.	-34	-37	0.2	5
Dec.	-52	-47	0.2	5
Year	3	-16	5.1	131

Barrow, Alaska
(71° 18' N, 156° 47' W; elevation 13 feet or 4 m)

	Temp.		Precip.	
	°F	°C	in.	mm
Jan.	-16	-27	0.2	5
Feb.	-18	-28	0.2	5
Mar.	-15	-26	0.1	3
Apr.	0	-18	0.1	3
May	18	-8	0.1	3
June	33	1	0.4	10
July	39	4	0.8	20
Aug.	37	3	0.9	23
Sept.	31	-1	0.6	15
Oct.	17	-8	0.5	13
Nov.	-1	-18	0.2	5
Dec.	-11	-24	0.2	5
Year	10	-12	4.3	110

Amundsen-Scott Station, Antarctica
(90° 00' S; elevation 9,200 feet or 2,800 m)

	Temp.		Precip.	
	°F	°C	in.	mm
Jan.	−20	−29		
Feb.	−40	−40		
Mar.	−66	−54		
Apr.	−76	−60		
May	−72	−58		
June	−71	−57		
July	−75	−59		
Aug.	−74	−59		
Sept.	−75	−59		
Oct.	−60	−51		
Nov.	−38	−39		
Dec.	−20	−29		
Year	−57	−49	0.1	3

the two polar categories does Amundsen-Scott Station fall?

If the particular place we are classifying does not belong to the polar *(E)* category, we next determine whether it has a dry *(B)* climate. In terms of vegetation, we have defined dry climates as associated with places where the precipitation is insufficient for tree growth. The climatic definition is "deficient precipitation." In a general way, we can consider that precipitation is deficient whenever it is less than a certain fraction of the potential evaporation from the surface and the potential transpiration from the plants. However, potential evaporation is difficult to measure, and data are available for only a very few places. We could use empiric formulas that give an approximation of potential evaporation based on other climatic factors, but unfortunately, potential evapotranspiration depends on a great many factors, such as temperature, insolation, length of day, wind, and kind of plant cover. The many formulas that have been developed to estimate potential evapotranspiration are more or less complicated and take into account all or some of the factors mentioned above. For the purpose of a world classification, however, we need a formula that utilizes only mean monthly temperatures and precipitation, since these are the only climatic data available on a world scale. The formula used here takes into account only the average annual temperature and the percentage of the total annual precipitation that falls in the six coldest months. It gives only a crude approximation of the potential evapotranspiration but has the advantage of being a simple formula, closely related to those used in the Köppen classification.[2] This formula reads as follows:

$$E = \frac{T}{2} - \frac{PPW}{4}$$

where E is the **water need** (a fraction of the potential evapotranspiration), expressed in inches of water, T is the average annual temperature in °F, and PPW is the percentage of precipitation that falls in the winter 6 months (for simplicity, October to March in the Northern Hemisphere and April to September in the Southern Hemisphere). To utilize this empiric relationship, we need only add up the precipitation of the 6 winter months and express this seasonal total as a percentage of the total annual precipitation. The rest is very simple arithmetic as the example below will show.

Ciudad, Juarez, Mexico
(31° 44' N, 106° 29' W; elevation 3,729 feet or 1,136 m)

	Temp.		Precip.	
	°F	°C	in.	mm
Jan.	42	6	0.2	6
Feb.	48	9	0.4	10
Mar.	54	12	0.3	8
Apr.	64	18	0.3	8
May	70	21	0.6	15
June	78	26	0.5	12
July	81	27	1.2	29
Aug.	79	26	1.3	33
Sept.	73	23	0.9	24
Oct.	63	17	0.9	24
Nov.	50	10	0.4	9
Dec.	43	6	0.4	9
Year	62	17	7.4	187

[2] This change from the original Köppen classification has the advantages of substituting one formula for three and of employing reasonably round numbers for the coefficients when the formula is expressed in English units. These round numbers are not only easier to deal with in a rapid classification of stations, but more importantly, they are much more in keeping with the intent of the Köppen formulas than the more complicated coefficients obtained by a senseless translation of the Köppen formulas into their literally exact equivalents in English units. The metric formula of which $E = \frac{T}{2} - \frac{PPW}{4}$ is an approximate counterpart is $e = 20t + 490 + 7PPW$, where e and t are in mm and °C, respectively. This formula gives $e = 20t$ for $PPW = 70\%$; $e = 20t + 140$ for $PPW = 50\%$; and $e = 20t + 280$ for $PPW = 30\%$; which are the three formulas that Köppen specified for climates with rain mostly in winter, evenly spread, and mostly in summer, respectively. Note that in these and succeeding expressions, T and P are in °F and inches; t and p in °C and mm; respectively.

Table 8-3. The Dry Climates

Climatic symbol	Climatic category	Definition based on vegetation	Climatic definition	Quantitative limit
BS	Steppe	Too little rain for trees, but enough for grass	Water need exceeds precipitation but is not twice as large	$\frac{T}{2} - \frac{PPW}{4} > P$; but $\frac{T}{2} - \frac{PPW}{4} < 2P$ $(20t + 490 - 7PPW > p$; but $< 2p)$
BW	Desert	Too little rain for any but specialized, drought-resistant plants	Water need is more than twice as great as precipitation	$\frac{T}{2} - \frac{PPW}{4} > 2P$ $(20t + 490 - 7PPW > 2p)$

The winter precipitation at Ciudad Juarez comes to 2.6 inches (66 mm), which is 35.1 percent of the total annual precipitation. Since the average annual temperature is 62°F (17°C), the water need can be estimated as $\frac{62}{2} - \frac{35.1}{4}$ or 22.2 inches or, using the metric formula, $e = (20 \times 17) + 490 - (7 \times 35.1)$ or $e = 584$ mm. The water need for Ciudad Juarez is thus greater than its precipitation (22.2 inches or 584 mm compared to 7.4 inches or 187 mm), and, therefore, it has a dry climate.

We have now defined dry (B) climates as those in which the water need (as established by the empiric formula) exceeds the precipitation. These dry (B) climates are further subdivided as in Table 8.3.

In the previous example, we found that the water need for Ciudad Juarez was 22.2 inches; its precipitation was 7.4 inches. The water need is not only more than the precipitation; it is more than twice as great. Thus, Ciudad Juarez not only has a dry (B) climate but also, within the (B) climates, is classified as desert (BW) rather than steppe (BS). The three examples below, for which the classification has been worked out, should make the procedure clear.

Minneapolis is not a dry station; estimated water need is less than the precipitation. Alexandria is a dry station; estimated water need is more than the precipitation. Water need is less than twice the precipitation. Therefore, Alexandria has a steppe (BS) climate. Alice

Minneapolis, Minn.
(44° 53′ N, 93° 13′ W; elevation 830 feet or 253 m)

	Temp.		Precip.	
	°F	°C	in.	mm
Jan.	14	−10	0.6	15
Feb.	18	−8	0.8	21
Mar.	29	−2	1.5	38
Apr.	46	8	1.8	45
May	58	14	3.0	76
June	68	20	4.4	112
July	73	23	3.8	96
Aug.	72	22	3.3	83
Sept.	61	16	2.4	62
Oct.	51	11	1.4	36
Nov.	33	1	1.4	35
Dec.	20	−7	0.8	20
Year	45	7	25.2	639

Precipitation in 6 winter months: 6.5 in.
Expressed as a percentage of total: $\frac{6.5}{25.2} \times 100 = 25.8$ percent
Estimate of water need $\left(E = \frac{T}{2} - \frac{PPW}{4}\right)$: $\frac{45}{2} - \frac{25.8}{4} = 16.1$ in.

Alexandria, Egypt
(31° 12′ N, 29° 53′ E; elevation 105 feet or 32 m)

	Temp.		Precip.	
	°F	°C	in.	mm
Jan.	57	14	1.9	49
Feb.	57	14	0.9	24
Mar.	60	16	0.4	10
Apr.	65	18	0.1	3
May	70	21	0.1	2
June	78	26	0.0	0
July	78	25	0.0	0
Aug.	79	26	0.0	0
Sept.	78	25	0.0	1
Oct.	74	23	0.2	6
Nov.	68	20	1.3	33
Dec.	60	16	2.2	56
Year	68	20	7.1	184

Precipitation in 6 winter months: 6.9 in.
Expressed as a percentage of total: $\frac{6.9}{7.1} \times 100 = 97.2$ percent
Estimate of water need $\left(E = \frac{T}{2} - \frac{PPW}{4}\right)$: $\frac{68}{2} - \frac{97.2}{4} = 9.7$ in.

Alice Springs, Australia
(23° 38' S, 133° 56' E; elevation 1,926 feet or 587 m)

	Temp.		Precip.	
	°F	°C	in.	mm
Jan.	83	28	1.7	43
Feb.	82	28	1.6	41
Mar.	77	25	1.2	30
Apr.	68	20	0.7	18
May	60	15	0.6	15
June	54	12	0.5	13
July	53	12	0.4	10
Aug.	58	14	0.4	10
Sept.	65	18	0.4	10
Oct.	73	23	0.7	18
Nov.	78	26	1.0	25
Dec.	81	27	1.5	38
Year	70	21	10.7	271

Precipitation in 6 winter months: 3.0 in.

Expressed as a percentage of total: $\frac{3.0}{10.7} \times 100 = 28.0$ percent

Estimate of water need $\left(E = \frac{T}{2} - \frac{PPW}{4}\right)$: $\frac{70}{2} - \frac{28.0}{4} = 28.0$ in.

Springs has a desert (BW) climate; estimated water need is not only more than the precipitation, it is more than twice as great.

So much for the dry climates. So far, in the application of our classification scheme, we have determined whether or not a station has a polar (E) climate or a dry (B) climate and, if the station is classified in either of these general categories, which subdivisions it falls into. Obviously, if the station is neither polar nor dry, it must fall into one of the remaining three major categories (A, C, or D), which are the climates that are neither too cold nor too dry for tree growth. What are the subdivisions of these major groups?

The tropical (A) climates have been defined as those without a true winter season (no mean monthly temperature below 64°F or 18°C). To be exact, we should say that the tropical (A) climates are those which are not dry (B) and which have no true winter. This is the reason for disposing, first, of the possibility that the water need exceeds the precipitation. The tropical (A) climates are subdivided according to their precipitation regimes. These divisions are shown in Table 8-4 below.

The next three examples show the procedure involved. You may either take it for granted that none of the three has a B climate, or you may estimate their respective water need to make sure.

In the first example, it is obvious that the station is tropical (A), since the temperature of the coldest month is well above 64°F (18°C). It is also obvious that it has a tropical rainforest (Af) climate, since the driest month has more than 2.4 inches (60 mm) of precipitation.

In the second example, it is again obvious that the station is tropical (A), but there is a marked dry season — the driest month, in fact, has only 0.2 inches (5 mm) of rainfall. It does not, therefore, have a tropical rainforest (Af) climate. Is the total annual precipitation sufficient for classification as a monsoon (Am) climate

Table 8-4. The Tropical Climates

Climatic symbol	Climatic category	Definition based on vegetation	Climatic definition	Quantitative limit
Af	Tropical rainforest	Rain and temperature adequate for growth of broadleaved evergreen tropical rainforest	No winter; no dry season	No month colder than 64°F (18°C); no month with less than 2.4 inches (60 mm) of rain
Am	Monsoon	Same as above	No winter; marked dry season, but compensated by high annual rainfall	No month colder than 64°F (18°C); total rainfall more than 100 inches (2,500 mm) less 25 times the rainfall of the driest month $P_{Total} > 100 - 25 P_{Driest}$ ($P_{Total} > 2{,}500 - 25 p_{Driest}$)
Aw, As	Tropical savanna	Climates conducive to growth of semideciduous forest	No winter; marked dry season with only moderate annual rainfall	No month colder than 64°F (18°C); does not qualify as either Af or Am Aw: winter dry savanna As: summer dry savanna

Kisangani (Stanleyville), Zaire
(0° 31' N, 25° 11' E; elevation 1,361 feet or 415 m)

	Temp.		Precip.	
	°F	°C	in.	mm
Jan.	77	25	3.3	84
Feb.	77	25	4.0	102
Mar.	77	25	6.9	175
Apr.	77	25	5.6	142
May	77	25	6.1	155
June	76	24	3.5	89
July	74	23	4.4	112
Aug.	74	23	8.9	226
Sept.	75	24	7.5	191
Oct.	76	24	9.7	246
Nov.	76	24	6.7	170
Dec.	76	24	2.8	71
Year	76	24	69.4	1,763

Rangoon, Burma
(16° 46' N, 96° 09' E; elevation 12 feet or 4 m)

	Temp.		Precip.	
	°F	°C	in.	mm
Jan.	77	25	0.2	5
Feb.	79	26	0.2	5
Mar.	84	29	0.3	8
Apr.	87	31	1.4	36
May	84	29	12.1	307
June	81	27	18.4	467
July	80	27	21.5	546
Aug.	80	27	19.7	500
Sept.	81	27	15.4	391
Oct.	82	28	7.3	185
Nov.	80	27	2.8	71
Dec.	77	25	0.3	8
Year	81	27	99.6	2,529

($P_{total} > 100 - 25 P_{driest}$)? or, in metric units, is $P_{total} > 2,500 - 25 P_{driest}$? $100 - (25 \times 0.2) = 95$ inches, or $2,500 - (25 \times 5) = 2,375$ mm. The average annual total, 99.6 inches (2,529 mm) exceeds this limiting value. Hence, Rangoon has an *Am* climate.

In the third example, the station is once again tropical (*A*), and there is a marked dry season; therefore, it does not qualify as tropical rainforest (*Af*). In fact, the driest month has no precipitation. The limiting value for the total precipitation is thus $100 - (25 \times 0)$, or simply 100 inches (2,500 mm). The annual total of 57.3 inches (1,453 mm) is not greater than 100 inches (2,500 mm), so Taxco does not qualify as having an *Am* climate. It is, therefore, an *Aw* station (*Aw* rather than *As* because the dry season is in winter rather than summer).

Those places whose climates do not fit any of the above categories must have either mesothermal (*C*) or microthermal (*D*) climates; that is, climates with either a moderate winter (coldest month between 27° F and 64° or between -3° C and 18° C) or a cold winter (coldest month below 27° F or -3° C). These two major groups are subdivided in accordance with two criteria: (1) the seasonal regime of their precipitation and (2) the warmth of their summers. (See Table 8-5).

Within each of the two climatic groups *C* and *D*, there are a number of possible subcategories, such as *Cfa, Cwb, Csa, Dfb, Dwd*.[3] For most purposes these possibilities are combined into a few categories as shown below.

Taxco, Mexico
(18° 33' N, 99° 36' W; elevation 5,756 feet or 1,754 m)

	Temp.		Precip.	
	°F	°C	in.	mm
Jan.	67	19	0.0	1
Feb.	69	21	0.2	5
Mar.	72	22	0.4	9
Apr.	75	24	0.9	23
May	76	24	3.0	77
June	72	22	10.6	269
July	70	21	11.9	301
Aug.	69	21	13.5	342
Sept.	69	21	13.0	329
Oct.	69	21	3.5	90
Nov.	68	20	0.2	5
Dec.	67	19	0.1	3
Year	70	21	57.3	1,454

Climatic symbol*	Climatic category
Cfa, Cwa	Humid mesothermal
Cfb, Cfc	Mid-latitude marine
Csa, Csb, Dsa, Dsb	Mediterranean
Dfa, Dfb, Dwa, Dwb	Humid continental
Dfc, Dwc, Dwd	Subarctic

*The combinations Csc, Cwc, Dfd, Dsc, Dsd are possible but rare enough to be excluded. Cwb is common in tropical highlands; it is excluded from the discussion and from the world map for the sake of simplicity.

---

[3] The combinations *Cfd, Cwd,* and *Csd* are logically impossible, since the definitions of *C* (coldest month between 27° F and 64° F) and of *d* (coldest month below −33° F) are incompatible.

Table 8-5. Subdivisions of the Mesothermal (C) and Microthermal (D) Climates

Climatic symbol	Climatic definition	Quantitative limit
f	No marked seasonality of precipitation	Six coldest months have between 30 percent and 70 percent of total annual rainfall
w	Dry season in winter	Six coldest months have less than 30 percent of total annual rainfall
s	Dry season in summer	Six coldest months have more than 70 percent of total annual rainfall
a	Hot summers	Warmest month warmer than 72°F (22°C)
b	Long, warm summers	Warmest month between 50°F (10°C) and 72°F (22°C), but at least four months warmer than 50°F (10°C)
c	Short, warm summers	Warmest month between 50°F (10°C) and 72°F (22°C); fewer than four months warmer than 50°F (10°C)
d	Very cold winters	Coldest winter month colder than –33°F (–36°C)

Other limits are also used in the Köppen system: a ratio of 3:1 between the wettest winter month and the driest summer month for the boundary between s and f climates; and a ratio of 10:1 between the wettest summer month and the driest winter month for the boundary between w and f climates. The quantitative limits based on percentages are preferable for several reasons: They use the same kind of data used to determine the boundary between dry and humid climates; and they do not lay too much stress on individual months, as do the ratios indicated above.

An example of each of the five categories is shown below. Classify each station by first noting whether it has summer drought, winter drought, or no seasonality of precipitation; then by deciding whether it has hot summers, long warm summers, short warm summers, or very cold winters. Fit each station into its proper group.

Urbana, Illinois
(40° 06′ N, 88° 14′ W; elevation 743 feet or 226 m)

	Temp.		Precip.	
	°F	°C	in.	mm
Jan.	27	–3	2.2	56
Feb.	30	–1	1.9	48
Mar.	40	4	3.2	81
Apr.	51	11	3.7	94
May	62	17	4.1	104
June	71	22	4.0	102
July	76	24	3.3	84
Aug.	74	23	3.3	84
Sept.	65	18	3.2	81
Oct.	55	13	2.8	71
Nov.	41	5	2.4	61
Dec.	30	–1	2.1	53
Year	52	11	36.2	919

Paris, France
(48° 49′ N, 2° 28′ E; elevation 164 feet or 50 m)

	Temp.		Precip.	
	°F	°C	in.	mm
Jan.	38	3	2.1	53
Feb.	40	4	1.7	43
Mar.	45	7	1.5	38
Apr.	51	11	2.0	51
May	57	14	2.2	56
June	63	17	2.0	51
July	67	19	2.3	58
Aug.	66	19	2.4	61
Sept.	61	16	1.9	48
Oct.	53	12	2.2	56
Nov.	44	7	2.2	56
Dec.	38	3	2.1	53
Year	52	11	24.6	624

132   Chapter 8

Eugene, Oregon
(44° 07′ N, 123° 15′ W; elevation 359 feet or 109 m)

	Temp.		Precip.	
	°F	°C	in.	mm
Jan.	39	4	6.3	160
Feb.	43	6	5.0	127
Mar.	46	8	4.3	109
Apr.	51	11	2.4	61
May	56	13	2.1	53
June	61	16	1.4	36
July	67	19	0.3	8
Aug.	66	19	0.4	10
Sept.	61	16	1.3	33
Oct.	53	12	3.8	97
Nov.	46	8	5.6	142
Dec.	42	6	6.6	168
Year	52	11	39.5	1,004

Vladivostok, U.S.S.R.
(43° 07′ N, 131° 55′ E; elevation 95 feet or 29 m)

	Temp.		Precip.	
	°F	°C	in.	mm
Jan.	7	−14	0.2	5
Feb.	14	−10	0.3	8
Mar.	27	−3	0.4	10
Apr.	40	4	1.3	33
May	49	9	2.1	53
June	56	13	2.6	66
July	65	18	3.0	76
Aug.	69	21	4.8	122
Sept.	62	17	4.0	102
Oct.	49	9	1.6	41
Nov.	31	−1	0.6	15
Dec.	15	−9	0.4	10
Year	40	5	21.3	541

Archangel, U.S.S.R.
(64° 35′ N, 40° 36′ E; elevation 20 feet or 6 m)

	Temp.		Precip.	
	°F	°C	in.	mm
Jan.	8	−13	0.8	20
Feb.	10	−12	0.8	20
Mar.	17	−8	0.9	23
Apr.	30	−1	0.8	20
May	41	5	1.3	33
June	53	12	1.8	46
July	60	16	2.8	71
Aug.	55	13	2.6	66
Sept.	46	8	2.2	56
Oct.	34	1	2.0	51
Nov.	23	−5	1.3	33
Dec.	12	−11	0.9	23
Year	32	0	18.2	462

Summary

In this chapter we have mentioned twelve climatic categories, as follows:

Climatic symbol	Climatic category
ET	Tundra
EF	Frost
BS	Steppe
BW	Desert
Af	Tropical rainforest
Am	Monsoon
Aw or As	Tropical savanna
Cfa or Cwa	Humid mesothermal
Cfb or Cfc	Mid-latitude marine
Csa, Csb, Dsa, or Dsb	Mediterranean
Dfa, Dfb, Dwa, or Dwb	Humid continental
Dfc, Dwc, or Dwd	Subarctic

The steps needed to categorize any place on the earth's surface are the same as those used to outline the system in this chapter. In other words, one determines, by using the appropriate quantitative criteria, whether:

(1) the station has a polar climate. If so, whether, it is tundra or frost. If not whether

(2) the station has a dry climate. If so, whether it is teppe or desert. If not, whether

(3) the station has a tropical climate. If so, whether it is rainforest, monsoon, or savanna. If not, whether

(4) the station has a mesothermal climate. If so, whether it is humid mesothermal, mid-latitude marine, or mediterranean. If not, whether

(5) the station has a microthermal climate, and one determines whether it is humid continental or subarctic.

This procedure is illustrated below, using data for Carrollton, Missouri.

(1) Is the station polar? Is the warmest month colder than 50° F (10° C)? Obviously not. Therefore, we proceed to the second question.

Carrollton, Missouri
(39° 22′ N, 93° 30′ W; elevation 809 feet or 247 m)

	Temp.		Precip.	
	°F	°C	in.	mm
Jan.	28	-2	1.5	37
Feb.	31	-1	2.0	51
Mar.	40	4	3.0	75
Apr.	55	13	3.3	84
May	64	18	4.8	122
June	74	23	4.5	114
July	78	25	4.8	122
Aug.	76	24	4.6	117
Sept.	69	20	3.3	85
Oct.	57	15	2.8	72
Nov.	43	6	1.7	42
Dec.	33	1	1.8	46
Year	54	12	38.1	967

(2) *Is the station dry? Does the water need exceed the precipitation?* We estimate the potential evapotranspiration from $E = (T/2) - (PPW/4)$ ($e = 20t + 490 - 7PPW$). T, the average annual temperature, is 54°F (12° C); PPW, the percentage of the precipitation that falls in the six winter months, is 33.6. E is therefore 18.6 inches (495 mm using the metric formula), which is less than the precipitation. The station is not dry.

(3) *Is the station tropical? Is the coldest month warmer than 64° F (18° C)?* Obviously not. We therefore proceed to the next question.

(4) *Is the station mesothermal? Is the coldest month between 27° F and 64° F (between -3° and 18° C)?* The answer is "yes." Therefore, the station is mesothermal. Since the percentage of precipitation that falls in winter is 33.6 percent, it has no marked seasonality of precipitation (f); and since the warmest month is warmer than 72° F (22° C), it fits into category a.

The station is thus classified as having a *Cfa* climate, and it fits into the humid mesothermal group. The classification of any other station follows the same steps.

This scheme of classification is not an end in itself but merely a useful way of grouping certain climatic characteristics, so that one can talk about them conveniently and, above all, so that one can map them in a way that is convenient for both the mapper and the map reader. The map (front end-paper) shows the resulting pattern, which is discussed in the next chapter.

Suggestions for Further Reading

Miller, A. Austin, *Climatology*, 8th ed., Methuen Co., Ltd., London, 1955, pp. 78-99.

Thornthwaite, C. W., "An Approach Toward a Rational Classification of Climate," *Geographical Review*, Vol. 38, No. 1 (1948), 55-94.

9 World Distribution of Climates

Introduction

The reasons for the occurrence and distribution of the individual climatic elements have been discussed in the earlier chapters. Some attention was given to combinations of climatic elements in the sections dealing with air masses; and the description of the distribution of precipitation over the world dealt with the interplay of a number of climatic phenomena. So far, however, we have not dealt with the interaction of all these elements at particular places. For obvious reasons, we can discuss neither every element nor every place; we must generalize therefore, on the basis of the classification described in Chapter 8.

The Köppen classification makes possible identification of twelve major climatic categories in terms of their average monthly and annual temperature and precipitation characteristics. In this chapter we shall describe, for each of the twelve categories, the distribution of the climatic type over the surface of the earth, the atmospheric processes that are dominant, and the climatic characteristics that result. Associated soil and vegetation will be mentioned briefly; fuller descriptions of these are found in later chapters. We begin with the treeless climates — first the polar then the dry climates; then we shall deal with the tropical, mesothermal, and microthermal climates in that order, which is roughly from the equator toward higher latitudes. Examples from each category are included, consisting of data tables followed by graphic representations. Each of these climatic stations is located on Fig. 9-1.

Tundra Climates

The tundra (ET) climates are found in high latitudes, over the arctic air-mass source region. They occupy the Arctic Ocean Basin and its coastal fringes. Although they prevail over all the Bering Sea and extend to 50°N latitude off the Labrador coast, their southern limit on the continents lies north of the Arctic Circle. In the Southern Hemisphere, they are almost completely restricted to the Antarctic Ocean between 50°S and 66½°S latitudes; their only continental location are the southwestern tip of South America and the northern portion of the Palmer Peninsula, in Antarctica.

The boundary between arctic and polar source regions is sometimes referred to as the arctic front, but cyclonic storms are only weakly developed. The more vigorous cyclonic precipitation associated with the polar front does not reach these high latitudes, even in summer. Furthermore, neither steep lapse rates nor

World Distribution of Climates 135

Fig. 9–1. Location map of climatic stations representative of the twelve categories of the Köppen classification.

high absolute humidities are possible under polar temperature conditions. Temperatures are low, even in summer when the sun is above the horizon for 24 hours, because of the low angle of solar incidence and the depletion of solar energy in the long traverse of the atmosphere by the oblique rays of the sun. Summer warmth is further reduced by the oceanic character of tundra regions. Winters, on the other hand, are continental in the arctic tundra, since the ocean is covered with ice and snow and since its surface behaves like any other snow-covered surface in high latitudes. The Antarctic Ocean remains unfrozen in winter, although its winter temperature is modified by the air drainage from the Antarctic land mass and by the significant amounts of ice that float out from the continent; the antarctic tundra is, therefore, much warmer in winter than its boreal counterpart.

The climatic characteristics that result from this interaction of atmospheric processes over the particular kinds of surfaces found in high latitudes can be summarized by saying that the tundra is a land of uniformly low temperatures in summer; low to bitterly cold temperatures in winter; and low evaporation, low moisture content, and low precipitation all year long. The variety within tundra climates can be inferred from the following data (also shown graphically in Fig. 9-2).

Coppermine, Northwest Territories (ET)
(67° 49′ N, 115° 15′ W; elevation 30 feet or 9 m)

	Temp.		Precip.	
	°F	°C	in.	mm
Jan.	-20	-29	0.5	13
Feb.	-22	-30	0.3	8
Mar.	-14	-26	0.6	15
Apr.	1	-17	0.6	15
May	22	-6	0.5	13
June	38	3	0.8	20
July	49	9	1.4	36
Aug.	47	8	1.8	46
Sept.	36	2	1.2	30
Oct.	20	-7	1.1	28
Nov.	-4	-20	0.7	18
Dec.	-15	-26	0.4	10
Year	12	-11	9.9	252

Coppermine, on the shores of the Arctic Ocean in the Northwest Territories of Canada, exemplifies the continental aspect of the tundra; the winter temperatures reveal the continental character of the ocean in winter, and the precipitation regime is distinctly continental (a

136 Chapter 9

Angmagssalik, Greenland (ET)
(65° 36′ N, 37° 33′ W; elevation 97 feet or 30 m)

	Temp. °F	Temp. °C	Precip. in.	Precip. mm
Jan.	16	−9	2.9	74
Feb.	14	−10	2.4	61
Mar.	18	−8	2.6	66
Apr.	25	−4	2.1	53
May	35	2	2.0	51
June	42	6	1.8	46
July	45	7	1.5	38
Aug.	44	7	2.1	53
Sept.	38	3	3.3	84
Oct.	30	−1	4.7	119
Nov.	22	−6	3.0	76
Dec.	19	−7	2.7	69
Year	29	−2	31.1	790

Orcadas, South Atlantic Ocean (ET)
(60° 44′ S, 44° 44′ W; elevation 12 feet or 4 m)

	Temp. °F	Temp. °C	Precip. in.	Precip. mm
Jan.	32	0	1.4	36
Feb.	33	1	1.6	41
Mar.	32	0	1.9	48
Apr.	27	−3	1.6	41
May	20	−7	1.2	30
June	14	−10	1.0	25
July	14	−10	1.3	33
Aug.	15	−9	1.2	30
Sept.	21	−6	1.2	30
Oct.	26	−3	1.2	30
Nov.	28	−2	1.3	33
Dec.	31	−1	1.2	30
Year	24	−4	16.1	407

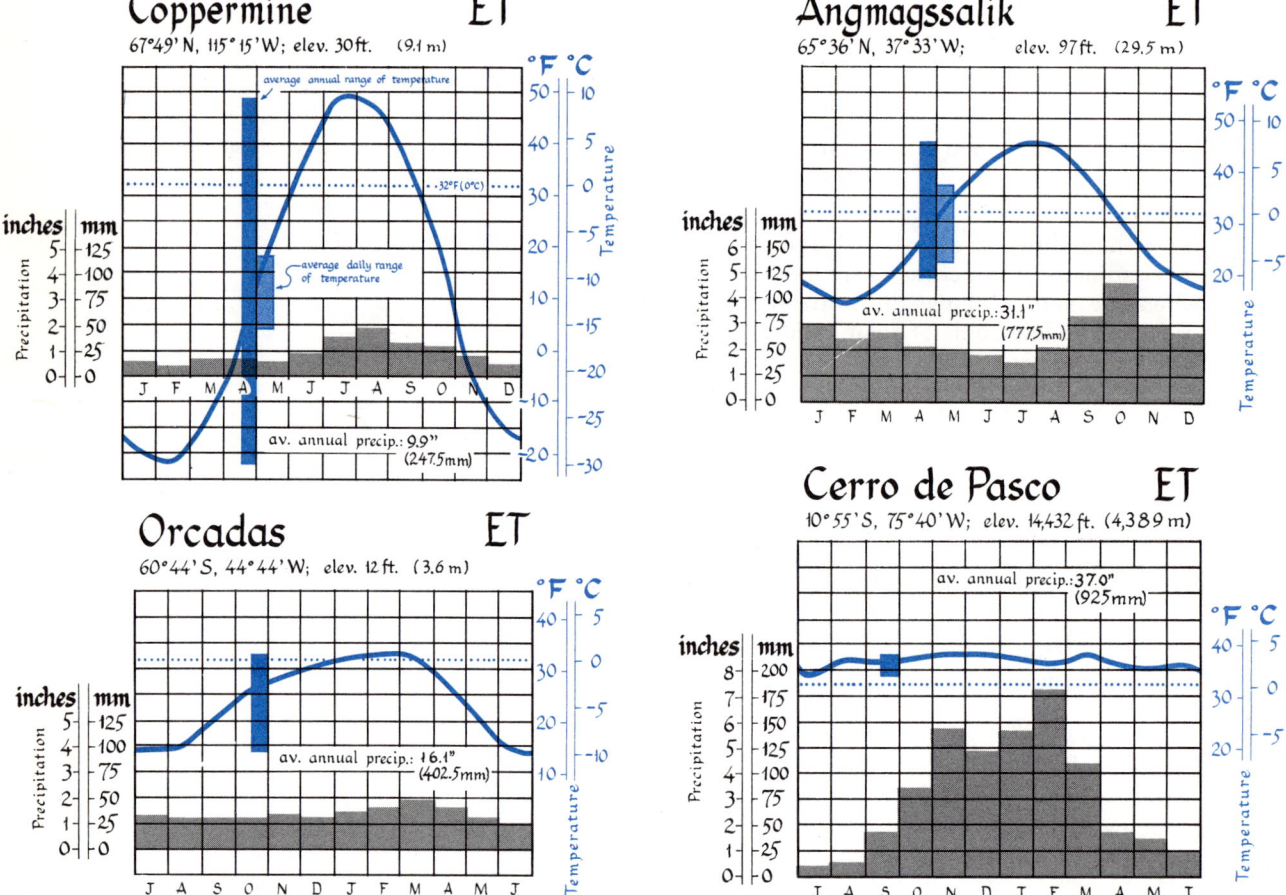

Fig. 9-2. Temperature and precipitation at selected *ET* stations. The dark blue vertical bar represents the annual range of temperature, the light blue bar the mean daily range (the average difference between the highest and lowest temperatures of each day of the year).

World Distribution of Climates 137

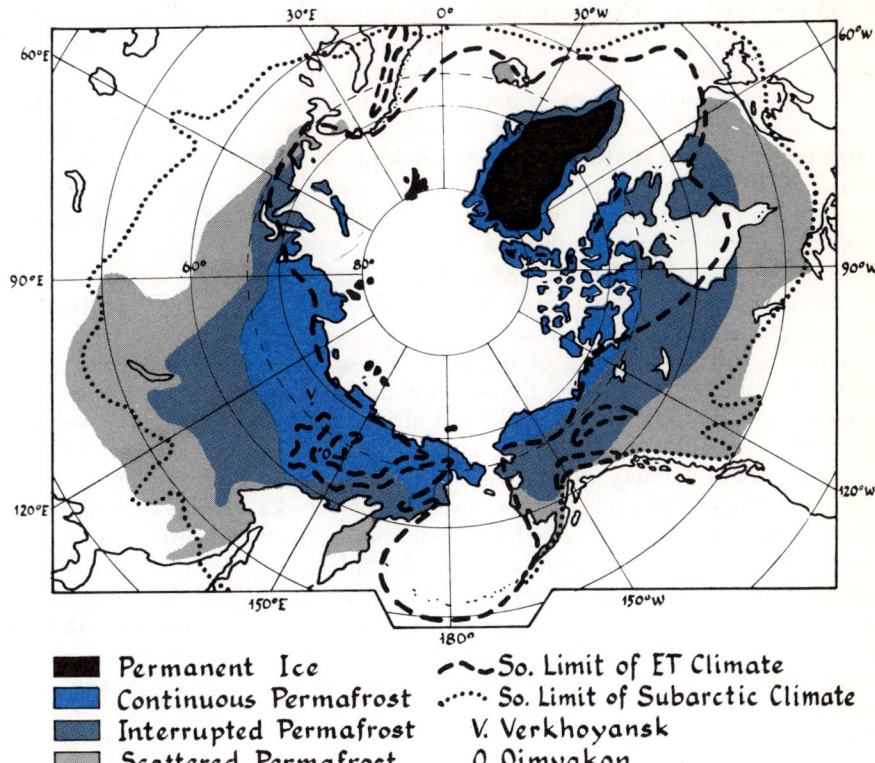

Fig. 9-3. The distribution of permafrost. Note the correspondence of the outer limits of permafrost with the southern boundary of subarctic climates.

low annual total, with a very modest maximum in summer).

Angmagssalik, on the eastern coast of Greenland, benefits from exposure to North Atlantic waters. The effect of the warmth of the West Wind Drift is particularly obvious in the relatively high winter temperatures and in the large amount of precipitation, whose source is warm *mP* air masses in the North Atlantic.

Orcadas, an Argentine naval station in the southernmost part of the Atlantic Ocean, reflects the fact that the neighboring Antarctic waters are cold, but not nearly as cold as the Arctic Ocean in winter; it also supplies more moisture than the Arctic Ocean, but considerably less than the North Atlantic.

The differences among the three stations reflect differences in winter temperatures and precipitation; summer conditions are much more similar. This fact is reflected in the annual ranges of temperature, which are 71° F (39° C) at Coppermine, 31° F (17° C) at Angmagssalik, and only 19° F (11° C) at Orcadas. Since almost all the tundra lies in high latitudes, seasonal differences in temperature are intensified by differences in the amount of daylight. The long winter nights are in marked contrast to the long summer days, and the transition between the two is rather abrupt in spring and fall. This feature of the climate has the additional result of reducing daily ranges of temperature to very low values, seldom larger than 4 or 5° F (2 or 3° C). The mid-latitude contrast between nighttime and daytime temperatures is replaced by a greater seasonal contrast and a much-reduced diurnal range.

Tundra climates are largely found over water. Where they do occur on land, they are associated with characteristic soils and natural vegetation, both of which bear the name "tundra." Tundra vegetation is an association of lichens, mosses, sedges, and dwarf birches and conifers on its southernmost fringes. Tundra soils result more from physical than from chemical processes. The air is not warm enough to melt the water in the subsoil, which remains permanently frozen. This condition, known as **permafrost,** not only reduces chemical reactions, but more importantly, it creates an impermeable layer at a depth that varies with summer temperatures. Tundra soils are, therefore, very poorly drained and may be quite boggy in summer. The distribution of permafrost is shown in Fig. 9-3. The zone of continuous permafrost corresponds to some degree to the distribution of *ET* climates over land, but extends into the subarctic climates. The frozen layer, perhaps representative of more severe climatic conditions in the Pleistocene, is unbroken and may extend to depths of

1,500 feet (460 m). It presents some curious problems for human occupancy, since buildings must be insulated not only against the air but also against the soil — loss of heat to the subsoil is bad enough, but structures have been known to melt the permafrost with disastrous consequences. Road building presents similar difficulties. Interrupted permafrost is a zone of nearly continuous frozen ground with frost-free patches under lakes and rivers. Southward, frost-free ground becomes the rule, but there are disconnected patches of permafrost. These last two zones fit the distribution of subarctic climates reasonably well.

There is another location of tundra climates on the earth. As one moves toward the equator the summer coolness of high latitudes is duplicated at increasingly high altitude. This high tundra is difficult to distinguish from the lowland polar variety when it occurs in high-latitude mountain ranges, such as the Alaska Range; even the frost climates have their counterparts over the mountain glaciers. In lower latitudes, the warmest month may still average below 50° F (10° C) if one goes high enough (about 12,000 feet or about 3,600 m at the equator), but winter temperatures become milder and less distinct from summer as one approaches the tropics. As the seasonal differences decrease, diurnal contrasts increase; the 6-month rhythm of the polar tundra is replaced by the 12-hour rhythm of equatorial latitudes. The consequent differences in amount and frequency of light and heat have profound significance for equatorial plant life and soil formation, which have only superficial similarities to the tundra vegetation and soils of high latitude.

Highland tundra climates are found in large, discontinuous patches in the Rocky Mountains, in a fairly continuous belt along the Andes, in a very large chunk over the Tibetan Plateau, and in smaller fragments over other high mountains.

Cerro de Pasco, in the Peruvian Andes, illustrates a tropical tundra climate (Fig. 9-2). The relatively high precipitation reflects its windward location. The Tibetan tundra is much drier, as befits its trans-Himalayan position with respect to the Indian Ocean.

Frost Climates

Even polar temperatures are sufficiently high to average one or more months above 32°F (0°C), at least near sea level and over ground that is free of snow or ice. Frost (EF) climates are, therefore, found almost entirely on the ice caps of Greenland and Antarctica. Elevation is obviously a factor in producing very low temperatures: Greenland rises to 10,000 feet (3,000 m), and much of Antarctica is higher. But the principal agent responsible, particularly in summer, is undoubtedly the permanent surface of fresh ice. As we have seen, ice has a very high albedo and may reflect up to 80 percent of the incoming solar radiation; much of the heat contained in overlying air masses is used to melt or evaporate the ice and is thus unavailable for raising the temperature of the atmosphere. Furthermore, since the ice cover is a poor conductor of heat, particularly when it is thousands of feet thick, no heat can be drawn from the ground beneath. For all these reasons, the already low temperatures of polar latitudes at high elevations are further reduced.

The ice caps have no rivals as far as average annual temperatures are concerned (-57°F or -49°C at the South Pole, -23°F or -31°C at Eismitte on the Greenland Ice Cap), nor in terms of summer cold (-20°F or -29°C in December at the South Pole, +10°F or -12°C in July at Eismitte). But even though the winter months are incredibly cold (-76°F or -60°C in July at the South Pole, -53°F or -47°C in February at Eismitte), there are lowland stations in the northeastern part of Siberia that are colder than Greenland — and near enough to the fantastic temperatures of Antarctica to make little difference to a reasonable person. Oimyakon has an average January temperature of -60°F (-51°C), Verkhoyansk of -58°F (-50°C)(see Fig. 9-3 for the locations of these two stations). Indeed, Verkhoyansk once held the record for the lowest temperature ever observed at the surface of the earth: -89.9°F (-67.7°C) on February 5 and 7, 1892.[1] This record has been surpassed by several Antarctic stations:

Cerro de Pasco, Peru (ET)
(10° 45′ S, 77° 16′ W; elevation 14,432 feet or 4,399 m)

	Temp.		Precip.	
	°F	°C	in.	mm
Jan.	37	3	5.6	142
Feb.	36	2	7.2	183
Mar.	38	3	4.4	112
Apr.	36	2	1.7	43
May	35	2	1.5	38
June	36	2	1.0	25
July	34	1	0.4	10
Aug.	37	3	0.5	13
Sept.	36	2	1.7	43
Oct.	37	3	3.4	86
Nov.	38	3	4.7	119
Dec.	38	3	4.9	124
Year	36	2	37.0	938

---

[1]The same temperature was recorded at Oimyakon on February 6, 1933

−102.1°F (−74.5°C) was observed at the South Pole on September 17, 1957, and −126.9°F (−88.3°C) on August 24, 1960, at Vostok, which is about 11,200 feet above sea level and near the pole of inaccessibility (78° 27′ S, 106° 52′ E). In any case, neither Verkhoyansk nor Oimyakon represents polar climates, for both have distinct summer seasons with temperatures well above 50°F (10°C) in July; they are part of the subarctic climatic zone *(Dwd)*, which is discussed later on. Monthly means of temperature and precipitation for Eismitte, Vostok (probably the coldest place occupied by man), and Jungfraujoch follow and are shown in Fig. 9-4; data for Oimyakon are shown for comparison.

All the means are for very short periods; the data for Eismitte represent one year's record during the period 1930–1931; those for Vostok are for the period 1961–1963. Even so, they show the extremely low temperatures at all seasons and an annual range of temperature that is high but not as high as in the Northern Hemisphere tundra or in continental subarctic climates. The data also indicate the low precipitation, especially over Antarctica; there is some precipitation in almost every month of record, but the average monthly totals are never over 0.3 inches (8 mm). The tables and graphs do not, of course, show the low rate of evaporation and melting or the tremendous contrast between the long polar night and day. Another noteworthy feature of ice cap climates is the occurrence

Eismitte, Greenland (*EF*)
(70° 54′ N, 40° 42′ W; elevation 9,940 feet or 3,029 m)

	Temp.		Precip.	
	°F	°C	in.	mm
Jan.	−43	−42	0.6	15
Feb.	−53	−47	0.2	5
Mar.	−40	−40	0.3	8
Apr.	−25	−32	0.2	5
May	−6	−21	0.1	3
June	2	−17	0.1	3
July	10	−12	0.1	3
Aug.	−1	−18	0.4	10
Sept.	−8	−22	0.3	8
Oct.	−32	−36	0.5	13
Nov.	−45	−43	0.5	13
Dec.	−37	−38	1.0	25
Year	−23	−31	4.3	111

Oimyakon, U.S.S.R. (*Dwd*)
(63° 28′ N, 142° 49′ E; elevation 2,381 feet or 726 m)

	Temp.		Precip.	
	°F	°C	in.	mm
Jan.	−60	−51	0.3	7
Feb.	−54	−48	0.2	6
Mar.	−24	−31	0.2	5
Apr.	1	−17	0.2	4
May	35	2	0.4	10
June	56	13	1.3	32
July	62	17	1.6	40
Aug.	53	12	1.5	37
Sept.	37	3	0.8	20
Oct.	4	−16	0.5	12
Nov.	−30	−34	0.4	11
Dec.	−53	−47	0.3	9
Year	2	−17	7.7	193

Vostok, Antarctica (*EF*)
(78° 27′ S, 106° 52′ E; elevation 11,200 feet or 3,400 m)

	Temp.		Precip.	
	°F	°C	in.	mm
Jan.	−29	−34	Unavailable	
Feb.	−48	−44		
Mar.	−74	−59		
Apr.	−88	−67		
May	−87	−66		
June	−88	−67		
July	−88	−67		
Aug.	−83	−64		
Sept.	−85	−65		
Oct.	−72	−58		
Nov.	−52	−47		
Dec.	−28	−33		
Year	−68	−56	1.9	48

Jungfraujoch, Switzerland (*EF*)
(46° 33′ N, 7° 59′ E; elevation 11,736 feet or 3,577 m)

	Temp.		Precip.	
	°F	°C	in.	mm
Jan.	6	−14	Unavailable	
Feb.	7	−14		
Mar.	10	−12		
Apr.	16	−9		
May	21	−6		
June	27	−3		
July	30	−1		
Aug.	31	−1		
Sept.	28	−2		
Oct.	21	−6		
Nov.	13	−11		
Dec.	8	−13		
Year	18	−8		

140 Chapter 9

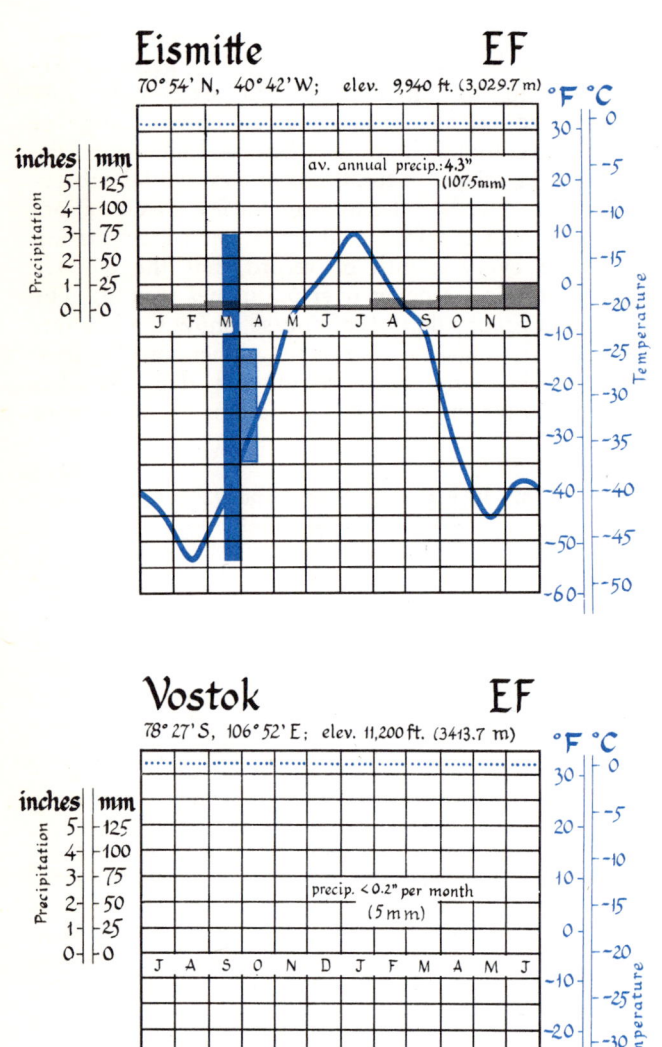

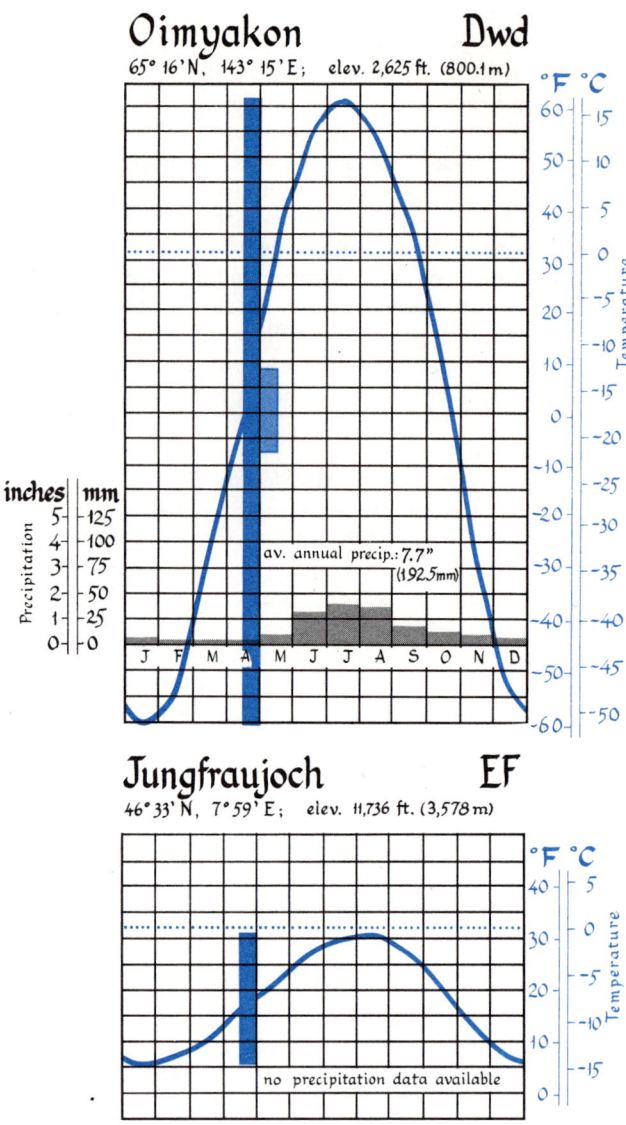

Fig. 9-4. Temperature and precipitation at selected *EF* stations. Vostok, on the antarctic plateau, is the coldest place on the surface of the earth for which temperatures are recorded. Eismitte is in a corresponding location on the Greenland Ice Cap, but is much warmer and somewhat wetter than Vostok. Jungfraujoch is a middle-latitude *ET* station at high altitude. Oimyakon is not *ET* (it has too much summer warmth), but is presented to show its very low winter temperatures.

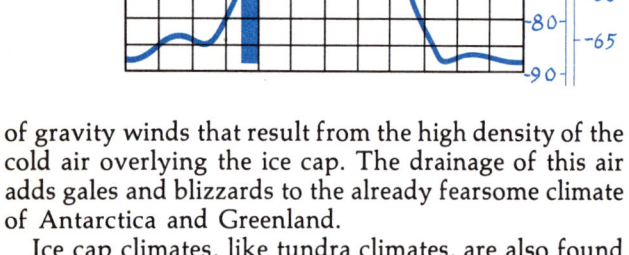

of gravity winds that result from the high density of the cold air overlying the ice cap. The drainage of this air adds gales and blizzards to the already fearsome climate of Antarctica and Greenland.

Ice cap climates, like tundra climates, are also found in lower latitudes, but at such high altitudes that their areal extent is negligible on a world scale. An example of a mid-latitude station with an *EF* climate is Jungfraujoch, in the Bernese Alps of Switzerland (Fig. 9-4).

The limits of such *EF* climates correspond fairly closely with the snow line, except in subtropical

World Distribution of Climates   141

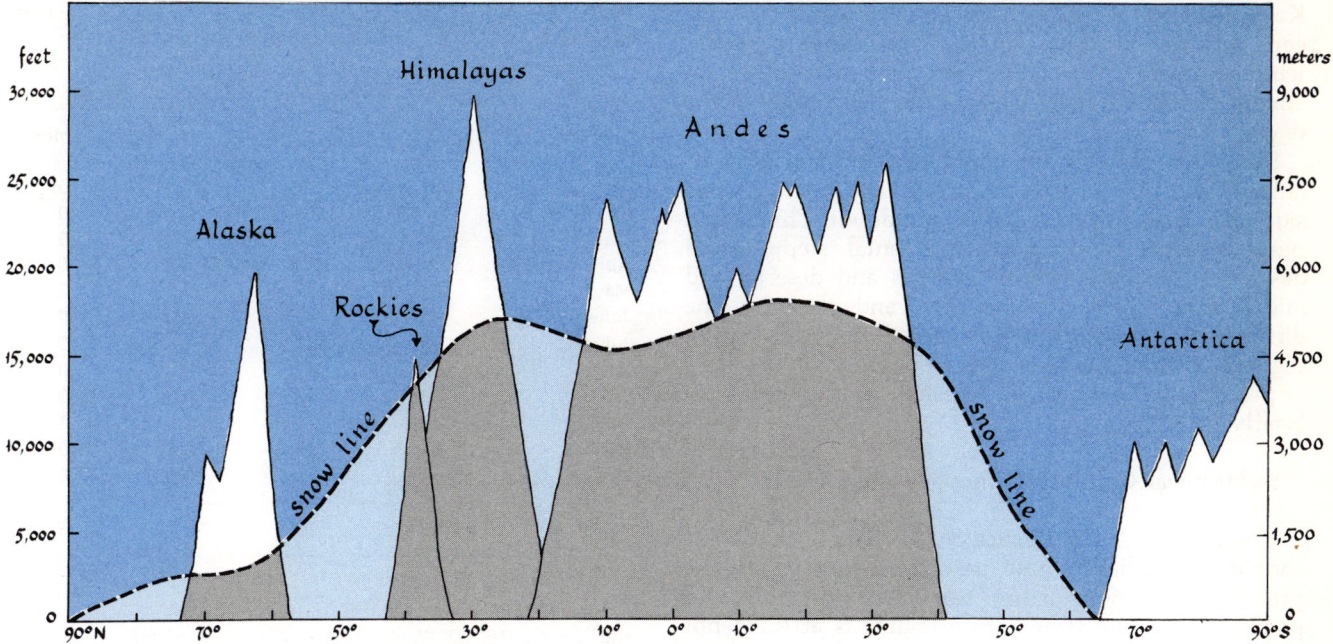

Fig. 9-5. The latitudinal variation in the elevation of the snow line. The elevation of the snow line decreases toward the poles, leaving much of Alaska and all of Antarctica above the limit of perpetual snow. The dip in the snow line, near the equator, results from higher precipitation, not lower temperatures. The various mountain ranges are shown schematically to indicate the approximate position of the snow line in various latitudes.

latitudes where the high elevation of the lower limit of perpetual snow reflects dearth of precipitation rather than excess of warmth (though subtropical summers may, in fact, be slightly warmer than the warmest months along the equator). A generalized relationship between latitude and height of the snow line is shown in Fig. 9-5.

Steppe and Desert Climates

The steppe *(BS)* and desert *(BW)* climates are considered together for convenience of description and understanding. The steppe climate is marginal to the desert climate, in both location and characteristics; it is, therefore, simplest to talk about core areas of drought and their climatic features and to treat the steppes as marginal and more humid variants of the deserts.

In Chapter 7, considerable attention was devoted to the causes and distribution of areas of low precipitation. Three types of locations with deficient precipitation were identified: a subtropical zone associated with the stable lapse rates of the subtropical high pressure belt, a mid-latitude zone associated with continental interiors, and a polar zone.

The definition of a desert climate, however, involves not only precipitation but also **water need. Potential evapotranspiration** (which determines water need) is very low in the polar regions; even small amounts of precipitation result in a surplus of snow or water, as evidenced by the continuing existence of the Antarctic Ice Cap despite loss of ice to the ocean along its edges and by the waterlogged soils of the continental tundra in summer. In subtropical and middle latitudes, however, 10 inches (250 mm) of precipitation is not enough to supply the water need, except where the average annual temperature drops below 40°F or 4°C (either because of altitude or latitude) or where the precipitation is unusually concentrated in winter.

The core areas of desert and steppe climates are thus the same as the centers of low precipitation in subtropical and mid-latitudes. The margins of the drylands do not, of course, correspond with the 10-inch (250-mm) isohyet, since temperature and hence water need vary considerably over the latitudinal range and diversity of surface involved. Nevertheless, the isohyet of 10 inches (250 mm) is roughly concordant with the outer limit of subtropical deserts and, because of the reduced water need, with the outer edge of the mid-latitude steppes. This correspondence in area is close enough that we need not repeat the discussion of causes and distribution of low precipitation given in Chap. 7.

Keep in mind, of course, that we have added the concept of water need, so that we must consider temperature and seasonal concentration of precipitation among the factors that influence the distribution of the dry climates.

Since dry climates are found over a wide range of latitude and over oceans as well as land, there is considerable variety of both steppe and desert. Three main types emerge: subtropical continental steppes and deserts, subtropical oceanic steppes and deserts, and mid-latitude continental steppes and deserts. The differences among them are based on temperature regimes; further subdivisions can be made on the basis of precipitation regimes, but we shall refer to these only briefly.

Subtropical Continental Steppes and Deserts

The general characteristics of this part of the world should be familiar by now, particularly in terms of the previous discussions of heat balance, temperature, and precipitation. The temperature regime is affected not only by the latitudinal position and continental character of the surface but also by the dryness of the air and land. The lack of moisture, as we have seen, means that very little heat is used in evaporation; radiational surpluses must be disposed of by heating the ground and the air. Furthermore, dry air allows for a maximum impact of solar energy; few clouds reflect incoming radiation, and little water vapor in the air scatters or absorbs insolation. On the other hand, radiational cooling is also relatively unobstructed. Deserts and, to a lesser degree, steppes have an annual temperature range that is relatively high for their latitudes and have the highest *diurnal* ranges in the world.[2] They also have the highest summer temperatures, both average monthly and extreme.[3]

The data to the right and Fig. 9-6 illustrate the annual march of temperature and precipitation in these climates.

Aswan, in upper Egypt, and Greenland Ranch, in Death Valley, are representative of subtropical continental desert climates, although Death Valley is in sufficiently high latitude that it has an annual temperature range more characteristic of mid-latitudes

---

[2]Except possibly for high mountain climates, for which data are lacking.

[3]The highest temperature ever recorded was 136.4°F (58.0°C) on September 13, 1922, at Azizia, about 50 miles (80km) south of Tripoli in Libya. Greenland Ranch, in Death Valley, California, has the highest average July temperature. Very high average annual temperatures, however, are not limited to deserts. This dubious honor is shared by wetter places on the north coast of the African Horn. Massawa, for instance, has an average annual temperature of 85°F (29°C), Djibouti of 86°F (30°C).

Aswan, Egypt (BW)
(24° 02' N, 32° 53' E; elevation 364 feet or 111 m)

	Temp.		Precip.	
	°F	°C	in.	mm
Jan.	62	17	0.0	0
Feb.	65	18	0.0	0
Mar.	72	22	0.0	0
Apr.	81	27	0.0	0
May	88	31	0.1	2
June	92	33	0.0	0
July	92	33	0.0	0
Aug.	92	33	0.0	0
Sept.	89	32	0.0	0
Oct.	84	29	0.0	1
Nov.	74	23	0.0	0
Dec.	64	18	0.0	0
Year	80	26	0.1	3

Greenland Ranch, California (BW)
(36° 27' N, 116° 52' W; elevation -168 feet or -51 m)

	Temp.		Precip.	
	°F	°C	in.	mm
Jan.	52	11	0.3	6
Feb.	58	14	0.2	6
Mar.	67	19	0.2	4
Apr.	76	24	0.1	3
May	84	29	0.0	2
June	94	34	0.0	1
July	102	39	0.1	3
Aug.	99	37	0.1	3
Sept.	90	32	0.1	3
Oct.	76	24	0.1	2
Nov.	61	16	0.2	4
Dec.	53	12	0.2	5
Year	76	24	1.6	42

than of the subtropics. Niamey, in the Niger Republic, represents steppe climates on the equatorial side of the deserts. It has a more moderate temperature regime than Aswan, largely due to the fact that the summer is a period of cloudiness and rainfall, associated with the northward advance of equatorial convergence precipitation in summer. Note the effect of the cloudiness on temperature. Three relatively distinct seasons result: a cool, dry season in winter; a hot, dry season in spring; and a cooler, wet season in summer and early fall. This pattern is also typical of savanna and monsoon climates, except that the wet season is much more pronounced in the wetter *A* climates.

The subtropical continental dry climates are associated with desert soils (including **lithosols** and

Niamey, Niger Republic (BS)
(13° 31′ N, 2° 06′ E; elevation 709 feet or 216 m)

	Temp.		Precip.	
	°F	°C	in.	mm
Jan.	76	24	0.0	0
Feb.	80	27	0.0	0
Mar.	87	31	0.1	2
Apr.	92	33	0.3	7
May	93	34	1.5	37
June	89	32	3.1	80
July	84	29	5.3	135
Aug.	81	27	8.6	219
Sept.	83	28	3.7	93
Oct.	87	31	0.8	20
Nov.	82	28	0.0	1
Dec.	76	24	0.0	0
Year	84	29	23.4	594

Fig. 9-6. Temperature and precipitation at selected subtropical continental *BS* and *BW* stations. Aswan illustrates desert conditions in the Sahara. Niamey, at the southern edge of the Sahara, has a steppe climate with summer rain. Greenland Ranch illustrates the very high summer temperatures that can be reached in continental interior deserts. Note also the high diurnal ranges of temperature.

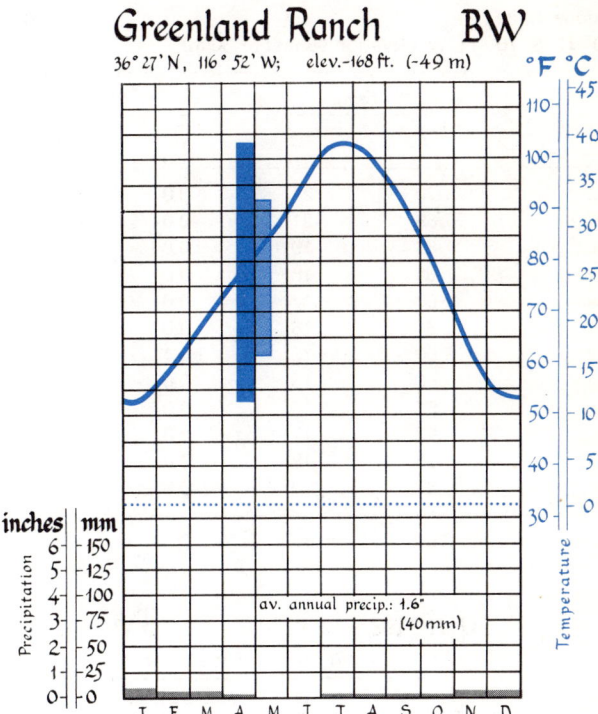

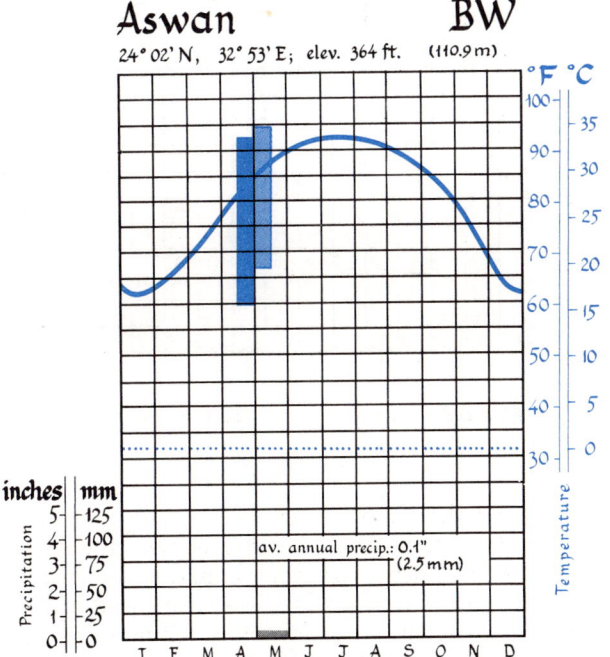

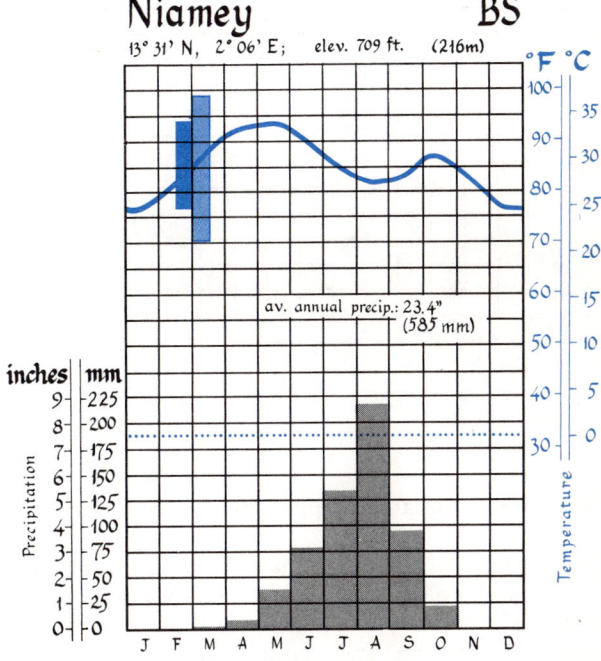

sandy soils) grading into **brown steppe soils** on the wetter margins, under *BS* climates. Vegetation varies from low patchy desert scrub to sparse tropical grasses in the steppe climates. There are scattered areas with no vegetation, but these are more closely associated with lithosols or sand than with low precipitation; complete absence of vegetation is the exception and not the rule in desert climates.

144 Chapter 9

Iquique, Chile (BW)
(20° 12' S, 70° 11' W; elevation 98 feet or 30 m)

	Temp.		Precip.	
	°F	°C	in.	mm
Jan.	70	21	0.0	0
Feb.	69	21	0.0	0
Mar.	67	19	0.0	0
Apr.	65	18	0.0	0
May	63	17	0.0	0
June	61	16	0.0	1
July	60	16	0.1	1
Aug.	60	16	0.0	0
Sept.	61	16	0.0	0
Oct.	63	17	0.0	0
Nov.	65	18	0.0	0
Dec.	68	20	0.0	0
Year	64	18	0.1	2

Agadir, Morocco (BS)
(30° 23' N, 9° 34' W; elevation 164 feet or 50 m)

	Temp.		Precip.	
	°F	°C	in.	mm
Jan.	57	14	1.5	39
Feb.	59	15	1.1	28
Mar.	62	17	0.9	24
Apr.	65	18	0.7	18
May	67	19	0.2	5
June	70	21	0.0	0
July	72	22	0.0	0
Aug.	73	23	0.1	2
Sept.	71	22	0.3	7
Oct.	69	21	0.8	20
Nov.	64	18	1.3	34
Dec.	58	14	1.6	41
Year	66	19	8.5	218

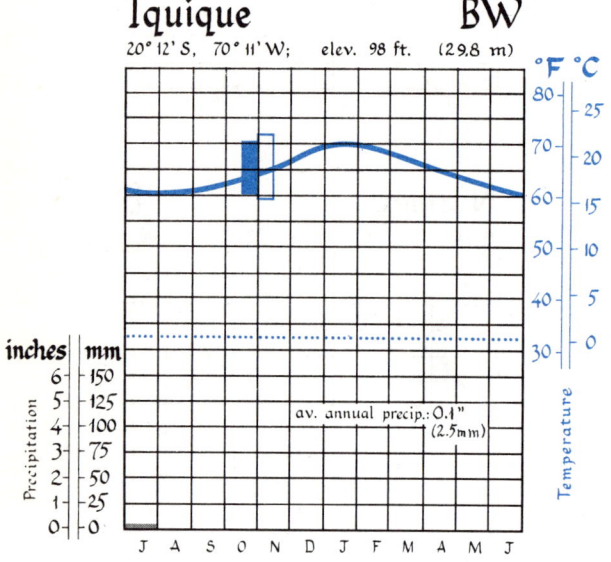
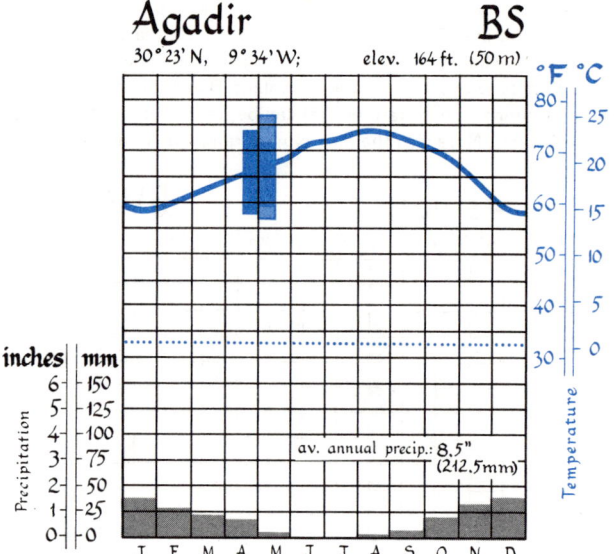

Fig. 9-7. Temperature and precipitation at selected subtropical oceanic BS and BW stations. Note the much smaller annual ranges of temperature than at Aswan or Niamey.

Subtropical Oceanic Steppes and Deserts

This variant of the dry climates is found along subtropical west coasts and extends westward over the oceans for considerable distances. The major difference between the oceanic and continental varieties of dry climates lies in their temperature regimes. The subtropical oceans though quite warm at all seasons, do have a moderating effect on the annual march of temperature. They also reduce summer temperatures by providing water for evaporation. Other peculiarities of this climatic type are introduced by the presence of cold currents immediately offshore. As we have seen, this feature adds to the stability of the air near the coast; despite the prevalence of humid marine air masses resulting from onshore winds, these west coasts are extremely dry. The air is humid, fog is common, but the ground remains dry.[4] These effects can be seen by com-

---

[4] Except where fog drips from vegetation that is nourished by the water thus extracted from the air. Such fogs are particularly common along the Peruvian coast, where the associated drizzle is known as *garúa*.

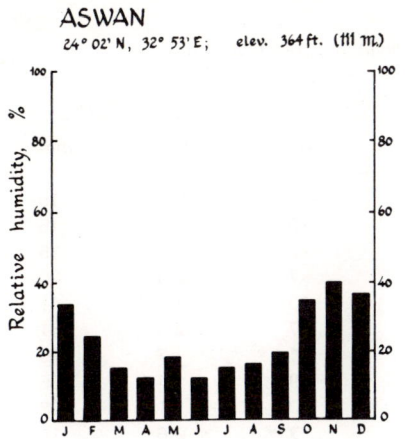
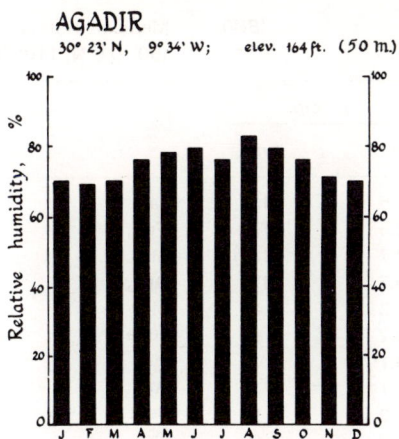
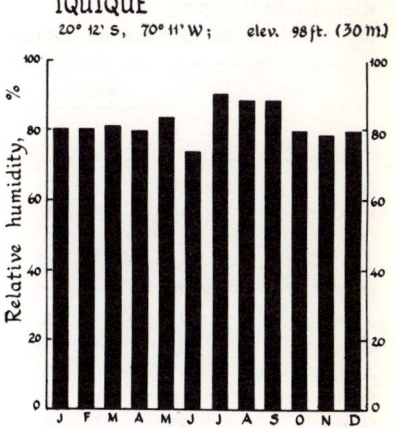

Fig. 9–8. Relative humidity at Aswan, Agadir, and Iquique. Aswan is clearly much drier than the two coastal stations.

paring the climatic data for Iquique and Agadir to that for Aswan (Figs. 9-6 and 9-7), and particularly by comparing relative humidities (Fig. 9-8).

Iquique represents the extraordinarily dry conditions of the Atacama Desert, which was discussed at some length in Chapter 7. Agadir is a coastal steppe station in southern Morocco. Note that the highest relative humidities occur from June to September, precisely when it rains the least. The stability of the subtropical high-pressure belt, accentuated by the stabilizing effects of the Canary Current, dominates the summer climate. Winter, on the other hand, is the rainy season, despite slightly lower relative humidity. Agadir in summer is under the influence of subtropical stability, which is replaced in winter by cyclonic precipitation associated with the polar front. A comparison of Agadir with Niamey (Fig. 9-6) reveals the basic differences between oceanic and continental subtropical steppe climates, as well as between steppes on the poleward side of the subtropical deserts and those on the equatorial margin. The poleward steppes not only are colder than the equatorward ones[5] but also have a distinctly different precipitation regime. Furthermore, the winter rains of Agadir, for example, are subject to much less evaporation than the summer rains of Niamey. Much less rain is needed to differentiate steppe from desert on the poleward side of the subtropical deserts than on the equatorward side.

Mid-latitude Steppes and Deserts

The locations and causes of these dry areas were discussed in Chapter 7. Since their major development is in about 40° latitude in the interior of continents, they do not occur in the Southern Hemisphere. Their distribution in North America is quite different from that in Asia. The predominant grain of the major mountain systems in North America is north-south; hence, the interior drylands are also broken into north-south fragments that merge without a break into the subtropical drylands to the south. Further, the desert portion of the mid-latitude drylands is very small. In Asia, on the other hand, the mountains run principally in an east-west direction. The drylands, although interrupted by mountains known as the Tien Shan, run principally from east to west and are clearly separated from the subtropical drylands of Iran and Western Pakistan by the Hindu Kush and other mountains. Furthermore, the desert area is very large.

Two stations illustrate this mid-latitude climate (Fig. 9-9). Krasnovodsk, just east of the Caspian Sea, reflects some of the expected characteristics of a mid-latitude continental interior. It has a fairly high annual range of temperature, with hot summers. The low precipitation is concentrated in winter and spring rather than in summer; eastward moving mP air masses bringing cyclonic precipitation can reach this far inland because of the east-west grain of the land and of the easy access along the Mediterranean. Bear in mind that Krasnovodsk is on the southwestern margin of the Old World mid-latitude drylands. To the east, the precipitation regime changes to one of winter drought and concentrated, though meager, summer rain. To the north, temperatures decrease. Ulan Bator, the capital of Mongolia, has an average January temperature of -24°F (-31°C). Even though Ulan Bator is 4,300 feet (1,300 m) above sea level, the gradient of winter temperature between it and Krasnovodsk is most impressive.

Medicine Hat, in Canada's Alberta Province, represents conditions at the northern limits of the mid-latitude drylands in North America. It is colder than Krasnovodsk and has a higher annual range of

---

[5] In summer, however, continental steppes north of the Sahara are just as hot as those south of the desert.

Krasnovodsk, U.S.S.R. (BW)
(40° 00' N, 52° 59' E; elevation −68 feet or −21 m)

	Temp. °F	Temp. °C	Precip. in.	Precip. mm
Jan.	36	2	0.5	13
Feb.	40	4	0.5	13
Mar.	47	8	0.7	18
Apr.	56	13	0.7	18
May	69	21	0.3	8
June	77	25	0.2	5
July	83	28	0.1	3
Aug.	83	28	0.1	3
Sept.	75	24	0.2	5
Oct.	63	17	0.3	8
Nov.	51	11	0.4	10
Dec.	43	6	0.6	15
Year	60	16	4.6	119

Medicine Hat, Alberta (BS)
(50° 01' N, 110° 37' W; elevation 2,144 feet of 653 m)

	Temp. °F	Temp. °C	Precip. in.	Precip. mm
Jan.	12	−11	0.6	15
Feb.	14	−10	0.6	15
Mar.	28	−2	0.6	15
Apr.	45	7	0.8	20
May	55	13	1.6	41
June	63	17	2.4	61
July	69	21	1.7	43
Aug.	66	19	1.4	36
Sept.	56	13	1.1	28
Oct.	45	7	0.6	15
Nov.	28	−2	0.7	18
Dec.	19	−7	0.7	18
Year	42	6	12.8	325

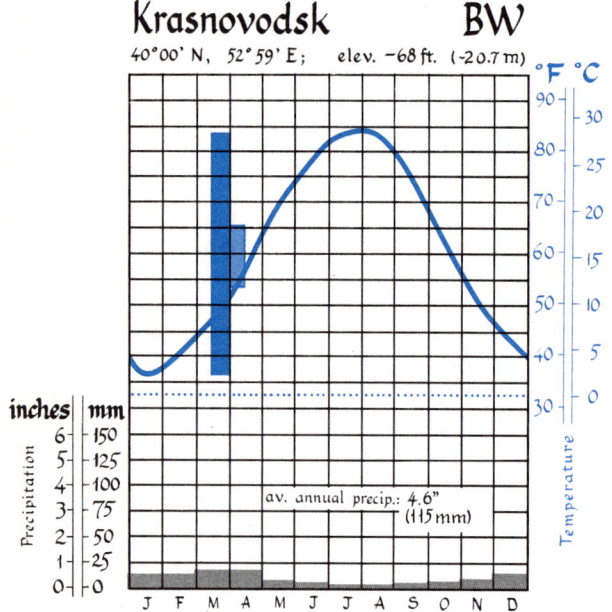

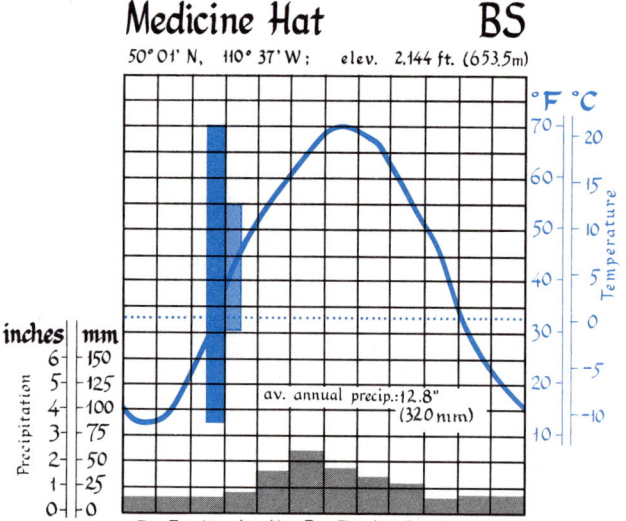

Fig. 9-9. Temperature and precipitation at selected mid-latitude BS and BW stations. High annual and moderately high diurnal ranges of temperature characterize these extratropical dry stations.

temperature. It is also wetter, as befits a steppe station, but its rainfall is mainly in summer. Medicine Hat is much closer to the North Pacific than Krasnovodsk is to the North Atlantic, but the Canadian Rockies effectively shut out winter cyclonic rains from the prairie provinces of Canada.

The soil and vegetation associations of the mid-latitude drylands are fairly similar to those of the subtropics. **Sierozems** are more prevalent, and **lithosols** less so; but on a world scale one can speak of desert soils in all of the drylands. There is, perhaps, more difference in the natural vegetation; but the term "desert scrub" will have to cover the variety for the purposes of our quick overview.

Rain-Shadow Deserts

There are many rain-shadow areas in the world. Most of them are either small or local accentuations of subtropical or mid-latitude drought. Patagonia in southeastern Argentina is distinctive in being a relatively large area of rain-shadow drought that is causally unrelated to the other major drylands. Nowhere else in the world is there a coastal desert in such high latitude nor with such mild temperatures. Sarmiento in Patagonia (Fig. 9-10) is slightly inland and illustrates this climate fairly well. It has a steppe climate but is near the desert boundary. The temperatures are moderate, and the annual range is small for an east coast.

Sarmiento, Argentina (BS)
(45° 36' S, 69° 05' W; elevation 879 feet or 268 m)

	Temp. °F	Temp. °C	Precip. in.	Precip. mm
Jan.	65	18	0.2	5
Feb.	64	18	0.3	8
Mar.	58	14	0.3	8
Apr.	52	11	0.4	10
May	45	7	0.8	20
June	38	3	0.8	20
July	37	3	0.6	15
Aug.	42	6	0.5	13
Sept.	46	8	0.4	10
Oct.	53	12	0.3	8
Nov.	58	14	0.2	5
Dec.	62	17	0.3	8
Year	51	11	5.1	130

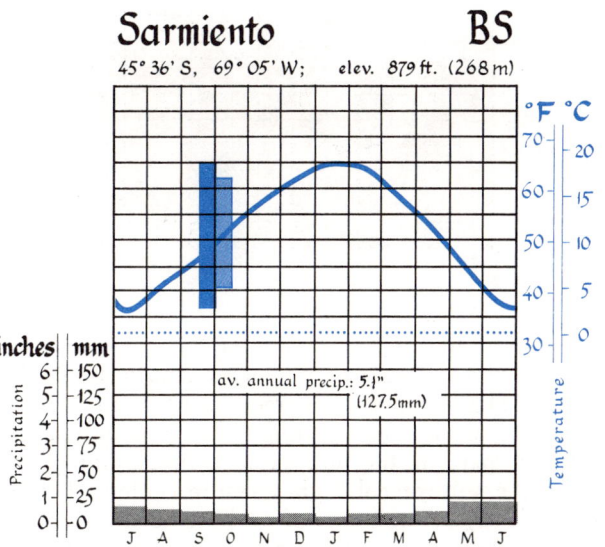

Fig. 9-10. Temperature and precipitation in a rain-shadow location. Note the mild temperatures for a dry climate in relatively high latitudes.

Tropical Rainforest Climates

The tropical rainforest (Af) climates are, by definition, constantly wet (the driest month must have more than 2.4 inches or 60 mm of precipitation) and warm (the coldest month must be warmer than 64°F or 18°C). They generally can be identified with that part of the world where the Intertropical Convergence Zone and its convergence phenomena dominate atmospheric processes all year. This zone, as we saw in Chapter 7, lies close to the equator, except that it is dislocated some 5° or 10° to the north over the Atlantic and the western Pacific Oceans. On land, the extent of Af climates is discontinuous and occupies only a part of the equatorial

Uaupés, Brazil (Af)
(0° 08' S, 67° 05' W; elevation 272 feet or 83 m)

	Temp. °F	Temp. °C	Precip. in.	Precip. mm
Jan.	80	27	10.3	262
Feb.	80	27	7.7	196
Mar.	80	27	10.0	254
Apr.	80	27	10.6	269
May	79	26	12.0	305
June	78	26	9.2	234
July	78	26	8.8	224
Aug.	79	26	7.2	183
Sept.	80	27	5.1	130
Oct.	80	27	6.9	175
Nov.	81	27	7.2	183
Dec.	80	27	10.4	264
Year	80	27	105.4	2,679

Mbandaka, Zaire (Af)
(0° 03' N, 18° 16' E; elevation 1,066 feet or 325 m)

	Temp. °F	Temp. °C	Precip. in.	Precip. mm
Jan.	76	24	3.8	97
Feb.	76	24	5.6	142
Mar.	76	24	5.0	127
Apr.	77	25	5.6	142
May	76	24	6.2	157
June	75	24	3.4	86
July	74	23	3.8	97
Aug.	74	23	3.9	99
Sept.	75	24	7.0	178
Oct.	75	24	8.9	226
Nov.	75	24	8.5	216
Dec.	76	24	5.4	137
Year	76	24	67.1	1,704

Singapore, Singapore (Af)
(1° 18′ N, 103° 50′ E; elevation 33 feet or 10 m)

	Temp.		Precip.	
	°F	°C	in.	mm
Jan.	79	26	9.9	251
Feb.	80	27	6.8	173
Mar.	81	27	7.6	193
Apr.	81	27	7.4	188
May	82	28	6.8	173
June	81	27	6.8	173
July	81	27	6.7	170
Aug.	81	27	7.7	196
Sept.	81	27	7.0	178
Oct.	80	27	8.2	208
Nov.	80	27	10.0	254
Dec.	80	27	10.1	257
Year	80	27	95.0	2,414

Pukapuka, South Pacific (Af)
(10° 55′ S, 165° 54′ W; elevation 7 feet or 2 m)

	Temp.		Precip.	
	°F	°C	in.	mm
Jan.	82	28	13.7	348
Feb.	82	28	16.1	409
Mar.	82	28	10.1	257
Apr.	83	28	8.3	211
May	83	28	8.1	206
June	82	28	6.3	160
July	81	27	6.1	155
Aug.	82	28	6.2	157
Sept.	82	28	5.7	145
Oct.	82	28	8.8	224
Nov.	82	28	12.7	323
Dec.	82	28	15.4	391
Year	82	28	117.5	2,986

Tamatave, Malagasy Republic (Af)
(18° 07′ S, 49° 24′ E; elevation 20 feet or 6 m)

	Temp.		Precip.	
	°F	°C	in.	mm
Jan.	80	27	14.4	366
Feb.	80	27	14.8	376
Mar.	79	26	17.8	452
Apr.	77	25	15.7	399
May	74	23	10.4	264
June	71	22	11.1	282
July	70	21	11.9	302
Aug.	70	21	8.0	203
Sept.	71	22	5.2	132
Oct.	74	23	3.9	99
Nov.	77	25	4.6	117
Dec.	79	26	10.3	262
Year	75	24	128.2	3,254

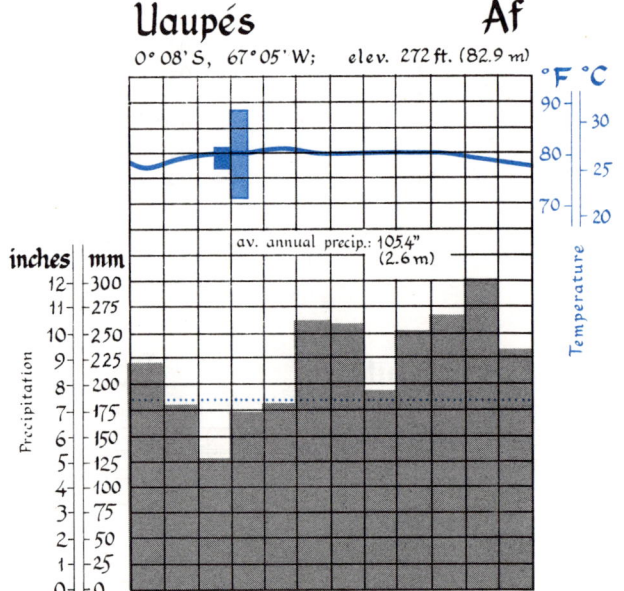

Uaupés Af
0° 08′ S, 67° 05′ W; elev. 272 ft. (82.9 m)
av. annual precip.: 105.4″ (2.6 m)

portions of the continents: the northwest coast of South America and the western part of the Amazon Basin, a section of the Congo Basin, and more extensively, the Indonesian archipelago. Beyond this equatorial zone, Af climates occur in four continental locations without any appreciable oceanic extension. These are the Caribbean coast of Central America, the central part of the east coast of Brazil, the east coast of Madagascar, and the east coast of the Philippine Islands. All four areas are east coasts in about 10° to 20° latitude. They owe their heavy year-round precipitation to the fact that mT air masses move onshore from the east or northeast (southeast in the Southern Hemisphere). This movement of air is often referred to as the *ending trades*, for it is indeed the equatorial part of the anticyclonic circulation of air around the subtropical high-pressure centers. This is not, however, a simple case of onshore northeast trades in the Northern Hemisphere and southeast trades in the Southern Hemisphere. Seasonal shifts in the pressure field introduce complications in the movement of the air masses; the situation with regard to Brazil and the

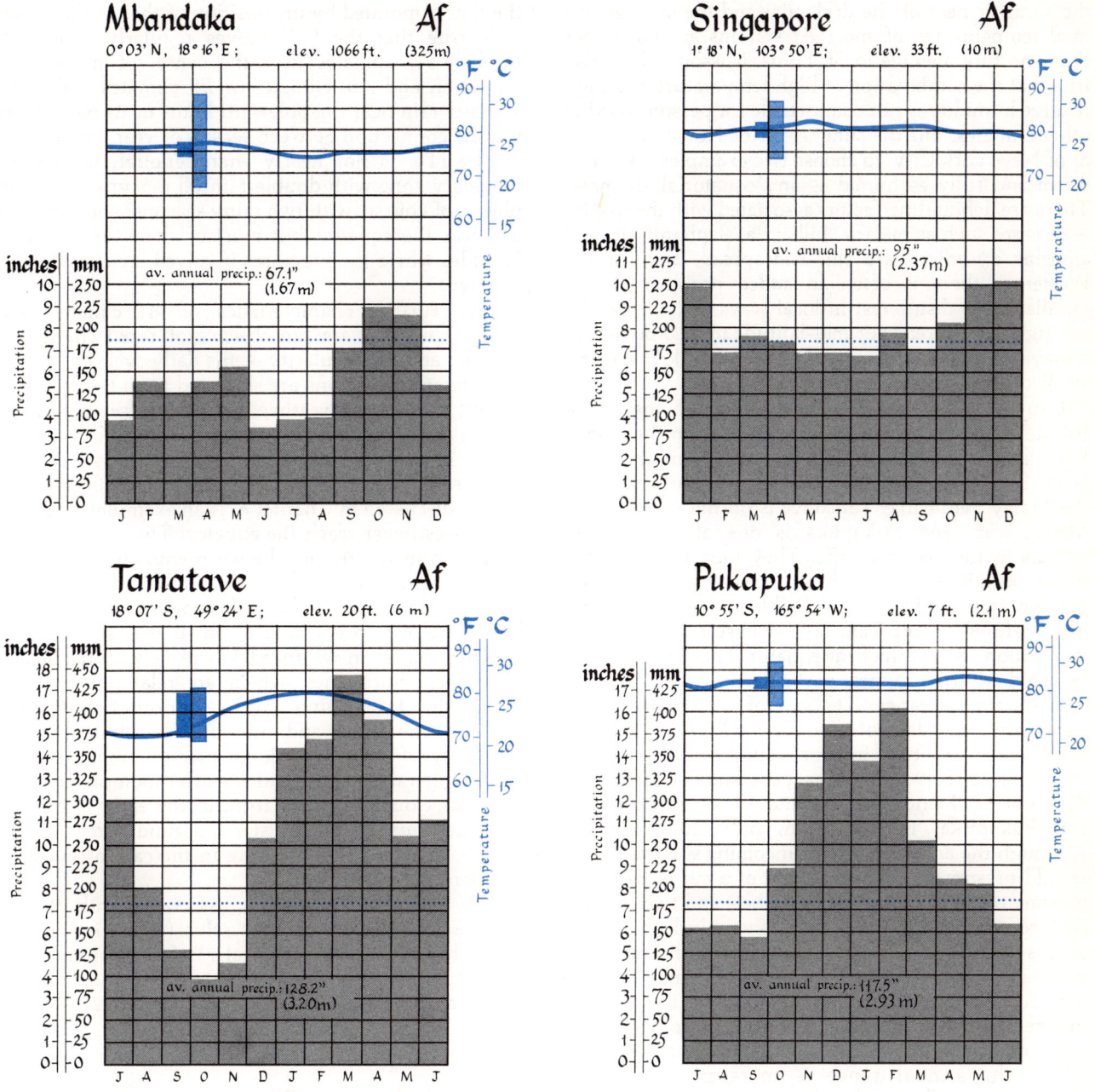

Fig. 9-11. Temperature and precipitation at selected *Af* stations. They all have relatively even rainfall all year; only Pukapuka and Tamatave have some seasonality; but even there the dry season has much more than 2.4 in. (60 mm) per month. At each station the daily range greatly exceeds the annual range of temperature, which is very small indeed.

Philippines is particularly complex. Nevertheless, these trade-wind coasts may be considered to have similar *Af* climates for roughly similar reasons.

Much has been said and written about the monotony and enervating quality of this climate, mostly by Europeans and North Americans. It is true, of course, that seasonal variation of temperature is very restricted. There is, on the other hand, considerable diurnal range — as much, on the average, as in the Mississippi Valley or along the west coast of North America. Temperatures are high, to be sure, but the warmest month in the tropical rainforest does not compare with the summer heat of the drylands; and the average annual temperatures of most *Af* stations are no higher than the July average in St. Louis, Missouri. It is also true that the combination of high temperature and high relative humidity is not comfortable for persons used to other climates, but the summer air of Houston, Texas, or of New York City, to choose two examples, can be as warm and fully as humid as any equatorial air mass. There are debilitating factors associated with the tropics — diseases such as malaria, bilharzia, elephantiasis, and amoebic dysentery are still widespread, and for the Westerner there is often an added problem of unfamiliarity or disinterest in local ways of living and a too-rigid transference of mid-latitude mores. Be that as it may, the objective climatic data for several *Af* stations are listed and shown in Fig. 9-11.

Uaupés is on the Rió Negro, one of the major tributaries of the Amazon; Mbandaka is on the Congo River and is about 1,000 feet (about 300 m) above sea level; Singapore is on an island at the southern tip of the Malay Peninsula; Tamatave is on the east coast of Madagascar; and Pukapuka is one of the Danger Islands in the South Pacific. They thus represent the major locations in which *Af* climates are found. All the stations have a negligible range of monthly temperatures except Tamatave, whose regime betrays its position 18° south of the equator. They all have fairly sizable diurnal ranges, particularly when compared to their annual range. The size of the diurnal range varies with the degree of continentality. Uaupés and Mbandaka, the two interior stations, have diurnal ranges of 17°F (9.4°C); Singapore and Tamatave, on the coast, 13°F and 12°F (7.2 and 6.7°C), respectively; and Pukapuka, in mid-ocean, only 10°F (5.6°C). Although the annual march of incoming solar radiation should presumably take the form of a curve with two maxima (in March and in September) and two minima (in December and in June), there is very little indication of this double periodicity in the temperature regime. A very faint trace of a double curve can be imagined in the data for Singapore and Mbandaka. The temperature differences from month to month are too small in equatorial climates to show any kind of clear trend; and where the annual range is more marked, as at Tamatave, it reflects other factors.

There is more variability in rainfall than in temperature, both during the year and among stations. The variation between stations reflects many factors, among which the generally drier situation of the elevated, plateau-like interior of Africa is evident. The annual variability has complex causes; the often-quoted relationship between high sun and rainy season is not really evident. The ITC is not a sharp line, nor does it move with clock-like precision across each latitude at the time appointed by the position of the overhead sun. It is true that the ITC moves regularly enough to produce *Af* climates near the center of its seasonal migration and *Aw* climates or *BS* climates with some summer rain near the poleward limits of its movement; but the connection between season of year and position of the ITC is not nearly sharp enough to create a planetary zone with double rainfall maxima. There are places, of course, with two rainy seasons; some of these are near the equator, but most of them are in higher latitudes where the processes involved are considerably different.

Two further characteristics of *Af* climates that deserve mention are the absence of rapid changes in weather and the regularity of the daily cycle of rainfall. The equatorial regions are sheltered from the relatively abrupt changes in weather that occur elsewhere when one air mass is replaced by another. Some indirect effect of the invasion of mid-latitudes by outbursts of polar air is sometimes felt in the tropics (the surge of the trades, discussed in Chapter 6, is an example), but polar air masses never reach the equator. The most common transition in the tropics, between equatorial and trade air, does not entail any noticeable difference in temperature at the surface; the pattern of clouds and likelihood of rain is changed, but that is all. The only striking change in weather takes place with the advent of tropical hurricanes, which are relatively common along the trade-wind coasts. Another difference between equatorial and mid-latitude climates is the regularity of the convectional precipitation. At many places, the rains are dependable afternoon phenomena; cumulus clouds begin to form in the early afternoon, and the convective rain starts around three to four o'clock or later in the evening in the case of maritime locations.

The soils and vegetation associated with the *Af* climate are **ferralites** and the **selva** (which is simply a Spanish word for the tropical rainforest, from which the *Af* climates get their name). These soils and the selva are more fully discussed in Chapters 20, 21, and 22.

Tropical Monsoon Climates

Tropical Monsoon *(Am)* climates owe their distinctiveness to the high amounts of annual precipitation and to the strong concentration of rainfall in three or four summer months. Because any all-inclusive scheme of classification must draw hairline boundaries between its categories and because the Köppen classification in particular uses certain kinds of definitions to delimit the *Am* climates, it follows that monsoon climates grade

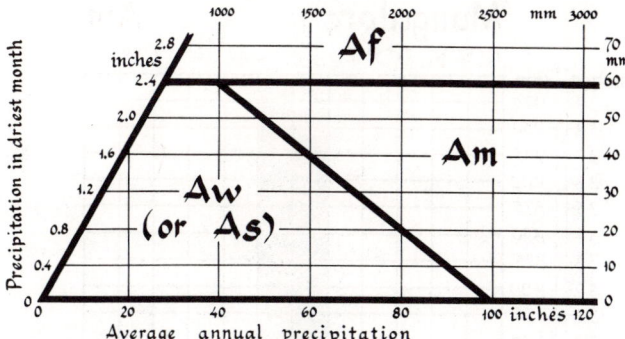

Fig. 9-12. A graphic representation of the boundaries between Af, Am, and Aw climates. Any place with more than 2.4 in. (60 mm) of rain in the driest month is classified as Af (above the horizontal line at 2.4 in. or 60 mm). The boundary between Am and Aw is at 40 in. (1000 mm) if the driest month has as much as 2.4 in. (60 mm). As the precipitation of the driest month decreases to zero (downward in the graph) the total annual rainfall needed increases to 100 in. (2500 mm), so that the boundary line slants downward and to the right. As a result of these definitions, we can see that Aw and Af share a very narrow boundary.

imperceptibly into Af or Aw climates.[6] Figure 9-12 shows in graphic form the boundaries set for the Af, Am, and Aw climates by the Köppen classification. As you can see, a direct transition between Af and Aw climates is possible, but it can take place only over a narrow range of annual precipitation (between 28.8 and 40 inches or between 720 and 1,000 mm using the metric formula). It follows that over most of the range of A climates, Af is separated from Aw by a belt of Am. This intermediate variety of Am climates is usually found in a very narrow belt and has neither very high rainfall nor a very pronounced dry season. To avoid overcomplicating the world map of climates (front endpaper), this narrow belt of relatively atypical Am climate has been omitted; Am climates are shown only where they cover a fairly wide area and have their typical characteristics of high annual precipitation and pronounced winter drought.

The location of Am climates can be seen on the front endpapers, and their causes have been discussed previously in connection with tropical atmospheric disturbances (Chapter 6) and with areas of high precipitation (Chapter 7). They are found on the northeast coast

---

[6] The Am climates also grade into C climates, as do also the Af climates where they impinge on tropical mountain ranges. The area of such Cf climates is small, but they are found on both flanks of the northern Andes, on the intermediate mountain slopes of Central America, in the Cameroon Highlands, and in the higher ranges of Malaya and Indonesia.

World Distribution of Climates

of South America, the coast of Africa from Sierra Leone to Liberia and from Nigeria to the Cameroons, the Malabar Coast of India, the west coasts of Burma and Malaya, and the Philippines. There is probably more diversity in the causal phenomena from one region of monsoon climate to another than in any of the other climatic categories devised by Köppen. Without describing in some detail the air-mass movement and other relevant atmospheric phenomena for each of the major areas which experience Am climate, we cannot really explain their distribution. We can say, however, that the areas mentioned are all tropical windward

Mangalore, India (Am)
(12° 52' N, 74° 51' E; elevation 72 feet or 22 m)

	Temp.		Precip.	
	°F	°C	in.	mm
Jan.	80	27	0.1	3
Feb.	80	27	0.1	3
Mar.	83	28	0.2	5
Apr.	85	29	1.5	38
May	85	29	6.2	157
June	80	27	37.1	942
July	79	26	38.9	988
Aug.	79	26	23.5	597
Sept.	79	26	10.5	267
Oct.	80	27	8.1	206
Nov.	81	27	2.9	74
Dec.	80	27	0.5	13
Year	81	27	129.6	3,293

Cairns, Australia (Am)
(16° 55' S, 145° 47' E; elevation 16 feet or 5 m)

	Temp.		Precip.	
	°F	°C	in.	mm
Jan.	82	28	16.6	422
Feb.	81	27	15.7	399
Mar.	80	27	18.1	460
Apr.	77	25	11.3	287
May	73	23	4.4	112
June	71	22	2.9	74
July	69	21	1.6	41
Aug.	71	22	1.7	43
Sept.	73	23	1.7	43
Oct.	77	25	2.1	53
Nov.	79	26	3.9	99
Dec.	81	27	8.7	221
Year	76	24	88.7	2,254

Cherrapunji, India (Cwb)
(25° 15' N, 91° 44' E; elevation 4,309 feet or 1,313 m)

	Temp.		Precip.	
	°F	°C	in.	mm
Jan.	53	12	0.7	18
Feb.	55	13	2.1	53
Mar.	62	17	7.3	185
Apr.	65	18	26.2	665
May	66	19	50.4	1,280
June	68	20	106.1	2,695
July	68	20	96.3	2,446
Aug.	69	21	70.1	1,781
Sept.	69	21	43.3	1,100
Oct.	66	19	19.4	493
Nov.	60	16	2.7	69
Dec.	55	13	0.5	13
Year	63	17	425.1	10,798

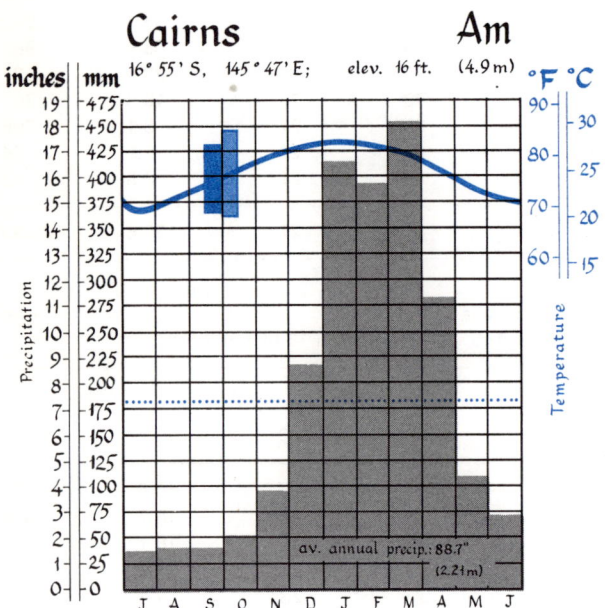

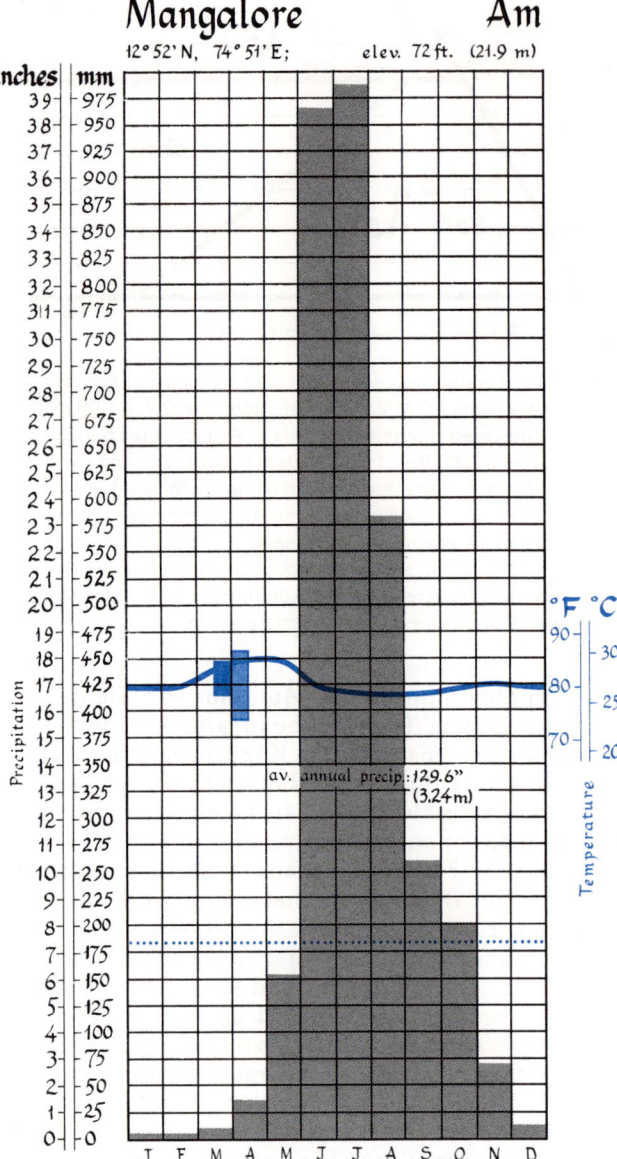

Fig. 9-13. Temperature and precipitation at selected *Am* stations. Mangalore is a typical Indian monsoon station. Cairns illustrates the much weaker Australian monsoon. Cherrapunji, although not warm enough to be classified as a tropical (*A*) station, has an incredible monsoonal regime of precipitation.

coasts onto which moist and unstable *mT* air masses move during the summer. The following stations, also shown in Fig. 9-13, are representative of *Am* climates.

Mangalore, on the Malabar Coast of India, illustrates both the most typical and the best-known area of monsoon climate. It is also representative of most other *Am* regions where the relief features are relatively modest (in this case the low mountains known as the Western Ghats) and the orographic accentuation of precipitation is not excessive. One noteworthy feature of the monsoon climate is the division of the year into the three seasons, already alluded to in our discussion of Niamey. The cool, dry season of winter is not very different; the hot, dry season of spring is, in fact, considerably cooler than at Niamey but no less dry and dusty, if shorter. The major difference is in the sudden onset of the mon-

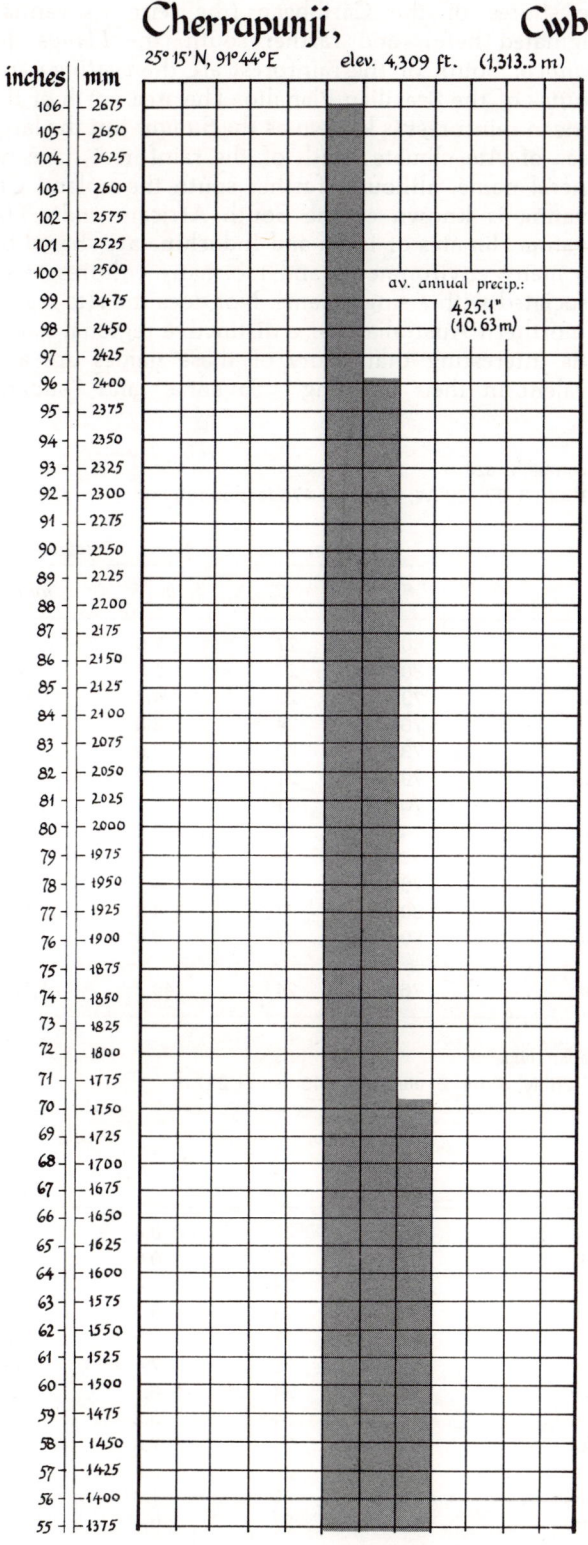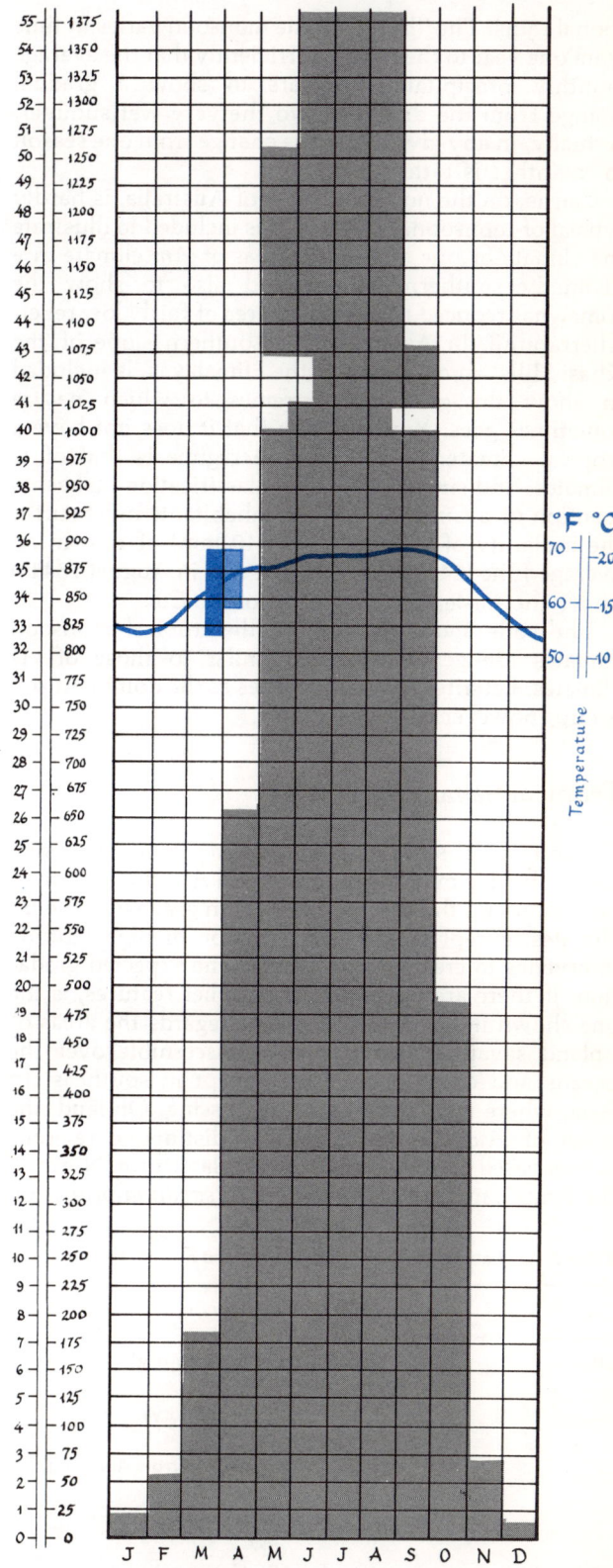

World Distribution of Climates 153

soonal rains. The "burst" of the monsoon varies in time from one year to the next — sufficiently that the average monthly precipitation appears to show a gradual change from the dry winter to the very wet summer. Actually, in any given year, the change from one season to the other is rather abrupt.

Cairns, on the northeast coast of Australia, is hardly typical of monsoonal climates. It is included to illustrate the climate of one of the few areas of Am climate in a distinctly southern latitude and also to show the somewhat reduced rainfall in an area of fairly low relief. Cherrapunji, in Assam on the southern slope of the Khasi Hills, an extension of the Himalayas, is included to show the grotesque extremes to which nature sometimes goes. You will note that it does not have a tropical climate; its temperature regime is that of C climates; and since the Köppen classification makes no mention of a Cm category, the inhabitants must suffer the indignity of 100 inches (2,540 mm) of rain in an average June (241 inches or 6,121 mm in August 1841!) under the misleading classification of Cw.

The soils and vegetation of the tropical monsoon climates are generally quite similar to those of Af climates. **Laterites** replace **ferralites** as the dominant soil group, however.

Tropical Savanna Climates

The tropical **savanna** (Aw) climates completely surround the dry margins of the Af and Am climates (over both land and the oceans) and in turn grade into the BS climates, except where high elevation or high latitude intervenes to create a Cw climate. The expected gradation, if there are no complicating relief features, is the one shown in Table 9-1. If one disregards the areas of upland savanna, this pattern is discernible over the oceans and all the continents except in southeastern Asia, where the dry climates are missing. On land, the areas of Aw climate often bear distinctive regional names based on either vegetation or landform. North of the American rainforest there is the *savanna* proper on the shores of the Caribbean (the term "savanna" originated here); and farther south, the *Llanos del Orinoco*. South of the rainforest are the *caatinga* and *campos* of the Brazilian Planalto. The area north of the African rainforest is known as the *Sudan*; but the large areas of Aw climate south of the rainforest have no general name, although farther south the upland Cw savanna is known as the South African *veldt*. The savanna climates of India and Indochina are shared by too many environments and too many cultures to be associated with a single name. Nor does the Aw region of northern Australia bear a distinctive regional name. It is interesting that many of these names are ambivalent in their meaning. "Savanna" and "sudan"

Table 9-1. Latitudinal Arrangement of A and B Climates

Latitude	Climate
30° N	BS (summer dry)
	BW
	BS (winter dry)
	Aw
0°	Af (with narrow Am margins)
	Aw
	BS (winter dry)
	BW
30° S	BS (summer dry)

Corumbá, Brazil (Aw)
(19° 00′ S, 57° 39′ W; elevation 476 feet or 145 m)

	Temp.		Precip.	
	°F	°C	in.	mm
Jan.	80	27	7.0	178
Feb.	80	27	5.8	147
Mar.	79	26	4.7	119
Apr.	76	24	3.3	84
May	73	23	2.6	66
June	70	21	1.3	33
July	70	21	0.7	18
Aug.	74	23	0.9	23
Sept.	77	25	2.6	66
Oct.	78	26	4.1	104
Nov.	80	27	4.8	122
Dec.	80	27	6.4	163
Year	76	24	44.2	1,123

Ibadan, Nigeria (Aw)
(7° 26′ N, 3° 54′ E; elevation 745 feet or 227 m)

	Temp.		Precip.	
	°F	°C	in.	mm
Jan.	80	27	0.4	10
Feb.	82	28	0.9	23
Mar.	83	28	3.6	91
Apr.	82	28	5.3	135
May	80	27	5.9	150
June	78	26	7.4	188
July	76	24	6.1	155
Aug.	76	24	3.3	84
Sept.	77	25	7.0	178
Oct.	78	26	6.2	157
Nov.	80	27	1.8	46
Dec.	80	27	0.4	10
Year	79	26	48.3	1,227

refer as much to flatness of terrain as to the distinctive grassland of the *Aw* climates. "Llanos," of course, refers merely to flat plains, but there is a distinct connotation of grass-covered lowland in the usage of the term. "Campos" and "caatinga" are basically vegetational terms referring to grassland and thorny scrub, respectively. "Veldt" also refers to grassy vegetation. But, in general, there is some confusion of "flat" and "grassy" in the regional names given to areas of *Aw* climate.

The climatic situation is represented by the graphs (Fig. 9-14) for Corumbá, Brazil, Ibadan, Nigeria, and Saigon, Viet Nam. Their precipitation regimes recall the *Am* climates, but the rainfall is much less.

Saigon, South Viet Nam (Aw)
(10° 47'N, 106° 42'E; elevation 30 feet or 9 m)

	Temp.		Precip.	
	°F	°C	in.	mm
Jan.	79	26	0.6	15
Feb.	81	27	0.1	3
Mar.	83	28	0.5	13
Apr.	85	29	1.7	43
May	84	29	8.7	221
June	82	28	13.0	330
July	81	27	12.4	315
Aug.	81	27	10.6	269
Sept.	81	27	13.2	335
Oct.	81	27	10.6	269
Nov.	80	27	4.5	114
Dec.	79	26	2.2	56
Year.	82	28	78.1	1,983

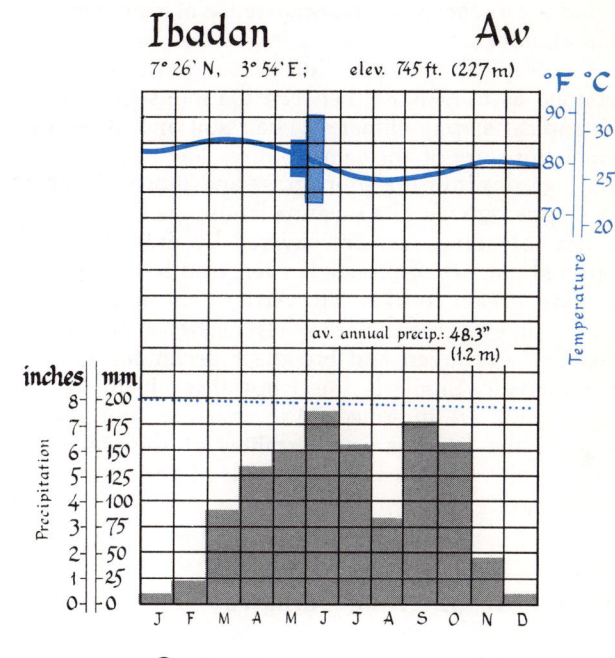

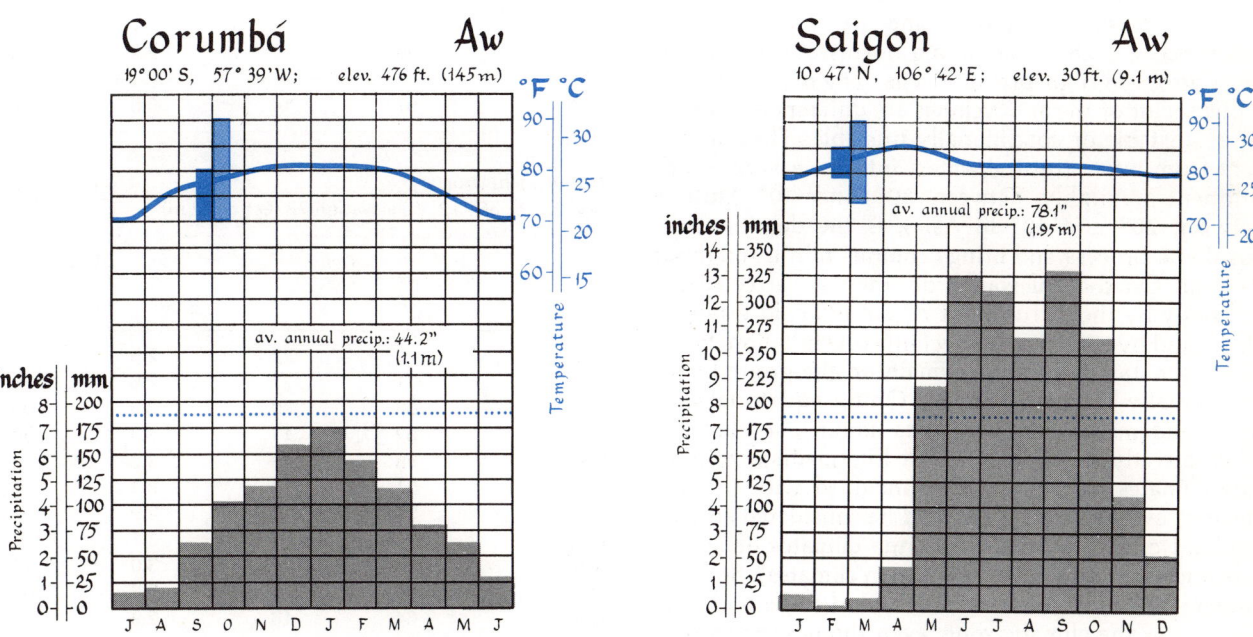

Fig. 9-14. Temperature and precipitation at selected *Aw* stations. Note the marked seasonality of the precipitation, and its effect on summer temperatures; as a result of depressed summer temperatures, Saigon and Ibadan have spring maxima.

Saigon represents the wetter portion of the range of possible annual precipation, Ibadan and Corumbá the drier portion. The temperature regimes are still markedly equatorial, even though the annual range is around 10°F (6° C) rather than the 2° or 3° of the *Af* stations. Generally speaking, the temperature regimes resemble those of *Am* climates and for roughly the same reasons. The potentially higher temperatures of the high-sun period are reduced by the occurrence of summer rains (and clouds).

The tropical savanna climates are most easily understood as transitional between the monsoon and the subtropical steppe climates. Their soil and vegetation associations are of the same nature. The tropical rainforest grades into the short grass and thorny scrub of the steppe by way of a savanna landscape that consists of flat to rolling surfaces covered by grass but with scattered patches of **broadleaf evergreen** trees, especially along stream courses. The African savannas consist largely of evergreen species; elsewhere, a mixture of broadleaf evergreen and **broadleaf deciduous** is more characteristic. Similarly, the **tropical red loam** soils of the savanna are intermediate in character and genesis between the **laterites** and **ferralites** of the equatorial climates and the **brown steppe soils** of the subtropical *BS* climates.

Humid Mesothermal Climates

Humid mesothermal *(Cfa* and *Cwa)* climates are associated with the western and northwestern parts of the subtropical high-pressure cells. The *mT* air masses in this sector of the anticyclones are quite moist and are either unstable or conditionally unstable. Their movement from warm tropical waters into the east coasts of continents, roughly between 30° and 40° latitude (between 20° and 40° in Asia), is the dominant atmospheric process and brings copious rain throughout the year. On the poleward side, they are affected occasionally by the intrusion of *cP* air and cold waves in winter and by considerable cyclonic precipitation. Some snow falls, but it does not remain on the ground very long, usually less than a month. On the equatorial side, the humid mesothermal climates adjoin *Aw* climates, either directly or by way of an intermediate zone of *Cwa* climates (which we have included with the *Cfa* climates, even though they could be considered either a highland extension of the savanna climates or an extratropical extension of the monsoon climates). Both the temperature and the precipitation tend to be higher in this part of the climatic zone, even though the winters are drier. On the west, the humid mesothermal climates become drier and grade into steppe climates of the continental interiors.

The major locations of *Cfa* and *Cwa* climates, as shown on the front endpaper, include most of the southeastern part of the United States; a considerable portion of southern Brazil, Uruguay, and the humid Pampa of Argentina; a narrow stretch of the Natal coast of South Africa and a much larger upland area in the interior; the northern and higher parts of India and Indochina, as well as parts of China; and most of the east coast of Australia. Since these are all in latitudes of fairly strong poleward temperature gradient in winter, they include a considerable range of winter temperatures; but the summer temperatures are everywhere high, ranging from the defined minimum of 72° F (22° C) to average monthly values greater than 85° F (29° C).

St. Louis, Missouri (Cfa)
(38° 38' N, 90° 12' W; elevation 465 feet or 142 m)

	Temp.		Precip.	
	°F	°C	in.	mm
Jan.	34	1	2.2	56
Feb.	36	2	2.2	56
Mar.	44	7	3.2	81
Apr.	56	13	3.6	91
May	67	19	3.8	97
June	77	25	3.9	99
July	81	27	3.3	84
Aug.	79	26	3.6	91
Sept.	72	22	3.0	76
Oct.	61	16	2.8	71
Nov.	46	8	2.7	69
Dec.	36	2	2.2	56
Year	57	14	36.5	927

Montevideo, Uruguay (Cfa)
(34° 52' S, 56° 12' W; elevation 72 feet or 22 m)

	Temp.		Precip.	
	°F	°C	in.	mm
Jan.	72	22	2.9	74
Feb.	71	22	2.6	66
Mar.	68	20	3.9	99
Apr.	62	17	3.9	99
May	56	13	3.3	84
June	51	11	3.2	81
July	50	10	2.9	74
Aug.	51	11	3.1	79
Sept.	54	12	3.0	76
Oct.	58	14	2.6	66
Nov.	64	18	2.9	74
Dec.	69	21	3.1	79
Year	61	16	37.4	951

Precipitation may be as low as 30 inches (762 mm) on the western margins or as high as 60 inches (1,524 mm) on the subtropical edges.

These climatic conditions are illustrated by St. Louis, Missouri, Kuang-Chou (Canton), China and Montevideo, Uruguay (Fig. 9-15).

St. Louis is fairly close to the northern border of *Cfa* climates in the United States and is somewhat interior in location. This position is reflected in the high annual range of 47° F (26° C), and the modest annual total of precipitation with its slight summer maximum. The data for Kuang-Chou, just south of the Tropic of Cancer on the southeast coast of China, reflect not only its southerly location but also the northward penetration of monsoonal influences. It is decidedly a *Cwa* station, with high temperatures in both the wet summer

Kuang-Chou (Canton), China (Cwa)
(23°06′ N, 113°18′ E; elevation 29 feet or 9 m)

	Temp.		Precip.	
	°F	°C	in.	mm
Jan.	56	13	1.8	46
Feb.	57	14	2.7	69
Mar.	62	17	3.6	91
Apr.	71	22	5.9	150
May	80	27	9.9	251
June	82	28	10.6	269
July	84	29	9.9	251
Aug.	84	29	9.6	244
Sept.	82	28	5.4	137
Oct.	75	24	2.3	58
Nov.	68	20	1.6	41
Dec.	61	16	1.4	36
Year	72	22	64.7	1,643

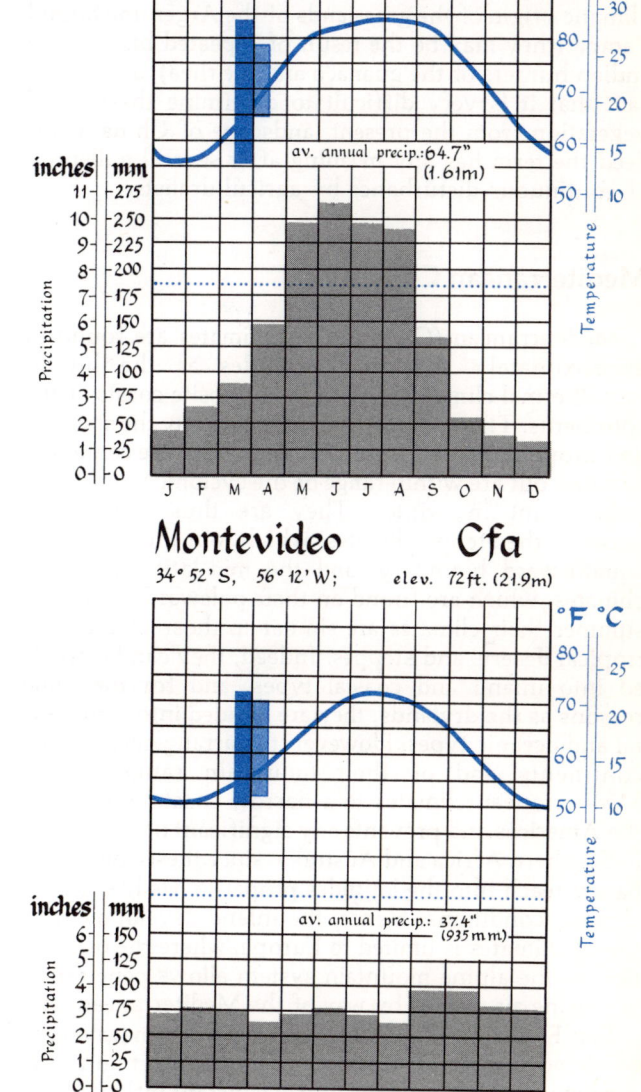

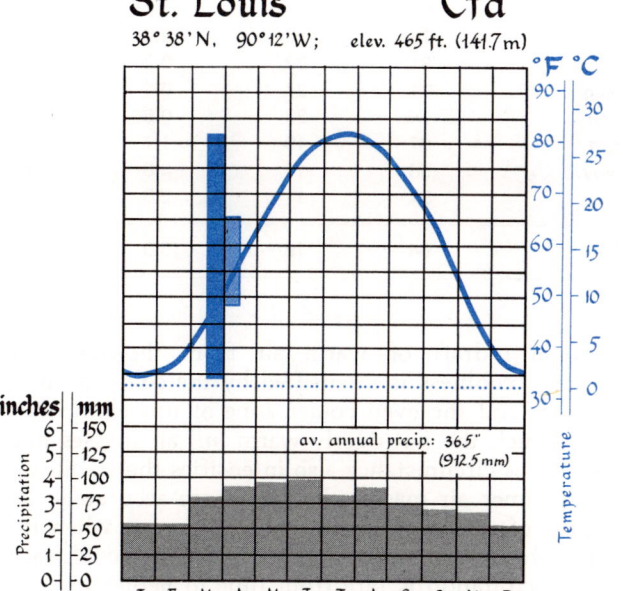

Fig. 9-15. Temperature and precipitation at selected *Cfa* and *Cwa* stations. Canton has considerable seasonality of rainfall, despite its *f* classification. Note the milder temperatures at Montevideo in the Southern Hemisphere as compared to St. Louis.

and the dry winter. Montevideo, the capital of Uruguay, is slightly more oceanic, reflecting the generally smaller seasonal contrasts of the Southern Hemisphere. It has milder temperatures than either St. Louis or Kuang-Chou, and its rainfall is very evenly distributed.

Humid mesothermal climates are generally associated with **subtropical red and yellow soils.** Vegetation associations are more complex, ranging from grasslands to **broadleaf evergreen** forests. **Broadleaf deciduous** and even **needleleaf** forests, as well as mixtures of all these types, are also found. The situation is even further complicated, since there is some question about the climatic origin of the grasslands of the Argentine humid Pampa (they may be the result of repeated burning by Indian hunters of the guanaco and the rhea); and by the fact that it is very difficult to determine the natural vegetation from the present landscape of China, if indeed the term has any meaning after several millennia of continuous disturbance by agriculturalists.

Mediterranean Climates

Mediterranean (*Csa* and *Csb*) climates are found in approximately the same latitudes as the humid mesothermal climates, but on the opposite coasts of the continents. These regions are dominated by the stability and drought of the subtropical high-pressure centers in summer but are within range of the cyclonic rains of the polar front in winter. They are thus transitional between the steppe climates, which join them on their equatorward boundary, and the mid-latitude marine climates, which are found on their poleward border. In summer their climates are similar to those of the subtropical deserts and steppes. Indeed, they can be divided into inland and coastal types, and for the same reasons as the drylands, they are divided into continental and oceanic types. However, the arrangement of the continents and of their mountain ranges limits Mediterranean climates to a narrow coastal section in the Americas and prevents any significant development in Southern Africa and Australia, since these continents barely reach into the latitudes proper to Mediterranean climatic conditions. Hence, the inland development of the *Cs* climates is limited to Europe, where the orientation of the alpine mountain system allows penetration of marine air masses by way of the Mediterranean Sea.

San Francisco (Fig. 9-16) represents the marine type of Mediterranean climate. The effect on the temperature regime of the nearly constant inflow of *mP* air masses is evident; winters are oceanically mild, and the summer temperature is not much higher. Average monthly temperatures do not rise above 60° F (16° C) until September and October, when there is an oc-

San Francisco, California (*Csb*)
(37° 47' N, 122° 25' W; elevation 52 feet or 16 m)

	Temp.		Precip.	
	°F	°C	in.	mm
Jan.	50	10	4.6	117
Feb.	53	12	3.7	94
Mar.	54	12	2.9	74
Apr.	55	13	1.4	36
May	57	14	0.6	15
June	59	15	0.1	3
July	59	15	0.0	0
Aug.	59	15	0.0	0
Sept.	62	17	0.2	5
Oct.	61	16	0.9	23
Nov.	57	14	2.0	51
Dec.	52	11	4.3	109
Year	57	14	20.7	527

Athínai (Athens), Greece (*Csa*)
(37° 58' N, 23° 43' E; elevation 351 feet or 107 m)

	Temp.		Precip.	
	°F	°C	in.	mm
Jan.	48	9	2.2	56
Feb.	49	9	1.6	41
Mar.	53	12	1.4	36
Apr.	59	15	0.8	20
May	68	20	0.8	20
June	76	24	0.6	15
July	81	27	0.2	5
Aug.	81	27	0.4	10
Sept.	74	23	0.6	15
Oct.	67	19	1.7	43
Nov.	58	14	2.8	71
Dec.	51	11	2.8	71
Year	64	18	15.9	403

casional indraft of warm air from the northeast. Summer coolness is intensified by the cold California Current and the even colder zone of upwelling water over which the not overly warm *mP* air masses must pass. This circumstance also intensifies the stability of the summer air masses. Fog or, more accurately, low stratus often forms in this moist air mass when it is cooled below its dew point by contact with cold water, but precipitation is rare from June to September. The summer fog has some independent effect on the temperature, but it is not the main cause of the summer coolness. Indeed, since fog is much more prevalent at night than during the middle of the day, it may keep as

Adelaide, Australia (Csa)
(34° 56′ S, 138° 35′ E; elevation 140 feet or 43 m)

	Temp. °F	Temp. °C	Precip. in.	Precip. mm
Jan.	73	23	0.8	20
Feb.	74	23	0.7	18
Mar.	70	21	1.0	25
Apr.	64	18	1.8	46
May	58	14	2.7	69
June	54	12	3.0	76
July	52	11	2.6	66
Aug.	54	12	2.6	66
Sept.	57	14	2.1	53
Oct.	62	17	1.7	43
Nov.	67	19	1.1	28
Dec.	71	22	1.0	25
Year	63	17	21.1	535

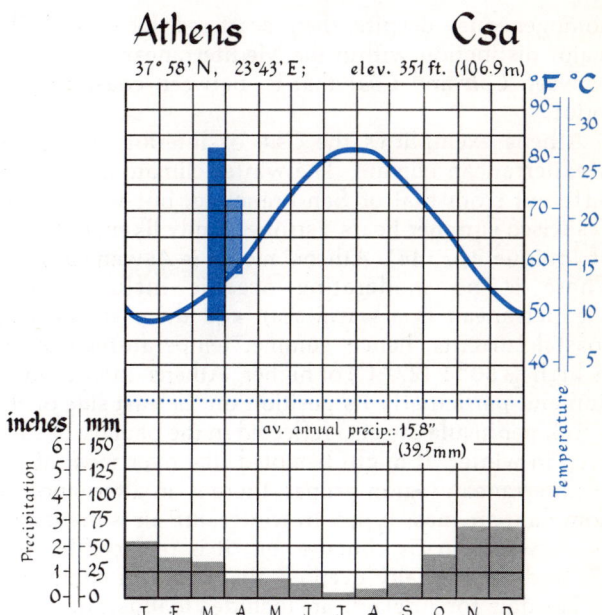

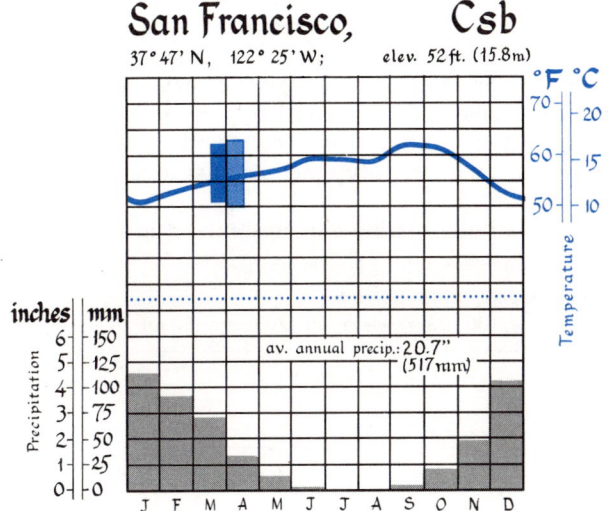

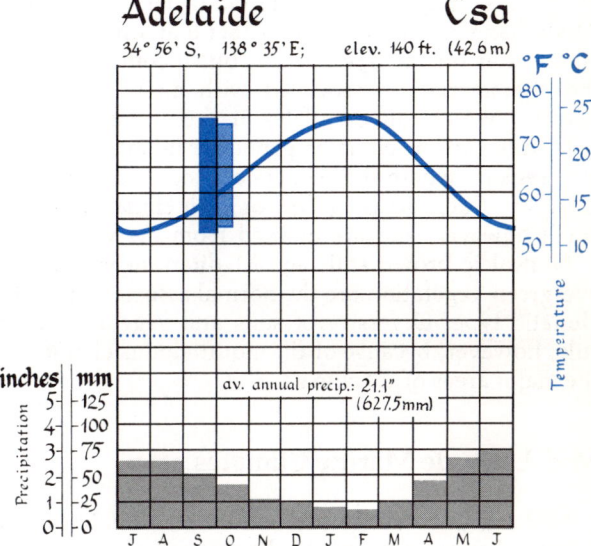

Fig. 9-16. Temperature and precipitation at selected *Csa* and *Csb* stations. Marked summer drought is what unifies the Mediterranean climates. Athens has continental temperatures; San Francisco oceanic ones; and Adelaide intermediate temperatures.

much heat in as it keeps out. Both fog and low summer temperature are caused by the advection of cooled *mP* air masses.

In winter, the subtropical anticyclone migrates far to the south, and the cold current disappears. The flow of air continues to be from the northwest; but these winter *mP* air masses are part of the polar-front zone of cyclonic storms, and precipitation reaches a maximum in January. The amount of rain decreases equatorward as one goes beyond the effective range of cyclonic storms; the Oregon coast has nearly 100 inches (2,540 mm) per year; Los Angeles, only 15 inches (380 mm).

Somewhat similar conditions prevail along other west coasts in the same latitude (roughly between 30° and 40°) although the effects of the cold currents are either weaker (in the case of the Benguela and West Australian Currents) or are displaced to the latitude of the desert coasts (in the case of the Humboldt or Canary Currents). Nevertheless, these *Csb* climates are quite homogeneous, despite their scattered locations. The major distinction within the Mediterranean climates is between *Csb* and *Csa;* that is, between coast and interior.

Athens exemplifies the *Csa*, or interior, variety of Mediterranean climate. The winter climate is not very different from that of San Francisco, but while a San Francisco summer bears a strong family likeness to that of Iquique, (p. 144), Athens' model is Aswan (p. 143). There is no moderating oceanic influence (the Mediterranean is a warm sea), and there are no cool coastal currents; hence, summer temperatures may go as high as 80° F (27° C) or higher. Athens' interior location and particularly its position on the east side of the Greek peninsula are also reflected in the paucity of rain, even in winter. It might be noted that over most of the Mediterranean region proper, there is an alternation of dominant air-mass types. In winter, *mP* air is prevalent, as elsewhere in *Cs* climates, but it is replaced by *mT*, *cT*, and occasionally even *cP* air masses in summer.

The data for Adelaide are included to illustrate both a Southern Hemisphere Mediterranean climate and a somewhat intermediate situation between the oceanic regime of San Francisco and the continental conditions of Athens. This intermediate character of the South Australian climate is due to the fact that, unlike Athens, Adelaide is on the open ocean; but it completely lacks the characteristics associated with an offshore cold current, which differentiates it from San Francisco.

Mediterranean climates penetrate inland or upslope into regions where the winters are too cold to fit the *C* category. Such areas of *Dsa* or *Dsb* climates are very small however, and are omitted from the world map.

Noncalcic brown soils and Mediterranean **broadleaf evergreen** vegetation are the normal associations of this climatic type. Diversity of soils and vegetation is the rule, however, because of the mountainous character of the major areas of *Cs* climate.

Mid-Latitude Marine Climates

Mid-latitude marine (*Cfb* and *Cfc*) climates are the poleward extensions of Mediterranean climates along west coasts and adjacent oceans. They are dominated by *mP* air masses during the year; but unlike the Mediterranean climates, they are not very strongly influenced by the subtropical anticyclones in summer. Nor are they influenced by cold currents in the Northern Hemisphere; as a matter of fact, these high-latitude west coasts are bathed by the warm waters of the West Wind Drift. As a result, summer air masses, though much more stable than the winter *mP* air, are potential purveyors of considerable precipitation. Characteristically, both the winter and the total annual precipitation increase poleward, while the time of maximum precipitation comes earlier in the year (see Fig. 7 - 4). Temperatures decrease, of course, but they retain the oceanic character of the coastal Mediterranean climates. The gentleness of the poleward temperature gradient along west coasts can be illustrated by some simple comparisons.

Between San Francisco (Fig. 9 - 15) and Prince Rupert, on the British Columbia coast (Fig. 9 - 16), there is a difference of 16½° of latitude, or about 1,100 miles (1,800 km). In that distance, the average January temperature decreases from 51°F to 35°F (from 11°C to 2°C), and the average July temperature from 59°F to 55°F (from 15°C to 13°C).

The smallness of this gradient can be appreciated if it is compared to its east-coast analog. From Richmond, Virginia, to Goose Bay, Labrador (whose latitudes correspond to those of San Francisco and Prince Rupert), the average January temperature drops from 39°F to 0°F (from 4°C to -18°C), and the average in July, from 78°F to 61°F (from 26°C to 16°C). On the other hand, precipitation gradients are more marked along the west coast. San Francisco has only 20.7 inches (526 mm); Prince Rupert, 95.3 inches (2,421 mm). The differences in amount and regime are apparent from Fig. 9 - 16. Between Richmond and Goose Bay, on the other hand, there is a decrease in precipitation from about 40 inches to approximately 30 inches (from about 1,000 to 750 mm) and the regime remains very similar, both stations having a remarkably even distribution of precipitation during the year, with only a slight predominance of July and August over the other months.

Mid-latitude marine climates extend inland only in

Prince Rupert, British Columbia (*Cfb*)
(54° 17' N, 130° 23' W; elevation 170 feet or 52 m)

	Temp.		Precip.	
	°F	°C	in.	mm
Jan.	35	2	9.8	249
Feb.	36	2	7.6	193
Mar.	39	4	8.4	213
Apr.	44	7	6.7	170
May	48	9	5.3	135
June	53	12	4.1	104
July	55	13	4.8	122
Aug.	57	14	5.1	130
Sept.	53	12	7.7	196
Oct.	47	8	12.2	310
Nov.	41	5	12.3	312
Dec.	36	2	11.3	287
Year	45	7	95.3	2,421

Europe. As in the case of Mediterranean climates, the east-west alignment of mountain ranges allows easy access to marine air masses. In North America, this development is blocked by the Canadian Rockies; in the Southern Hemisphere, the zone of these marine climates barely reaches far enough north to impinge on the attenuated southern tips of the continents. Hence, mid-latitude marine climates have a large oceanic domain in both hemispheres, but their continental sphere is limited to northwestern and central Europe. The relatively sharp inland gradients, particularly of rainfall, are illustrated by the data for Valentia, Eire, and for London (Fig. 9-17).

Valentia is on the southwestern coast of Ireland, and London is only 450 miles (720 km) to the east as the crow flies. But there is sufficient relief between the two to make London 5° F (3° C) colder in winter and 5° F warmer in summer. More significantly, the precipitation drops from 56 inches (1,420 mm) to 23 inches (584 mm). Farther east, winter temperatures continue to decrease inland, but summer temperatures hardly change. Precipitation remains between 20 and 25 inches (between 500 and 625 mm), except where relief is sufficiently marked to add its orographic impetus to winter cyclonic storms and summer thundershowers.

Mid-latitude marine climates over the oceans are

Valentia, Eire (Cfb)
(51° 56' N, 10° 15' W; elevation 30 feet or 9 m)

	Temp.		Precip.	
	°F	°C	in.	mm
Jan.	45	7	7.0	178
Feb.	45	7	4.7	119
Mar.	46	8	3.8	97
Apr.	48	9	3.1	79
May	52	11	3.2	81
June	57	14	3.2	81
July	59	15	4.3	109
Aug.	59	15	4.6	117
Sept.	57	14	4.3	109
Oct.	52	11	5.1	130
Nov.	48	9	6.3	160
Dec.	46	8	6.3	160
Year	51	11	55.9	1,420

Invercargill, New Zealand (Cfb)
(46° 29' S, 168° 21' E; elevation 12 feet or 4 m)

	Temp.		Precip.	
	°F	°C	in.	mm
Jan.	57	14	4.2	107
Feb.	57	14	3.3	84
Mar.	55	13	4.0	102
Apr.	51	11	4.1	104
May	46	8	4.4	112
June	43	6	3.6	91
July	41	5	3.2	81
Aug.	44	7	3.2	81
Sept.	47	8	3.2	81
Oct.	51	11	4.1	104
Nov.	52	11	4.2	107
Dec.	55	13	4.0	102
Year	50	10	45.5	1,156

London, United Kingdom (Cfb)
(51° 29' N, 0° 00'; elevation 149 feet or 45 m)

	Temp.		Precip.	
	°F	°C	in.	mm
Jan.	39	4	2.0	51
Feb.	40	4	1.5	38
Mar.	44	7	1.4	36
Apr.	48	9	1.8	46
May	54	12	1.8	46
June	60	16	1.6	41
July	64	18	2.0	51
Aug.	63	17	2.2	56
Sept.	59	15	1.8	46
Oct.	51	11	2.3	58
Nov.	44	7	2.5	64
Dec.	40	4	2.0	51
Year	50	10	22.9	584

North Pacific (Cfb)
(45°–50° N, 135°–140° W)

	Temp.		Precip.	
	°F	°C	in.	mm
Jan.	43	6	Unavailable	
Feb.	43	6		
Mar.	42	6		
Apr.	43	6		
May	46	8		
June	50	10		
July	54	12		
Aug.	58	14		
Sept.	57	14		
Oct.	54	12		
Nov.	48	9		
Dec.	45	7		
Year	49	9		

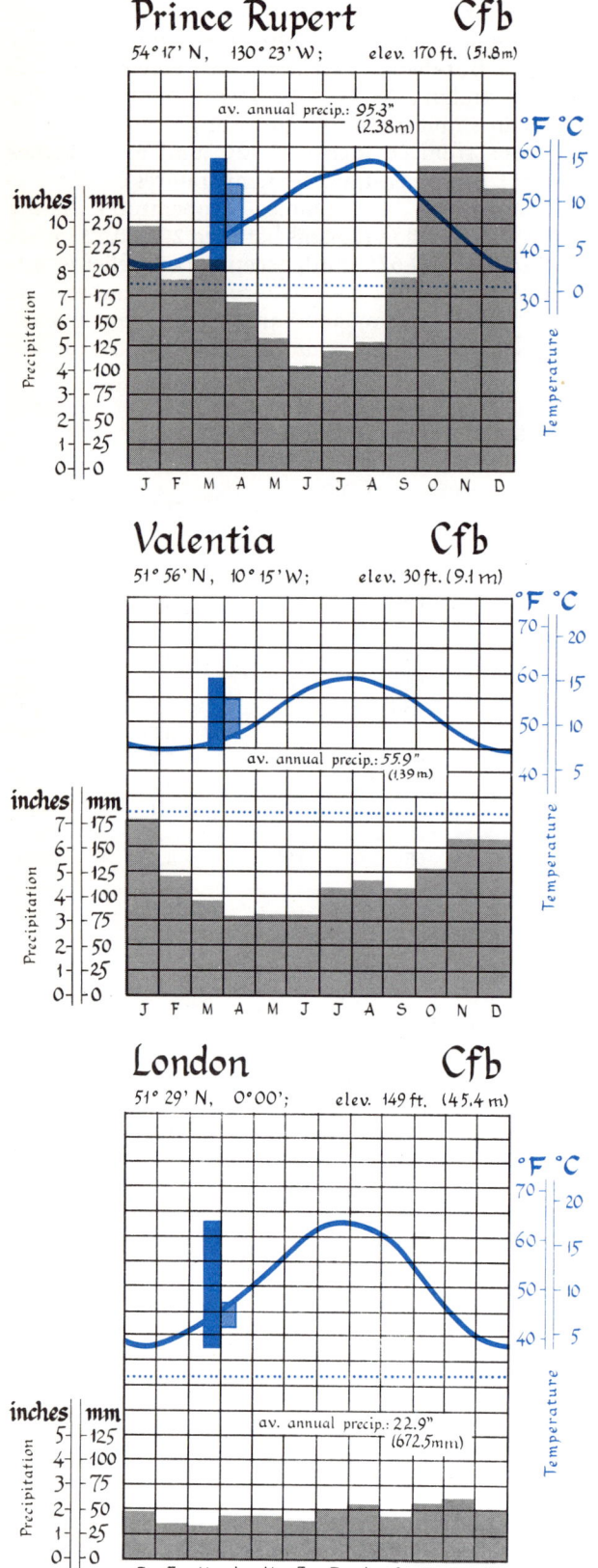
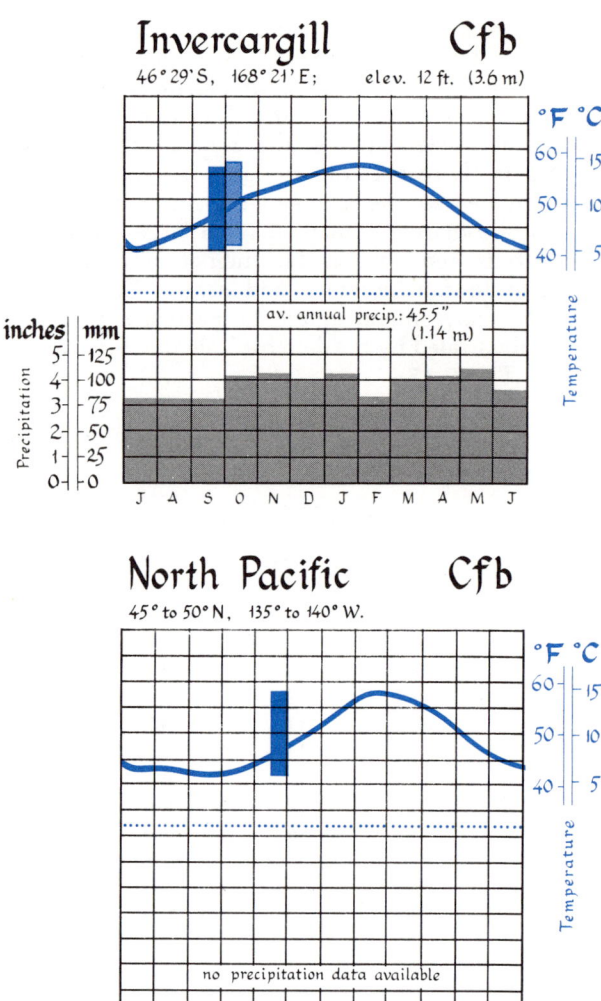

Fig. 9–17. Temperature and precipitation at selected *Cfb* stations. Mildness and humidity characterize all of these stations despite their different locations.

represented by the data for Invercargill on the south coast of South Island in New Zealand, and for the North Pacific (Fig. 9–17). Invercargill is perhaps representative of the nearly circumplanetary belt of *Cfb* and *Cfc* climates in the Southern Hemisphere, for which climatic data are hard to come by. This supposition is strengthened by comparing its temperature regime with that for the quadrangle in the North Pacific. These two stations do not differ very much from the coastal climates of Prince Rupert or Valentia, although the absence of orographic turbulence may be seen in the total precipitation.

Mid-latitude marine climates are associated largely with oceanic surfaces, on which neither soils nor vegetation are very well developed. On land, there is a diversity of soils as well as of vegetation. In general, the gray-brown **podzolic** and **podzolic soils** predominate, though the diverse mountain environment and the small size of geomorphic units in Europe go a long way toward obscuring any zonal arrangement of soils. Natural vegetation is similarly heterogeneous. In North America, a **coniferous** forest is dominant and relatively untouched by man; in Europe, a **broadleaf deciduous** forest gives way inland to a mixed forest of broadleaf deciduous trees and **evergreen conifers;** while in New Zealand and in southern Chile, a broadleaf deciduous forest grades northward into a mixed forest.

Humid Continental Climates

The humid continental (*Dfa, Dfb, Dwa, Dwb*) climates occupy the central and eastern portions of North America and Eurasia in higher mid-latitudes; they are practically confined to continents in the Northern Hemisphere and are completely absent from the Southern Hemisphere, since the cold winters implied by *D* are not found in conjunction with the warm summers implied by *a* or *b*, in the oceanic Southern Hemisphere. The polar front dominates the climates of these continental areas. On the one hand, it makes this zone a battleground between polar and tropical air masses, thus emphasizing extremes of temperature and precipitation — heat waves, cold waves, blizzards, and convective downpours are all annual phenomena in the humid continental climates (parts of the state of Illinois experienced all these, not to mention hail and destructive winds, within the single month of January 1967). On the other hand, this position under the polar front implies some cyclonic precipitation during most of the year, even though winter is relatively dry in the western part of the North American area and is decidedly dry in the eastern part of the Eurasian area (see Chapter 7). Despite the tendency toward winter drought, snow is a salient feature of these climates; whether snowfall is light or heavy, the long cold winter allows the snow to lie on the ground for three, four, or even five months.

The range of temperature and precipitation regimes in the humid continental climates is illustrated by the data for Ann Arbor, Michigan, Moscow, U.S.S.R.,[7] and Harbin, Manchuria (Fig. 9-18).

Ann Arbor is a *Dfa* station on the southern border of *D* climates in North America. As you can see from its winter temperatures, it is close to the border of the *C* climates, and it is barely warm enough in summer to qualify as an *a* climate. In other words, the northern boundary of the hot summer (*a*) climates is very close to the southern boundary of the cold winter (*D*) climates in this part of the United States. However, with increasing continentality, the area of *Dfa* climates broadens out to the west. Ann Arbor's summer maximum of precipitation is fairly well marked, but there is a fair amount of precipitation in even the driest months (in the same latitude but at the western edge of the *Dfa* climates, O'Neil, Nebraska, averages 3.9 inches or 99 mm in June but only 0.4 inches or 10 mm in January).

Relatively moderate summer maxima of precipitation, in contrast to the very heavy summer rain of monsoon or savanna climates, may give a misleading idea of the seasonal trend of the water balance. Let us compare the monthly averages of precipitation at Ann Arbor

Ann Arbor, Michigan (*Dfb*)
(42° 17' N, 83° 44' W; elevation 871 feet or 265 m)

	Temp.		Precip.	
	°F	°C	in.	mm
Jan.	24	-4	1.9	47
Feb.	24	-4	1.8	47
Mar.	34	1	2.3	58
Apr.	46	8	2.8	71
May	58	14	3.5	88
June	68	20	3.4	87
July	73	23	2.8	72
Aug.	70	21	2.7	68
Sept.	63	17	2.8	70
Oct.	52	11	2.6	66
Nov.	38	3	2.4	60
Dec.	28	-2	2.0	52
Year	48	9	31.0	786

Moskva (Moscow), U.S.S.R. (*Dfb*)
(55° 46' N, 37° 0' E; elevation 505 feet or 154 m)

	Temp.		Precip.	
	°F	°C	in.	mm
Jan.	15	-9	1.5	38
Feb.	16	-9	1.4	36
Mar.	24	-4	1.1	28
Apr.	39	4	1.9	48
May	54	12	2.2	56
June	62	17	2.9	74
July	65	18	3.0	76
Aug.	62	17	2.9	74
Sept.	52	11	1.9	48
Oct.	40	4	2.7	69
Nov.	27	-3	1.7	43
Dec.	18	-8	1.6	41
Year	39	4	24.8	631

---

[7]Moscow, Idaho, has a *Dfb* climate, as does its larger Soviet namesake, but is not nearly cold enough in winter to exemplify the Eurasian *D* climates.

Harbin (Haerphin), China (Dwa)
(45° 43' N, 126° 40' E; elevation 526 feet or 160 m)

	Temp.		Precip.	
	°F	°C	in.	mm
Jan.	-1	-18	0.2	5
Feb.	5	-15	0.2	5
Mar.	23	-5	0.4	10
Apr.	42	6	0.9	23
May	56	13	1.7	43
June	66	19	3.7	94
July	72	22	4.4	112
Aug.	70	21	4.1	104
Sept.	58	14	1.8	46
Oct.	40	4	1.3	33
Nov.	21	-6	0.3	8
Dec.	4	-16	0.2	5
Year	38	3	19.2	488

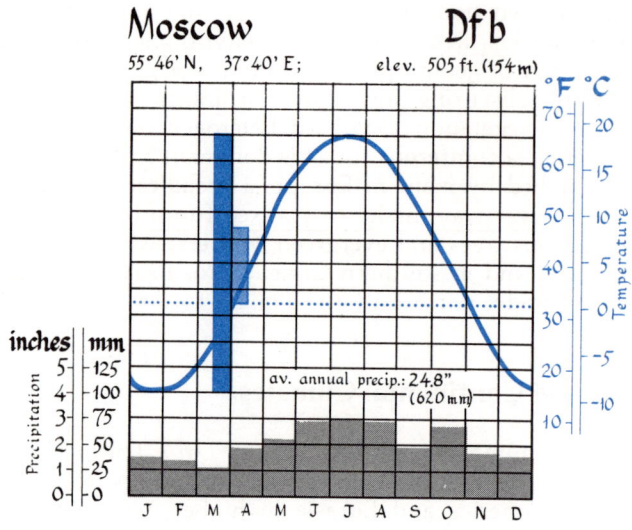

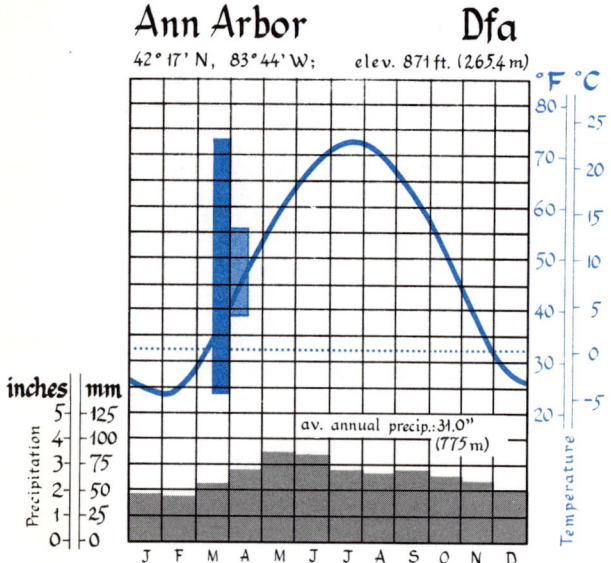

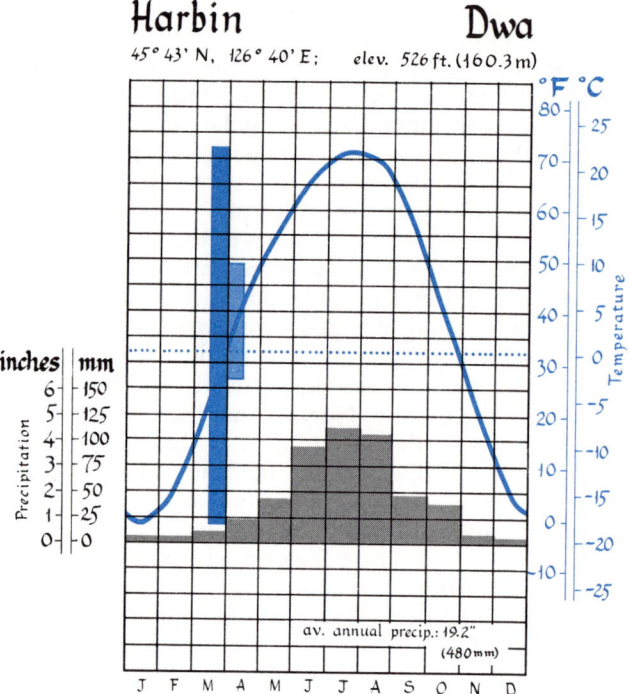

Fig. 9-18. Temperature and precipitation of selected *Dfa*, *Dfb*, and *Dwa* stations. Annual ranges of temperature are among the highest in the world. Precipitation is usually highest in summer. The winter temperatures illustrated here range from 24°F (-4°C) at Ann Arbor to -1°F (-18°C) at Harbin.

with its monthly averages of potential evapotranspiration (Fig. 9-19). Precipitation represents supply of water, and potential evapotranspiration represents the demand; the differences between them indicate the average surplus or deficit of water in any given month (Fig. 9-19)[8]. It is obvious from the figure that, for Ann Arbor, supply is unable to meet demand in summer, whereas the lower winter precipitation is more than adequate to meet winter needs, which in any case are nonexistent during the months of below-freezing weather.

The resulting march of the water balance gives a very

---

[8]The scheme used here to describe the water balance by comparing water supply and demand is that of C.W. Thornthwaite.

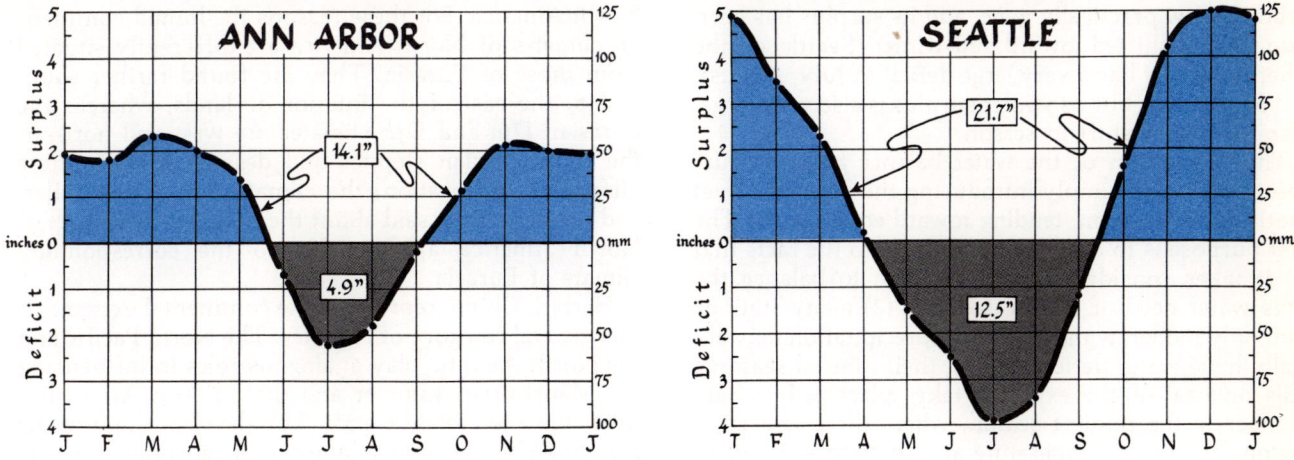

Fig. 9-19. Water deficit (gray) and water surplus (blue) at Ann Arbor and at Seattle. Note that Ann Arbor has a water deficit in summer even though summer has more precipitation than winter.

different picture of soil and plant moisture conditions than does the precipitation alone. Viewed in this way, Ann Arbor has a summer-dry climate, despite the opposite seasonality of its precipitation regime, which does not mean that its climate is similar to that of Mediterranean places with nearly rainless summers. Seattle, Washington, for example, has the same average annual water surplus as Ann Arbor; both places get about 9 inches (230 mm) more precipitation than is needed in order to meet the demands of potential evapotranspiration. Both have a surplus in winter and a deficit in summer; but surplus and deficit are both much larger at Seattle than at Ann Arbor. The significance of this difference to soil and plants becomes clearer if we take into account one additional fact. Soils are capable of storing moisture during periods of surplus, and this stored moisture can be utilized by plants during periods of deficit. The moisture-holding capacity of soils varies with their texture, their structure, and their depth. On the average, however, a soil can absorb the equivalent of 4 inches (100 mm) of rainfall before becoming waterlogged. It follows that the first 4 inches (100 mm) of surplus in the fall at Ann Arbor will be absorbed by the soil and the remaining surplus must run off either on the surface or into the groundwater table. By the same token, the first 4 inches (100 mm) of deficit in the summer can be compensated for by the stored moisture in the soil; the remaining deficit will, of course, affect the plants. The same reasoning applies to Seattle. If we take this additional factor into consideration, however, a different picture emerges. Figure 9-20 represents the net water surplus and deficit after deducting the contribution of stored soil moisture from the deficit and after deducting the absorption of water by the soil from the surplus. As you can see, the summer deficit of Ann

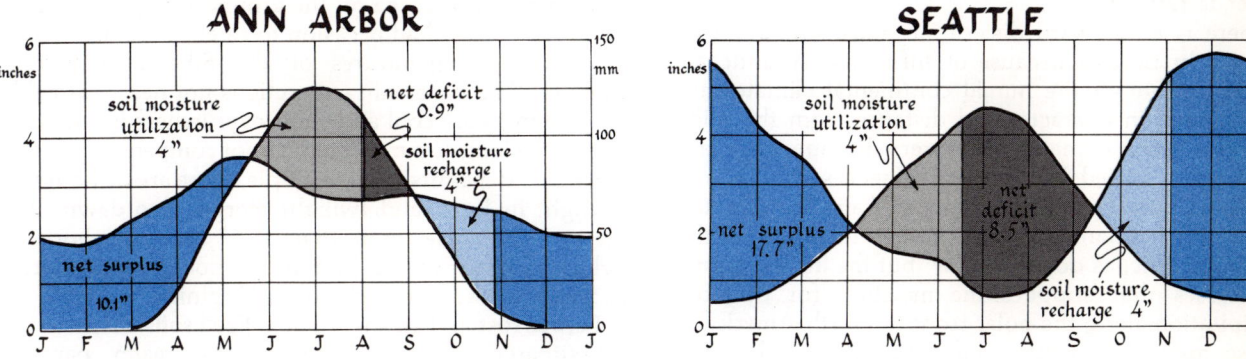

Fig. 9-20. The water balance at Ann Arbor and at Seattle. The net deficit is very small at Ann Arbor, after the soil moisture has been utilized to alleviate most of the summer deficit.

Arbor is now practically gone, and its surplus has been reduced and shifted toward mid-winter; Seattle, on the other hand still has a very large deficit in July, August, and September; but its winter surplus is also very large, despite a slight shift in season.

This discussion of the water balance has two purposes other than simply introducing the concept of yet another open system, tending toward equilibrium. The first purpose is to draw your attention to the facts that (1) it takes considerable summer rain to balance the large water need of that season and (2) many stations with only moderate maximums of precipitation may actually have water deficits during their rainiest seasons. This reversal of the expected takes place only where summer is the rainiest season, where there is a large seasonal range in temperature and, therefore, in water need, and where the summer maximum of precipitation is relatively modest. Elsewhere, the rainiest season is also the season of water surplus.

The second purpose of this discussion is to point out that despite the reversal of seasonality introduced by considering both supply and demand of water, summer rains do not produce the same kind of water balance as winter rains. Furthermore, since the reversal is limited to certain situations only, the distinctions made in the Köppen classification among *w*, *f*, and *s* climates are valid.

Moscow (Fig. 9-18) has a *Dfb* climate and illustrates the humid continental climates of eastern Europe and central Asia. Its high annual temperature range, cold winters, and relatively rainy summers all point to the continental character of this inland climate. But all things are relative. The latitude of Moscow places it more than 900 miles (1,450 km) north of Ann Arbor. To bring out the relative mildness of the European Russian climate, we should compare it with that of Goose Bay, Labrador, which is in the same latitude. As you may recall, January at that Labrador station averages a very cold 0° F (-18° C); in comparison, the Muscovite winter is fairly mild.

There is a poleward dislocation and moderation of climates in Eurasia because of inflowing Atlantic air masses. Although the humid continental climates are far removed in character and distance from the midlatitude marine climates of western Europe, they are distinctly modified by oceanic influences. The North American humid continental climates are also affected by the North Atlantic. Ann Arbor gets its rain from the North Atlantic, as does Moscow; but the inflow of *mT* air masses that supplies the moisture for cyclonic precipitation along the polar front in North America is not so marked in winter and does not extend so far north as the eastward penetration of marine air into Eurasia. Furthermore, the tempering effect of Pacific air masses are blocked by the western cordilleras of North America. For these reasons the humid continental climates of North America are differently situated from those of Eurasia. They are found farther south and to the east of the interior drylands, whereas the Eurasian *Dfa* and *Dfb* climates are west and north of the central Asian steppes and deserts. Despite these differences in location, the climates are quite similar, and what we have said about the climate of this part of North America applies also to the corresponding climate of Eurasia.

Harbin, China, represents the continental extreme of this general category of climates. The North Pacific and the North Atlantic play analogous roles in influencing the Manchurian summer and that of Ann Arbor, but there the resemblance ends. Low-level outflow of air from the Siberian anticyclone completely dominates the winter climate around Manchuria. Subzero temperatures and negligible precipitation are the inevitable results.

The humid continental climates in the southern portion of the area are covered by **podzol soils,** but their continental margins impinge on **prairie soils** and **chernozems.** The natural vegetation is mainly a mixed forest of **broadleaf deciduous** trees and **evergreen conifers,** though it also includes the southern reaches of the **taiga.**

Subarctic Climates

Subarctic *Dfc*, *Dwc*, and *Dwd* climates are found between humid continental and tundra climates in the source region of *cP* air masses. This large area is influenced very little by the ocean. Churchill, on Hudson Bay, and Yakutsk, in northeastern Siberia, are representative stations (Fig. 9-20).

Bitter winter cold is followed by a very rapid warming in summer, tempered only by the high latitude. Winter is very dry, although the small amounts of snowfall are preserved on the ground for half the year. We have already had occasion to refer to the incredibly low winter temperatures of the Siberian subarctic climates. The American subarctic is not nearly as cold, but summers are cool and short everywhere in the subarctic; the long summer days cannot compensate for the low sun. At Yakutsk, for example, there are 20 hours of daylight in June, and twilight merges into dawn; but the sun never rises more than 51½° above the horizon (which is about the position of the noon sun in March at San Francisco or Richmond, Virginia).

Permafrost and strongly **podzolized soils** characterize the subarctic climates. They are the realm, par excellence, of the **taiga,** the largest stretch of unbroken though relatively scrawny forest on the surface of the earth.

Churchill, Manitoba (Dfc)
(58° 47′ N, 94° 11′ W; elevation 43 feet or 13 m)

	Temp.		Precip.	
	°F	°C	in.	mm
Jan.	−19	−28	0.5	13
Feb.	−16	−27	0.6	15
Mar.	−6	−21	0.9	23
Apr.	14	−10	0.9	23
May	30	−1	0.9	23
June	43	6	1.9	48
July	53	12	2.2	56
Aug.	52	11	2.7	69
Sept.	41	5	2.3	58
Oct.	27	−3	1.4	36
Nov.	5	−15	1.0	25
Dec.	−11	−24	0.7	18
Year	17	−8	16.0	407

Yakutsk, U.S.S.R. (Dwd)
(62° 01′ N, 129° 43′ E; elevation 535 feet or 163 m)

	Temp.		Precip.	
	°F	°C	in.	mm
Jan.	−49	−45	0.3	8
Feb.	−33	−36	0.2	5
Mar.	−10	−23	0.1	3
Apr.	16	−9	0.3	8
May	40	4	0.4	10
June	57	14	1.1	28
July	63	17	1.6	41
Aug.	57	14	1.3	33
Sept.	42	6	1.0	25
Oct.	17	−8	0.5	13
Nov.	−19	−28	0.4	10
Dec.	−42	−41	0.3	8
Year	11	−12	7.5	192

Fig. 9-21. Temperature and precipitation at selected *Dfc* and *Dwd* stations. The continental characteristics of these climates are more marked in Asia than in North America, but cold dry winters and extremely large annual ranges are evident at Churchill as well as at Yakutsk. See also Oimyakon (Fig. 9-4).

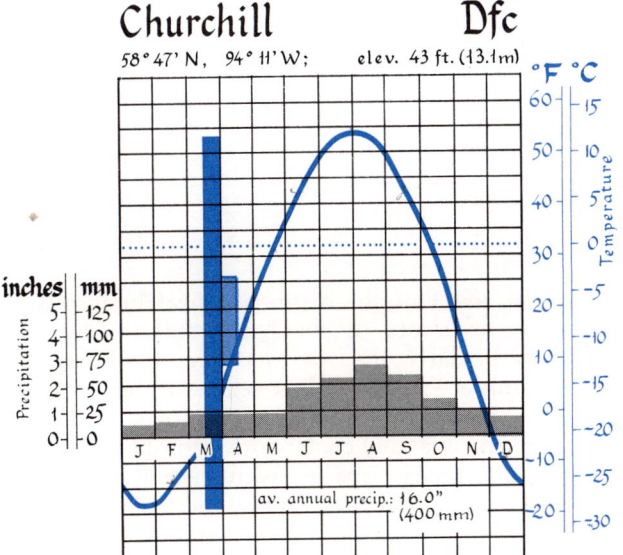

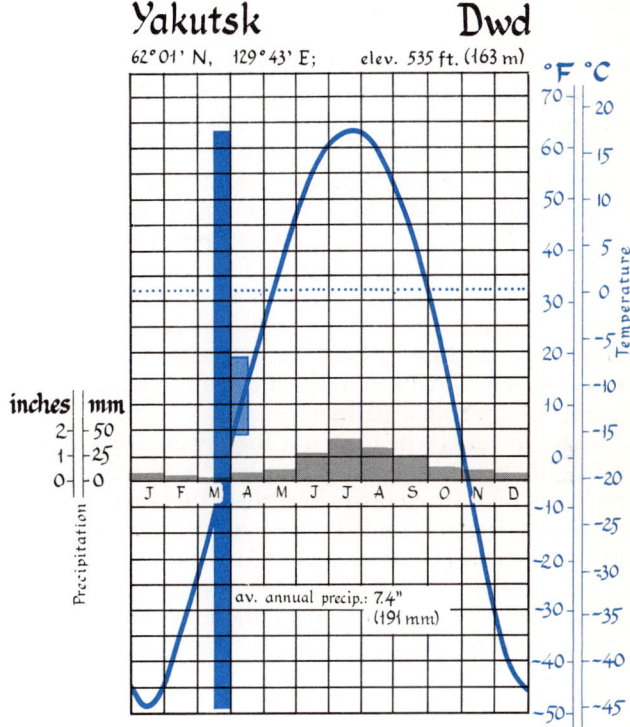

Special Climates of a Non-Zonal Nature

The climates we have described so far are grouped together on the basis of broad morphological similarities. It is not surprising that similar climates occupy similar locations; identical latitudes have similar regimes of insolation, and equivalent location with respect to continental and oceanic borders results in nearness to similar centers of atmospheric action and hence to dominance by comparable air-mass regimes. Insolation and air-mass movement are not only fundamental climatic factors, but they also result in a zonal distribution of climatic types.

Some differences in climate are the result of different

locations, but not in the broadly zonal sense of the word. We have already spoken of the difference between north- and south-facing slopes (Chap. 1 and Chap. 2) with regard to climate and vegetation. The impact of such locational differences within a mountain landform cannot be overstressed. For instance the average annual precipitation on the island of Kauai, in Hawaii, varies from 53 inches (1,345 mm) at Kilauea Point on the windward northeast coast, by way of 476 inches (12,090 mm) on the windward side of Mount Waialeale, to 23 inches (586 mm) at Makaweli on the southwest leewardside, over a total distance of less than 30 miles (50 km). The highest volcanic peaks of Central Mexico, notably Citlaltepetl and Popocatepetl, have glaciers down to about 14,000 feet (4,300 m) on their north-facing slopes, but only bare rock on their south-facing slopes where radiation is much higher.

Half Moon Bay, on the California coast, has an average July temperature of 57.9°F (14.4°C) while Redwood City less than 15 miles (24 km) away but on the east side of the Santa Cruz Mountains is 10°F (5½°C) warmer. In addition, the warmest month along the coast is September (when the ocean is at its warmest) whereas a few miles inland, in the shelter of the mountains, the maximum is in July (which is normal for continental stations).

Other elements of the climate are not so closely tied to location, however. The most significant variable is the character of the surface, whose importance we have already stressed. The basic distinction, between land and water, can best be thought of as part of the zonal pattern; likewise, the presence or absence of snow on the land or ice in the sea is another difference in the nature of surface which can produce significant differences in climate and which can also be thought of as part of the zonal pattern.

Some surfaces which are not zonally distributed, but are highly localized may nevertheless have such a striking effect on the air masses over them that they produce climates different from those of surrounding areas of distinct surface characteristics.[9] Many such surfaces are sufficiently widespread and have strong enough impacts on the climate to merit attention. We shall discuss only two, the forest and the industrial city[10].

___

[9] An even stronger distinction is made between zonal and non-zonal soils (Chap. 22), for the non-zonal soils bear very little imprint of the zonal factors of soil formation, whereas the most strikingly local climates almost invariably resemble the zonal climates around them more than they do other local climates.

[10] Instead of thinking in terms of zonal and non-zonal climates, one can think of the localized phenomena as climatic sub-systems within the more generalized overall atmospheric system. One might, of course, prefer the simpler labels — generalities and local details.

Forest Climates

The distinctive feature of the forest is that it distributes the surface which is active with respect to heat and water budgets over the whole height of the forest, instead of concentrating the climatic action near the ground. Much of the surface heat exchange is shifted from the ground to the tops of the trees which form the crown of the forest. This transfer is most complete where there is a fairly continuous canopy. There are, therefore, strong contrasts between the seasons in a deciduous forest as well as strong contrasts between a dense forest with trees of uniform height (characteristic of temperate deciduous forests) and a sparse forest (of Mediterranean or Savanna climates) or one with a multi-storied canopy (of the tropical rainforest).

The distinctiveness of the forest with respect to the heat budget and differences among various kinds of trees is reflected by differences in albedo, as given in Table 9-2. The generally low albedos mean that the canopy traps a much larger portion of the insolation than other surfaces, both vegetated and bare. But little of this insolation reaches the surface, except when deciduous forests are leafless.

Table 9-2. Albedos of Various Kinds of Vegetation

Desert shrub	30–40%
Chaparral	20–25%
Temperate deciduous forest in leaf	10–20%
Young oak in leaf	18%
Temperate coniferous forest	1–15%
Young pine	14%
Crown of old pine forest	1%

The net effect of the distinctive heat budget is to reduce the extremes of the temperature within the forest, particularly the maxima. The most extreme temperatures are found in the treetops, while the forest floor may be one or two degrees warmer during the night, and considerably cooler during the day (although some Mediterranean species with very low rates of summer transpiration may reverse the trend). In mid-latitude forests the maximum daily temperature may be as much as 5°F (about 3°C) cooler than that over bare ground outside the forest, and on hot days the difference may be twice as great. Similarly, maximum soil temperatures can be as much as 10°F (about 6°C) cooler in the forest than outside it.

Water exchanges, in contrast to heat transfers, take place closer to the ground. This is particularly true since evaporation, which occurs in every part of the forest, is reduced by the generally lower temperatures under the canopy, whereas transpiration is obviously intense and draws most of its moisture from the ground. As with heat budgets, there are differences between species, and especially between closely spaced and widely separated trees. Relative humidities are, however, 5 to 10 percent higher inside the forest than outside it. This general surplus comes from the combination of high rates of transpiration, low ambient temperature, and relative absence of wind. The first factor supplies the absolute humidity, the second increases the relative humidity, and the third allows the humidity to accumulate without being dispersed.

A third climatic particularity of the forest is its impact on air movement. As can easily be imagined, the denser the forest the greater its braking power, but even a single row of Lombardy poplars will serve to shelter farmsteads from the full impact of flatland winds. But within a dense stand the reduction in speed is spectacular. A wind of 15 miles per hour (about 7 meters per second) near the top of a pine forest drops rapidly to half that speed as it enters the frictional layer near the tree tops and is slowed to about 2 miles per hour (1 meter per second) in the forest itself. Even winds of hurricane force are unnoticed some miles inside a dense tropical forest.

City Climates

City climates have recently become the focus of considerable attention, for the special nature of the manmade surface has striking climatic implications and not all of them happy ones. We shall briefly summarize the principal distinctions of the city surface, and its climatic consequences. We can begin with a near-truism — the city is the focus of man's modification of the physical environment. While there have been much more extensive changes elsewhere, in terms of forest destruction, replacement of wild vegetation by crops, or removal of the surface in strip mines, nowhere has the change been as intense and had more implications for the climate than in the city of the industrial age.

Lowry has grouped the climatically significant aspects of the structure of the city into several sets of "basic influences that set a city's climate apart from that of the surrounding area."[11] These include the surface materials of the city, the shapes and orientations of the structures, and the industrial city air.

The stone and concrete of city buildings have a higher thermal conductivity and a higher specific heat than the soils of surrounding countryside. City surfaces thus accept heat about three times as fast as soils, even though it takes more energy to raise the temperature of stone, brick or concrete by the same amount as a soil. Thus the city surface may not be as warm as that of the countryside during the warmest part of the day, but it will have stored much more heat below the surface. The stored heat is therefore available to warm the city at night.

The roughness of the city structures, particularly the complexity of its surfaces, both vertical and horizontal, acts as a barrier or baffle to wind and insolation alike. Wind speeds are therefore reduced by small but noticeable amounts, perhaps as much as 25 percent; and strong winds are slowed down even more (the observed annual frequency of winds exceeding 10.5 m/sec. or about 25 mph is about 370 hours at Croydon, outside the London agglomeration, but only 13 hours inside the city). The reduction in wind speed has obvious consequences for the concentration of dust and other pollutants in the air and for the transfer of heat out of the city. At the same time it causes turbulence and vertical mixing in the air layer near the ground which may remove pollutants from the lowest strata of air, but only when some wind is blowing, of course. The baffle also traps radiation, which is why the complicated surface of the city which has a larger area than that of the simpler surfaces of the countryside, can store more heat.

The city generates heat at a prodigious rate compared to the country. Industrial processes, domestic heating (as well as the heat transfer by air conditioning in summer), automobile exhausts, and human metabolism add up to an enormous production of heat. On Manhattan Island, for example, industrial activity and house warming alone produce more than twice as much heat as does the January insolation (but only one sixth as much in midsummer). Even a smaller city, such as Denver, produces about 90 langleys of industrial and municipal heat per day, which is almost as much as its average insolation on a cloudless day in winter. An automobile burns fuel at about the same rate as a home furnace, and even humans produce heat at a rate that varies between 100 watts when they are at rest and 300 watts when they work. The 36 million people living in the giant urban sprawl from Boston to Washington in 1970 produced about 70 ly per day, which is nearly one sixth of the energy received from the sun over this surface of almost 7,000 square miles (about 17,000 km²).

This artificial production of heat, when combined with the greater storage capacity of city structures, produces higher temperatures over the city than above surrounding countryside. The effect is so striking that one speaks of the **urban heat island**. Temperatures in the city have thus risen by 1° or 2° over the past hundred years (the annual mean by nearly 1°C and the average of the winter minima by nearly 2°C). T.J. Chandler has calculated temperatures for London and its surroundings during the period 1931-60.[12] These

---

[11] Lowry, William P., "The Climate of Cities," *Scientific American*, Vol. 217, No. 2 (August 1967), 16.

[12] T.J. Chandler, "London's urban climate," *Geographical Journal*, Vol. 127, March 1962, pp. 279-302.

Table 9-3. Average Temperatures of London and Surroundings (1931–60)

	City		Suburbs		Countryside	
Average maximum	58.3°F	(14.6°C)	57.6°F	(14.2°C)	57.2°F	(14.0°C)
Mean annual	51.8°F	(11.0°C)	50.5°F	(10.3°C)	49.2°F	(9.6°C)
Average minimum	45.2°F	(7.3°C)	43.1°F	(6.2°C)	41.8°F	(5.4°C)

temperatures, shown in Table 9-3, indicate that the city is a little warmer than the countryside during the warmest part of the day, but considerably warmer at the time of minimum temperatures.

The difference between average maximum temperatures is only 1.1°F (0.6°C), but that between the average minimum temperatures is 3.4°F (1.9°C). These differences may be even greater on particular occasions when the city center may be 10°F (5½°C) warmer than its surroundings. It should be obvious that differences in temperature exist within the city as well. As Bryson and Ross point out, "Heavily traveled streets in most cities are 2° or 3° warmer than side streets in winter. The areas around stop lights are usually 2° or 3° warmer than the areas between stop lights — not because of the lights but because cars idle there."[13] The heat island corresponds very roughly in size and intensity with the size of the city, but there appears to be a maximum effect (of about 15°F or 8°C) regardless of city size. Furthermore, the heat island is present even over very small cities. Corvallis, Oregon which has fewer than 30,000 inhabitants, for example, has nevertheless developed city-center temperatures at least 10°F (5½°C) warmer than those of the surrounding Willamette Valley countryside.

The urban heat island is not a permanent phenomenon. It is much weaker during the day, and may indeed disappear if insolation is high, since the dirty air over the city will then affect incoming solar energy much more than the cleaner country atmosphere. It will also disappear if the wind reaches significant velocities such as 25 mph (11 m/sec.) for a city as large as London or as small as 5 mph (2.2 m/sec.) for small towns. In short, the heat island will be best developed under conditions of low insolation and weak winds, conditions that are most commonly found on winter nights; the heat island will be least significant on hot midsummer days.

The peculiar hydrologic arrangement of the city is yet another factor that differentiates urban from rural surfaces. Except for the precipitation that falls on parks or gardens, all other rainfall is removed as quickly as possible by drains and sewers, and the snowfall is often

carried away. As a result the city surface is drier. Aside from the obvious difference between paved street and muddy field, the drier city has much less water to evaporate and hence more of the energy trapped in the heat island is used to warm the city air, thus adding to the warmness of the city. Despite these differences in wetness the city air seems to have only slightly smaller relative humidities and about the same absolute humidity as country air.

The last but certainly not the least of the factors differentiating the city surface from its country cousin is its potential as a source of air pollutants. Even small cities without much industry produce almost as much pollution per inhabitant as very large ones. This is due, in part, to the large role of the automobile in creating atmospheric pollution. About 90 percent of the carbon monoxide, about 80 percent of the hydrocarbons, about 25 percent of the nitrogen oxides and a smaller fraction of sulfur oxides and other particulates are produced by the internal combustion engine. No wonder, then, that even completely residential sections of cities contribute heavily to air pollution.

Air pollution is not new (John Evelyn commented sadly on the smells, smoke and soot of London in 1661), but it has become aggravated in modern times with the expansion of industry and the increasing use of automobiles. The result is the development over almost all western cities of an **urban dust dome,** with ten times as many condensation nuclei, particulates and gaseous admixtures as in country air.

The consequences of this dust dome are many. It reduces incoming insolation by 15 to 20 percent (and ultra-violet radiation by considerably more). But the dust dome also cuts down outgoing radiation from the city, so that the net effect may be to raise temperatures slightly. The dust brought into the city by air movements from the cooler country toward the warmer city (movements similar to the land breeze discussed on p. 45) circulates within the dome, and is lifted by turbulence over the city itself. At some distance aloft the air circulation is outward from the city center and the pollutants move outward also, but since vertical movement decreases, the pollutants often sink over the outer city. The concentration of dust and other particles at the top of the dome has other climatic consequences. The blanket of pollutants absorbs radiation from above and also from below; it thus creates a layer of warm air

---

[13]Reid A. Bryson and John E. Ross, "The Climate of the City," in *Urbanization and Environment,* Duxbury Press, North Scituate, Massachusetts, 1972, pp. 61-62.

some distance aloft. This stratification of temperature creates an inversion or intensifies the stability of already existing inversions. In any case it serves to place a lid on the heat and pollution of the city by limiting upward movement to the turbulence that takes place within the dome itself. A city that experiences inversions is thus particularly subject to air pollution. If horizontal air movement is likewise restricted by mountain barriers, as in the case of Los Angeles, the situation may become critical, especially to those who suffer breathing disorders.

The consequences of the appearance of a dust dome and heat island for precipitation are not clear. The increased concentration of condensation nuclei and the turbulence over the rough city surface should lead to more condensation within the pollution dome. The stability of the air must work to reduce precipitation potential, however. Fog would thus appear to be a common condensation form over cities, despite the relative warmth which may keep city temperatures above the dew point when the general ambient temperature is not very low. Indeed, some studies indicate that in winter fog is twice as frequent over cities than over the country, but only one-third more frequent in summer. The precipitation contrasts between city and country are less clear, although most observers believe that there is more precipitation over cities, and that it is spread over a larger number of rainy days.

More obvious consequences of the presence of a dust dome include its effects on the eyes, noses and throats of the population. The most easily quantifiable of such effects of the dust dome are clearly demonstrated by the reduction in visibility observed over most cities in the past decades. For instance, since the end of World War II, Tokyo has had a threefold increase in the number of days during which horizontal visibility is less than 2,000 m (1¼ miles). Similar increases have also taken place in much smaller places. In the southern end of the Willamette Valley, in Oregon, the incidence of days with visibility exceeding a mile (1,600 m) has been catastrophically reduced in the last decade alone.

These then are some of the peculiarities of city surfaces and their climatic consequences. One last remark may serve to emphasize the powerful impact of human institutions on the physical environment. Studies in both the United States and Great Britain indicate that there is more precipitation over industrial cities during weekdays than on weekends. The same is true of temperatures; Sunday is more like a day in the country than the weekdays when industrial and other urban activities are at full throttle. How strange that the biblical injunction to rest once every seven days in an originally pastoral society, should eventually trigger important climatic changes in cities of an industrial era, thirty centuries later.

Summary

This chapter is, in itself, a summary. It has focused on the combined effects of the various atmospheric processes we previously considered separately. In one way, this is the only chapter that deals with climate in its full connotation. Even so, it is a summary in another sense, for we have really talked about the climate of fairly large areas and used particular places only to illustrate major climatic regions. The interaction of atmospheric processes and the variegated surface of the earth gives every place a distinctive, even unique, climate. Thus the heat balance at some specific place depends not only on its income of solar radiation as determined by latitude, time of year, and condition of the atmosphere, but also on the energy loss from the surface as determined largely by the outgo of terrestrial and atmospheric radiation and in part by the amount of heat used to evaporate water and to warm the atmosphere by turbulent convection. Both income and outgo therefore depend in part on the condition of the troposphere, particularly with regard to moisture content and cloudiness. These two items are constituent elements of the water balance at this same place — a balance which is simply described, yet quite complex. There is a supply of water to the surface and a demand or water need; the difference between the two determines the amount and seasonality of the water surplus or deficit. But the supply consists not just of precipitation, whose amount and regime depends on complex processes ranging from evaporation and condensation to air-mass characteristics and movement, but also on moisture storage in the soil and its transfer to and from the surface of the earth. At the same time, the water need is mainly determined by the amount of energy available to evaporate surface water and to produce transpiration in plants; the water need is also a function of the humidity of the air, of the strength of the wind, and of several other characteristics of the lower troposphere.

The interrelation of all of these processes, which taken together create climate, cannot be described simply, but the whole is another example of an open system with interacting inputs and consequences to the system. And yet this climatic system is incomplete unless we also consider the impact of these atmospheric processes on particular kinds of surfaces. Water behaves differently from land in transferring heat to and from the surface; and land surfaces are themselves sufficiently varied to produce important climatic differences. More basic yet is the significance of the general arrangement of landforms and water bodies — their consequences for the heat and water budgets of the surface and their impact on air-mass formation and movement.

A general knowledge of landforms is therefore necessary in order to completely develop our understanding of climate; likewise the development of landforms cannot be understood unless we know something about climate. But geomorphology, or the study of the earth's relief, is also a separate and equal part of our knowledge of the earth's surface, including the relief features of the ocean basins. The next section of this book therefore focuses on landforms. As in the discussion of climates we will first explain the most important tectonic or land-forming processes. We will then proceed to a description of the distribution of relief forms as they appear over the surface of the earth.

Suggestions for Further Reading

Kendrew, W.C., *Climates of the Continents*, 4th ed., Oxford University Press, Oxford, 1953, pp. 29-575.

Lowry, William P., "The Climate of Cities," *Scientific American*, Vol. 217, No. 2 (August 1967).

Marcus, M. et al, *Urbanization and Environment*, Duxbury Press, North Scituate, Massachusetts, 1972.

Miller, A. Austin, *Climatology*, 8th ed., Methuen & Company, Ltd., London, 1955, pp. 100-280.

Trewartha, G.T., *The Earth's Problem Climates*, The University of Wisconsin Press, Madison, 1961.

10 Introduction to Landforms

Introduction

The shape and contour of the land surface, or landforms,[1] are of fundamental importance for several reasons. Most obviously and literally, landforms are the fundamental upon which all terrestrial events, both cultural and natural, take place. Furthermore, the slope of the land is an important factor in the evolution of soils; the elevation of a mountain range can critically influence the amount and distribution of rainfall, and mountain barriers may check or channel the migration of plants. Thus, the form of the land plays an active role in the unfolding of natural events. It also plays a passive, though important, role with respect to human events. The significance to man to the form of the land is subtle, changing, and passive: subtle, because man's reasons for doing things in certain ways are frequently not clear even to himself; changing, because different people at different times react to similar land surfaces in very different ways; and passive, because man and not the land is the agent of any change in human events. In any case, knowledge of the nature and distribution of

---

[1]Many related terms are often used to refer to the shape of the land surface. Since they are not completely synonymous, the differences in their meanings should be kept in mind. The most general term is probably *topography* (Greek *topos* place + *graphein* to describe), which refers to description of anything that characterizes a place, whether it be a man-made feature such as a road or a house or the portrayal of the form of the land. *Physiography* (Greek *physis* nature + *graphein*) is an equally broad term, referring to a description of any natural phenomenon; but it is often used in the more narrow sense of a descriptive and genetic interpretation of the relief features of the earth. *Geomorphology* (Greek *ge-* earth + *morphology* the study of shape) is more to the point; it is the study of the land and sub-marine relief features of the earth's surface. It is probably the best term to describe the subject matter of this part of the text.

The words "landform," "terrain," and "relief" refer more specifically to the objects themselves than to their study or description. The differences among their meanings are subtle. "Landform" is perhaps more all-inclusive of every aspect of the shape and appearance of part of the land, although, strictly, it applies to individual features and not to their totality. "Terrain" carries mainly the message of area, though it too can be defined as the physical features of a tract of land. "Relief" or "relief features" differs somewhat from the other two in laying more stress on elevation, particularly on the difference in elevation between the highest and lowest parts of a piece of land.

landforms is fundamental to the study of most patterns of physical and cultural geography.

Aside from its close connection with other distributions on the surface of the earth, geomorphology may legitimately be studied for its own sake. In such a study one should appreciate not only the nature and distribution of landforms but also the processes responsible for their origin. Thus, geographers focus their attention upon landforms as both elements of the landscape and a topic for special study.

The Geographic View of Landforms

Geographic study of landforms is explanatory as well as descriptive. Some geographers in the course of their studies may be primarily concerned with description while others may focus on how landforms develop. For example, an agricultural geographer would be most interested in degree of slope, length of slope, and orientation of slope with respect to sun and wind. These aspects of slope are significant in terms of runoff and the receipt of sunlight and rain. The shady slope may have a growing season different from that on the sunny side, while the degree and length of slope influence the rate of soil erosion and the feasibility of mechanization. An urban geographer might describe a potential site in terms of the suitability of the landforms for various urban uses. His concern might be with slope or the kind of local rock underlying the surface. On the other hand, he might describe an already existing city as a manmade landform with its peculiar impact on runoff and air movement.

Geographers need a systematic, descriptive analysis or classification of landforms, one that is flexible and easy to apply and that is based on useful and meaningful terrain qualities. One of the aims of this text is to discuss landforms on the basis of such a classification.

To answer the questions "What," "Where," and "How much" with minimal ambiguity, the descriptive terms used in a classification of landforms should be standardized. Some are terms of common usage, but their meanings differ, depending on individual interpretations reflecting the backgrounds of users. Such elementary and fundamental terms as "plains," "hills," and "mountains" are used quite ambiguously in common parlance. Hence, we shall define these terms in such a way that they are meaningful in a descriptive classification.

But geographers are also interested in studying the process of landform development — that is, in determining how particular landforms came to be as they are. Consequently, another aim of this section will be to examine some of the processes involved in landform development. Brief definition of these processes are sufficient at this time. Three surface-molding processes operate on the solid exterior of the earth — **weathering, erosion,** and **deposition.** Weathering is the decay and collapse of rock as the result of chemical and mechanical changes in the rock. Erosion involves the pickup and transport of weathered materials. Deposition is the process whereby transported material is dropped as sediment.

Landforming Factors

The landform qualities resulting from weathering, erosion, and deposition are strongly conditioned by *rock composition, rock structure, climate, vegetation,* and *time*. We may call these "landforming factors" because they profoundly affect the end result of the landforming processes. A change in any one of the factors will result in a change in the landform.

To take a simple case, areas of hard rock, which are resistant to weathering and stream erosion, will form uplands or ridges, while areas of rock vulnerable to weathering will be eroded into lowlands and valleys if all the other factors are equal (Fig. 10-1). By the same token, rocks of the same type will produce different landforms if their structure (their position and arrangement) differs (Fig. 10-2).

Climate is equally important in the development of relief features. For the moment let us consider only one aspect of climate: precipitation. Warm, dry areas have intermittent streams and infrequent **sheet wash,** and relatively little chemical weathering occurs because of the shortage of surface moisture. Large differences in precipitation obviously produce different landform developments.

Vegetation has an obvious relationship to the process of weathering — particularly to erosion. A dense, deeply-rooted cover protects the land, even from the

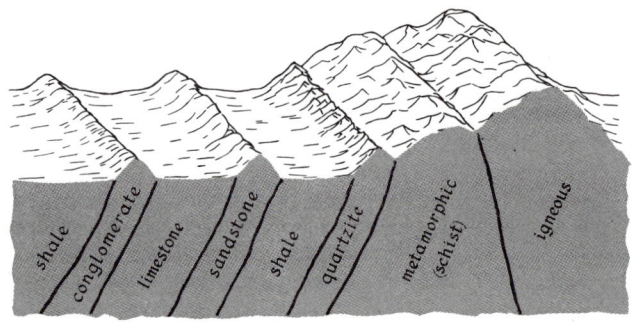

Fig. 10-1. The general relationship between rock type and landform. The resistant conglomerate, sandstone, and quartzite strata form linear ridges, while the less resistant shale and limestone strata have been eroded into valleys. The more resistant and massive schist and igneous rocks underlie the mountain ranges.

Introduction to Landforms 175

Fig. 10-2. The tilting of these originally horizontal sedimentary strata in the Uinta Mountains of Utah produces different landforms than if the beds were horizontal. The tilting is also evidence of large-scale movement in the earth's crust. *Photograph by W.R. Hansen, U.S. Geological Survey.*

impact of torrential downpours, and retards the downslope movement of weathered particles. A sparse or shallow-rooted cover has entirely different consequences for land formation.

Time is also of great importance to landform development; in some cases, it is the most important factor. Landforms long exposed to erosion are very different from those that have existed for only a short time, even when all the other factors are the same. Any study that tries to explain how a particular landform evolved must carefully consider all five landforming factors.

The Hydrologic Cycle

The movement of water from the surface of land and sea to the atmosphere and then back to the earth and its dispersal in surface and subsurface runoff are of critical importance to the landforming processes. A description of the principal components of the hydrologic cycle must, therefore, precede any discussion of landforms, even though the elements of the cycle involve both the atmosphere and the **hydrosphere** or the water of the earth.

In the simplest sense, the hydrologic cycle is a cycle of removal of water from the surface of the earth through evaporation and its return to the surface through condensation and precipitation (Fig. 10-3). Water is evaporated from the soil, from the surfaces of plants, and from the surfaces of streams, lakes, and oceans; it is also transpired by plants. The oceans, covering more than two-thirds of the earth's surface, provide most of the water vapor to the atmosphere. The vapor-laden air that moves onto the land may become involved in convectional systems or storms in which condensation occurs and water falls on the land as rain or snow. Precipitation over the oceans returns water directly to the primary reservoir. A portion of the water falling on the land is re-evaporated, some runs off the surface and collects into streams (permanent and intermittent), and further portion seeps into the ground.

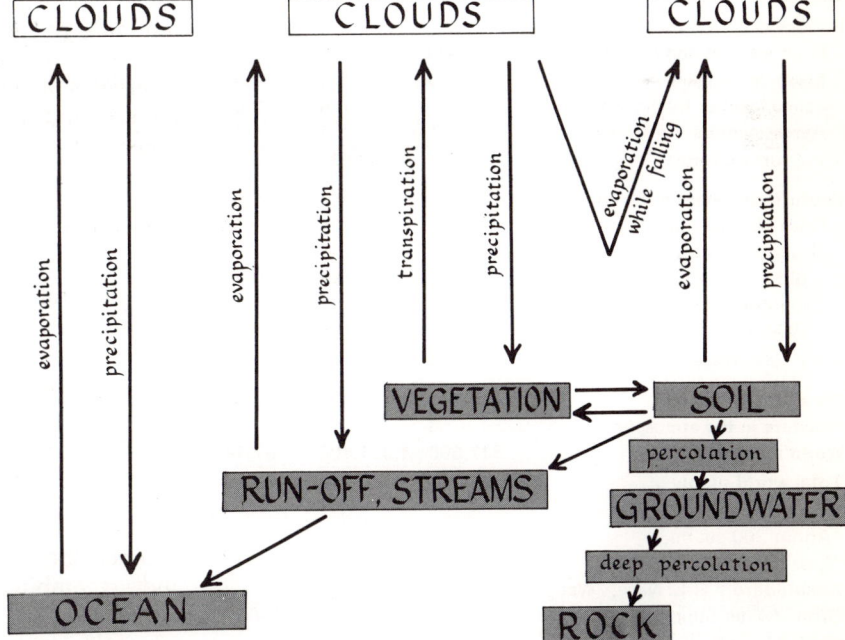

Fig. 10-3. The hydrologic cycle. Water is evaporated from the ocean and released on the surface of the sea or land by condensation and precipitation. On land, some water is lost through evaporation, some runs off in streams, and some seeps into the soil, becoming part of the groundwater, which in many instances is a source of water for springs and streams.

Here, some of the water is absorbed by plant roots, carried up through the stems, and evaporated from the leaves. Most of the remainder continues downward and becomes a part of the groundwater — the water saturating the weathered, cracked, and jointed zone of the surface rock. Some of the groundwater may move laterally and appear at the surface as seepages or springs.

The estimated distribution of the total volume of water involved in the hydrologic cycle is shown in Table 10-1. One striking feature is the surprisingly small portion of the water in the atmosphere that supplies the moisture to the land. Not so surprisingly, in view of the small percentages of water in the air, is the very small amount of water in streams.

Nevertheless, the returning surface flow of the hydrologic cycle is the primary agent of erosion. In humid areas, the surface flow operates continuously as streams and periodically, during heavy rain, as sheet wash and swollen streams. This running water is capable of removing enormous amounts of material from the land.

Two distinct sources of energy are involved in the hydrologic cycle: *solar energy*, which evaporates the water and provides, indirectly, the power to move the vapor-bearing air laterally into the land and vertically in the convective systems and storms; and *gravity*, which creates the kinetic energy that results in flow downslope at a rate proportional to the steepness of the slope and the volume of water.

The hydrologic cycle does not operate everywhere over the earth with the same intensity. More evaporation takes place over low-latitude oceans than over the interior of a high-latitude continent because of greater amounts of available energy and available moisture. The amount of condensation and rainfall varies from very small amounts, practically zero in some places, to hundreds of inches per year (Chapter 7). Consequently, stream flow and surface wash are infrequent in some areas while in other areas stream flow is continuous and surface wash is frequent. Directly and indirectly these contrasts cause significant differences in landforms.

Earth Materials

The lithosphere (Greek *lithos* stone + sphere), or outer crust of the earth, is the building material from which the landforms are carved. Water, ice, and wind move across the surface of the land and engrave a multiplicity of designs, reflecting the varying character and structure of the underlying rock. The character of rocks — that is, their degree of weakness or resistance to the agents of weathering and erosion — is determined by their composition and origin. Since rocks are a physical union of minerals and since minerals are a chemical combination of elements it follows that there are many kinds of rocks whose characteristics depend on their mineral composition and on the size and distribution of the mineral grains. A short review of the major kinds of minerals is therefore in order.

Minerals

A mineral is the product of a chemical union of two or more elements forming a natural inorganic substance, usually having characteristic and consistent natural properties of color, hardness, luster, and manner of fracturing. For example, the two elements oxygen and silicon can combine to form the mineral **silica** (SiO_2). A common form of silica is quartz, which has a hexagonal crystalline form, light color, marked hardness, glassy luster, and ragged fracture.

Since the elements of the lithosphere can combine and recombine in many ways there are literally thousands of minerals. However, relatively few minerals are sufficiently abundant to be considered

Table 10-1. Approximate Distribution of Water in the Hydrosphere

	Volume of water		
Location	Cubic miles (1,000's)	Km^3 (1,000's)	Percentage of total
Water on land:			
Surface water on the continents			
Polar ice caps and glaciers	7,300	30,429	2.24
Fresh-water lakes	30	125	0.01
Saline lakes or inland seas	25	104	0.01
Average amount in streams	*	1	**
Total surface water	7,355	30,659	2.26
Subsurface water on the continents			
Root zone of soils	6	25	**
Groundwater less than ½ mi. (800 m) deep	1,000	4,168	0.30
Groundwater more than ½ mi. (800 m) deep	1,000	4,168	0.30
Total subsurface water	2,006	8,361	0.60
Total water on land	9,361	39,020	2.86
Moisture in the atmosphere	3	13	**
World's oceans	317,000	1,321,400	97.14
Total world supply	326,364	1,360,433	100.00

*About 300 cu. mi.
**Less than 0.005 percent
Adapted from R.L. Nace, "Water Management, Agriculture and Groundwater Supplies," U.S. Geological Survey Circular 415, 1960, pp. 1-11, Table 2.

major rock-forming types, and of these, four merit particular attention:

1. Of the **silica** minerals, the most important is quartz (SiO_2), which is very resistant to both erosion and weathering. Rocks composed primarily of quartz are extremely durable.
2. The **feldspars**, a very important family of minerals, contain silica, aluminum, and varying amounts of potassium, calcium, and sodium. In general, the feldspars or aluminosilicate minerals are light to medium-dark and hard; they break along two cleavage or splitting planes at right angles or nearly so. Despite their hardness, the feldspars weather rapidly in humid climates, eventually forming clay. The important feldspars are:
    a) Potassium feldspar ($KAlSi_3O_8$), in which pink is the most prominent color; orthoclase and microcline are the principal varieties.
    b) Sodium feldspar ($NaAlSiO_3O_8$), which are white to light gray; albite is the principal variety.
    c) Calcium feldspar ($CaAl_2Si_2O_8$), which are gray to dark gray; labradorite is the common variety.
   A distinction is commonly made between the potassium feldspars and the other two; in such a case they are referred to as orthoclase and plagioclase, respectively.
3. The **ferromagnesian** minerals are, as the name suggests, composed primarily of iron and magnesium. They also contain aluminum, sodium, and silica, together with other minerals of small importance to rock formation but vital to plant growth. Rocks formed from these minerals are dark and relatively easily weathered. Among the most important ferromagnesian silicates are:
    a) Hornblende ($NaCaFeMg + Si_4O_{11}$)
    b) Augite ($CaMgFe + Si_4O_{11}$)
    c) Biotite ($KFeMg + Si_4O_{10}$)
    d) Olivine ($MgFe + SiO_4$)
   The first three are black; olivine is light green.
4. The fourth family of minerals, the **carbonates**, is composed mainly of two minerals — calcite and dolomite. Calcite ($CaCO_3$) is a combination of carbon dioxide (CO_2) and calcium; dolomite ($CaMg(CO_3)_2$), of CO_2 with both calcium and magnesium. Both of these rocks, which often occur in large masses of such single-mineral rock as limestone and dolomite, are vulnerable to solution or dissolving by groundwater, and they break down readily in humid climates. However, both are relatively resistant in arid climates, where solution weathering is greatly reduced.

Rocks

The form and properties of a rock are dependent upon the character and appearance of its physically combined minerals, which, in turn, depend on the physical and chemical properties of the individual minerals and the environment in which the combining occurred. Since there are many minerals and several environments in which minerals can combine, there are many types of rock. We shall describe only those most common in the lithosphere.

Three primary classes of rock types can be distinguished on the basis of their origin. These are **igneous rock** (Latin *ignis* fire), or rock resulting from the solidification of a liquid mass, as of lava; **sedimentary rock** (Latin *sedire* to sink down), which, as the name implies, is a rock composed of the debris of other rocks deposited by wind or water and then consolidated; and **metamorphic rock** (Greek *meta* change + *morphe* form), the product of changes in rock character brought about at or near the surface of the earth. Each of these primary types of rock can, in turn, be further subdivided.

Igneous rocks. Igneous rocks are subdivided according to two important factors: (1) the rate at which the mass cooled, since this controls the size and arrangement of the mineral grains, and (2) the composition of the original liquid mass. The latter determines what minerals will be formed during cooling and solidification. Lava exposed on the surface cools very quickly and offers little time for the growth of mineral crystals. Consequently, the crystals are small, and the rock has a fine-grained texture. Should the molten rock fail to reach the surface, however, the overlying rock retards heat loss, increases the time of crystal growth, and allows the formation of larger mineral crystals. Thus, there are **extrusive** (fine-grained) and **intrusive** (coarse-grained) **igneous rocks**.

Varieties of intrusive and extrusive rocks are differentiated by their mineral composition, color, and crystal size (Table 10-2). *Granite* is one of the best-known and most abundant of the intrusive rocks. It is a coarse-grained mixture of quartz and feldspars with a small amount of ferromagnesian minerals. Because of its preponderance of light-colored quartz and feldspar, granite is usually light gray in color, and the scattered dark mineral crystals give it a speckled appearance. *Gabbro* lies at the other end of the scale of intrusive rocks. It is a heavy, dark-colored rock composed almost entirely of ferromagnesian minerals and dark feldspars. Like granite, gabbro occurs in large masses and is a common rock in the lithosphere.

Table 10–2. Igneous rocks

Intrusive	Extrusive	
Granite	Rhyolite	Light color—Acidic
Diorite	Andesite	
Gabbro	Basalt	Dark color—Basic
Coarse-grained	Fine-grained	

Fig. 10-4. Columnar jointing in basalt, Devils Postpile National Monument, California. *Photography courtesy of the U.S. Department of the Interior, Geological Survey.*

Many varieties of intrusive rock have characteristics intermediate between granite and gabbro — in the main, the difference between them is the result of differences in content of dark minerals. The dark minerals have a higher specific gravity than the light minerals; for example, the specific gravity of granite is 2.6 and that of gabbro, 3.2.

The fine-grained, extrusive equivalents of granite and gabbro are *rhyolite* and *basalt*. Rhyolite is a light-gray rock with a dull surface, because the crystals are too fine to be seen by the naked eye. Basalt is a dense, black rock that contracts somewhat upon cooling and frequently forms into five- and six-sided vertical columns (Fig. 10-4). Basalt is the most widespread of the lavas, occasionally occurring in flows covering thousands of square miles.

Igneous rocks composed mainly of light-colored minerals of lower density belong to the granite groups. Because the rocks of this group are high in silica content, they are also called *acidic* rocks, since silicon, when combined with water, forms silicic acid. These rocks are often referred to as **sial** (for *si*lica and *al*umina). The dark, basaltic group of igneous rocks, low in silica and rich in iron and magnesium, are rich in bases. These *basic* rocks are referred to as **sima** (*si*lica and *ma*gnesium).

Sedimentary Rocks. Sediments can derive from any kind of existing rock and can be deposited on land (by streams, wind, and glaciers) or in shallow, continental seas. Rocks formed from these deposits are called **clastic** (Greek *klastos* broken) **sedimentary rock** and are subdivided according to their composition and texture (Table 10-3). Sediment may also be derived from the remains of organic material or may result from chemical reaction and precipitation. However, the majority of sedimentary rocks were deposited on ocean bottoms by waves and marine currents. Originally then these rocks consisted of unconsolidated sediment.

Table 10-3. Clastic Rocks

Size of particles (inches)	(mm)	Unconsolidated sediment	Consolidated rock
>0.04	>1.0	Boulders, cobbles gravel	Conglomerate
0.004–0.04	0.1–1.0	Sand	Sandstone
0.00008–0.004	0.002–0.1	Silt	Siltstone
<0.00008	<0.002	Clay	Shale

Sediment carried by moving water is usually deposited in a selective manner. If the transporting current is strong enough, it may carry coarse gravel as well as finer particles. If the current weakens, the larger fragments will be laid down first as a layer of gravel. Further slackening of the current may produce a second layer, composed of sand. As the speed of the water changes, dissimilar layers will be deposited. These beds, or *strata*, differ from one another in texture and composition, since the sorting process of running water not only separates the sediment by density but by size as well.

Marine sedimentary rock deposited in shallow seas occurs in successive strata of various kinds, separated

Fig. 10-5. Horizontal sedimentary strata near Mitchell, Montana. The strata are composed of banded sandy shale. The well-defined contacts between each layer are bedding planes. *Photograph by C.D. Walcott, U.S. Geological Survey.*

by *bedding planes*. In their initial state the rock layers and bedding planes are nearly horizontal (Fig. 10-5).

Consolidation of sediments to form rocks is chiefly the result of compaction and cementation, with cementation playing the most important role. Gradually, desposits of silica, calcium carbonate, and iron oxide become established in spaces between the particles and cement them into a solid rock. This process of change from sediments to rocks is called **lithification**.

The lithified sediments form several kinds of rocks based primarily upon sediment size (Table 10-3). The coarsest, *conglomerate* is composed of rounded pebbles, cobbles and small boulders, along with a mix of sand and silt. Rounding of the rocks in the conglomerate is the result of abrasion during periods of turbulent water transporation. *Sandstones* can be divided into several types on the basis of the kind of cement or composition of the sand grains. Normally the grains are of quartz, but under certain circumstances they may also be formed from feldspars or from broken pieces of shell and coral. In this latter instance the distinction between sandstone and limestone is difficult to make. This is occasionally also true in the case of *siltstone*, the product of compaction and cementation of silt. Siltstone frequently has the feel of very fine sand. This is not surprising since the distinction between coarse silt and very fine sand is quite arbitrary (Table 10-3). When thoroughly compacted by pressure, clay forms a laminated (very fine layered) rock called *shale* that easily breaks up into small flakes and plates.

The organic and chemical sedimentary rocks are composed of material produced by the growth of plants and animals or by chemical precipitation. For example, coal is an important sedimentary rock of organic origin. The one of greatest importance is *limestone*, which is most frequently formed from the cemented remains of coral, shellfish, and small, lime-fixing plants. The cement, calcium carbonate, is taken from the organic remains by solution and then redeposited around them. Inorganic limestone may result from direct chemical precipitation in waters heavily charged with lime in solution. The capacity of sea water to hold lime in solution is proportional to its temperature. One way to bring about direct lime precipitation is to cool warm, lime-rich sea water. This is most frequently brought about when cool and warm ocean currents meet and mix.

Metamorphic Rocks. As the name suggests, these are rocks that have undergone change, through either heat, pressure, the addition of a new mineral, or a combination of all three processes. In some instances, the change may be so complete that the form of the original rock is destroyed. In general, the result of metamorphism is to make the rock harder and more compact than the original, thus increasing its resistance to weathering.

Metamorphism also tends to change sedimentary rock to crystalline forms. For example, limestone, through the processes of metamorphism, becomes *marble*. Upon being subjected to intense pressure and heat, the calcite in the limestone forms and reforms new and larger crystals, while the evidence of the organic life in the original rock is obscured or altered beyond recognition.

Shale, upon being metamorphosed, turns into *slate*.

In the original shale, the rock is soft and weak, and the layers of sedimentation can be readily observed as either fine lines or thin strata. As the result of great pressure and heat, new planes of parting, or cleavage, are formed. These new cleavage planes are imposed upon the slate without regard to the original direction of stratification, or bedding. Continued pressure and internal shearing cause continued changes, and the slate becomes *schist*. Schist differs from slate in that it contains an abundance of mica and other new materials that have formed during the metamorphic process. Schist can also be derived from other rocks, such as granite, by a similar process of intense metamorphosis.

Another kind of metamorphic process is the deposition of silica by underground water around the particles in siltstone, sandstone, and conglomerate. Since the particles in these rocks are also normally quartz the result is the extraordinarily durable rock *quartzite*. Quartzite is one of the most resistant rocks found in nature; it invariably forms high, craggy ridges that stand well above the surrounding terrain.

Igneous rocks are also subject to metamorphism. For example, we already mentioned that granite may be changed to schist. Granite may also be changed to *gneiss*. The original crystalline form is not seriously altered, but the dark and light minerals are arranged in streaks or bands. Some gneisses may have been derived from sedimentary rock subjected to great change. Gneiss ranks with granite in resistance to weathering and erosion.

Resistance of Rocks to Erosion. Each rock has its own peculiar resistance to weathering and erosion, based on its composition. In the remarks that follow, it is assumed that only differences in rock composition are in question, as changes in the other landforming factors may completely alter the relative resistance of any given rock.

Generally, igneous rocks are more resistant than sedimentary rocks, while metamorphic rocks lie somewhere in between (Fig. 10-1). Granite and gneiss usually form higher mountains, while slates and schist tend to form uplands and hills. Quartzite is extremely durable and, given equal bulk, would probably be a more common highland rock than granite.

Within the sedimentary rocks there are considerable differences in realtive resistance, shale and limestone usually being less resistant than sandstone and conglomerate. Not all sandstones and conglomerates (or other sedimentary rocks) are of the same strength; variations in the kind and degree of cementation cause differences in resistance to weathering and erosion.

Climate affects the relative resistance of some of the rock types. For example, in mid-latitude humid climates, granite is usually stronger than limestone.

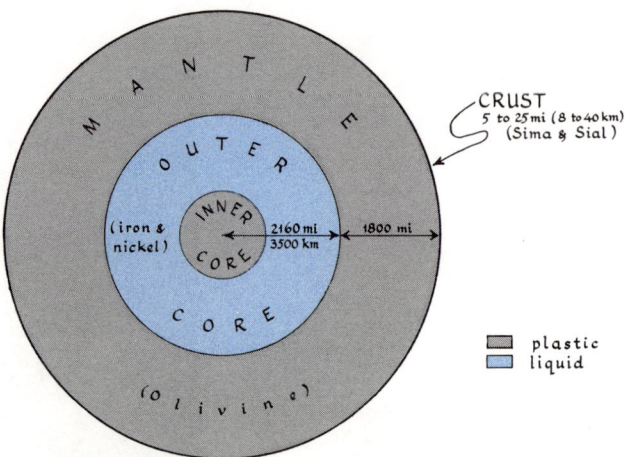

Fig. 10-6. The earth's interior.

However, in arid climates the reverse is true; limestone is more resistant than granite or other crystalline rocks. It is difficult to say whether this reversal stems from decreased resistance of granite or increased resistance of limestone due to the drastic reduction of solution weathering in arid lands.

The Form of the Earth's Crust

Our discussion so far has told us something about the kinds of materials that make up the outer crust of the earth. Let us now look at the form, or arrangement, of the material. The center of the earth consists of a solid *inner core* and a molten *outer core*, made of iron and nickel (Fig. 10-6). These cores, about five times denser than the material of the surface, are probably hotter than 4,000° F (about 2,5000° K), and have a radius of about 2,000 miles (3,200 km). Most of the remaining volume of the earth, a layer about 1,800 miles (2,900 km) thick, is composed of solid olivine minerals. This solid **mantle** reacts as a high-density and rigid medium when earthquake waves pass through it. Outermost and thinnest of all these layers is the **lithosphere,** or earth's crust, which varies in thickness from 5 to 25 miles (8 to 40 km). It has enough plasticity to adjust to some long-term pressures (which take place

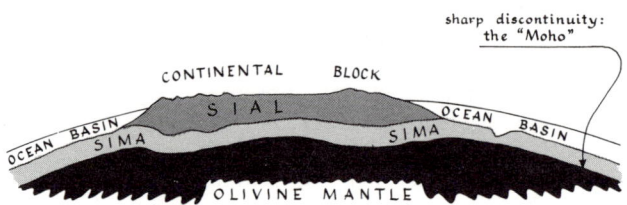

Fig. 10-7. The sima and the sial.

over millions of years) by flowing without breaking, but it is also brittle enough to crack when subjected to sudden stresses, such as faulting or mountain building.

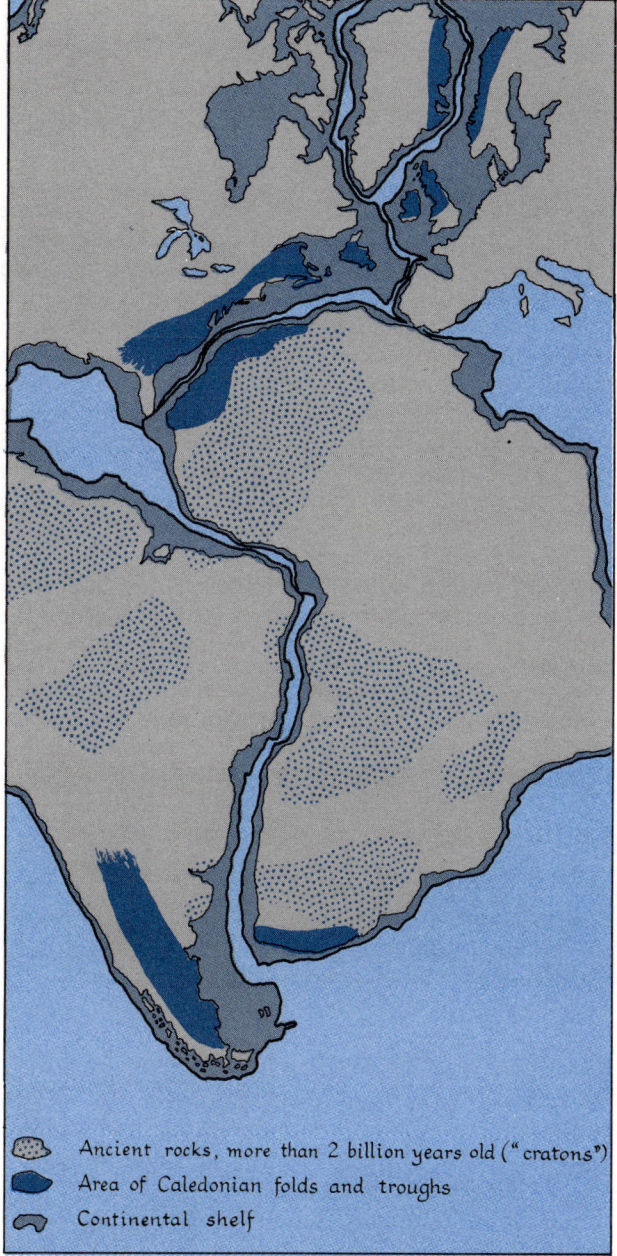

Fig. 10–8. Hypothetical reconstruction of the continental blocks before drift. The shapes of the blocks permit a very close fit at the edges of the continental shelves. In addition, both old and newer rock types and structures are arranged in very plausible formations across the contacts between blocks. Modified from P. M. Hurley, "The Confirmation of Continental Drift," *Scientific American*, April 1968, pp. 56, 60.

The earth's crust consists of two separate layers of different composition and elevation (Fig. 10-7). The discontinuous upper layer, the **sial,** makes up the continental blocks. It is composed of relatively light-weight and light-colored granitic minerals. The lower layer, the **sima,** is continuous and forms the floor of the ocean basins. It is composed of somewhat denser and darker basaltic rock, dominated by ferromagnesian minerals. The arrangement of the crust is, therefore, similar to a set of continental blocks floating on a heavier layer of basement rock.

Considerable evidence exists that the continental blocks are indeed floating. Apparently, they were once joined together and have been forced apart by the extrusion of material on the ocean floors which has caused them to drift to their present locations (Fig. 10-8). The close correspondence of the shapes of the edges of the continental shelves, the similarity in rock type and biota between continental margins now separated by wide expanses of ocean, the puzzling evidence of enormous changes in past climates, and the peculiar orientation of magnetic minerals in solidified rock that do not point to the present position of the magnetic poles are all explainable in terms of continental drift.

The principal levels of the continental blocks or platforms and oceanic basins can be seen in Fig. 10-9, which shows the relative areal extent of categories of elevation at 6,000-foot (about 1,800 m) intervals. Even this crude interval cannot obscure the fact that two categories stand out. Twenty-six percent of the earth's surface lies between sea level and 6,000 feet (1,800 m) of elevation. Another 39 percent lies between 12,000 and 18,000 feet (3,600 and 5,500 m) below the sea. Surprisingly little land lies between these two major levels; the contrast between them is relatively sharp.

If we use categories less crude than 6,000-foot intervals, the distinction between continental sial and oceanic sima becomes even more marked (Fig. 10-10). The margin of the continental blocks is not at sea level but near the 500-foot (150 m) **isobath.**[2] The continental platform, partly covered by water, thus constitutes about 35 percent of the total surface, and two-thirds of that lies in the narrow range between 500 feet (150 m) below and 3,000 feet (900 m) above sea level. The portion of the platform that is covered by water is called the **continental shelf.** The slope of the shelf depends on the general inclination of the continental margins, but its outer edge is almost always near the 500-foot (150 m) isobath. Consequently, the width of the shelf

---

[2]An isobath is a line of equal depth below sea level (Greek *isos* equal + *bathos* depth). The equivalent word for a contour line above sea level is "isohypse."

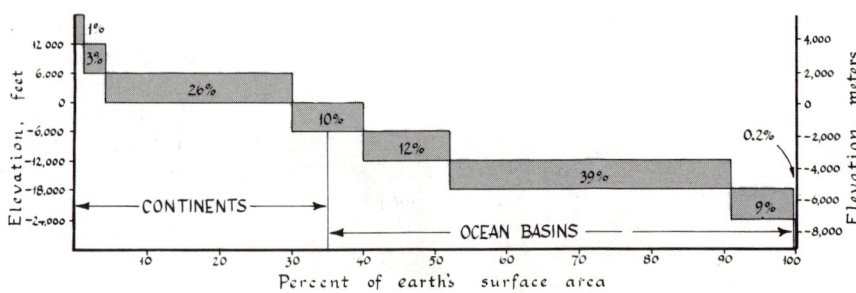

Fig. 10-9. The relationship between surface area and elevation. Each block of the diagram represents a 6,000-foot (about 1800m) interval of elevation; the length of the block is proportional to the area that falls between the various 6,000-foot elevation limits.

varies greatly from place to place. Beyond the outer edge of the shelf lies the **continental slope,** a relatively steep drop of more than 12,000 feet (3,600 m) to the abyssal plains of the oceanic floor. The continental blocks and oceanic basins are often called the *first-order relief forms.*

In the new language of continental drift the earth's crust is divided into several plates. The plates, some with continents imbedded in them, are the basic structural units of the earth's crust and can be thought of as moving more or less independently of one another. Figure 10-11 shows the arrangement of these crustal blocks in the past and at present. Note the wide range of sizes and the imperfect correspondence between the basic continental building blocks and the continents as we normally think of them. One of the smallest, the California Plate for example, is shown as an appen-dage to the main part of North America; an appendage which is believed to be sliding northwestward along the zone of faults and fractures — a zone in which earthquakes often make headline news in California.

The back endpaper shows the major zones in which crustal material is being brought to the surface along the floors of the ocean basins. The situation in the Atlantic Ocean is perhaps the most clearcut. Notice that the Mid-Atlantic Ridge, the zone of extrusion from below, crosses Iceland. The consequences for that volcanic island were unfortunately demonstrated in January 1973 (and on numerous previous instances) by destructive eruptions on the small off-shore island of Heymaey which almost obliterated the important fishing port of Vestmannaeyjar. The breaks in the ocean floor, resulting from the continuous outpouring of molten rock are also shown on the back endpaper.

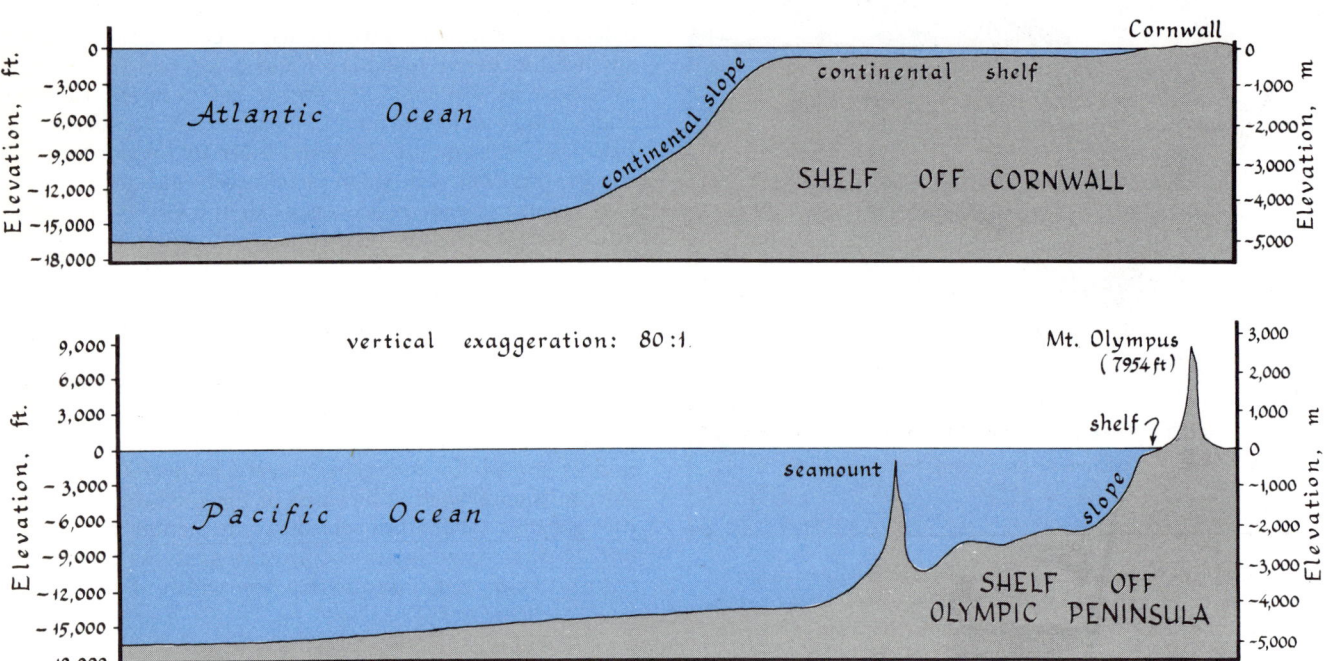

Fig. 10-10. The continental shelf and slope. The top diagram illustrates the wide continental shelf and relatively smooth slope along the coast of northwestern Europe. The shelf in the western part of the United States is much narrower, and the slope much less regular.

Introduction to Landforms 183

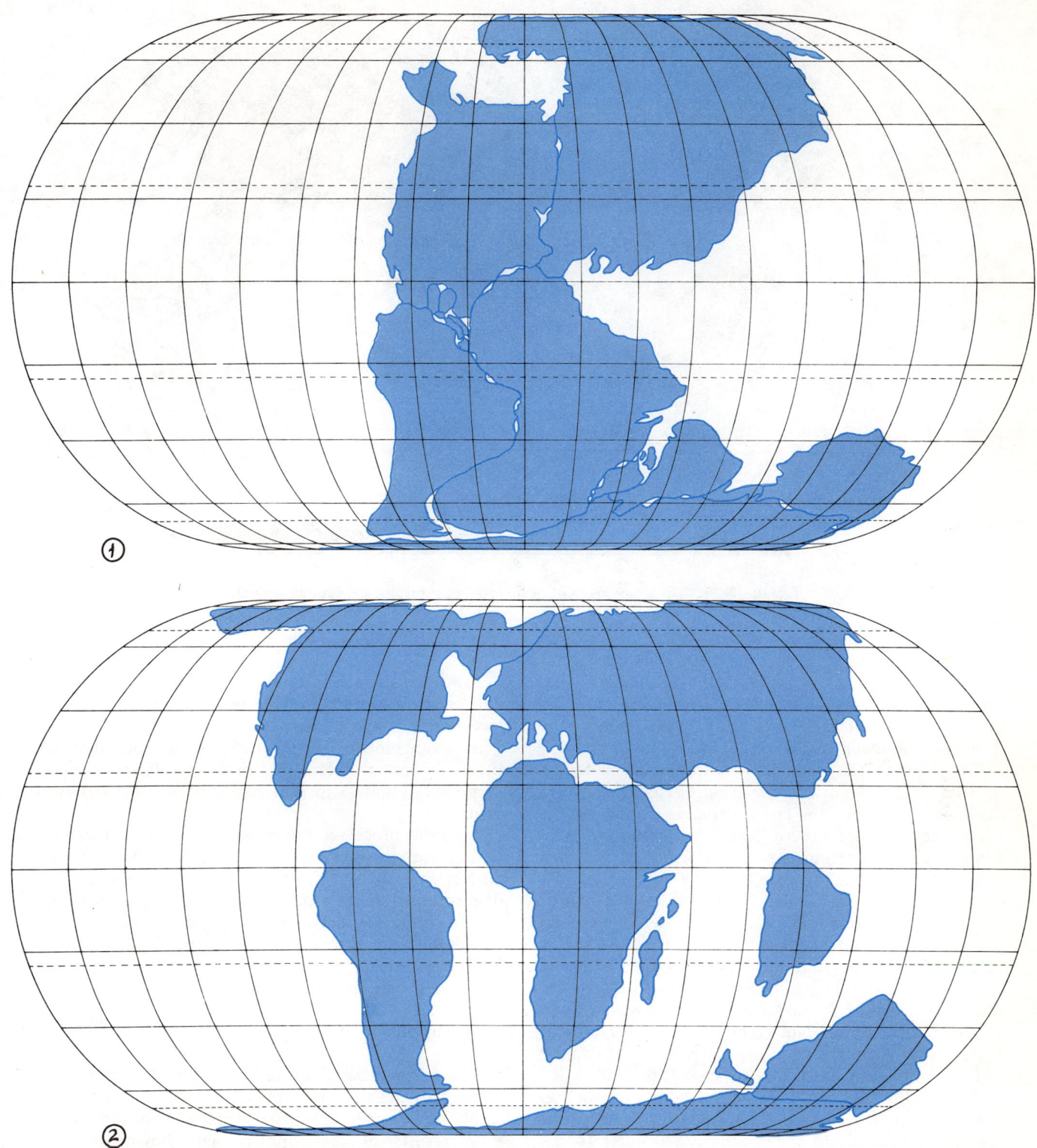

Fig. 10–11. The arrangement of tectonic plates in the past and at present. (1) Mesozoic. (2) Cenozoic (3) Present and Future. In (3) the plates, as they may appear 50 million years from now, are shown in blue.

184 Chapter 10

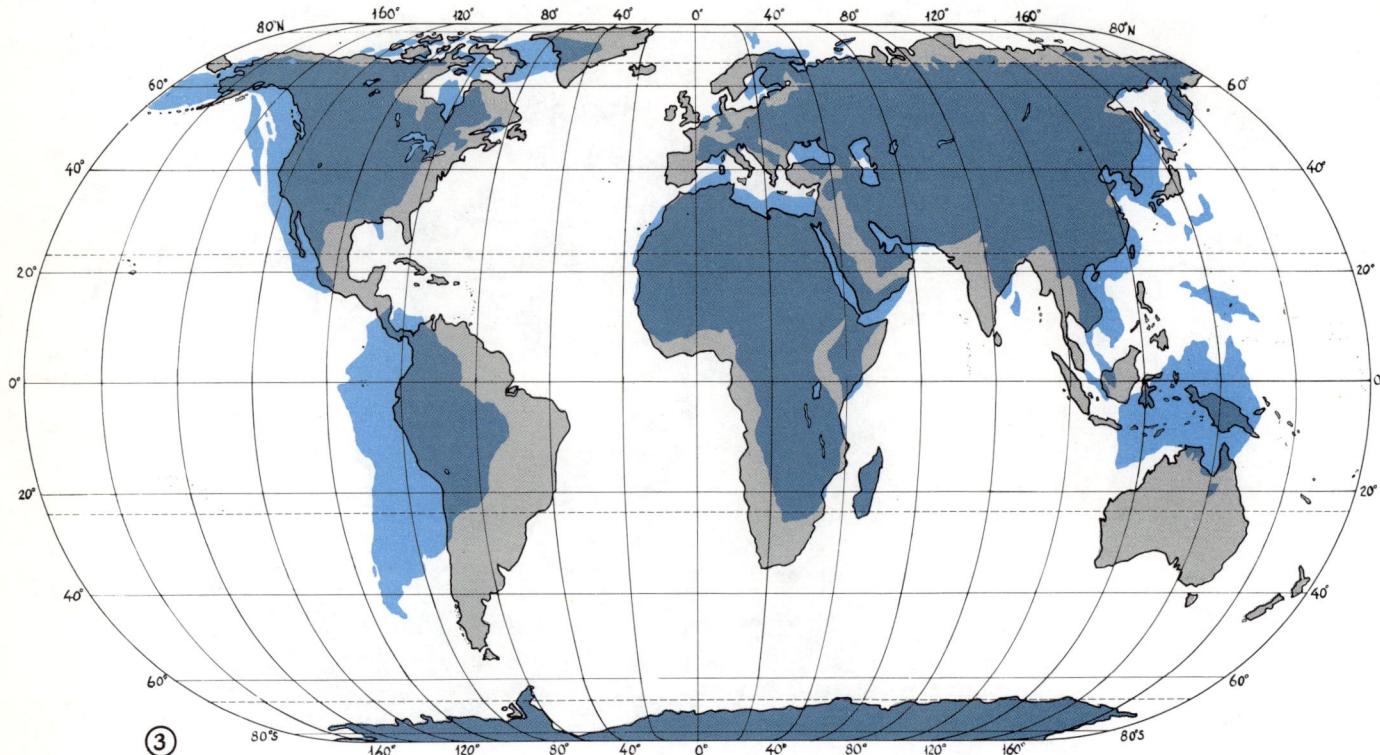

Fig. 10–13 (3). The arrangement of tectonic plates at present and future (in blue).

New crust is thus being created from sub-crustal materials mainly in the ocean basins. The resultant spreading of the ocean floor is the mechanism that causes the continental plates to drift apart from an original single land mass. The drift is illustrated by the three sets of maps in Fig. 10-11. Not shown are the areas where crustal material is being returned to the mantle. The creation of new crust obviously poses a dilemma — either the crust expands continuously, which seems unlikely (even though the pattern of fractures in the ocean floor is said to resemble that which occurs in the cover of an expanding sphere), or else there is a return of crustal material to the mantle whence the new material originally came. The second alternative is the most plausible, since it allows for an equilibrium in the long run. Furthermore, zones in which the crust is being shoved downward have been discovered. One of the major ones is in the western part of the Pacific Basin. Not coincidentally, this area of large-scale downwarping is also the zone of the deepest holes in the ocean floor (see Chap. 19, p. 290).

The landforms superimposed on the plates, such as major mountain systems and lowlands or oceanic ridges and deeps, constitute the *second-order relief* forms. They are largely the product of tectonic forces and can best be described in terms of those forces and the processes they set in motion.

Tectonic Processes

Landforms are conditioned not only by the chemical and physical character of the rock but also by the way in which the rock is arranged at the surface of the earth. This arrangement, or structure, of the rock depends primarily on the **tectonic processes** (Greek *tekton* carpenter, builder) to which the rock has been subjected.

Tectonic processes are those processes that deform and rupture the earth's crust using energy derived from within the earth. They fall into two major groups: **diastrophism** (Greek *diastrophe* twisting, distortion), and breaking, bending, or warping of the earth's crust called **faulting,** and **vulcanism** (Latin *Vulcan* god of fire), the movement of molten rock from within the earth onto or into the earth's crust.

Diastrophism

The tilting, bending, or breaking of sedimentary strata provides the most obvious evidence of movement in the earth's crust (Fig. 10-2). Some of the more common results of diastrophism are shown diagrammatically in Fig. 10-12. It represents two types of faulting; tensional and compressional (a fault being a fracture in the crust along which a slipping movement

Introduction to Landforms 185

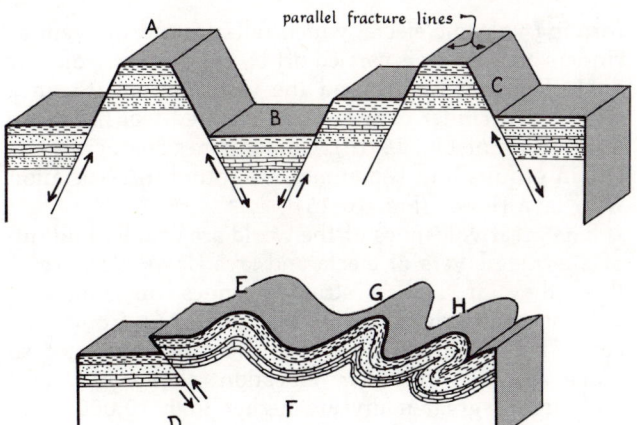

Fig. 10-12. Common faults and folds. *A* is a horst; *B* is a graben; *C* is a normal fault; *D* is a reverse fault; *E* is an anticline; *F* is a syncline; and *G* and *H* are asymmetrical anticlines.

has occurred). *A* represents a **horst** (German *horst* eyrie); parallel or near-parallel fractures have developed in the earth's crust because of tensional stresses. As the outer blocks slip downward along the fractures, the central block remains in place or is pushed up, creating a block mountain. Movement along the fault planes is not accomplished in one convulsive upheaval; many small movements over a long period are required to raise the block relative to the surrounding land surface.

Tensional stress may also produce a **graben** (German *graben* ditch); in Fig. 10-12 *B*, a block flanked by parallel fractures, has dropped downward, sharing the fate of the keystone when the sides of an arch are pulled apart. Grabens are of variable size. They may be quite small, only a few miles in length and only a few hundred yards wide, or they may extend for hundreds of miles. One of the largest extends from Israel to Tanzania; it forms the Red Sea Basin and extends southward through the highlands of Ethiopia and into East Africa, where its course is evidenced by long, narrow lakes. This enormous feature was not created at one time but was the result of repeated small movements occurring over millions of years.

A **normal fault** (*C* in Fig. 10-12), occurs when tension in the earth's crust is relieved by subsidence along a single fracture, forming a steep fault scarp, or cliff. Fault scarps vary in height from a few feet to several thousand feet (Fig. 10-13). Weathering and erosion will soon obliterate the traces of a low fault scarp, but the more massive scarps may be carved into rugged mountain fronts. Normal faults rarely exist as isolated features but usually occur in a series creating a sequence of steep mountain scarps. For example, normal faulting is believed to be the structural cause of many of the mountain ranges in Nevada, Utah, and Arizona.

Compression, or crustal shortening, may result in a **reverse fault**, in which one block slides up over the other (*D* in Fig. 10-12). In some cases opposing reverse faults may also cause graben valleys. Some geomorphologists believe that compressional faulting is the best explanation for the Rift Valley of East Africa. Under compressional stress rocks frequently yield gracefully, bending into large upfolds and downfolds, especially noticeable in sedimentary rocks. These up-

Fig. 10-13. A small normal fault with a displacement of about 25 feet (8m), on a creek a mile north of Bluff Springs, Texas. The vertical displacement of sedimentary strata on each side of the fault shows that the block on the right has dropped downward relative to the other. *Photograph by L.W. Stephenson, U.S. Geological Survey.*

Fig. 10-14. Strong folding in Fish Creek Canyon, California.

folds (E in Fig. 10-12), are called **anticlines** (Greek *anti* against + *klinein* to lean); the downfolds (F in Fig. 10-12) are **synclines** (Greek *syn* together + *klinein*). Folds can be open and symmetrical, or they can be closed and asymmetrical or even overturned (G and H in Fig. 10-12). The degree of folding depends upon the strength of the rocks and the duration and intensity of the stress applied to them (Fig. 10-14). Folding is an important form of rock structure because the folds upend strata of varying resistance, and subsequent landforms have a characteristic linear pattern of ridges and valleys.

Very gentle folding is called warping. It involves a slight distortion of the earth's crust over a rather large area. Warping is frequently difficult to trace, even in sedimentary strata, because the deformation of the rock is slight. The peninsulas of Florida and Yucatán are examples of warping. They have been raised slightly above the sea by a broad, gentle arching of the earth's crust. Hudson Bay is an area of active warping; the bottom of the bay is rising very slowly, and its water is being spilled into the Arctic and North Atlantic Oceans.

Vulcanism

The second major group of tectonic processes involves the movement of molten rock within the earth's crust or its extrusion onto the surface.

Extrusive Vulcanism. The forms of **extrusive vulcanism** are controlled to a large degree by the manner of extrusion. In some circumstances, hot, liquid rock or **lava** and ash and cinder explosions or **volcanic ejecta** are forced from vents or small pipes in the earth's crust because the molten rock contains a high proportion of gas under great pressure. Some of the lava explodes into small fragments that solidify in the air forming volcanic ejecta, which falls around the vent as cinders and ash or is carried off by the wind as volcanic dust. The material around the vent accumulates in a steep-sided **cinder cone** several hundred feet high (top left portion of Fig. 10-16). Cinder cones frequently occur in groups and sometimes are found in association with lava flows. (Fig. 10-15)

The great volcanoes of the world are usually built up of alternate layers of ejecta and lava flows. Because of this, they are called **stratovolcanoes** or **composite volcanoes** (top right portion of Fig. 10-16). Composite cones are striking features of the landscape, not so much because of their elevation above sea level (although a great many are higher than 10,000 feet), but because they often stand at considerable heights above the surrounding land and are characterized by a marked symmetry of their gently concave slopes (Fig. 10-17). Their steep-sided, gently curving form is the result of alternating explosive expulsions of ejecta and gently flowing lava. The ejecta tends to build steep slopes; while the lava breaks through the side of the volcano, thus flattening the lower slopes of the cone. Most of the composite cones lie in a great ring around the margins of the Pacific Ocean, where, in places, they are the most important relief features in the landscape, as along the west coast of Central America, in central Mexico, in Japan, and in Java. Composite volcanoes also occur in the Mediterranean area, especially in Italy and Sicily (back endpaper). Some of the best-known volcanoes are Popocatepetl in Mexico, Fujiyama in Japan, Mayon in the Philippines, Vesuvius in Italy and Rainier, Hood, and Shasta in the United States.

Occasionally lava wells up from a vent under low

Fig. 10-15. Mount Capulin, a cinder cone in New Mexico. The steep, straight slopes and central depression are characteristic features of cinder cones. *Photograph by W.T. Lee, U.S. Geological Survey.*

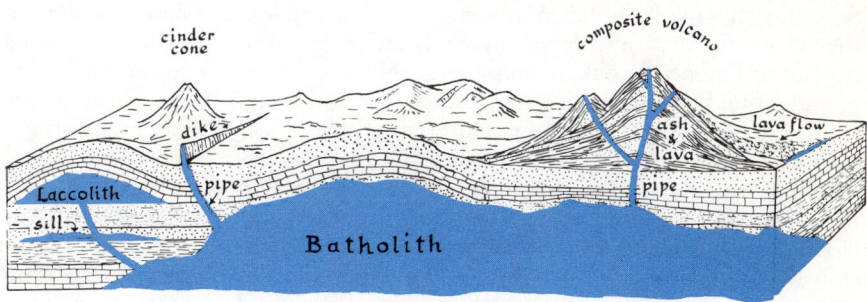

Fig. 10-16. Intrusive and extrusive forms of vulcanism.

pressure so that explosions and the emission of ejecta are infrequent. Instead, the lava, usually a dark, highly fluid basalt, spreads out in thin sheets and forms a gently sloping dome usually rising from the ocean floor. Volcanoes of this type are called **shield volcanoes** or **lava domes.** The major area of occurrence is the Hawaiian Archipelago, whose islands are all lava domes, although some have been greatly modified by erosion.

Other forms of extrusive vulcanism result from the emission of lava from long fissures or cracks. The lava spreads out from the fissures as a viscous or sticky liquid and gradually buries the surrounding terrain. Repeated flows produce a lava plateau. In some places the fissures are widely distributed, and lava flows cover thousands of square miles. For example, the Columbia Plateau in the Pacific Northwest of the United States is the result of repeated flows of lava over a long period. Other fissure flows occurred in Ethiopia, in the Deccan Plateau of India, and in southern Brazil.

Intrusive Vulcanism. Intrusive vulcanism occurs when molten rock is injected in to the earth's crust. There are several types of intrusions, which are differentiated by size and mode of occurrence (lower portion of Fig. 10-16). The largest form, the **batholith** (Greek *bathys* deep + *lithos* rock), is an irregular mass of igneous rock of great size that extends deep into the crust. It was once a great molten mass of rock, or **magma,** that rose through the surrounding rock by melting and incorporating that rock into its own body. As the pressures that initiated the rise of the magma were relieved, the molten rock stagnated at some distance beneath the crust of the earth and slowly cooled and solidified into crystalline rock. Most batholiths, such as the Idaho and Sierra Nevada batholiths, do not consist of a single intrusive mass but are a complex of igneous intrusions of different ages. Batholith rock, frequently granite, is usually more resistant than the rock into which it is intruded.

The hard mass of a crystalline batholilth appears at the surface of the earth only when overlying rock has been stripped away, commonly as the hard core of an eroded mountain surrounded by partially removed sediments. Frequently, batholithic masses are cut by faults, the faulting occasionally raising the batholith even further above the surrounding land. Batholiths are the core of many of the world's more important mountain ranges. The Sierra Nevada of California, composed mainly of a granitic batholith, is about 450 miles (over 700 km) long and about 120 miles (almost 200 km) wide. Other mountain ranges developed on a batholith core are the Northern Rockies in Wyoming, and the Uinta Mountains in northern Utah.

Frequently associated with the batholiths are laccoliths, dikes, and sills. **Laccoliths** (Greek *lakos* cistern + *lithos*) often occur in conjunction with batholiths as small offshoot masses of crystalline rock connected to the parent magma by subterranean conduits (left portion of Fig. 10-16). Hot gases and liquids escaping

Fig. 10-17. Shishaldin Volcano, Aleutian Islands. Note the concave curvature of the mountain sides—especially on the left. The steep but gently curving form is characteristic of composite volcanoes. *Photograph courtesy of the U.S. Dept. of the Interior, Geological Survey.*

from the magma have taken advantage of fractures in the overlying rock and have spread laterally between beds of sedimentary rock, forming large, blister-like accumulations between the strata. In cross section a laccolith looks something like a mushroom, with the laccolithic mass resembling the top, and the feeder pipe, through which material came from below, resembling the stem.

Dikes and **sills** are minor features of volcanic intrusion (left portion of Fig. 10-16). As the magma forces its way into the rock near the surface, it may penetrate along vertical joints and form buried walls of igneous rock called dikes. Sills are the result of the spreading of liquid rock sheets between bedding planes of sedimentary rock prior to solidification.

Igneous intrusions result in the accumulation near the surface of the earth's crust of large and small masses of rock that may have a greater resistance to weathering and erosion than the surrounding rock. Upon removal of the overlying rock, the exposed intrusions eventually form highland areas. Laccoliths may become the basis for hill lands or small mountain ranges, while batholiths are the foundation of most of the world's major mountain ranges.

Orogenic Diastrophism

Folding, faulting, and vulcanism have occurred repeatedly and over wide areas throughout the history of the earth. The raising of great mountain ranges, including the injection of enormous granitic intrusions into the cores of many ranges, took place during relatively short periods separated by longer eras of erosion and denudation. The word orogeny (Greek *oros* mountain + *genesis*) is used to refer to these convulsive mountain-building periods. Table 10-4 indicates the approximate time schedule. Long periods of sedimentary deposition in each of the geologic eras are separated by relatively brief periods of orogeny. The names of the orogenies, derived from mountain ranges they created, vary from one region to another; only the most important names are indicated. Notice also the apparent connection between the orogenic upheavals and changes in the dominant life forms. The revolutionary changes in the environment associated with an orogenic period may very well have spelled the doom of species too well adjusted by overspecialization to the previous environmental conditions, thus allowing less specialized forms to take over.

The Tectonic Map of the Continents

Perhaps the easiest way to describe the major lineaments of the continents is to divide the surface into broad categories based on its tectonic history. The orogenic events in Table 10-4 can be simplified into three major periods of mountain building and, therefore, three kinds of second-order relief features. These are (1) the very old, worn-down mountains of Precambrian age; (2) the middle-aged mountains formed during the Caledonian and Hercynian orogenies of the middle and late Paleozoic; and (3) the young mountains of the Alpine orogenies in the Tertiary. Tables, basins, and masses represent the nonorogenic categories and refer to surfaces partly or completely covered by nearly horizontal sedimentary rock.

Table 10-4. Geologic time table

Eras	Periods (in Cenozoic only)	Epochs (in Quaternary only)	Principal orogenies	Age (in years before present)	Dominant life form
Cenozoic	Quaternary	Recent		15,000	Man
		Pleistocene	Cascadian, Alpine	2,000,000	
	Tertiary				Mammals
			Laramide	60,000,000	
Mesozoic			Beginning of continental drift	85,000,000	Reptiles
			Appalachian, Hercynian	230,000,000	
Paleozoic					Amphibians
			Acadian	350,000,000	
					Fishes
			Taconian, Caledonian	425,000,000	
				600,000,000	Invertebrates
					Few known fossils
Precambrian			Many orogenies	4-5,000,000,000	

In addition, major areas of extensive faulting are categorized as rift zones. These categories and their distribution are shown on the back endpaper.

Precambrian Shields. The earliest orogenies are Precambrian in age, but the mountain systems that were formed more than 600 million years ago have been almost completely worn down by erosion, and their rocks have been strongly altered by metamorphism. These nearly horizontal surfaces of little relief, composed of granitic rocks and metamorphosed sediments, are called **Precambrian shields.** They are the oldest and most stable parts of the earth's crust. Four major areas are shown on the map: the Laurentian Shield in eastern Canada, the Baltic Shield in northern Europe, the Angara Shield in Siberia, and the Antarctic Shield.

Paleozoic Mountains. The next oldest orogenies are those that occurred during the Paleozoic. The middle-aged **Paleozoic mountains** have not been planed down by a billion years of erosion, but they have been worn down into low-rounded mountains. Included in this category are also certain Paleozoic and older rocks broken into tabular blocks with intervening basins.

The major systems of Caledonian and Hercynian age include such important systems as the Appalachian ranges in North America, the coastal mountain systems of northwestern Europe, the Urals, the mountain ranges of eastern Australia, parts of Antarctica, and large expanses of mountain ranges intermixed with tabular masses and basins in east-central Asia and southern China. Note that the Paleozoic mountains of northwestern Europe, northwestern Africa, and eastern North America were apparently created in a single block before the drifting apart of the separate tectonic plates (see Figs. 10-8 and 10-11).

Tertiary, or Alpine Mountains. **Tertiary,** or **alpine mountains,** products of the latest orogenies, are still high and rugged, their ruggedness intensified, in some cases to spectacular proportions, by glacial action. Alpine mountains are often found in association with recent vulcanism and zones of fracture. They occur in two main belts. One circles the Pacific Ocean and includes the western cordilleras of North and South America and the island archipelagoes that form the eastern limit of the Asian continent. The other system extends from the Atlantic to the Pacific and includes the Alps, the Caucasus, and the Himalayas among its most important ranges. Many of these young mountains may owe their genesis to the folding of the crust which took place when one tectonic plate was pushed into another. The Himalayas, for example, are along the contact between the Australian and Eurasian plates. The western cordilleras of North America are similarly located with respect to the Pacific and North American plates.

Basins, Tables, and Masses. These categories include a variety of underlying surfaces or **basement rock** covered by sediments. The underlying rock may be horizontal shield material; some of it may be very old such as the cratons shown in Fig. 10-11, but its ruggedness has been preserved by the cover of sedimentary strata; and, in some cases, both the basement rock and the overlying material are young. **Basins** are surfaces almost completely covered by nearly horizontal sedimentary strata. **Tables** are areas that have been partly stripped of their sedimentary cover. **Masses** are complex surfaces whose origin is not very clear. Many of them are Southern Hemisphere shields with some sedimentary cover. They also include areas of horizontal basaltic lava flows.

Other Features of the Tectonic Map. In addition to these categories, the black endpaper shows certain other features. It indicates the extent of the continental shelf where it is wide enough to be shown at the scale of the world map. The map also identifies two major areas of block faulting, the great Rift Valley of Africa and a similar zone extending from the Gulf of California into the faulted mountains of the southwestern United States (Fig. 10-18). Finally, it locates the major areas of recent volcanic activity.

Fig. 10-18. The trace of the San Andreas fault in the central valley of California. The movement along this fault is horizontal. The lateral displacement of some valleys and the cutting off of others is clear evidence of this. The San Andreas fault is probably a part of the crustal movement related to plate tectonics. *Photograph courtesy of the U.S. Dept. of the Interior, Geological Survey.*

Summary

In their study of differences from place to place on the surface of the earth, geographers are concerned with both the description of those differences and the processes that create them. Thus, we began this chapter by indicating the importance of description in the study of landforms. This description is useful for the understanding of other areal distributions related to the form of the land as well as for the study of the processes that create the differences among landforms.

Weathering, erosion, and deposition, the major processes affecting the development of landforms, are treated in the next chapter, but their effect is strongly conditioned by rock composition, rock structure, climate, vegetation, and time. These important variables are, therefore, called the landforming factors. Equally pertinent is the hydrologic cycle. The movement of water from the surface of the land and sea to the atmosphere and then back to the earth, though it involves only a tiny fraction of all the water in the lithosphere, is of paramount significance to land formation because the action of water is one of the principal components of weathering, erosion, and deposition.

The possible combinations of minerals that make up rocks are extremely numerous, but the most important groups are the durable silica minerals, the more easily weathered feldspars and ferromagnesian minerals, and the soluble carbonates. These, separately or in combination, make up most of the rocks. The great variety of resulting rocks can also be sorted out into a limited number of categories. The igneous rocks, formed by the solidification of the liquid outer crust, are either extrusive or intrusive. The former are extruded to the surface in liquid form and cool quickly on exposure; the latter cool below the surface and are only subsequently exposed. These rocks have various characteristics, depending on their rate of cooling and on their constituent minerals. Sedimentary rocks are the result of deposition of weathered portions of rocks. Their qualities vary according to the size of the particles involved and the degree of cementation. Metamorphic rocks are igneous or sedimentary materials that have been transformed by pressure or heat. The character of these rocks depends on that of the original material and on the strength and duration of the metamorphic action.

All these rocks are part of the outer crust of the earth, a thin layer less than 25 miles (40 km) thick and composed of relatively light blocks (the sial) floating on a more continuous layer of denser rock (the sima). The former represent the continental masses, the latter the floors of the ocean basins. This outer crust is subject to a number of deformations, the agents of which are called the tectonic processes. Tectonic deformation includes diastrophism, which involves the bending and breaking of sedimentary layers of other rock structures, and vulcanism, which involves the addition of material near the surface by extrusion or intrusion.

Some of the diastrophism is orogenic; it is so widespread and of such strength and duration that it creates mountain systems over very large areas. Such diastrophism has occurred repeatedly in the past. The oldest orogenies, Precambrian in age, created very large mountain systems, but they have been reduced to low, nearly level features such as the Laurentian, Baltic, Angara, and Antarctic Shields. The Paleozoic orogenies produced ranges that have not been completely worn down by erosion but are generally low and rounded in outline. The Appalachians. the Urals, and the coastal ranges of northwestern Europe are examples of such middle-aged mountains. The Tertiary orogenies produced mountains whose relative youth is seen in their greater height and ruggedness. The cordilleras of western North and South America and the Himalayan ranges are typical of such mountains. The distribution of these tectonic features is shown on the tectonic map, whose generalized lineaments bring considerable order out of what might otherwise appear to be a chaotic mixture of landforms of different types and ages. Even more order emerges if one remembers the pattern of continental drifting, which has separated once contiguous areas of similar orogeny.

Suggestions for Further Reading

Continents Adrift, Readings from *Scientific American*, W. H. Freeman and Company, San Francisco, 1972.

Emmons, W. H., Ira S. Allison, Clinton R. Stauffer, and George A. Thiel, *Geology: Principles and Processes*, McGraw-Hill Book Company, Inc., New York, 1960.

Hammond, Edwin H., "Small-Scale Continental Landform Maps," *Annals of the Association of American Geographers*, Vol. 44, No. 1 (1954), 33-42.

Holmes, Arthur, *Principles of Physical Geology*, 2nd ed., The Ronald Press Company, New York, 1963.

Putnam, W. C., *Geology*, Oxford University Press, New York, 1964.

Strahler, A. H., *Physical Geography*, 3rd ed., John Wiley & Sons, Inc., New York, 1969.

11 Weathering and Mass Wastage

Introduction

Gradational processes, as the name implies, reduce topographic inequalities by wearing down the highlands and filling in the lowlands. The gradational agents are streams, glaciers, and wind; in addition, gravity acts directly upon loose material lying on slopes. The gradational agents and gravity are effective only if the rock at the surface has been decomposed and fragmented through weathering. Gradation can thus be viewed as a two-stage process: (1) the preparation of material for removal (or **weathering**), and (2) its removal by erosion (or **degradation**) and subsequent deposition (or **aggradation**).

Weathering processes tend to differ much less from place to place than does rock composition. Generally speaking, therefore, in any given locality, the *rate* of weathering depends largely upon the kind of rock being weathered (provided that structure, climate, vegetation, and time are reasonably constant). Some rocks break down more readily than others; this variation in the rate at which rocks are prepared for removal is called *differential weathering*. Differential weathering prepares rocks for *differential erosion*, which in turn frequently carves topographic designs patterned after the type and structure of the underlying rock.

The processes of erosion and deposition will be discussed in subsequent chapters when we consider the agents that accomplish this work. The emphasis here is on weathering.

Weathering

Weathering, the prerequisite for effective gradational action, is the internal decay and collapse of solid rock brought about by chemical and physical processes that eventually reduce the most massive rock to very small fragments. Physical and chemical processes operate together; and while it is difficult to distinguish between them in nature, the two types of weathering are discussed separately for convenience of exposition.

Chemical weathering alters minerals or removes them from rocks. Rock minerals are usually chemically stable in the environment under which the rocks were formed. But tectonic processes lift up the rocks and erosion exposes them to the atmosphere — a new environment where their minerals are subject to change through contact with abundant oxygen, carbon dioxide, and water. The ensuing chemical changes result in rock decomposition. *Physical* weathering, on the other

Table 11-1. Examples of the More Common Forms of Chemical Weathering.

Oxidation:	$4\,FeO$ ferrous iron oxide, partly soluble in water	+	O_2 oxygen	=	$2\,Fe_2O_3$ ferric iron oxide relatively insoluble in water	
Hydration:	$\boxed{K\,AlSi_3O_8}$ orthoclase	+	H_2O water	=	$\boxed{K\,AlSi_3O_8}$ H_2O layer of water molecules adhering to crystal	
Hydrolysis:	$\boxed{K\,AlSiO_3O_8}$ orthoclase crystal	+	$H^+\,OH^-$ disassociative water	=	$\boxed{H\,AlSiO_3O_8}$ hydrogen cation displacing potassium	+ KOH in solution

Once the potassium cations are replaced by hydrogen, the orthoclase crystal breaks down, eventually into kaolinite clay whereas plagioclase tends to form montmorillonite clay.

Carbonation:	$CaCO_3$ calcite	+	H_2CO_3 carbonic acid $(H_2O + CO_2)$	=	$Ca(HCO_3)_2$ sodium bicarbonate in solution
Solution:	$NaCl$ table salt	+	H_2O	=	$Na^+ + Cl^-$ sodium and chlorine ions enclosed in water molecules

Evaporation reverses this process exactly, so that the original salt can be recovered unchanged in quality and amount.

hand, fragments rocks without chemical change, although it too is activated by the exposure of rocks at the surface. Physical weathering results in rock disintegration.

Chemical Weathering

Rock decomposition is brought about by **oxidation, hydration, carbonation,** and **solution** (see Table 11-1). In weathering processes, oxidation usually means the union of oxygen with other elements or minerals to form oxides, the oxidized matter losing electrons to the oxygen. The formation of ferric oxide is most effective in the presence of water on rocks containing iron compounds. Oxidation of ferromagnesian silicates causes the silicate structure to break down; in the case of primary iron or iron oxides, oxidation results in bulkier and softer minerals.

Hydration refers to the absorption of water by minerals through the adhesion of water molecules to the surfaces of the mineral crystals. Common among the feldspars, hydration leads to the eventual formation of clay by *hydrolysis*.[1] Hydration not only weakens the mineral structure but greatly increases its volume. Hydration also accompanies the oxidation of iron. The rust that collects on iron implements is a combination of iron oxide and absorbed water and is a new and softer compound with a greater volume than the original iron.

Carbonation occurs when calcium or potassium from the feldspars combines with carbon dioxide from the groundwater to form calcium or potassium carbonates. Because these new carbonate substances occupy more space than the original minerals, they exert pressure on the rock minerals from which they were derived. The new substances are also more susceptible to removal through solution by groundwater.

Solution is another means of chemical weathering. Some minerals, such as sodium chloride (table salt), are

---

[1]Hydrolysis is defined as chemical weathering that involves the addition of the elements of water to the weathered substance.

soluble in fresh water, while others, such as calcium carbonate, can be dissolved in a weak solution of carbonic acid formed when carbon dioxide (derived from the air or from organic matter in the soil) mixes with water. Acting on calcium carbonate, carbonic acid produces a salt, calcium bicarbonate, which is readily soluble in water. Taken into solution, the dissolved minerals can be carried off by groundwater and streams. This process of dissolution and removal is called **leaching.** Leaching weakens rocks by removing minerals; rocks composed entirely of soluble minerals may disappear completely.

As is evident, chemical weathering requires significant moisture to operate most effectively. High temperatures are equally important, since the rate of chemical reaction increases sharply with increasing warmth. Consequently, variations in climate produce significant variations in the rate of chemical weathering. It is most rapid and complete in the warm and humid tropics (Fig. 11-1). With decreasing temperature or decreasing rainfall, it becomes less effective, and physical weathering becomes more important.

Physical Weathering

Rock disintegration results from rock expansion through removal of pressure by *unloading*, by *thermal expansion* and *contraction*, by *crystal growth*, or by *organic activity*. Unloading takes place when rocks are uplifted and eroded so that the confining pressure of a thick layer of overlying materials is removed. The exposed rock expands slightly in volume, develops laminated joints, and separates into spherical layers, much like those in an onion. These joints are most evident in massive crystalline rock where they are roughly parallel to the ground and become more widely spaced with depth. The breaking away and removal of the outer rock layers from the main mass is called **exfoliation.** When this process is associated with unloading, the resulting landforms are called exfoliation domes[2] (Fig. 11-2).

Thermal expansion and contraction resulting from solar heating and nocturnal cooling is believed to cause granular disintegration and small-scale exfoliation of crystalline rock. Solar heating is most effective in deserts, where there are few clouds and little water vapor in the air and very little vegetation to shield the rocks from the sun's rays. Since the ground is usually dry, very little solar energy is used to evaporate water,

---

[2]Half Dome and the Royal Arches in the Yosemite Valley of California and Sugar Loaf Mountain in the Bay of Rio de Janeiro are three famous examples.

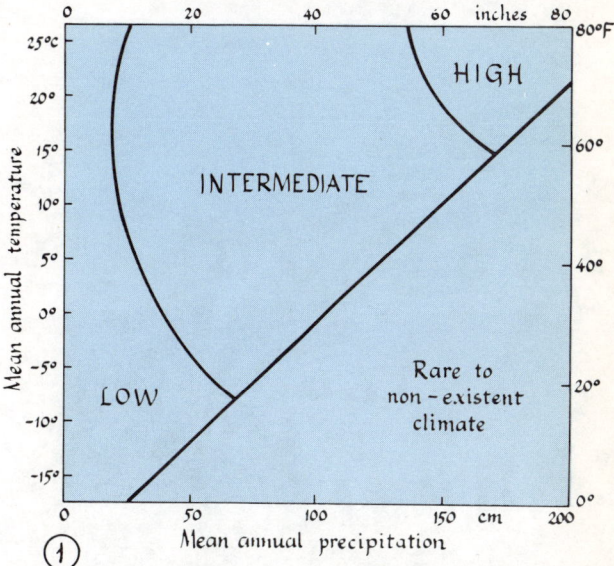

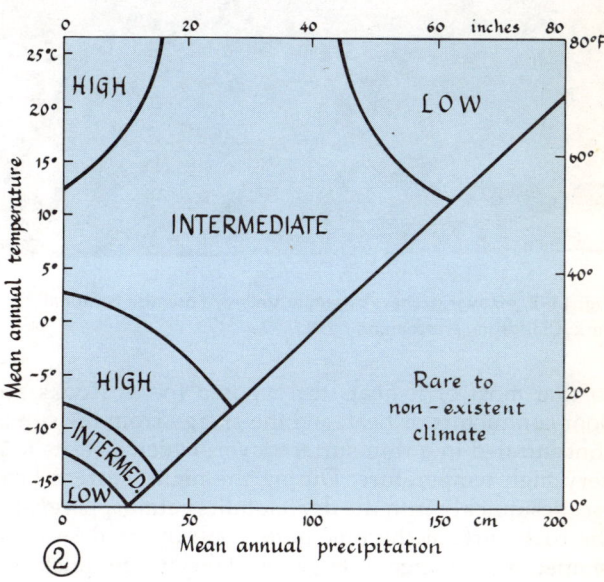

Fig. 11-1. A diagrammatic relationship between weathering and climate. The intensity of chemical weathering is simply related to climate, as shown in (1). The highest intensity of weathering occurs when both rainfall and temperature are high; as rainfall decreases, chemical weathering becomes less important than physical weathering. The relative intensities of physical weathering are shown in (2) and reveal a more complicated pattern. The highest intensities are found when rainfall is relatively low, but the lowest intensities are at two climatic extremes—hot and humid on the one hand, and cold and dry on the other. Modified from L.B. Leopold, M.G. Wolman, and T.P. Miller, *Fluvial Processes in Geomorphology*, W.H. Freeman and Co., San Francisco, 1964, p. 42.

194 Chapter 11

Fig. 11-2. Royal Arches, Yosemite Valley, Yosemite National Park, California. *Photograph by R.L. Day.*

so that most of it heats the exposed rocks. Rocks are poor conductors of heat, and the energy from the sun is concentrated in a thin surface layer, which it heats to a very high temperature. During the night, surface heat escapes quickly into the dry, cloudless atmosphere, and the rock surfaces become quite cool. In the desert the temperature change is large and takes place quickly.

As the rock heats during the day and cools during the night, differential expansion and contraction of the various mineral crystals set up internal stresses between minerals near the surface. The re-creation of these internal stresses day after day, for long periods, is thought by some authorities to cause the rock to crumble away, grain by grain. Stress due to the large temperature difference between the rock surface and rock interior eventually may cause the surface to scale off in thin layers. The mineral grains and thin bits of exfoliation shell, called *scree*, collect around the base of the rocks, where they remain for some time because further breakdown and removal are retarded by the shortage of moisture.

Even the driest desert has occasional rain; enough moisture is present to permit hydration of some of the feldspars. Hydration weakens the crystals and causes them to expand. This too can cause mechanical granular disintegration and small-scale exfoliation. Some authorities believe that hydration causes much more granular disintegration than does thermal expansion and contraction.

Expansion as the result of growth of ice and salt crystals also causes rocks to break and disintegrate. The growth of ice crystals is an effective means of **wedging**, especially in rocks having numerous fractures and joints. Water penetrates into the fractures and expands upon freezing, exerting considerable pressure on the rock and enlarging the fractures. On subsequent melting, the water runs farther into the rock. Refreezing renews the pressure on the rock at a greater depth, eventually breaking it apart. Frost wedging is most frequent in middle and high latitudes, where repeated freezing and thawing occur throughout the winter. In low latitudes, frost does not occur except at high elevations, while in high latitudes alternations of freezing and thawing are limited to brief periods in spring and fall. However, frost wedging is very significant at high elevations and in high latitudes because of the relative ineffectiveness of chemical weathering with low temperatures and frozen water (Fig. 11-3).

Rock disintegration resulting from growth of salt crystals is also common in arid and semiarid climates. During periods of drought, capillary action brings water to the surface of sandstone and other porous rocks. As the water evaporates, small crystals of salt develop which, in time, pry grains of sand from the face of the rock.

Organic activity is a form of physical weathering brought about by the wedging action of plant roots and the burrowing of animals. Small plants send tiny rootlets into small cracks and joints; as the plants grow, the roots expand and exert pressure on the rock. Under certain favorable circumstances, pressure of root growth may aid in the widening of rock fractures. Growth of the rootlets is undoubtedly effective in prying many small pieces from rock weakened by chemical weathering. The burrowing and tunneling of animals and insects also can dislodge fragments from partially weathered rock. In any individual instance, the amount displaced may be small, but the great number of workers results in a large total dislodgment.

Diastrophism also produces physical weathering. The forces of uplift may cause solid rock to develop joints and cracks. Because of this fragmentation of the originally massive rock, diastrophic jointing can be

Fig. 11-3. View from Muir Pass on the Whitney Trail, Mt. Whitney, California. The coarse rock fragments in the talus cones at the base of the escarpment are the product of frost wedging. *Photograph by James Bier.*

classified as a gross form of mechanical weathering.

The foregoing discussion has stressed the effect of climate, but other factors also exert considerable influence upon the rate of weathering. For example, the steeper the slope, the greater the run-off and the smaller the amount of rain water available for chemical weathering. Growing plants and burrowing animals aid in the mechanical breakup of rocks, and both are also indirectly involved in rock weathering. To a considerable degree, it is the amount and kind of decayed plant and animal matter on the surface of the ground that controls the amount of carbon dioxide which can combine with water to form carbonic acid, a most important agent in rock weathering.

Rock composition also influences the rate at which weathering processes operate. Quartzite, for example, will expand and contract rather uniformly when it is heated and cooled since it is composed of a single mineral, quartz. It is also relatively impervious to all forms of chemical weathering. Granite, on the other hand, consists of several minerals including feldspars. Consequently, it weathers sooner than quartzite, whether in an arid or a humid climate. Finally it should be emphasized that while climate, relief, biota, and rock composition combine to condition the rate at which rock weathers at a given place, it is the length of time over which the various processes operate that determines how much weathering occurs and hence the quantity of weathered material produced.

Regolith or Mantle Rock

The end product of the weathering process is a layer of decomposed rock and soil (see Chap. 21 for a discussion of soil formation), called **regolith** or **mantle rock**, which covers the surface of the land to a variable depth depending upon differences in the landforming factors, particularly differences in the character of the rock and the slope of the land. Through chemical and physical weathering, rock is prepared for movement downslope to the valley floors, where it can be carried off by streams.

Mass Wastage

By itself, weathering does not produce significant landforms. Important relief features develop only when regolith is removed from its place of origin. The moving force is the pull of gravity acting directly or indirectly through the agency of streams, winds, and glaciers. Direct gravitational pull results in downslope migration wherever weathered material is available and the slope is reasonably steep, say greater than 5°. This migration is called **mass wastage** or **mass movement**.

More material is moved and at a faster rate on steep slopes than on gentle slopes, because the steeper the slope the more nearly the direction of movement approaches the vertical line of maximum gravitational pull. On very steep slopes even large masses of bedrock may break free under favorable conditions and fall down with catastrophic suddenness. Because the regolith moves more readily on steep slopes, it is likely to be thinner than on gentle slopes. On slopes of lesser inclination, the regolith moves more slowly but more persistently; a small amount of downhill migration is going on all the time.

The persistence of mass wastage and its presence on all slopes except those of very low gradient make it as significant as stream erosion to degradation. It is certainly the most effective way of wearing away the interstream ridges, which are generally beyond the reach of stream erosion. However, downslope migration begins only when resistance to movement has been overcome. What causes this resistance, and how is it overcome?

Processes Affecting Mass Wastage

The causes of resistance on all slopes are *cohesion*, *friction*, and the restraining action of *plant roots*. The intensity of resistance varies according to slope. On steep slopes resistance is low, and weathered material is readily started on a downslope migration. Resistance increases with decreasing slope until so low a gradient is reached that mass wastage stops altogether.

Cohesion, the property of sticking together, increases as the size of the weathered particles decreases; it reaches a maximum in clay. Clay has unusual importance in mass movement because it can act as either a cohesive agent or a lubricating agent. Cohesion also varies according to the quantity and kinds of clay minerals present. Consequently, regolith rich in certain kinds of clay is highly resistant to the onset of movement as long as the clay contains less than a certain minimum of water. Once the clay is thoroughly wet, its reduced friction more than offsets its cohesiveness. In any case, cohesion in the regolith is a force that resists the beginning of mass wastage but has little retarding effect once movement is underway.

Friction, in contrast to cohesion, not only resists the beginning of mass wastage but also retards the movement once it has started. Friction depends on the pressure within the weathered material and on the size and roughness of the particles. On gently slopes, material may move fairly readily if it is only lightly compacted; but if the weathered layer is thick, the increased pressure causes the particles to press more tightly upon one another, thereby increasing the friction within certain limits. Friction also increases with the area of contact within the regolith; if the particles are large, the number of contacts and area of contact are small. Conversely, if the particles are small, the number of contacts is large; hence the area of rubbing increases, and so does friction as long as the particles remain dry.

But as the size of particles decreases, the water-holding capacity increases, and the greater moisture content of the soil may then reduce friction.

Mass wastage is also hindered by the frictional action of roots upon the regolith. If they are long enough, roots can even bind the regolith to the relatively unweathered rock beneath. However, even on heavily forested slopes, roots do not completely prevent mass wastage. Most tree roots spread out fairly close to the surface, and only a few penetrate beyond 4 or 5 feet. Downslope migration can take place beneath the surface layer of roots and between the more deeply penetrating roots. In some instances, as in the tropics, the regolith is so thick that even the deep taproots of trees do not reach through the entire layer of weathered material. In any case, forests are commonly associated with high rainfall, and the well-soaked regolith moves relatively easily. Perhaps the most important effect of plant roots, especially those near the surface, is that they act as a coarse sieve, permitting finer particles to move while blocking or slowing down the larger fragments.

Even though gravity exerts its pull at all times, some triggering action is required to overcome strong cohesiveness, friction, or the binding action of roots. Possible triggering actions include changes in volume of the regolith, decreases in cohesion, and decreases in friction.

Changes in volume bring about repeated, small, downslope movements. These changes in volume may have several causes. Changes in temperature from day to night cause repeated expansions and contractions. In humid and subhumid regions, soils swell as they absorb moisture and contract and crack as they dry. In mid- and high latitudes, alternate freezing and thawing of soil moisture expands and contracts the regolith. All these changes in volume result in a steady downslope migration of the regolith

Addition of water to the weathered layer causes a decrease in cohesion of the clay within the layer. The change from the coherent to the noncoherent state in clay may occur suddenly once a certain water content is exceeded. For example, when montmorillonite clay is saturated, a slight jar will change it from a binding agent to an effective liquid lubricant through the release of absorbed water. Kaolinite clay is a much less effective lubricating agent even when saturated, since it absorbs relatively small amounts of water.

Friction can also be reduced by soaking the weathered layer with water. An abundance of water not only causes an increase in weight but also acts as a lubricant, permitting particles in the regolith to rotate against each other more freely. Trampling by large animals, shaking by earthquakes, or the passage of trains and other heavy vehicles may also trigger mass wastage in the regolith.

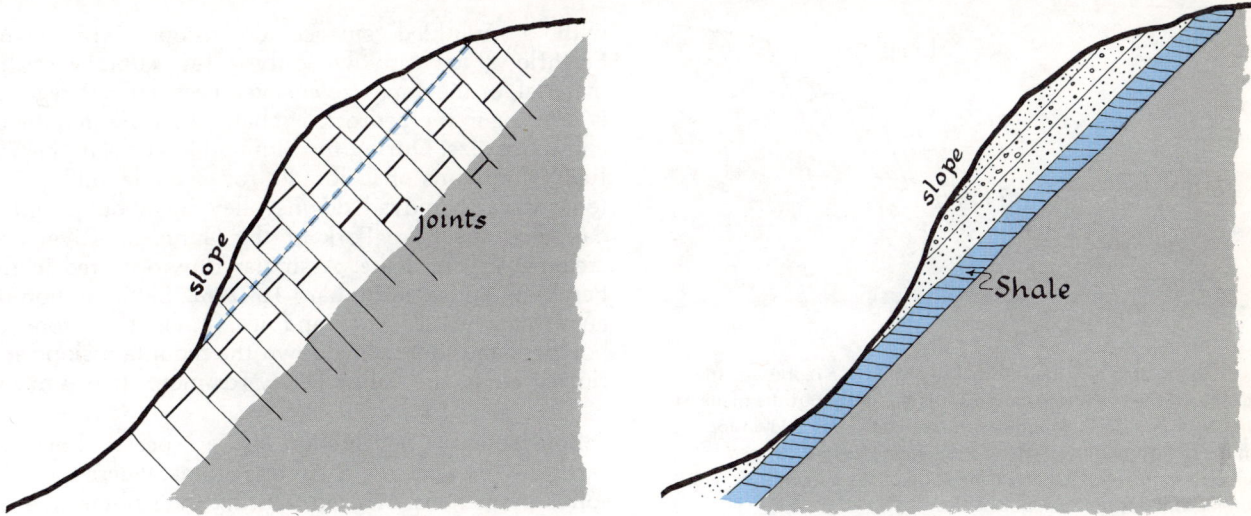

Fig. 11-4. Circumstances of rock structure and slope conducive to landslides. In the left diagram, the valley slopes are steeper than the rock joints. At the right, the slopes exceed the dip of rock strata. Once the shale layer is thoroughly soaked, a landslide is very likely to occur.

Under natural conditions the modification of vegetation so as to change the binding action of plant roots upon the regolith probably occurs so slowly that it does not become a significant factor in accelerating mass wastage. However, deforestation or overgrazing on grassy slopes may play an important (but as yet undetermined) role in speeding the rate of downslope migration of the regolith.

Forms of Mass Wastage

Material moves downslope by falling and rolling, by sliding, by flowing, or by any combination of the three. Some of the more common kinds of mass movement are described below.

Rockfall. The simplest downslope movement is **rockfall**, where individual particles tumble downhill. The moving fragments vary in size from sand grains to huge boulders. Rockfall is especially active on steep slopes of cliffs with exposures of bare rock, where weathering loosens the rock particles. Eventually the fragments form a *talus cone* or a *talus slope* at the base of the cliff (Fig. 11-3). Features such as these are very likely to be found along all steep and rocky slopes.

Landslides. A **landslide** is a spectacular kind of mass wastage in which large amounts of material slide and tumble rapidly downslope. Landslides may occur on any steep mountain slope or on slopes undercut by streams or excavations made for roads or railroads. Landslides also occur on slopes underlain by dipping strata or jointed rock where the dip-angle of the strata or joint is approximately the same as the inclination of the slope (Fig. 11-4). Dipping strata are especially conducive to landslides if the strata contain layers of weaker rock, such as thin layers of shale interbedded with thick layers of sandstone. If the shale is weakened by weathering or lubricated by water, the weight of the overlying sandstone may cause the whole mass to break away and slide downslope. It is a well-known fact that landslides occur more frequently during rainy seasons, undoubtedly an indication that reduced cohesion and a decrease in friction are probably the motivating factors.

Slumps. A **slump** is a form of landslide in which rock breaks along the edge of a cliff and slips downhill with a rotary movement (Fig. 11-5). Slumping usually occurs whenever massive strata rest upon shale or other clay-rich sedimentary rock. When the underlying rock

Fig. 11-5. Slumping along an escarpment, showing the characteristic rotary movement.

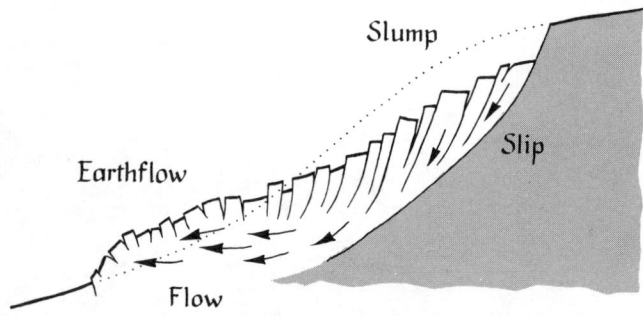

Fig. 11-6. Sketch of a mud flow. The dotted line represents the original surface. The initial movement is slumping, but the material at the lower end flows. Muddy earth frequently flows downslope from the toe of the flow. After C.F.S. Sharpe, *Landslides and Related Phenomena*, 1938, by permission of the author and Columbia University Press, publishers.

is weakened by the addition of moisture or is removed by erosion, the cap rock breaks, and a large block slides down and tilts backward along the curving surface of the slip. Small-scale slumping also occurs along stream channels and sea cliffs, where the regolith and bedrock are constantly being eroded.

Earthflows. Earthflows are characteristic of steep slopes saturated with moisture. Masses of water-soaked regolith slide and flow downslope over a period lasting from several minutes to a few hours. Materials slump or slip at the top, creating a spoon-shaped scar upslope, and flow downhill to form an irregular bulge with a wrinkled surface downslope (Fig. 11-6). Earthflows may involve only a few cubic yards of material, or they may cover several acres. If the regolith is deep or if the bedrock is shale, the flow may be of enormous size. During the Slumgullion flow in the San Juan Mountains of Colorado, for example, millions of tons of earth moved down-valley for about 6 miles, damming the Lake Fork of the Gunnison River and creating Crystal Lake. A similar flow occurred in the Peruvian Andes on January 10, 1962. Set in motion by an avalanche of snow and ice, millions of tons of boulders and mud came down the mountain slope and flowed along the valley below for more than 9 miles.

Solifluction. Solifluction is a special kind of earthflow that occurs in Arctic regions, where the subsoil is permanently frozen. During the summer the ice in the soil melts to a shallow depth, and the melted layer becomes saturated with water that cannot drain downward because of the frozen subsoil. This saturated layer slowly moves downslope as a unit, but some parts move faster than others, thus causing the surface to develop a wrinkled appearance (Fig. 11-7). Solifluction can also occur outside the Arctic over a subsoil rich in clay. Here the clay layer prevents the downward drainage of soil water resulting in a saturated layer that moves downslope in a manner similar to its Arctic counterpart.

Soil Creep. The mass movements described above are usually well defined and fairly rapid. **Soil creep,** on the other hand, refers to the slow, downslope move-

Fig. 11-7. Solifluction lobes in the Seward Peninsula of Alaska. The lobes, or solifluction terraces, more on the order of an inch or two per year by creep and solifluction. *Photograph courtesy of the U.S. Geological Survey.*

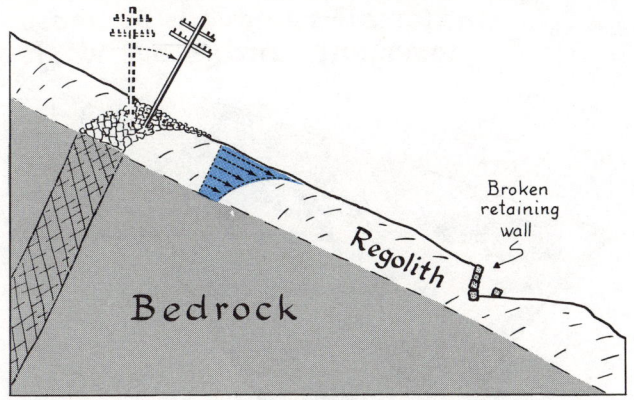

Fig. 11-8. Evidence of the variable movement of soil creep. Because the surface layers of the regolith move more rapidly than the subsurface layers, vertical structures are soon rotated out of plumb. After C.F.S. Sharpe, *Landslides and Related Phenomena*, 1938, by permission of the author and Columbia University Press, publishers.

ment of the entire regolith. It is active on most slopes covered by soil, even where gradients are as low as 5°. Since the whole layer of regolith is involved (though at a decreasing rate with increasing depth), and since creep is in action continuously (though at a different rate in different seasons) the total volume of material moved over a long period is very large and exceeds the combined effect of all other means of producing mass wastage.

Creep results from changes in volume brought about by alternate drying and wetting, heating and cooling, or freezing and thawing. Alternate freezing and thawing produce an additional effect called *frost heaving*. Soil particles are lifted at right angles to the slope when they freeze and are dropped vertically when they melt. This causes an additonal, gradual shift of surface material downslope.

The forces that favor creep work most efficiently at the surface of soil. Since soil is a poor conductor of heat, expansion and contraction of the regolith through solar heating and nocturnal cooling are greatest at the surface. For the same reason, in mid-latitudes at least, freezing and thawing and the attendant expansion and contraction are also most marked at the surface because heat stored in summer is conserved only in the lower layers of the regolith during the winter. The surface soil dries more completely than the subsoil and so contracts further during drought, which permits it to expand more during rain. Because the motivating forces are strongest at the surface, creep moves the top of the regolith more rapidly than the bottom. Proof of this variable movement is clearly demonstrated by the tilted fences and telegraph poles and the bulging and rup- tured retaining walls that can be seen on most moderate-to-steep slopes (Fig. 11-8).

Variations in the Rate of Mass Wastage

Mass wastage occurs in all climatic regions and on all slopes where the gradient exceeds a certain minimum value, but the rate of movement is highly variable. In high latitudes, for instance, solifluction comes to a complete stop when the temperature is below freezing, so that the amount of mass wastage is obviously greater in summer than in winter. In the wet tropics, on the other hand, rock reduction is going on and friction is being reduced year-round by the addition of moisture to the weathered layer. Mass wastage is probably going on at much the same rate all year long. In Mediterranean regions, where winters are wet and summers are dry, mass wastage is greatest during the winter, when there is a great deal more moisture in the regolith. By contrast, in the winter-dry tropics the period of maximum mass wastage is during the rainy summer months.

While the period of maximum movement clearly varies with climate, differences in the *kind* of mass wastage are not as easily related to climate. We may get some clue to this relation by examining the effect of climate upon the clay content of the regolith and then by considering the influence of clay content upon mass wastage. To keep things simple, we will discuss only the influence of montmorillonite and kaolinite clays upon a few aspects of mass wastage. Other forms of clay are found in the regolith, of course, but a discussion of the kinds and relative abundance of these clays, let alone their effect on mass wastage under different climates, is beyond the scope of this book.

The kind and amount of clay present in a given layer of regolith are affected not only by climate but also by the nature of the parent rock, the degree of slope on which the regolith lies, the kind of vegetation cover and the duration of weathering. To isolate the effect of climate upon clay formations, we must examine samples of soils formed under similar conditions of slope, slope aspect, vegetation, parent rock, and time.

Samples gathered in this manner show that changes in climate have a profound effect on the amount and kind of clay in the regolith. Soil samples collected in an east-west line across the United States — that is along a line of nearly equal annual temperature but of varying annual rainfall — show a marked increase in clay content with increasing moisture. At the same time, there appears to be a change in the kind of clay — at least between the dry and humid climates, with montmorillonite clay becoming important in soils of drier climates. Soil samples collected along a line of nearly constant moisture but varying annual temperature (north-south across the eastern United States) show a significant increase in clay content with increasing temperature. There is, in addition, a predominance of kaolinite clay at the warmer end in the southeast. If we increase both annual rainfall and temperature to the values of the rainy tropics, the kaolinite content of the soil drops sharply, because leaching becomes more intense.

What bearing do these circumstances have upon mass wastage? As we have seen, montmorillonite clay absorbs much more water than kaolinite clay and so acts as a more effective cohesive agent in the regolith. When montmorillonite becomes saturated, however, it quickly changes from a binding to a lubricating agent. This sudden change may bring about catastrophic landslides and earthflows. Since kaolinite absorbs relatively small amounts of water it is not an efficient lubricating agent when saturated. Consequently, if all other factors are equal, the kaolinite-rich regolith in the warm humid climate of the southeastern United States tends to have less spectacular landslides and earthflows than regions of montmorillonite-rich regolith. Differences in climate, at least insofar as they affect the clay content of the regolith, result in changes in the degree of movement rather than changes in the kind of movement.

The emphasis on climate in the last few paragraphs should not obscure the fact that other factors of the physical landscape also influence the rate of mass wastage. The earlier discussion of processes stressed the influences of the degree of slope, rock composition, rock structure and vegetation as well as climate upon the nature and rate of mass wastage.

Importance of Mass Wastage

Since mass wastage is operative at nearly all times, it is an important means of wearing down interstream areas; it transports weathered debris from the slopes to the valley bottoms, where it is picked up by streams and removed. Thus, it does a job that a stream cannot do. Streams can erode only within their wetted perimeter. Within this zone they cut down their channels through abrasion and scour. In certain kinds of soft rock they may erode laterally, but in consolidated rock, lateral erosion is unimportant compared to downward erosion. If lateral erosion is a relatively minor agent in reducing the landscape the stream alone merely cuts a deep, narrow slot into the terrain (Fig. 11-9). The removal of wedge-shaped rock portions on each side of the slot to give the stream valley its V-shaped cross section is accomplished mainly by mass wastage. To be sure, sheet wash, which is a special kind of stream flow, operates in conjunction with mass wastage

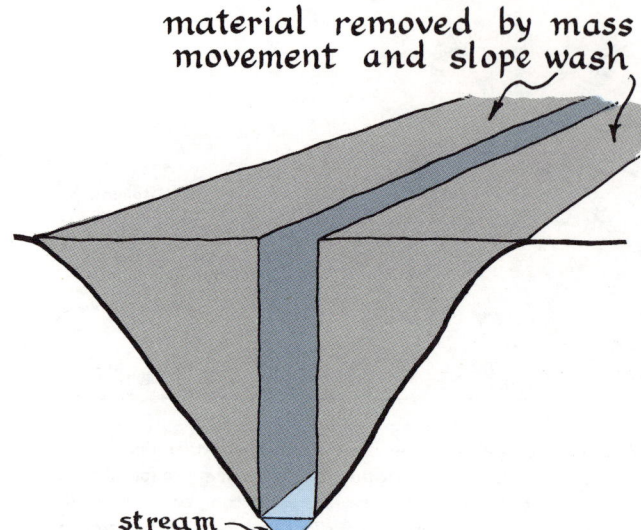

Fig. 11-9. Vertical and lateral stream erosion. The size of the shaded portion of the diagram indicates the importance of mass wastage in the erosional process.

to reduce slopes, but it occurs only during periods of heavy rain and is probably much less important than mass wastage.

Summary

Gradation is a two-stage process consisting of the preparation of material for transport and its subsequent removal and deposition elsewhere. This chapter was mainly concerned with the preparation of material, or weathering, and with one kind of removal called mass wastage, or mass movement.

Weathering proceeds at different rates under different conditions and thus leads to differences in landforms. Chemical weathering involves the processes of oxidation, hydration, carbonation, and solution. These processes are facilitated by the presence of adequate water, and the rate at which they take place is closely related to the temperature. As a general rule, therefore, chemical weathering is most marked in hot and humid climates and much less significant in cold and dry regions. Physical weathering, on the other hand, is much less tied to climate. It may take the form of exfoliation when the removal of the pressure of overlying material causes rock to peel along joints. Or it can involve thermal expansion and contraction, which are important agents of weathering wherever the temperature changes rapidly within short periods, as in deserts; or where the temperature repeatedly crosses some critical value, as in areas where the temperature hovers about the freezing point of water. Crystal growth and organic activity are other forms of physical weathering that are only indirectly related to climate.

Mass wastage is the name given to the localized, often unspectacular, but highly significant movement of material by the direct action of gravity. The factors that affect the rate of mass wastage include cohesion, friction, and root action. Cohesion, which depends on the kinds and amounts of clay and the water content of a rock acts only to retard the start of mass wastage. Friction depends on the internal arrangement of the weathered material, particularly the size and roughness of the particles. It not only affects the start of downslope migration but has a continuing effect once movement is underway. Friction is strongly conditioned by the presence or absence of water. The depth, consistency, and density of plant roots also affect the amount of mass wastage. Normally, some triggering action is needed to get the motion started. Changes in volume can do this, but the most important agent is water, since its presence decreases both cohesion and friction.

Mass wastage may take the simple form of rockfall or it may involve the spectacular motion of large landslides over long distances. It can consist of small slumps, with their characteristic rotary motion; or earthflow down steep slopes; or solifluction over the permanently frozen subsoil of polar regions; or slow and unspectacular but effective movement of large masses of materials by soil creep. Whatever form it takes, mass wastage is a significant element in the study of landforms because it transports material from the slopes down to the valley bottoms, where it can be picked up by streams and removed. It is therefore a necessary complement to stream action with respect to the transportaltion of weathered material.

Suggestions for Further Reading

Donahue, Roy L., *Soils: An Introduction to Soils and Plant Growth*, Prentice-Hall, Inc., Englewood Cliffs, N. J., 1965.

Illinois Highway Research Board, *Landslides and Engineering Practice*, Special Report 29, 1958

Jenny, Hans, *Factors of Soil Formation*, McGraw-Hill Book Company, Inc., New York, 1941.

Penck, Walther, *Morphological Analysis of Land Forms*, trans. Hella Czech and Katharine C. Boswell, St. Martin's Press, Inc., New York, 1941.

Reich, Parry, *A Survey of Weathering Processes and Products*, University of New Mexico, Pub. in Geology, Albuquerque, N. M., 1950.

Sharpe, C. F. S., *Landslides and Related Phenomena*, Columbia University Press, New York, 1938.

12 Gradation by Streams

Introduction

Streams are fundamental means of landform sculpture. They not only carry off the material brought to them by mass wastage but also create the slopes necessary to initiate mass wastage. Stream gradation is accomplished by three closely related processes: erosion, transportation, and deposition. Stream erosion is the picking up, breaking away, or dissolving of material in the stream bed. Stream transportation is the removal of the collected or dissolved material by stream currents. These processes are referred to as stream **degradation,** in contrast to stream **aggradation,** or **deposition.** Deposition of sediment results from a reduction in the velocity of stream flow and may take place upon the stream bed, along the valley floor, or at the bottom of a body of water into which the stream flows.

The result of stream gradation is to remove the products of weathering from the higher landforms and to transport them to lower locations. Gradation is accomplished by a great network of streams that can be divided into many *stream systems*. A stream system consists of a major stream and all its tributaries. The area drained by a stream system is a **drainage basin,** and the ridge between two individual streams is an **interfluvial ridge.** As stream systems entrench themselves into the land, relief grows more complex.

To understand the various aspects of stream gradation, we need some idea of the nature and properties of a stream. There are two major kinds of streams: **perennial streams,** which maintain their flow between rains, and **intermittent streams,** most common in semiarid and arid regions, whose flow is limited to a rainy season or even to a single rainstorm. Perennial and intermittent streams both flow in well-defined channels.

Less commonly observed is the intermittent flow that is not clearly channeled and occurs as **rill flow** and **sheet wash.** When the soil eventually becomes saturated during long heavy rains, rainwater can no longer filter into the subsoil, and the excess must flow downslope. Where the surface is bare and smooth, this slope runoff may be a continuous thin film, or sheet wash. On grass-covered slopes, the flow is divided by the grass stems into many tiny streams of water, while on rough slopes the water may collect into rills, or small temporary streams. Flowing over the soil surface, slope runoff exerts a dragging effect, the force of which depends on the thickness of the film of water and its speed of flow. On slopes protected by vegetation, sheet wash and rill flow pick up particles ranging in size from fine silt to coarse sand and gradually transport them downslope. This slow removal occurs on slopes everywhere. In humid climates, under natural conditions, the rate of removal is slow enough that soil horizons can be maintained; that is, soil-forming processes can keep up with the rate of erosion.

When water flows in a stream channel, the same processes are at work, but the relative importance of the various components is different. Gravity acting on water in a confined channel has two major components.

One component, acting downward, causes the water to exert pressure against the stream bed — the pressure varying according to the slope of the stream surface (or stream gradient) and the depth of the water (Fig. 12-1(1)). The other major component, the one that causes stream flow, is directed downstream, parallel to the water surface. The steeper the gradient and the greater the depth of a stream, the greater is the component of gravity causing stream flow and the greater is the average speed of that flow (Fig. 12-1(2)). With greater depth the average rate of stream flow increases because the frictional resistance of the bed decreases as the water deepens.[1] The rate of stream flow can therefore be changed by altering its gradient or its depth. Changes in gradient obviously come about through differences in the slope of the stream bed; changes in depth result from changes in the volume of flow, if the width of the stream bed is assumed to remain approximately constant.

The velocity of stream flow is not constant throughout any cross section of the stream. Friction between the water and the stream bed reduces the rate of flow along the bottom and along the sides of the stream. Away from the bed, the average rate of flow increases until it reaches a maximum in the center of the stream a short distance below the surface. These differences in speed are difficult to calculate. The slower-moving water nearest the bed exerts a dragging effect upon the more rapidly moving water near the surface. The resulting shear between the slower moving and the faster-moving water creates a disorderly condition of eddies and rotating masses of water, with upward, downward, backward, forward, and lateral movements. This turbulence makes it difficult to establish the velocity at any given point, let alone the average speed of the stream as a whole.

Stream Erosion

Relatively small increases in the average rate of stream flow result in large increases in turbulence, which is very important to the erosional process.

---

[1]This principle can be easily demonstrated. Assume a rectangular channel of great depth and a constant width of 10 feet. When, for example, a 2-foot length of the channel is filled by a stream to a depth of 10 feet, the wetted perimeter of the channel segment has an area of 60 square feet, whereas the volume is 200 cubic feet. If the depth of the stream doubles to 20 feet, the area increases to 100 square feet, and the volume becomes 400 cubic feet. Further doubling of stream depth makes it clear that stream volume increases at a much faster rate (geometric versus arithmetic) than the area of the wetted perimeter. Consequently, we can expect the volume also to increase faster than bed friction. Thus when stream depth becomes greater, the average speed of stream flow will increase to the point where it is checked by greater turbulence.

Through turbulence, with its partial upward movement, streams dislodge sand and silt from poorly consolidated materials in the stream bed and support them on the rising currents.

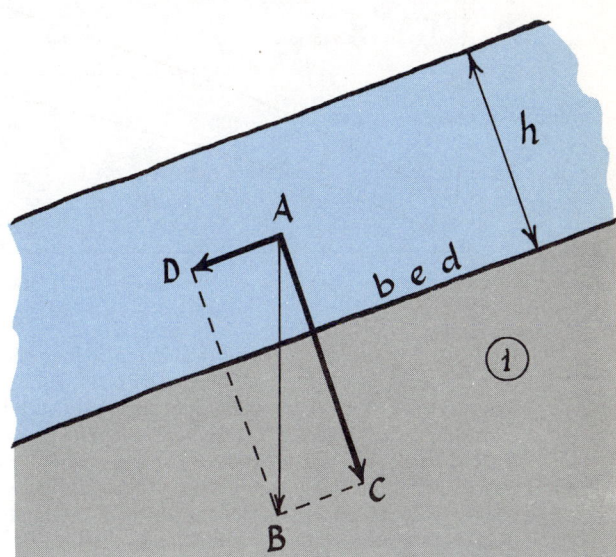

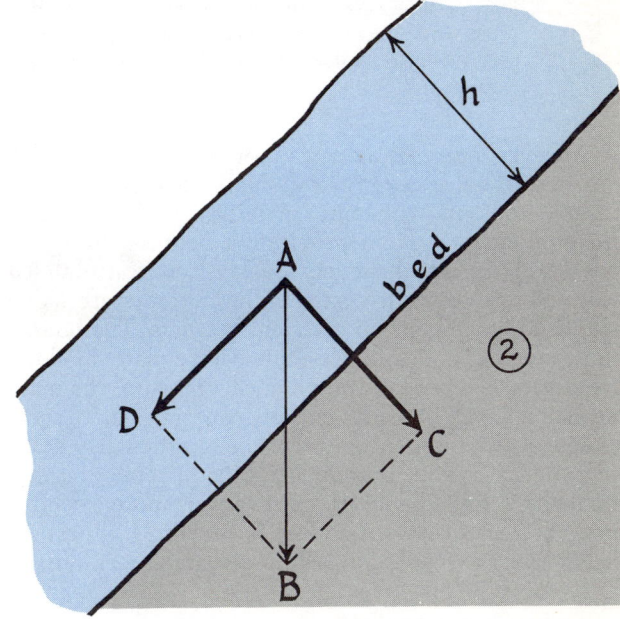

Fig. 12-1. (1) Vector diagram, disregarding friction, of forces operating in a stream channel. The length of line *AB*, representing the downward pull of gravity, is proportional to the weight of a layer of water of thickness *h*; line *AC* represents the component of gravitational force exerted against the stream bed, and line *AD* represents the component directed downstream. (2) Vector diagram showing the increase in the downstream component of gravitational force with an increase in stream gradient. The lines have the same meaning as in (1).

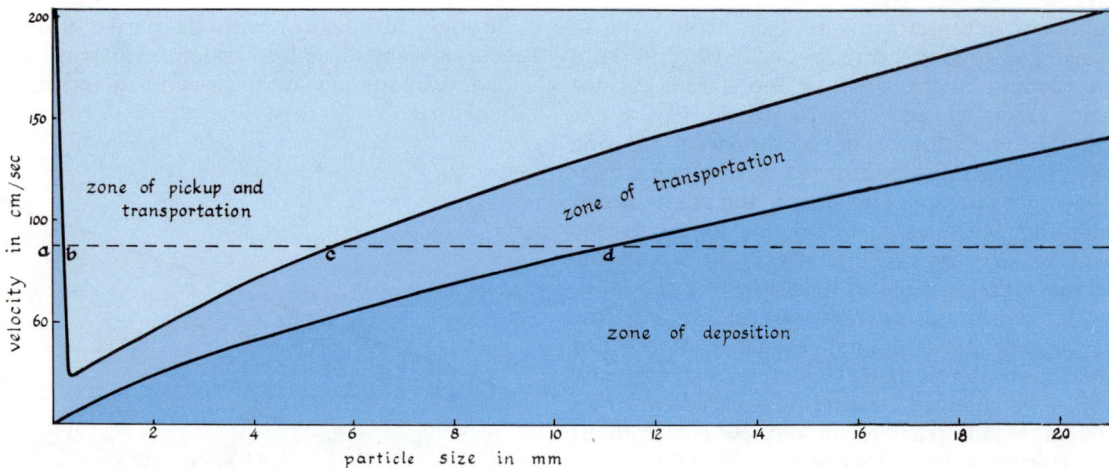

Fig. 12-2. The stream velocities required to pick up and transport various-sized particles. A stream with a velocity of 90 cm/sec, for example, cannot pick up particles in the size range shown by the dashed line between a and b (particles less than 0.2 mm in diameter), but it can transport such fine material if it has been contributed to the stream by mass wastage and sheet wash. The stream can both pick up and transport particles in the range indicated between b and c (particles between 0.2 and 5.5 mm in diameter). Larger particles cannot be picked up, although they can be transported (if they have been picked up upstream where the stream velocity was higher, either because of a steeper gradient or greater flow). The ability to transport without being able to pick up occurs in the size range between c and d (particles between 5.5 and 11 mm in diameter). Particles larger than 11 mm cannot be transported and are deposited. Note that for a slower stream, with a velocity of 60 cm/sec, for instance, the zone of transport and pickup is reduced (only particles between 0.3 and 2.4 mm can be picked up and carried), and particles as small as 5.0 mm are deposited. A fast-moving stream can pick up and transport material having a wide range of sizes, and deposits only the very largest particles. A slow-moving stream has a much-reduced capacity for pickup and transport, and deposits finer particles.

However, currents alone cannot effectively dislodge large particles or even very small ones (Fig. 12-2). Tightly packed coherent material, such as clay, prevents currents from penetrating between tiny particles and prying them loose. Clay and consolidated bedrock can be eroded only if the currents are laden with sand, gravel, or, on occasion, boulders. The bouncing, dragging, and rolling of this coarse material along the stream bed result in mechanical *abrasion*, the main mechanism of stream erosion on resistant rock. Chemical weathering, or *corrosion*,[2] particularly **solution** and **hydration,** is another means of removing material from the stream bed. Corrosion is more effective on some rocks than on others; for example, limestone yields readily to solution, while quartzite does not.

Stream Transportation

To erode its bed effectively, a stream must be capable of transporting not only the products of its own abrasion and corrosion but also the debris supplied to it by mass wastage and sheet wash (Fig. 12-2). Stream transportation is also largely dependent upon turbulence. Streams carry material by **suspension, traction,** and **solution.** Suspension occurs in a stream when the vertical currents of turbulence have an upward velocity greater than the sinking velocity of the particles in the water. The more violent the turbulence, the larger the individual particles and the greater the mass of sediment that can be maintained in suspension. There is, of course, a limit to the size of the suspended particles. The coarse material that cannot be transported by suspension remains on the stream bed. These coarse particles are in turn dragged, rolled, or bounced along the bottom of the stream by turbulence and the drag of stream currents; this process is called traction.

A significant portion of a stream's load is carried in solution, the dissolved matter becoming part of the moving water. Samples taken from rivers across the United States indicate that, on an average, about 20 percent of the total measured load is carried in solution. As one might naturally expect, the dissolved load is usually much lighter in streams flowing through arid regions.

---

[2]Corrosion (Latin *corrodere* to gnaw), or chemical weathering, should not be confused with corrasion (Latin *corradere* to rub), which is often used as a synonym for mechanical weathering, or abrasion.

Rate of Degradation

Streams do not erode and transport with equal efficiency. The rates of erosion and transportation depend principally on the velocity of stream flow. Accelerated stream flow brings about a rapid increase in turbulence and a coincident expansion in the capacity to erode and transport. Roughly speaking, an arithmetic increase in speed of flow brings about a geometric increase in the stream's ability to degrade.

The volume of stream flow also affects erosion and transportation. The effects of volume and velocity are difficult to separate because a change in volume inevitably results in a change in velocity as long as a stream is confined to its channel or to a narrow valley.

The variable resistance of rocks to weathering and erosion obviously influences the rate at which streams deepen and widen their valleys. Rocks resistant to abrasion and weathering retard the attack of a stream on its bed and reduce the production of weathered material on the valley sides, which retards both valley deepening and valley widening.

Streams do not deepen their channels uniformly. Observations of both straight and *meandering*[3] streams indicate that shallows and deeps occur along their channel floors. These are not spaced randomly; they occur at more or less regular distances of five to seven times the stream width. In meandering streams, the deeps are associated with the bends; thus, their spacing is one-half the length of a complete meander wave. The similarity of the spacing of deeps in straight and in meandering stream channels suggests that the tendency to meander is present in relatively straight stream channels in resistant rock. The tendency of streams to meander is possibly related somehow to the mode of stream flow. One explanation of meandering arises from observations that water in a channel appears to flow in a helical spiral; in other words it follows a path similar to that traced by the threads of a screw. This being the case, the meandering pattern may simply be a means of allowing a uniform rate of energy expenditure along a given length of stream channel.

Stream Deposition

Velocity of stream flow not only influences the capacity of a stream to degrade its bed but is also the primary variable in stream deposition. A decrease in velocity brings about a reduction in the quantity and vigor of turbulence; consequently, the ability of a stream to transport material also declines, and some of the particles in motion come to rest. The larger particles carried along by the stream are deposited first; as a stream continues to slow down, increasingly small particles come to rest until even the very finest settle out in nearly still water. In this process, as in the case of marine sedimentation, sorting by size occurs during deposition.

Landforms Resulting from the Work of Streams

So far, we have been studying how streams work; let us now investigate their role in the development of landforms. Stream action is both destructive and constructive with respect to landforms. Degradation initially roughens the land surface by the erosion of many valleys, while aggradation creates smooth, nearly flat surfaces by the deposition of stream sediment. Continued degradation by streams, together with mass wastage, if not offset by tectonic processes, produces ever widening stream valleys with gentler and lower valley slopes. As these progressively lower slopes intersect at the crests of interfluvial ridges, the initially rough terrain becomes increasingly smooth; the resulting, nearly level surface is often called a **peneplain** (Latin *pene* almost + plain). Thus, the long-term result of stream gradation is nearly flat land surfaces; but before this possible end result is attained, highly varied landforms can be created.

The landforms created by stream degradation consist of an assemblage of slopes, curved and straight, that are mainly shaped by stream erosion and mass wastage operating together. Several theories have been advanced to explain how such slopes develop and change. One idea places major emphasis on *slope decline*; slopes retreat with diminishing gradient (Fig. 12-3(1)). A second theory features *parallel retreat*; once a slope has been created, its form and angle are perpetuated as it moves back (Fig. 12-3(2)). A third idea, and the one presented here, is that of *parallel retreat and slope replacement*. This theory explains slope development by an initial retreat of the slope, parallel to its original inclination; but as the slope retreats further, it is consumed and replaced from below by successively lower and gentler slopes, or by successively lower and steeper slopes, depending on the rate of stream incision (Fig. 12-3 (3) and (4)).

Rates of Tectonic Uplift

Landforms resulting from stream degradation require some kind of tectonic deformation of the earth's crust to provide the potential relief from which the landforms are carved. If this raised surface is a former

---

[3] A meander (Greek *Maiandros* a river in Asia Minor proverbial winding course) is a sharp bend or turn of a stream.

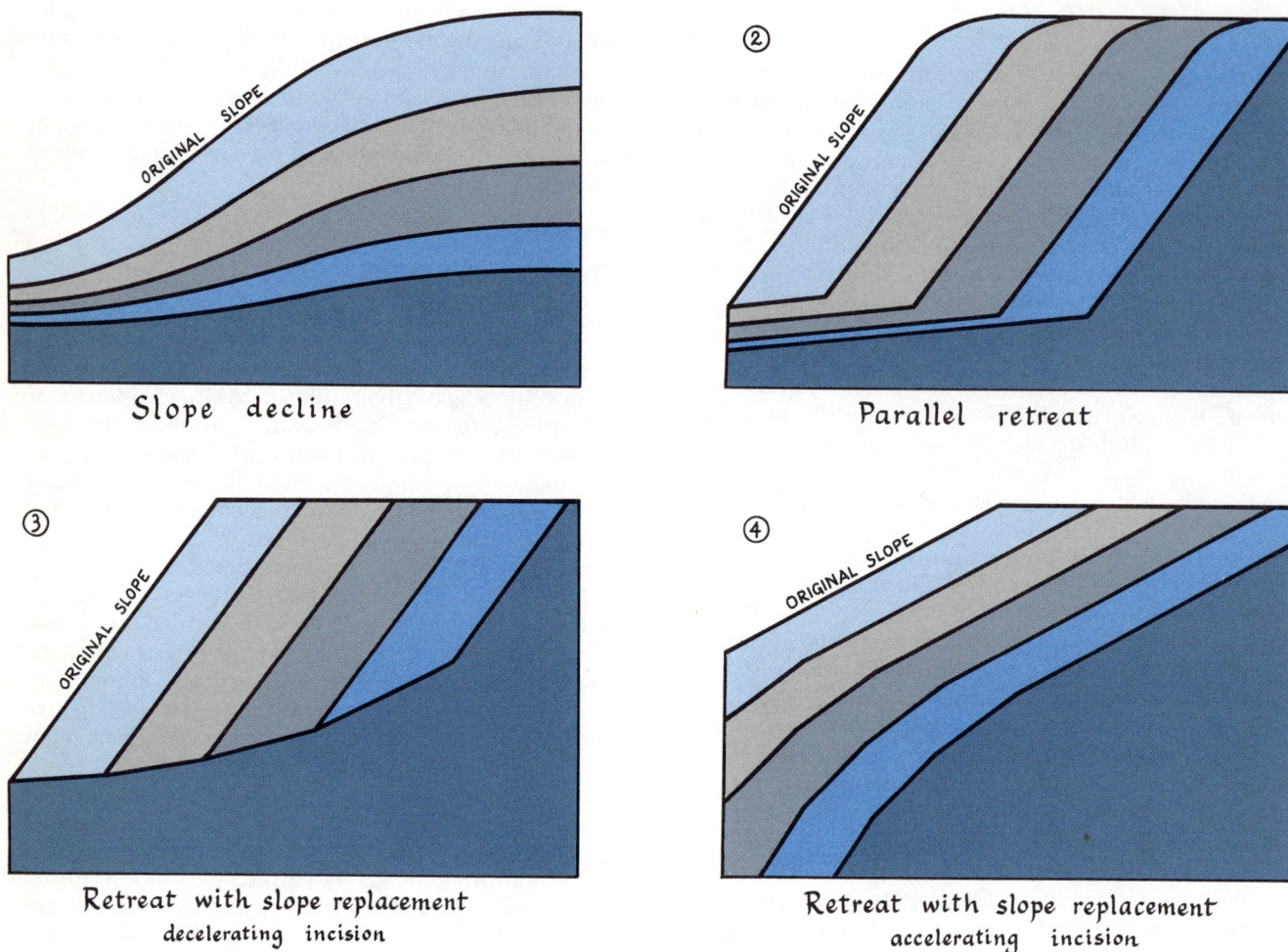

Fig. 12-3. Diagrammatic representation of theories of slope development. In (1) the upper slopes are flattened more than the lower, giving the valley side a convex shape. In (2) the slopes retreat parallel to their original inclination, so that the valley side remains straight. In (3) decelerating erosion causes gentler slopes to form in the lower part, thus giving the valley side a concave shape. In (4) accelerating erosion causes progressive steepening of the slopes with a convex shape, as in (1) but with straight slopes at the top of the valley side rather than at the bottom.

sea bottom, new streams are created as the land is uplifted. It is more likely that the raised surface consists of already existing landforms, on which new relief features are superimposed. In the latter case, the degrading processes of the preexisting streams accelerate or slow down, depending on whether their gradients and volumes increase or decrease.

Relief is created at varying rates. Land movement may start slowly, accelerate, then return to a slow rate. It may start quickly, then gradually slow down, stop, and start again. Other combinations of uplift rates may also occur. All these movements are important because of their effect on stream gradients. Stream volume may also change as the land surface is raised. The combination of altered gradient and volume produces varied and changing rates of stream flow and, consequently, varied and changing rates of stream incision or erosion. Let us consider three long-term conditions of degradation that illustrate the major kinds of significant valley development: a steady rate, an accelerating rate, and a decreasing rate. Each has a distinctive effect upon the development of stream valley slopes. To study the effect of different and changing rates of stream incision upon the development of vally cross sections, we will focus on a small segment of a stream valley and eliminate the effect of factors other than stream action by assuming for all of our examples that rock type and structure are uniform and that climate and vegetation are similar.

Valley Cross Sections

A steady rate of stream incision implies a balance between the rate of downcutting or downward erosion and the rate of mass wastage on valley sides. That is to say, the rate of incision is sufficient to remove all the weathered material brought down to the stream by mass wastage, and valley deepening proceeds at a rate proportional to valley widening. Figures 12-4(1) and 12-4(2) are cross-sections showing the valley sides of streams that are incising with the intensity required to remove the material from the valley sides through mass wastage and sheet wash. Because of the decreased influence of gravity upon the processes, the rate of mass wastage and sheet wash is less in valleys with gently sloping sides than in valleys with more steeply sloping sides; consequently, less intense downcutting is required to keep the gentle slopes straight. As long as the underlying rock remains the same, uniform rates of erosion produce valleys with straight, uniform slopes (as represented by the parallel retreat shown in Fig. 12-3 (2)). The gradient of the slopes depends on the intensity of erosion — the greater the intensity, the steeper the slopes.

To clarify the effects of accelerating stream gradation, let us assume that the increase in gradational force occurs in a sequence of distinct steps, each step representing an interval during which the rate of downcutting remains the same. During the initial period, intensity is such that it maintains relatively gentle valley slopes (Fig. 12-5). Subsequently, incision increases, causing steeper valley slopes to be established. During a third period, a still greater rate of downcutting results in even steeper slopes. This sequence of events creates angular convex slopes. If the intervals of increased downcutting are shortened and their numbers increased, the length of the individual slopes decreases, and the angular convexity approaches a smooth curve. A valley thus formed would have steep side slopes at the bottom and gradually gentler slopes near the top of the valley sides.

If downcutting continues to increase, valley slopes steepen until the limit of strength of the particular rock underlying the valley is reached. Beyond this point, no steeper slopes develop. These maximally steep slopes gradually replace all other slopes as the valley sides continue to retreat (Fig. 12-5). Both steady and increasing rates of downcutting produce V-shaped valleys, but the increasing rate *usually* results in steeper slopes. At this stage of accelerating erosion the valley sides would be represented by straight steep slopes extending all the way from the valley floor to the top of the interstream ridges.

As we consider decreasing rates of stream incision, let us again assume that the decrease occurs in a sequence of distinct steps. During an initial period, inten-

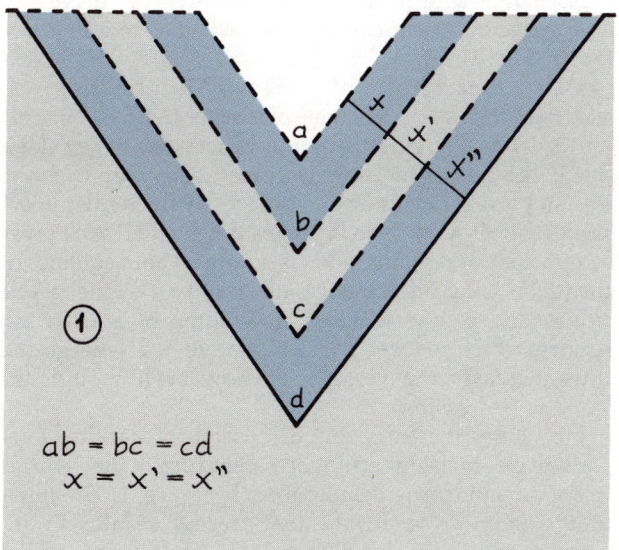
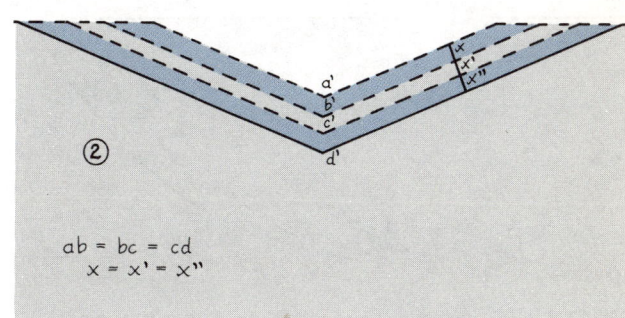

Fig. 12-4. Valley cross sections resulting from two distinct rates of stream incision. Underlying rock and other factors are assumed to be the same. In both (1) and (2), the rate of incision during each unit of time is such that it prevents slopes of inclination lower than the original from developing.

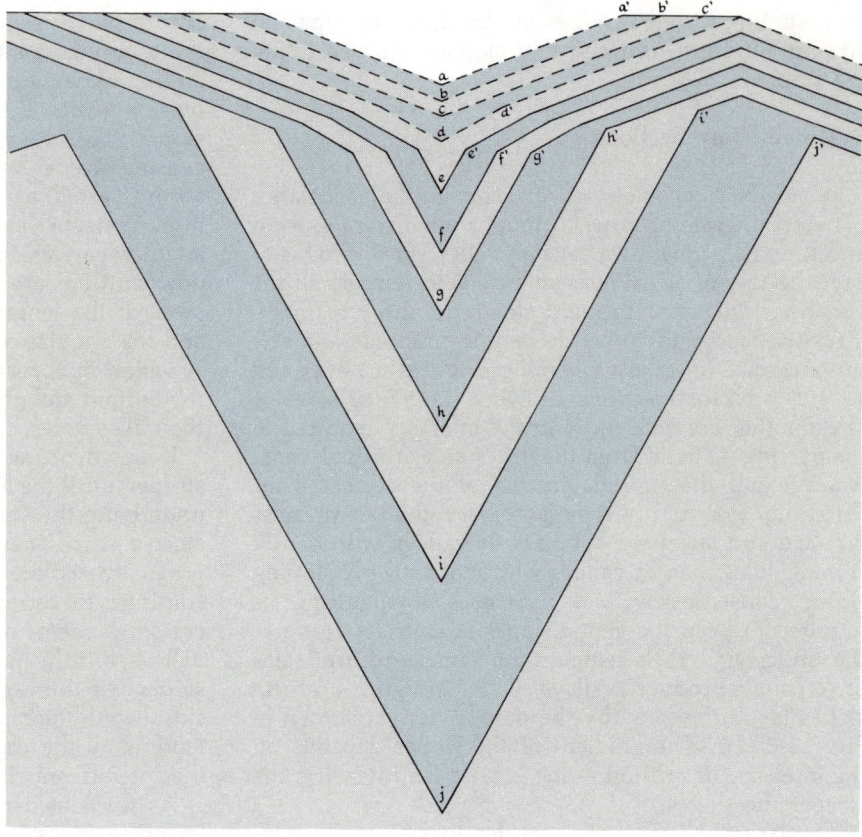

Fig. 12-5. Valley cross sections formed by accelerating rates of downcutting. If the downcutting remains constant, it maintains the relatively gentle valley slopes (*a-a'*, *b-b'*, *c-c'*); if during subsequent time units the rate of incision accelerates, it produces successively steeper slopes (*d-d'*, *e-e'*, *f-f'*, etc.) A very large number of such changes will produce a convex valley slope once the slopes reach their limit of steepness (determined by the internal strength of the rock), further acceleration of the rate of downcutting yields parallel slopes (*f-f'*, *g-g'*, *h-h'* etc.), whose increasing lengths gradually obliterate all other slopes (*j-j'*).

sity of incision maintains relatively steep valley slopes (Fig. 12-6). In subsequent periods, incision decreases, permitting progressively gentler slopes to develop during each period. This sequence of events creates angular concave valley slopes. Again, if the intervals of decreased erosion are shortened and their numbers increased, the length of the individual slopes decreases, and the angular concavity approaches a smooth curve (*xy* in Fig. 12-6).

When stream incision is steady or is increasing, it is assumed that mass wastage performs the most important role in removing the regolith from the valley sides. But if the rate of incision decreases, soil creep becomes less and less efficient on the progressively gentler lower slopes; sheet wash then becomes the most effective way to remove the regolith. This is a natural consequence of the decreasing importance of gravity on the gentle slopes and of the increase in volume of sheet-wash runoff, which reduces friction relative to discharge and thus increases the velocity of flow even though the slope is decreasing.

How does the foregoing discussion of varying rates of stream incision and corresponding valley forms relate to landform development? Figure 12-7 shows that steady downcutting by streams results in increasing local relief until the slopes of adjacent valleys intersect to form a ridge. From then on, local relief remains constant. The amount of local relief and the degree of slope of the valley sides depend upon the rate of the steady downcutting, assuming equal initial relief in all instances. If such factors as rock type, rock struc-

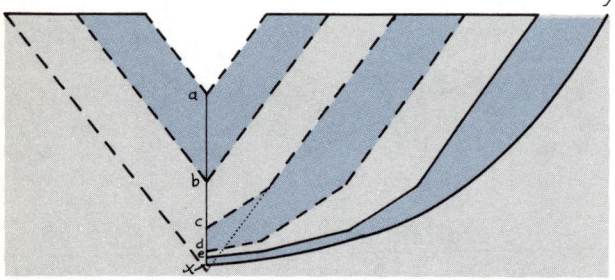

Fig. 12-6. Valley cross section formed by a decreasing rate of stream incision. During the first unit of time, erosion from *a* to *b* preserves the initial straight slope. In subsequent units of time, the successively lower rates of downcutting permit progressively gentler slopes to develop. A very large number of such changes will produce the concave valley slope *xy*.

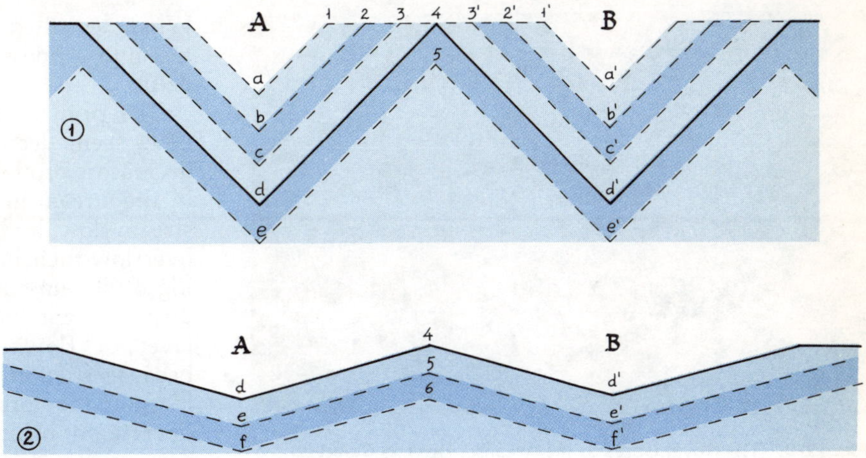

Gradation by Streams   209

Fig. 12-7. Development of constant local relief. In (1), steady and identical rates of downcutting by streams A and B result in increasing local relief until the beginning of time unit 4. Equal downcutting and slope retreat in subsequent time units will result in constant local relief. (2) indicates that if rock type and other variables are the same, the amount of local relief and the degree of slope depend upon the rate of incision.

ture, climate, and vegetation are not considered, a slow steady rate of erosion results in gentle slopes and low local relief, while a fast steady rate produces steep slopes and high local relief.

A long duration of a steady rate of stream incision eventually produces a state of *dynamic equilibrium* in the configuration of the land surface. Dynamic equilibrium is a condition in which all elements of the terrain are mutually adjusted, so they are being lowered at the same rate by stream erosion and mass wastage. Terrain in dynamic equilibrium is an example of an open system through which both mass (weathered rock and water) and energy are passing at a uniform rate. Eventually, the volume of stream flow changes, or the rate of uplift is altered, which upsets the balance by changing the rate of stream incision. When this happens the processes of stream erosion and mass wastage would eventually reach a new state of equilibrium.

An increasing rate of stream incision eventually creates greater local relief and steeper valley slopes than those produced by the steady rates. Once the limit of rock strength is exceeded, however, a new dynamic equilibrium can be established if the rock type, rock structure, climate, and vegetation remain the same.

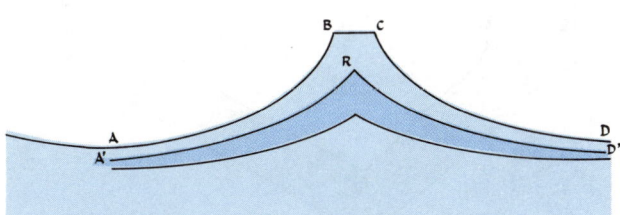

Fig. 12-8. Decreasing local relief. After slopes AB and CD intersect, continued stream downcutting at a decreasing rate will result in decreasing local relief as progressively less steep slopes intersect at the ridge (A'RD').

If the rate of downcutting slows, local relief increases, although less rapidly, until the slopes of adjacent valleys intersect. From then on, an important change takes place — local relief starts to decrease as progressively gentler slopes intersect at the ridge (Fig. 12-8). Eventually, this continued decrease leads to a nearly flat plain of erosion unless the land is once again uplifted and stream gradients are steepened.

In humid climates the valley sides usually intersect in rounded summits, not in sharp-crested ridges as shown in Figs. 12-4 through 12-6. This phenomenon of summit rounding is rather difficult to explain, but it is thought that the increasing curvature of slope away from the crest is necessary to accommodate the accelerated rates of creep required to remove increasing amounts of weathered rock downslope from the summit.

In an arid climate, where slopes on similar rock types tend to be steeper than in humid regions, similar processes yield slightly different results. It is felt that in arid regions, stream incision is relatively faster than weathering and mass wastage on valley slopes. In any case, the pace of events is probably slower in arid regions, since the rate of stream incision and of weathering and mass wastage are all slower than in humid regions. This interpretation should be applied with caution to such extremely arid regions as the central Sahara, since the extreme aridity may preserve forms dating from much earlier periods of greater rainfall.

Landforms of Stream Degradation

When tectonic processes produce rapid uplift, steep gradients allow streams to erode their valleys quickly. The depth of the valleys, or local relief, will depend not only on the rate of stream erosion but also on the amount of potential relief; deep valleys can be eroded

Fig. 12-9. The Royal Gorge of the Arkansas River in Colorado. The steep valley slopes indicate a rapid rate of downcutting. Slope irregularities are usual in nature, since uniform rock type and exposure are rare. The valley walls, however, do show a grossly convex form.

only into high initial relief. To simplify the description of the relation of cross-sectional valley development to rates of stream erosion, we have neglected all the other landforming factors. If one considers these other factors (for example, changing rock type and rock structure), the processes of valley growth would be as described, but their effects would bring about diversity in the patterns of the valleys and would prevent the development of straight valley slopes (Fig. 12-9). Through their influence upon the nature and capacity of erosion by surface flow, variations in climate and vegetation, for example, could further change the details of slope form.

The form and pattern of the ridges, hills, mountains, and peaks created by stream gradation are determined by the nature of all the local landforming factors. Eventually, however, time will become the paramount factor unless renewed earth movements occur. With time, stream gradients become less steep, and the rate of erosion decreases. Eventually, even the most massive mountain ranges, worn down by mass movement and sheet and rill wash, will be replaced by a surface of low local relief.

Landforms of Stream Deposition

Deposited stream sediment, regardless of its form of deposition or its character — whether coarse or fine — is called **alluvium**. Depending on where it is laid down, alluvium is deposited in three topographic forms. Deposits at the base of the hills or mountains, especially in arid regions, are called **alluvial fans**. On the middle and lower courses of stream valleys, alluvial deposits occur along the sides of the streams as **floodplains**.

When deposited at the mouth of a stream where it empties into a body of calm water, the alluvium forms a **delta**.

The primary cause of stream deposition is, as we have seen, decreasing velocity of stream flow. This decrease is brought about in several ways: by a decrease in the stream gradient, by a decrease in the depth of stream flow, by lateral dispersal of water when streams overflow their banks during floods, and by the checking of stream currents in a body of still water. Lateral dispersal reduces the rate of flow, not only because the layer of flowing water becomes thinner but also because friction with the surface increases. With the foregoing factors in mind, let us examine the way they operate to bring about the three forms of alluvial deposition.

Alluvial Fans. An alluvial fan is a low cone of gravel, sand, and finer sediment that resembles an unfolded Oriental fan in outline. The apex of the fan lies at the mouth of the valley or ravine where the stream emerges from a highland, while the main body of the fan is spread out over the adjacent lowlands (Fig. 12-10 and 12-11). Alluvial fans may be very small or cover many square miles. They are deposited by intermittent streams, especially those that flow only during rainstorms. Since intermittent streams (some of large size) are common in semiarid and arid regions, those

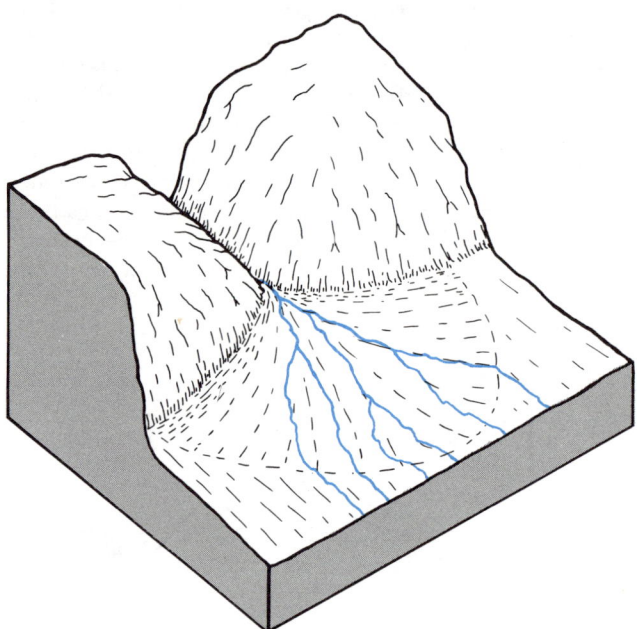

Fig. 12-10. An alluvial fan. The spread of flowing water once it leaves the narrow valley causes the fan shape. The distributary pattern of stream flow is characteristic of many fan surfaces.

Fig. 12-11. Alluvial fans in the Mojave Desert, California. The spread of water once it leaves the narrow valleys causes the fan shape. The distributary pattern of stream flow, characteristic of fans, is shown here by the numerous gullies on the fan surfaces. *Photograph by J.R. Balsley, U.S. Geological Survey.*

areas have an abundance of large alluvial fans. Mountainous regions in high latitudes also have many alluvial fans. Winter frost causes the streams to be intermittent during that season, and even in summer there is a large diurnal range in stream flow because of nocturnal freezing. In bringing about deposition of fans, the diurnal fluctuations in flow are probably more important than the annual variations. Cold-climate fans are numerous but are usually smaller than those found in arid climates. Fans are also small in humid climates because only small tributary streams are intermittent.

The loss of stream velocity which causes alluvial-fan deposition results from the combined action of lateral dispersal of floodwaters and loss of volume that occurs when a stream enters the lowlands at the base of the mountain.[4] In the highlands, during periods of flood, a swollen stream moves swiftly because its flow is confined by narrow valley walls; the current of water deepens as slope runoff increases. When this water flows onto the open surface of the plain at the foot of the mountain, it is no longer confined by the steep valley sides of the highland and spreads out quickly, dumping large amounts of sediment as it slows down.

The water, laden with sediment, occasionally spreads as a thin sheet over the entire surface of the growing fan, especially if the fan is small. More commonly, the water is dispersed by many small distributaries that fan out from the mouth of the valley. Since the individual distributary channels contain less water than the original channel, deposition of sediment takes place. This spreading and fanning out of flowing water from a fixed exit results in the characteristic outline of alluvial-fan deposits. The cone-shaped cross section is due to the fact that the greatest rate of deposition per unit area is near the apex of the fan. Toward the margins of the fan, the rate of deposition per unit area decreases as the surface area of deposition increases. From the apex of the fan to the outer margin, there is also a decrease in the size of deposited particles, ranging from very coarse gravel to fine silt and clay.

The relatively short duration and quick cessation of flow in intermittent streams confine the major deposition of alluvium to the immediate vicinity of the highland base. Even in arid climates, where scanty vegetation favors heavy slope runoff and strong erosion by that runoff, the alluvium is usually deposited within a few miles of the foot of the highlands. Perennial streams flowing out of highlands in humid climates usually cross the lowlands in valleys large enough to contain their flood flow. In these, the floodwater can spread only from one side of the valley, or bluff, to the other, but because of the relatively long duration of high water, the bluff-to-bluff flood may extend far downstream. Under these circumstances, alluvium is laid down as a long ribbon of deposits that eventually form a floodplain.

Floodplains. The primary cause of alluvial deposition on a floodplain is the reduction in stream velocity accompanying lateral dispersal of floodwater. A floodplain is usually flooded once a year, during the season of greatest runoff. At this time, streams overflow their channel banks, the overflow spreading out as a shallow sheet of slowly moving water. As the flood subsides, the receding water leaves behind a thin film of alluvium. Deposition is concentrated immediately along the stream-channel banks, where the overflowing water experiences its greatest loss of velocity as it leaves the relatively swift currents within the channel.

The greater deposition along the channel banks eventually builds up **natural levees,** which slope gently away from the edges of the channel (Fig. 12-12). Once natural levees develop, they prevent some of the flood water from returning to the main channel as floods subside. This water, trapped between the bluffs and natural levees, forms temporary ponds and

---

[4]This loss of velocity is often attributed to a decrease in gradient, but profiles across fan surfaces and up the confined channel show no such decrease. In fact, the gradient across the fan is frequently greater than that in the confined channel farther upstream.

Fig. 12-12. A natural levee deposited along a meander-scroll channel during the spring flood in 1964. Photograph taken on the Black Bottom, a portion of the Ohio River floodplain, in southern Illinois.

swamps where it deposits the finest sediments and the precipitates of any dissolved materials carried by the stream. This very fine alluvium, augmented by considerable organic debris from the swamp vegetation, forms the *backswamp*, or *levee-backswamp*, deposits.

Eventually the floodplain is built up above the level of all but the highest floods, and the rate of overbank deposition decreases to a very slow pace. Deposition then takes another major form in sandbars (*channel*, or *point-bar*, deposits) that accumulate along the inside of a river-channel **meander** during floods. The sandbar is deposited in a curve that approximates the shape of the meander. Eventually the point-bar deposit is raised above the general level of the river and is held in place by permanent vegetation. The river continues to erode laterally on the convex side of the channel, forming a second point-bar deposit which separates the first bar from the river. The first bar becomes part of the floodplain and is subsequently partly covered by alluvium deposited during floods until it reaches the general level of the flood plain. As a consequence of multiple point-bar deposition, the floodplain is crossed by curvilinear ridges and swales called *meander scrolls*. As time passes, point-bar deposits gradually cover the valley floor, forming a gently undulating surface on the floodplain (Fig. 12-14).

Streams flowing over floodplains frequently develop meandering, or smoothly looping, channels, which are exaggerations of the tendency of all streams to develop channels that twist and turn to some degree. In the easily eroded alluvium of floodplains, the turns become broad, smooth bends. Once formed, meander loops are frequently tightened at their bases by erosion of the bank on the outside of the stream curve. This bank erosion is usually more rapid on the downstream side of the bends. As a result, the downstream banks catch up with the succeeding upstream banks. Eventually, the meander is cut off as the two banks come together (Fig. 12-13). The cutoff meander becomes a *horseshoe*, or *oxbow lake* as the ends of the abandoned channel are blocked with sediment. Such lakes are soon filled with plant debris and with alluvium brought in by floods.

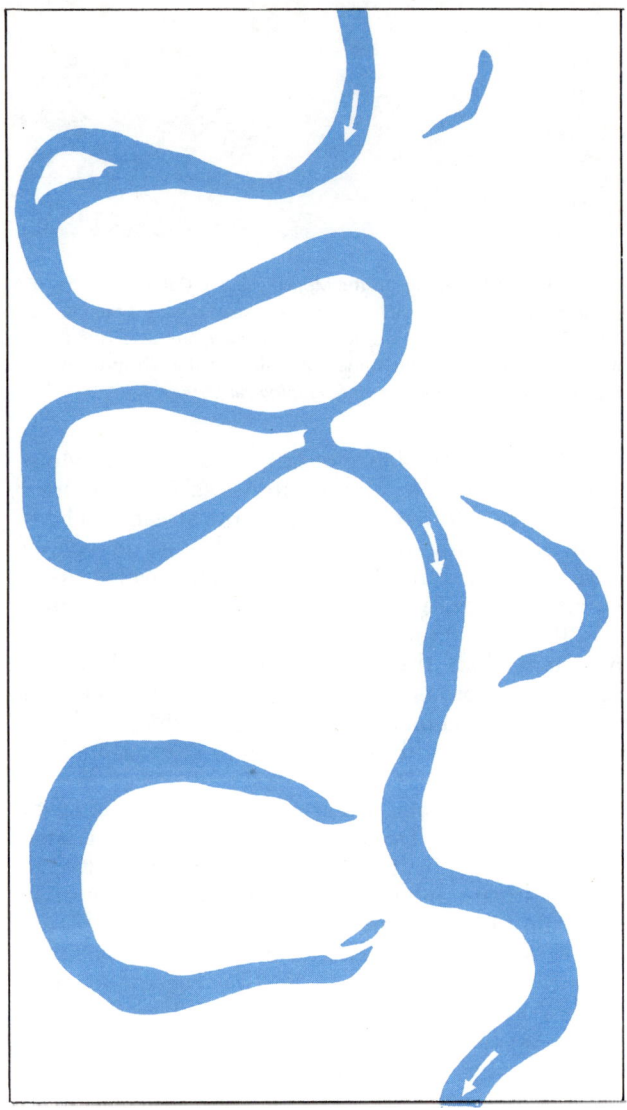

Fig. 12-13. Development of a cutoff meander as the result of lateral erosion primarily along the downstream bank of the meander. At the top the meander is still intact. In the middle the two ends of the meander have been joined by a new channel. At the bottom the river has abandoned the meander and flows through the cutoff channel.

The processes that result in meander formation and in the frequent cutoff of individual meander loops lead to a downvalley shift of the entire train of meanders. As these meanders migrate downstream, the entire floodplain is worked over by the river. Consequently, the surface is covered with a complex patchwork of natural levees, point-bar deposits, oxbow lakes, meander scrolls, and bits and pieces of backswamps (Fig. 12-14).

The survival of these features on many floodplains suggests that build-up of plain surfaces by overbank deposition has virtually ceased. Instead sediment is being temporarily stored at some places and, at the same time, an equal amount is being removed from others as the river meanders migrate across the floodplain. Consequently, in any given valley section, the floodplain can be considered as an example of an open system tending closely towards a steady state. The inflow of mass and energy (moving water and its load of sediment) at the upstream end of the section is balanced by an equal outflow at the downstream end. It is evident that while large rivers flow through areas of diverse climate, topography, surface rocks, rock structure and vegetation, they will approach a condition of dynamic equilibrium in their regime that reflects the influence of all the geographic factors.

As a consequence, man-induced changes in one or more of the geographic factors will upset the delicate balance in a stream system (more readily in small than in large stream systems) and create a condition of disequilibrium throughout the system. This will bring on either renewed build-up of the floodplain or its removal by erosion depending upon the nature of the change.

An example of the quick response of small stream systems to changes in the vegetation cover in their drainage basins comes from the tropical highlands of Papua in eastern New Guinea. Several decades ago steel axes were introduced to the Stone Age cultivators of the region. Armed with a more efficient instrument, the people cleared more forest for agriculture and soon expanded their cultivation to the relatively steep valley sides. Because of extensive forest removal and the accompanying acceleration of slope erosion, more sediment than before was dumped into main stream chan-

Fig. 12-14. Air photograph of part of the Willamette River floodplain, north of Salem, Oregon. Note the meander scars.

nels, which quickly became clogged and less capable of removing the increased runoff. As a result, the floodplains became excessively swampy; malaria increased, and the once productive valley flats had to be abandoned. The instability in the stream systems, the result of forest destruction in the local drainage basins, quickly produced increases in the accumulation of floodplain sediment. Once this new steady state is approximated for the local floodplains, presuming continued cultivation of the hill slopes, it will probably be characterized by more abundant and more rapid passage of water and sediment than formerly — and, in agricultural terms, the sediment, being removed faster than the soil-forming processes can generate new soil, will be of much poorer quality. Eventually, the accelerated erosion will cause the slope to become less steep or bedrock will be reached, thus reducing the sediment yield. Land abandonment and re-vegetation will accomplish the same results. Reduced sediment yield from the hill slopes will permit the stream to flow as before.

Floodplain deposition can take place only in valleys that allow an appreciable spread of floodwater across a wide valley floor created by stream erosion or by the deposition of other sediment. A wide-bottomed valley can result from a long period of decreasing stream erosion (Fig. 12-8). At some time in the course of the valley history when the lower slopes have become very gentle, floodplain alluvium may start to accumulate. Eventually, alluvium may be deposited to moderate depth as the floodplain broadens. Until recently, stream erosion was regarded as the primary agent of valley-floor widening. However, recent discoveries indicate that, in some large valleys at least, floodplain alluvium has been deposited upon preexisting delta sediments. In all probability, both valley widening and delta deposition preceded floodplain deposition, valley widening being more prevalent along the interior reaches of streams, while delta deposition is more common along the lower stream courses.

If floodplains are underlain by considerable depths of alluvium, prior delta deposits probably provided the wide floor necessary for flood deposition. For example, cores from wells drilled through the alluvium of the Mississippi River floodplain south of Cairo, Illinois, show an old stream valley buried by several hundred feet of sediment. Delta sediments form the lower layers of the valley fill, while the upper layers are composed of floodplain deposits (Fig. 12-15).

Similar evidence of deeply alluviated valleys comes from river floodplains all over the world. Near coasts, the buried valley bottoms are 300 to 450 feet (90 to 135 m) below present sea level. This worldwide phenomenon of deeply alluviated valleys has been in-

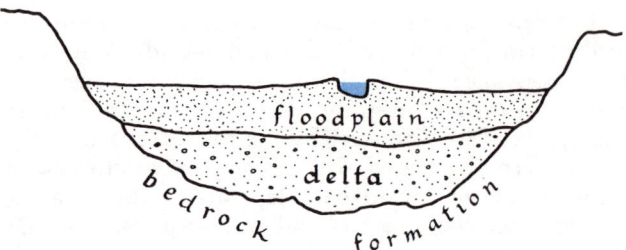

Fig. 12-15. Delta and floodplain alluvium partially filling a previously stream-eroded valley.

terpreted as evidence of a change in sea level that occurred about 15,000 years ago, at the close of the Wisconsin glacial stage, when the sea rose as water from the melting glaciers returned to the ocean basins (see Chapter 13). During the glacial period itself, sea level was about 450 feet (135m) lower than at present, and streams deepened their valleys, especially in their lower courses. Once sea level started to rise, the deepened valleys were flooded and became long, narrow bays, or estuaries. Stream sediment was then deposited at the heads of the bays in deltas that gradually expanded seaward. As sea level continued to rise, the bays were filled, first by delta deposits then by

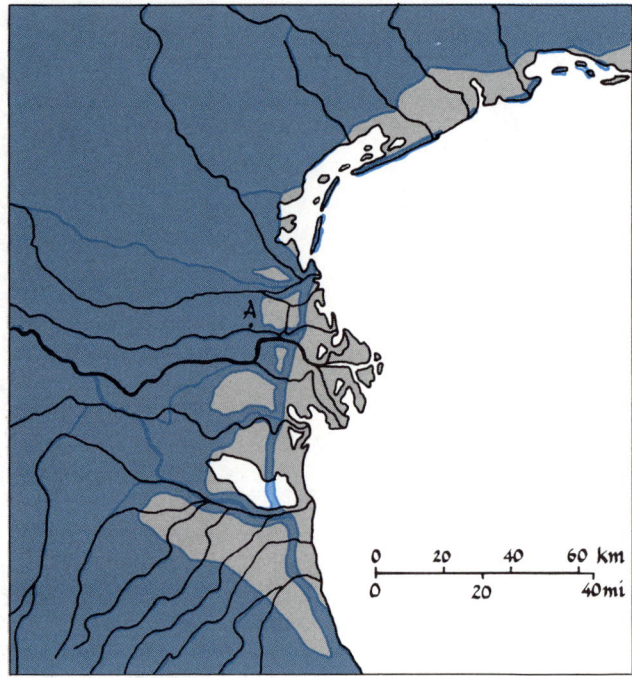

Fig. 12-16. The historical development of the Po Delta. Note the gradual seaward extension of the deltaic deposits and its consequence for the location of coastal cities such as Adria (A).

a covering floodplain sediment. At present, the sea has apparently reached a relatively stable high level. Floodplain deposition and erosion are continuing throughout the length of the filled estuaries in a state of approximate balance, whereas deltas are extending out from the old land into sheltered seas and gulfs (Fig. 12-16).

Deltas. Not all streams that have floodplains build deltas. On exposed coasts, with strong ocean currents immediately offshore, stream sediment is carried away and spread along the ocean floor. Most of the great deltas of the world are located along the shores of gulfs and seas where marine currents are weak or nonexistent and stream velocities are quickly checked by the surrounding masses of calm lake or seawater (Fig. 12-17).

As the stream currents are checked, the fresh water, loaded with sediment, disperses through the seawater, and sediment settles to the bottom. Eventually, a shallow submarine platform of alluvium is built up, across which the stream maintains its channel.

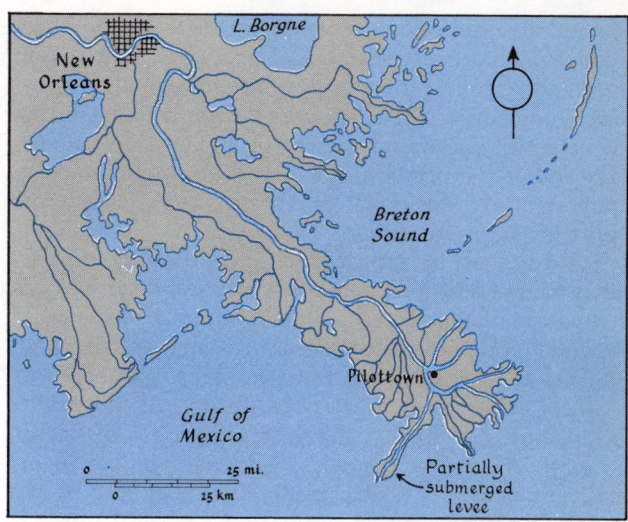

Fig. 12-18. Partially submerged natural levees along the distributaries of the Mississippi River at its mouth. These distributaries are recent; several older sets occur inland.

Submerged natural levees develop along both sides of the channel when floodwaters leave the relatively swift channel current and disperse laterally into the seawater over the submarine platform (Fig. 12-18).[5]

Deposition of a bar at the mouth of the channel causes the stream current to fork. This often results in an expansion of the submarine platform into two new channels, which in turn develop natural levees. As the new channels lengthen, the process is repeated whenever other channel-mouth bars are deposited. In this fashion, the stream acquires a number of branching outlets, or distributaries (Fig 12-18). In time, perhaps through deposition during periods of exceptionally high tides or great floods, natural levees along the distributaries are built up above the normal level of the sea. Continued deposition of floodwater sediment in the shallow water between the natural levees eventually raises the entire delta platform above the surface of the sea.

Summary

This chapter described the processes of stream gradation and discussed some of the results of these processes. In discussions of erosion and transportation,

Fig. 12-17. The Nile Delta looking eastward. The Mediterranean Sea is on the left, and the two northern arms of the Red Sea can be seen in the right background. The delta itself is sharply differentiated from its surroundings by the darkness of its irrigated fields. This photograph was taken from an altitude of more than 100 miles (160 km) in June 1965. *Gemini IV photograph courtesy of NASA.*

---

[5] Abundant suspended sediment in fresh water gives it a greater density than seawater. Thus, the fresh water can displace and move through the salt water.

considerable attention was paid to rill flow and sheet wash, which are less spectacular processes than stream flow and stream transportation but are equally important in forming the land between rivers. Understanding the effect of speed and volume of stream flow on erosion and transportation is a necessary preliminary to understanding the effect of various rates of tectonic uplift on slope formation.

Steady rates of stream incision produce straight-sided valleys and relatively constant local relief, but the inclination of the slopes and the amounts of local relief depend upon the rate of incision. If the land is being uplifted at an accelerating rate, the slopes and amounts of local relief tend to become greater. If the uplift diminishes in intensity, the decreasing rates of stream incision, in combination with mass wastage, can eventually reduce terrain of considerable relief to a nearly featureless plain.

Stream aggradation tends to build up flat surfaces in the form of alluvial fans, floodplains, and deltas. The end result of uninterrupted deposition is the reduction of the land surface to a flat plain. This origin of flat terrain is probably more common than the production of peneplains by erosion. In both cases, however, tectonic processes do not remain constant during the long period required to flatten out the landforms of the earth. Uplift or downwarping alters the rock type and its structure. These changes may modify the climate and other factors that can start a new phase of stream erosion under conditions very different from those of the earlier cycle.

Man-induced changes to vegetation can also start new phases of stream erosion and deposition. As we have seen, the removal of vegetation from fairly steep slopes for cultivation can greatly increase the rate of soil removal. This results in impoverished soil on the hill sides, clogged stream channels and wet, boggy floodplains. Accelerated soil discharge from the bare slopes will continue until bedrock is reached or the slope angle is lowered sufficiently to slow the rate of soil removal. A more likely event would be land abandonment and a slow reestablishment of vegetation. In any case, once the sediment input to the river is reduced, the river, no longer overloaded, can clean its channel and return to more normal conditions of over-bank flooding and meander development. The events leading to a reduction of sediment discharge are good examples of negative feedback in a natural system, a means whereby the system exercises a kind of self control.

Suggestions for Further Reading

Dury, G., *The Face of the Earth*, Penguin Books, Inc., Baltimore, 1959.

Leopold, Luna B., M. Gordon Wolman, and John P. Miller, *Fluvial Processes in Geomorphology*, W. H. Freeman & Company, San Francisco, 1964.

Penck, Walther, *Morphological Analysis of Land Forms*, trans. Hella Czech and Katherine C. Boswell, St. Martin's Press, Inc., New York, 1953.

13 Gradation by Glaciers

Introduction

Gradation by glaciers is presently limited to Greenland, Antarctica, and a few highland and mountainous areas (Fig. 13-1). But in recent geologic time — that is, during the past million or more years — glaciers were active over very large areas. Great sheets of ice developed repeatedly over northeastern North America and northwestern Europe. There were also accumulations of glacial ice in mountain valleys of most high mountain ranges over the world (Fig. 13-1). These glaciers of the past substantially modified the land over which they flowed.

Glaciers are large masses of ice on land that are caused to flow outward or downslope by their own weight. When more than 200 or 300 feet deep (60 or 90 m), glacial ice starts to move in a manner that is not entirely clear. Once in motion the ice seems to flow; it molds itself to the shape of the valley or other relief feature through or over which it is flowing. Measurements made on the surface of valley glaciers indicate that the center moves more rapidly than the sides, and it is believed that the surface moves faster than the bottom. Tributary glaciers join the main valley glacier accordingly,[1] in the same manner that tributary streams join a master stream (Fig. 13-2). However, despite these similarities to streams, glaciers do not flow as if they were liquid.

Glaciers seem to behave as a plastic substance; when they move, they undergo permanent deformations that do not allow the glacial ice to resume its original form once the deformational stress is withdrawn. Part of the movement may be the product of ice-crystal slippage. Under the stress of the overlying weight, many of the ice crystals in the lower parts of the glacier may become oriented in a fashion conducive to slippage. This movement is very roughly analogous to the movement in a pack of flat-lying cards when they are pressed from above at a slight angle to the vertical. Each card slides slightly on the card below in the direction of the pressure. Ice crystals may move in a similar manner. The individual movement along each crystal is slight, but the total movement of many crystals in the glacier may contribute substantially to ice movement. Pressure melting of ice and subsequent refreezing within a glacier may also create internal instability leading to some movement. Because glacial ice molds itself to the configuration of the land surface, resulting plastic deformation may be yet another means of ice movement.

Glacial ice moves at an exceedingly slow pace, ranging from a fraction of an inch to several feet per day. Nevertheless, glaciers are capable of considerable erosion and deposition, although the

---

[1] The top of the tributary glacier is always at the same level as the top of the master glacier.

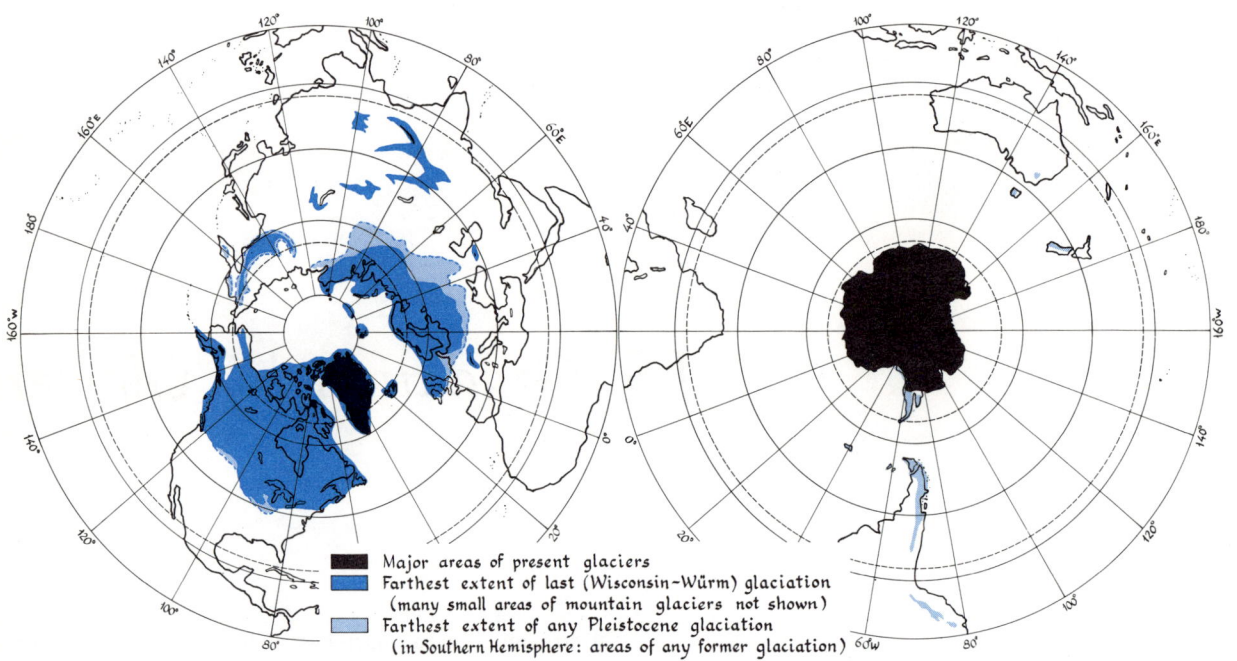

Fig. 13-1. The distribution of modern ice sheets. The farthest advance of the ice and the boundary of the last (Wisconsin) glaciation are also shown.

resulting glacial landforms are distinctly different from those caused by stream erosion. In addition, the glaciers themselves are landforms that exhibit considerable variety.

Glaciers are classified according to their place of occurrence and thus fall into two principal types: **valley, or alpine, glaciers** and **continental ice sheets.** Because the source of alpine glaciers is in high mountainous areas, they are confined mainly to the valleys, and their direction of flow is controlled by the configuration of the land. (Fig. 13-2). Continental ice sheets or glaciers are large masses of ice ranging in depth from a few hundred to several thousand feet and in area from about 100 square miles to the size of Greenland (720,000 square miles or 1,850,000 km²) or Antarctica (5 million square miles or 13 million km²). The large ice sheets do not necessarily flow in directions determined by the slope of the land. They usually move outward from the zone of thickest ice, flowing over and burying most of the landforms in their path. **Piedmont glaciers,** intermediate in location between valley glaciers and ice sheets, occur whenever alpine glaciers flow out of their valleys and coalesce at the base of mountains to form broad foothill ice sheets. Examples of present-day piedmont glaciers are found along the southern coast of Alaska and along the coasts of Greenland and Antarctica.

Glacial ice is quite unlike the ice of normal, everyday experience. It develops mainly from snow that is gradually compacted by complicated processes. During the summer, portions of the previous winter's snowfall are evaporated and melted. The remaining snowflakes form small grains by coalescence. At the same time, a certain amount of meltwater from above refreezes around the grains, and they eventually become rounded granules of recrystallized snow called **firn.** The change from snow to firn may require one or more seasons, depending on the temperature. Subsequent snowfall covers the firn with an ever deepening blanket of snow and of snow changing to firn. The resulting pressure packs the firn — a process that is speeded during summer days, when meltwater seeps through the firn and refreezes. Firn is gradually changed to ice by this packing, thawing, and refreezing. Once the mass of snow, firn, and ice accumulates to a certain critical thickness, it starts to move and can be called a glacier. As flow begins, the ice grows more dense because trapped air is expelled or compressed and because the ice crystals increase in size. As flow continues, the ice crystals become distorted;

some of the crystals continue to grow in size at the expense of surrounding crystals. This process of deformation changes ordinary ice to glacial ice, which bears the same resemblance to ordinary ice as metamorphic rock does to unaltered sedimentary rock.

Every glacier has an area of accumulation and an area of wastage. Each glacier can be considered an example of an open system with an input of snow and debris (mass) and potential energy of elevation in one climatic environment, and an outflow of ice to an area of different elevation and climate where the output from the system takes the form of debris deposition, water, water vapor and heat. The zone of accumulation on valley glaciers is the headward part of the glacier, while on ice sheets, accumulation occurs generally on the interior surface. All types of glaciers have a peripheral zone of **ablation,** where melting and evaporation take place (Fig. 13-3). As long as more snow is added in the region of accumulation than is lost in the zone of ablation, the glacier expands, becoming thicker and longer. Ice moves from the area of accumulation to the area of ablation, the excess causing the glacial front to advance. Once ablation balances accumulation, the glacial system temporarily attains equilibrium with ice arriving at the glacial front at the same rate it is being converted to water, water vapor and heat. If ablation exceeds accumulation the glacial front retreats, although forward motion within the ice sheet itself need not cease. Once ablation reduces the mass of ice to a critical thickness (about 300 feet or 100 m), motion ceases, and the glacier then becomes a mound of stagnant ice.

Crevasses, or tension cracks, in the surface of the ice are also common to all glaciers. These may be up to 30 feet (9 m) wide and 100 feet (30 m) deep and several hundred feet long. Crevasses result from tensional forces due to glacier motion and are more visible near the terminus or sides of a glacier than toward the center. They can also develop on the glacier in places where the slope of the valley bed increases. Crevasses that occur near the margin of the glacier provide a greater surface for ablation and also serve as repositories for the accumulations of sand and gravel that eventually result in characteristic glacial landforms when the ice melts completely.

Past Glaciations

Great though the modern ice sheets of Greenland and Antarctica may be, they are much smaller than the enormous continental expanses of ice that repeatedly covered much of North America, Europe, northern Asia, and southern South America during the

Fig. 13–2. Yentna Glacier in south-central Alaska. A large valley glacier fed by tributary glaciers. The dark lines represent zones of abundant debris incorporated into the body of the main glacier when it is joined by its tributaries. Note the dark bands squeezed into the flowing ice by the tributaries in the lower left of the photograph. *Photograph courtesy of the U.S. Dept. of the Interior, Geological Survey.*

Fig. 13-3. Glacial terminus at the head of Blue River on Mount Lewis Cass, Alaska. The glacier has been melting back in recent years, exposing the valley sides and leaving behind the rough-surfaced moraine in the foreground. The ice blocks in the lake, presuming they will be buried by sediment washed from the ice, may eventually form kettle holes as they melt. The sinuous ridge emerging from the glacier will eventually form a medial moraine. *Photograph courtesy of the U.S. Dept. of Agriculture.*

Pleistocene epoch[2] (Fig. 13-1). At least four major glacial stages occurred during the 1.5 million years of the Pleistocene Ice Age, each stage followed by an interglacial period during which most of the continental ice sheets melted away (Table 13-1).[3] The last glacial stage, which lasted for 100,000 years or more, ended 10,000 to 15,000 years ago with a rapid melting of the ice sheet.

Evidence for these multiple glaciations comes mainly from glacial deposits or glacial **drift.** Careful studies of these deposits reveal that the surface of each deposit was exposed to a long period of weathering before the next deposition (Fig. 13-4). Older deposits have well-developed soil profiles on them, and their forms are indistinct when compared with younger drift. These studies also show that the glacial advance of each stage reached a different maximum position and that some deposits of the later stages overlap those of earlier glaciations (Fig. 13-4). The last glacial stage, because of its superposition upon the earlier stages, has produced most of the existing glacial landforms. Because the last ice sheet retreated so recently, landforms resulting from it have been little modified by mass wastage and stream erosion.

During the maximum advance, the Illinoian continental ice sheet covered North America as far south as

Table 13-1.

North America		Europe	
Glacial stages	Interglacial periods	Glacial stages	Interglacial periods
Wisconsin		Würm	
	Sangamon		Eem
Illinoian		Riss	
	Yarmouth		Holstein
Kansan		Mindel	
	Aftonian		Cromerian
Nebraskan		Günz	

[2] Opinions about the duration of the Pleistocene vary. Earlier estimates, based on the rates of weathering of various glacial deposits, placed the beginning of the period 600,000 to 700,000 years ago. Extrapolations from carbon-14 dating of the youngest Pleistocene sea-floor deposits give estimates ranging from 300,000 to 600,000 years. Potassium-argon dating of lacustrine Pleistocene deposits in East Africa suggests that the epoch lasted for more than 2 million years. Among those who have written most recently on the subject, there is some consensus that the Pleistocene started at least 1.5 million years ago.

[3] The table suggests a contemporaneity between American and European events that is probable but not certain. Names for European glacial stages are from the Alps; those for the interglacial periods are from northern Europe.

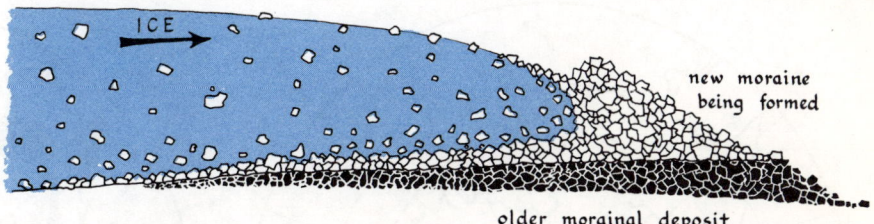

Fig. 13–4. A glacier that has overrun older glacial deposits. Note the new moraine being formed on top of the older morainal deposit. A weathered top of such buried moraines indicates that a long time has elapsed between successive glacial deposits.

the Missouri and Ohio Rivers, including northern Pennsylvania, New York, and New England. In Europe, the ice sheets extended over most of Great Britain, northern Germany, most of Poland, and the greater part of European Russia. At the same time, mountain areas experienced a great expansion of alpine glaciers. Large piedmont glaciers also developed on both sides of the Canadian Rockies and along the east side of the southern Andes. Few of these alpine and piedmont glaciers remain today. Only very high mountains and mountainous regions in mid-latitudes to high latitudes still have alpine glaciers, while the remaining small piedmont glaciers are mainly concentrated in southern Alaska and along the coasts of Greenland and Antarctica.

Causes of Glacial Advance and Retreat

What caused this well-documented cycle of glacial expansion and contraction? There may be a simple answer, but it has yet to be discovered. Instead, we have a number of partial explanations, many of which have serious flaws and none of which explains the totality of the phenomenon. To be completely satisfactory, a theory must account for all the following:

1. *The existence of large, thick ice sheets centered near both poles and extending over adjacent waters.*
2. *Repetitive advances and retreats of the ice and snow lines on mountains, plains, and oceans.*
3. *Advances and retreats that were contemporaneous not only within the Northern Hemisphere but between the Northern and Southern Hemispheres as well.*
4. *A series of small-scale variations in ice extent superimposed on larger-scale fluctuations (some ice advances and retreats are measured in decades; others accommodate these small fluctuations within a longer cycle measured in centuries; these in turn are part of even longer cycles, until we come to the major periods shown in Table 13-1).*

No existing theory explains all these facts satisfactorily. All the theories postulate that glaciers develop and grow whenever more snow accumulates in winter than is melted during the summer and that existing glaciers become smaller whenever ablation exceeds snowfall. Beyond these postulations, there is not even agreement on whether the disappearance of glaciers even from the centers of the Pleistocene ice sheets is the result of a warmer climate or a colder one! For example, it is obvious that immense amounts of water in the form of ice were stored on the land during the periods of glacial growth. Consequently, great quantities of water vapor must have been transported from the oceans to northeastern North America, to northwestern Europe, and to the Antarctic, where it fell as snow. Since cold air contains little water vapor even when it is saturated, the transfer of such large quantities of water vapor would seem to require winters substantially warmer than those presently observed in North America and Antarctica (but not necessarily in western Europe, where a slight *decrease* in temperature would probably favor greater accumulation of snow). In general, however, colder weather retards ablation in summer but also reduces snowfall in winter. Conversely, warmer weather, of course, increases ablation in summer but also results in heavier snowfall during the winter.

There is some evidence that warmer climates could not have caused the expansion of ice sheets. Between 1900 and 1940, for example, glaciers retreated rapidly all over the world, during a period in which the mean winter temperatures in many parts of the world increased by 1° or 2° over previous averages. This evidence is not very convincing, however. A small rise in temperature over a short period is a weak argument for postulating any kind of long-term relationship between climate and glaciation in the more distant past. Furthermore, the 20th-century warming may be more apparent than real. During the period in question, most urban meteorological stations were located in downtown areas. The taller buildings erected during this period cut down breezes and absorbed more heat than the smaller buildings of the previous era. The concentration of smoke, exhaust, dust, and other pollutants, which increased enormously during the same time, retarded the heat loss from the city. **Heat domes** have developed over most middle-sized and large cities. The observed average increase in temperature may reflect no more than this interesting example of man's ability to change the physical character of his environment. This hypothesis is strengthened by the fact that

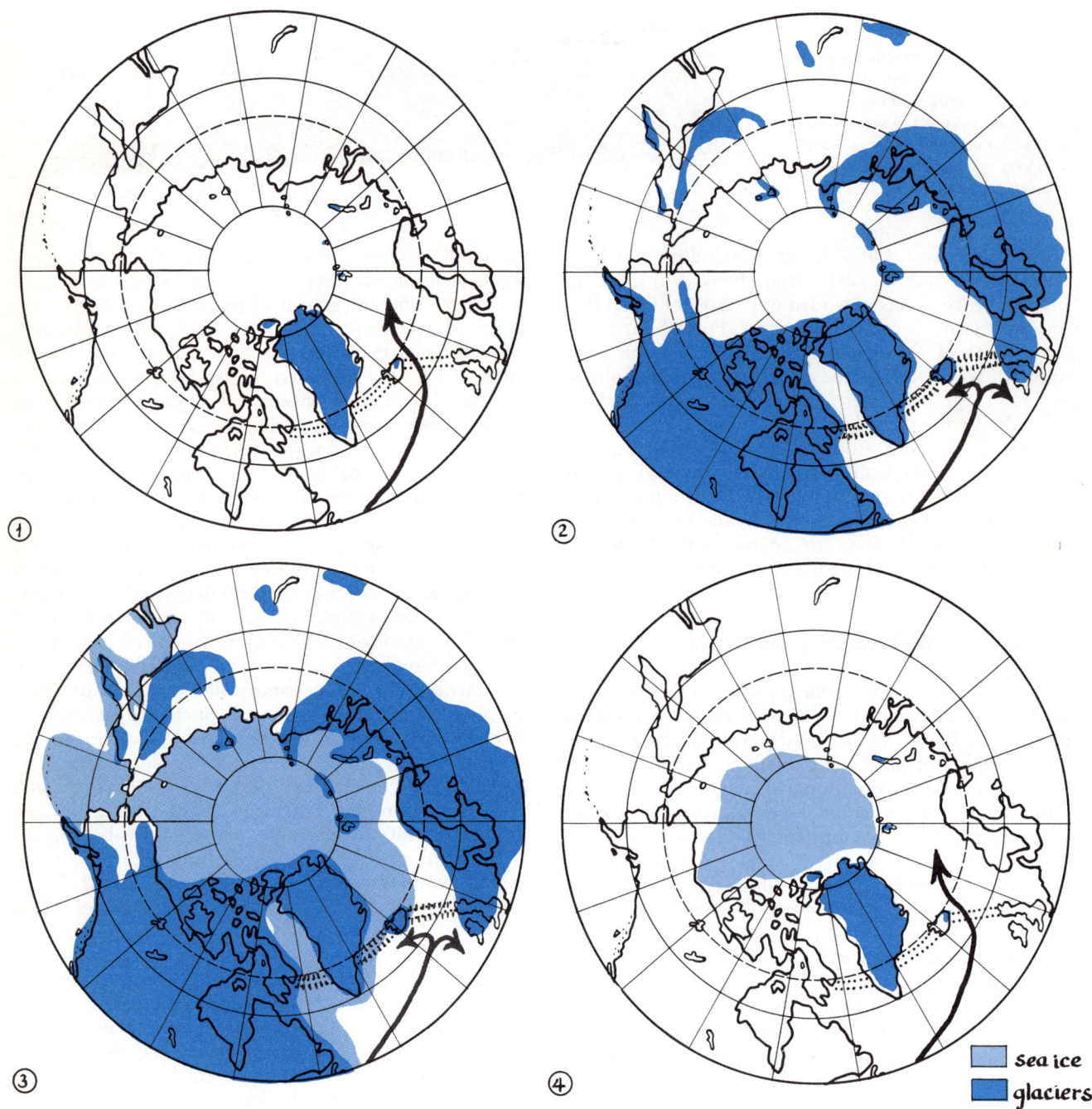

Fig. 13-5. Rise and fall of sea level in the Arctic Ocean and its effects on the growth of the continental ice sheet.

average temperatures leveled off or declined after 1940, when many recording stations were moved out of the heat domes of the cities to nearby airports. In any case, on the basis of the available evidence, it would be difficult to decide whether the recent world-wide wasting of glaciers is due to atmospheric warming or to some more subtle climatic changes that produced a reduction in accumulation rather than an increase in ablation.

At this point, we should briefly examine some of the theories currently advanced to explain glacial expansion and contraction. One theory suggests that ice advances may result from periodic increases in the temperature of the Arctic Ocean, produced when warm ocean currents from the south can penetrate the Arctic Ocean Basin and melt its sea ice (Fig. 13-1). The warmer winter climate and the increased evaporation from ice-free seawater combine to increase the amount of water vapor in the air. This water vapor in turn results in greatly increased snowfall on the continental fringe of the Arctic Ocean, eventually giving rise to expanding glaciers (Fig. 13-5(2)). The ice sheets expand, but, according to the theory, they cause their own destruction; for as the ice sheets accumulate on land the sea level drops, and the relatively shallow sill between the North Atlantic and the Arctic Ocean Basin reduces the exchange of water between the polar and the warmer seas. Consequently, the Arctic Ocean cools and freezes over, which reduces evaporation and, of course, snowfall (Fig. 13-5(3)). The glaciers, no longer adequately nourished, gradually waste away. As they melt, the sea level rises until it is once again deep enough to permit an inflow of warm North Atlantic water. The resulting warmth of the Arctic Ocean again melts the sea ice, which initiates another glacial stage (Fig. 13-5(4)). This theory calls for cyclical recurrence of continental glaciation throughout geologic history. Since there is no evidence for such a regular cycle before the onset of the Pleistocene glaciation the theory assumes that the cycle began when, according to paleo-magnetic evidence,[4] the North Pole migrated to its present position at the beginning of the Pleistocene. This assumption aside, the theory leaves a number of questions unanswered. How can the warming and cooling of the Arctic Ocean account for the growth of mountain glaciers in mid-latitudes and low latitudes? Where does the increased snowfall for the expanding Antarctic ice sheet come from? What accounts for the minor advances and retreats that are superimposed on the major stage of glacial expansion and contraction? Alternate thawing and freezing of the Arctic Ocean may be one factor aiding the development of alternate glaciation and deglaciation in the Northern Hemisphere but it cannot account for all the aspects of worldwide Pleistocene glaciation.

Another cause of the ice ages may have been periodic variations in the kind and amount of solar radiation. There is evidence that an increase in ultraviolet radiation from the sun creates climatic conditions unfavorable to glaciation, while a reduction in such radiation is favorable. Ultraviolet radiation from the sun increases with increasing numbers of sunspots on the solar surface, and there is clear evidence of 11-year, 44-year, and even longer cycles of sunspot frequency.

The amount and distribution of solar radiation that reaches the surface of the earth can vary, either because of changes in the total output of the sun or because of slight changes in the earth's positional relationship to the sun. Observations during the past 40 years show that the amount of solar radiation received at the top of the atmosphere does vary slightly and irregularly. Whether these slight variations can be related to changes in weather is not clear. At any rate, the changes in temperature and atmospheric circulation that accompany changing solar activity appear to be too small to start the growth of continental glaciers. We are free to assume, of course, that sometime in the past ultraviolet radiation may have been much less than at present and that periods of lower radiation may have lasted long enough to bring about the conditions necessary for glaciation.

Astronomic theories use the cyclic changes in the earth's orbit about the sun to account for changes in the amount and distribution of solar radiation received by the earth. The inclination of the earth's axis to the plane of the ecliptic (*obliquity of the ecliptic*) has varied between 21° 39' and 24° 36'. A decrease in obliquity decreases seasonal contrasts, while an increase in inclination increases seasonal differences. The *eccentricity of the earth's orbit* also changes over time. The sun occupies one of the foci of the earth's elliptical orbit; hence the earth is currently about 3 percent closer to the sun at its closest approach than when it is farthest away. At present, the earth is closest to the sun in the Northern Hemisphere winter. Small changes in the eccentricity of the earth's orbit bring about small variations in the receipt of solar radiation at any given place on the earth's surface. Finally, there is the wobbling of the earth's axis, which produces a slow change in the points of the earth's orbit that delimit the seasons. The slight conical movement of the axis is called the *precession of the equinoxes*. The obliquity of the ecliptic goes through a complete cycle (from 21° 39' to 24° 36' and back again) in about 41,000 years; the eccentricity of the earth's orbit varies over a period of 92,000 years; and the precession of the equinoxes has a periodicity of about 22,000 years. However, they do not combine in any simple cyclic fashion. Computations of their assumed combined effects on temperature seem to correlate with variations in the temperature of deep cores in the North Atlantic, but the major drawback of the astronomic theory is that the temperature changes are small, and those of the Northern Hemisphere are out of phase with those of the Southern Hemisphere. Furthermore, these perturbations must be presumed to have existed well before the beginning of the Pleistocene. They cannot be the causes of the ice age, but their effects may have contributed to the four major advances of the ice during the Pleistocene.

Theories that combine a number of different mechanisms provide more realistic explanations for the

---

[4] When certain rocks which are rich in iron solidify from the molten state, the iron fragments are "frozen" in a position that reflects the earth's magnetic field at that time. By plotting the changing orientation over a very long time, the approximate location of the North Pole can be identified. This kind of evidence suggests the North Pole entered the Arctic Ocean several million years ago.

causes of glaciation. For example, a theory combining changes in insolation with changes in the average elevation of the continents can account for the irregular occurrence of ice ages, as well as for the development of glaciers in low-latitude mountains during the Pleistocene and earlier in the history of the earth. For this theory, two kinds of supporting evidence seem reasonably certain. (1) The Pleistocene was a time of unusual crustal activity, and (2) changes in solar energy appear to correlate with advances and retreats of modern glaciers. There is abundant evidence of intensified uplift of highland areas during the Pleistocene. At that time, most of the major mountain ranges, including the western cordilleras in the Americas and the Alps in Europe, attained their present heights. The cordilleras cut across the zones of westerly winds and may have played a significant role in the growth of glaciers in those zones. The significance of the Alps, other than in the local development of mountain glaciers, is unknown. During the general uplift, the presumed centers of initial continental glaciations — the Laurentian Uplands of Canada and the highlands of western Scandinavia — were also rising. The temperature of these uplands dropped by about 3.6° F per 1,000 feet (0.65° C per 100 m) of uplift. This reduction of temperature, when combined with periodic changes in the quantity or character of insolation, could initiate an ice age. The periodicity of climatic change required for interglacial periods is assumed to come from the increase in temperature as the level of the highlands is reduced by erosion. This theory could be used to explain low-latitude mountain glaciation and the growth of the Antarctic ice cap. Alternate freezing and melting of sea ice in the Arctic Ocean could function to intensify the growth and decay of the glaciers in northeastern North America and Northwestern Europe.

H. C. Willet has proposed a theory that attempts to explain all of the previously listed facts. He begins by noting that when the sun emits more ultraviolet radiation than normally, the zonal (east-west) flow of air in the troposphere becomes dominant over the meridional (north-south) flow. During periods of low ultraviolet emission by the sun, meridional air flow is dominant. Zonal patterns tend to reduce the transport of tropical air toward the poles, and thus favor a glacial advance, particularly by reducing summer warmth. At the same time the zonal westerlies are intensified and bring increased storminess and snowfall. Meridional patterns, on the other hand, provide for a great deal of interchange of air between arctic and tropical regions. This situation may favor glaciation on the mid-latitude west coasts of continents, but the very warm summers in the continental interiors result in a retreat of continental ice sheets. In short, high ultraviolet radiation creates a zonal circulation favorable to glaciation, and low ultraviolet radiation favors a meridional circulation that causes glacial retreat.

This change in atmospheric circulation has been verified on a time scale of days. Sudden increases in ultraviolet emission have been followed after a few days by a shift of atmospheric circulation toward a glacial pattern, and decreases in the ultraviolet radiation by a shift to an interglacial circulation pattern. Many climatologists, meteorologists, geologists, and astronomers have accepted Willet's theory, even though there are some who object to extending a theory which has validity on a time scale of days to a scale of millennia or millions of years. They part company with Willet when he postulates that long-period changes in ultraviolet emission (which unfortunately cannot now be proven to have existed) produced the long cycles of glacial and interglacial periods, and that variations taking place in shorter periods and superimposed on the longer cycles are responsible for the smaller advances and retreats of the ice that have been measured in millennia; and so on down to the observed waxing and waning of small mountain glaciers that occur in cycles of several decades. Although there is doubt about some of Willet's assumptions, his theory has the merit of starting from a known variation in ultraviolet radiation by the sun (rather than an unprovable change in the total energy radiated) and the even greater advantage of being able, at least in theory, to account for the cycles within cycles that characterize the fitful starts and stops of glacial advance and retreat.

Other hypotheses have been proposed to account for the changing climates associated with glacial stages. The major theories are described above, and they illustrate the uncertainty of our knowledge about the causes of glaciation.

Glacial Gradation

Glacial Erosion

Glaciers are and were effective agents of erosion and transportation. We assume that, in their lower parts, the Pleistocene glaciers were crammed with erosional debris of all sizes, ranging from clay to huge boulders (Fig. 13-4), which in the main was picked up from the surface upon which the ice moved. The exact manner in which erosion is accomplished by modern glaciers is not clear, but probably the job is done by **plucking** and **abrasion.**

According to our postulations, the plastic glacial ice flows around particles in the regolith and loose blocks in the bedrock; and as the glacier continues to move, it

Gradation by Glaciers

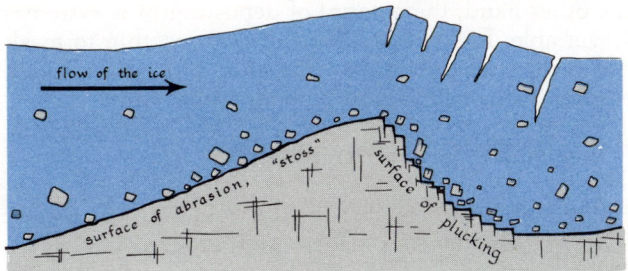

Fig. 13-6. Abrasion and plucking. The light lines represent cracks in the ice. The gentle stoss side has been smoothed by the abrading ice, whereas glacial plucking has steepened the lee side of the hill.

lifts and pries the material away. The plucked, or quarried, material then serves to abrade the surface over which the ice moves, scouring it to produce a gently rounded and smoothed topography (Fig. 13-6). The action is roughly analogous to that of very coarse sandpaper — sizable irregularities are smoothed away, but smaller gouges and **striations**, or scratches, are created at the same time (Figs. 13-7 and 13-8).

While the erosion rate of a glacier is related to the thickness of the glacial ice and its rate of movement, other factors, such as abundance and nature of drift and the resistance of the rock under the glacier, also come into play. It is not clear whether depth or speed is the most important factor; some glacial valleys deepen abruptly downvalley from the junction of tributary glacial valleys where both the depth and rate of glacial flow increased. It is difficult to make any quantitative statement concerning the degree to which erosion increases with increasing thickness and rate of movement, but it appears to be appreciable.

Glacial Transportation

Once the material is incorporated into a glacier, it is transported mainly near the bottom, where it was plucked. The upper layers of ice are relatively free of debris. However, there may be some material on top of a mountain glacier; from the sidewalls of the valley above the ice, rocks can be pried loose by frost weathering and roll down onto the surface of the glacier, and finer material may be washed down onto the ice. Glaciers transport both small and large particles with equal facility because of the viscosity[5] and great mass of the ice.

---

[5]Viscosity in a solid is the quality that allows the body to yield to stresses placed on it without breaking. In a glacier it allows the ice to surround any object in its path without causing a break-up of the ice.

Glacial Deposition

Glaciers deposit their load directly through ablation or lodgment and indirectly by streams of meltwater coming from the ice. When the rate of ablation is about the same as the rate of glacial movement, debris piles up around the margin of the glacier. If the rate of ablation exceeds glacial movement, the ice front retreats and leaves behind the material that the melting water can no longer carry. Lodgment is another form of direct deposition. If glaciers become clogged with debris in their lower parts, layers of drift may be plastered on the subglacial surface as the ice moves along. No sorting by water is involved in either form of direct glacial deposition; consequently, the material is usually composed of unstratified mixtures of various-sized particles. Material carried away from the ice by meltwater is laid down in stratified deposits in front of the ice or along stream channels leading away from the glacier.

Landforms Resulting from the Work of Glaciers

Glaciers produce distinctive erosional and depositional landforms. Often, however, glacial grada-

Fig. 13-7. Glacial striations on bedrock in Alaska that were formed by rock particles held in the moving ice. *Photograph courtesy of the U.S. Dept. of the Interior, Geological Survey.*

226 Chapter 13

Fig. 13–8. Glacial groove cut in bedrock near Grinnell Glacier, Glacier National Park. *Photography by James Bier.*

tion has merely modified preexisting landforms that were created by stream or wave erosion. To be sure, glacial erosion was deep in mountainous areas, where the ice flow was relatively fast and narrowly channeled by the high relief; but even here, evidence of former stream erosion is readily apparent. In areas of low relief, ice-sheet erosion was only skin-deep, smoothing out the valleys and rounding off the high points. On the other hand, the amount of deposition was extremely variable. In some areas the drift is too thin to mask the earlier relief features; in other places it is thick enough to develop its own topographic form.

Erosional Effects of Ice Sheets

The slowly moving continental ice sheets, completely covering the land, plucked and scoured away the regolith and some of the solid bedrock, leaving behind smooth, rounded surfaces. The uplands became rounded and polished knobs or hills of bare rock; valleys were opened and broadened as many small obstructions on their flanks were removed. Here and there, ice erosion produced asymmetric hills with gentle, smooth slopes on one side and angular, steep slopes on the other. The gentle, or *stoss*, sides of the hills, which face in the direction of ice approach, were smoothed by scour, while the steep, or *lee*, sides were roughened by the plucking action of the glacier (Fig. 13-6). Such glaciated hillocks are called **roches moutonnées** (French, meaning roughly sheep-like rocks). Little or no regolith has developed on these recently ice-scoured forms. The exposed rock still reveals the glacial grooves and striations scratched into the roches moutonnées by the drag of strong, sharp rocks that were firmly held by the ice (Figs. 13-7 and 13-8).

The erosion by continental glaciers radically changed

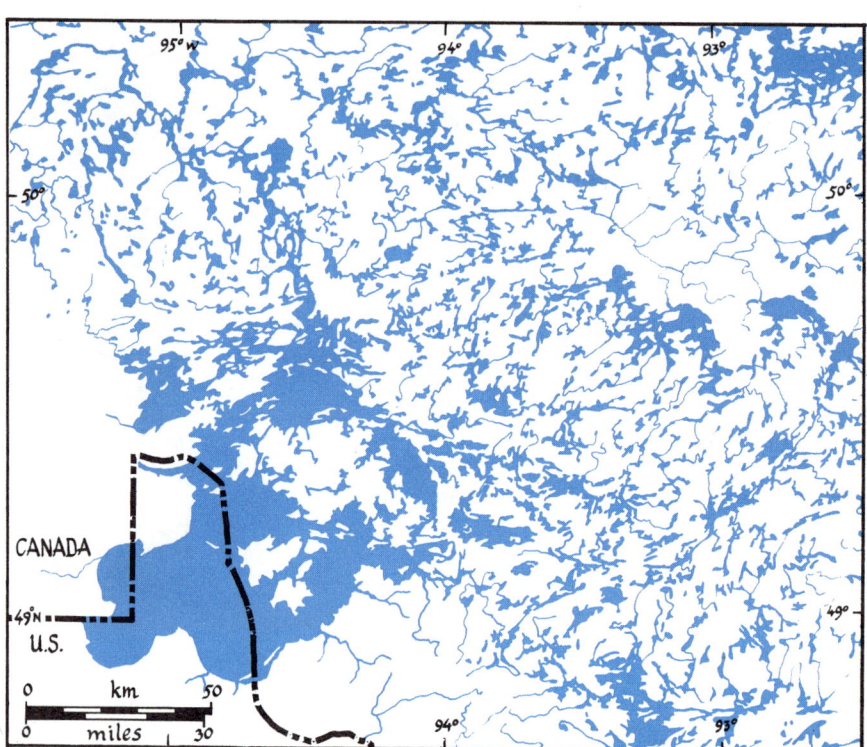

Fig. 13–9. Glacial lakes in a region of intense ice scour, northeast of Lake of the Woods, Minnesota.

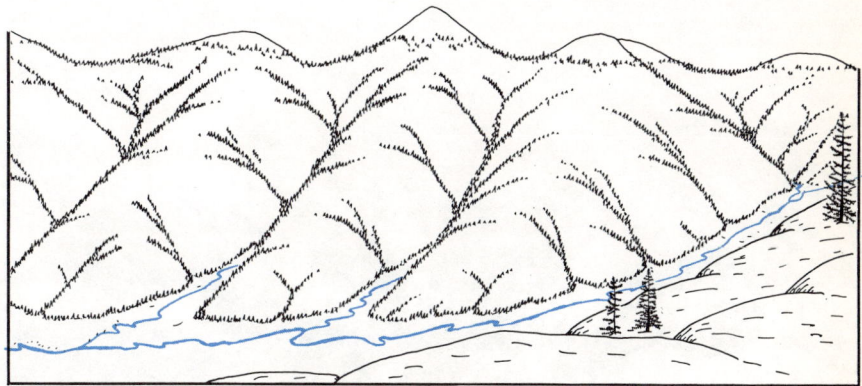

Fig. 13–10. (1) Alpine landforms before glaciation.

Fig. 13–10. (2) Alpine landforms during glaciation.

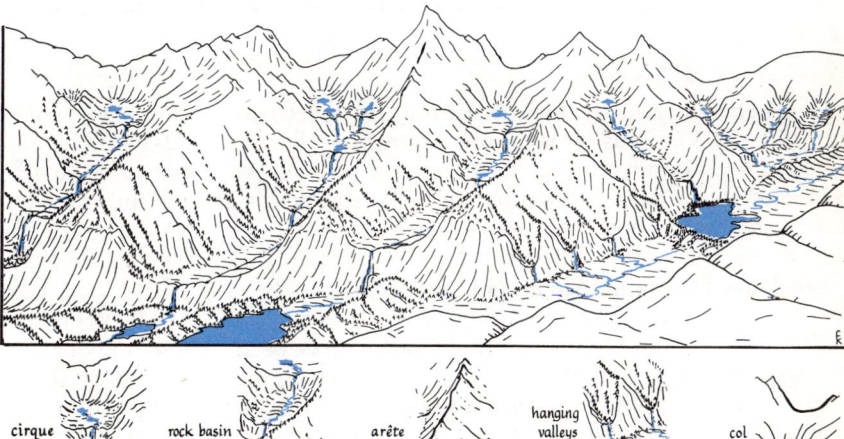

Fig. 13–10. (3) Alpine landforms after glaciation.

preexisting drainage patterns. Some old stream valleys were deepened by ice scour or partially filled by drift. Where the bedrock was weak or jointed, the ice plucked and scoured out closed basins in the valley floors, which were occupied by lakes after the glaciers melted back. Valley-floor gradients were changed to a series of basins or were steepened or flattened according to local circumstances of ice erosion and deposition. Following deglaciation, the streams adjusted to the disorganized pattern of slope gradients created by the glacial scour. The abrupt breaks in slope or localized reversals of gradient created a complicated drainage network interrupted by swamps, lakes, rapids and falls. In particular, numerous lakes of irregular shapes and varying

Fig. 13-11. A glacially eroded U-Shaped trough in the Tongass National Forest, Alaska. The small masses of ice along the right-hand crest are cirque glaciers separated by arêtes. In the left background is a good example of a triangular-shaped horn. *Photograph courtesy of the U.S. Dept. of the Interior, Geological Survey.*

sizes are characteristic features in areas of ice-sheet erosion (Fig. 13-9). Thus, erosion by continental glaciers subdued and rounded off uplands, opened up valleys, excavated rock basins for occupation by lakes, and disturbed the established drainage systems.

Erosional Effects of Alpine Glaciers

The form of most high mountain ranges has been strongly modified by the erosive action of alpine glaciers, even though the preexisting river pattern controlled the direction of ice flow. The characteristic system of master glaciers in larger valleys joined by smaller tributary glaciers (Fig. 13-10 (2)) was also a consequence of the channeling action of preglacial drainage networks (Fig. 13-10 (1)). Glacial ice filled the valleys to considerable depth. As the mass flowed along, it scoured and abraded the sides and bottoms of the valleys, gradually changing the V-shaped cross sections to U-shaped troughs (Figs. 13-10 (3), 13-11 and 13-12). The U-shape of the valley cross section was partly created by the dependence of glacial erosion on the thickness of the ice. The greatest erosion took place on the bottom of the stream valley, where the overlying ice was deepest. Erosion decreased up the valley sides, where the ice was shallower.[6]

The volume of ice and the rate of flow were normally much greater in the main trough than in the tributary valleys, so that the major valley was eroded much more rapidly than the tributary channels. Consequently, the floors of the side valleys were considerably higher than

---

[6] U-shaped valleys can develop by means other than glacial erosion. The presence of such valleys alone is not sufficient evidence for assuming glaciation of a mountainous region, since a simple decrease in the rate of stream incision will also produce a U-shape (see Fig. 13-6).

the floor of the main valley. Deglaciation thus exposed a series of **hanging valleys,** so named because of their perched location, which is nonaccordant with the main valley and is often occupied by a water fall (Fig. 13-12).

Glaciated valleys frequently terminate at their upper end in deep anphitheaters with high, steep sides and broad floors that frequently contain a rock basin. These **cirques** (French, *cirque* circus, amphitheater) (Figs. 13-10 (3), 13-11 and 13-13), where snow collects and it converted to glacial ice, are probably cut into the mountain slopes by a combination of frost wedging, mass wastage, and glacial erosion. Repeated thawing and freezing of water under the snow and firn, wedges out rock fragments from the mountain flanks. These fragments move downslope between the firn and the rock surface and gradually are incorporated into the evolving glacier. As the glacier flows out of the growing cirque, it scours and plucks the cirque floor, eventually forming a rock basin.

Continued glacial erosion on mountain flanks produces a variety of features. Glaciers may erode mountain peaks into **horns** by gouging out cirques in the mountain sides. As the cirques are cut back, their intersecting sides create roughly pyramidal peaks, which are characteristic skyline features of mountains subjected to alpine glaciation. The ridges between cirques frequently have sharp, ragged, knife-edge crests caused by frost wedging. In this form, the ridges are called **arêtes** (French, *arête* spine, razor) (Figs. 13-10 (3) and 13-11). Often cirques cut back enough that the ridge between them is partly removed, leaving a sharp-edged gap, or **col,** with a saddle-like profile (Fig. 13-10 (3)).

The impact of alpine glaciation on the landscape is twofold. Preexisting relief forms buried by the ice are modified by smoothing and rounding; the landforms above the ice develop sharp, angular forms. The total effect is one of marked contrast between the sandpapered smoothness of the lower slopes and valley flats and the quarried angularity of the higher peaks and ridges. In addition, Alpine glaciers "open up" mountains, permitting one to see the majestic scale of the ice-scoured topography. Ice-eroded mountains are among the most scenic of all natural features.

Depositional Effects of Ice Sheets

Glacial deposits, or drift, are of two major kinds: **till** and **stratified drift.** Till is a direct glacial deposit and is composed of an unstratified, heterogeneous mixture of rock fragments of all sizes. Stratified drift consists of sediment deposited by meltwater streams or in temporary lakes near the edge of the ice. These indirect deposits consists of layers of sorted and stratified gravels, sand, silt, or clay.

Fig. 13-12. The valley walls of a U-shaped trough in the Bernese Alps, Switzerland. The waterfalls near the village of Lauterbrunnen indicate the presence of hanging valleys. *Photograph courtesy of the Swiss National Tourist Office.*

Fig. 13-13. Snow-filled cirques in the Fiescher Horns. These cirques are tributary to the Aletsch Glacier, whose dark medial moraine can be seen in the foreground. *Photograph courtesy of the Swiss National Tourist Office.*

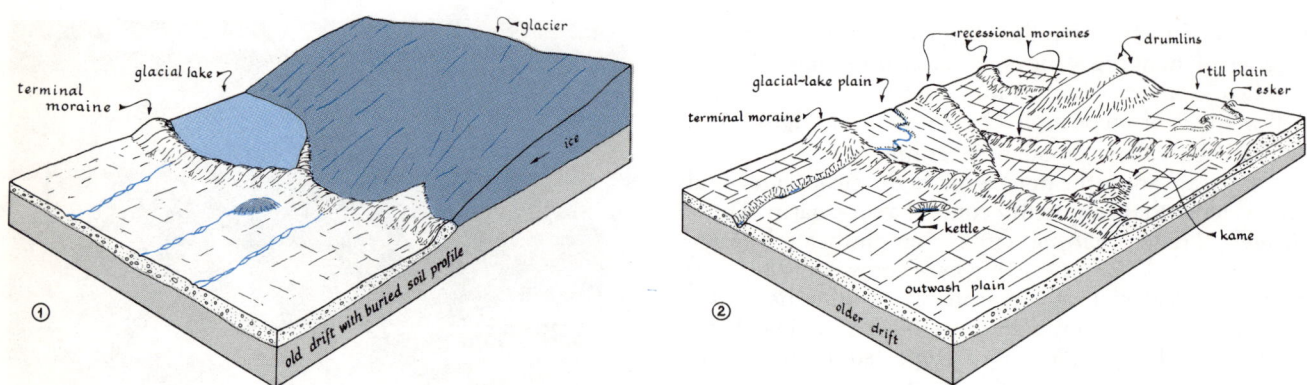

Fig. 13-14. Features of glacial and outwash deposition.

Alpine and continental glaciers produce similar depositional forms, except that the extent of continental deposition was vastly greater and the individual forms of glacial aggradation are frequently much larger. Consequently, continental glacial deposits are far more significant in the present landscape than are deposits of mountain glaciers.

Direct Glacial Deposition. The nature of the direct glacial deposits is a reflection of the state of the glacial system. Debris collects around the ice margin when the system is in a state of dynamic equilibrium. When the system is out of balance material is deposited beneath the glacier either as the ice front advances or as it retreats. When the rate of ice movement is balanced by the rate of ablation, the ice front remains stationary; but material is continually brought forward and accumulated along the ice front (Fig. 13-14 (1)). This heterogeneous debris piles up in the stagnant ice, and when the glacier melts back it appears on the ground as an unsorted deposit called a **moraine** (Figs. 13-14 (2) and 13-15). The moraine at the outermost edge of a glacier is called a *terminal* moraine. Pauses in the recession of the ice front cause the formation of a series of *recessional* moraines. After the ice has disappeared, the moraines appear as low hills and hollows that extend across the land in gently curved belts of rough topography.

The surface between moraines is often covered by sheets of till that were deposited beneath the ice to form

Fig. 13-15. Recent glacial moraine topography in the District of Mackenzie, N.W.T., Canada. The irregular topography associated with glacial moraines is clearly evident here. *Reproduced with the permission of the Geological Survey of Canada.*

Fig. 13-16. A view of drumlins. The smooth egg-shaped hills from a drumlin field behind a terminal moraine in Saskatchewan. *Photograph courtesy of the Geological Survey of Canada.*

a **till plain** (Fig. 13-14 (2)). Part of the till was deposited when the lower layers of moving ice were so heavily loaded with debris that the excess lodged on the bedrock and was left behind. Superimposed on this lodged till is an uneven layer of more till, left behind as the ice melted away. Because of this uneven deposition, till plains usually have gently rolling surfaces of wide, low rises and broad, shallow depressions. In areas where there was considerable preglacial relief or where the till sheet is thin, the original terrain configuration is still visible; but elsewhere, the morainic hills and intervening till deposits completely mask the preglacial form of the land.

Groups of smooth, spoon-shaped hills of till called **drumlins** are found on certain till plains, generally in a zone behind a recessional or terminal moraine (Fig. 13-14 (2) and 13-16). Characteristically, they are 0.5 to 1 mile (about 1 km) in length, a few hundred yards (a few hundred meters) wide, and about 50 to 100 feet (15 to 30 m) high. Drumlins normally occur in clusters of several hundred or more and are aligned so that the steep face of each drumlin faces in the direction from which the ice came, while their long axes are parallel to that direction. Drumlins were deposited beneath the ice sheets on their margins and appear to be streamlined accumulations of till that may have lodged on some sort of preglacial surface irregularity.

Indirect Glacial Deposition. Stratified drift deposits, often called **glaciofluvial deposits,** are classified according to their form, their composition, and their position in relation to the ice front. Smooth, gently sloping **outwash plains**, composed of crudely stratified deposits of sand and gravel, may occur for long distances beyond the outer margin of morainic deposits (Fig. 13-14 (2)). These outwash plains are large, nearly flat alluvial fans that were built up by sediment washed from the ice by meltwater. They are occasionally pitted by **kettle holes,** sometimes filled by lakes. These depressions mark the position of blocks of ice left behind the main ice sheet when it retreated. Outwash material was then deposited around and over the block. When the block of stagnant ice finally melted, its disappearance left a hole in the otherwise smooth surface of the outwash plain.

Other deposits of stratified drift along the ice margin are **kames** and **kame terraces.** Kame terraces are accumulations of sand and gravel deposited by streams between the glacier and the adjacent valley wall or hillside, which appear as valley-side terraces after the ice is gone. Kames are mound-like hillocks of sand and gravel that develop initially as small deltas or steep alluvial fans built out against the flank of the ice (Fig. 13-14 (1)); after the ice retreats, these deposits eventually collapse and form irregular mounds (Fig. 13-14 (2)). Kame-forming sediment can also collect in openings or depressions in the ice, to be dumped on the ground during glacial retreat.

Stratified drift also occurs in long, sinuous ridges of sand and gravel, which are called **eskers** (Fig. 13-14 (2)). These may have originated as deposits of temporary meltwater streams flowing through tunnels in the bottom of the ice. Some esker deposits, often scores of miles long, show little sign of having been disturbed, indicating that they were probably deposited under stagnant ice that melted away without subsequent motion. Other eskers, for example the ones currently being formed by the retreat of Alaskan glaciers, appear to originate in the middle rather than at the bottom of the glacier and collapse onto the surface of the ground when the glacier retreats.

Well-stratified deposits of drift were laid down on the floors of temporary glacial lakes (Fig. 13-14 (1)), forming **glacial-lake plains** when the lakes ceased to exist. Glacial lakes developed where the land sloped downward toward the ice front, forming an enclosed basin between the sloping ground and the ice margin, or where water was impounded between the ice front and a relief barrier. Meltwater deposited an abundance of fine sediment on the lake floor in stratified layers of silt and clay. These fine-textured strata, called *varves*, commonly have a banded appearance due to the alternation of light and dark layers. Each pair of layers is thought to represent one year's deposits. The relative turbulence of water in summer allowed only the light-colored layer of coarser material to be deposited; in winter, the lake was frozen over and still, and the dark clay layer could settle out. Because of this feature, varves have been used to date the occurrence of glacial-lake deposits in Scandinavia and elsewhere. The deposition of layer upon layer of silt and clay on the bottoms of these marginal lakes produced the extremely level surface of glacial-lake plains (Fig. 13-14 (2)).

All these different types of drift combine to form complex land surfaces, varying from rough moraines to flat lake plains. The diverse composition of the drift adds to this complexity. Clay-rich, poorly drained till; sandy, well-drained outwash; and deep clay or silt lake deposits all have different surface expressions. It should also be remembered that there once existed a series of landforms associated with the drift of glaciations earlier than the Wisconsin stage. These have been so altered by stream erosion and mass wastage and by later glacial action that they have lost much of their special drift characteristics. Many of the Wisconsin and other glacial deposits have been the basis for exceptionally productive soils. For example, the great corn belt of the mid-western part of the United States coincides remarkably well with the mid-continent area of continental glacial deposition.

Summary

The world-wide advances and retreats of glacial ice are responsible for many of the landforms in high and middle latitudes. In fact, some of the most spectacular scenery in the world is associated with the present and past locations of Alpine or mountain glaciers. Various theories, none of which completely explain all of the facts, have been advanced to account for the waxing and waning of the great Pleistocene ice sheets. Fluctuations of solar output, changes in sea level, changes in land elevation, continental drift, variations in the earth's position with respect to the sun, and changes in the predominant flow of air in the troposphere have all been suggested as causes of climatic change that alternately favored and hindered the formations of continental ice sheets.

Glacial gradation consists of the various mechanisms of glacial erosion, glacial transportation, and glacial deposition. Erosion by ice is mainly a function of the thickness of the glacier and its rate of movement. Glaciers transport both small and large particles which are concentrated near the bottom of the ice, although some material may tumble onto the ice surface and very fine particles may penetrate into the ice itself. Deposition occurs mainly in the zone of ablation or melting, where the material is dumped without being sorted by size. Occasionally material is lodged directly on the subglacial surface if the ice is clogged with debris and the subsurface is rough.

Glacial landforms are the result of either deposition or erosion. Erosional forms associated with continental ice sheets range from mere gouges or striations to a complete re-ordering of the surface so that previous stream networks are completely disarranged, giving way to a multitude of lakes or marshes in the hollowed-out basins and a confused pattern of drainage. Alpine glaciers have sculptured mountain peaks into horns by gouging out cirques in the mountain sides; they have also widened preexisting stream valleys into U-shaped troughs with spectacular hanging valleys at the entry point of small side glaciers into the much deeper main valley. Depositional forms are either direct, mainly in the form of unstratified moraines at the edges or front of the glacier and till plain deposits where the glacier receded too slowly to leave a sharply defined morainal deposit, or they are indirect. Indirect deposits, often called glaciofluvial, are the stratified or sorted deposits carried beyond the edge of the ice by glacial meltwater. Outwash plains pitted by kettle holes, kames, eskers, and glacial-lake plains are the most important of these depositional landforms.

Glacial landforms are the youngest features of the earth's surface, except for the most recent faulting and volcanic activity. They cover enormous areas in the middle and high latitudes, including all of Antarctica, and have created incredibly beautiful peaks and valleys wherever alpine glaciers were or are still active. The outwash deposits of the continental ice sheets extended the effect of the ice beyond the margin of the ice itself and shaped a large area of irregular plains that still bear the imprint of their glaciofluvial origin.

Suggestions for Further Reading

Beckinsale, R.P., "Climatic change: A critique of modern theories" in *Essays in Geography for Austin Miller*, Ed. by J.B. Whittow and P.D. Wood, University of Reading, 1965.

Flint, Richard F., *Glacial and Pleistocene Geology*, John Wiley & Sons, Inc., New York, 1957.

Longwell, Chester R., and Richard F. Flint, *Introduction to Physical Geology*, John Wiley & Sons, Inc., New York, 1965, Chapt. 13.

Putnam, William C., *Geology*, Oxford University Press, New York, 1964, Chapt. 13.

Schultz, Gwen, *Glaciers and the Ice Age*, Holt, Rinehart and Winston, Inc., New York, 1963.

14 Gradation by Wind

Introduction

Wind, usually a minor factor in landform development, is an important agent of gradation in some parts of the world. It is less significant as a gradational agent than mass wastage, streams, or glaciers; but it can, under favorable conditions, produce unique and important landforms of erosion and deposition.

Air movement is similar to the flow of water. For example, over a smooth surface, wind speed varies from zero, in a thin layer next to the surface, to progressively higher values aloft. The resulting shear between the various layers of air moving at different speeds causes turbulence similar to that in moving water. High wind velocity substantially increases the shear between the moving layers of air, which intensifies turbulence. Wind turbulence may also be increased by heating from below. If the air is warmed at the surface, it expands and becomes less dense, leading to much greater vertical turbulence than in air that is being cooled by the surface (see Chapter 6). Vertical temperature differences also exist in flowing water but the contrasts are weaker, and their influence upon stream turbulence is negligible. Turbulence in streams is largely mechanically induced, whereas turbulence in wind is both dynamic and thermal in origin. In general, however, the motion of air passing over the ground is similar to the motion of water flowing over a stream bed.

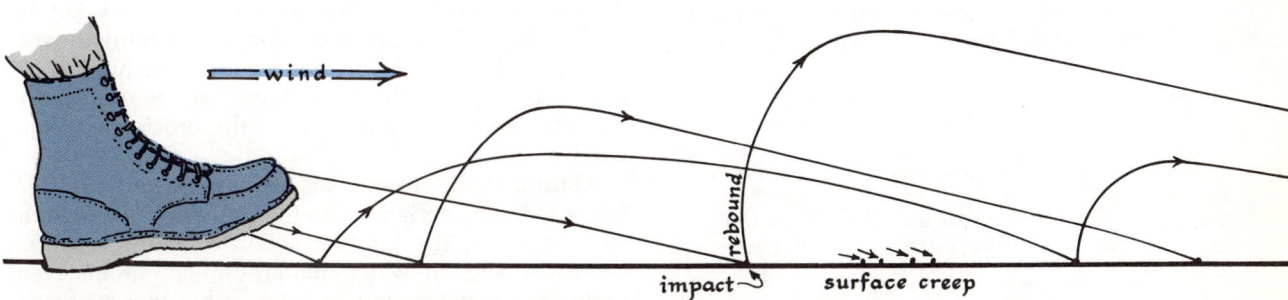

Fig. 14-1. Saltation, suspension, and surface creep.

Wind Erosion

Degradation by the wind consists of two processes, **deflation** (Latin *deflatus* blown off) and **abrasion** (Latin *abrasus* scraped off). Deflation is analogous to the removal of material from a stream bed by the direct action of the water or to the plucking and quarrying of glacial ice. Abrasion by wind is directly comparable to abrasion by water or ice.

Deflation

Turbulent whirls of air remove materials from the ground in much the same way as a stream removes material by the action of turbulent water. There are, of course, limits to the size of particle that can be removed from the surface. Experiments have shown that fine and medium sand[1] is most easily moved. This sand is large enough to protrude into the moving air yet light enough to be picked up. It begins to move when the wind reaches a speed of about 10 miles per hour (about 5 m per second), 6 inches (about 150 mm) above the ground. Fine silt particles (smaller than 0.001 inches (.025 mm) in diameter) are very resistant to movement because they do not extend above the still layer of air close to the ground. These fine particles can be removed, but only through the impact of larger grains which bounce along the surface and splash some of the fine-textured dust into the airstream. Very coarse material, such as gravel, is usually too heavy to be moved even by winds as strong as 35 miles per hour (16 m per second).

Once material is picked up, it is transported by **suspension, saltation** (Latin *saltatus* jumped, leaped), and **surface creep.** Suspension is possible whenever the upward velocity of turbulent wind currents is greater than the rate of fall of the particles being carried. Thus borne aloft, the dust may be carried great distances by the wind. Saltation is a means of transportation whereby sand particles leap, jump, or bounce along the surface. The individual grains fly through the air in a curving path, rising abruptly, then falling in a long, gentle arc, striking the surface at a low angle. Upon landing, the grains may bounce again and once more be caught up by the wind, or they may force other particles upward into the airstream to repeat the process (Fig. 14-1). As a result of the constant impact of bouncing grains, the surface layer of sand creeps downwind. In addition, large particles, too big to be involved in saltation, may be moved downwind by the pounding action of innumerable particles of sand. Surface creep may, in time, move large sheets of coarse sand and fine gravel for considerable distances.

Abrasion

Wind abrasion, like that of water, requires cutting tools to transmit the wind energy to the surface being eroded. Sand and smaller particles driven by the wind against an exposed surface wear the surface away by mechanical impact. A mechanical sandblaster imitates this effect. Although wind abrasion is similar to stream abrasion, it usually produces only minor geomorphic features.

Optimum Conditions for Wind Erosion

Degradation by the wind may occur anywhere on earth, but its rate and magnitude depend on a number of factors. The grain of the land can be an important factor. Rough ground increases friction and thus reduces wind speed. Some landforms channel the wind and accelerate it locally. We have just seen the importance of wind speed to deflation, but the rate of deflation also depends on the kinds of material provided by weathering, on the amount of vegetation cover, on the texture and moisture content of the material available for removal, and, to a lesser degree, on the surface temperature.

Although these factors interact in a complex way, their individual effects can easily be described. For example, vegetation reduces erosion by shielding the surface from the force of the wind and by the binding action of its roots. If the vegetation is destroyed, the soil particles become vulnerable to the erosive effect of wind.

Abundant moisture binds particles together, effectively reducing deflation. As the amount of water in the soil decreases, the binding action of surface tension decreases, and the water film around soil grains eventually becomes too thin to act as a binding agent.

It follows from the above that optimum conditions for wind erosion are found in those areas where the

---

[1]The descriptive categories of "clay," "silt," "sand," "gravel," and so on can be precisely, if arbitrarily, delimited by the diameters of the particles involved. The U.S. Department of Agriculture uses the following definitions for purposes of soil-texture identification:

Material	Diameter	
	(inches)	(mm)
Gravel	>0.04	>1
Coarse sand	0.02-0.04	0.5-1.0
Medium sand	0.01-0.02	0.25-0.5
Fine sand	0.004-0.01	0.1-0.25
Silt	0.00008-0.004	0.002-0.1
Clay	<0.00008	<0.002

We shall use these definitions wherever we refer to size of particles.

vegetation cover is thin, where the surface material is of intermediate size or has a proper combination of sizes, and where the material frequently dries out for some time. A heated surface plays a lesser role, although it does favor deflation by increasing turbulence. In general, plowed and tilled fields or areas of recent deposition, such as dry river courses, beaches, and areas of recent glacial drift, are particularly susceptible to wind erosion. Arid lands are also susceptible to deflation, since the ground is dry, frequently quite hot, and relatively bare of vegetation.

The concentration of sand dunes in arid regions (compare Fig. 16-3 and the front endpaper) would indicate a close association between deflation and aridity (since dune deposition presupposes strong wind degradation). The association is not perfect, however. Dunes appear to be well developed in the dry lands of Africa and Asia, but the deserts of North and South America have few and small areas of sand dunes. These phenomena suggest the importance of factors other than climate in wind degradation. The sparse plant cover may account for the relative abundance of sand dunes in the Old World, in whose deserts nomadic herders have been grazing their flocks for several millennia, with the result that vegetation is usually very sparse. The deserts of North America and much of South America have a better-developed plant cover, even where rainfall is less than in Old World deserts. Nearly bare ground is found only in small areas, such as parts of the delta of the Colorado River, of central New Mexico, and of the coast of Peru and northern Chile. Significantly, these are also areas of sand-dune accumulation in the Western Hemisphere. Evidently, vegetation and climate are both important factors in the stimulation of wind erosion. At present, we have insufficient data for evaluation of their relative importance to deflation. It is clear, however, that wind becomes a significant agent of erosion and deposition in arid regions with sparse vegetation.

Landforms of Wind Gradation

Forms of Wind Erosion

Features produced by wind abrasion are relatively insignificant. On clay-bound sand or other soft but cohesive sediments erosion results in minor forms, such as **yardangs** (Turkish *yardang* steep bank, precipice), which develop on bare surfaces exposed to erosion by continuous, steady winds that produce long, shallow grooves and sharp-crested ridges in the surface. Resistant rock surfaces are only occasionally etched by the wind. Other minor forms include **ventifacts** (Latin *ventus* wind + *factum* something done), which are cobbles whose sides have acquired an angular quality because of sand abrasion.

The principal forms of deflation occur in loosely consolidated material. **Blowouts** are deflation hollows, usually only a few feet deep and up to a mile in diameter, that frequently develop in semiarid regions where the grass cover has been destroyed. Another common form of blowout occurs on sandy beaches where sand has been piled up by wind action and then partially anchored by vegetation. In exposed places, renewed deflation gradually forms elongated, spoon-shaped depressions with crests of moving sand on the downwind ends. The deflation hollow is obviously an erosional feature; the downwind deposition feature is called a blowout dune, which is discussed in the section "sand dunes."

Active deflation on unconsolidated deposits of mixed sizes, as in alluvial fans, produces a gravel cover called **reg** or desert pavement (Fig. 14-2). As the wind blows across the fan surface, the fine particles are winnowed out and carried off in suspension while the sand particles are removed by saltation and surface creep. The remaining gravel eventually forms a continuous cover of pebbles. Once reg is established, wind erosion is greatly reduced or may cease altogether. Such reg surfaces are found in most desert areas, but they are best developed in the western Sahara, where they cover tens of thousands of square miles.

Forms of Wind Deposition

Sand and silt set in motion by wind eventually come to rest. The form assumed by these wind deposits depends on the size of the particles and on the cir-

Fig. 14-2. Desert pavement or reg. Wind has removed everything finer than the pebbles. The scale is given by the pocket watch. *Photograph by J. Gilluly, U.S. Geological Survey.*

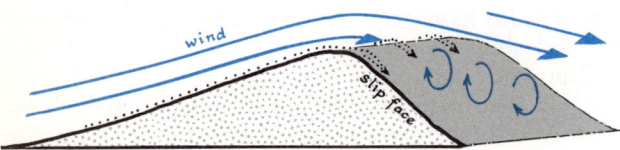

Fig. 14-3. Cross section of a transverse dune. The wind-driven sand forms a gentle windward slope; the lee slope is as steep as the angle of repose of the sand.

cumstance under which they are deposited. Coarse material is normally deposited locally, but silt can be deposited far downwind in thin layers of dust. The fine dust is usually an insignificant fraction of the total surface material. In the past, however, silt particles accumulated to such depths in certain favorable localities, that they formed significant deposits of fine-textured material called **loess** (Swiss German *lösch* loose) (Figs. 14-6 and 14-7).

Larger particles are deposited near their source in hills of sand which are shaped by the wind. The shape and prevalence of these dunes are controlled to a large degree by the amount of sand available for deposition and by the character and direction of the prevailing winds. Occasionally, preexisting irregularities in the surface or patches of vegetation are also important factors in determining dune shapes.

Sand Dunes. Dunes may be active, changing their form and position under the force of the wind, or they may be fixed where the sand is held down by vegetation. Both active and passive dunes can be classified according to their shape. The most important types are **tranverse, longitudinal, barchan** (Russian *barkhan*, from a Kirghiz word meaning "moving sand dune"), and **blowout** dunes.

Transverse dunes are associated with moderate wind velocity and an abundant supply of sand. They have the form of steep-faced waves and are called transverse because the general alignment of their crests is at right angles to the direction of the prevailing wind. The upwind side of the dune has a long, gentle slope, while the lee side is relatively steep (Figs. 14-3 and 14-4). The lee slope, or *slip face*, has an angle of about 35°, the maximum angle of repose for sand; steeper slopes cause the sand to slide down the face as it is blown over the crest of the dune. When viewed from above, the sharp-crested ridges are discontinuous and have an irregular outline.

Another variety is the longitudinal dune, whose crests are parallel to the predominant wind direction. One of the more important is the **seif** dune, which occurs in the deserts of northern Africa, Arabia, and southwestern Asia. Seifs are great sand ridges, several hundred feet high and many miles long, that have uneven crests with slip faces first on one side, then on the other. They appear to be elongated by major prevailing winds, while periodic crosswinds of shorter duration clean up the sand between ridges and account for slip faces on their side slopes. These dunes occur in groups a few hundred yards to several miles apart, and the ground between dunes is relatively free of sand. Despite their large size, they apparently require only a meager supply of sand. Longitudinal sand ridges are also prominent in the desert of central Australia. Here the ridges are lower and have irregular courses across the landscape.

Barchans are crescent-shaped dunes that occur as isolated sand hills (Fig. 14-5). The maximum dune length and width is about 400 yards (400 m), and the greatest height is about 100 feet (30 m). The lee face is on the concave side; the horns point downward and indicate the direction of dune migration. Barchans seem to be the result of winds of consistent direction and

Fig. 14-4. Transverse dunes in the Great Sand Dunes of New Mexico. The prevailing wind is from the left. *Photograph by James Bier.*

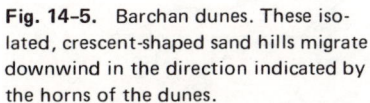

Fig. 14-5. Barchan dunes. These isolated, crescent-shaped sand hills migrate downwind in the direction indicated by the horns of the dunes.

moderate velocity and of limited-to-moderate supply of sand. The sand may initially collect around some obstacle, or in some instances it may accumulate through some random chance of wind deposition. Once a sufficient mass of sand has accumulated, it begins to move downwind, and it gradually takes a crescent form because sand particles move faster around the ends of the dune. Other barchans frequently develop off the horn tips of upwind dunes, forming a succession of dunes extending downwind.

Blowout dunes, in which the convex face of the dune crest faces downwind, result from incomplete vegetation cover and from wind that blows primarily from one direction. The most frequent dune of this type is the coastal blowout, which forms where abundant sand is available. An elongated depression is created by wind erosion, and the sand is piled up on the downwind side, forming a parabolic dune. The slip face, on the inland side, slowly advances landward.

Loess. In several parts of the world, especially in the mid-latitude belts of westerly winds (see Fig. 4-14), there are extensive surface deposits of fine-grained material called loess. It is usually pale yellow or buff and has a tendency to maintain vertical cliffs along road cuts or stream embankments (Fig. 14-6).

In the United States there are large loess areas in the Upper Mississippi-Missouri River Basin and on the east side of the Mississippi River south of its junction with the Ohio River. These layers are especially thick on the eastern sides of the valleys, which generally trend north-south. Loess areas also exist in southeastern Washington and western Idaho. Other extensive deposits occur in central Europe, the Pampa of Argentina, and central Asia. The greatest area of loess is in northern China, where it covers hundreds of thousands of square miles and where thicknesses of more than 100 feet (30 m) are common. Smaller deposits are found on South Island, New Zealand (Fig. 14-7).

The mineral composition of the loess particles varies slightly from region to region, but the particles are always small and usually consist of undecomposed minerals. Typical loess is rich in calcium and is partially consolidated in thick layers by a calcium-carbonate cement. The aeolian, or wind-borne, origin of loess is substantiated by the partially weathered appearance of loess particles and by the location of some large areas of loess on the downwind sides of deserts in central Asia, northern China, and the Argentina Pampa. Loess deposition is not occurring to a noticeable degree at present; but it is believed that in the past (probably during the Pleistocene epoch), conditions were favorable for intensive deflation in the deserts of interior Asia and western Argentina. The silt-sized products of arid weathering were borne eastward in great dust storms by the westerly winds and were deposited in subhumid areas east of the deserts, where the silt was compacted into loess and partially cemented by its own lime content.

This theory of origin does not entirely explain the

Fig. 14-6. A steep bluff along a road cut in a loess deposit. *Photograph by E.W. Shane, U.S. Geological Survey.*

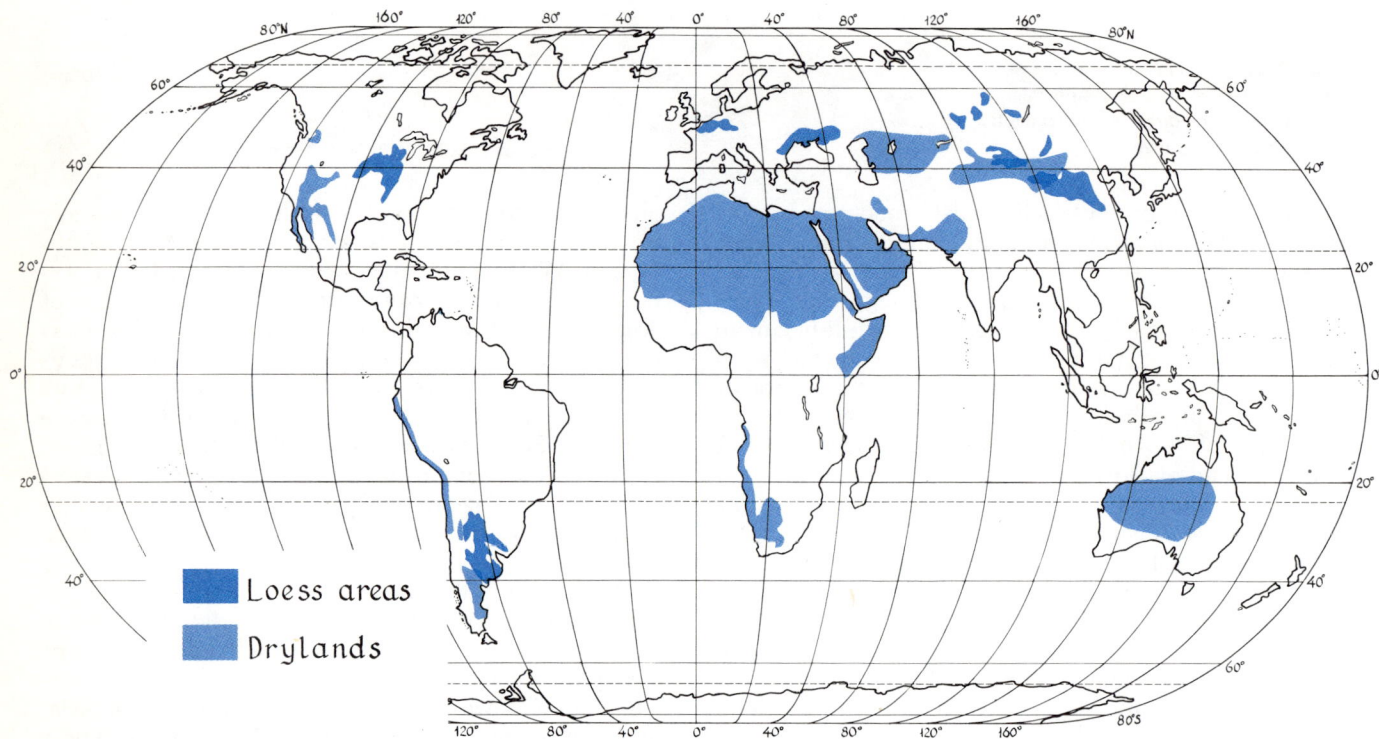

Fig. 14-7. Major loess areas and dry lands of the world.

American and European loess deposits. In particular, some modification of the notion of desert origin is required to account for the thick loess deposits immediately on the eastern sides of river valleys in the central part of the United States. According to one theory, optimum conditions for deflation probably prevailed around the continental glaciers in Europe and North America during the waning stages of the last glaciation. The combination of exposed, unconsolidated till and huge amounts of meltwater caused the streams that were discharging from the waning glaciers to clog their valleys with great amounts of sediment. In winter, the till and valley train deposits dried out, and their bare surfaces became prime targets for deflation. The dust picked up by the prevailing westerly winds accumulated mostly on the eastern sides of the valleys.

Summary

Wind erosion removes and transports loose material in a manner very similar to stream erosion, though the action is not confined to a channel and the particles removed have a smaller range in size. Erosion by deflation involves the removal of small particles by turbulent air, while abrasion implies a blasting action by wind-drawn particles. The most important factors affecting the rate and degree of erosion are wind velocity, the nature of the surface material, vegetation, surface moisture, and topography. Optimum areas for wind erosion are those where the vegetation cover is thin, the texture of the surface material is of a proper combination of sizes, and the material is frequently dry for considerable periods. The forms created by wind erosion are usually small and of minor importance. Present-day deposition by the wind is largely in the form of dunes, whose shape and movement depend on the kinds and quantity of sand and on wind velocity. Another landform of wind deposition is loess. The large-scale deposition of silt at relatively great distances from its source was prevalent during the Pleistocene but is probably not occurring at a significant rate today.

Suggestions for Further Reading

Bagnold, R.A., *The Physics of Blown Sand and Desert Dunes*, Methuen & Company, Ltd., London, 1941.

Dury, G., *The Face of the Earth*, Penguin Books, Inc., Baltimore, 1964, Chapt. 16.

Longwell, Chester R., and Richard F. Flint, *Introduction to Physical Geology*, John Wiley & Sons, Inc., New York, 1965, Chapt. 14.

Putnam, William C., *Geology*, Oxford University Press, New York, 1964, Chapt. 12.

15 The Classification of Landforms

Introduction

The character and location of landforms have a powerful influence upon the climatic patterns which modify the distribution of soils, vegetation, wild animals and to a considerable degree, the economic activities of people. The constituent materials of landforms are an important factor in the development of soil and vegetation patterns, and play a significant role in animal and human health, although this relationship is not yet completely understood or appreciated. Obviously, landforms are a most important ecologic element in determining the nature and distribution of many properties of the natural environment.

Landforms may modify the regional climate in a number of ways. For example, mountains which lie across the path of moisture-bearing air masses will promote wet climates on the windward slopes and dry climates on the lee slopes. The transition may be abrupt, as along the crest of the Sierra Nevada Mountains in California where, in a space of two or three miles, one may pass from a cool, humid climate to a desert. Equally dramatic changes in soil, vegetation and animal population often accompany such abrupt differences in climate.

Mountainous areas with large local relief, such as the isolated mountains which rise several thousand feet above the deserts of south central Arizona, also provide widely varied habitats for plants and animals. Orographically accentuated precipitation increases up the windward slopes, the temperature decreases, evaporation is reduced and consequently, the effectiveness of the snow and rain that do fall is increased. The combined gradients of water supply and water need bring about rapid changes in vegetation and animal population with changing elevation. In one hour, traveling by car on a desert highway, one may rise from an arid to a near arctic environment, crossing as many environmental zones as may be found in a trip from Arizona to central Canada.

Smaller landform types can also affect local climates. Hills and interspersed hollows may bring about changes in soils and vegetation patterns by dispersing and concentrating rainfall runoff even where the local relief is too small to cause significant changes in temperature. In addition, as we have seen in Chap. 3, north-facing slopes are usually cooler and more moist than south-facing slopes. One impact of slope orientation upon vegetation has been mentioned in Chap. 1. In that example from southern Illinois we saw that the trees found on the north-facing slope were the same as those found on flat ground 400 miles farther north.

Constituent material, especially rock type, can exercise a strong influence upon the development of soils and vegetation even though other factors of the natural setting remain unchanged. Among some of the noteworthy cases are the effects of various types of limestone upon soils and vegetation. Soft limestones and marls (limestones rich in clay) will develop into grass-covered, shallow, dark soils (called **rendzinas**), except in arid and very cold, humid climates. On the other hand, hard limestones with high clay content almost always develop into deep, red, clay-rich soils (often called **tierra rojas**) that support forests.

The effect of rock type upon animals is more subtle and indirect. The presence or absence of particular minerals in the parent rock is reflected in the soil layer and in the vegetation growing in the soil. Animals eating the vegetation may be correspondingly affected. For example, solenium, a substance toxic to animals, sometimes is consumed by them in the form of plants generally called loco weeds. These same grasses make perfectly good fodder if they grow on soils whose parent material does not contain solenium. The absence of some vital element in the parent material may also eventually affect the animals' diet. For example, the early settlers on the high highlands of Kenya were mystified by very low reproductive rates of their cattle in an environment that appeared to be ideal for livestock raising. Eventually the cause was traced to male sterility because of a lack of cobalt in the soils derived from the basaltic rocks characteristic of the highlands. Once traces of cobalt were added to salt licks, the reproductive rate increased to a normal level.

While additional effects of mineral deficiencies or abundancies have been noted in studies of deer and other animal populations, similar investigations relating human health to soil or rock quality have only begun. What little work has been done suggests there may be an important relationship between the quality of human health and mineral deficiencies stemming from the innate characteristics of soil parent material.

Only a few examples of the ecologic effect of landforms have been mentioned. These, however, seem to indicate that the role of the various landform properties in determining the nature and quality of the physical environment is varied, complex and frequently quite subtle. Effective analysis will require a detailed knowledge and understanding of the nature of landform properties as they are distributed across the surface of the earth.

Landforms as Part of a Dynamic System

The discussions of landforming processes in the previous chapters suggest that landforms and their

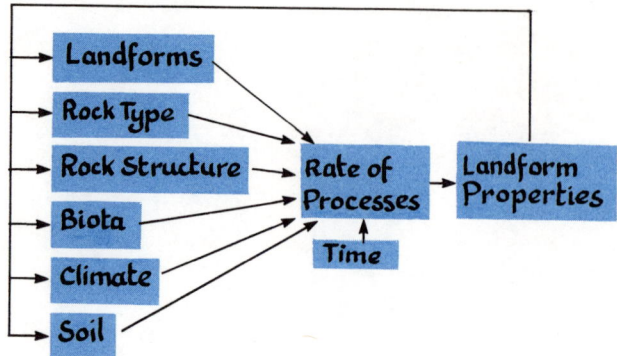

Fig. 15–1. The landform system: a schematic arrangement illustrating how the properties of the system are a function of the interaction between landforming factors and processes.

properties are part of a complex system of interrelated geographic factors and processes. Figure 15-1, a variation of Fig. 1-1, shows how landforms are treated as a dynamic system, meaning that changes in landform properties will take place through time. The factors condition the rate at which the processes operate; the processes, in turn, through their ability to change the form or nature of the physical factors, modify the influences of the factors upon processes. This means that the properties of the landform system, such as slope angle, slope form, local relief, and surface expression, are a function of the landforming factors interacting with the landforming processes.

Several simplified models of the interrelationships between some of the geographic factors and the processes operating on slopes, for example, will illustrate this concept. Figure 15-2 shows the functional relationship between slope angle, a landform property, and some of the geographic factors. The plus sign in the diagram indicates a direct relationship between

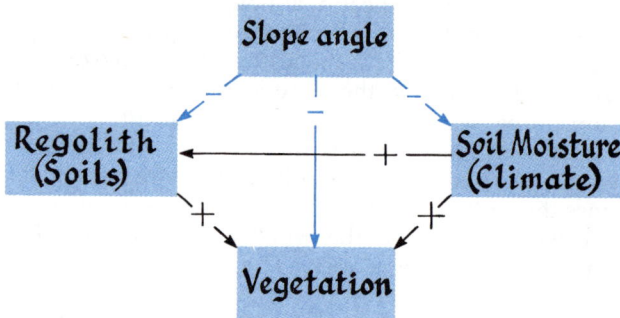

Fig. 15–2. The functional relationship between slope and the factors of soil moisture, vegetation and regolith. Plus signs indicate a positive correlation, minus signs a negative correlation.

slope and the other factors, whereas the minus sign indicates an inverse relationship. Thus, the diagram tells us that, as slope angles increase, soil moisture decreases, the thickness of the regolith decreases and the degree of vegetation cover becomes less. On the other hand, the factors other than slope angle are in direct relationship; as soil moisture decreases the vegetation cover also decreases and the regolith becomes thinner.

Figure 15-3 represents a diagram of the relationships between slope angles and some of the landforming processes. The sketch indicates that as slope angles increase the rates of mass wastage and sheet wash increase. But, as these operate more rapidly they tend to reduce the slope angle, providing a kind of self regulation known as a negative feedback loop in the jargon of systems analysis. A more complicated negative feedback is illustrated in Fig. 15-4 where a new variable, stream channel erosion, is introduced. As channel erosion increases, the slope angle also increases, thus leading to an acceleration in the rates of mass wastage and sheet wash. The resulting increase of slope debris that is added to the stream slows the rate of channel incision and, in time, leads to a decrease in slope angle of the valley sides. As a consequence, a condition of dynamic equilibrium will be established, in which the slope angles no longer change because they are in reciprocal adjustment with the other factors.

Although the discussion here has been limited to slopes, it is apparent from the study of the processes of land formation that there is no specific topographic form associated with the concept of dynamic equilibrium. Any landform may be in equilibrium or closely approaching it, but recognition of this condition in actual landforms may not always be easy. In any case, the theoretical intervention of negative feedback between processes and form insures that all landforms will eventually tend towards equilibrium.

Dynamic equilibrium is an important ecologic concept with respect to landforms as well as other elements of the physical environment. Interference with the factors of the natural setting by man may throw the physical environment out of balance by creating negative feedback[1] that operates at wildly fluctuating rates. Thus, if we look back at Fig. 15-3, we will be reminded that for any given slope angle the rates of mass wastage and sheet wash will also be strongly influenced by the nature and degree of the vegetation cover. The destruction of the vegetation cover will

---

[1] Positive feedback mechanisms may also be set in motion. Because their cumulative effect is often explosive, these processes cannot operate for long before they destroy the whole system and thus choke themselves off.

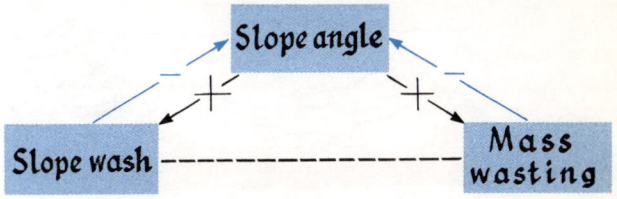

Fig. 15-3. The relationships between slope angle and slope eroding processes. The meaning of the algebraic signs is the same as in Fig. 15-2. The dashed line indicates an inferred relationship.

remove the retarding frictional effect of plant roots on mass wastage and will lay the soil bare to the direct stresses of sheet wash. Debris may be added to the streams faster than it can be removed, clogging stream channels and burying flood plains. The negative feedback of sediment upon stream processes becomes disastrous and local equilibrium may be upset for some time. You will recall that this is what happened in eastern New Guinea when the steel ax was introduced to the mountain cultivators of that area.

Up to this point the emphasis has been upon landforming processes and, to a lesser degree, upon the important geographic factors that influence the rate at which the processes operate. Since landforms are also one of the factors, a system of landform categories will be useful in studying the impact of the land surface upon other elements of the physical environment.

To identify, study, and interpret differences and similarities between the landform areas that make up the land surface of the earth, we need a simple scheme that allows for meaningful comparative description. In a world study of any phenomenon as complex as landforms, regional comparisons can be obtained only by generalizing the landform patterns and by ignoring or abstracting a certain amount of detail. One set of landform properties that can be used to generalize the complexity of the earth's relief features into meaningful

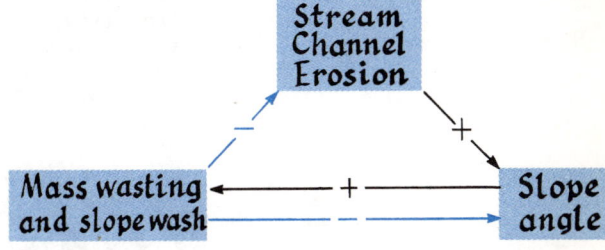

Fig. 15-4. The association between stream channel erosion, slope angle and slope erosion. The meaning of the algebraic signs is the same as in Fig. 15-2.

Fig. 15-5. The importance of slope location. In (1), the steep slopes rise above the general level of the land, forming basin and range plains. In (2), the steep slopes decline below the general level, forming plateaus.

categories is (1) *slope*, (2) *local relief*, (3) *constituent material*, and (4) *surface expression*.

Landform Properties

Two features of slope are important in terrain description: slope inclination and slope form. Slope inclination, or gradient, can be expressed as an angle or as a percentage. A slope of 10 percent, for example, represents a change in elevation of 10 feet over a horizontal distance of 100 feet; a slope of 100 percent implies a rise of 100 feet for every 100 feet of horizontal distance; a completely vertical wall has an infinitely great slope if expressed in percentages. In nature, slopes may range from nearly flat to vertical, but for purposes of general description three slope classes — steep (greater than 18 percent), moderate (between 6 and 18 percent), and gentle (less than 6 percent) — will be sufficient.

The form of a slope refers to its straightness, concavity, convexity, or combinations thereof. It also refers to the length of slope, to its orientation, and to its location relative to the surrounding land. For example, two slopes of equal inclination and straightness may have quite different significance if one leads upward from the general level of the terrain and the other inclines below this level (Fig. 15-5).

Local relief is the difference in elevation between the highest and lowest points in a given area. It gives some indication of the *degree* of slope and of the *texture* of the landform being considered. The texture of a landform refers to the spacing of the landform elements. Many closely spaced ridges and valleys, for instance, determine a fine or close-knit texture; wider spacing of the same units represents a coarse texture. Thus, if the local relief is small, either the texture is close-knit (Fig. 15-6 (1)) or the slopes are low (Fig. 15-6 (2)). Greater local relief indicates steeper slope angles (Fig. 15-6 (3)) or coarser texture (Fig. 15-6 (4)). Local relief is an important descriptive element and is frequently used to distinguish major terrain types.

To be meaningful and useful for comparison, landform properties must refer to areas of similar size. It would be misleading to compare the local relief within a square mile to the local relief of an area ten times larger. Furthermore, an areal unit of 1 square mile (or, say, 1 km²) might encompass a small plain in a mountainous region but would yield no indication of the properties of the surrounding mountains. On the other hand, if we use a larger unit area, say 100 square miles (or, say, 100 km²), we may overestimate the local relief. For instance, a nearly flat but gently sloping plain with a gradient of only 1 percent exhibits a difference in elevation of 528 feet in 10 miles (or 100 m in 10 km); therefore, the local relief in a 10-mile square will be more than 500 feet (or more than 100 m in a 10-km

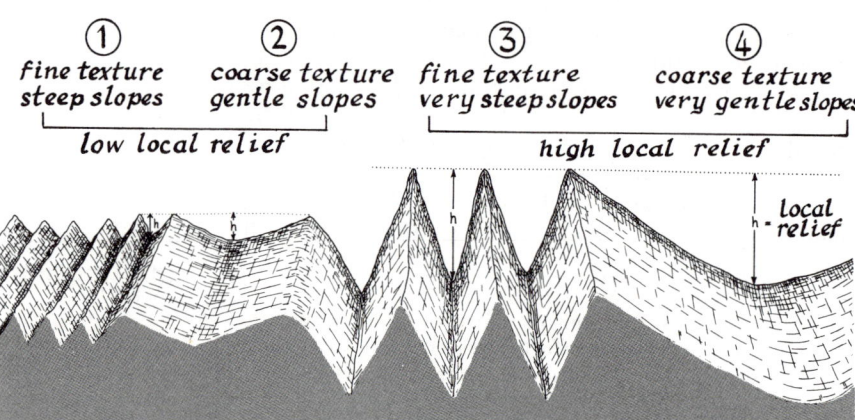

Fig. 15-6. The relationships between slope angles, landform texture, and local relief.

square), despite the essential flatness of the surface. To avoid these problems, we take the local relief to be the difference in elevation between a ridge and its adjacent valley, no matter how far apart the high and low points might be.

Constituent material includes both rock type and structure. These two items are, of course, landforming factors in themselves. It would seem inappropriate, therefore, to include them as elements of terrain description. There are circumstances, however, in which similar landforms may develop on unlike types of material as the result of entirely different processes; but knowledge of the constituent material would permit the recognition of these distinct processes. Furthermore, differences in bedrock and structures may result in markedly different soil types, vegetation cover, and availability of surface water. Failure to include underlying material in landform description, then, may lead to erroneous conclusions about the processes operating upon a given landform unit and about its further development.

Surface expression has two important facets in landform description: the *pattern* of the landforms as seen from above and the horizontal spacing, or *texture*, of these forms. Pattern is difficult to define quantitatively. We normally use descriptive terms, such as "linear," "dendritic," "trellis-shaped," or "randomly distributed," to refer to the patterns of landforms, especially of drainage systems. The texture of the landform, which is more easily quantified, may be defined either as the number of ridges or valleys or as the number of significant slope changes that occur along a line of a given length.

Landform Categories

There are six major categories of landforms: high mountains, low mountains, hills, plains with high relief features, plains of moderate relief, and plains of slight relief. These six categories can be delineated by certain descriptive criteria, which are empirical and arbitrary, since they can be varied according to the objectives of the user and since the categories are essentially subjective concepts. After all, one man's mountain may be another man's hill. Nevertheless, the terrain types identified by this system are established by a uniform set of descriptive properties. Consequently, units of terrain from widely separated areas can be compared objectively. Figure 15-7 shows one quantification of the limits of the six categories; their world distribution is shown on Figs. 16-3 and 17-3.

From Fig. 15-7, we see that plains of slight and moderate relief have low local relief (less than 300 feet or about 100 m), and more than half of their surface is

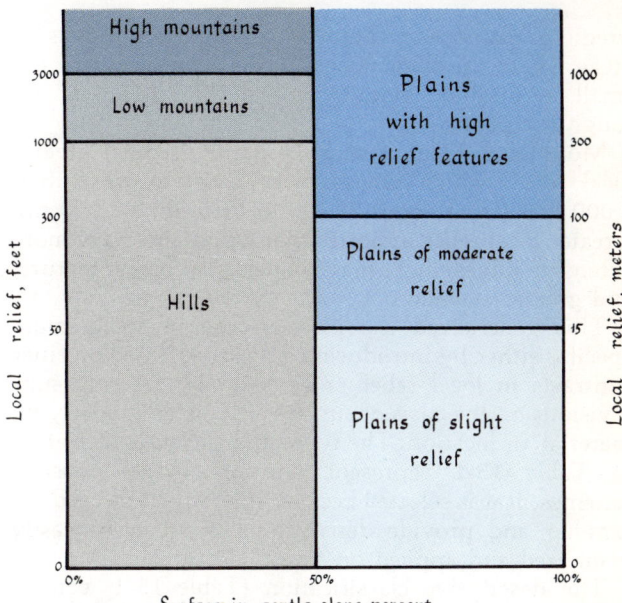

Fig. 15-7. Graphic identification of the six major landform orders. The horizontal axis represents the percentage of the total surface that is in gentle slope. Its values increase from 0 at the left to 100 percent at the right. The boundary between hills and mountains on the one hand, and plains on the other, lies along the value of 50 percent. The vertical axis represents local relief, running from 0 at the bottom to values of more than 2000 feet at the top. Plains with slight relief, for example, have less than 50 feet of local relief and are thus shown below the value of 50 feet of local relief.

in gentle slope. They are land units with low slope angles and with a surface expression that ranges from nearly flat to rolling and irregular. They can occur at any elevation. Altitude is irrelevant to their identification; it is their smooth, relatively flat surface that is important.

Plains with high relief features are of two kinds: (1) plateaus and (2) plains with basin and range topography. Plateaus resemble plains in their surface form, but steep slopes are present in the deep valleys or canyons that divide the land surface into tablelands of varying size. In plateaus, therefore, the high relief features are below the general level of the plain. Moreover, plateaus are usually set off from land of lower elevation by one or more escarpments. Basin and range plains consist of flat lowlands of various sizes that are separated by ranges of hills or mountains. In this case, the high relief features lie above the plains.

When the area in gentle slope drops below 50 percent of the total surface, we no longer speak of plains but rather of hills or mountains. Hills are areas with moderate-to-steep slopes with a local relief of less than 1,000 feet (about 300 m). It should be emphasized that hills are not merely transitional, between rolling and irregular plains and mountains. There are extensive areas where the dimensions of the landforms are too small to produce mountains and the surface is too rough for plains.

Mountains differ from hills mainly in their greater local relief, which ranges from 1,000 to more than 6,000 feet (from about 300 m to more than 1,800 m). Greater local relief indicates that mountains have more extensive ridges, more massive peaks, a coarser texture, and greater elevation.

This generalized classification can be made more specific either by introducing subclasses based on finer contrasts in local relief or slope or by creating subdivisions on the basis of surface expression, underlying material, or location. The types and subtypes identified in Table 15-1 represent one of several possible schemes; it was selected because the properties used are familiar and provide descriptions that can be easily visualized and applied on a world scale.

The descriptive classification (Table 15-1) can be used either to categorize an individual landform or to map the surface configuration of all the world (see Figs. 16-3 and 17-3). Its usefulness can be greatly enhanced by combining its descriptive quality with our knowledge of process and of landforming factors. For example, plains have certain characteristic features that are explicitly stated in the classification (e.g., "most of the land in gentle slopes," "local relief less than 300 feet"). These features raise certain questions about process and origin: Are the nearly flat surfaces the result of deposition under water, are they young features on which stream erosion has not yet carved irregularities, or are they flat because of the resistance of an underlying horizontal layer of sediment or basalt? These kinds of questions can be answered only if we know something about the constituent material and the tectonic process operating on the landform in question.

A first approximation to the necessary tectonic information is provided by the back endpaper which shows three general kinds of information: orogenic belts, shield areas, and areas of relatively undisturbed sedimentary rock. The sedimentary rock, including alluvium as well as marine deposits, has been little involved in tectonic deformation except for relatively minor warping. The shield regions are areas of ancient rock (metamorphic and igneous) that have been relatively stable and quiet for a long time. Long-continued erosion has reduced many of the shields to areas of low relief. Some, however, have undergone repeated broad-scale uplift or sinking, with little visible disturbance of their constituent rock.

A comparison of the back endpaper and Figs. 16-3 and 17-3 shows a clear general relationship between major landform regions and major tectonic regions. Plains coincide with areas of undisturbed sedimentary rock; hills and mountains follow the orogenic belts, although hills, as well as plains, occur in some of the shield areas. The purpose of the next two chapters is to expand upon these general relationships, using both descriptive and genetic material to provide a more detailed survey of world landform regions.

Table 15-1. Classification of Landforms

I. Plains of slight relief
The local relief is less than 50 feet (about 15 m); more than half of the surface is in gentle slopes (of less than 6 percent). These plains can be subdivided on the basis of different constituent materials and landforming processes.

II. Plains of moderate relief
The local relief is between 50 and 300 feet (between 15 and 100 m); more than half the surface is in gentle slopes.
They can be subdivided, according to surface expression, into
A. Smooth plains,
B. Irregular plains,
C. Pitted plains.

III. Plains with high relief features
The plain surface itself has a local relief of less than 300 feet (100 m), but the high relief features within the plain have a local relief greater than 300 feet (100 m); more than half the surface is in gentle slopes.
They can be subdivided, according to surface expression, into
A. Basin and range plains,
B. Cuestaform plains,
C. Plateaus or tableland plains.

IV. Hills and low mountains
Hills have a local relief less than 1,000 feet (300 m), low mountains between 1,000 and 3,000 feet (between 300 and 1,000 m); less than half of the surface is in gentle slopes. They can be subdivided, on the basis of surface expression, into
A. Linear hills and low mountains;
B. Nonlinear hills and low mountains, which take three forms:
1. Nonlinear hills or low mountains with rough slopes,
2. Nonlinear hills or low mountains with rounded slopes,
3. Nonlinear hills or low mountains with compact, cone-shaped slopes.

V. High mountains
Local relief is greater than 3,000 feet (1,000 m); less than half of the surface is in gentle slopes.
They can be subdivided, according to surface expression, into
A. Volcanic mountains,
B. Glaciated mountains,
C. Nonglaciated mountains.

Summary

Landforms can be viewed in different ways and are therefore studied with various purposes in mind. One can think of the surface forms of the earth as an ecological element that determines the nature and distribution of many properties of the natural environment (such as climate, soils, and vegetation), and which, in itself, is an important element of any habitat.

Landforms may also be viewed as part of a dynamic geographic system in which the component parts are interrelated by many feedback mechanisms. The relationships between slope angle and other landform properties, or between it and landforming processes illustrate the essentials of this approach.

Whatever view is taken, it is useful to begin with a system of landform categories based upon the major landform properties of slope, local relief, constituent material, and surface expression. These categories, together with our knowledge of processes, will allow us to appraise the nature and distribution of the surface configuration of the earth.

Suggestions for Further Reading

Hammond, E.H., "Analysis of Properties in Landform Geography: An Application to Broad-Scale Landform Mapping," *New Approaches to the Geography of the United States, Annals of the Association of American Geographers*, Vol. 54, No. 1 (March 1964), 11-19.

———, "Small-Scale Continental Landform Maps," *Annals of the Association of American Geographers*, Vol. 44, No. 1 (March 1954), 33-42.

16 Plains

Introduction

The purpose of this chapter and the next is to describe the distribution of the various kinds of landforms and to call attention, in a general way, to the relationship between the character of landforms and the factors that have affected their development. The landforms will be described according to the classification established in Chapter 15; the resulting types will then be appraised, where possible, with respect to the primary factors in their development.

A word about map scale and levels of generalization is in order. A small map scale requires cartographic or conceptual generalization, and usually both (see Appendix). Since the maps showing the distribution of landforms have a very small scale, they cannot be very detailed. This lack of detail is compensated for, in part, by the fact that we are using a classification that is itself fairly general; but even so it has often been necessary to resort to considerable cartographic generalization where landforms of a particular type cover very small areas. The descriptions and the maps should not be interpreted to mean that a particular landform covers a region to the exclusion of all others; it simply means that the described landform is dominant.

Let us now turn to the landform categories. We begin with those of least relief, starting with nearly flat plains and working our way toward the more rugged landforms.

Plains of Slight Relief

These flat plains have less than 50 feet (15 m) of local relief and a predominance of nearly level land. The critical feature is the smooth surface; consequently, subdivisions of flat plains are made primarily on the basis of different constituent material and not according to surface expression. The smooth surfaces of these plains are usually the result of the even distribution of sediment by water. Flat plains, therefore, include recent marine plains, lake plains, floodplains, delta plains, and piedmont alluvial plains.

Recent Marine Plains

Recent marine plains consist of portions of the **continental shelf** that were exposed by a relative rise of the land within the last quarter of the Pleistocene epoch. Most of the submerged shelf is very smooth, with the exception of occasional shallow basins. When exposed, the shelf becomes a flat plain, which is underlain by loosely consolidated sedimentary rock, primarily soft sandstone, shale, and limestone. These strata have very slight seaward slopes and are frequently veneered by a deposit of sand.

Recent marine plains have nearly flat surfaces because of the previous even distribution of sediment by waves and marine currents (Fig. 16-1). As the plains are exposed by gentle uplift, the processes of stream

Plains 247

Fig. 16-1. Salisbury Plain, a recent marine plain located along the coast of Dorsetshire, England.

gradation modify their surfaces. Rivers, originating in the old land, extend their courses across the newly exposed, seaward-sloping platform, and new tributary streams develop locally (Fig. 16-2). The network of tributary streams develops slowly in shallow valleys, and large areas of the plain remain undrained because of the extremely low gradient.

The water table is high because the sea is close. Swamps are frequent in the floodplains of the larger streams and on poorly drained, flat interstream surfaces. Some **interfluves** are so wide and their slopes so gentle that hundreds of square miles may be covered by swamp, as in the Everglades of Florida, the Okefenokee Swamp astride the Georgia-Florida boundary, and the Dismal Swamp of Virginia and North Carolina. Yet wherever there are slight rises in the land as a consequence of small, initial undulations of the continental platform, interfluve swamps are small, and the plain is usually well drained, since the surface material is sandy.

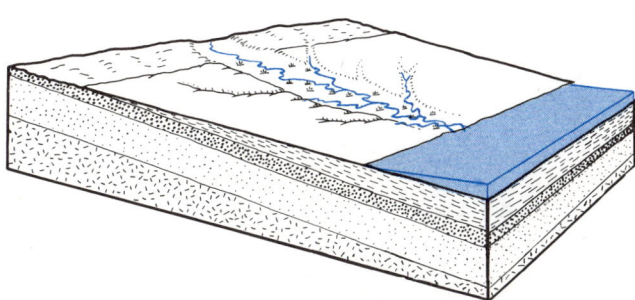

Fig. 16-2. A recent marine plain. The tilt of the underlying strata has been exaggerated in order to simplify the drawing.

Recent marine plains occur along the Atlantic and Gulf coasts of North America from New Jersey to Mexico, in eastern Nicaragua, along the northeastern coast of South America, and in patches along the Arctic coast of North America (Fig. 16-3).

Lake Plains

Lake plains are the most nearly flat of all the landform features and have the fewest surface interruptions. They are underlain by very fine, flat-bedded sediments deposited in lakes. Lake plains of various origins occur in widely separated areas. The plains around the margins of the Great Lakes, the Lake Agassiz Plain, and the Ontario clay belt in Canada are the products of sediment deposited in former large Pleistocene glacial lakes. Some of these plains were formed on the edges of the continental ice sheets, while others were created by deposits from major streams blocked by the glacial ice.

Other lake plains were not associated with glaciers. The Congo Lake Plain and that of the south-central Sudan, both in Africa, are the products of enormous lakes that existed in the middle and late Pleistocene. Wide, flat sheets of fine sediment were laid down in broad, shallow basins that were apparently formed through differential uplift. Parts of the continent lagged behind the general uplift and became depressions without outlets to the sea. When the Nile and the Congo Rivers eventually eroded their valleys through the marginal rim of the continent, the lakes were drained; and large expanses of nearly flat terrain were exposed.

Large lake plains also developed in areas of interior

248   Chapter 16

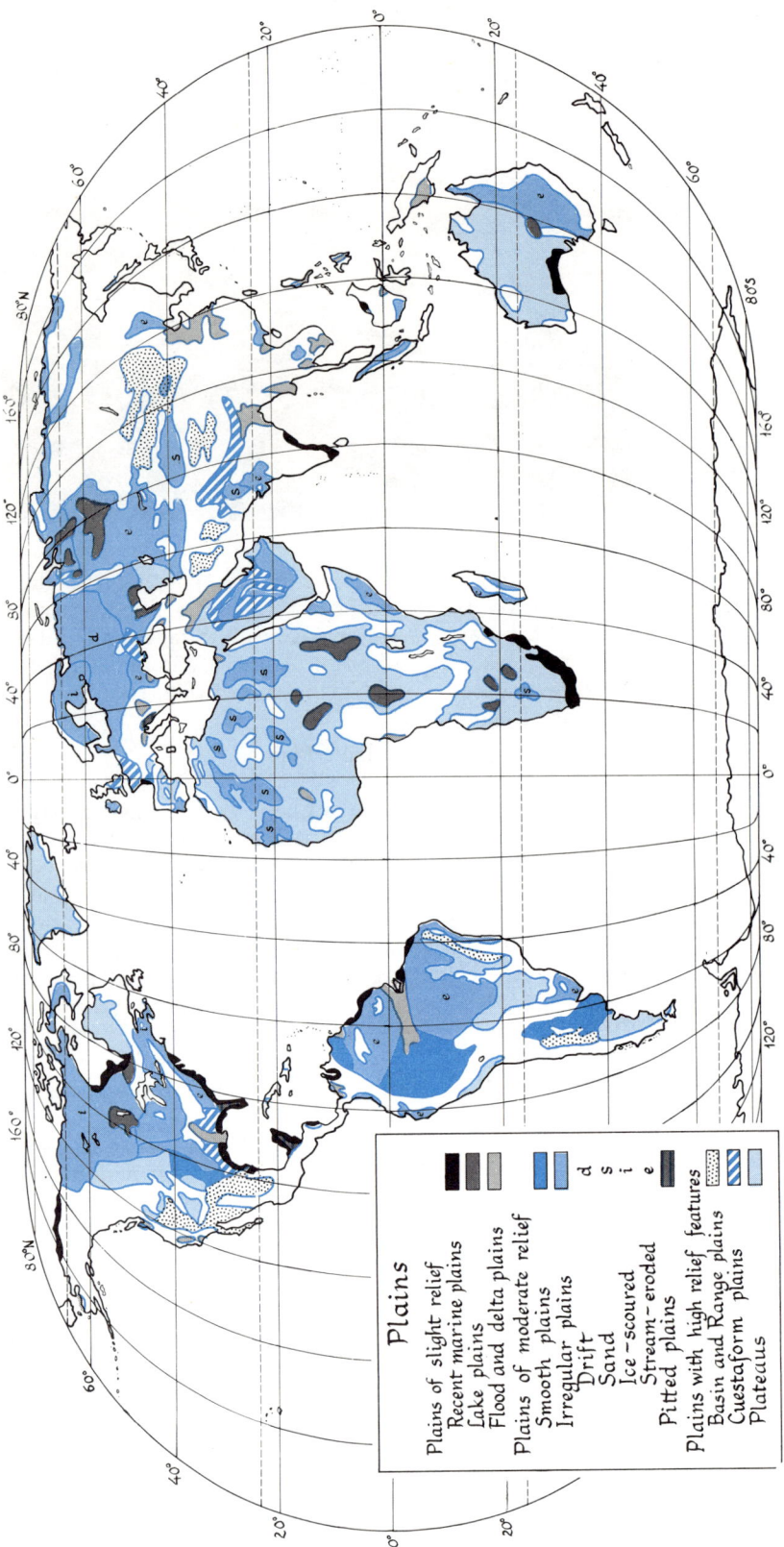

Fig. 16-3. The distribution of plains.

Fig. 16-4. A lake plain surface near Salt Lake City, Utah. The straight, horizontal lines on the hill slope in the background are traces of old lake shores.

drainage during the last glacial period, when increased rainfall and reduced evaporation impounded large bodies of water in now-arid surroundings (Fig. 16-4). A return to more arid conditions at the end of the Pleistocene deprived these lakes of inflowing water, and continued evaporation dried them up. Their flat surfaces are now preserved by the current aridity. Two of the most important examples in North America are the plains that were formed by Lake Lahontan in western Nevada and Lake Bonneville in western Utah. Other noteworthy examples are the Chad Basin Plain in Africa, the Lake Eyre Plain in central Australia, and the plains around the Caspian Sea in the U.S.S.R. (Fig. 16-3).

The surfaces of all three types of lake plains have yet to be seriously modified by streams and mass wastage. Their lack of relief is a direct consequence of the recent exposure of their lake deposits.

Floodplains

Floodplains range in size from a few acres to thousands of square miles. They occur wherever streams deposit significant amounts of alluvium along their valley floors (see Chapter 12). Differences in sediment composition are more significant in differentiating the valley floor than the minor relief provided by natural levees and point-bar deposits. The better-drained sandy loam deposits of the natural levees, the clay of the oxbow-lake deposits, and the silt clay in the levee backswamp result in soils of quite varied textures. These soils are diverse, not only because of their distinctive textures and arrangements within the river valley, but also because of the varied origin of the sediments concentrated in a single floodplain by the tributaries of a major river system. Despite their diversity, these soils are often more amenable to agriculture than those on neighboring uplands because of their greater youthfulness, depth, and friability, or tendency to crumble easily.

Many floodplains are areas of intensive agriculture

Fig. 16-5. A portion of the floodplain of the Ohio River. The ridge and swale terrain is characteristic of meander-point deposits.

and high population because of the availability of water and the fertility of the soil. The major agricultural floodplains of the world include those of such rivers as the Sacramento-San Joaquin, the Mississippi and its tributaries (Fig. 16-5), the Paraná in South America, the Nile, the Tigris and Euphrates in the Near East, the Amu Darya and Syr Darya in central Asia, the Indus and Ganges in India, and the Yangtze and Hwang (Yellow) in China (Fig. 16-3). Even minor floodplains may be of great agricultural productivity. The river plains of Japan, although they are small, provide food for an enormous population. Exotic rivers that have their source in well-watered regions but flow through deserts are particularly striking features of the landscape, because the intensity of human activity and the density of population in the floodplain contrast so strongly with the unused emptiness of the surrounding desert.

The agricultural value of floodplain alluvium is often reduced by flood hazard. Artificial levees may simply raise the level of the river bed without reducing the threat of flooding. Where flooding has been controlled, as in Egypt, agricultural productivity has been decreased because the soil minerals are no longer being replenished by annual deposits of silt. A number of major rivers have fairly sizable floodplains that are not intensively used for reasons of climate, poor soil, or simply lack of development. Among these are the Amazon in South America, the upper Niger in Africa, and the Lena, Ob, and Amur Rivers in Eurasia.

Delta Plains

Delta plains differ from floodplains principally in that they have natural distributary channels, a higher water table, and an expanding coastal margin. Most of the rivers that have important floodplains also have major deltas, and the transition from floodplains to deltas is gradual. However, some streams have floodplains and no deltas, and some delta-building streams do not have significant floodplains.

The distributary channels of delta-building streams are similar to floodplain streams, but they have two distinctive features: They branch in a downstream direction; and as long as they are active, they do not drain the delta, since their stream beds are often at least as high as the rest of the delta (see Chapter 12). Distributaries simply convey river water to the sea; drainage is provided by abandoned, sealed-off distributaries and by interlevee streams (Fig. 16-6). The natural levees of delta channels are high and broad upstream and become narrower and lower downstream, eventually disappearing in the swampy seaward margin of the delta. The land between the levees is usually wet, since it is only very slightly above sea level. Salt

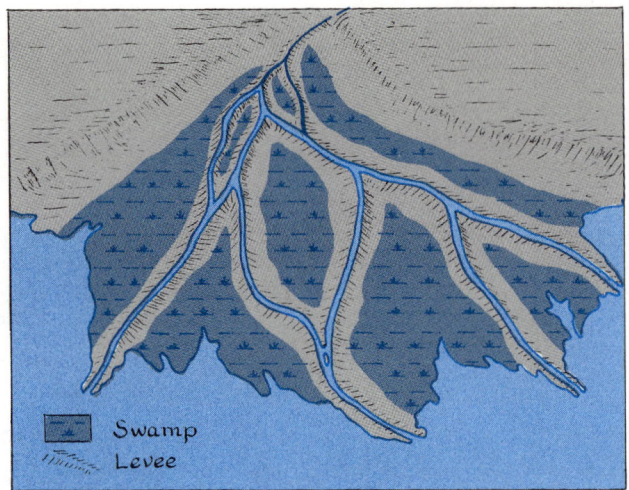

Fig. 16-6. Interstream swamps and drainage in a delta.

marshes, which prevail near the sea, give way to fresh-water swamps inland, above the reach of the tide. As a consequence of the wet ground, the higher, natural levees are the important sites for settlement, roads, and cultivation.

Despite marshiness and flooding, deltas are frequently areas of dense agricultural population. Their surfaces are crisscrossed by drainage ditches and artificial levees, and, in some instances, they are bounded seaward by dikes. One of the earliest intensively used deltas was that of the Nile — the Lower Kingdom of ancient Egypt has been continuously tilled for nearly 4,000 years.

The Netherlands is composed almost entirely of the deltas of the Rhine, Maas (Meuse), and Scheldt Rivers (Fig. 16-7). One of the more noteworthy aspects of this delta plain is the fact that, for the past several centuries, land has been reclaimed from the sea. Dikes have been built beyond the seaward margin of the delta, salt water has been pumped out, and the salt has been removed from the alluvium by providing good drainage through ditches and additional pumping. Each enclosed area, called a *polder*, then becomes new farmland. Presently, there is even greater need for fresh-water storage; and the Dutch plan to close off most of the delta from the North Sea to create a lake of fresh water.

Some deltas are noteworthy for reasons other than supporting large populations. For example, the Hwang River in northern China is building a double delta north and south of the Shantung Peninsula. The Hwang, heavily charged with silt from the loess highlands it drains, has built up its channel to such a degree that the occurrence of exceptionally high floods occasionally causes it to change its course from one side of the Shantung Peninsula to the other. Such changes are accompanied by great flooding and by an enormous loss of life.

The Colorado River has formed a deltaic plain at the head of the Gulf of California. In so doing, it has dammed off the Salton Basin, which is 275 feet (84 m) below sea level. Water from the Colorado River is used to irrigate the piedmont alluvial plain on the edges of the basin. In 1907, the Colorado broke through one of the canals and flooded a large part of the basin, leaving the Salton Sea as the remnant of the overflow.

There are other famous delta plains of considerable size, such as the deltas of the Po, Indus, Ganges, and Mississippi Rivers. Delta plains of equal size but of less fame are those of the Volga, Niger, Danube, Rio Grande (Rio Bravo in Mexico), and the Orinoco (Fig. 16-3). Most of these delta plains have fairly dense populations; the Mississippi and Orinoco Deltas are exceptions. Much of the alluvium in the Mississippi Delta is deposited into a trough, which is gradually sinking, in part because of the enormous weight of the accumulated sediment. Deposition has shifted from one side of the delta to the other because of changes in the location of the distributaries. Today, the delta is growing only in the area of most active deposition, while elsewhere the shore line is moving inland very slowly as the delta surface gradually sinks. This subsidence may be thought of as a kind of negative feedback that keeps the delta from expanding beyond a certain size; there is an equilibrium of sorts between expansion by sedimentation and contraction by sinking. As a result, the Mississippi Delta is excessively wet, and settlement is confined almost entirely to the natural levees. The lowland swamps and marshes have not been exploited except for fishing, trapping and exploitation of oil deposits.

The Orinoco Delta is also characterized by poor drainage and widespread flooding. Until recently, little capital has been invested in developing this large delta plain, but the Venezuelan government has now begun the construction of levees and dikes in an effort to control flooding and encourage settlement of the delta.

Piedmont Alluvial Plains

The largest piedmont alluvial plains occur at the foot of highlands in arid and semiarid climates. Their undulating surface results from the overlapping convex slopes of individual alluvial fans (see Chapter 12). The underlying materials are usually coarse gravel near the base of the mountains and fine silt and clay near the toe of the fan.

The ability of fans to absorb and store water makes piedmont alluvial plains an especially useful agricultural landform. Water seeps into the coarse material of the radial slope at the apex of the fan and gravitates through the alluvium until its further penetration is checked by the fine material near the toe. This buried water can be used for irrigation on the finer alluvium in the lower portions of the fan, to which it is easily distributed by gravity (Fig 16-8).

Piedmont alluvial plains vary in size from small, relatively isolated plains of a few square miles (a few square km) in area to large, continuous features 50 to 100 miles (80 to 160 km) wide.[1] Major examples are found in California, on both the eastern and western sides of the Central Valley and in the Los Angeles-San Bernardino Lowland; in Chile, on the eastern edge of the Central Valley; and in many places in central Asia, such as the Tarim Basin and the basins of Turkestan. Smaller piedmont alluvial plains are part of the basin and range plains of many dry regions, such as the Iranian Plateau.

These piedmont plains have been significant areas of human activity and settlement. The oases on the fans of the northern and southern edges of the Tarim Basin in central Asia mark the site of the famous silk and jade caravan routes of antiquity. The piedmont alluvial plains in the Fergana Valley southeast of Tashkent and those west of Samarkand in Russian Turkestan may

Fig. 16-7. The Rhine delta. The complicated pattern of streams and distributaries has been given individual names for each significant stretch of river. The streams flowing in from the east are the combined channels of the Rhine and Maas. The largest stream from the south is the Scheldt.

---

[1] But even the largest are too small to show at the scale of the world map (Fig. 16-3).

252 Chapter 16

Fig. 16-8. The surface of the Cedar Creek alluvial fan in southwestern Montana. *From the Ennis Quadrangle, United States Geological Survey.*

have been early centers of plant domestication. The smaller, higher alluvial plains in the basins of the central Mexican highlands were the sites of an early evolution of agriculture in pre-Columbian Mexico and later provided the base for the expansion of the Aztec Empire.

Plains of Moderate Relief

Plains of moderate local relief, between 50 and 300 feet (between 15 and 100 m), may have extensive areas of nearly flat terrain, or they may have rolling and irregular surfaces; they are of varied origin and are composed of diverse materials. These plains can be divided into three types of increasing roughness of surface: smooth, irregular, and pitted plains.

Smooth Plains

The most common type of smooth plain with moderate relief is an old piedmont alluvial plain. It consists of segments of flat terrain separated by flat-bottomed, steep-walled valleys that are most deeply entrenched near the mountain base. The interfluves, except for the valley margins, have remarkably smooth surfaces (Fig. 16-9).

Old piedmont alluvial plains occur along the bases of large mountain systems and were deposited by streams flowing from the highlands. The original surfaces of these plains were very similar to those of the piedmont alluvial plains that are still in the process of formation. At some time, however, and for reasons that are not entirely clear, stream activity changed from deposition to erosion. Perhaps this change was due to uplift or to a change in climate or to both. In any event, the streams that originally formed the plains eroded valleys into the fans, depressions that are now separated by flat-topped interfluves. In time the streams developed floodplains, some several miles wide. In some places, smaller tributary streams have eroded belts of rough terrain located along the margins of the major valleys (Fig. 16-10).

Among the better-known examples of smooth plains of moderate relief are the Pampa (Quechua *pampa* plain) in Argentina and the Llanos (Spanish *llano* plain) in Venezuela. Similar plains occur on the northern flank of the Pyrenees, the southern edge of the Alps in the Po Valley, along the southern slope of the Himalayas in the middle and upper parts of the valleys of the Ganges and Indus Rivers, and along the northern slopes of the Caucasus Mountains. The High Plains, east of the Rocky Mountains in the United States, are also old piedmont alluvial plains. They extend from the Llano Estacado in Texas to the northern boundary of Nebraska (Fig. 16-3).

Fig. 16–9. A piedmont alluvial plain. The smoothly rounded contours of the interfluves in the high plans of Kansas are clearly evident under the cover of prairie grass plains. *Photograph courtesy of the U.S. National Park Service.*

Irregular Plains

Irregular plains have a moderate local relief, between 100 and 300 feet (between 15 and 100 m), but the surface is not smooth; short steep slopes are often present. This general category includes drift, sand, ice-scoured, and stream-eroded plains. Each has a distinguishing set of surface characteristics that are related to the material composing it and to the agent that molded its surface.

Drift Plains. The surfaces of drift plains may be nearly flat or rolling, and irregular with fairly steep, short slopes; local relief is greater than 100 feet (30 m). Their detailed surface expression is quite varied, due to the contrasting kinds of deposition that took place in, under, and around the continental ice sheet. Nearly flat areas of lake and outwash deposition are often interspersed with till deposits, which are smooth to rolling, and with morainic belts of rough terrain (Chapter 13, Fig. 13-13).

In many areas the drift buried the preexisting terrain to a depth of scores or even hundreds of feet; the present landform is entirely the result of glacial deposition. The till was deposited partly by lodgment and partly by being dumped at the retreating edge of the ice. Consequently, the till plain has a gently rolling surface of scattered hillocks and shallow depressions, frequently without an outlet. Because of poorly integrated stream patterns, till plains are often swampy and require extensive drainage before they can be permanently settled.

254 Chapter 16

Fig. 16-10. A portion of the High Plains north of Dodge City, Kansas. *From the Lake McBride Quadrangle, United States Geological Survey.*

The line of farthest advance of the continental glacier and the several stopping places during its retreat (with many brief readvances) are marked by belts of rougher terrain that reflect the dumping by the melting ice of material at the shifting margin. Some moraines are low and smooth and are difficult to identify at a casual glance; others are well-defined zones of rough land. These differences in surface configuration result from variations in the duration of ice-front stability and from differences in the age and constituent material of the moraines.

The large amount of meltwater at the ice front frequently resulted in smooth surfaces of water-deposited sediment — glacial-lake plains and outwash plains. The flat surfaces of some outwash plains are pitted with kettle holes, which probably resulted from the melting of stranded blocks of ice buried by the outwash deposit. Outwash plains are frequently covered with thin sheets of loess. The silt is usually less than 30 feet (9 m) thick; consequently, the loess surface reflects the preexisting drift topography.

Drift plains occur in two major areas, the central United States and northwestern Europe (Fig. 16-3). In general, drift plains are found only in association with the last (Wisconsin) glaciation. The characteristic features of drift plains that were formed during earlier glacial periods have been nearly obliterated by subsequent stream erosion and mass wastage. Climate and age are thus both important factors in the determination of the presence of drift plains. Particular conditions were required to form the glaciers necessary for the deposition of drift, while its preservation is due to the recency of its origin.

Sand Plains. The surface configuration of sand plains is largely the product of wind gradation and consists mostly of dune deposits. As we have already seen, the occurrence and the form of the major kinds of depositional features, such as transverse, seif, or barchan dunes, are conditioned by the direction and strength of the wind and by the availability of sand (see Chapter 14).

Large sand plains with seif dunes include the Rub'al Khali in southern Arabia and the Libyan Desert of eastern Libya and western Egypt (Fig. 16-11). Transverse dunes prevail in the sand plains of the Great Erg in the western Sahara. Sand plains also occur east of the Caspian Sea and south of the Aral Sea in the Caspian-Turan Lowlands, where the sandy surface is partly anchored by vegetation. This is also true in north-central Nebraska, where an extensive region of old dunes is now anchored by grass. Sand plains of limited extent occur in areas of Australia and, to a small degree, in the Atacama and Mojave Deserts (Fig. 16-3).

The concentration of sand plains in the Old World, particularly in areas of ancient pastoralism, has led to some speculation that extensive sand plains are not natural features — that they exist only to a limited degree in regions of vegetative and climatic equilibrium, even in deserts. It is thought that much arid-land vegetation has been destroyed in the Old World as a result of prolonged overgrazing. As a consequence, the sand in alluvial fans and the sand derived from bedrock were exposed to wind erosion and eventually accumulated to form sand plains.

Evidence from the southern margin of the Sahara Desert in West Africa seems to support the theory. Small but growing sand plains are now advancing southward into subhumid regions. Areas that should normally support grass, or even forests, have been buried under moving sand. The possibility that a considerable part of the subhumid portion of West Africa may be engulfed by this southward expansion of sand is a serious problem facing the new nations in that part of the continent.

Man-induced wind erosion in West Africa is a good example of positive feedback in a natural system. Once the sand dunes form and begin to move they bury and kill the vegetation, thus increasing the potential for wind erosion. The addition of more sand to the dune system expands the area of destruction and further increases the supply of sand — so it moves at an ever-increasing rate. However, in West Africa, the dunes are moving south in the direction of higher rainfall. Eventually the higher rainfall will act as a negative feedback by providing conditions favorable to vegetation growth sufficient to stabilize the dunes. But before this can occur, thousands of square miles of grassland will be destroyed. Although this is an extreme case it does illustrate an important principle. In unmolested nature the operation of relatively slow negative feedback provides adequate protection to natural systems because the changes in the system are also relatively slow. Rapid, man-induced changes, on the other hand, may result in great damage before the negative feedback can catch up.

Sand and aridity seem to go together, but sand plains are even more closely associated with a reduction in vegetation — a reduction that can be brought about by increased aridity, but also by overgrazing or by deposition in fans, floodplains, deltas, or beaches. The critical role of the plant cover in determining the amount of suitable material available for wind erosion cannot be overstated.

Ice-scoured Plains. The configuration of ice-scoured plains consists of smoothly rounded surfaces with long gentle slopes, open valleys, and abraded ridge crests. The drainage pattern is poorly integrated and is characterized by numerous lakes, swamps, and rapids. The soil is thin and patchy; large areas of bare, polished rock are common (Fig. 16-12). Such plains are chiefly characteristic of moderately high-latitude regions in the Northern Hemisphere. They occur in much of Canada east of the Rocky Mountains, on the Scandinavian Peninsula, in central and northern Finland, and in those parts of European Russia adjacent to Finland (Fig. 16-3). Their appearance and location are clear indications of their origin by ice scour.

Some rolling plains with smoothly rounded outlines occur in nonglaciated areas. They are usually associated with crystalline rock, mainly granite, and have been eroded under varied climates. The rolling surface of the Guiana Uplands in northern South America is an example of this kind of plain.

Stream-eroded Plains. All plains are presently subject to stream erosion. In some, however, the erosion has proceeded to the point where other, earlier surface-molding processes are of secondary topographic importance. The landscape normally consists of smooth or gently rolling interstream surfaces where traces of earlier landform processes are preserved (Fig. 16-13). The valleys have moderate, steeply sloping sides, and their floors are narrow, although some of the larger valleys may contain small floodplains.

Fig. 16-11. Longitudinal seif dunes in southern Arabia looking toward the Hadramawt Plateau. This photograph was taken from an altitude of more than 100 miles (160 km) in August, 1965. *Gemini V photograph courtesy of NASA.*

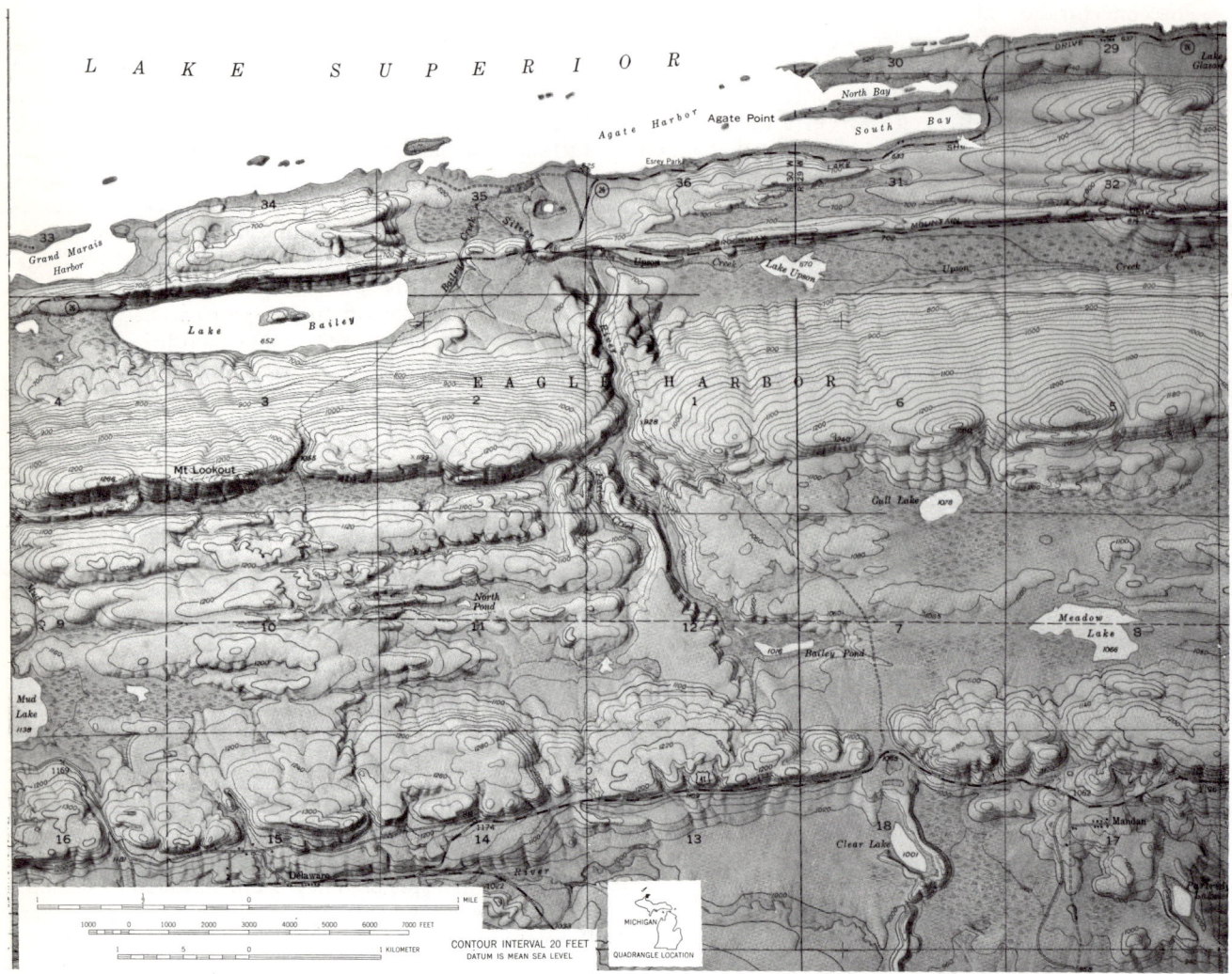

Fig. 16-12. The ice-scoured cuestaform plain along the south coast of Lake Superior in the Upper Peninsula of Michigan. The rounded form, numerous small lakes, and swamps are characteristic of the many areas of ice erosion. The sinuous ridge east of Clear Lake is probably an esker (see Chap. 13). *From the Delaware Quandrangle, United States Geological Survey.*

The older and higher marine plains along the Atlantic and Gulf coasts of the United States are stream-eroded plains. They have been dissected by streams so that the original level surfaces have been changed from nearly flat to irregular plains. In the central part of the United States, the drift that lay south of the maximum advance of the Wisconsin ice sheet has been changed from a drift plain to a stream-eroded plain. These two examples represent relatively recent change by stream gradation. Other, older and more irregular plains of stream erosion occur in the Amazon Basin and cover a major portion of the southern part of the U.S.S.R. (Fig. 16-3).

Pitted Plains

A characteristic feature of pitted plains is the presence of enclosed depressions of varying size and number. In some instances, the sides of the depressions are steep; in others, the sides are relatively gentle, and the terrain has a rolling quality. The most common type of pitted plain is the **karst** plain[2] (Fig. 16-14); small

---

[2]The term "karst" is derived from the Serbo-Croatian word, *kras* (stone), by way of the Italian name "Carso," given to the region along the northeastern coast of the Adriatic Sea, where karst landforms are well developed.

Fig. 16-13. The plains of Picardy. A rolling, stream-eroded plain in northern France. *Photograph by J.A. Russell.*

areas of pitted outwash plains also fit this description.

The constituent material of karst plains is invariably limestone. In some cases, the limestone occurs at the surface; in others, it is buried by layers of other rock or by a deep, residual soil. In every instance the depression results from the removal of limestone, below the surface, by solution and erosion. Pits, or **sinks**, are found in areas where the limestone occurs at the surface (Fig. 16-15). Where the limestone is buried by other material, the outline of the depression is softened by the overlying material. In areas where the limestone has only recently been exposed to solution weathering, as in western Cuba, pitting of the surface is only partly developed. In other places, as in central Jamaica, the limestone has been subject to solution weathering for some time, so that the surface is extremely rough. Weathering by solution is not very pronounced in dry climates, but the characteristic pitted landforms of karst plains develop even in arid areas, although more slowly than in humid lands.

The best-known and best-described area of karst is the Adriatic coast of Yugoslavia. Another notable area of limestone plains is the zone between Florida and Yucatán, a low, broad limestone platform that separates the Gulf of Mexico from the Caribbean Sea and the Atlantic Ocean. To the north is the karst plain of central Florida; in the center, the limestone-covered island of Cuba; and to the south, the karst plain of Yucatán.

Fig. 16-14. A karst plain in New Mexico. *Photograph by Jerome Wyckoff.*

258   Chapter 16

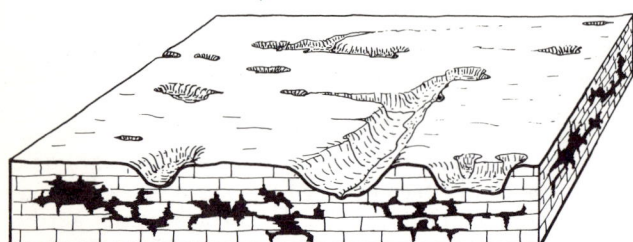

Fig. 16-15. Karst forms. The limestone on which the karst develops has been dissolved into linear valleys, circular sinkholes, and underground caverns (shown in black).

Fig. 16-16. A bolson near Summer Lake, Oregon.

Other karst areas include southern Indiana, the Mammoth Cave region of Kentucky, Apulia on the heel of the Italian boot, and the Causses region of France; in fact, karst plains are evident wherever limestone is found at or near the surface (Fig. 16-3).

Plains with High Relief Features

Some plain surfaces are interrupted at intervals by features of high local relief. If the interruptions are hills or mountains, the nearly level land is in the lower part of the landform unit; if the interruptions are valleys, the nearly level surface is in the upper part of the landform.

Plains with nearly level land in the lower elevation range are of two kinds. These can be distinguished by surface expression and constituent material. One is characteristic of arid regions and is represented by the basin and range plain or the alluvial basin plain. The other is the cuestaform plain, and its particular surface expression is the product of rock type and structure.

Basin and Range Plains

Basin and range plains are a series of alluvial basins, or **bolsons** (Spanish *bolsón* a large purse), enclosed by widely separated hills or mountains (Fig. 16-16). The highlands were produced by folding or faulting or both, and the intervening structural depressions were partly filled with the alluvium deposited by intermittent streams from the highlands. Each basin is a local pocket of centripetal interior drainage, which is characteristic of arid areas. The inflow of water from surrounding highlands hardly ever equals the potential evaporation from the basin floor, and consequently, water in the bolson cannot overflow even the lowest pass to cut a break in the mountain rim.

The detailed configuration of the basin floor reflects differences in depositional form. A belt of alluvial fans rims the bolson at the foot of the enclosing highlands. In the center is a flat expanse of fine lake sediments, which were deposited in temporary lakes formed during periods of higher-than-normal rainfall. The lake floor, when dry, is called a **playa** (Spanish *playa* beach). The lake sediments are rich in soluble bases — so much so that the dry playa floor is frequently white with alkali (Fig. 16-17). Some of these deposits may contain salts of commercial value, such as the borax salts of the southwestern United States and the sodium nitrate deposits of northern Chile.

In some bolsons, the alluvial fans are replaced by **pediments**, which have the same surface appearance but are gently sloping rock surfaces covered with a thin veneer of sand and gravel. The pediments that are underlain by soft rock are probably the product of lateral stream erosion; the origin of hard-rock pediments is a subject of considerable debate. Some geomorphologists

Fig. 16-17. A playa in Death Valley California. The roughness of the salt pools indicates that this lake bottom has not been covered even by temporary lakes since the Pleistocene. *Photograph by Jerome Wyckoff.*

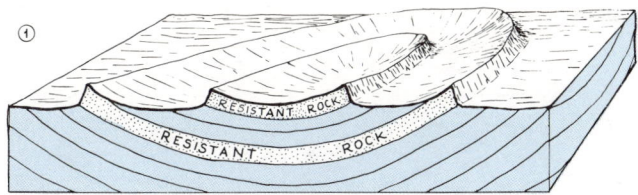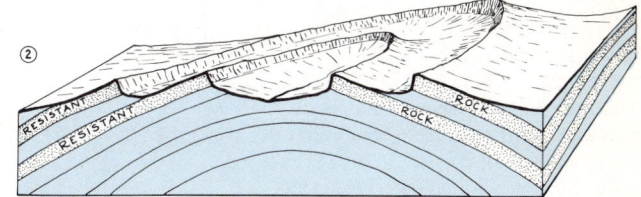

Fig. 16–18. Cross section of cuestas developed on a basin (1) and a dome (2). In the basin, or syncline, the steep sides of the cuestas face outward. In the dome, or anticline, the escarpments face inward.

are of the opinion that pediments formed on resistant rock occur in humid areas as well as in arid regions but are obscured by a thick cover of regolith.

Basin and range plains are conspicuous in the Great Basin of the United States and southward into central Mexico. They are also found in the arid region of northwestern Argentina and on the dry plateau of Iran. Some very large bolsons occur in Sinkiang and western Mongolia; the biggest is the Tarim Basin. Some large basins of interior drainage contain sizable salt lakes, such as the Great Salt Lake in Utah and Lake Eyre in Australia (Fig. 16-3). Some of these lakes are remnants of the much larger Pleistocene water bodies, which were discussed in connection with lake plains.

Cuestaform Plains

Cuestaform plains develop on sedimentary strata of varying degrees of resistance that have been tilted slightly and later eroded. Small, local streams that follow the weak strata erode broad, open valleys; the more resistant strata stand out as interstream ridges. The ridges, or **cuestas** (Spanish *cuesta* from Latin *costa* rib or side), which give the plain its name, are asymmetric in cross section. The relatively steep escarpment on one side results from the sapping of a resistant layer by the removal of weaker underlying sediments; the gentle *dip slopes* on the other side testify to the slow erosion of the top of a resistant layer, which is nearly horizontal.

Sedimentary rocks that have been tilted, without warping, along a single axis show up as a series of long, parallel cuestas with broad intervening lowlands. This type of cuestaform plain is found in central Alabama and eastern Texas. A more intricate pattern develops when the strata are not only tilted but also gently warped into a series of domes and basins. The escarpments around a dome are infacing; the cuestas around eroded basins are outfacing (Fig. 16-18). Such a plain has developed in the lowlands surrounding the Great Lakes. The cuestas, formed by resistant limestone, dolomite, and sandstone, comprise a large system of nearly concentric ridges and lowlands that extend for long distances across the northern part of the Middle West. The Niagara cuesta, for example, can be traced from eastern Iowa to the south shore of Lake Ontario (Fig. 16-19).

Most of the uplands of southeastern England are cuestaform ridges, developed upon sedimentary strata that have been warped into a major basin and several minor domes. The Paris Basin has a similar landscape of outfacing cuesta ridges, which partly enclose the basin to the east (Fig. 16-20). Much of the Ukraine is a cuestaform plain, in which the major rivers flow at the foot of a series of northeast-facing escarpments. The world distribution of cuestaform plains is shown in Fig. 16-3.

Plateaus

Tableland plains, or plateaus, are the product of deep erosion of plain surfaces by widely spaced streams (Chapter 15, Fig. 15-5). The plains may be underlain by nearly horizontal sedimentary strata, lava flows, or by a complex mass of rock flattened by long erosion. The surface must have been uplifted sufficiently to allow

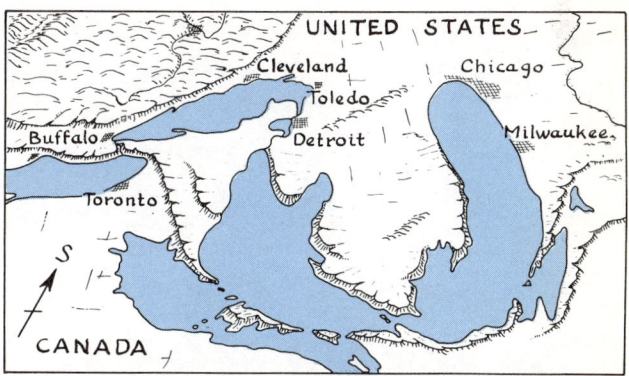

Fig. 16–19. Cuestas around the Great Lakes. Various sets of cuestas are shown, including those whose escarpments mark the southern limit of Lakes Ontario and Erie as well as parts of Lakes Huron and Michigan. Another set crosses Lake Huron and forms the islands that separate Georgian Bay from the main part of the lake. This escarpment continues westward along the north shore of Lake Michigan.

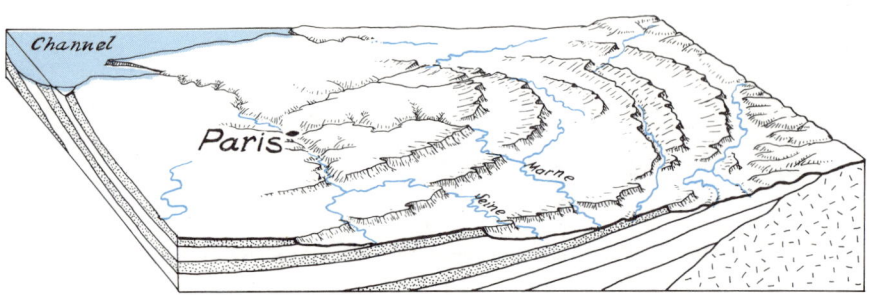

Fig. 16-20. Cuestas in the Paris Basin. These are outward facing escarpments that extend eastward from the lowest (but most recent) deposits around Paris toward the higher (and older) sediments that form cuestas along which the Meuse (Maas) and Moselle Rivers flow in the easternmost part of the diagram.

streams to erode deep, relatively narrow valleys. The large extent of undissected surface and the concentration of water in a few streams are due to aridity or to a resistant cap rock. In humid areas, the stream pattern is close-knit; and weak rocks are dissected into rough hills, such as the inappropriately named Cumberland "Plateau" of western Appalachia. The distribution of plateaus is shown in Fig. 16-3.

Plateaus formed on horizontal sedimentary rock in arid climates are composed of broad tabular uplands edged by abrupt escarpments. The drainage network of deeply incised streams is coarse and preserves the wide, flat interfluves. The plateau surface may be further interrupted by mesas and buttes, which are the remnants of an older and higher tableland surface (Fig. 16-21). The Colorado Plateau is a well-developed example of such a tableland (Fig. 16-22).

Lava plateaus are built by repeated lava flows, usually basaltic, that cover large areas. The horizontal structure of the basaltic flows makes for a rough similarity with sedimentary tablelands. However, the volcanic plateaus have particular qualities that stem from their porosity and the chemical reaction of the constituent rock. Andesite lavas are acidic; but basalt often weathers into soils rich in bases, which can be made into valuable agricultural land. The semiarid Columbia Plateau of the Pacific Northwest is highly productive where irrigated. In wetter climates, the basalt may weather into rich soils, even under tropical temperatures, as in the central highlands of Kenya in Africa, and the Deccan Plateau of India.

The plateaus of western Australia and of Africa south of the Sahara are the result of uplift of surfaces that had been previously eroded to near flatness. The primary constituent material of these high tableland plains is a complex mixture of ancient metamorphic, igneous, and sedimentary rock. The landform consists of flat surfaces interrupted both by valleys and by isolated

Fig. 16-21. Partial view of a mesa north of Fossil Butte, Wyoming. Notice the horizontal strata exposed in the cliff face. *Reproduced with permission of the National Park Service, U.S. Dept. of the Interior.*

Plains 261

Fig. 16-22. The gorge of the Colorado River. Notice the deep, narrow canyon which has been cut into the horizontal sedimentary rocks that form the Colorado Plateau. The upper part of the gorge is broken up into many mesas. *Photograph by James L. Bier.*

hills and mountains (Fig. 16-23). The canyons are the result of recent erosion, while the hills and mountains appear to be residual features that survived the ancient phase of erosion. Tableland plains of such continental proportions may include quite different landforms that cover very large areas; the lake plains of the Congo and the Sudan, for example, are part of the larger complex of African tablelands.

Aridity is probably the critical factor in preserving the plateau of western Australia; but it cannot explain the African tablelands, since they are found under a variety of climates, including that of the tropical rainforest. However, if we examine the rivers on the plateau surface, we note that they flow for long distances in broad, shallow valleys. Near the margin of the continent, they descend abruptly from the plateau to the sea by way of waterfalls and rapids in deep, narrow gorges (Fig. 16-24). The fact that these streams, many of which are very large, have not eroded valleys of more uniform gradient suggests that the African Plateau has only recently been uplifted. Indeed, studies of marine terraces along the coast of Tanzania in East Africa indicate that part of the continent is still rising. The continued presence of the African Plateau can thus be ascribed to its relative youth.

Summary

This chapter was devoted to the several kinds of plains — to their appearance, constituent material, origin, and distribution. Plains of slight relief have flat

Fig. 16-23. The flat expanses of the East African Plateau. The platform of the plateau is part of the African Table, an old complex surface whose original relief features were worn down by the long period of erosion since the pre-Cambrian. The relatively deep canyons that drain the surface are not visible except from their very rims.

Fig. 16-24. Victoria Falls of the Zambezi River, Rhodesia. One of the major waterfalls of the world, it is located at the edge of the East African Plateau. Its presence indicates the recency of formation of the plateau; otherwise stream erosion would have reduced the edge of the plateau to a more regular and gentle slope.

surfaces that result from the uniform distribution of sediment in the sea, in lakes, or in floodplains and from the recency of their origin. The development of piedmont alluvial plains is strongly affected by climate, since they attain significant size only in arid regions. Plains of moderate relief are conditioned by a number of factors, including process, rock type and structure, age, vegetation, and climate. In general, they owe their slightly more irregular aspect to deposition by glaciers or wind, which is less uniform than deposition by water, or to the roughening effect of glacial scouring, stream erosion, or solution weathering. Plains with high relief features are associated with particular conditions of climate or rock type and structure; certain plateaus owe their existence to the recency of their fomation since erosion will destroy their tableland character if other factors do not intervene.

Suggestions for Further Reading

Hunt, Charles B., *Physiography of the United States*, W.H. Freeman and Company, San Francisco, 1967.

King, Lester C., *The Morphology of the Earth*, Hafner Publishing Company, Inc., New York, 1962.

Thornbury, William D., *Regional Geomorphology of the United States*, John Wiley & Sons, Inc., New York, 1965.

17 Hills and Mountains

Introduction

Hills and mountains differ from plains not in terms of local relief, but in that most of the land is not in gentle slopes. The values of local relief range from very small to well over 3,000 feet (1,000 m). Within this wide range, we can distinguish between hills (local relief less than 1,000 feet or 300 m), low mountains (local relief between 1,000 and 3,000 feet or between 300 and 1,000 m), and high mountains (local relief greater than 3,000 feet or 1,000 m). High mountains, in addition to their greater local relief, possess certain characteristics that derive from the climate at their high altitude rather than from their ruggedness. No such distinction exists between hills and low mountains. The surface expression of these two forms is very similar, except for the size of the blocks that make them up. Hill lands do occur as separate landform regions; but where they have local relief close to 1,000 feet (300 m), they merge imperceptibly into low mountains. Because of their similar surface expression, hills and low mountains are discussed together.

Hills and Low Mountains

Hill lands and low mountains are areas of rough terrain with more than half of their surface in moderate or steep slopes and with local relief of less than 3,000 feet (1,000 m). Several kinds of hill and low-mountain landscapes may be distinguished by pattern, texture, and slope profile. Hills and low mountains may have either linear or nonlinear patterns. Nonlinear hills can have rough, rounded, or conic slope profiles, with associated differences in the texture of the landforms. All these differences are related to variations in the landforming factors.

Hills and Low Mountains with Linear Relief

Hills and low mountains with linear relief were evolved principally from stream erosion on sedimentary rock that had been uplifted and warped into anticlinal and synclinal structures similar to those shown in Chapter 16, Fig. 16-18, except that the strata are more strongly bent. The valleys outline the outcrops of softer sediments, while the uplands, called **hogback ridges,** are composed of more resistant rock. As Fig. 16-18 demonstrates, the top of anticlines do not necessarily become highlands, nor do synclines always form valleys; the contrasting erodability of the individual sedimentary strata controls the location of ridges or valleys.

Where the horizontal axis of the anticlines and synclines is straight and parallel to the general level of the land, the ridges and their intervening valleys are also straight and parallel; where the axes dip with respect to the surface, the ridges and valleys exhibit very tight

264 Chapter 17

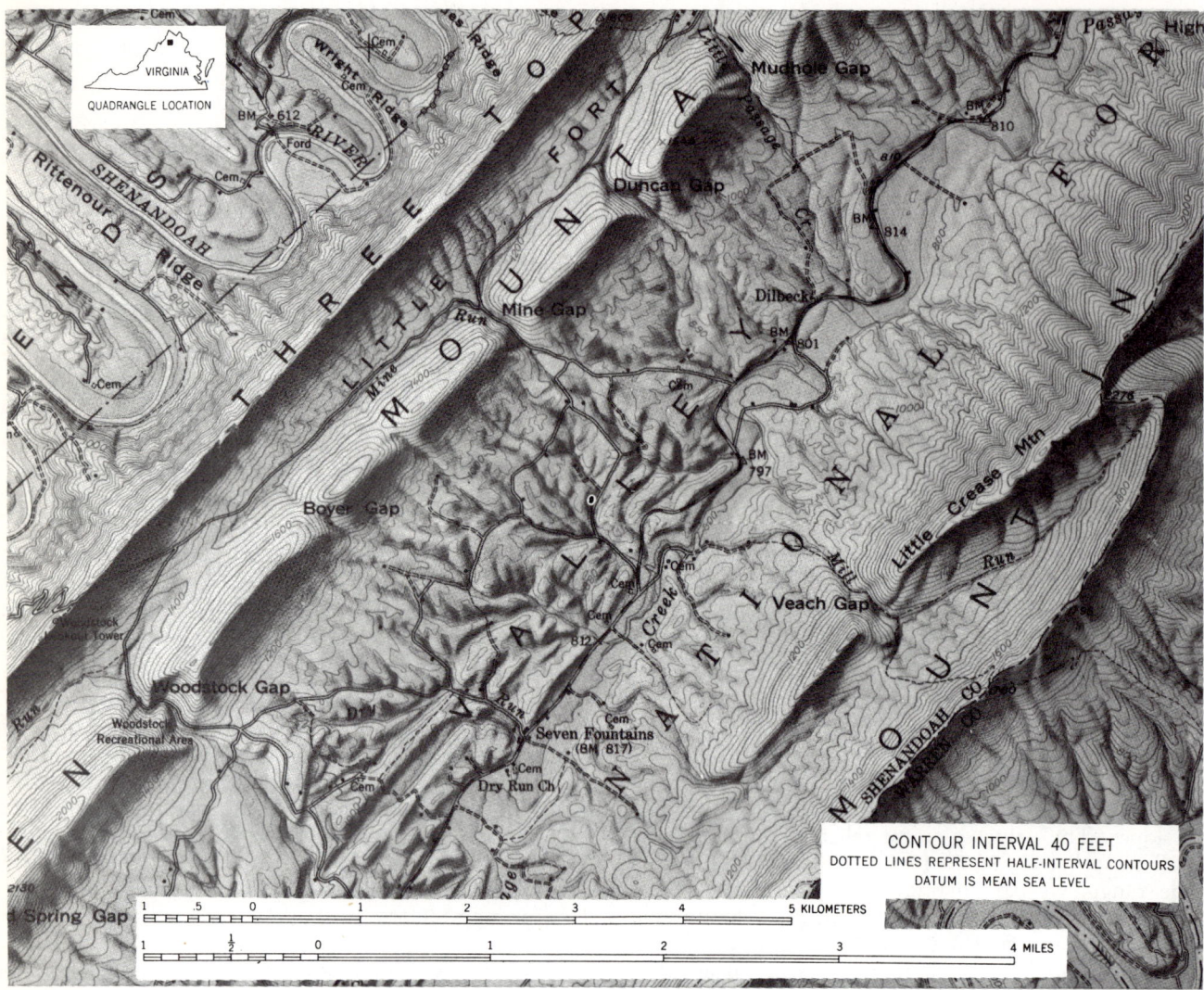

Fig. 17-1. Linear hills and valleys paralleling an anticlinal axis in the Shenandoah Valley of Virginia. The steep sides of the cuestas face southeastward toward the axis of the anticline. *From the Strasburg Quadrangle, United States Geological Survey.*

turns. Occasionally, they sprawl across the land in broad twists and loops, reflecting the particular geometric relationship between the land surface and the twisted axes of the folds (Fig. 17-1).

Hogback ridges are cuestas formed on steeply inclined strata. They are much more symmetric in outline than ordinary cuestas, since they have no gently tilted backslope. All slopes are fairly steep, and the crests are continuous and nearly unbroken; hence, travel across the grain of the land, a matter of crossing ridge after ridge, is difficult. The valleys, on the other hand, are usually broad and have gently sloping bottoms that permit easy passage along their floors.

The major regional streams in the valleys probably existed before the onset of orogeny and have maintained their courses across the hogbacks as the land was uplifted. Consequently, the ridges are occasionally cut by narrow stream valleys, called **water gaps** (Fig. 17-2). **Wind gaps** are higher passes; they represent river cuts abandoned during the course of uplift and erosion. The drainage pattern is trellis-shaped; most of the stream courses follow the axis of the structure, but a number of right-angle turns occur where water gaps open onto the valley floor.

The eastern part of Appalachia is a striking example of local variations on a regional theme of linear ridges and valleys, produced by small differences in pattern, texture, and process. On the east is the Shenandoah Valley, a long, straight, and continuous lowland set off by linear ridges, which extends southward from the Potomac River for more than 100 miles (160 km) between the Blue Ridge on the east and the linear landscape of ridges and valleys on the west. To the west, the valleys are not so wide, but their open quality persists. Still farther west, in West Virginia and in western Virginia, the ridges are even closer and the valleys are narrow; the nearly level land occupies less than 20 percent of the total surface. In addition to these detailed differences in texture, part of this area has been slightly affected by glaciation. The linear pattern remains virtually unchanged, however; the principal modification by glacial abrasion is a thin regolith and a more rounded outline of the ridges.

Other regions of similar landforms are the Ouachita Mountains of western Arkansas and eastern Oklahoma; the concentric, curvilinear ridges of the Black Hills of western South Dakota; the linear Jura Mountains between France and Switzerland; and the low-mountain country of the Atlas Range in northern Africa. (Fig. 17-3).

Linear landforms may also result from faulting. In the Coast Ranges of California, for example, zones of badly fractured rock are bounded by long, parallel faults. By removing the crushed material, the streams have eroded long, straight valleys and given a crudely linear appearance to the low mountains.

Hills and Low Mountains with Nonlinear Relief

Nonlinear hills and low mountains of rough relief. This landform is characterized by V-shaped valleys, angular slopes, narrow ridges, a dendritic drainage pattern, and a fine or moderately coarse texture. Local relief can be quite low; it is the high percentage of surface in appreciable slope that gives this landform its rough appearance and fine texture.

The Allegheny-Cumberland region, south of the maximum advance of the continental glaciers in the eastern United States, is a good example of rough, hilly terrain that, in places, grades into low mountains. This

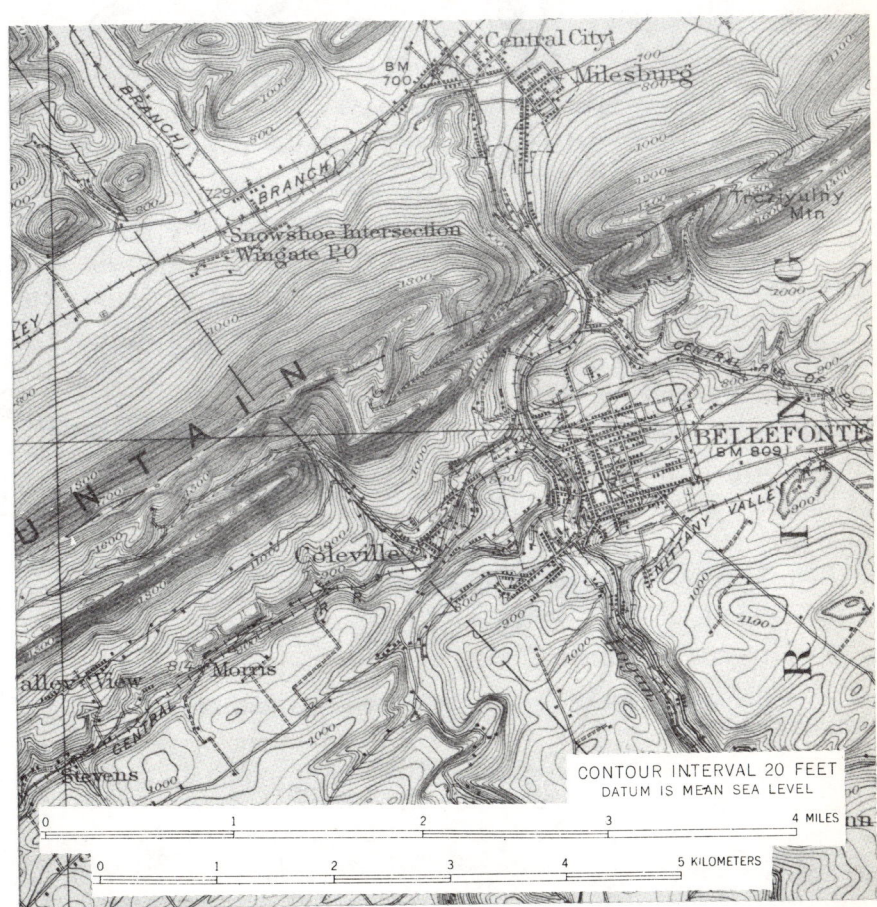

Fig. 17-2. The importance of water gaps to transportation routes in the linear mountains of Pennsylvania. Note the railroads and roads converging on the water gap north of the town of Bellefonte. *From the Bellefonte Quadrangle, United States Geological Survey.*

266 Chapter 17

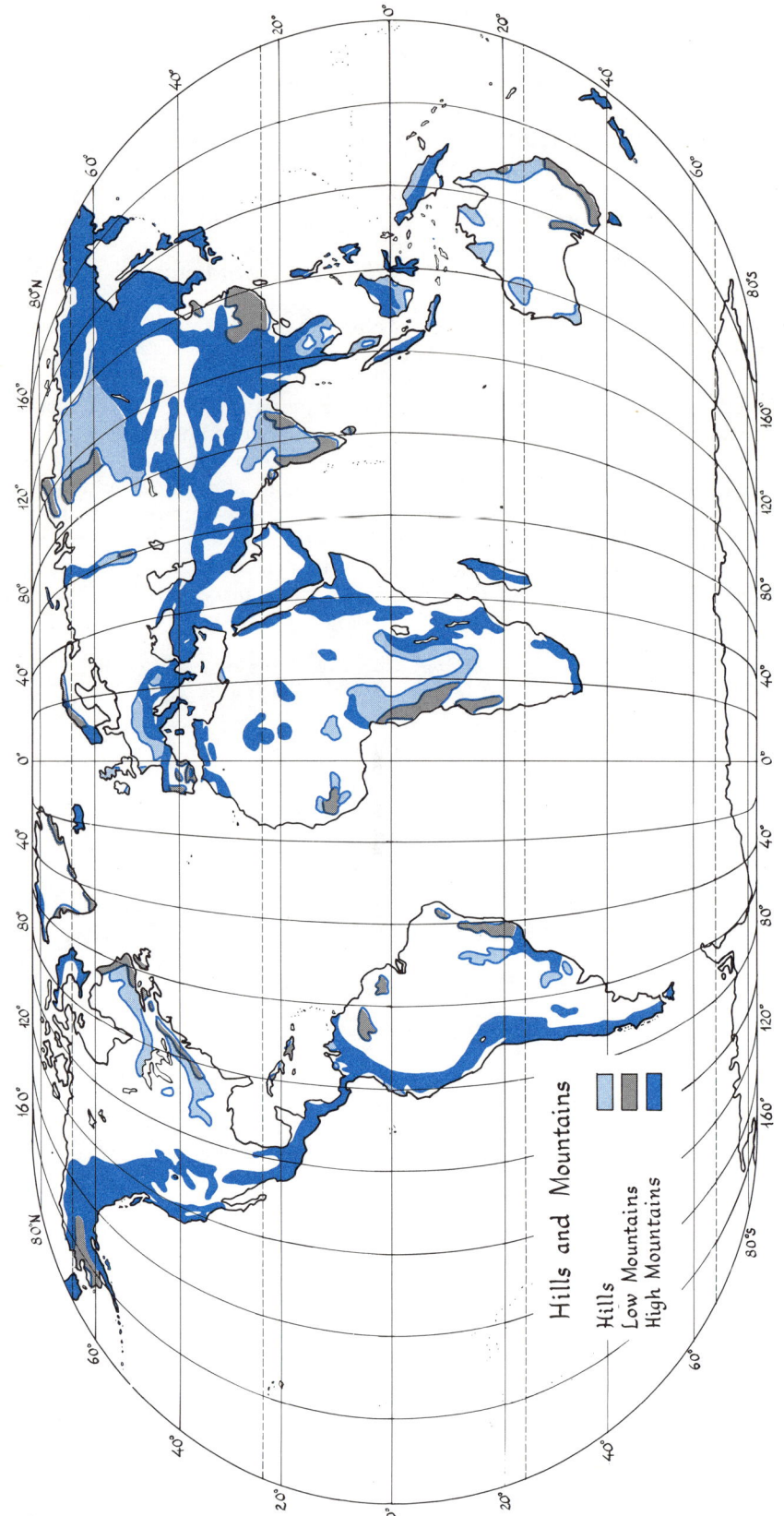

Fig. 17-3. World map of hills and mountains.

Hills and Mountains 267

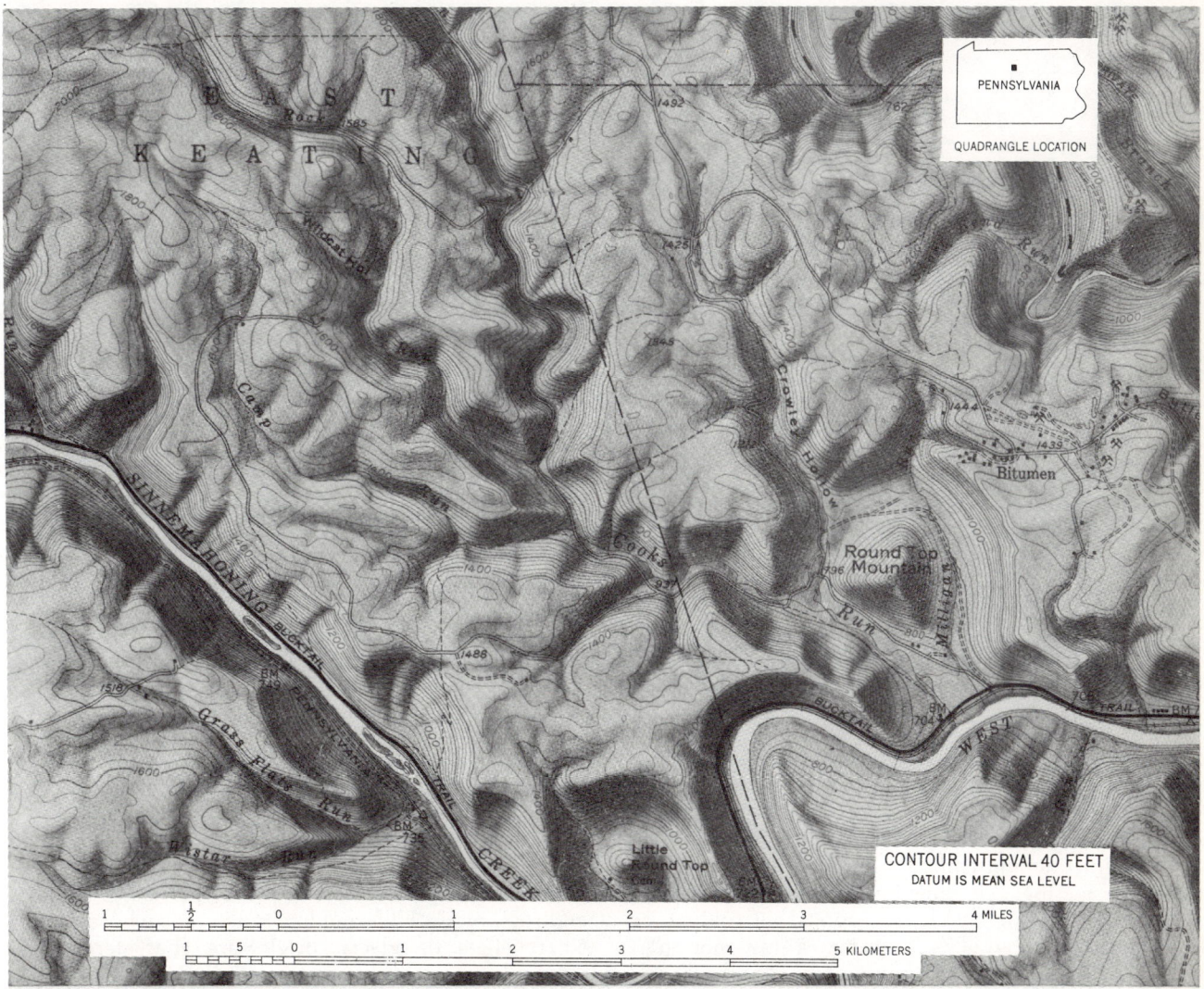

Fig. 17-4. A section of rough hills in the Allegheny-Cumberland region of north-central Pennsylvania. Notice the similarity in elevation of many of the ridge crests. *From the Renovo West Quadrangle, United States Geological Survey.*

area is underlain by gently warped sedimentary strata in the form of a shallow basin. The eastern margin is formed by the Allegheny Front, a southeast-facing, cuestaform escarpment of resistant cap rock that slopes gently westward. It extends for a distance of 700 miles (1,100 km) from northern Pennsylvania to Alabama.

In the east, near the Allegheny Front, streams are generally closely spaced, separated by sharp-crested ridges of uniform elevation. These low, rough mountains are often close to 3,000 feet (1,000 m) above the valleys; only the larger streams have floodplains, and these are small (Fig. 17-4). Westward, in Ohio and central Kentucky, there is less relief and the valleys are more spacious. Even the uplands are relatively broad and rolling, especially where resistant cap rocks are present (Fig. 17-5).

Other important regions of rough uplands underlain by horizontal sedimentary strata are the Ozark Mountains in Arkansas and southern Missouri and the hills and scattered low mountains on the well-consolidated sedimentary strata of the central Siberian highlands north of the Angara Shield (Fig. 17-3).

Rough uplands are also found on some shield areas of ancient crystalline rock and on ancient rock complexes of mixed origin. These areas have gone through several periods of folding, faulting, intrusion, and metamorphism. The structure is so complex that there is little systematic structural control of stream

Fig. 17-5. The western part of the folded Appalachians in West Virginia. The valleys are less closely spaced than farther east. Note the broad sweep of the uplands. *Photograph by Jerome Wyckoff.*

networks. Dendritic drainage patterns have developed wherever there is sufficient initial relief and stream flow. This landform occurs in several places. One is in southern China, a region of sediments overlying an igneous and metamorphic basement rock of such venerable age that it is sometimes classified as a shield. A large-scale uplift, without appreciable folding or faulting, caused streams to carve a low-mountain landscape of considerable roughness. Another such region, the Blue Ridge Mountains, extends from northern New Jersey into northern Georgia. It is scarcely more than a few miles wide in the north but broadens toward its southern limits, where it is more than 50 miles (80 km) wide and has a local relief of 2,000 or 3,000 feet (600 or 1,000 m). The mountains are composed of igneous and metamorphic rocks of extreme age and structural complexity. In the absence of a systematic structural arrangement, streams developed a dendritic pattern and eroded a low, mountainous landform of irregular summits with rounded crests.

Regional variations in roughness are undoubtedly related to both the type of rock and the kind of climate. For example, extremely rough surfaces develop on poorly consolidated, fine-grained sedimentary rock in semiarid climates. The complicated ridges and ravines of the badlands in South Dakota and other steppe and desert regions are examples of this kind of terrain (Fig. 17-6). The same kind of rock in a humid climate would be quickly eroded to a much smoother surface.

Nonlinear Hills and Low Mountains of Rounded Relief. This landform type is the result of either glacial erosion or the interaction of particular climates and rock types. Hills and low mountains whose rounded relief is of glacial origin are characterized by smoothly curving slopes, open valleys, and land surfaces predominantly in moderate-to-steep slopes (Fig. 17-7). The regolith is thin or nonexistent, and outcrops of glacially polished rock may occur at the surface. Lakes and swamps, linked by rapids and waterfalls, are typical features of the drainage. The lack of regolith, the rounded surface, and the disorganized drainage are evidence of recent erosion by continental glaciers. Examples of such rounded topography are the Scottish Highlands, the Scandes of Norway, the uplands of northern and southeastern Canada, the highlands of Labrador, the White and Green Mountains in New England (including their extension into the Gaspé Peninsula), and the Adirondack Mountains in New York (Fig. 17-3). In all these examples, the preglacial landform was the product of stream erosion and mass wastage upon constituent materials, usually without significant structural pattern.

All rounded hills and low mountains are not the

Fig. 17-6. The badlands of South Dakota. Notice the many bedding plains and the fine-grained nature of the sediments shown in this exposure. *Reproduced with permission of the United States Department of Agriculture.*

product of glacial erosion. Deep regolith, valleys with convex walls and relatively open floors, or valleys with narrow, V-shaped cross sections are often found in regions of granitic rock and humid climates. The rounded hills of the western Blue Ridge and the low hills in the western part of the Maryland Piedmont are examples. Rounded forms may also develop in subhumid climates on fine-grained sandstone rock. Such rounded hills occur in the Coast Ranges of California on relatively soft sandstones.

Conic Hills. These consist primarily of cinder cones formed by volcanic eruptions of lava that contained large amounts of gas. The gas turned the lava into a bubbly mass which exploded into small fragments as it solidified. These small, solid particles of volcanic ash formed a compact, cone-shaped pile of debris around the vent (Fig. 17-8). The unconsolidated ash is eroded easily by streams; hence the preservation of the conic shape is probably due to either arid climate or, in humid climates, the recency of cone formation. Cinder cones usually rise 500 to 1,000 feet (150 to 300 m) above the surrounding land. While they frequently occur in clusters, they do not cover large areas and are often superimposed on mountains of other origin. Cinder cones are numerous around the margins of the Pacific Ocean; noteworthy concentrations are found on the island of Java and in the state of Michoacán in Mexico.

One final word about hills and low mountains. Comparisons of the world distribution of hills and low mountains (Fig. 17-3) with the tectonic map (back endpaper) suggest an areal coincidence between these landforms and areas of old folding. It is tempting to conclude that hills and low mountains are much older forms than the higher mountains. To be sure, the underlying structure of hills and low mountains is very old; but their form, which resulted from erosion on recently uplifted land, is probably much younger. The lower local relief of hills and low mountains cannot, therefore, be ascribed to old age; it is probably due to a rate of uplift slower than the rapid orogeny that produced the high mountains.

High Mountains

High mountains are an obvious feature of the landscape because of their elevation and great local relief. The spectacular features of the landform — the coarse texture of the terrain, the large size of the individual ridges and valleys, the absence of flatland, and the steepness of the slopes — are all associated with its tectonic youth. The orogenic processes of folding, faulting, and vulcanism, which created these moun-

270 Chapter 17

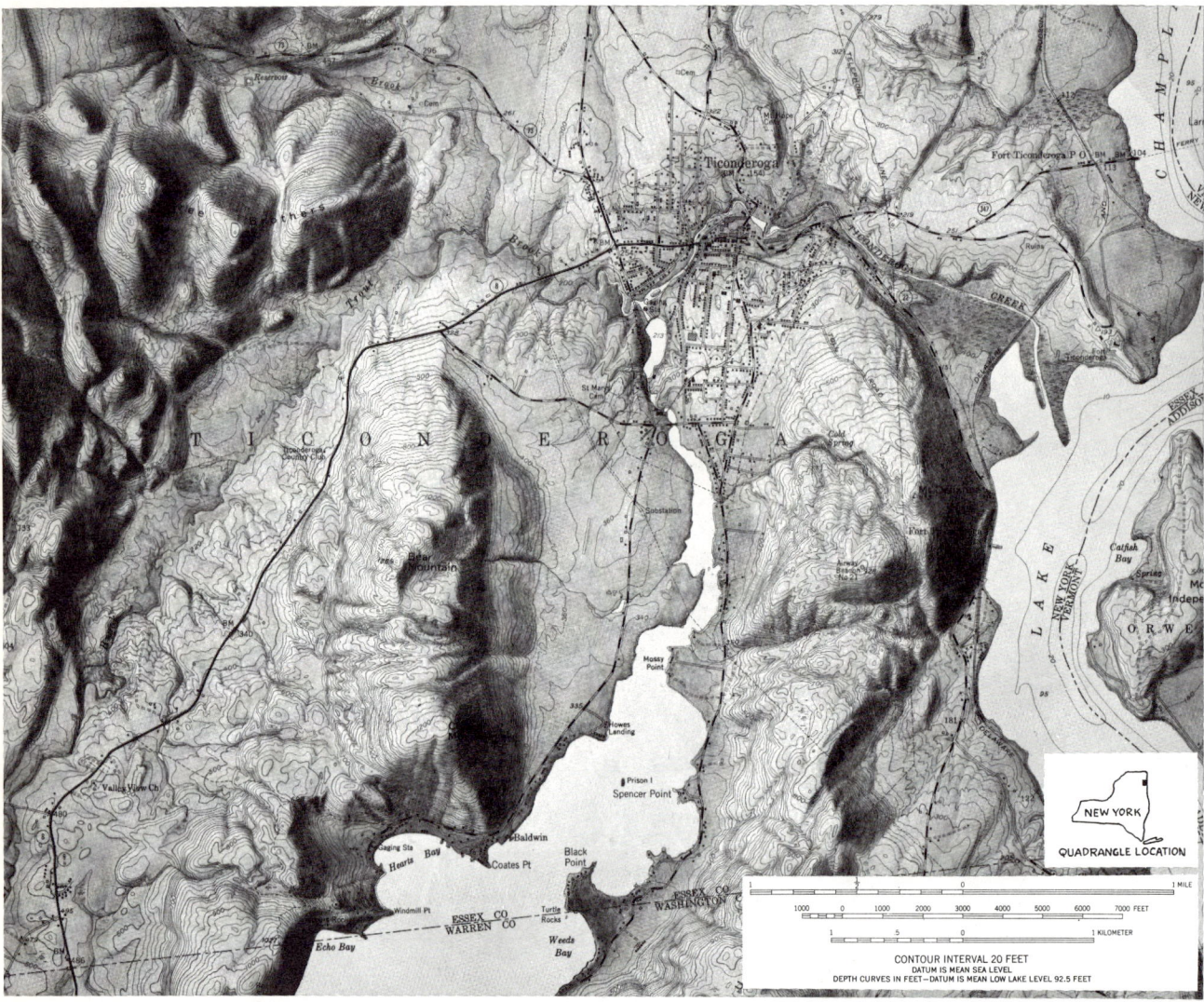

Fig. 17-7. A section of the Adirondack Mountains just west of Ticonderoga, New York. The rounded form and locally steepened slopes are not unusual in ice-scoured uplands. *From the Ticonderoga Quadrangle, United States Geological Survey.*

tains, all started in the recent past and appear to be continuing into the present. The explosive quality of the rockfalls down the steep bare-rock valley sides in the Himalayas suggests that the current rate of stream incision in those mountains is continuing to create valley slopes too steep to be maintained by the strength of the underlying rock. In less spectacular areas the steep narrow valleys suggest a very high rate of stream incision while the frequency of landslides and rockfalls is evidence of the rapid rate of mass wastage.

A distinction between linearity and nonlinearity is not very useful in high mountains. As a consequence of folding and faulting, many mountain ranges have a grossly linear trend similar to the pattern of some hills and low mountains. However, when examined in detail the linear quality is frequently diluted by other features. Large volcanoes may dot the surface; glacial erosion may have created local patterns often at right angles to the linear trend; deep valleys may be eroded into the mountain flanks, and many cross-ridges, related to local structure may be superimposed on the larger linear form.

A more useful classification distinguishes between volcanic, glaciated, and nonglaciated mountains. The

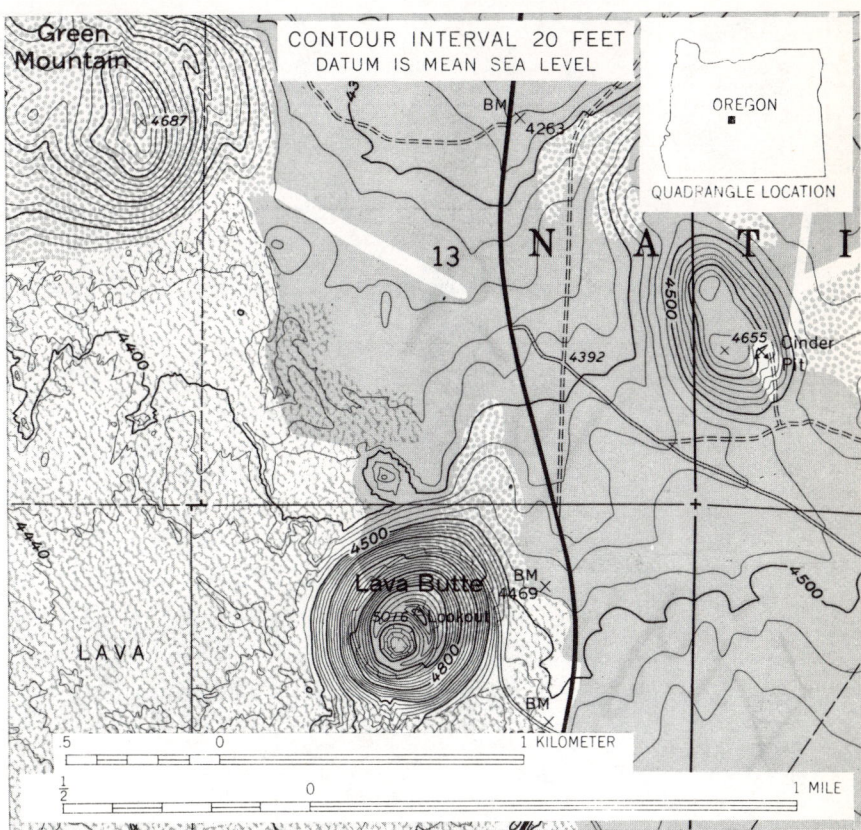

Fig. 17-8. Cinder cones in central Oregon. There is a small crater on top of Lava Butte (the depression is shown by the hatched lines). The lava flows from this Butte have displaced the Deschutes River several miles westward from its former course near the Butte. *A portion of the Lava Butte Quadrangle, United States Geological Survey.*

differences in form resulting from the differences in processes are sufficiently striking to warrant the threefold division.

Volcanic Mountains

Volcanic mountains differ from conic hills in size, local relief, and origin. Conic hills consist of cinder cones seldom more than 1,000 feet (300 m) high. Volcanic mountains, by definition, are at least 3,000 feet (1,000 m) above their surroundings. The steeper, more spectacular mountains are composite volcanoes, resulting from repeated depositions of ash and lava. They have symmetrically curving flanks, although stream and glacial erosion may have roughened the sides of the cones (Fig. 17-9). Shield volcanoes are gentler in outline, but have equally great relief because of their enormous size. Volcanoes may occur singly or in clusters. They may be the main mountain-forming element, as are Fujiyama and the shield volcanoes of Hawaii; or they may be superimposed upon ranges of earlier and different origin, as are the volcanoes of Central America, of the Andes of Colombia and Chile, and of the high Cascades in the Pacific Northwest.

The distribution of volcanic mountains (back endpaper) is revealing; most of them occur in two major zones (1) around the Pacific Ocean and within its basins (many of the islands in the Pacific are volcanic in origin), and (2) along a line from the Azores Islands eastward into Turkey. This concentration is related to recent and even contemporaneous crustal disturbances in these areas. The frequent earthquakes in the Mediterranean region and around the Pacific margin are evidence of current orogenic activity. These zones are also among the areas of the highest, therefore presumably the most recent, mountains of the world. During mountain building, joints, cracks, and fissures in the crust are developed by folding, faulting, and magmatic activity. These zones of weakness allow the lavas and gases, which are under great pressure, to escape to the surface and build volcanoes.

Indeed, the folding and uplift characteristics of mountain regions may, at times, be the fundamental cause of both intrusive and extrusive vulcanism. Most major mountain ranges have batholithic cores against which lie strata of sedimentary and metamorphic rock (Fig. 17-10). The structural relationship suggests that the distortion of the overlying stratified rock was due

272   Chapter 17

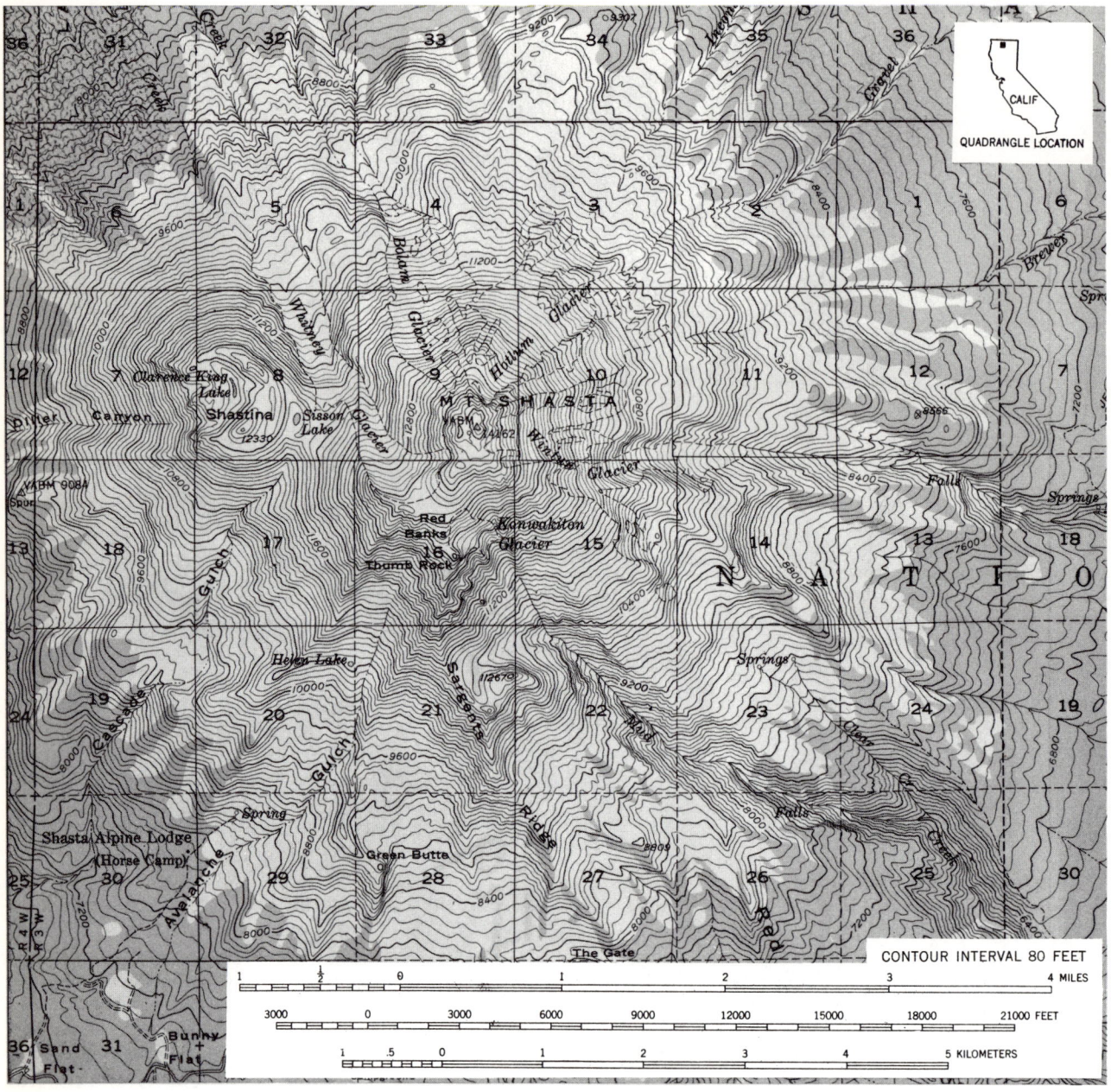

Fig. 17-9. Mt. Shasta, California. Note the generally symmetrical flanks of the cone, and also the variety introduced by secondary cones, such as Shastina. *A portion of the Mt. Shasta Quadrangle, United States Geological Survey.*

to the upthrust of the batholith. It is possible, however, that the upfolding came first and initiated the development of the magma, which then expanded against and into the rising overlying rock. At depths of 10 or more miles (15 or more km) below the surface, magmatic material has a very high temperature but is in a solid state because of the great pressure of the overlying material. Upfolding, or doming, of the surface rock is accompanied by a reduction of pressure and a consequent liquefaction of the rock. Continued folding and the expansion of the magma might permit the molten rock to approach the surface. Where the overlying material is weak, some of the magma would escape to the surface to form lava flows and volcanoes.

Hills and Mountains 273

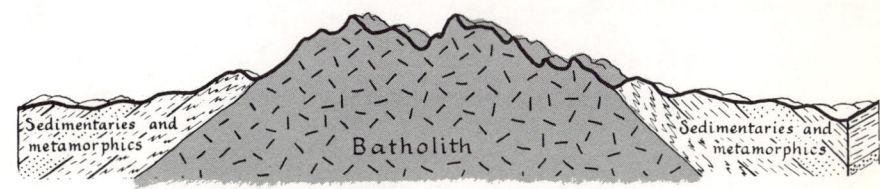

Fig. 17-10. Diagram of a batholith and overlying sedimentary strata. The batholith was formed at depth, but when it was uplifted it tilted the sedimentary and metamorphic rocks on its flanks; the previous cover, on top of the batholith, has been completely eroded away.

Glaciated Mountains

Snowfields and glaciers are significant features on many mountains in high latitudes or in higher mountains in mid-latitudes. On some, the snow is a winter phenomenon; on others, even summer temperatures are insufficient to melt the snowfield. The lower limit of permanent snow decreases from about 15,000 feet (4,500 m) near the equator to 1,000 or 2,000 feet (300 or 600 m) in northern Norway and southern Alaska. Obviously, this latitudinal decrease cannot be uniform, since the climatic factors controlling supply and melting of snow do not have a uniform latitudinal gradient; in any case, they vary from place to place, even in the same latitude. During the Pleistocene glaciations, the snow line was 3,000 to 4,000 feet (1,000 to 1,300 m) lower than it is today.

In the mountains of the western continental United States, the modern glaciers are very small, rarely more than a mile long. In the Alps and the Canadian Rockies, alpine glaciers 5 to 10 miles (8 to 16 km) long are not unusual. The largest valley glaciers occur in the Himalayas, in the mountains of southeastern Alaska, and in the southern Andes in Chile, where they attain lengths of 20 to 40 miles (30 to 60 km).

Snow and glaciers are contemporary features of many alpine mountains, but the Pleistocene glaciers were much larger and left an imprint upon the mountains that is clearly evident today. The cross-valley profiles were frequently changed from narrow V-shapes with overlapping lateral spurs to relatively open U-shapes. The open form of the glacial valleys permits an unobscured view of the impressive landscape of cirques and horns, hanging valleys and waterfalls, tarn lakes and alpine meadows, and remnant glaciers and morainal deposits (Fig. 17-11).

Fig. 17-11. Aerial view of the Bernese Oberland. The Brienz Range, shown in the middle ground, is mostly above 7,000 feet (or 2100 m). The Bernese Alps are on the skyline. The Finsteraarhorn (14,032 feet or 4277 m) is the sharp peak third from the left. The peak at the extreme right is the Jungfrau (13,653 feet or 4161 m). These magnificent peaks are examples of mountains sculptured into sharp-sided horns by alpine glaciation. *Photograph courtesy of the Swiss National Tourist Office.*

Fig. 17-12. Glaciated mountains in Switzerland. The highest peak is the Finsteraarhorn of the Bernese Alps. To the right are the Fiescher Horns. *Photograph courtesy of the Swiss National Tourist Office.*

Glaciated mountains occur in most of the major alpine cordilleras (Fig. 17-12). The northern Rocky Mountains of the United States and Canada, the Sierra Nevada of California, the Cascades, the central and southern Andes, the Pyrenees, the Alps, the Caucasus, and the mountains of central Asia and northeastern Siberia all have extensive areas of glaciated mountain topography.

Nonglaciated Mountains

The rough, extremely rugged mountains are characterized by broken, angular slopes and craggy ridges or crests. The rapidly incising streams create deep canyons with steep slopes and overlapping spurs on the valley sides. Because of the rapid rate of mass wastage along the steep slopes, the regolith is thin and rocky. For this reason, and because of aridity in many localities, the vegetation on crests and slopes is thin or absent. These angular mountains are associated with stream erosion on folded and faulted structures or on rocks of even more complex form. Notable examples include the rugged highlands of western China, the rough mountains of the western United States, and the Sierra Madre Occidental and Oriental in Mexico (Fig. 17-3). There are other areas of rugged mountainous landscape that occur in parts of Turkey and northwestern Iran, southeastern Siberia, the Balkans of southern Europe, and the Andes of northern Venezuela.

Summary

This chapter has considered a wide variety of hill and mountain landscapes. Hills and low mountains with linear relief are most often associated with sedimentary layers of fairly steep inclination that are arranged in synclinal and anticlinal structures. Nonlinear hills with rough relief are closely associated with erosion of elevated surfaces that are composed of either homogeneous surface materials or very complex materials without obvious structural trends. In either case, structure did not control drainage, so that dendritically arranged streams have incised a very rough pattern into the land. Nonlinear hills with smoothly rounded or conical profiles are related to particular combinations of climate and rock type or to their volcanic origin.

High mountains are the product of recent orogenic processes of folding, faulting, and vulcanism. The spectacular appearance and symmetric beauty of high volcanic cones are a distinctive feature of certain mountain ranges or isolated volcanic highlands. No less spectacular and equally distinctive are the glaciated mountain landscapes. Nonglaciated mountains are perhaps less scenic; but they have equally great relief, and their angular and rugged forms have an attraction of their own.

Suggestions for Further Reading

Bullard, F.M., *Volcanoes in History, in Theory, in Eruption*, University of Texas Press, Austin, 1962.

Hunt, Charles B., *Physiography of the United States*, W.H. Freeman and Company, San Francisco, 1967.

King, Lester C., *The Morphology of the Earth: A Study and Synthesis of World Scenery*, Hafner Publishing Company, Inc., New York, 1962.

Thornbury, William D., *Regional Geomorphology of the United States*, John Wiley & Sons, Inc., New York, 1965.

18 The Continental Margins and the Ocean Basins

Introduction

The seaward margins of the continents often consist of three distinct elements. The *landward element*, which bears the evidence of past oceanic transgressions and of the resultant marine erosion and deposition, is the zone of *marine terraces* (steplike benches along the coast of marine origin). The *middle element* is the shore itself, the zone of abrupt transition between land and sea, where marine gradation is presently concentrated. The *seaward element* is the *continental shelf*, that part of the continental surface which is presently covered by the sea.

Beyond the outer edge of the continental shelves lies the ocean basins, consisting of the *continental slopes* and the ocean floors. The ocean floor is not, as was once thought, a nearly featureless plain, interrupted here and there by occasional mountain masses. The recent widespread use of echo-sounding equipment to measure the depth of the sea has shown that the ocean floor is as complex as the continental surfaces.

Sea level has fluctuated considerably. Deep trenches and high ranges on the ocean floor indicate that tectonic processes have raised and lowered the ocean floor over wide areas, thus changing the water-holding capacity of the basins. In addition, during the Pleistocene, sea level changed repeatedly as a result of the growth and melting of continental glaciers. Evidence indicates that sea level fell by nearly 450 feet (140 m) below its present level as land ice expanded during the last stage of the Wisconsin glaciation. When the glaciers began to melt, 15,000 to 18,000 years ago, the sea began to rise. About 6,000 years ago, it reached approximately its present level. Since then, only small changes have occurred; the net result has probably been a small rise in sea level.

To complicate matters, the continental margins of the oceans also have been subject to change in elevation. Tectonic processes operating in the Pleistocene and since then have both uplifted and depressed shore zones along many coasts.

Marine terraces, when they occur, provide evidence of the changing relative position of land and sea. They are terrace features of marine erosion or deposition (or

The Continental Margins and the Ocean Basins 277

Fig. 18-1. Marine terraces on the northeastern side of Curaçao, Netherlands Antilles. The flat surface is a platform of former marine erosion. At least two fossil sea cliffs can be seen in the background.

both) that are found on the landward side of the shore zone (Fig. 18-1). Very probably both land and sea movement were involved in their formation and preservation — the uplift of the terraces above sea level preserving them from destruction by subsequent marine erosion. Continued study of marine terraces may provide conclusive evidence that both of these movements occurred; and, with luck, the individual role of each in terrace formation may be isolated. In any case, evidence accumulating from many coastal zones relates some marine terraces to Pleistocene interglacial high sea levels.

Factors of Shore Formation

The shore can be divided into a number of fairly distinct zones (Fig. 18-2). Proceeding seaward, the following zones are distinguishable: *coast*, *shore zone* (consisting of *backshore* and *foreshore*), *nearshore*, and *offshore*. The coast is the area inland from the sea cliff or beach ridge, extending landward to the point where some major change in landforms occurs — such as a change from a coastal plain to a hill zone. The backshore is the zone from the sea cliff to the shore line, or the line of high tide. The shore line is often marked by a low ridge, called a berm, which represents an accumulation of sand at the inland limit of high tide. The foreshore lies between the shore line and the line of low tide. Beyond the foreshore is the nearshore, which covers the part of the continental shelf that is less than 6 fathoms deep at low water. The rest of the continental shelf is called the offshore (Fig. 18-2).

The shore zone is one of the most active areas of erosion and deposition on the earth's surface. Here, in some localities, the surf batters the land with immense energy, cutting wide platforms and producing high cliffs. Elsewhere, the action of the surf may be predominantly one of construction, forming long barrier islands of sand. In yet other regions, coral organisms may construct shore features of a special type. These and other kinds of shores are the products of a number of factors. The active factors in shore formation include changes in sea level as well as wave action, each of which can produce both erosion and deposition. The passive factors in shore formation include the slope of the submarine platform adjacent to the shore, the hinterland (inland) topography, and the character of the rock of the shore-zone topography.

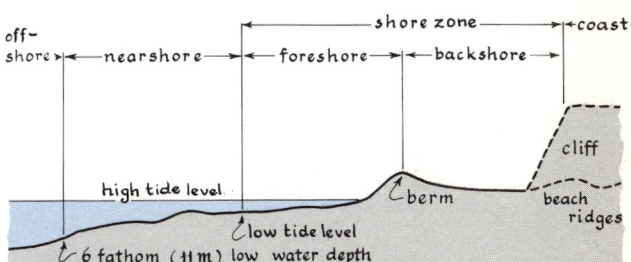

Fig. 18-2. A normal shore profile. Note that the various categories of shoreland are separated by four significant features—the sea cliff or beach ridges, the berm at the point of high tide, the limit of low tide, and the point at which the sea reaches a depth of about 6 fathoms (36 feet or about 11 m).

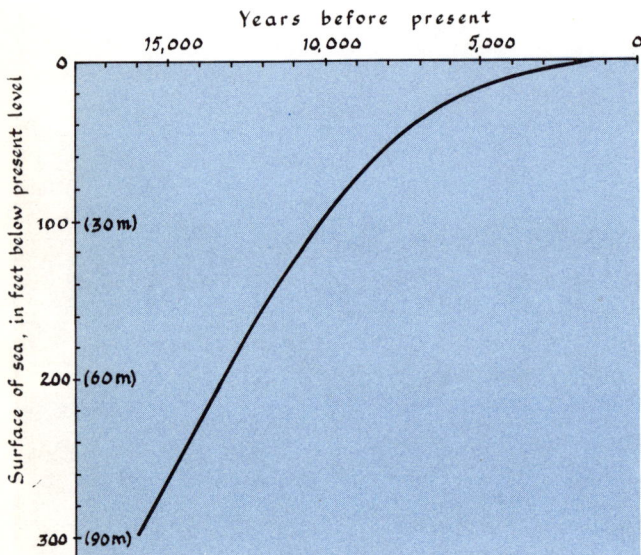

Fig. 18-3. Postglacial rise of sea level. Estimated levels at individual dates have been omitted; the curve is an approximation of the general trend. From F. P. Shepard, "Sea Level Changes in the Past 6,000 Years: Possible Archaeological Significance," *Science*, Vol. 143 (February 7, 1964), 574–576. (Copyright 1964 by the American Association for the Advancement of Science.)

Active Factors in Shore Formation

Changes in Sea Level. The most important sea-level change affecting the development of present shore lines is the postglacial rise associated with the last deglaciation. Carbon-14 dates of submerged shell heaps and of vegetation of known depth below the present sea level indicate that the sea rose relatively quickly until about 6,000 years ago. Since that time, the rate has been much slower (Fig. 18-3), although even now most coasts are gradually being submerged.

Waves and Wave Action. Surf consists of wind-induced waves of oscillation in the open sea that are modified to breakers, or waves of translation, as they approach the shore. The relation of waves to wind and the mechanics of wave motion are complex and difficult topics, but we shall try to summarize the most important aspects of wave formation.

Friction between moving air and water transmits the motion of the air to the water surface. At very low wind speeds, the air flow is laminar or nonturbulent, and the surface of the water remains undisturbed. As the wind speed increases, air flow becomes turbulent, and small disturbances are set up on the surface of the water. Once the water surface becomes irregular, friction between the wind and water increases; and if the wind persists, the waves become larger.

The time available for wave growth depends upon the fetch, or the length of open water over which the wind is blowing. As long as the wind is moving faster in the direction of wave movement than are the waves, each wave acts as a small barrier to wind, and an eddy forms in the lee of the wave. As a result, the wind pressure is greater on the windward side of the wave than on the lee. This pressure difference allows the wave to increase in speed and height, but not indefinitely. For any given wind speed, the height of the wave has an upper limit beyond which increasing fetch produces no noticeable result.

Theoretically, in the open sea, waves have a symmetrical cross section and occur in a series, each wave having nearly the same amplitude and length as its neighbor (Fig. 18-4). But the character of sea waves is, in fact, variable because of the varied conditions of wind velocity and of turbulence in areas of wave generation. Further irregularities result from the superimposition of wave patterns from different directions.

When waves move from their place of generation, those of originally different amplitude and length may combine to form larger waves or may partially cancel each other. This process of intensification or destruction reduces the number of irregularities, and the waves

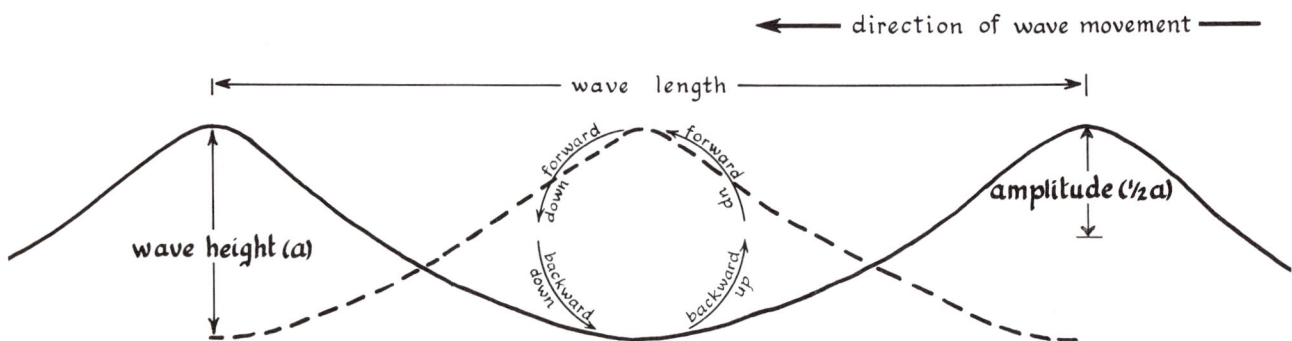

Fig. 18-4. The amplitude and length of a sea wave. The circle shows the path of a water particle at the surface during the passage of a sea wave.

become the smooth and even-crested feature called **swell**.

Swell in the open sea produces very little horizontal movement in the water; the wave shape moves across the surface of the water in the same way that a ripple moves across a rug that is flipped up and down. There is a difference, however, since friction between the wind and the surface of the water produces a rotating movement of water particles near the surface instead of purely vertical movement like that of the rug. As the wave form passes, the water is moving forward in the crest, then down to form a trough, and backward to rise as the next crest approaches (Fig. 18-4). Each particle, at the end of its turn, is only very slightly downwind from its original position. This rotary motion without much horizontal displacement accounts for the fact that objects floating on the sea are not pushed along appreciably by the waves. Frequently, logs and other debris that lie too deep in the water to be affected by the wind move in the direction of water currents that are independent of the wind direction and not in the direction of wave movement.

At the surface, the diameter of the path of the water particle in its rotary motion is equal to the amplitude of the wave. But with depth, the diameter of rotation decreases, so that at a depth approximately equal to one-half the wave length, the motion of the water particle virtually ceases.

The importance of waves to our discussion lies in their behavior as they approach the land — that is, in the transformation of waves of oscillation into surf, or waves of translation. Near the shore, waves are affected by the shallowness of the water. At a water depth equal to about one-half the wave length, the approaching waves begin to shorten and their velocity decreases. At the same time, the wave steepens on its landward side until the wave crest breaks and spills forward. The break probably occurs when the speed of the rotating water particles at the top of the crest exceeds the speed of the slowing wave form. The average water depth at which the break occurs is about the same as the wave height.

When the wave crest breaks, the energy of the wave can be used in eroding, transporting, and depositing coastal material. The prevalence of erosion or deposition depends, in part, on local weather conditions such as storminess and changes in wind direction. In mid-latitudes, summer and winter are often contrasting seasons in this respect; deposition prevails in summer, erosion in winter. The seasonal contrast is probably directly related to differences in wave type. Winter is the stormy season in mid-latitudes; consequently the frequency of steep-crested storm waves increases. Observations suggest that storm waves have a weak surge or swash across the beach relative to the return flow or backwash, and they are therefore conducive to erosion. During the summer the frequency of long, relatively flat waves or swell increases. These have a stronger swash than backwash and the sand they move up the beach face builds up.

Passive Factors in Shore Formation

Slope of the Nearshore Platform. The east coast of the United States is characterized by deposition along its shores; the west coast, by erosion. The main reason for this regional contrast appears to be the slope of the nearshore part of the continental shelf. The average slope of the west coast shelf is about 5 percent; that of the east coast is only 2 percent. Steep slopes mean deep water near the shore; hence the wave crests do not break until they are very close to the land, and their energy is concentrated in a narrow zone. This condition, favorable to shore-zone erosion, is characteristic of the west coast of the United States. The gentler slopes of the east coast cause waves of similar amplitude to break farther offshore. The breaking waves pick up sediment; but as the surf moves across the gently rising bottom, it slows down and deposition takes place, usually in the nearshore. In time, this action produces submerged sandbars.

Hinterland Topography. An important passive factor that determines the local character of most shores is the nature of the land adjacent to the coast. The horizontal plan, or outline, of a shore is largely the result of the action of the rising sea on the coastal terrain during the last deglaciation. For example, a rising sea along a flat coastal plain produces a straight shore line, whereas a sea rising against linear hills perpendicular to the advancing shore produces an irregular shore line. The shape of the resulting bays depends upon the character of the ridges and valleys.

Shore-Zone Rock Types. Among the many rock types whose individual characteristics affect the development of the shore zone, coral limestone deserves special mention. Corals are tiny animals that secrete lime to form an external wall which partially encloses their fragile bodies. They usually live in large colonies, in which enormous numbers of their small honeycomb structures coalesce to form large masses of coral limestone. The coalescence is furthered by several species of tiny lime-absorbing plants, whose tissues accumulate on the coral reef and help to cement it.

The important reef-forming corals need special environments to survive. They require clear, constantly moving water that is warmer than 68° F (20° C). Sunlight is conducive to the growth of coral and is particularly necessary for the growth of the lime-absorbing marine plants. These requirements limit the depth of growing coral to 200 or 300 feet (about 75 m).

Coral reefs are thus restricted to certain tropical shores, where warm water and abundant sunshine are present. Under these conditions, coral growth normally produces land-tied **fringing reefs** and off-shore **barrier reefs** that greatly modify the original shore-zone topography. They are common along the shores of the Caribbean Sea and the southern part of the Gulf of Mexico, around the islands of the tropical part of the Pacific Ocean, including Indonesia, and along the tropical shores of the Indian Ocean.

Coral formations along tropical mainland shores are usually a combination of fringing and barrier reefs enclosing a lagoon (Fig. 18-5). Frequently the base of the coral on the ocean side of the barrier reef is below the level at which reef coral can grow; and yet the barrier reef surface, if not exposed at low tide, is veneered with living coral. How did this coral at the bottom of the barrier reef become established at depths it cannot presently tolerate? The simplest explanation is that the coral became established as a fringing reef during the last glacial period, when sea level was lower. Subsequently, the sea began to rise and the lower levels of the colony died; but new coral continued to grow on top of the old, thus keeping pace with the rising sea. When the sea eventually reached its present high level, the old, land-tied fringing reef became an offshore barrier reef; and a new fringing reef developed around whatever remnant of the island mass still remained above water. This simple sequence of events was complicated by tectonic movement along some of the coasts and by the fact that sea level rose and fell repeatedly, even during a single glacial period.

Coral atolls, which occur in the open ocean away from continental shores, are also difficult to account for. They consist of barrier reefs enclosing lagoons without a central island. The lagoons range in diameter from a few hundred yards to as much as 15 miles (25 km). The lagoons are often partly filled with living coral, and their depth is usually restricted to the narrow range between 200 and 300 feet (between 50 and 100 m). Most of the atolls occur in the Pacific and Indian Oceans, frequently in association with volcanic peaks. Some of these peaks are only small islands rising above the lagoon surface; some have only a narrow band of fringing reef around their base; and some have both fringing and barrier reefs.

Charles Darwin arranged the fringing reefs, the barrier reefs, and the decreasing size of the central islands into a temporal sequence and theorized that this succession could produce an atoll (Fig. 18-6). He believed atolls to be the result of the upward growth of coral upon subsiding volcanoes, wherein the coral grew as rapidly as the land sank so that the living coral was never much below sea level.

A different theory argues that the volcanic foundation of the atoll did not sink but that sea level changed. When the sea was high, the volcanic island basement was planed off by marine erosion to form a broad platform on which coral became established. During glaciation, sea level dropped; and the corals may have perished because of exposure and the lower temperature of the sea. The platform was again planed off at this lower level. When the sea rose during an interglacial period, coral reefs were reestablished around the margins of the platform and grew upward with the rising sea. Once the platform was established, this sequence could recur during each cycle of changing sea level.

This theory explains why atoll lagoons have such a narrow range in depth. It also provides for a thin layer of coral over the platform. However, test holes that were bored through the coral cap rock on Eniwetok Atoll in the Marshall Islands showed that 4,000 feet (1,200m) of reef limestone, which must have developed in shallow water, overlie a core of volcanic rock. In this instance, Darwin's theory of subsidence was substantiated. It does not necessarily mean that all atolls are the product of subsidence, nor does it mean that the role of Pleistocene sea-level changes in the origin of atolls can be ignored. Continuing studies indicate that atolls have complex origins and that both changing sea level and land subsidence were involved in their development.

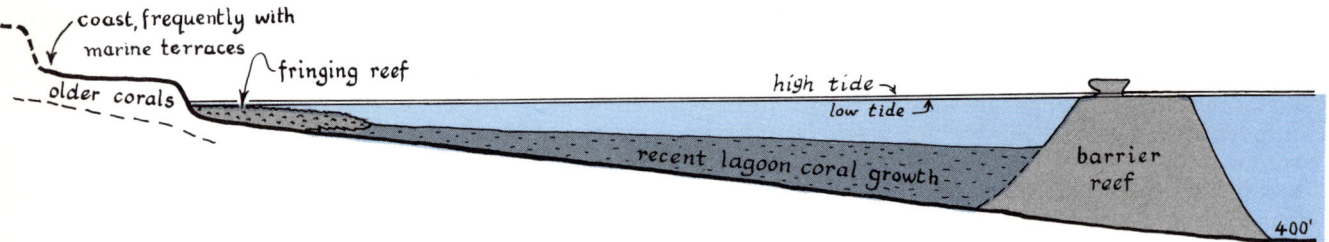

Fig. 18-5. An idealized sketch of fringing and barrier coral reefs as they commonly occur along the coast of East Africa. The fringing reef is tied to the shore, and is separated from the barrier reef by a lagoon whose floor consists of new coral deposits.

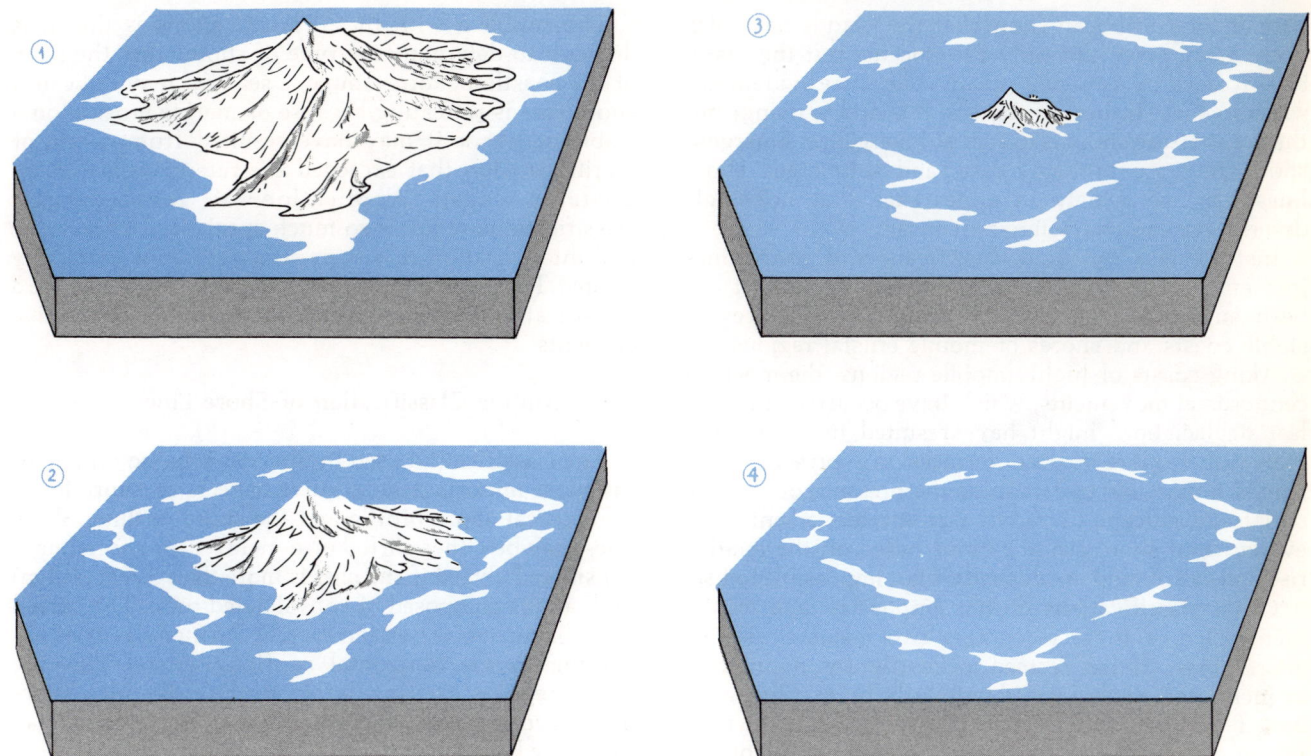

Fig. 18–6. The development of an atoll according to the sequence proposed by Charles Darwin. (1) A volcanic peak projects above sea level and is flanked by a fringing reef; (2) as it subsides, the old fringing reef continues to grow but is separated from the island core by a narrow lagoon; (3) the lagoon widens; and (4) the island core has completely disappeared, leaving only the barrier reef built upon the former fringing coral.

Classifications of Shore Lines

Genetic Classifications

One of the earliest shore-line classifications was proposed by William Morris Davis and subsequently modified by Douglas Johnson. Their arrangement was the essence of simplicity. Shores were classified according to whether their features suggested emergence, submergence, or neither. The three types were assumed to undergo modification by wave erosion in a series of sequential changes.

The Davis-Johnson classification is noncommittal about whether the land or the sea moved. However, the notion of Pleistocene sea-level fluctuations makes the scheme difficult to apply. It is reasonably certain that sea level has risen nearly 450 feet (140 m) during the past 18,000 years; so it is logical to conclude that most coasts should have submerged shores. To be sure, there may be some areas where recent rapid uplift of land has surpassed the rate and amount of sea-level rise, but they are probably rare. We can safely assume that most coasts do have shore lines of submergence. As a consequence, much of the Davis-Johnson classification is meaningless.

Shore-zone processes are probably much more complex than any simple genetic classification would suggest. For example, evidence from marine terraces on the central California coast indicates that the shore line is emerging. The emergence was probably discontinuous and the terraces were eroded while the land stood still. The deformation of the terraces through crustal movement indicates that the coast has been uplifted at least as recently as the last half of the Pleistocene. However, records of wells drilled into the floodplain alluvium of local valleys show that during the last deglaciation the sea rose at least 200 feet (60 m) with respect to the land, which would indicate that the sea rose faster than the land. Thus, despite the fact that the coast of central California has been rising, it has a submerged shore line. The embayments usually associated with a rising sea level are absent because the drowned river valleys have been filled with alluvium. Marine terraces on some of the islands in the southern part of the Caribbean show that the islands have also been rising since the early Pleistocene, but the many bays formed by drowned stream valleys are evidence of submergence. In both instances, the land is rising; but the form of the shore is that of submergence. Emergent shore lines are probably rare, and submerged shore lines vary so widely in appearance that no single description can characterize them all.

Instead of a threefold classification of shore lines into emergent, submergent, and neutral coasts, it has been suggested that they be divided into shores of stable coasts and shores of mobile crustal regions.

Along coasts of highly mobile regions, the most recent crustal movements, which have occurred since the last deglaciation, might have resulted in emergence. However, unless marine terraces are present, it is almost impossible to determine the presence or absence of late Pleistocene or recent tectonic movement. It is evident that shore classifications based on the relative rise and fall of land are difficult if not impossible to use.

Other classifications do not emphasize land movement. One of these recognizes two major classes of shore lines: those formed primarily by nonmarine agencies and those formed primarily by marine agencies. These are further subdivided according to the type of nonmarine and marine agencies responsible for the configuration of a particular shore.

Another twofold classification distinguishes between advancing and retreating shores. The advance of the shore seaward represents the prevalence of emergence and deposition over submergence and erosion; shore retreat represents the opposite. Each primary class is divisible into two subgroups: the advancing class into emerged coasts and outbuilt coasts; the retreating class into submerged coasts and eroded coasts. These subgroups may be further subdivided according to the processes or agents involved; for example, an outbuilt coast may be one formed by the seaward growth of a delta.

The problem with these classifications is that it is difficult to determine the type of shore from the form of the coastline. For example, a sea cliff may indicate a shore that is retreating because of marine erosion on a stable coast, or it may have resulted from a rate of marine erosion that exceeded the rate of uplift on an unstable coast. Application of these genetic classifications requires so much specialized knowledge and interpretation that their effective use is extremely limited. For general purposes, it is more useful to have a classification based on readily recognizable descriptive elements.

Descriptive Classification of Shore Lines

The classification used in this text is based on two of the most important descriptive aspects of shore lines: the vertical shore profile and the shore outline. Shore lines may be either cliffed or noncliffed. A shore line is considered cliffed if the cliff is more than 5 feet (1.5 m) high and is composed of consolidated rock. The cliffed and noncliffed classes can each be further classified according to the horizontal configuration — they are either regular or irregular in outline. Regular shore lines have generally smooth outlines, although in detail they may have small irregularities, or **crenate bays,** that are open to the sea. Irregular shores are those having numerous bays of varying size and shape that are sheltered from the sea. It is evident that the boundary between the regular and irregular types is related to the size of the bays. If, for each unit of bay depth (inland penetration), there are at least five units of unembayed shore line between bays, the shore belongs to the regular class; otherwise, it is classified as irregular.

The regular and irregular classes are broad types that can be subdivided according to local conditions. Regular shores may include crenate shores, smooth and nearly straight shores, and lagoon shores with barrier islands of sand or coral. The barrier islands are in-

Table 18-1. Descriptive Classification of Shore Lines*

	Regular			Irregular				
	Smooth	Crenate bays	Lagoons	Cone-shaped bays	Pouch-shaped bays	Rectangular bays	Elongated bays	Composite lagoons
Cliffed shore lines	18-7	18-8	18-9 †	18-10	18-11		18-12	†
Noncliffed shore lines		18-13		18-15 ‡				18-16

*The numbers refer to the figures illustrating the major types.
†Rare, since barrier islands are usually coral reefs, which are noncliffed.
‡Sand and coral-reef barrier islands.

terrupted by tidal inlets which are most widely spaced where the tidal range is greatest. All of the regular subtypes can be found in association with both the cliffed and the noncliffed shores, although lagoons are rarely found along cliffed shores. The types of irregular shores are identified by the shape and dimensions of the bays in a given area or by the presence of composite lagoons, irregular shores with barrier islands. Some of the more common bay types are long, relatively narrow, cone-shaped bays called **ria bays** (sometimes with branching tributaries); pouch-shaped bays; rectangular bays; and narrow, extended estuaries, frequently branched, which may be called **elongate bays.** The classification is shown in Table 18-1.

Types of Shores

We now turn to the study of selected shore types identified by the descriptive classification, beginning with the primary distinction between cliffed and noncliffed coasts. In general, cliffed shores are associated with highlands and relatively steeply sloping nearshore submerged platforms. They are characterized by steep, wave-cut cliffs and by a general absence of features of wave deposition. Noncliffed coasts are usually associated with gently sloping nearshore submerged platforms and with low hinterlands. They are characterized by general prevalence of features of wave deposition, such as wide beaches, offshore bars, and sand barrier islands. Cliffed coasts can have features of deposition, and noncliffed coasts can have features of

Fig. 18-7. A cliffed shore on the north coast of Curaçao, Netherlands Antilles. The coral limestone, undercut by wave abrasion and solution, has broken off in large blocks and fallen into the sea. The blocks check further wave erosion of the cliff until they have been worn away by the sea.

Fig. 18-8. The cliffed regular coast of northern California in the vicinity of Point Delgada. The coastline is parallel to the trend of the coastal mountains, hence the sea finds few weak or low spots to attack; instead the whole shore is eroded into a straight coast with cliffs, as the ocean wears away the mountain ranges.

marine erosion; however, long-term deposition and conditions favorable to cliffing tend to be mutually exclusive.

Coral features can become established on both cliffed and noncliffed shores wherever the conditions for reef coral growth are favorable and the sea waves are not too turbulent. Even in the case of cliffed shores, features of coral origin can occur as both barrier and fringing reefs at the same time that a moderate amount of mechanical and solution erosion takes place at the foot of the cliff.

Cliffed Shores

Regular Cliffed Shores. Cliffed shores are sometimes smooth, especially in regions of uniform rock or where the supply of beach sand is relatively abundant (Fig. 18-7). More frequently, they are crenate with few headlands or promontories and with embayments that are open to the sea. Beaches are restricted to the embayments or occur near the mouths of streams. The hinterland usually rises rather steeply from the shore to form hills, mountains, or plateaus, while the grain of the hinterland is parallel to the shore (Fig. 18-8). Examples of this type of shore line are found along the west coast of North America and South America between 45° N and 42° S latitudes. There are few natural harbors, and nearly everywhere the highlands are at the edge of the sea; seaward, the continental shelf is narrow and has a relatively steep slope. Cliffed regular lagoon coasts are found in restricted areas where coral barrier reefs can grow. Examples occur along most of the tropical coast of East Africa and along the shores of many of the islands in the Philippines and Indonesia (Fig. 18-9).

Fig. 18-9. A cliffed shore formed in coral limestone on the mainland coast of Tanzania. Cliffing has occurred despite the presence of coral-reef barrier islands, 3 to 5 miles (5 to 8 km) offshore.

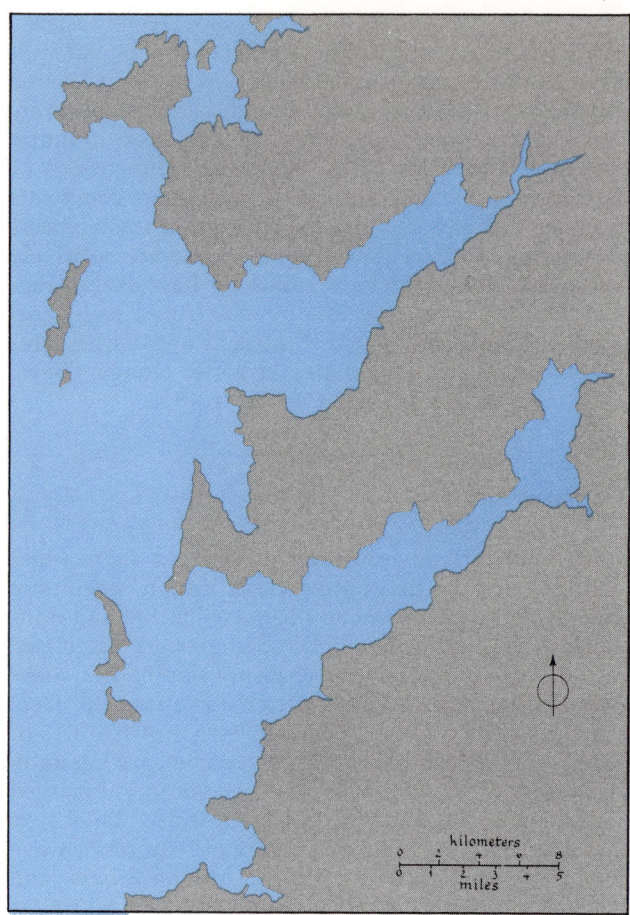

Fig. 18-10. A portion of the cliffed ria coast of northwestern Spain. Here the grain of the land is at right angles to the coast. Hence there are cliffs only at the headlands between the long, relatively shallow bays which were flooded by rising interglacial sea levels.

Irregular Cliffed Shores. Irregular cliffed shores are subdivided according to bay shape, which in turn depends upon the nature and kind of valley found in the hinterland. The first prerequisite for irregularity is that the grain of the land must intersect the coast at an angle. If the valleys are relatively narrow with gentle gradients, the rising interglacial sea flooded them to form long, relatively shallow bays, and the shore became irregular and cliffed with cone-shaped bays. Cliffing occurs only on the headlands between the bays; there is little or no erosion within the bays because of the loss of energy by the wave front as it spreads out upon the bay shores. Classic examples of this shore are found in northwestern Spain, where cone-shaped bays or ria bays commonly penetrate deeply into the land (Fig. 18-10). Similar shores are also

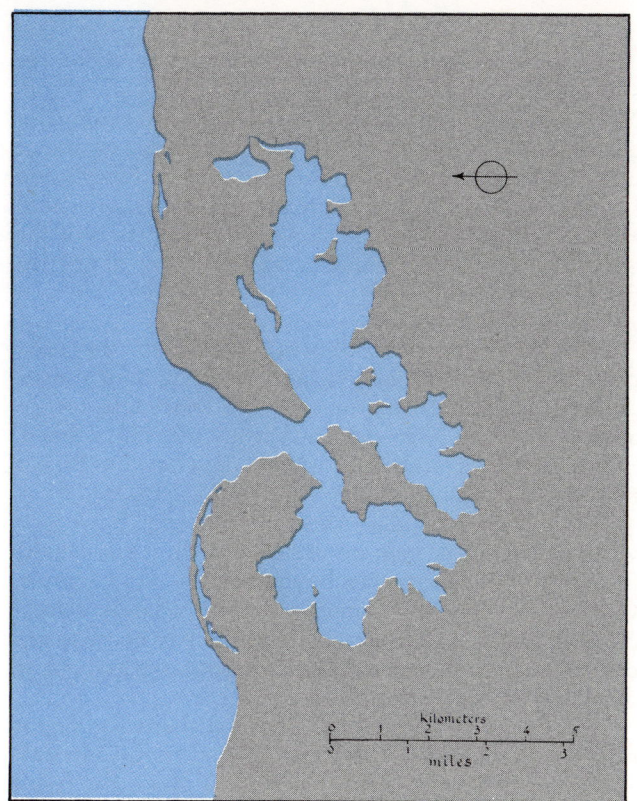

Fig. 18-11. Pouch-shaped bays on the north coast of Cuba west of Havana. A rim of coral protects the shore and keeps the entrance to the bay narrow.

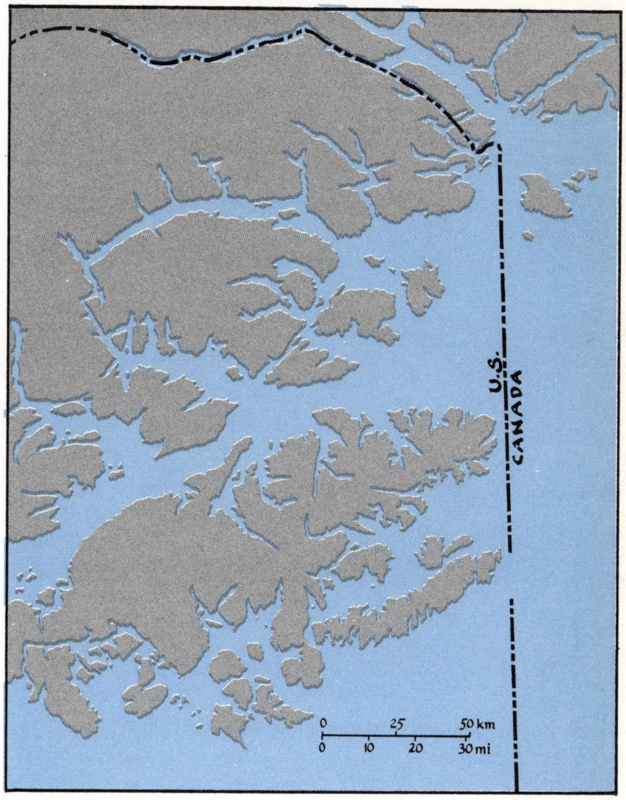

Fig. 18-12. A portion of fjord coast of British Columbia. Glaciers scoured deep, narrow and steep-sided valleys which were subsequently invaded by rising seas during the present post-glacial period.

found on the coasts of Greece, western Turkey, and western Korea.

Irregular cliffed shores with rectangular and pouch-shaped bays are also the result of the interaction of landforms and rising sea level. The coast of China south of the Yangtze River is a good example of a rectangular embayed shore. This is an estuarine shore consisting of drowned stream valleys, but the bays have blunt heads and an angular form. The shore of northern Cuba, where a rim of coral limestone borders the coast, has pouch-like bays with narrow entrances and wide embayments behind the limestone rim (Fig. 18-11). This type is common along tropical shores bordered by elevated coral reefs.

Irregular cliffed shores may also have elongated bays that are deep and wind inland for many miles as narrow inlets. The inlets are bordered by highlands that drop steeply to the shore. The deep trough-like character of the bays and their generally smooth sides, together with the alpine topography of the surrounding highlands, suggest that the bays are drowned glacial valleys. During the Pleistocene, glaciers scoured deep U-shaped valleys in the highland margins of the sea, often below the present sea level and even below the lower sea levels of the glacial periods. When the ice melted, the sea rose and invaded the valleys, producing **fjords** (Fig. 18-12), which are often deeper immediately beyond their mouths than the ocean. Fjords occur mainly along west coasts in high mid-latitudes — in British Columbia and southern Alaska, in Norway, along the southern coast of Chile, and on the west coast of South Island in New Zealand. Similar shores also occur on the coast of Labrador and in Greenland, where some of the valleys are still filled with ice.

Noncliffed Shores

The noncliffed shores can also be divided into regular and irregular types. While the various subdivisions of these two types of noncliffed coasts (Table 18-1) do appear in nature, the most important examples involve regular lagoon shores and irregular composite lagoon shores. The major emphasis of the following discussion will, therefore, be on these two kinds of

Fig. 18-13. A noncliffed smooth shore along the central coast of mainland Tanzania. Note the relatively straight but low beach ridge.

shore lines; less attention is given to other kinds of noncliffed coasts.

Noncliffed Regular Shores. Regular shores are either smooth or lagoonal. The smooth shore consists of gently sweeping curves and is composed mainly of sandy beaches with a backing of sand ridges or sand dunes (Fig. 18-13). These shores are rare because the gently sloping nearshore platform that leads to an absence of cliffs also encourages the development of sandbars that eventually form barrier islands. This development results in a lagoon coast rather than a smooth one.

In the development of lagoon coasts, the transition from a sandbar that is still submerged to a sand barrier island may be accounted for by a slight drop in sea level. The transition may also be brought about by a temporary rise in sea level during a period of strong onshore wind. The wind not only raises the level of the sea slightly, but also increases wave activity, so that sandbars may be built up enough to stand above normal sea level. But strong winds and waves also frequently destroy sandbars, so it is only transitory storms of exactly the right intensity that can transform sandbars into offshore islands. In any case, sand barrier islands lead a precarious existence; many are formed but few survive. Those that do survive soon acquire wave-built beach ridges and sand dunes that gradually widen the island (Fig. 18-14).

Noncliffed regular lagoon shores are usually found in areas of gently sloping nearshore platform, low hinterland, and abundant sediment. These occur for short distances along the Atlantic and Gulf coasts from New Jersey to Yucatán (Fig. 18-15). They are also characteristic of the south shore of the Baltic Sea and of much of the coast of West Africa between Liberia and the Niger Delta. Noncliffed regular lagoon shores with coral barrier islands occur along low-lying parts of the east coast of Central America and on some of the Indonesian islands.

Noncliffed Irregular Shores. Many of the noncliffed irregular coasts have lagoons along their shores. Along parts of the Atlantic and Gulf shore line, for example, the rising sea drowned a gently rolling plain. Subsequently, sand barrier islands developed to form a noncliffed composite lagoon shore (Fig. 18-16).

Noncliffed irregular shores with ria bays and without barrier islands do occur; but they are usually found in rather large embayments, as, for example, Chesapeake Bay. Presumably, barrier islands cannot develop in the protected waters of the bay because under such conditions the waves are not large enough to deposit sandbars.

Another type of irregular shore without lagoons is called a **fjard coast.** It is characterized by an intricate shore line with many long, branching bays, the result of a sea rising against a plain of glacial erosion. It may be cliffed or noncliffed and the headlands between bays are low and rounded. Usually, fjard coasts are associated with gently sloping continental shelves and low hinterlands; offshore islands are plentiful, but offshore bars are rare. Good samples of fjard shores occur on the coasts of Maine and southern Sweden.

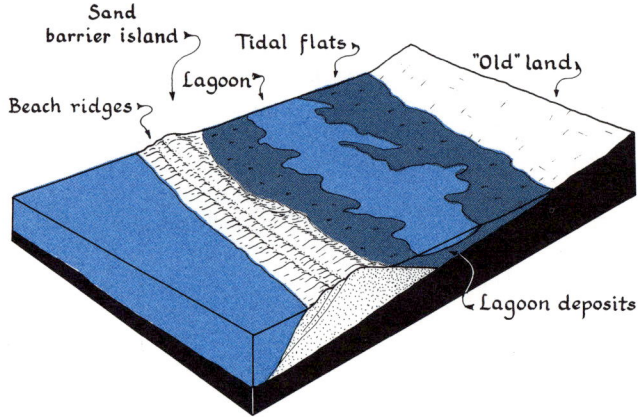

Fig. 18-14. A sand-barrier island with several wave-constructed beach ridges. The offshore island is easily moved or destroyed by wind and tide, since it consists only of loose sand. The lagoon behind the barrier island is also an ephemeral landform.

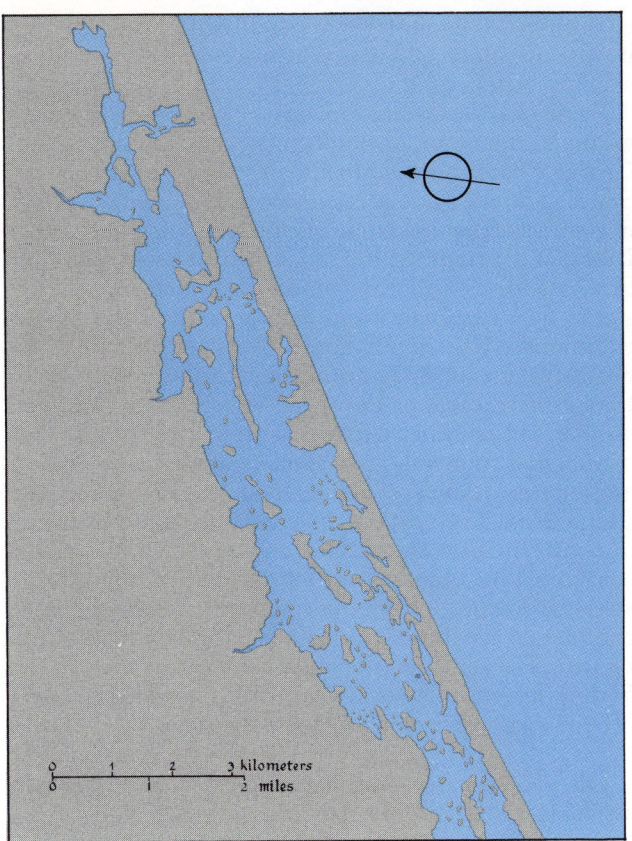

Fig. 18-15. A noncliffed regular lagoon shore on the coast of North Carolina. A low hinterland and many rivers contributing sediment created this complex of lagoon and barrier island. Sediment from the continent keeps the barrier island alive, while ocean currents create its smooth seaward shoreline.

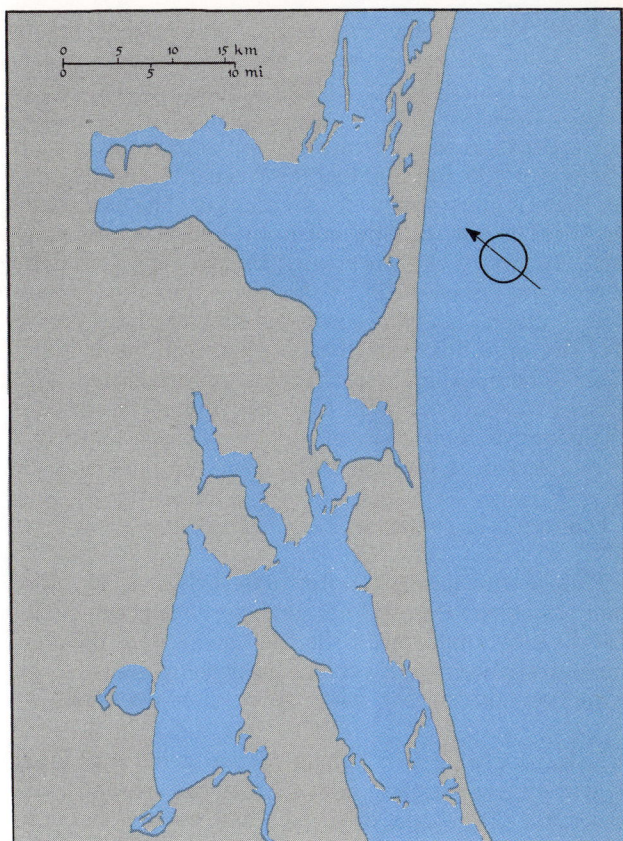

Fig. 18-16. A noncliffed lagoon ria shore on the coast of South Carolina. Rising seas drowning a rolling plain have created this irregular lagoon and ria coast. The features are much coarser than in Fig. 18-15, particularly if we keep in mind the difference in scale between the two figures.

The Continental Shelf

We have discussed two of the elements of the margins of the continents — the landward element or marine terraces, and the middle element or shore. The continental shelf is the third major element of the ocean-continent margin. The landward margin of the shelf plays a critical role in the development of the shore line, as we have seen. Seaward, the shelf is an extension of the continent and, in some instances, is an unbroken continuation of the adjacent mainland coastal plain. The continental shelf ranges in width from less than 5 miles along portions of the Pacific coast of the Americas to a width of 240 miles (390 km) off Newfoundland and almost 200 miles (320 km) off the Amazon River. The tectonic map (back endpaper) shows that the shelf is present along nearly all coasts of the world, but with considerable variations in width. In general, the lands bordering the Indian Ocean and the eastern margins of the Atlantic and Pacific Oceans have relatively narrow continental shelves, whereas the lands on the western borders of these oceans have wide shelves. The Arctic Ocean and the neighboring North Sea and Bering Sea also have relatively wide continental shelves, as do the northern and southern margins of Australia.

A relationship appears to exist between narrow shelves and tectonic instability, on one hand, and wide shelves and tectonic stability and stream deposition, on the other. Narrow continental shelves frequently border regions of high mountains in zones of contemporary mountain building. The steep slope of the continental shelf in these tectonically active zones accounts for its narrowness. On the other hand, shelves off large rivers, especially in closed seas where deltaic deposition is occuring, are quite wide; examples are found in the

Gulf of Siam, in the Yellow Sea, and in the Persian Gulf. The shelf off the relatively stable eastern coasts of North America and South America is also quite wide.

The depth of the continental shelf at its outer margin is between 400 and 500 feet (between 120 and 150 m), but there is considerable local variation. The surface of the shelf is not smooth and unbroken but has many small irregularities; it is crossed by low ridges, shallow basins, and occasionally by shallow, submerged stream valleys. Such valleys occur in the shelves of the North Sea, the Java Sea, and off the mouth of the Hudson River. These submerged channels are associated with present-day rivers on the land and represent seaward extensions of their channels that were eroded during a glacial period of low sea level.

Economic Importance

The distribution of continental shelves is of more than academic interest. First, they are the setting of the world's most important fisheries; and second, the continental shelves have become important areas of oil extraction during the past two or three decades.

Fish. Oceanic fish belong to two general groups: (1) demersal, or bottom fish, such as codfish and haddock; and (2) pelagic, or surface fish, such as herring, pilchard, and tuna. Demersal fish are confined to regions of the continental shelf or reefs, while pelagic fish can live in the open sea wherever there is adequate food. Neither type is randomly distributed in the ocean; the greatest number are found where there is abundant food.

Fish, like land animals, depend on plants to transform minerals into food. The mineral food is taken up by microscopic plants. Small crabs, barnacles, sea worms, and tiny fish all graze on these many groups of algae. In turn, countless numbers of small flesh eaters prey on the grazers. This great complex of tiny plants and animals is called plankton. Such fish as herring, mackerel, and sardines eat the plankton and, in turn, are preyed on by larger fish. The flesh eaters die, their bodies decay, their minerals return to the sea, and the cycle starts again.

Food-rich regions in the open oceans are those with abundant minerals and oxygen. Such areas tend to occur in mid-latitude seas, where there is seasonal overturn of water associated with winter chill, so that the consequent mixing brings a renewal of minerals to the surface. These regions are also found where ocean currents of unlike temperature mix with considerable turbulence. Upwelling water along certain coasts (see p. 298) also provides an abundance of mineral food and marine life. The seas on continental shelves are provided with great amounts of mineral food by streams and by the marine vegetation that thrives in the relatively shallow waters of the shelf.

Some shelf areas are especially rich in minerals. For example, the Grand Banks off Newfoundland and George's Bank off the northeastern coast of the United States combine wide areas of shelf with a large supply of minerals from the land. The juxtaposition of the cold Labrador Current with the warm waters of the Gulf Stream adds to the local abundance of minerals and oxygen. This fortunate combination is the basis for an enormous fish population. The rich fisheries of northwestern Europe around the North Sea and along the northern coast of European Russia are also located on a broad continental shelf supplied by rivers with continental minerals. A similar situation is found around the islands of Japan and in the Yellow and China Seas. The northern part of this region is a zone of contrasting ocean currents on a broad shelf. Off the coast of China, minerals are supplied by the many large streams flowing out of the continent. There are other continental shelf regions which are not yet fully exploited. They include the great continental shelf in the Bering Sea and the relatively unused continental shelf off the east coast of southern Argentina.

Oil. Oil in the Continental Shelf has become an available resource only during the last 20 years or so because drilling beneath several hundred feet of water involved new techniques. But today, oil and natural gas trapped in the sediments of the continental shelf are attracting worldwide economic interest. In the United States, the intensity of the debate between the individual states and the federal government about jurisdiction over the offshore zone is largely based on its oil wealth. Oil-bearing geological structures occur in the shelves off the southern California coast and off the Gulf coasts of Texas, Louisiana, and Alabama. Potential production may be great. For example, the area of the continental shelf along the Texas coast where conditions are favorable for oil accumulation is as large as the state of Illinois. Similar attention has recently been focused on the North Sea, particularly with respect to discoveries of natural gas. Other shelf areas have not yet attracted as much attention, but there are some authorities who believe that the offshore waters will become the most significant areas of oil extraction in the future.

Continental shelves raise many questions of interstate and even international sovereignty, since the limits of national jurisdiction over offshore waters were established prior to the realization of their economic importance. The continental shelf may extend as much as 100 miles (160 km) beyond the limits presently claimed by most governments. The question of jurisdiction over these potential sources of oil may be difficult to adjudicate.

Fishing in international waters on the shelf has also created problems in the past and may become even more troublesome as the world population increases. The legal questions about this activity are complex and, in fact, may be insoluble within the framework of our present international laws. For instance, is it proper for Soviet fishermen to appear in large numbers off the coasts of Oregon and Washington, even if they observe our 3-mile limit? What about American fishermen off the coasts of Mexico? Is it proper for a nation to drill oil wells in an internationalized section of a continental shelf that is, from a physical standpoint, a part of another nation's mainland? Such problems may become matters of urgent international concern in the future.

The Ocean Basins

The ocean basins begin at the edge of the continental shelf and are composed of two parts: the continental slope and the ocean floor. Whereas the continental shelf is merely a part of the continental land mass which happens to be currently flooded by the sea, the continental slope is the rim of the ocean basins and as such belongs to the truly oceanic world. The proper boundary between continents and oceans is thus the outer edge of the shelf, or the top of the basin rim. The ocean floor lies at an average depth of about 12,000 feet (3,600 m) below sea level, but has a complicated surface of submarine troughs and crests.

The Continental Slope

The average height of the continental slope above the ocean bottom is about 12,000 feet (3,600 m), and its greatest height is about 30,000 feet (9,000 m). The inclination of the slope is gentle, rarely exceeding 6 percent, but it does not represent a smooth decline to the ocean depth, since the surface is broken by numerous ravines and submarine canyons.

These very large valleys, or submarine canyons, are difficult to explain. They are found all over the world and, in some instances, are cut into relatively recent sediment. Consequently, the canyons cannot all be ancient features preserved by the covering sea; some of them are apparently still in the process of formation.

There are two major types of submarine valleys on the continental slope: **shelf notches** and **California canyons.** The notches consist of large slots that have been eroded into the outer edge of the shelf for a few miles but extend down the slope to great depths. They have steep walls, as much as 8,000 feet (nearly 2,500 m) high, and their bottom gradients are similar to those of subaerial canyons. A few notches have shallow headward extensions across the shelf to the vicinity of large river mouths.

Other submarine canyons, first described from the California coast, have a branching pattern of tributaries. They extend shoreward into the shelf and seaward to depths of 8,000 feet or more (2,500 m). Canyons of this type have gradients similar to youthful streams, and their branching tributaries make accordant junctions with the main canyon. In addition, there are no breaks in gradient where the canyons cross active fault zones. They appear to be experiencing continuing formation.

The processes that create the notches and the California canyons are not entirely understood, although there have been some interesting speculations. Lowering sea level by a "mere" 10,000 feet (3,000 m) would, of course, allow for the possibility that the canyons were cut by subaerial streams. Such a theory, however, needs to account for the removal of nearly a third of the water in the ocean basins! Another explanation hypothesizes an enormous upwarping of the continental margin. Erosion on the seaward side of the uplifted margin could, of course, carve out canyons, but there is no evidence of the erosion and deposition which would also occur on the landward edge of the upturned continental margin.

Turbidity currents are perhaps the most satisfactory explanation of submarine canyons. The downslope flow of water that is denser than its surroundings is analogous to nocturnal valley breezes of chilled highland air that displace the lighter air of the valleys and flow into the lowlands. A turbidity current is set in motion when a mass of water is colder or has more sediment than the surrounding water and is, therefore, denser. The unclogged character of the canyon floors seems to indicate that turbidity currents are presently sweeping the submarine valleys clean, but they may have been even more effective during the Pleistocene.

During the Pleistocene, some streams transported enormous amounts of debris from the glacial meltwaters, so that pools of turbid water collected on certain parts of the continental shelf, especially in the depressions. The water in these pools perhaps eventually flowed through depressions in the edge of the continental shelf and rushed down the continental slope with considerable erosive force. Since the turbidity currents were concentrated at certain places (wherever there was a low point in the continental shelf), renewed action of such currents in the same channel might eventually have cut the deep submarine canyons and their accordant tributaries. Continued flow since the Pleistocene would account for the unclogged floors and the lack of concordance between canyons and fault zones.

The canyon that extends from the estuary of the Congo River across the narrow West African shelf is of unusual interest in determining the age of submarine canyons. The mouth of the Congo, in the not-too-distant past, may have been several hundred miles to the north. Uplift of the western edge of the African land mass dammed the Congo, forming a large lake. The Congo then spilled across a low point in this natural dam and so took its present course to the sea. In the process, the river cut through the rim and drained the lake. The exposed lake sediments are not very old, since they contain records of early Pleistocene human settlement. Therefore, if the Congo did change its course, the event occurred relatively recently in geologic time, and the great slot that extends into the Congo estuary must also be recent.

The Ocean Floor

The ocean floor is probably as diverse a surface as the land; it is marked by many ridges, peaks, seamounts, deeps, and basins. Some of the major features deserve special mention. The Mid-Atlantic Ridge is a long, curving mountain range extending from the North Atlantic floor to the continent of Antarctica. This mountain range has a local relief of 5,000 to 10,000 feet (1,500 to 3,000 km) with respect to the ocean bottom and is one of the world's great mountain ranges. This underwater cordillera and other similar, submerged ridges are the result of the appearance of new crustal material on the ocean floor by the mechanism of thermal convection in the earth's outer crust (see Chapter 10).

Another spectacular feature of the ocean floor is the ocean deeps. The Bronson Deep off the north coast of Puerto Rico has a depth of 28,500 feet (8,700 m) below sea level and nearly 16,000 feet (4,900 m) below the general surface of the ocean floor. Other deeps occur on the rim of the Pacific Ocean, off the west coast of South America, off the east coast of the Philippines, and off the east coast of Japan. They are not limited to the ocean margins but are also found in the western floor of the Pacific Basin. Some of these have depths of 20,000 to 35,000 feet (6,000 to 10,000 m). The deepest yet discovered is the Marianas Trench of the South Pacific, 36,198 feet (11,033 m) below sea level. The role of deeps in providing troughs into which old crustal material is disposed of downward into the earth's core is briefly discussed in Chapter 10.

In addition to these deeps on the Pacific floor, numerous small basins are set apart by undersea ridges. Recent studies off the coast of California reveal a series of parallel escarpments running almost due east and west. Each escarpment is about 500 to 600 miles (800 to 1,000 km) from its nearest neighbor and appears to be the result of undersea faulting.

Other features of interest are **seamounts,** or flat-topped mountain peaks. Some are several thousand feet beneath the surface, while others are only a few scores of feet from the surface. The truncated seamounts have the same shape as volcanic cones whose tops have been planed off by wave action. These and other features, most of which have been discovered recently, give the ocean basins a diverse topography suggesting that tectonic processes have been as active there as they have been on the continents.

Summary

In this chapter we discussed the components of the continental margins and the ocean basins. The continental margins consist of three areas: the marine terrace zone, the shore zone, and the continental shelf. The terrace zone, where present, reflects the changing relative position of land and sea. The shore zone is the area of present marine erosion and deposition, which, because of its dynamic quality, received the greatest emphasis in the discussion. The factors of shore formation which influence the rate of marine erosion are of two kinds — passive and active. The passive factors are the slope of the submarine platform, the hinterland topography and the nature of the rock of the shore zone. The active factors are wave action and changing sea level. The third area of the continental margins, the continental shelf, is narrow where tectonic action is strong and wide where tectonic action is weak. It is associated with jurisdictional problems arising because of the economic importance of these offshore waters with respect to fishing and oil deposits.

The ocean basins consist of two parts: the continental slope and the ocean floor. Tectonic and gradational processes have given the ocean basin a highly diversified form, including the presence of submerged mountain ranges and ocean deeps, both of which are associated with the formation and destruction of new crustal material.

Suggestions for Further Reading

Guilcher, A., *Coastal and Submarine Morphology*, John Wiley & Sons, Inc., New York, 1958.

Johnson, D. W., *Shore Processes and Shoreline Development*, John Wiley & Sons, Inc., New York, 1919.

King, C. A. M., *Beaches and Coasts*, St. Martin's Press, Inc., New York, 1960.

Shepard, F. P., *Submarine Geology*, 2nd ed., Harper & Row, Inc., New York, 1963.

19 The Oceans

Introduction

For a long time, man's view of the sea was quite simple. He thought of it as a source of food, as a means of transportation, or as a barrier. In the past century, appreciation and understanding of the complexity of the seas have greatly increased. For example, the impact of the oceans on weather and climate is a subject of great importance and interest. The oceans are enormous reservoirs of heat, moderating both seasonal and latitudinal contrasts of air temperature. At the same time, they are the fundamental source of all the moisture that falls on the land. The oceans also contain vast supplies of minerals; nearly all the minerals available on the land exist in appreciable quantities in solution in the seas. Man has extracted salt from the ocean by evaporation for thousands of years. Recently, chemical techniques have been devised to extract magnesium from seawater, and many other minerals no doubt will be similarly extracted in the future.

Potentially, the most important product of the ocean is fresh water, which is being distilled today, on a small scale, in many arid parts of the world. However, man is on the threshold of discovering an economical method of large-scale extraction, which may be a fortunate development considering the present rates of fresh-water pollution and population increase.

The ocean has long been a source of food in the form of fish, shellfish, and sea mammals. The sea may also become important for the production of algae and seaweed, which can be processed into a variety of foods. The Japanese have already made remarkable progress in this field. With further refinement of processing techniques, the oceans will become an even greater source of food.

The seas provide an excellent means of transporting bulk cargo — it is much cheaper to ship wheat, iron ore, coal, and oil by water than by land. In the future, transportation methods for bulk cargo may change. For instance, cargo submarines with atomic propulsion systems could haul cargo rapidly and smoothly — well below the potentially stormy surface. And great plastic bags floating partly submerged and towed by tugs could be used to transport oil.

It is evident that the oceans are of great importance to man and, as such, merit the careful consideration of geographers. More must be learned of their composition, surface characteristics, movement, and animal and mineral resources. It is ironic that we have recently been learning more about the surface of the moon than we know about the ocean.

Characteristics of Ocean Waters

The water surface of the earth can be divided into regions in the same way as the land surface, although the criteria establishing the regions are obviously different. Regional oceanic distinctions can be made on the basis of temperature, salt content, and movement.

These are criteria applied to the ocean-surface layer, which rarely exceeds a depth of 1,000 feet (300 m). This layer has the greatest influence upon weather and climate and is the part of the ocean in which most marine life is found. As nearly as we can determine, deep-water characteristics have little immediate influence on the atmosphere or the land, except where water wells up from medium depths. Nor do we know much about the deeper parts of the oceans. The emphasis in this chapter, therefore, will be on the surface layers rather than on the oceanic depths.

Physical Characteristics

Composition of Seawater. The oceans are not just salty water; they are a complex chemical solution of many minerals derived from the earth's crust and of organic salts derived from the decay of plants and animals. These materials amount to about 3.5 percent of the total weight of the ocean water. Sodium and chlorine, which produce the common salt of the sea, are the most important elements, comprising approximately 86 percent of the total weight exclusive of hydrogen and oxygen. Next in importance are magnesium (3.7 percent), sulfur (2.6 percent), calcium (1.2 percent), potassium (1.1 percent), and bromine (0.2 percent). Note that these percentages refer only to the solid substances found in ocean water. They indicate a very small concentration of solids, since the dissolved matter amounts to only 3.5 percent of the weight of the water. Other elements occur in even lesser amounts. Interestingly, sodium, calcium, and potassium are present in seawater in almost the same proportions as in human blood, an indication perhaps that enclosed within man's skin is a marine solution that nourishes him as his pre-terrestrial ancestors were nourished in the sea. Nutrient salts are also present in seawater — the phosphates, nitrates, and nitrites which are used directly by plants and indirectly by animals for nourishment and growth.

Salinity. Salinity refers to the concentration of all these dissolved solids in a given sample of seawater and is expressed as so many parts of salts per thousand parts of water (written as $^0/_{00}$). This concentration varies with time and place because of differences in evaporation and in the amount of fresh water added by rain and streams. The normal salinity of the surface layers in the open ocean is between 33 and 37 parts per thousand. It is relatively low in high latitudes, increases to a maximum near 30° latitude, and decreases from there to the equator (Fig. 19-1).

This latitudinal arrangement of salinity is related to the gross distribution of evaporation and precipitation (Fig. 19-2). In high latitudes, the rate of precipitation and ice melt is relatively high compared with the low

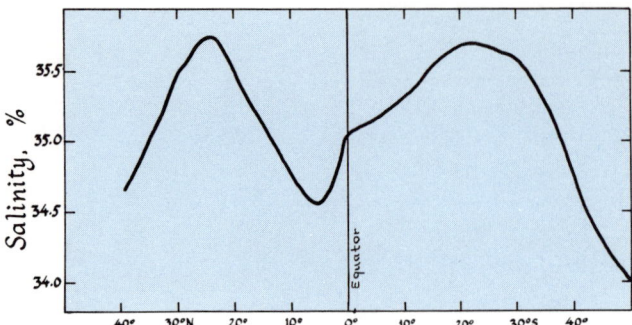

Fig. 19-1. The variation of the salinity of surface water with latitude. Modified from Sverdrup, Fleming, and Johnson, *The Oceans*.

rate of evaporation. The Arctic Ocean has particularly low salinity because it is fed by numerous large rivers. The high salinity of the lower mid-latitudes results from the high rates of evaporation and the low rates of precipitation that are associated with the belt of subtropical high pressure. Enclosed seas in this latitude, such as the Mediterranean and Red Seas, are particularly saline because their relative shallowness results in high surface temperatures (and evaporation) and a greater concentration of salts than in a deeper, cooler ocean. The general dryness of this zone has an additional effect in that stream flow is small compared with that in other latitudes. The equatorial seas, on the other hand, are rainier and cloudier and are fed by large rivers. The combination of a larger supply of fresh water, a higher rate of precipitation, and a lower rate of evaporation makes these waters much less saline than the neighboring subtropical oceans.

Superimposed on the latitudinal distribution of surface salinity is the fairly strong contrast between the Atlantic and other oceans. The smaller Atlantic is considerably saltier than the Pacific Ocean. It is also, of course, much more saline than any of the polar seas.

Temperature. Temperature is one of the critical characteristics that distinguish one water mass from another. Differences in temperature cause differences in water density, since warm water usually has a lower density than cold water. If the surface water becomes cooled, the dense layer sinks; and warmer water rises to replace it in a convectional overturn of water. Unlike fresh water, seawater continues to grow denser until the freezing point is reached; upon freezing, freshwater ice is formed, intensifying the local salinity and causing the local seawater to become even more dense.

In general, surface temperatures decrease with increasing latitude, ranging from about 81° F (27° C) at the equator to about 29° F (-2° C) in the Arctic and Antarctic Oceans (Fig. 19-3). However, the western

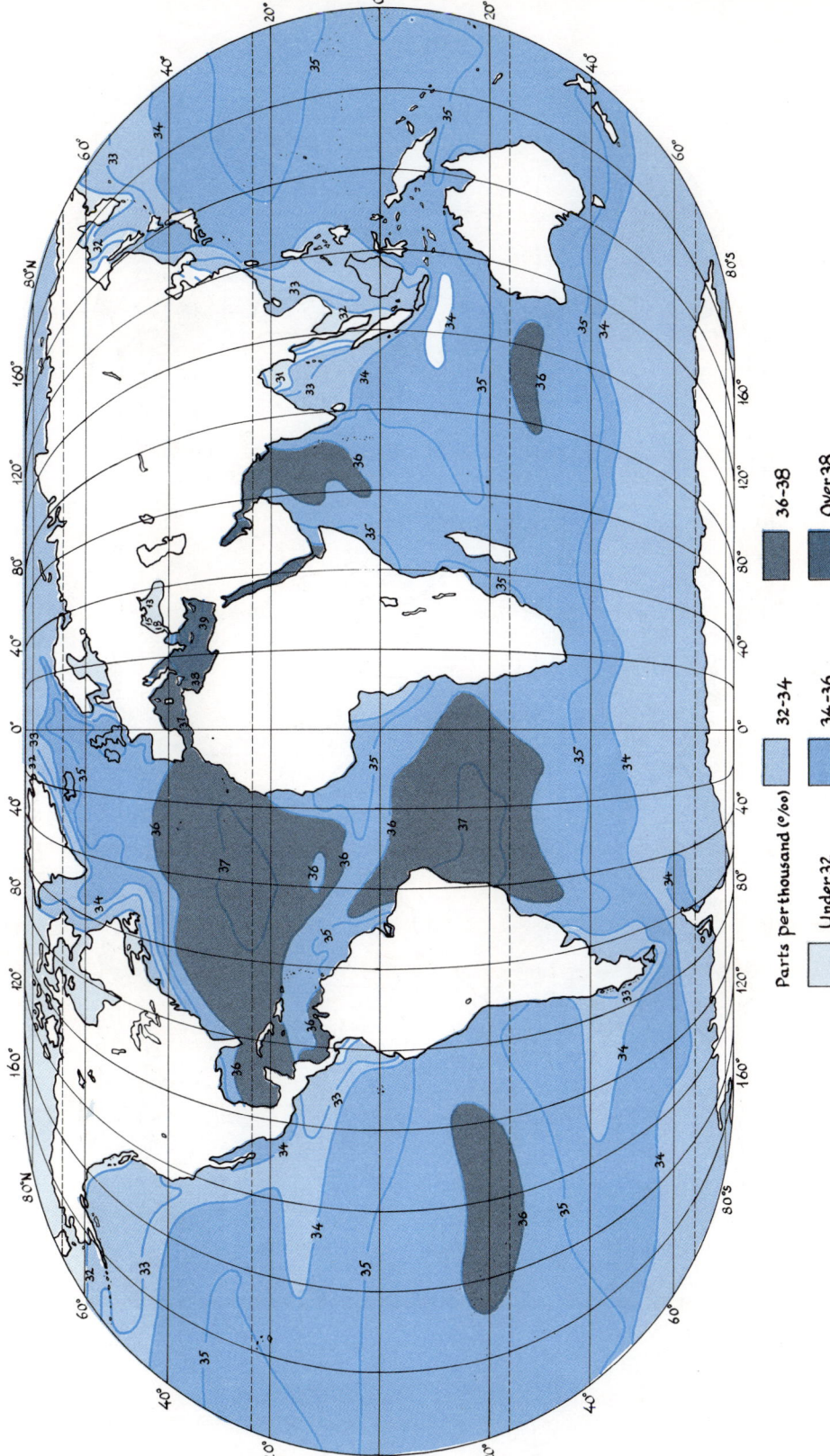

Fig. 19-2. The average distribution of surface salinity, in parts per thousand. Note the high salinity in subtropical latitudes where precipitation is low and evaporation is high. There is a relative minimum in equatorial latitudes, along the ITC, where precipitation is much higher than in the subtropics. The lowest salinity is found in polar oceans where low evaporation and melting ice combine to form sea water with very low salinity.

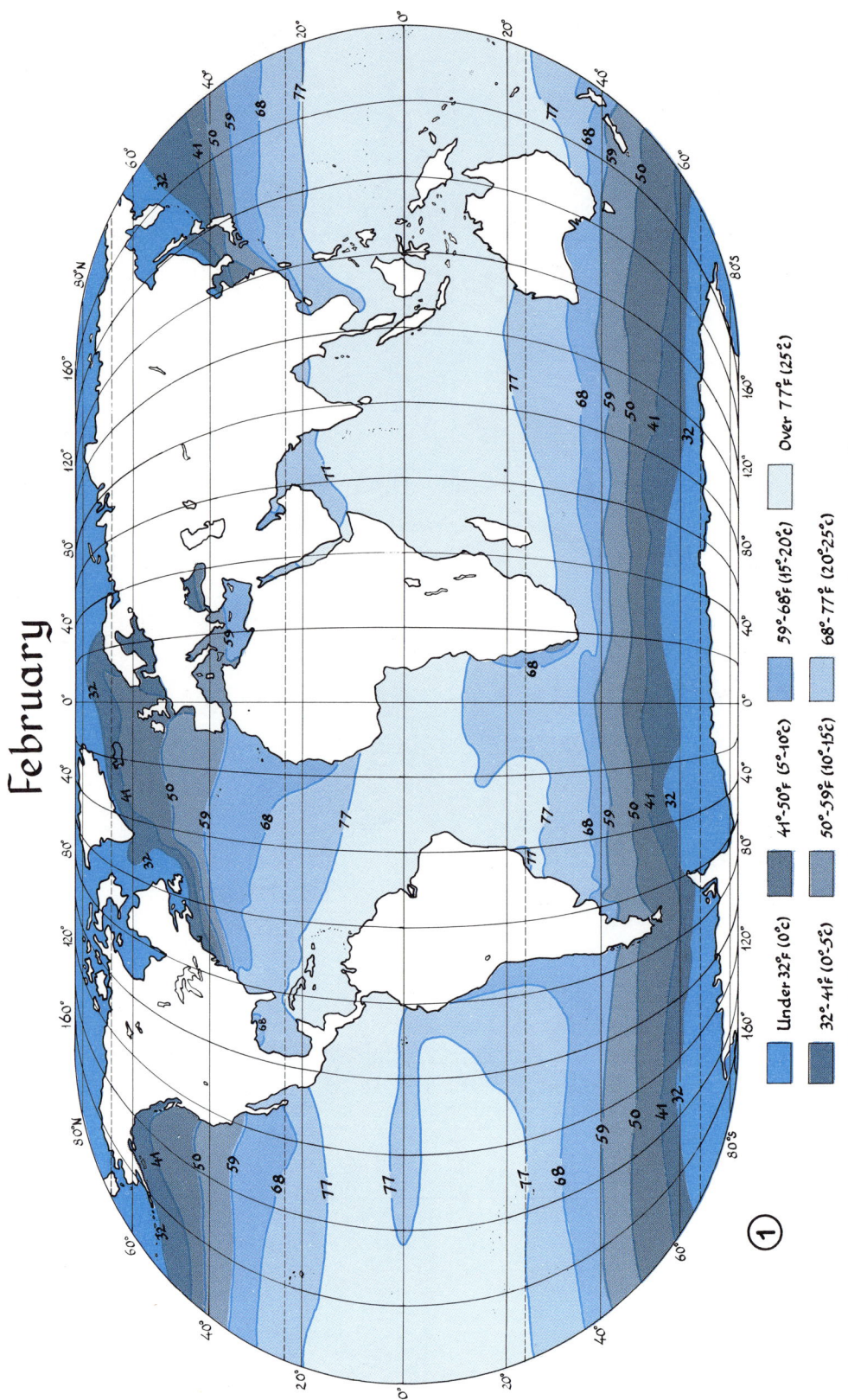

Fig. 19-3. Average surface temperature of the oceans in February (1) in °F.

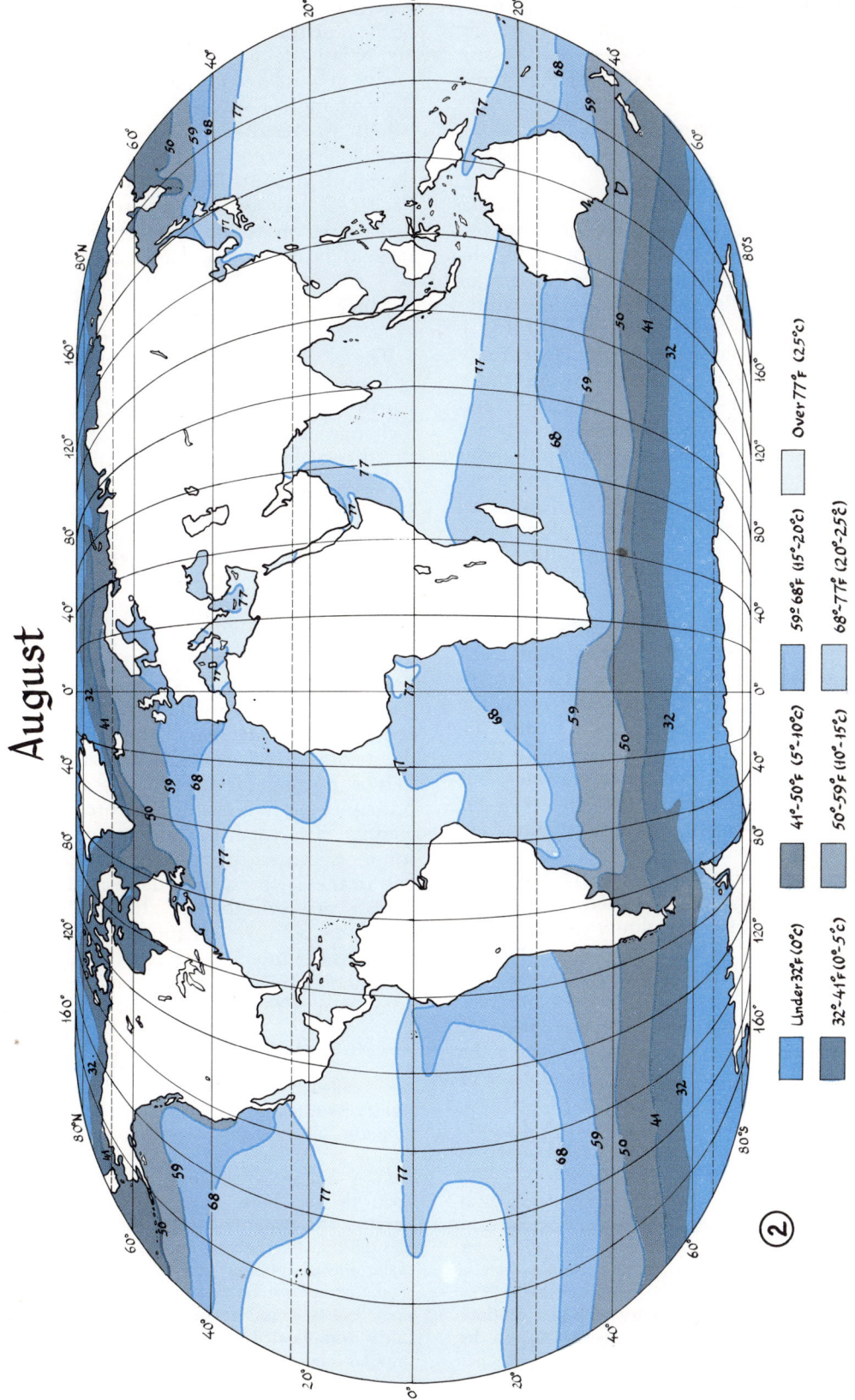

Fig. 19-3. Average surface temperature of the oceans in February (1) and in August (2), in °F. The isotherms follow parallels of latitude, except where ocean currents distort the pattern. Note the poleward displacement of isotherms in winter in regions of warm currents (as in the North Atlantic in February or along the eastern coast of South America in August). Cold currents create an equatorward bulge in isotherms, which is most marked in summer (as along the west coast of North America in August or along the west coast of South America in February).

parts of the oceans are warmer than the eastern parts because of the contrasting effects of poleward-moving and equatorward-moving currents (see pp. 34-36). The Northern Hemisphere seas are warmer than those of the Southern Hemisphere; the average surface temperature of the northern seas is about 67° F (19° C), while that of the southern oceans is about 61° F (16° C).

A seasonal variation also exists in the surface temperature of the oceans, at least in high latitudes. This variation is most clearly demonstrated by the seasonal shift in position of the sea-surface isotherm of 32° F (0° C) in the North Pacific Ocean. From a winter position south of the Aleutian Islands (Fig. 19-3 (1)), this isotherm moves poleward until it lies entirely within the Arctic Ocean in midsummer (Fig. 19-3 (2)). A similar movement occurs in the North Atlantic and Antarctic Oceans. The seasonal contrast of surface temperature decreases toward the equator until about 30° latitude. Nearer the equator, seasonal temperature changes occur only on the east side of the oceans, where cold currents flowing from polar regions carry some of the high-latitude seasonal temperature contrasts with them.

Density. Seawater with a salinity of 35 parts per thousand is about 3 percent more dense than fresh water at a temperature of 39° F (4° C).[1] The density varies with the degree of salinity and with the temperature: the greater the salinity, the greater the density; the higher the temperature, the lower the density. Pressure also affects the density of seawater; but since we are mainly interested in the surface-water layer, density changes due to differences in pressure are small enough to be ignored.

In areas of high precipitation or where fresh water is supplied by rivers or melting ice, the density of ocean water is low because surface salinity is less than average. In some regions of low temperature and low evaporation, the surface density may also be low because of low salinity. For example, in the Baltic Sea, where evaporation is low and there is a large inflow of fresh water, the specific gravity is very low, because the average salinity is only 10 parts per thousand. Seas of high evaporation and little inflow, such as the Red Sea, are very dense, despite their warmth, because the surface salinity may reach 40 parts per thousand or more.

Open oceans may also be quite dense. The central part of the North Atlantic Ocean, the Sargasso Sea, in about 30° N latitude, has a high density because it lies in the center of the North Atlantic high-pressure cell, where the exchange of surface water is small and evaporation is high as a result of the bright sky and dry air. Similar areas of high density, though less developed, occur in the South Atlantic Ocean and in the North and South Pacific Oceans. In the Indian Ocean, a surface high-density region extends south from the Persian Gulf almost to the equator, and a second lies west of Australia.

Movement in the Oceans

The ocean is full of movement. The restless motion of waves and swell has already been described, but it is worth repeating that waves and swell represent mainly a change in the form of the ocean surface rather than the displacement of water vertically or horizontally. There is, however, considerable vertical and lateral displacement of water near the surface and at depth. Subsidence, upwelling, and currents (both surface and subsurface) are very important; but by far the greatest vertical disturbance of the water is due to tidal movement.[2] We will begin our discussion with tidal movement.

Tides. Nearly everywhere, the surface of the sea rises and falls twice each day. This tidal movement results from the gravitational attraction of the moon and, to a lesser degree, of the sun. If we consider only the moon's effect, which is more than twice as great as that of the more distant sun, we can explain the general principles of tide-producing forces in a fairly simple manner. The earth and the moon revolve about their common center of gravity, which actually lies just within the earth on the side nearest the moon. Viewed from a very great distance, the small swing of the earth about this point would appear insignificant when compared to the much greater swing of the moon; so it is reasonable, but not quite accurate, to speak of the revolution of the moon around the earth.[3] Even though

---

[1] The density of any substance is the mass of a unit volume at a given temperature. The density of pure water is greatest at 39° F (4° C), at which point it is 1 gram per cubic centimeter (about 62.4 pounds per cubic foot) and that of seawater is 1.028 grams per cubic centimeter. The specific gravity of a substance is the ratio between its density and that of pure water. Thus, the specific gravity of pure water is 1.0, and the specific gravity of seawater (with a salt content of 35 parts per thousand) is 1.028.

[2] So-called tidal waves are not tidal movements but are the result of submarine earthquakes. These tidal waves are more properly called *tsunamis*, even though this word refers to any kind of storm wave in Japanese.

[3] The same principle applies to the revolution of the earth about the sun. Both are revolving about a common center of gravity. The disparity in mass between earth and sun is great enough that this point is so close to the center of the sun as to make the distinction unnecessary in most contexts. These relationships are further complicated by the fact that every body exerts a gravitational pull on every other; it becomes fairly complicated to describe very accurately the mutual gravitational effects of three celestial bodies, let alone four or more.

the earth's revolution about the common center of gravity is much smaller than that of the moon, it is, nevertheless, responsible for the existing form of the tides. At the center of the earth, the centrifugal force generated by this revolution exactly balances the gravitational attraction between earth and moon; otherwise, the two bodies would crash together. The centrifugal force remains nearly constant everywhere on the surface of the earth and is always directed away from the moon; but the gravitational force, directed toward the moon, decreases very rapidly as the distance from the moon increases. As a result, on the side of the earth closest to the moon, the gravitational force of the moon is much greater than at the center of the earth. Gravitational pull exceeds centrifugal force, and the oceans are pulled toward the moon in a tidal bulge several feet thick. On the side away from the moon, the gravitational attraction of the moon is less than the centrifugal force, and the oceans are pulled away from the moon in a similar tidal bulge. In between these two antipodal points, on both sides of the earth, the opposing forces are about equal (as they are at the center of the earth); but the flow of water toward the tidal bulges is strong enough to create lower water than would be the case in a tideless world. These relationships are shown in Fig. 19-4.

The earth rotates within this tidal bulge once every 24 hours, so that the common center of gravity between earth and moon also rotates. Because the moon is also moving, a complete rotation of the center of gravity takes about 24 hours and 50 minutes. As a result, the tide — like the moon — rises about 50 minutes later each day. If the surface of the earth were covered by a layer of water of uniform thickness, the tidal bulges would pass any particular meridian twice during a period of 24 hours and 50 minutes. During that time, two high, or **flood tides,** and two low, or **ebb tides,** would occur. Figure 19-4 indicates that the high water and the time of the moon's transit of the meridian are simultaneous. On the actual earth, the tide lags behind the moon's passage by several hours. This lag, called the lunitidal interval, varies according to location and season.

Even on a completely oceanic earth, the tides would vary in height from day to day because of the complicating action of the sun. The tide-producing effect of the sun reinforces that of the moon whenever the sun, moon, and earth are in line — that is, when the moon is either new or full. At this time, the flood tides are at their highest, and the ebb tides are at their lowest. These are the **spring tides.** In between new and full moon, at the first and third quarters of the moon, the sun, moon, and earth form a right angle. The tide-producing forces of the sun and the moon are opposed to one another. The greater effect of the moon still results in a tidal bulge, but the flood tides are much lower than during spring tides, and the ebb tides are

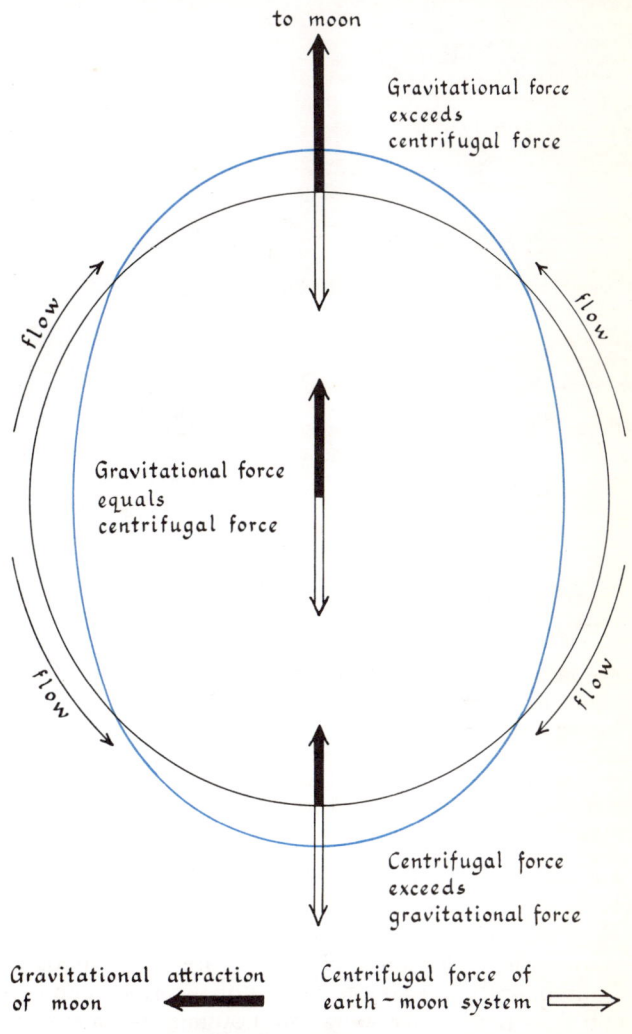

Fig. 19-4. The gravitational and centrifugal forces of the earth-moon system and their relationship to tides.

correspondingly higher. These are the **neap tides.** These relationships are shown in Fig. 19-5.

The earth is not uniformly covered by water, of course, so that the local tidal range and period may vary greatly from place to place according to local conditions of coastal morphology, wind, atmospheric pressure, and configuration of the ocean floor. The declination[4] of the sun and moon, the elliptical orbit of the earth around the sun, and that of the moon about the earth

---

[4]Declination of a celestial body is the angle on the earth between the body and the plane of the equator measured along the celestial meridian of the body. For example, the declination of the sun varies between 23½° N and 23½° S.

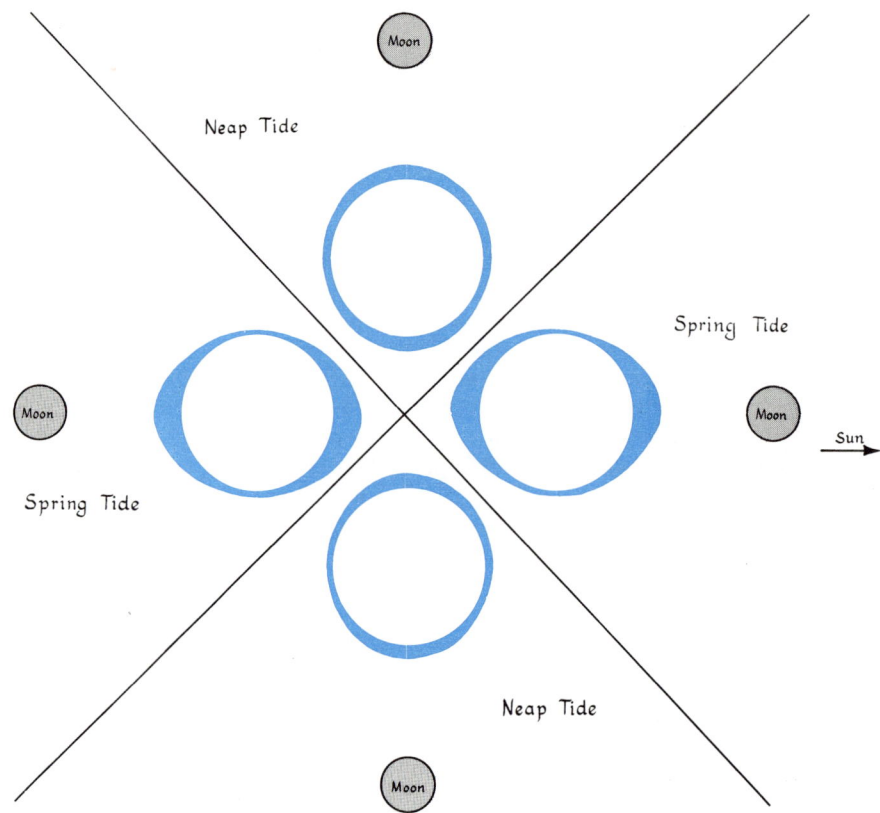

Fig. 19-5. Spring and neap tides. The highest tides, the spring tides, are formed when the sun, moon and earth are in line. The neap tides, formed when the sun and the moon's gravitational pull are at right angles, are low tides.

are additional variables that affect the magnitude and period of the tidal bulge. These factors may cause complex rhythmic tidal movements that partially cancel the main lunar tides on which they are superimposed. Consequently, the simple pattern of two flood and two ebb tides per day, arriving about 50 minutes later each day, may be so modified by local conditions as to be hardly recognizable.

In general, there are three main types of tides: *diurnal*, *semidiurnal*, and *mixed*. The diurnal tide, with only one flood and one ebb per day, is uncommon but does occur in the Gulf of Mexico and in other partly enclosed seas. Not all enclosed seas have this tidal regime. For example, the North Sea, despite its generally enclosed character, has northern and southern connections with the Atlantic Ocean that create a very complex tidal pattern. The semidiurnal tide has two full cycles of movements in a 24-hour period, and its floods and ebbs are about equal in magnitude. This is the common form of tide in the Atlantic Ocean. The mixed tide is more common in the Pacific and Indian Oceans, where two flood tides occur every 24 hours, but one of the flood tides may be so low that the next ebb tide may scarcely show any fall; or one of the low tides may be quite high, so that the succeeding flood tide scarcely brings the water surface above mean sea level. The mixed tides seem to be combinations of daily and semidaily tides, with local harmonic movements controlled by basin shape and coastal outline (Fig. 19-6).

Vertical Movement and Deep Ocean Currents. Vertical and horizontal currents are set in motion by surface contrasts in temperature and in salinity and by the force of the wind. One characteristic vertical exchange involves large amounts of water but takes place within a relatively shallow surface layer. It occurs in mid- and high-latitude oceans, where there is a pronounced winter season. During this cold period, the relatively warm surface water, cooled by radiation and by conduction to the atmosphere, becomes dense and sinks; it is replaced at the surface by relatively warm and lighter water from below. This layer, in turn, is cooled and is replaced from below. The convectional overturn of water occurs throughout the winter season and results in a thorough mixing of the surface layer of water.

Another form of vertical movement involves persistent upwelling and is found off the west coasts of continents in mid-latitudes (Fig. 19-7). The reason for the rising movement along these coasts is the prevailing wind, which blows parallel to the shore or at a slight angle to it. These winds generate local, shallow surface currents that, because of the deflective effect of the earth's rotation (see Chapter 4), move at an angle of about 45° from the wind direction. The prevailing winds along these west coasts are equatorward, so that clockwise deflection in the Northern Hemisphere causes the surface water to move away from the shore (Fig. 19-7). The surface deficit is replaced from below. The upwelling water is colder than the surface water it replaces, and it intensifies the coldness of the major currents moving from high latitudes toward the equator.

There are also regions of perennial subsidence. High salinity produces density subsidence in the centers of the North Atlantic eddy and similar eddies in the South Atlantic and in the Pacific and Indian Oceans. Because of high evaporation and low precipitation in the center of these great eddies, surface density is high; the water sinks and is replaced by a slow inflow of water from the inner margins of the eddies. The density induced by high salinity is offset to a considerable degree by the warmth of these subtropical waters; consequently, the water subsides only to a depth of about 1,000 feet (about 300 m).

Perennial water subsidence also takes place in high latitudes. Poleward-moving ocean currents are warm and relatively saline. As they move poleward, their water loses heat to adjacent waters and particularly to the air. The high salinity of these waters makes them denser than the local water masses, and subsidence results. This subsidence is especially well developed in the West Wind Drift of the North Atlantic. The warm, saline water is subjected to intense cooling in winter. The inflowing, sinking water causes Arctic Water to flow into the North Atlantic. A substantial part of the West Wind Drift, however, subsides to the bottom of the North Atlantic and spreads southward as *North Atlantic Deep Water*.

There are two general areas of sinking water in the Antarctic Ocean. One occurs on the borders of Antarctica, in the Weddell and Ross Seas. Here, the water is cooled sufficiently to sink, then flow northward as *Antarctic Bottom Water* at the bottom of the Indian, South Pacific, and South Atlantic Oceans. The second area of subsidence occurs in the West Wind Drift zone that girdles the Antarctic Continent. This water, the *Antarctic Intermediate Water*, which is quite cold but only moderately saline, sinks to a general level of about 3,000 feet (1,000 m) and also flows northward.

The subsidence of these great quantities of ocean water is a component of deep-water circulation in all oceans. It is more vigorous in the Atlantic Ocean, partly because of strong subsidence of water in both the extreme north and the extreme south. There are, as we

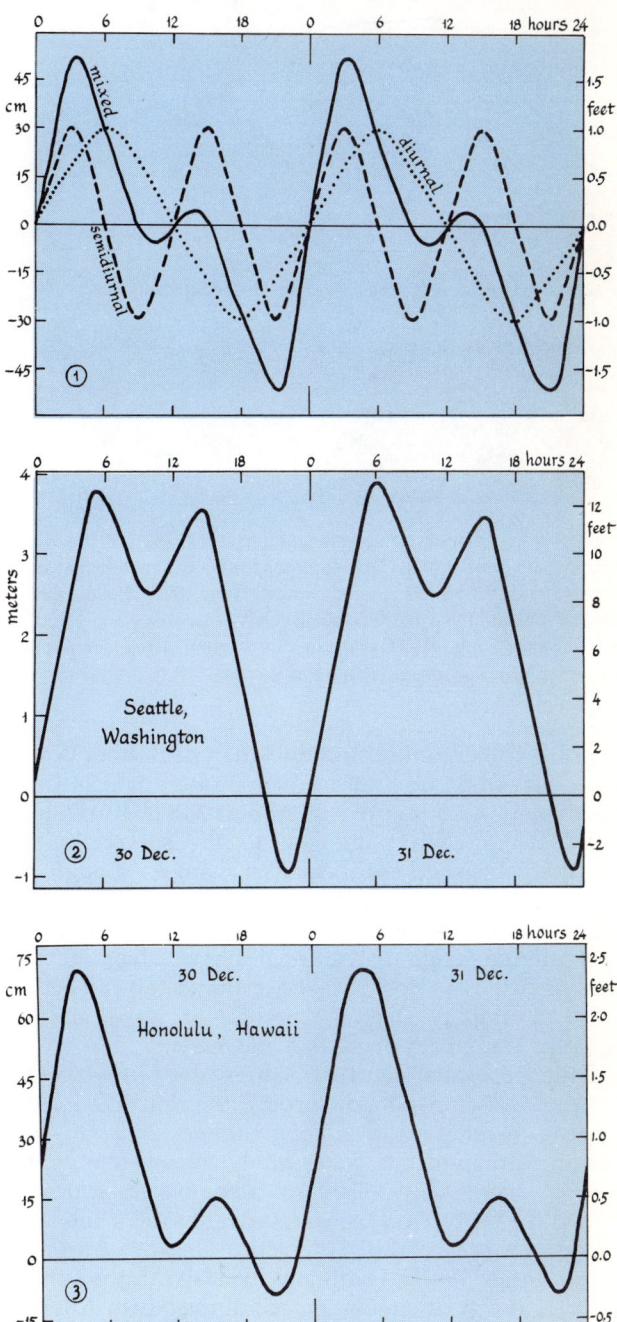

Fig. 19-6. Tidal curves. (1) represents idealized examples of the three tidal types. (2) and (3) are examples of mixed tides.

have seen, three general layers of movement in the Atlantic: the Antarctic Bottom Water, which flows northward to at least 30° N latitude; the North Atlantic Deep Water which flows southward gradually rising

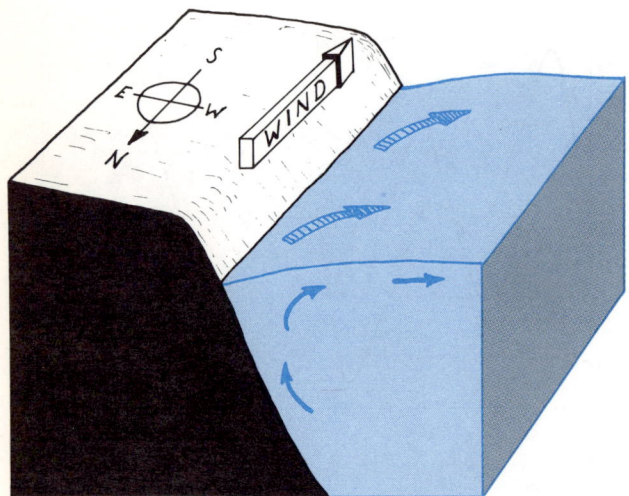

Fig. 19-7. The origin of upwelling water as the result of the offshore displacement of surface water by the wind. The block arrow shows the direction of the wind parallel to the coast. The hatched arrows indicate the offshore displacement of surface water resulting from the deflective effect of the earth's rotation. The black arrows show the resulting upwelling in cross section.

over the colder, underrunning Antarctic Bottom Water; and the Antarctic Intermediate Water, which flows northward at a depth of about 3,000 feet (about 1,000 m), gradually moving to the surface in the Northern Hemisphere. Lateral exchanges, though not so easily identified, also occur at various depths in other oceans, completing the cycle of motion set up by subsidence and upwelling. These deep lateral movements are important, since they keep the oceans from developing stagnant pockets of water and are necessary adjuncts of surface movement.

The deep water returns to the surface in ways that are not well understood. In part, the return is by upwelling in mid-latitudes along the west sides of continents, although this water rarely comes from depths greater than 900 feet (300 m). Some waters appear to rise to the surface on the equatorward side of the West Wind Drift in the Antarctic Ocean. Antarctic Intermediate Water eventually blends with the surface layer of the great circulation cell in the North Atlantic. Once the deep water rises to the surface, it becomes a part of the surface currents that eventually take the water back to high latitudes. This is a very simplified presentation of a most complicated, little-known system.

As nearly as can be determined, the lateral exchange of deep ocean water due to density differences is extremely slow; a single cycle may take several hundred years. The slowly moving portion of the cycle is in the deep flow. Surface currents have a relatively high velocity (3 to 6 miles or 5 to 10 km per hour), but because they are relatively shallow and because they move both poleward and equatorward, (for example *some* surface water is not returning to high latitudes), the return at the surface of the immense volumes of Arctic and Antarctic Bottom Water from the equatorial regions to their polar source regions is obviously indirect and therefore slow.

Surface Currents. The large, relatively fast surface currents have already been discussed in Chapter 4. It should be stressed that these surface circulations are not closed; water is exchanged with adjacent seas. For example, the West Wind Drift in the Antarctic Ocean forms a continuous band of water that moves from west to east. However, some of this cold water becomes part of the Peru, Benguela, and West Australian Currents. At the same time, there is replacement of this loss, probably from the Brazil and Mozambique Currents.

It is evident from the foregoing discussion of oceanic circulation that the oceans form a gigantic open system with a complex array of energy and mass exchanges. In general, heat energy is gained in the tropics and released in high latitudes. The pattern of energy input from the winds in the atmosphere is more complex. In simplest terms, maximum energy input occurs where the winds blow steadily in the same direction as the earth is rotating (principally in the zone of westerlies), and energy is lost where the air flow is opposed to the rotation (as in the zone of easterly trades). In addition, some energy is gained from the gravitational force acting upon large volumes of water with different densities.

Mass exchange between the oceans and the atmosphere occurs in areas where rainfall exceeds evaporation or where evaporation exceeds precipitation. The world ocean system would have a virtually perfect steady state or balance between inflows and outflows of mass were it not for periodic formation of continental glaciers.[5] During stages of glacial maxima the oceans suffered a mass loss proportional to the degree of glaciation. During subsequent deglaciation the water locked in the ice eventually returned to the sea. Though these exchanges represented less than 5 percent of the total oceanic mass, they were, nonetheless, periods of substantial disequilibrium to the ocean system.

Water Masses

Water masses are large areas of surface water about 1,000 feet (300 m) thick, of nearly uniform temperature

---

[5] The addition of juvenile water (water which has been derived from the interior of the earth) to the oceans has probably been very slight since early geologic time.

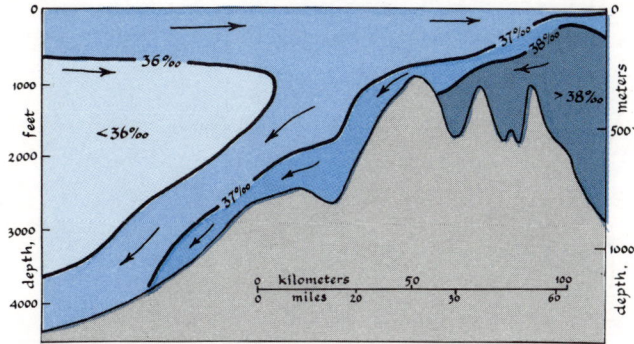

Fig. 19-8. Cross section of the flow of water across the sill at Gibraltar. Fresher and lighter Atlantic water flows into the Mediterranean at the surface and replaces water lost by evaporation. The denser, more saline Mediterranean water flows out below the lighter surface water and sinks to a depth where it meets similar densities. Modified from Sverdrup, Fleming, and Johnson, *The Oceans*.

and salinity and thus also of nearly uniform density. Surface salinity and density vary both seasonally and from place to place because of differences in evaporation and in addition of fresh water by rain and streams. Surface temperatures also differ because of regional and seasonal differences in the absorption of solar radiation, in radiational cooling and in the conduction of surface heat to and from the atmosphere. These differences are not randomly distributed but result in fairly well-defined regional water masses.

A few examples will illustrate the unique qualities of certain well-defined regional water masses. The Mediterranean Sea, an area of high evaporation, is a source region for high-density, saline water that spills out of the Mediterranean Basin across the sill at Gibraltar and flows out into the Atlantic Ocean at a depth appropriate to its density (Fig. 19-8). The water has been identified by its temperature-salinity characteristics, and its movements have been traced for long distances. The Gulf Stream also has characteristic values of salinity and temperature that set it apart from the adjacent waters. The Labrador Current, flowing southward along the Labrador coast, is easily distinguished from adjacent waters of the North Atlantic by its lower temperature and salinity (Fig. 3-9).

Water-Mass Regions in the Open Ocean

Oceanic water-mass regions can be distinguished on the basis of their characteristic temperature, salinity, and velocity. To be sure, the boundaries between regions, at the surface of the water, are broad zones rather than sharp lines, since water mixes readily and the currents frequently swing to and fro over wide areas. Nevertheless, temperature and salinity values tend to be uniform over fairly large areas, and the changes from one water mass to another are sufficiently abrupt to allow us to distinguish the major regions, which are discussed below and are shown in Fig. 19-9.

Tropical Oceanic Regions. These regions occur in low latitudes on both sides of the equator in the Pacific and Indian Oceans. The *Pacific Equatorial Water* is a uniform water mass with surface temperatures ranging from about 75° to 85° F (24° to 29° C) and with values of salinity around 35 parts per thousand. The water region forms a westward-pointing wedge that virtually crosses the Pacific. The surface temperature is relatively high, and the annual range is low. The salinity is moderate. The surface movement is generally westward, although the Equatorial Countercurrent forms a narrow, eastward-flowing thread of water just north of the equator. On the eastern side of the Pacific, the westward-flowing currents converge, while on the western side, they diverge.

This tropical oceanic water mass is characterized by a relatively sparse flora and fauna and a deep blue color. The principal fish are pelagic or surface such as tuna, and are mainly concentrated in the east. Fish are abundant and varied in the upwelling waters off the west coasts of North America and South America. These upwelling areas are transition zones between Antarctic Water, Central Oceanic Water, and Pacific Equatorial Water (Fig. 19-9).

The *Indian Equatorial Water* is very similar to its Pacific counterpart insofar as surface temperature and salinity are concerned. But it does differ in two respects. The Red Sea supplies very saline water at depth, and the direction of the North Equatorial Current becomes northeasterly during the summer monsoon.

Central Oceanic Regions. These regions are much larger than the tropical areas just discussed (Fig. 19-9). They cover most of the Atlantic Ocean even in low latitudes, since the water temperatures, even near the equator, are too low to produce an equatorial water mass. *Central Oceanic Water* also occupies large parts of the Pacific Ocean on either side of the Pacific Equatorial Water and a broad belt south of the Indian Equatorial Water. Temperature and salinity characteristics of these water masses are given in Table 19-1.

The greater salinity of the Atlantic Ocean is obvious. The table obscures the fact that Southern Hemisphere waters are generally cooler than those of the Northern Hemisphere in the same latitudes. North Atlantic and North Pacific Central Waters extend far to the north and include cool waters, which in the Southern Hemisphere would be classified as Subantarctic. Hence, these northern waters appear to be cooler than their southern counterparts when, in fact, they are actually warmer if compared latitude by latitude.

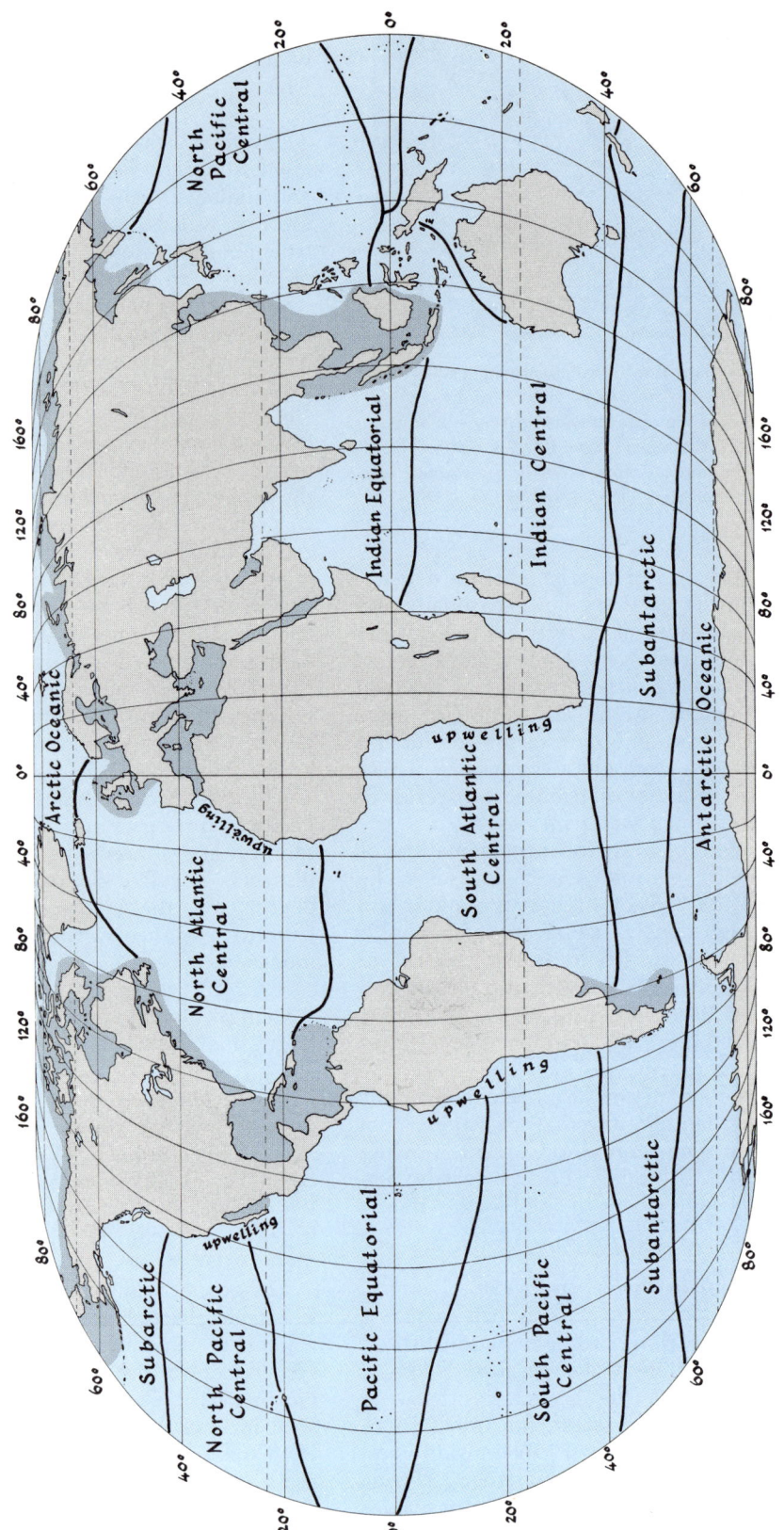

Fig. 19-9. The distribution of water-mass regions. Continental shelf waters are shown in dark blue.

Table 19-1. Temperature and Salinity Characteristics of Central Oceanic Water Masses

Water mass	Latitudinal temperature variation (°F)		Range of salinity (0/00)
	In January	In July	
North Atlantic Central Water	40-80	50-85	34½-37½
South Atlantic Central Water	60-80	50-85	35-37½
North Pacific Central Water	50-85	70-85	34½-35½
South Pacific Central Water	60-85	50-80	34½-36½
Indian Central Water	60-80	50-75	35-36

In general, the central oceanic regions coincide with the great circulation eddies north and south of the equator. On their eastern sides are the cold, equatorward-flowing currents, on the west the warm, poleward-flowing currents. On the poleward margins there are considerable ranges in seasonal temperatures. Because of these changes in temperature, the central oceanic regions experience considerable seasonal overturn of surface water. In the central parts of the rotating eddies, the water contains few nutrients and a sparse flora and fauna, with a consequent deep blue color.

Subarctic and Subantarctic Regions. The *Subarctic Water* has low surface temperatures and salinities. For example, its average surface temperature at latitude 50° N is between 35° and 40° F (between 2° and 4° C). The surface salinity is about 32 parts per thousand. On the eastern side of the region, temperature and salinity are slightly higher. The *Subantarctic Water* is transitional between the Central Oceanic and Antarctic Waters. It has low but variable salinities and temperatures. The region lies in the belt of the West Wind Drift and experiences considerable mixing, both horizontally and vertically.

Polar Oceanic Regions. The southern limit of the Arctic Oceanic Water is the same as the limit of winter ice. A good part of the surface is covered by ice throughout the year, and even when it is not, surface water temperatures are near the freezing point. The annual range in temperature is, therefore, small. Surface salinities are usually low, except in those areas of inflowing water from lower latitudes.

The *Antarctic Oceanic Water* occupies a belt around the Antarctic continent and extends northward to between latitudes 52° and 54° S. Its temperature is only a degree or so above freezing, and its salinity is about 30 to 35 parts per thousand, with local variations related to inflow of warmer water from the north.

Continental Shelf Water

In addition to the water in the open ocean, there are water masses that are greatly modified by the underlying presence of the continental shelf and by the addition of fresh water from rivers. Three general subtypes result from these modifications: polar, mid-latitude, and tropical continental shelf water masses.

Polar Continental Shelf Regions. Polar Continental Shelf Waters occur mainly along the margins of the continental shelf of the Arctic Ocean; they are much less extensive in the Southern Hemisphere because of the limited continental shelf around the Antarctic Continent. *Arctic Continental Shelf Water* is usually frozen in winter but open in summer. The annual range, however, is slight, and the salinity is lower than in the water of the open Arctic Ocean because of the addition of fresh river water during the summer season. The formation of sea ice is one of the characteristic features of polar shelf water. The temperature at which sea ice forms depends upon the salinity; the higher the salinity, the lower the freezing point. The water, therefore, first freezes near the land,[6] in the form of small, needle-shaped crystals which are perpendicular to the sea surface. This crystal position keeps the ice flexible and able to dampen the waves and ripples, eventually permitting the development of a solid ice crust, which gradually extends seaward. The ice cover originating in this fashion is called *land ice* or *shore ice*. It should not be confused with true *sea ice*, which develops on the open ocean and is often called *pack ice* in the Arctic.

Icebergs are another phenomenon typical of polar shelf waters. In the Northern Hemisphere, they are mainly formed by the breaking off, or calving, of large chunks of the tidewater glaciers in Greenland. They are also formed along the Alaskan coast, but to a much lesser extent. The icebergs are transported equatorward into the open ocean by currents. Southern Hemisphere icebergs are generated by the *shelf ice* of Antarctica, which is a seaward extension of the Antarctic ice cap onto the continental shelf. The shelf ice is flat. Consequently, the icebergs are more tabular and elongated than their rugged, Northern Hemisphere counterparts; some of them are dozens of miles long.

Mid-Latitude Continental Shelf Regions. *Mid-Latitude Continental Shelf Waters*, with the exception of water off Argentina, are restricted to the Northern Hemisphere. They are distinguished from the Central

---

[6]Sea ice consists only of pure water; the salts in the freezing water are precipitated into the rest of the ocean, which increases its salinity. As a result of this increased salinity, a lower temperature is required for the formation of additional ice.

Oceanic Waters by their shallow depth and the presence of cold currents from higher latitudes that hug the mid- and high-latitude east-coast continental shelves. These regions are also characterized by fresh-water inflow from the land and an excess of precipitation over evaporation. These continental waters contain the world's greatest fisheries, partly because organic and mineral matter is supplied in great abundance by rivers and by polar ocean currents close to the continental shelves.

Tropical Continental Shelf Regions. *Tropical Continental Shelf Waters* around the islands of Indonesia and in the South China Sea are found in the zone of monsoonal circulation. The annual variation of temperature is small, but there are marked seasonal changes in salinity and in the direction of the currents because of the shifting monsoon winds. A similar region lies between Australia and New Guinea, in the Coral and Arafura Seas, but the surface salinity is higher and shows much less seasonal variation.

Enclosed Seas

Because of their enclosed nature, these seas often reflect the qualities of the surrounding land. Their temperature-salinity characteristics are frequently very different from those in the neighboring open ocean.

Enclosed Seas of the Old World. The Mediterranean Sea represents a mid-latitude continental modification of oceanic water. The streams tributary to the Mediterranean have small drainage basins, and the climate is characterized by hot, dry summers. Consequently, evaporation exceeds precipitation, and stream flow is low. The loss of water is balanced by inflowing surface water from the Atlantic, which enters the basin through the Straits of Gibraltar. Since evaporation is a surface phenomenon, the inflowing Atlantic surface water soon becomes saline, sinks, and is mixed by convection currents. The basin is, therefore, filled with water of uniform temperature and high salinity (38 parts per thousand). The outflow of saline water from the Mediterranean Sea leads eventually to a total exchange of water between the basin and the Atlantic Ocean. The time required for this exchange has been calculated to be about 75 years.

The Red Sea and the Persian Gulf are similar to the Mediterranean, although they are hotter and more saline. Air and sea temperatures are among the highest in the world, precipitation is very low, and they scarcely receive any fresh-water runoff from the surrounding deserts. The Black Sea represents an extreme case of an enclosed water body. The surface layer has a low salinity because of the many large rivers flowing into it and is, therefore, light. Winter chilling is not enough to overcome this low surface density; convective currents do not develop, and vertical mixing does not take place. As a result, only surface water can flow out by way of the shallow Bosporus and Dardanelles; the deep water remains trapped in the basin. The Black Sea is, therefore, stagnant below a depth of 450 feet (140 m). The deeper water contains little or no oxygen, thus inhibiting marine life and leading to the generation of hydrogen sulphide, a characteristic feature of basins of stagnant water.

The Baltic Sea is very similar to the Black Sea in its vertical stratification. Surface salinity is low; there is no convective overturn in winter; and there is a series of shallow sills across the exit to the deep water of the Skagerrak. However, the Baltic Basin is occasionally ventilated during periods of strong west winds when the low-salinity surface is pushed eastward, allowing the deep water to come to the surface.

The Caribbean Sea and the Gulf of Mexico. The Caribbean and the Gulf are warm seas whose average salinity is 35 parts per thousand in the upper levels. Little convective mixing occurs, since their surface waters are not seasonally cooled. Consequently, at the level of sill depth,[7] the water in the Caribbean is less dense than that in the neighboring ocean. Atlantic water thus spreads across the basin floors of the Caribbean Sea and the Gulf of Mexico. This movement at depth is little disturbed by the relatively shallow surface current that enters the Caribbean from the east, crosses the Gulf of Mexico, and exits between Florida and Cuba as the Gulf Stream.

Summary

The surface of the ocean, like the surface of the land, can be divided into readily distinguishable regions. Each region has its own combination of temperature, salinity, movement, and marine life. Horizontal variations in temperature and salinity lead to regional variations in water density, which in turn set great convective and deep oceanic-water currents in motion. The surface currents of the oceans are thought to result from well-defined terrestrial wind systems. These surface currents, common to all oceans, are remarkably similar in pattern and distribution. The patterns of

---

[7] Anegada Passage, east of Puerto Rico, and the Windward Passage between Haiti and Cuba have sill depths of approximately 600 feet (180 m).

salinity and temperature also have a certain degree of regularity over the surfaces of the oceans. Using the patterns of salinity and temperature, and, to a lesser degree, movement, oceanic regions of similar characteristics can be identified and mapped. Certain marginal seas, because they are surrounded by land or underlain by broad continental shelves, are set apart from the open oceans by their modified temperatures and degrees of salinity.

Suggestions for Further Reading

Cotter, Charles H., *The Physical Geography of the Oceans*, Hollis & Carter, Ltd., London, 1965.

Dietrich, Günter, *General Oceanography: An Introduction*, trans. Feodor Ostapoff, Intersciences Publishers, Inc., New York, 1963.

King, Cuchlain A. M., *An Introduction to Oceanography*, McGraw-Hill Book Company, Inc., New York, 1965.

Munk, Walter, "The Circulation of the Oceans," *Scientific American*, Vol. 193, No. 3 (September 1955), 96-104.

Sverdrup, H. U., R. H. Fleming, and M. W. Johnson, *The Oceans: Their Physics, Chemistry, and General Biology*, Prentice-Hall, Inc., Englewood Cliffs, N. J., 1942.

Williams, Jerome, *Oceanography: An Introduction to the Marine Sciences*, Little, Brown and Company, Boston, 1962.

20 The Environment and Distribution of Vegetation

Introduction

The study of vegetation, as also of soils, transcends the arbitrary line between physical and cultural geography in a more obvious way than does either geomorphology or climatology. Geographers are concerned with the origin and distribution of vegetative patterns in the plant cover of the world. There are at least three different kinds of general plant groups that can be separated according to the degree of modification by man. There are, first of all, the cultivated plants of the fields and the ornamentals of the garden, which together make up the bulk of domesticated plants. Secondly, there are the nondomesticated plants whose distribution has been more or less strongly modified by man. This modification may be in the form of harvest, as in the case of timber trees; or replacement, when vegetation is cleared for cultivation; or even eradication, which may be conscious, as in the case of noxious weeds, or unconscious, as in the case of many useful species that have been thoughtlessly destroyed. Third, are those assemblages of undomesticated plants whose composition may have experienced some slight modification that is difficult to perceive.

The first two categories are part of the material studied by cultural geographers, since the composition and distribution of these plants (and for the domesticates, even their origin) are the result of cultural rather than physical processes. Only the third category is properly discussed as an aspect of physical geography. Hence, the focus of this chapter is upon "natural" vegetation — vegetation that has had little interference from man and whose form and composition vary in a systematic way with other elements of the physical environment, such as landforms, or climate.

However, very little vegetation is completely natural, particularly in the mid-latitudes of the Northern Hemisphere. The widespread and long-continued activity of man as farmer and herdsman has profoundly modified the original vegetation over much of the world. Nevertheless, large areas have been only peripherally or only slightly touched by human activity; and there are enough remnants of the original vegetation cover, even when they occur only in small and isolated stands, that we can theoretically reconstruct the natural cover over larger areas. It is obviously difficult, in some instances, to decide how much alteration of the natural vegetation has taken place or how far extrapolation from existing "natural" assemblages may safely be carried because plant communities are dynamic. They are subject to continuing change as the result of disease, storms, droughts, freezes, and the like.

There are two reasons for discussing the composition and distribution of natural vegetation, despite its absence through destruction over large parts of the world and the conjectural nature of its natural character in other parts. First, the form of individual plants and their composition into characteristic plant communities, even when slightly modified by man, are primarily the result of the interaction of physical processes. To be sure, a plant is neither a rain gauge nor a thermometer, but its density, its form, and its growth habits tend to reflect the climatic, edaphic, and geomorphic conditions of its immediate environment. In a sense, therefore, the plant cover, more than any other aspect of nature, integrates the totality of the physical environment into a visible element of the landscape. Second, natural vegetation is not only scenery but has considerable impact upon the scene itself. The influence of plants on geomorphic processes has already been briefly discussed in Chapter 11; their influence on the development of soils is well known and is discussed in Chapter 21; and their influence on climate, although overstated in the past, is nevertheless significant. Thus, an appraisal of the nature and distribution of the natural vegetation is an appropriate part of physical geography.

At a world scale, biogeographers do not focus their attention upon the floristic composition of plant assemblages but rather upon the structure, pattern and distribution of those communities. The largest and most generalized units, such as forest, desert shrub, tundra, and grassland, are called *plant formations*. Within these formations are found *plant communities* or *associations*, such as broadleaf evergreen tropical forests, broadleaf deciduous mid-latitude forests, and prairies. These associations occur repeatedly in similar environments, where they have the same general appearance even though the species involved may be completely different from one location to another. The similarity of these associations is thought to result from common physical needs, since they cover extensive areas and occupy characteristic physical environments. For these reasons vegetation associations are the particular focus of this chapter.

According to one theory, a plant community on well-drained, relatively level surfaces, under conditions of constant climate, and without interference by man, animals, or fire, gradually changes in character. The first community is composed of the simplest types of plants, while the last is the most complex both in regard to the number and nature of plant species. Each community, starting with the first, creates an environment that is favorable to the creation of a subsequent, more complex, and more demanding community. The ultimate stage, called the *climax community*, is characterized by an equilibrium between the plants and their environments and persists as long as the environment is not changed by agents outside the plant assemblage. That is to say, the development of the last plant community no longer creates an environment in which a still more complex community can develop. But nature and man are not unchanging. Lava flows occur, floods take place, man starts fires and introduces new plants, animals and insects, climate changes, and diseases spread; any of these events may upset the orderly progression to a vegetation climax.

Plant Communities and Factors Affecting Their Distribution

Even though the concept of a vegetational climax must be modified because of natural and man-induced interferences with its orderly development, the idea that plant communities are subject to change is a real and useful concept. Because of the problems associated with the application of this concept to the study of vegetation, other ideas have been developed to avoid those difficulties. One of these ideas is the concept of an **ecosystem,** a concept which can be applied to the study of all living things, even though we will confine its application to vegetation. An ecosystem is a basic functional unit of nature consisting of all the organisms in the community and their non-living environment that are linked by flows of nutrients and energy. A plant community and its associated animal population is such an ecosystem. Because the community exchanges mass and energy with its surroundings it is also an open system. In order to see how an ecosystem operates we will examine, in a general way, the flow of nutrients and energy through a particular ecosystem, a mid-latitude broadleaf deciduous forest that is expanding. In our example the forest has nearly reestablished itself in a formerly deforested area. The general environmental circumstances under which this has taken place include a temperate climate with adequate year-round precipitation, warm summers and cool winters, a gently rolling terrain, and a moderately fertile soil with a good water-holding capacity.

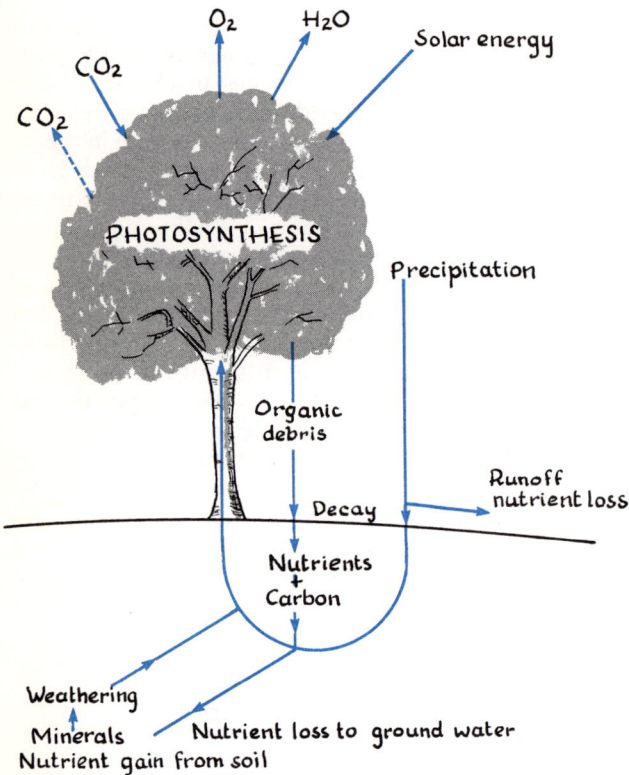

Fig. 20-1. A diagrammatic representation of the nutrient cycle in a forest.

Nutrient flows through all plant systems are cyclic in character. Nutrients are taken up from the soil and used in conjunction with photosynthesis[1] to replenish losses of plant tissue and to produce plant growth. In a mid-latitude forest, plant tissue losses take place during the annual leaf-shedding in the fall, and as a lesser but more or less continuous rain of debris in the form of dead branches, bits of bark, and fruit husks. Eventually, the trees themselves will fall. As this debris breaks down by decay and by the action of soil organisms, the contained nutrients are released to the soil where they become available to the plants once again. In the soil, the nutrients are taken into solution by soil water and conveyed to the plant by its roots (Fig. 20-1). Once the cycle is well established the loss of nutrients by run-off or to ground water is relatively slight. Such losses that do occur are offset by flows of nutrients from weathered minerals in the soil. The rate of flow of nutrients through plants, other things being equal,

---

[1] A process in which chlorophyll absorbs light energy and catalyzes the formation of sugar and starch from water and the carbonic acid gases in the plants.

depends upon the amount of water and energy available to the plants.

The flow of energy through plant communities has been measured with sufficient accuracy to treat the flow in terms of a budget, balancing the inflow of energy against the outflow. Presuming an adequate supply of moisture, if more energy enters a plant community than leaves, the surplus energy is reflected in plant growth, and the community expands. If more energy leaves than enters, then death and decay exceed regeneration and growth, and the community contracts.

Figure 20-2 is a diagrammatic representation of energy flow through the forest selected for this discussion. For the sake of simplicity, it is assumed that the animal population in the forest remains constant in numbers and composition. Of any 100 units of energy that enter the community in a given period of time, 55 are used to maintain the plant matter already in existence, one unit supports the animal life and 24 units are consumed in the decay that breaks down the organic debris and humus. All these processes account for 80 of the incoming units. The remaining 20 units are stored in new tissue or plant growth, 18 in growth above ground and 2 in root growth.

As we have already seen, open systems tend toward a steady state where the inflow of energy and mass is balanced by outflow. In the case of the plant community under discussion, the forest is approaching a condition where plant growth is balanced by decay, that is, a climax condition. However, as was suggested earlier, plant communities are subject to continuous change because of fluctuations in weather, introduction of plant diseases, or perhaps changes in numbers or kinds of animals. For example, several years of drought will slow the growth of mature trees and kill many seedling trees, thus decreasing the stored energy. In contrast, a lengthy period of increased precipitation will speed the growth of mature trees and assure the survival of many seedlings and so greatly increase the stored energy in the forest. With respect to disease, the Dutch elm disease in the northeastern United States is well known for its profound and far reaching effects upon the composition of the forest of that area. As another example, the recent dramatic reduction in the rabbit population in Great Britain due to the viral disease *myxomatosis* resulted in a remarkable increase in plant growth on that island.

The ecosystem we have been analyzing, an expanding broadleaf deciduous forest as well as all other ecosystems, will change as local circumstances change. Similarly, other ecosystems will develop in different localities because of the varied influence of a number of factors. These factors, other than time (which is implicit in the climax theory) include climate, landforms, soils and organisms.

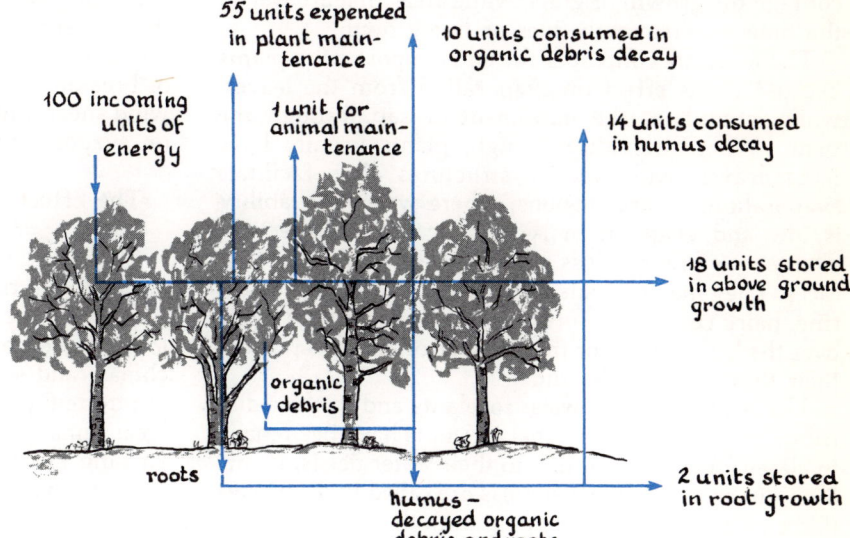

Fig. 20-2. A generalized diagram of energy flow through an expanding forest community. The flow lines have been simplified to keep the diagram from becoming cluttered. For example, the 55 units of energy expended in plant maintenance involve all the trees in the community, but for simplicity's sake, the energy is shown emerging from a single source. Adapted from G.M. Woodwell, "The Energy Cycle of the Biosphere," *Scientific American*, Vol. 223, No. 3, Sept. 1970. pp. 70–71.

The Effect of Climate

The climatic factor is perhaps the most important. Temperature, precipitation, humidity, light, and wind are particularly critical to the distribution of plants. Temperature is a rough indication of the amount of heat in a system and, as such, serves as an approximate index to the amount of solar energy available to the system. Temperature is thus a convenient substitute for more precise ways of describing the level of energy flow through a plant community. The effect of temperature cannot easily be evaluated from existing records. Temperature values are published mainly as monthly or annual averages, but the temperatures critical to plant growth are frequently not averages or even momentary values, but rather the duration of temperatures above or below certain critical extremes.

Most of the known relationships between temperature and plant growth are empirical and tell us little about the physiological effect of heat. We know, for instance, that a great many physiological processes in plants slow rapidly if the temperature drops below 45°F (7°C). But this phenomenon does not explain why, for certain plants, a brief and intense frost has no effect, while much milder frosts can be absolutely killing if they last more than a few days. Nor does it help us to understand, except in a most general way, why the poleward limit of tree growth in Canada and Eurasia coincides so closely with the isotherm of 50°F (10°C) in July.

Duration of specified temperatures seems to be more important than the mere occurrence of an extreme. This observation is true with respect to not only frosts but higher temperatures as well. The poleward limit of many tropical plants, for example, appears to be established by varying durations of temperatures below 65°F (18°C).

A change in temperature often plays an important role in plant growth. Certain physiological reactions in plants require particular combinations of variations in weather. One Mediterranean plant requires a period of increasing temperatures and decreasing moisture availability before it will begin to flower. Other Mediterranean plants require decreasing temperature and increasing moisture content of the soil for the germination of their seeds.

In general, then, the life activities of plants are conditioned by temperature, which sometimes activates or terminates certain reactions and sometimes determines the speed at which these activities take place. Different plants are variously adapted with respect to minimum, optimum, and maximum temperatures for their physiological functions.

Precipitation and humidity are scarcely less important than temperature in affecting the distribution of plant life. Obviously, the rainfall in a given area is a factor of outstanding importance to plant life, since the difference between precipitation and evaporation determines the amount of water available for plant growth. This amount can be modified locally, since the moisture-holding capacity of a soil is affected by the degree of slope and the permeability of the soil. Excessive soil moisture content is as important as deficient moisture in restricting or favoring certain plants. The annual cycle is also important. Periods of excessive wetness followed by periods of drought seem to encourage the growth of grass, while areas that are wet all the time are frequently sites of tree growth.

The moisture content of the air is important to plants because of its effect on evaporation from the leaves, which stimulates the movement of sap through the plant. Where humidity is high, plants usually have large leaves with special structures that facilitate evaporation. In arid regions, where water availability is low and evaporation is high, leaves are specially adapted to prevent loss of moisture from their surfaces. They may be coated with a film of wax or have a fine, hairy covering to prevent the direct passage of air over the leaf surface, or they may be oriented to present their thin edges to the sun.

The relationship of water to plants and to their distribution is so important that attempts have been made to classify plants according to their water needs. Plants living in excessively moist areas are called **hydrophytes** (Greek *hydros* water +*phyton* plant), those that tolerate drought are called **xerophytes** (Greek *xeros* dry + *phyton*), while those living under conditions of alternate deficit and surplus are called **mesophytes** (Greek *meso* middle + *phyton*). Woody hydrophytes are characterized by relatively shallow root systems, long slender trunks, thin bark, and large leaves with restricted means of preventing water losses from their surfaces. Woody xerophytes, on the other hand, tend to have short trunks, thick bark, deep roots, and a variety of mechanisms to prevent the loss of moisture from the leaf surfaces. Succulent xerophytes store moisture in fleshy tissues for use during periods of drought. This adaptation is characteristic of cactus in the New World and some euphorbias in the Old World.

Light is important to plants, since it is essential for photosynthesis. Photoperiodism, the response of plants to the duration of daylight, is an important conditioner of plant growth, since many plants require a rather specific duration of daylight for successful flowering. The common ragweed, for example, is a short-day plant. It flowers only when there is less than 12 hours of daylight. In low latitudes, it would bloom too early, before its growth processes were well under way. In high latitudes, it would bloom too late in the year to escape the harmful effects of frost. Other plants are adapted to long-day flowering. This difference in light-duration requirements may be a genetically fixed characteristic; many long-day plants will not bloom in the tropics, and short-day plants will not bloom when taken to high latitudes. Photoperiodism may be one of the factors limiting plant migration.

Wind is a fourth climatic element that affects plant growth because it increases water loss by constantly bringing unsaturated air into contact with the leaves and young stems. Because of the greater evaporation, trees and shrubs grow away from the direction of the prevailing wind, leading to the peculiar form of plants that are subjected to considerable wind. Wind can also affect vegetation mechanically, through uprooting trees or breaking off branches or other parts of the plant by wind shear, but the greatest impact is through the effect on evaporation.

The Effect of Landforms

Elevation, slope orientation, degree of slope, and variations in rock type significantly affect the distribution of plant communities. Generally, landforms affect local vegetation indirectly through their influence on climate and soils.

Strong topographic contrasts tend to produce distinct local climates. The effect of elevation in arid regions is a striking example. Even a gradual ascent over a short distance may take one from the desert vegetation of the lowlands to the forested slopes of higher elevation. This rapid change results from the fact that small increases in elevation may lead to considerable increase in precipitation and to significant decrease in temperature and evaporation.

Slope itself may have an important effect on local climate and thus on vegetation. Equatorward-facing slopes, for example, receive a great deal more sunlight than slopes facing poleward. As a result, the soil and air temperatures are lower on the poleward-facing slopes, evaporation is less, and more water is available to the plants. The consequent difference in appearance between the **ubac** and the **adret** depends on whether temperature or moisture is the limiting factor in plant growth. In very cold areas, most of the vegetation may be concentrated on the warmer, equatorward-facing slopes. In warmer areas, the vegetation may be more lush on the moister, poleward-facing sides of a valley.

Another effect of slope is to concentrate or to disperse surface water. Precipitation may be relatively uniform over a sizable area; but on the slopes, a large proportion of the falling water runs off rapidly, so that only a small part of the precipitation infiltrates the soil. On the lowlands, on the other hand, the collected runoff results in more infiltration than the rainfall would indicate. This contrast is complicated by the fact that more snow falls on the ridges than in the valleys. The snow does not infiltrate the soil until it melts, thus creating both temporal and spatial differences in infiltration rates. These microclimatic differences greatly affect the local distribution of plants.

Rock type is important in controlling the distribution of certain kinds of plants. For example, recent lava flows have an obvious impact upon the local vegetation, since they destroy the previous cover and create a new surface, where water and organic matter are not easily available to incipient vegetation because of the porous nature of the cooled lava. Faulting may expose rock strata of varying composition, which may become the setting for different plant types. Frequently, these types may be related to water conditions in the upturned strata or to differences in the chemical composition of the constituent minerals in each rock layer. Landslides, stream erosion, and deposition or the development and gradual movement of sand dunes create local but very important differences in the plant environment.

The Effect of Soils

Soils influence plant growth through their chemical constituents, chemical reaction, and capacity to hold water or air. The chemical properties of soils are frequently very important in determining the local character of vegetation. Limestone soils, for example, can often be mapped by the type of plant covering. Certain kinds of limestone rock are frequently covered by grass because grasses flourish in calcium-rich soils. The chemical reaction of soil is also important. The hydrogen-ion concentration may affect soil fertility and so modify the plant cover. Since more exchangeable hydrogen is attached to soil particles because of leaching, there is less room for plant food. It follows that acid soils, with a high concentration of hydrogen, have a lower natural fertility than basic soils, with their greater concentration of basic ions necessary for plant growth. Excessive concentration of basic ions, however, also limits the kinds and amounts of plants that can grow in a given soil. A balance between acidic and basic ions is most conducive to vigorous plant growth.

Physical properties of the soil affect water availability. In sandy soils, water drains readily, and the vegetation may be strongly xerophytic because water retention by the soil is low. If the soil contains less sand and more clay, the water-holding capacity is greater, and a different and denser cover can take root. Too much clay creates an impermeable layer in the soil. Above the layer, the soil may become waterlogged; below it, there will be a deficiency of soil moisture. Under such conditions, deeply rooted plants will fare very poorly in competition with shallow-rooted, water-tolerant vegetation.

Biotic Factors

The biotic factors, in the widest sense, include the effects on plant growth of all living organisms, plant or animal. Some biotic factors, such as the competition for a limited amount of soil water, affect the whole plant community. The established plants may use so much of the available water as to exclude young plants; the whole plant community is thereby affected. The shading effect of taller species also affects the entire community. Some plants react adversely to shade and cannot tolerate it, while others require it as a necessity for survival; but all are affected in some way. Another general effect is produced by earthworms, which change the character of the soil by bringing mineral matter from below and carrying organic matter from the surface downward. In so doing, they create circumstances favorable for the growth and perpetuation of a number of plants.

Other biotic relationships affect only individuals within the plant community. For example, several plants of the genus *Yucca* in the western United States can be pollinated only by yucca moths *(Pronuba);* their natural distributions, therefore, are determined by the range of the moths. When transplanted beyond this range by man, the plants bloom profusely year in and year out but produce no seed. There are other interesting examples of plant dependence on a single insect. Some orchids of the tropical rainforest depend on specific insects for pollination. The persistence of the blossoms depends on the time of arrival of the insects, for once they are fertilized, the flowers fade quickly.

Epidemics bring about great changes in the natural vegetation. As previously mentioned, the Dutch elm disease has virtually eliminated elms from many parts of the northeastern United States. And the white pine blister rust has killed much of the white pine that was not previously destroyed by lumbering in the forests of northeastern North America.

Changes in the animal population can also result in changes in the number and composition of the plant population. The saguaro cactus in Arizona may soon be eliminated from many areas as a consequence of the extermination of the coyote by man. The disappearance of the coyote has allowed pack rats and jack rabbits to increase enormously, to the detriment of the cactus — the pack rats consume the young saguaros, while the rabbits eat the plants that shade and protect the cactus (see Fig. 20-17). Because of this twofold attack, the saguaro cactus is unable to reproduce in numbers sufficient to maintain its population. The change in animal population need not be brought about by man, as in this example; it may be due to natural causes.

Man is undoubtedly the most influential biotic agent. Cutting, grazing by domesticated animals, tilling, and using fire have profoundly altered the natural vegetation over wide areas. Repeated burning by man is believed by some to be responsible for the extensive distribution of grasslands in the areas of tropical savanna climate. As technology becomes more complex, mankind becomes able to change the vegetation in other ways: by irrigating, by draining swamps, by damming streams, by building cities and highways, by cultivating, and by mining. All kinds of enterprises

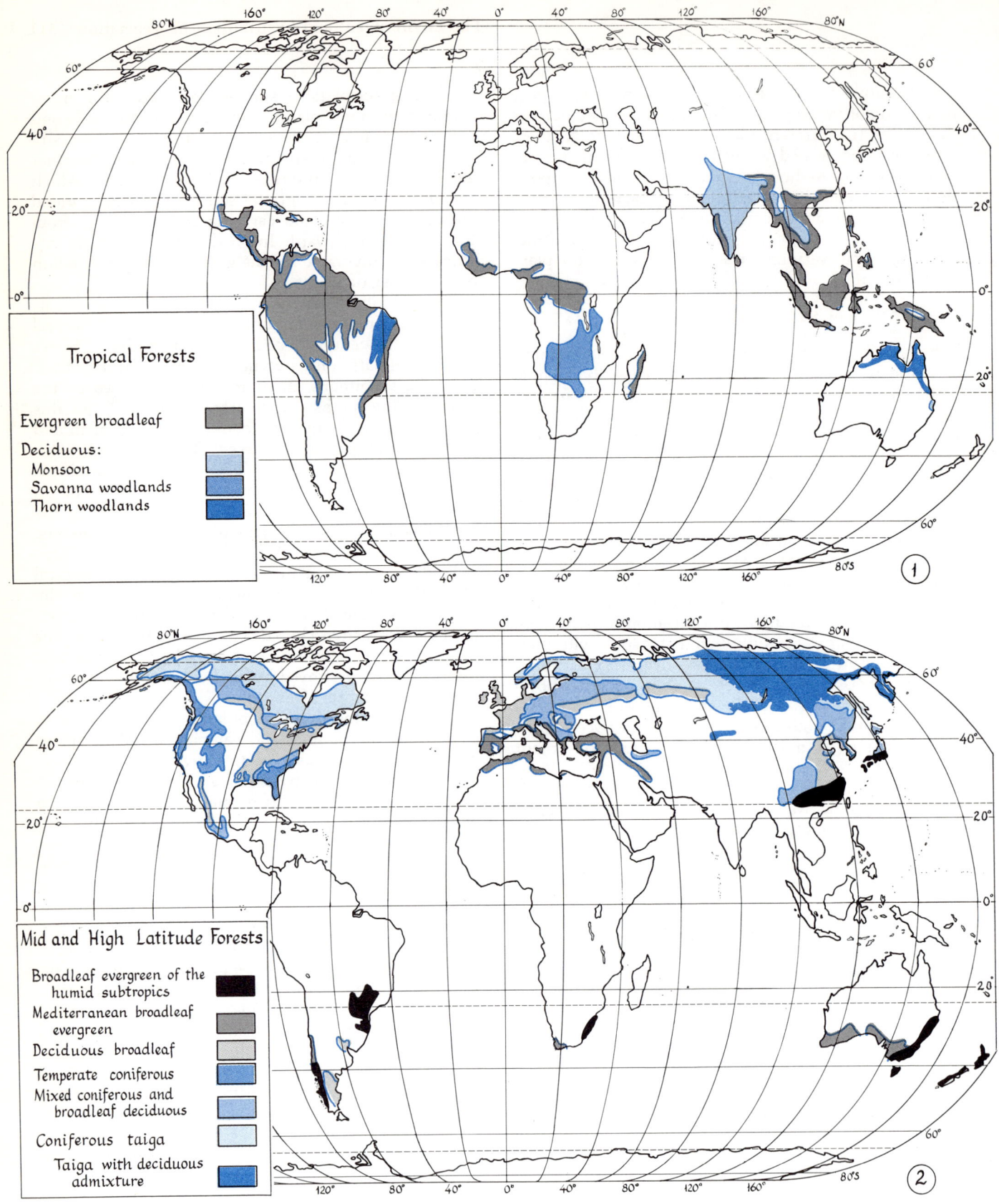

Fig. 20-3. The world distribution of vegetation; (1) tropical forests, (2) mid-and high-latitude forests, (3) grasslands, and (4) deserts, tunda, and highland vegetation.

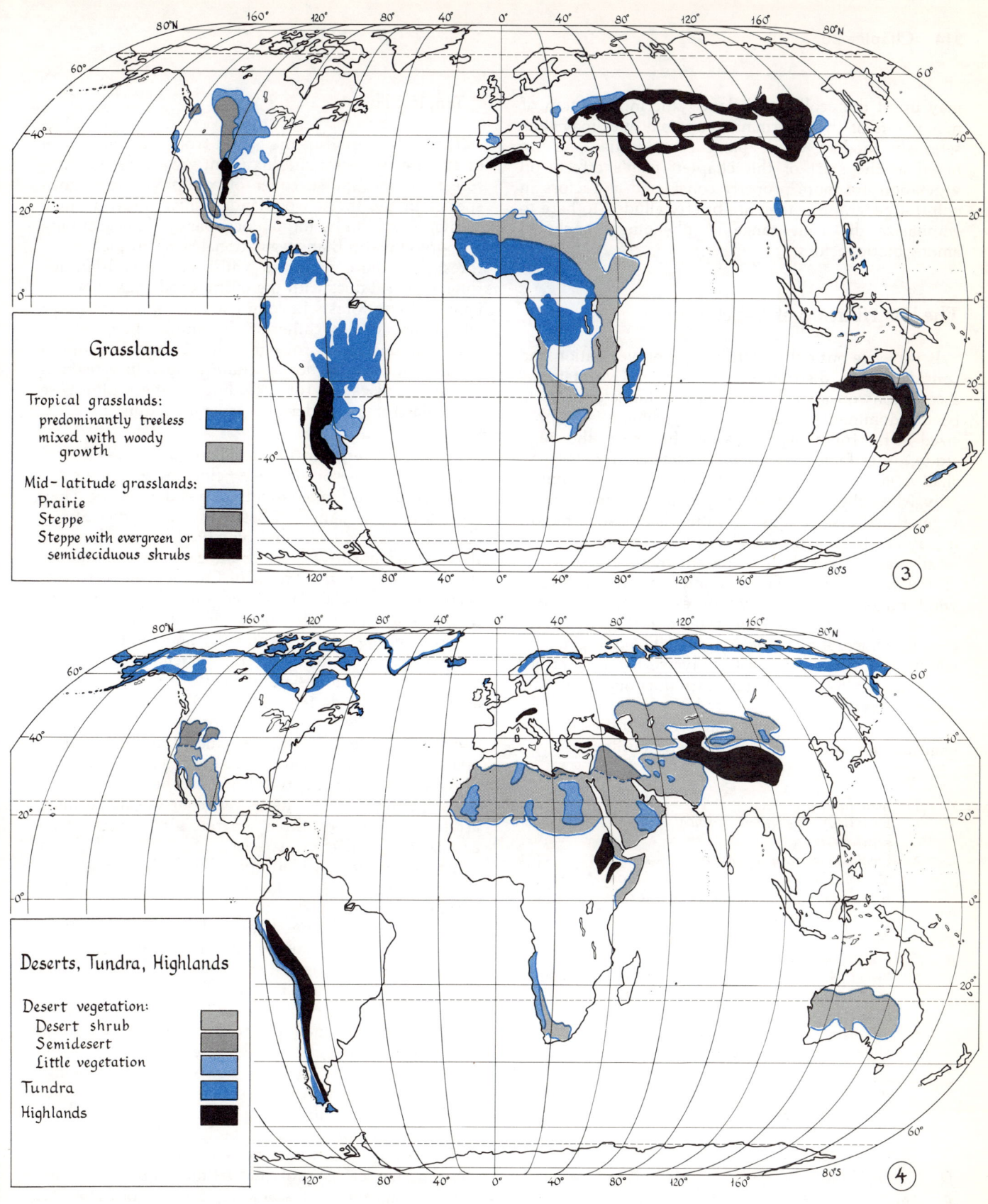

may upset the ecological balance; the introduction of new plants or the extermination of old ones may thus seriously modify whole plant communities. As we noted at the start of this chapter, man's effects on vegetation are more properly considered as factors in cultural geography; they are briefly alluded to here to emphasize the wide range of phenomena included among biotic factors.

The Distribution of Vegetation

If we leave out cultural influences on vegetation and consider the distribution of major vegetational associations on a world-wide basis, climate appears to be the dominant influence; hence, climate is emphasized in the following discussion. In a more thorough study, other factors would be considered in detail. We shall refer to the influence of factors other than climate only when they result in fairly sizable occurrences of vegetational associations. Elevation is one such factor; it may create an extreme diversity of vegetation within small areas, particularly in mountainous areas, where elevation changes very rapidly and exposure to sun and wind differs markedly within extremely short distances. One of the maps of vegetation (Fig. 20-3 (4)) therefore includes a category of highland vegetation, which is, however, much too diverse to be described adequately in a general introduction to plant associations.

The categories of major plant communities that we shall use are shown in the following outline.

A. *Tropical forests*
  1. *Broadleaf evergreen tropical forests, or tropical rainforests*
  2. *Deciduous tropical forests*
     a. *Monsoon forests, or semideciduous forests*
     b. *Savanna woodlands, or dry forests*
     c. *Thorn woodlands*
  3. *Mangrove forests*
B. *Mid- and high-latitude forests*
  1. *Broadleaf evergreen forests of the humid subtropics*
  2. *Mediterranean broadleaf evergreen forests*
  3. *Broadleaf deciduous forests*
  4. *Temperate coniferous forests*
  5. *High-latitude coniferous forests, or taiga*
C. *Grasslands*
  1. *Tropical grasslands, or savanna grasslands*
  2. *Mid-latitude grasslands*
     a. *Prairie*
     b. *Steppe*
D. *Desert shrub*
E. *Tundra*

Tropical Forests

Tropical forests run the gamut from the multiplicity of evergreen species and luxuriant growth of the rainforest, to the sparse cover of xerophytic deciduous species in the thorn woodlands, which show a much greater alteration by human occupancy. This grouping is more climatic than vegetational. The tropical rainforest has more affinities with certain mid-latitude forests than it does with the thorn woodlands. The one item all these forests have in common is tropical heat; all plants are megatherms, meaning they require high temperatures for growth. Common tropical location and similar low-latitude temperature regimes are the justification for including all these otherwise dissimilar forests under one general heading.

Broadleaf Evergreen Tropical Forests. These forests occur on suitable soils in the *Af* climates, where the dry season, if any, is limited in duration and intensity. They are found in the Amazon Basin of South America and along the Pacific coast of Colombia and Ecuador, and they extend northward along the east coast of Central America as far as the Tropic of Cancer. Similar forests occur in equatorial Africa and on the east coast of Madagascar. They also occur in western India and Ceylon, in the Malay Peninsula, and on the Burma coast and interior lowlands; they extend southward to the islands of Indonesia and eastward to New Guinea and the Fiji Islands (Fig. 20-3 (1)).

In general, these tropical rainforests have four major components: trees and shrubs, vines, epiphytes (a plant attached to another plant above ground and sometimes drawing nourishment from it), and, where light conditions permit, coarse herbs. The trees are usually arranged in two or three separate strata, each characterized by different species. The highest layer is composed of rather widely spaced trees that rarely form a continuous canopy. The second layer usually has a continuous mass of tree crowns. The trees in the third layer thus grow in the shade and usually have small, slender trunks and narrow, tapering crowns. In some instances, the two highest strata are both open, so that it is the third layer that provides the dense canopy, but the more common circumstance is for the upper layer to be open, the second layer to be dense, and the third to be open (Fig. 20-4). It should be emphasized that the trees in the lowest layer are not only young saplings waiting for an opportunity to grow into the next higher layer but are also adults adapted to the particular light and humidity conditions of the layer they inhabit.

The tropical rainforest does not have the seasonal rhythm characteristic of mid-latitude forests. Flowering, fruiting, and the loss and replacement of leaves can

The Environment and Distribution of Vegetation 315

Fig. 20-4. An evergreen broadleaf tropical forest.

they send roots down to the soil and become independent of their host. The vines then form an interlocking web of growing roots around the trunk that gradually kills the host tree by strangulation.

Some epiphytes (Greek *epi* on, upon + *phyton*) do not have any ill effects upon the host, although they get their nourishment from the organic matter contained within the tree bark. They often have special growth forms to trap rainwater, since they do not send roots into the tissue of the host plant. The number and diversity of epiphytes is great, and their presence especially distinguishes the tropical rainforest from the temperate forest communities. Among some of the more important epiphytes are the bromelias and orchids.

Herbaceous or herb-like growth is not plentiful on the shaded floor of the forest. There is competition for light, for root space, and even for soil moisture in a typically dense tropical rainforest; for example, some of the undergrowth has purple leaves, especially adapted to use the ultraviolet light that filters through the upper canopies. Giant herbs and coarse grasses grow in clearings where tree growth is checked, perhaps by periodic flooding. Various wild species of the banana family are among the more common inhabitants of these openings.

The gross form of the tropical rainforest, its year-round growth, and its dense foliage are dependent on the warm, wet tropical climate. The rainforest provides the special environments for many epiphytic, parasitic, and other herbaceous plants that can exist nowhere else.

Deciduous Tropical Forests. These forests occur where there is a seasonal alternation of water surplus and deficit. Several kinds of tropical forests reflect this seasonal rhythm. In order of decreasing rainfall in the environment, they are the monsoon forests, the savanna woodlands, and the thorn forests.

Monsoon forests are associated with the *Am* climates of India, southeast Asia, and Australia. They are also found on the margins of the tropical rain-forest in Africa, Madagascar, Indonesia, in South America, and in Central America (Fig. 20-3 (1)). The vegetation is less luxuriant than in the tropical rainforest; it is more open and not so clearly separated into layers. Many of the trees, such as teak, shed their leaves during the dry season, the degree of leaf fall depending upon the severity of the drought. On the other hand, some trees retain their leaves even in the dry period. Furthermore the dry season is usually the time of flowering, among other plant activities. The monsoon forest is clearly not **deciduous** in the sense that leaf fall marks the end of an active season and the beginning of a dormant season. Hence, it is often referred to as being **semideciduous**.

In the semideciduous forests, the trees are relatively

take place at any time of the year. Most trees lose and replace their leaves continuously through the year, so that the individual tree is evergreen. Other species renew their leaves in a shorter period, so that some bare trees can be observed in the forest, even though the forest as a whole remains evergreen. The trees of the tropical rainforest are characteristically hydrophytes, meaning that they have slender trunks, thin bark, broad leaves, and buttressed roots. Many of them produce very hard, heavy wood.

The numerous woody vines of the tropical rainforest are called *lianas*. They are dependent on the external support of trees to reach the light at the top of the forest. The vines climb to the crowns and branch profusely from tree to tree, thus linking the crowns so firmly that if a tree is cut at the base, it often remains standing. Some vines cling to the trunks by small, rootlike tendrils. Others scramble upward by the use of spines or wide, angular branches. Some climbing vines are best identified as stranglers. These plants begin their lives as epiphytic growths on tree branches. Later,

Fig. 20-5. Mango trees in Tanzania. The round crown and short trunk of the mango are characteristic of the trees in the semi-deciduous monsoon forest. *Photograph by D.R. Altschul.*

short and have thick bark and round, large crowns. The mango tree shown in Fig. 20-5 has the characteristic shape of this type of vegetation. The forest may consist primarily of trees with touching crowns, or the canopy may be interrupted. In the latter case, there is a lower layer of dense undergrowth. The leaves of the trees are usually large; and because they function mainly during the rainy season, they show little xerophytic adaptation. Climbers are fewer and smaller than in the tropical rainforest and are frequently herbaceous rather than woody. Generally speaking, the vegetation is less luxuriant than in the rainforest, but the multitude of species is almost as great, despite a history of much more intensive human occupancy.

The savanna woodlands, as the name suggests, occur in regions of *Aw* climate, where the dry period is longer and the total annual rainfall less than in the monsoon forest. They occupy much of the Brazilian Plateau, large areas of eastern and central Africa, much of central India, and the margins of the Caribbean (Fig. 20-3 (1)).

The savanna woodland, or dry forest, varies in density from an open, park-like assemblage to a relatively dense stand. Even where it is densest, however, visibility is never much less than 100 yards (about 100 m), in great contrast to the limited view afforded by the rainforest. The trees are normally 40 to 60 feet (12 to 18 m) high, have thick bark, and usually develop an umbrella-shaped crown (Fig. 20-6). The leaves are small and provide little shade. The species comprising this forest are limited in number and consist mainly of acacia, mimosa, and other members of the legume family. The forest floor is covered with grass; there are few lianas and epiphytes. In winter, the trees are leafless, and the whole woodland assemblage is dry. Fires commonly sweep through the grass, and only trees large enough to endure this periodic burning are able to survive.

In some places, the forest appears to be giving way to grass because of periodic killing of saplings by fire, often deliberately set by man. In other areas, fire-resistant trees have been able to survive in a fire-climax woodland. The respective roles of climate, natural burning, and deliberate firing in creating particular kinds of savanna woodlands are not easy to separate. Furthermore, the close connection between large, level stretches of land and the occurrence of savanna vegetation suggests an additional complication in assigning causes to the present distribution of this deciduous tropical woodland.

Thorn woodlands are the most xerophytic of the tropical forests, since they occur on the dry margins of the savanna bordering the BS climates. They are found in northeastern Brazil (where they are commonly referred to as *caatingas*), on the low islands in the Caribbean, in southwestern Africa, in the central part of northeastern India, and in the west and center of Australia (Fig. 20-3 (1)). They also occur on well-drained, sandy soils or on clay-rich soils in the more humid areas of the savanna climate. This woodland is difficult to map because it intergrades with the savanna grasslands. It is, therefore, often combined with the latter as a single category on small-scale maps.

Many of the trees of the thorn forest belong to the legume family. They are thorny and show markedly xerophytic traits; their leaves are small, and the trees have deep roots. Other species store water in swollen or

Fig. 20-6. The savanna woodland, or dry forest, of south-central Tanzania. The grass-covered forest floor is characteristic of this vegetation type.

The Environment and Distribution of Vegetation 317

Fig. 20-7. A baobab tree in eastern Tanzania. In this moist savanna climate, patches of thorn forest are associated with soils that are rich in clay and sometimes saline.

large trunks; among the latter is the African baobab, with its great, thick trunk, long roots, and water-holding tissues (Fig. 20-7). Cactus and other succulents are scattered throughout the thorn forest of the Western Hemisphere, whereas the giant euphorbias represent this niche in the Eastern Hemisphere.

Mangrove Forests. The most characteristic and important vegetational type along tropical shores is the mangrove forest. The mangrove has a number of unusual characteristics that enable it to thrive in the saline environment of tropical tidewater. The plants are salt-tolerant, and some have a leathery or fleshy foliage with a glossy surface that protects the leaves against excessive transpiration. The root system consists either of strut-like members (stilt roots) that are above ground (and above the water) and are provided with breathing pores or of a series of conical offshoots from the buried roots that stick out of the surface like growing asparagus shoots. These aerating roots conduct oxygen into the underground parts of the root system. In some species, the seeds, which often germinate while still attached to the parent tree, drop roots downward into the mud. Or if the tide is in, the seeds can be transported by water to another locale suitable for growth.

As a result of these special adaptations, mangrove forests are able to develop on mud or sand flats that are exposed at low tide but are otherwise covered by salty water (Fig. 20-8). Especially favorable conditions for the growth of mangroves occur in creeks or lagoons and in quiet bays. The waterborne seeds or seedlings can root on the tidal flats and grow into a dense forest.

Mangroves often spread inland along brackish swamps and in lagoons, in a succession of mangrove forests that differ in appearance. The contrast between the forest that grows in the open water and that which develops on the inland margin, where seawater comes for only a brief period during high tide, is often quite marked. Some mangrove species may even grow in fresh water or in areas which are dry for months at a time.

Mangrove forests grow along the southern and eastern coasts of Asia, in northern Australia, and in the Melanesian Islands. They also occur in Central America and along the tropical shores of South America. Some degree of mangrove growth usually takes place along arid shores in tropical or subtropical environments where the water is warm, but not along the west coasts of continents washed by cold ocean currents. In general, the mangrove forests extend only marginally into the subtropics; they belong in the special habitat provided by warm-water tropical coasts. Although mangrove forests are an important plant association, they do not appear in Fig. 20-3 (1) because their narrow linear distribution cannot easily be shown at the scale of the world maps of vegetation.

Mid-Latitude and High-Latitude Forests

This group of forests spans the climatic range from subtropic to subarctic and thus involves a correspondingly wide variety of vegetation associations. In rough order of increasing latitude, broadleaf evergreen forests occur in the humid subtropical and Mediterranean climates; broadleaf deciduous forests, in the humid continental climates; and coniferous evergreen forests, in the mid-latitude marine and subarctic climates — in places of particularly severe climate, deciduous conifers

Fig. 20-8. Mangrove trees growing in a lagoon beach north of Dar-es-Salaam, Tanzania.

may be found. At the humid subtropical margin of these associations, the forest is strongly reminiscent of the tropical rainforest, but poleward, the seasonal rhythm of a short, active summer and a long, dormant winter becomes more and more pronounced. The winter is cold and the water frozen, so that the forest species become increasingly xerophytic in order to survive.

Broadleaf Evergreen Forests of the Humid Subtropics. These forests are found in association with the humid subtropical climates *(Cfa* and *Cwa)*. They are lower and less luxuriant than the tropical rainforest; they have fewer vines, epiphytes, and buttressed tree trunks; they also reflect the greater seasonality of the climate by the admixture of deciduous trees. Such forests occur locally along the Gulf coast of the United States and more extensively in southern Japan and China, southwestern South Africa, southern Chile, and in New Zealand. They also occur in southeastern Australia (Fig. 20-3 (2)).

In the forests of the southeastern part of the United States, live oak predominates, although evergreen magnolia is common. The forest distribution is limited by several physical factors, such as sandy soils, poor drainage, and fire. In regions of poor drainage, the vegetation is dominated by bald cypress, while in sandy areas frequently swept by fires, pine replaces the oak and magnolia.

In other areas of temperate rainforest in the Northern Hemisphere, oaks, laurels, and magnolias are dominant. Numerous shrubs form a dense undergrowth; woody climbers, epiphytes, and ferns are plentiful. In New Zealand, the broadleaf evergreen forest includes many large tree ferns, as well as conifers and some species of the Southern Hemisphere beech. The forest in Australia is dominated by the southern beech and the eucalyptus.

Mediterranean Broadleaf Evergreen Forests. The dry summers and mild, moist winters of the Mediterranean climates *(Csa* or *Csb)* are associated with a broadleaf evergreen forest with some admixture of coniferous species (Fig. 20-3 (2)). This Mediterranean woodland, where it is least disturbed by man, consists of a mixed assemblage of brush 3 to 10 feet (1 to 3 m) high and trees. The dominant broadleaf trees are low, rounded evergreen oaks, although in Australia the vegetation consists primarily of eucalyptus. The trees and shrubs are well adapted to withstand summer drought, since the small, hard, and waxy leaves retard the loss of moisture from their surfaces. The short tree trunks are covered with thick bark, and the root systems are deep. Many of the grassy and herbaceous plants are conditioned to flower and set seed by decreasing moisture and increasing temperature in

Fig. 20-9. Oak-sagebrush chaparral in the Santa Monica Mountains, California. *Photograph by R.L. Day.*

spring. On the other hand, germination is encouraged by increasing moisture and decreasing temperature in late fall.

Most of the Mediterranean woodlands have been strongly modified by man. The present vegetation consists mostly of brushy thickets of mixed deciduous and evergreen bushes, known as **chaparral** in California and *maquis* or *garigue* in the Mediterranean (Fig. 20-9). This brush may represent the original vegetation cover in some regions; but more likely it is secondary growth, replacing a mixed assemblage of woody shrubs and chestnut-oak forest that was probably destroyed by man. The destruction of the Mediterranean forest in the Cape region of South Africa and its transformation into a low cover of brush has been well documented. In the Mediterranean Basin, the same vegetational change probably took place as early as the first millennium B.C.

Broadleaf Deciduous Forests. These forests are composed mainly of broadleaf trees that lose their leaves in winter. They constitute the main vegetational community that once covered much of temperate Europe, eastern Asia, and the eastern United States (Fig. 20-3 (2)). These areas have cool to cold winters and mild to hot summers, with adequate year-round precipitation. From the physiological standpoint, winter is the dry period, either because low temperatures hinder the absorption of water by the roots or because the water is frozen for periods of time. To prevent excess loss of moisture, which is chiefly from the leaves, the plants are leafless in winter.

The deciduous mid-latitude forest has a single stratum of tall trees, with shrubs and young trees forming a partial substrate. The forest floor is usually open and receives considerably more summer sunlight than the floor of the tropical rainforest (Fig. 20-10).

Small perennial herbs send up shoots and leaves early in the spring, flowering and fruiting rapidly before the full-leaf canopy of the forest develops. During the summer, these small plants die or remain green but inactive.

All mid-latitude broadleaf deciduous forests are structurally similar and follow the same leafing habit, but they have different floristic compositions. In western and central Europe during the Pleistocene, the forest was trapped between the continental ice sheet and the piedmont glaciers descending from the Alps. As a result, many of its tree species perished. In the United States and in eastern China, there was ample room for southward migration as the ice sheets expanded. Consequently, the American and Asian forests are presently more varied and luxuriant than their European counterparts. Many species of oak, beech, birch, hickory, walnut, maple, elm, ash, tulip trees as well as pines and other conifers abound in North America. In eastern China, however, very little of the original forest has survived the thousands of years of human occupancy.

Temperate Coniferous Forests. Conifers have narrow, needle-like, or sometimes scale-like leaves that are replaced continuously during the year. They frequently have gummy sap, thick bark, and well-developed root systems. Most conifers are evergreen, and all of them bear seeds in the form of a cone — hence their name.[2] The dominant coniferous trees are spruce, pine, fir, larch, and hemlock; but some coniferous forests also include deciduous broadleaf trees, such as birch, beech, and maple.

Coniferous forests occur in mid-latitudes under a variety of precipitation and temperature regimes and occasionally intermixed with broadleaf deciduous trees (Fig. 20-3(2)). Indeed, they extend well into the tropics, for the pine is one of the most widespread of all tree genera and is found on sandy soils or on highland slopes almost everywhere in the Northern Hemisphere. But expanses of coniferous forests large enough to show on a small-scale map of the world are rare in temperate latitudes. Small remnants of what may once have been large stretches of coniferous forest occur in Europe and the Far East, but the major areas are all in North America. These include the coniferous forest of the Pacific Northwest, the pine barrens of the southeast, and the remnants of a pine-hemlock forest around the Great Lakes.

The moderate temperatures and abundant winter precipitation of the Pacific Northwest support a dense coniferous forest that extends from southern British Columbia to northern California. This forest includes some of the world's largest and tallest trees, such as the coastal redwood and the Douglas fir, which reach heights of more than 300 feet (90 m) and girths greater than 60 feet (18 m) (Fig. 20-11). Other conifers, such as spruce, fir, and pine, are also present, as well as a considerable admixture of deciduous trees, such as bigleaf maples and both deciduous and evergreen oaks.

The southern pine barrens is an extensive coniferous forest region that occupies the Atlantic and Gulf coastal plains from New Jersey to eastern Texas. It is composed of pine with angular branches, a thick, somewhat reddish bark, and a fairly sparse canopy (Fig. 20-12). Among the primary species in the forest are short leaf, slash, loblolly and long leaf pine. This coniferous forest

Fig. 20–10. Deciduous broadleaf forest of Indiana. The relatively open floor of the forest encourages the growth of shrubs and herbaceous plants. Several small trees and herbaceous plants are clearly visible in the foreground. *Photograph courtesy of the U.S. Department of Agriculture.*

---

[2]The major deciduous conifer is the larch, which grows in the mountains of northwestern North America, in Quebec and maritime province of Canada, on the humid margins of the northern Great Plains, and in the highlands of central Siberia. It is more common in the taiga than in the temperate coniferous forest. Bald cypress and Sabino are also deciduous conifers (or near-conifers).

Fig. 20-11. Redwoods in Muir Woods National Monument, north of San Francisco. Note the lush undergrowth in this "temperate rainforest". Northward the redwoods give way to firs and spruces. *Photograph by George A. Grant, National Park Service.*

Fig. 20-12. Typical pine forest of the Atlantic coastal plain northeast of Gainesville, Florida. Periodic, controlled burning is necessary to suppress the growth of palmetto shown here on the forest floor. *Photograph by R.L. Day.*

is somewhat out of place in an area of high rainfall and long growing season. To be sure, the soils are sandy and evaporation is high, so that the root zone of the trees is quite dry; these factors might account for the presence of a xerophytic pine forest. In recent years, however, it has been observed that wherever the vegetation was protected from burning, broadleaf trees such as magnolia and sweet gum were able to take over the forest in a matter of 40 or 50 years. If periodic burning is resumed, the broadleaf species are virtually eliminated, and the thick-barked, fire-tolerant pines reestablish their dominance. The existence of the southern pine forest is probably related more to the control of fire by man than to soil and climate.

Another great coniferous forest once occupied the southern shores of the Great Lakes and extended from northern Minnesota eastward into Maine. This forest association, in a region of continental climate with cold winters, cool summers, and moderate precipitation, was dominated by white pine, red pine, and hemlock. It probably did not represent a climatic climax, since it was best developed on sandy outwash plains of glacial origin. In any case, little of the forest remains standing because the timber has been cut down or destroyed by fire.

High-Latitude Coniferous Forests. "Taiga" is the name given to the coniferous forest that occurs in the subarctic climates of North America and in Eurasia, where it forms the largest single forested area on earth (Fig. 20-3 (2)). Conifers, such as larch, spruce, fir, and pine, predominate, although willow, birch, and beech also occur. The xerophytic character of the conifers has survival value in the subarctic because the soil is frozen much of the year; moisture is available only during the short summer season of 1 to 3 months. Even in summer, water absorption by roots is retarded by the coolness and the acidity of the soil. Since growth is slow because of the long winters and cool, short summers, the trees are relatively small. The trunks are usually not more than a foot in diameter, and tree height is frequently less than 50 feet (15 m). Trees are larger on the southern margins of the subarctic forest, where the summers are longer and warmer; but the northern margin of the forest is at the climatic limit of tree growth, and the growing season is extremely short and very cool. The trees are very small; even 50-year-old trees are scarcely more than 5 feet (1.5 m) high.

In central and eastern Siberia, where the winters are not only extremely cold but also extremely dry, the loss of moisture from plant surfaces is excessive. The deciduous habit of some conifers is essential for survival under such circumstances. The Dahurian larch, for example, is the dominant species in the forest of north-central Siberia. The climate of this part of the world not only restricts growth because of winter drought and short summers, but it also creates a permanently frozen subsoil. As a consequence, trees are small, shallow rooted, and extremely slow growing.

Grasslands

The presence of grasslands, whether they be savanna grasslands in the tropics or prairie and steppe grasslands in mid-latitudes, has long been related to the occurrence of subhumid and semiarid climates. Today, other causal factors are beginning to receive considerable attention. Even though the evidence is incomplete and somewhat contradictory, it is now obvious that climate is only one of several factors that affect the distribution of grasslands. Landforms, soils, and the actions of man, particularly with regard to fire, must all be taken into consideration.

Let us begin our discussion by emphasizing three special characteristics of grasses:

1. *Grasses need large amounts of calcium and flourish in areas where there is an abundance of lime in the soil.*
2. *Grasses are characteristically quite adaptable to periods of prolonged wetness followed by periods of intense drought.*
3. *Grasses have an annual growth habit and so are quite resistant to fire. The roots survive the destruction of the leaves and are available for growth at the beginning of the next rainy season.*

From these characteristics, we can deduce that grasses should grow better than trees in areas characterized by alternate wet and dry seasons and by repeated burning over a long period of time.

Savanna Grasslands. The tropical savanna grasslands present a varied appearance ranging from an almost complete grass cover to a mixture of grass and scattered trees that occasionally occur in small clumps (Fig. 20-13). These tropical grasses are found in both the tropical savanna and the steppe climates (Fig. 20-3 (3)). On the humid edge of the *Aw* climate, the dominant grasses may be 12 feet (3.6 m) high; on the arid margin of the *Aw* or in the *BS* climate, they may be less than 2 feet (60 cm) high. This variation in height depends not only on rainfall but also on location. The grasses of Africa and southern Asia are quite tall, whereas South American grasses rarely exceed 6 feet (1.8 m) in height and may be much shorter.

In general, savanna grasses grow in bunches, and bare ground is commonly visible between clumps. The continuous sod characteristic of mid-latitude grasslands does not occur in the tropics. The leaves of the savanna grasses are frequently stiff and leathery; only the fresh shoots are truly suitable for livestock grazing. As a

Fig. 20-13. Palm savanna grassland, Tanzania. The branching trees are fire-resistant Doum palms.

result, the grasses are burned in the dry season to make room for new growth at the beginning of the rainy seasons. This repeated burning accounts, in part, for the lack of sod and for the bare ground between the clumps of grasses.

For some time, the savanna grasslands, found in areas of seasonal rain and drought, were thought to be a transitional vegetation type between the tropical broadleaf forests and the deserts. In recent years, this theory has been seriously questioned on a number of grounds. For one thing, the characteristic vegetation in many parts of the *Aw* climates is a savanna forest and not a grassland. Furthermore, woodland growth occurs in the tropical steppes, which are even drier than the *Aw* grasslands. Woodlands are thus found in areas that are as dry as or drier than grasslands. Hence, it appears that the original vegetation of many of the savanna grasslands may have been a dry tropical forest and that the grasses have taken over for nonclimatic reasons.

Tropical savanna grasses are frequently associated with nearly level land that has fairly good drainage. This grass vegetation may be the result of repeated burning by man. The grass survives on the fire-swept flats, but the perennial trees can survive only in protected declivities or on higher ground. In some areas of grass, as in the pine savannas of Honduras, the soils are very sandy. Since pines and grass are better able to survive the low moisture content of the soil than other species, it seems reasonable to ascribe the presence of these grasses to the natural edaphic or soil conditions. There is evidence, however, that the undisturbed original vegetation was brush and conifers, which were destroyed by the use of fire to improve grazing. Very small differences in elevation may exert a considerable influence in areas of seasonal flooding. In the llanos of Venezuela and the *pantanal* of southwestern Brazil, the grass vegetation may result

Fig. 20-14. Coastal Plain Savanna, Tanzania. The small clump of trees left of center is growing on a termite mound.

from alternate drought and flooding by vast, shallow sheets of water. Grass grows on the flat surfaces; clumps of trees or small forests patches occur on areas of well-drained higher ground. A similar arrangement is found in some of the savannas along the coastal plain of East Africa, except that the tree clumps are on termite mounds which are akin to very large ant hills, and the grass grows on the nearly flat surface between the mounds (Fig. 20-14).

Savanna grasslands cover enormous areas (Fig. 20-3 (3)). In addition to the llanos and pantanal, South America has a third major tropical grassland region in the *campos* of south-central Brazil. In Africa, the savanna grasses extend in a broad band across the Sudan from the Atlantic Ocean to the eastern highlands and southward into Tanzania and Mozambique. Savanna grass also occurs in western Madagascar and in the interior of Australia around the central desert, especially along its eastern and western margins.

Mid-Latitude Grasslands. The tropical grasslands, whatever their origin, typically include scattered trees or shrubs. The mid-latitude grasslands are usually treeless except along water courses (Fig. 20-15). All the mid-latitude grasslands are dominated by extremely hardy, perennial grasses with narrow leaves. Usually, the rooting is shallow and forms a matted turf, which holds the rainfall and prevents erosion. It is believed by some that the sod of old, dead leaves may prevent the germination and growth of trees. There are, however, considerable areas of steppe grasslands in which shrubs, both evergreen and deciduous, dot the landscape (Fig. 20-3 (3)).

Apparent climax grasslands occur in mid-latitude plains with an average annual precipitation of 10 to 30 inches (250 to 750 mm). In general, these areas have cold and fairly dry winters and warm or hot summers; precipitation is at a maximum in late spring and early summer. The mid-latitude grasslands thus occur mainly in the interior of continents and are more extensively developed in the Northern Hemisphere than in the Southern Hemisphere. There are four major types of grasslands, structurally similar but floristically diverse; (1) the prairies of central and western North America, (2) the steppes of the Soviet Union, (3) the Pampa of Argentina, and (4) the scattered grasslands in the high veldt of South Africa, in southern Australia, and on the east coast of South Island, New Zealand (Fig. 20-3 (3)). In all these places, the original grasses have been greatly changed or completely destroyed by grazing, by cultivation, or by the introduction of alien plants.

The original prairies of central and western North America are almost completely gone. When European farmers first settled the area, there were three major types. The tall-grass prairie was dominated by tall, luxuriant, and deep-rooted grass; forests were confined to river valleys. These prairie grasses were more than 5 feet (1.5 m) tall, and under favorable conditions, grew to twice that height. Westward, with decreasing precipitation, the shorter mixed-grass prairie, characteristic of pioneer Nebraska, was found. Shrubs and trees occurred only in depressions and along water courses. Farther west, up to the base of the Rocky Mountains, was the broad belt of short-grass prairie, with grass less than 12 inches (30 cm) tall. The steady decrease in height toward the west was accompanied by a change from a complete cover of sod to scattered bunches of grasses on relatively bare ground. Much of this prairie is believed to be the product of repeated burning by the Indians, who used fire to drive game

Fig. 20-15. Short steppe grass on the smooth, rolling hills of the inner Coast Ranges east of Oakland, California. *Photograph by R.L. Day.*

and perhaps to preserve the grass against encroachment by the forest.

The steppes of the Soviet Union cover large areas south of the coniferous forest and north of the interior desert. They extend from eastern Europe deep into eastern Asia, with large, isolated patches occurring farther east and south. They too have been greatly reduced by cultivation. These steppe grasslands are similar to those of North America in their structural appearance, although they differ greatly in floristic composition. By and large, the grasses are shorter and are mostly composed of hardy perennials.

The grasslands of the Argentine Pampa may have been very similar to the prairie grasslands in the midwestern United States. They occurred on flat to gently rolling plains and probably were not a climatic climax. Edaphic factors and fire played important roles in the origin of the grassland. In any case, most of the present grasses were introduced from western Europe; as in North America, the original sod has been almost completely eradicated.

The grasslands of the veldt of South Africa and those of southern Australia are also comparable to the northern steppe and prairie of the central United States. The grasses are confined to the rolling or nearly flat interfluve surfaces, and the trees are confined to stream courses or ravines.

Desert Vegetation

Desert vegetation is closely associated with *BW* climates (Fig. 20-3 (4)). It consists of a scattered growth of plants with strong xerophytic characteristics, separated by considerable areas of bare ground or ground occupied only by very small plants (Fig. 20-16). The plants are adapted to withstand severe drought in many ways. The woody shrubs have long roots, en-

Fig. 20-17. The thick, fleshy stem of the saguaro is characteristic of many cacti in the American deserts. Here a palo verde in the Organ Pipe Cactus National Monument is serving as the nurse plant. *Photograph by R.L. Day.*

Fig. 20-16. Scattered, drought-resistant plant growth characteristic of the desert vegetation of southwestern Arizona. *Photograph by R.L. Day.*

abling them to obtain moisture well below the surface. They show other xerophytic characteristics, such as short trunks, thick bark, and evergreen leaves. Some desert plants endure drought by drying up almost entirely without harm to themselves. These brown, dusty-looking objects appear dead, but immediately after a rain they become active. Other plants, such as cacti, store water in their stems or other organs (Fig. 20-17). These succulents usually have well-developed root systems near the surface, enabling them to gather soil moisture whenever it is available. Other desert plants are able to survive because of the extraordinarily resistant quality of their seeds. These may lie on the ground for several years until the proper amount of moisture is absorbed by the soil. When the soil contains sufficient moisture for the plants to fulfill their life cycles, the seeds germinate; and the plants grow and produce a new set of seeds in a very short time. Other plants survive by virtue of their smallness and modest requirements for water. Some are so small they cannot readily be seen without a hand lens.

While desert vegetation exhibits some variation in detail from continent to continent, it is nearly the same from a structural standpoint, at least in those areas where it has not been seriously affected by man's activities. In general, the New World deserts are more spectacular than the Saharan and central Asian drylands. The large, tree-like plants of the Sonoran and Mohave Deserts are a case in point. Saguaro, opuntia cacti, ocotillos, and Joshua trees give these deserts the appearance of colorful, prickly woodlands. Even where the vegetation is low and gray, as in the sagebrush and shad-scale desert of the northern part of the Great Basin, it covers a considerable portion of the ground, although it consists of distinctly spaced bushes (Fig. 20-18). Similar stands of aromatic, woody brush from 1 to 6 feet (30 to 180 cm) high are found in other areas that have 5 to 10 inches (13 to 25 cm) of precipitation and rather cold winters with slight snowfall, such as the desert-steppes of the Soviet Union between the grasslands and the true desert, the deserts of central Asia, the eastern edge of the Sahara, and the periphery of the South African desert.

The most arid areas have very little vegetation; the plants are small and widely scattered. Some of the deserts, particularly those of the Old World, have extensive sand-dune regions. In some cases, the dunes may be partially fixed by shrubby growth; in other areas, they are on the move and free of vegetation. Permanently frozen ground, glaciers, and moving dunes are about the only completely barren environments.

Fig. 20-18. Sagebrush semidesert in south-central Oregon. The foreground is a blowout.

Tundra

The tundra lies poleward of the tree line or above it, in close association with the *ET* climates (Fig. 20-3 (4)); that is, where the average temperature of the warmest month is below 50°F (10°C) but above freezing — a considerable range of possible summer temperatures and an even greater variation in winter (see Fig. 9-2). A detailed study of the tundra vegetation would reveal numerous subtypes that reflect the wide range of climatic conditions. In general, however, differences in vegetation within the tundra are related to latitude (or altitude), and we can describe the gradation of vegeta-

Fig. 20-19. A tundra landscape near Cape Dorset in the Northwest territories of Canada. This summer view shows a nearly continuous cover of low growth. The color of the ephemeral plants gives way to the purely mineral world of the uplands whose dark color is relieved only by occasional patches of snow. *Photograph by J. Feeney, Information Canada Photothèque.*

tion as a relatively simple latitudinal (or altitudinal) arrangement.

The vegetation of the tundra is composed of a few dwarfed, woody plants, ephemeral herbaceous plants, sedges (grass-like plants with solid stems), mosses, and lichens (Fig. 20-19). The warmest part of the tundra includes a number of stunted and misshapen willows and birches scattered among the sedges and the ephemeral plants that give the tundra a brief period of color when they bloom. Poleward, the sedges and herbaceous growth give way to the duller tones of mosses and of lichens. The lichens dominate the landscape where temperatures become extreme. Eventually, even they give way to a mineral landscape of rock and snow, in which only the most favorable microhabitats harbor a small number of especially hardy plants.

This simple description does not imply that the tundra is less varied than other vegetational associations. Altitude, for example, is as important in creating local contrasts in the tundra as it is elsewhere. Microhabitats of varying attractiveness to plants abound. One example is the south-facing slope of the arctic tundra. It not only receives more insolation than a horizontal surface, but it also protects the vegetation from north winds. In addition, **solifluction** occurs in the slopes, and the drainage, although poor, is infinitely better than on the waterlogged permafrost of the flat bottom lands. As a result, the plant community found on such a slope is distinct from that of the surrounding flatland.

Summary

In this chapter we have studied the distribution of the world's natural vegetation and some of the factors that affect it. Almost all the physical processes we have studied play a role in determining the composition and distribution of natural vegetation. But man's role in modifying and removing vegetation is so important that we could not limit the discussion to physical processes, even by excluding domesticated plants and vegetation obviously modified by human action. It is apparent that the distinction between physical and cultural geography is arbitrary, particularly in relation to plant distributions on the surface of the earth.

Aside from man, the single most important factor affecting the distribution of plants is climate. The major plant associations that were used as a basis for the description of the distribution of vegetation on a world scale are clearly related to climatic conditions. In many instances, the relation between a particular association and a specific climate is very close. This is not surprising, given the importance of heat, light, and moisture to plant growth and the fact that the climatic classification used in this text was elaborated from the evidence of plant distributions.

Other factors were more briefly dealt with, but the importance of landforms and soils, of other vegetation, and of time could not be excluded even in the highly generalized account of the distribution of vegetation. It should be obvious that the physical landscape is conditioned by all the physical processes we have been talking about and that the various elements interact in an open system, in which a change in a single element not only affects every other element but is in turn itself affected by the changes in the other elements.

Suggestions for Further Reading

Dansereau, Pierre, *Biogeography: An Ecological Perspective*, The Ronald Press Company, New York, 1957.

Eyre, S. R., *Vegetation and Soils*, Aldine Publishing Company, London, 1963.

Newbigin, Marion I., *Plant and Animal Geography*, 2nd ed., E. P. Dutton & Company, Inc., New York, 1948.

Polunin, Nicholas, *Introduction to Plant Geography and Some Related Sciences*, McGraw-Hill Book Company, Inc., New York, 1960.

21 The Nature and Formation of Soils

Introduction

Soils, of fundamental importance because they support plants which provide food for animals and men, are defined as a natural system at the surface of the regolith that is suitable for plant growth. This system, which is divided into horizons or layers of mineral and organic constituents that are usually unconsolidated and of variable depth, differs from the parent material below in physical properties and constitution, in chemical properties and composition, and in biological characteristics. The following discussion of the physical and chemical properties should provide a better understanding of the term "soil." The physical properties with which we shall be concerned are soil profile, color, texture, structure, porosity, and water. The chemical properties include mineral constituents, organic content, and chemical reaction.

The Nature of Soils

Physical Properties

Soil profile. The profile of a soil refers to its vertical zonation, as a result of soil-forming processes, into horizons of varying color, texture, structure, consistency, and chemical properties. The rock from which a soil is formed is vertically homogeneous with respect to its physical and chemical properties, but weathering processes operate most effectively at the surface of the soil and decrease in intensity with increasing depth. Other processes transfer material downward (and occasionally even upward) into the weathered layer. New material is added or existent material is removed to produce layers that are generally parallel to the surface. These horizons are most obvious in middle-latitude mature soils in which three main layers can usually be distinguished.

Soil horizons are identified alphabetically. The A horizon[1] in humid climates is the chemically leached and physically eluviated layer. **Leaching** is the removal of soluble material by percolating water; **eluviation** (Latin *ex* out of + *luere* to wash) is the downward removal of clay by suspension in the percolating water. The A horizon is often rich in organic material and is the

---

[1]The A horizon can be separated into sublayers by subscripts. A_{00} and A_0 refer to the horizon of organic debris lodged on top of the soil proper; in the A_{00} horizon, the debris is mainly undecomposed, and in the A_0 layer, it is partly decomposed. Below this in the soil the horizon may be further divided into the A_1 and A_2 horizons based upon the amount of humus each contains.

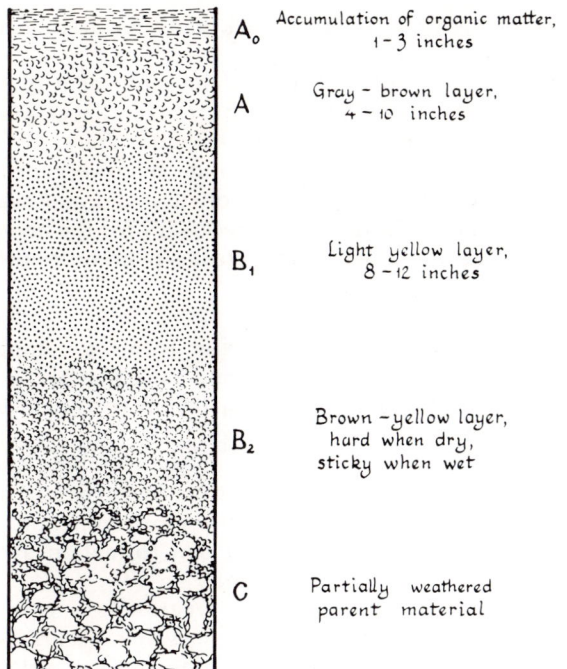

Fig. 21-1. A forest soil profile. No A_{00} horizon is shown, although there may be a thin layer of undecomposed organic matter present. The B horizon is divided, in this example, into B_1 and B_2 horizons, which differ in the amount of clay accumulation and in color.

layer of maximum biologic activity. Its depth, as well as that of lower layers, varies greatly from soil to soil.

Immediately below the *A* horizon is a zone of accumulation, the *B* horizon. Here, various leached and eluviated substances are accumulated. **Illuviation** (Latin *in* into + *luere*), the deposition of clay within the interstices of the coarser material, may cause the zone to become increasingly impenetrable to water.

The *C* horizon is the partially weathered, little-changed **parent material** from which the soil is derived. The transition from the *C* horizon to bedrock is gradual and often imperceptible (Fig. 21-1).

Soil horizons are important in distinguishing one kind of soil from another. The essential difference between two unlike soils may be not so much the difference in the physical or chemical properties derived from their parent materials as the contrast between the distributions of those properties in the soil profiles.

Soil Color. One of the most recognizable properties of soil profiles is color. The surface color changes from place to place. For instance, there is a marked regional gradation from the red soils of the southeastern United States to the black soils of the midwest. Even within a given soil, color changes with depth. For example, some mid-latitude forest soils have gray-brown *A* horizons and orange *B* horizons. The color of the *C* horizon is variable, since it depends on the local rock type from which the soil developed.

Soil color is a rough indicator of the organic and inorganic qualities of a soil. Black color in the *A* horizon is usually associated with a large amount of organic material and carbonates. In some instances, however, black color indicates undesirable soil properties, such as the presence of alkali in arid climates or the predominance of reduction[2] as a soil-forming process in wetter areas. Red indicates a concentration of iron oxide; yellow indicates the presence of limonite; and white may reflect the occurrence of lime, sand, kaolin, or gypsum.

Texture. Texture refers to the relative abundance of particles of various sizes. Standard categories of texture, such as clay, silt, and various sizes of sand, are quantitatively defined (see footnote 1, Chapter 14). The relative frequency of different particle sizes in a particular soil is recognized by such terms as "clay soils," "sandy clay loam soils," "sandy loams," "sand," and so forth. The major categories are shown in Fig. 21-2.

Textural quality is an important soil characteristic, since it partly determines the water capacity of the soil. It also affects the rate at which soils are capable of absorbing rainwater, a process called infiltration. For example, sand permits water to enter and, in some instances, pass through the soil profile rapidly, while abundant clay content may retard the rate at which the water enters the soil. The difference or similarity in texture between one horizon and another is a critical feature of any given soil.

Structure. Soil structure refers to **aggregates**, or the arrangement and grouping of soil particles into aggregates of characteristic and readily recognizable form. Structure is better developed in the *B* horizon than at the surface because the fine particles tend to move out of the *A* horizon to the *B* horizon and because the agents responsible for forming the structures are usually concentrated in the lower layer. The structure varies from granular, crumb like, or platy near the top of the *B* horizon to prismatic or columnar in the lower parts of the horizon. Occasionally, the structure may indicate conditions detrimental to soil water and aeration. For example, clay particles may be so structured that they lie one over the other much as bricks in a wall, thus preventing the entrance of air and water into the soil. In the presence of bases such as ions of calcium, magnesium, and potassium, the clay particles unite in small aggregates or clusters called **floccules** (Latin *floc-*

---

[2] Reduction, the opposite of oxidation, is the removal of oxygen from chemical compounds.

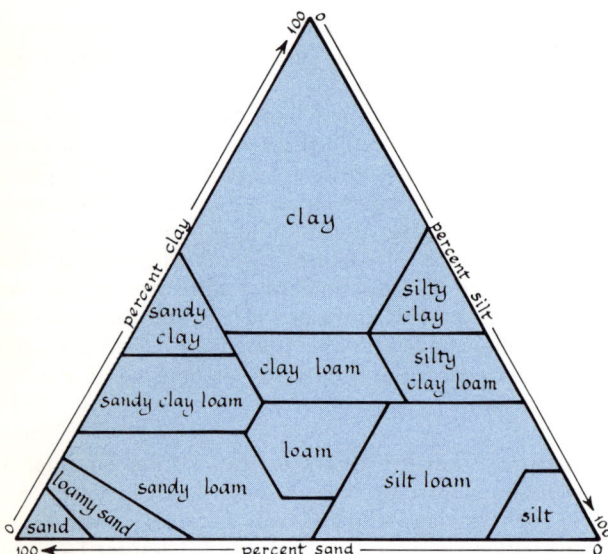

Fig. 21-2. Textural categories. Any soil sample can be placed in its proper category if we know the relative proportion of clay, silt, and sand in the soil. In the triangle, percentages of clay range from zero along the bottom line to 100 percent at the apex of the triangle; sand percentages go from zero along the right-hand side to 100 percent at the left-hand corner of the triangle; and silt percentages, from zero along the left-hand side to 100 percent at the right-hand corner of the triangle. Thus a soil containing 50 percent clay and 50 percent sand (hence no silt) belongs on the left-hand side (0 percent silt), half-way between the bottom side and the top corner (50 percent clay) and also half-way between the right-hand side and left-hand corner of the triangle (50 percent sand). This location as you can see is appropriately categorized as "sandy clay." The small triangle representing this category includes all soils with 45 to 65 percent sand (right-hand edge to left-hand corner); 35 to 55 percent clay (bottom edge to top corner); and hence 0 to 20 percent silt (left-hand edge to right-hand corner).

cus flock of wool). These aggregates give the soil a more open quality, thereby allowing penetration by water and air. Changes in soil structure may, therefore, have an important influence on the rate at which water is absorbed and retained by soil.

Soil Porosity. Soil porosity can be defined as the average amount of interstitial space that can be occupied by air and water per unit volume of soil. Porosity is a reflection of soil texture and structure. It is evident that in a soil of low porosity, plant roots have access to very little oxygen and not much water. Even with adequate or excessive rainfall, not much water can penetrate the soil, and most runs off at the surface. In a soil of great porosity, rainwater infiltrates rapidly into the large interstitial spaces and may run through the horizons quickly. Both high and low porosity may thus cause plants to suffer from lack of water despite adequate rainfall.

Porosity not only influences the movement of soil water and its availability to plant roots but also affects the amount of air found in the soil and, therefore, the respiration of plant roots and micro-organisms. Furthermore, the oxygen, carbon dioxide, and water vapor contained in the air are necessary for the organic and inorganic weathering that creates plant nutrients in the soil.

Soil Water. Water in the soil forms a complex, dilute solution of acids, bases, and salts that also contains minute bits of soil in suspension (between .002 and .2 mm in size) called colloidal soil particles. Water usually enters the soil from the top, but it may enter from below if the groundwater level rises. Water charged with acids or bases serves two functions; it becomes an important agent of soil formation and a source of mineral food for plants.

Soil water takes three forms: *hygroscopic*, *gravitational*, and *capillary*. Hygroscopic water occurs as a thin film on soil particles and colloids. It is more plentiful in humid climates than in arid ones and in soils rich in colloids rather than in those low in colloids. Hygroscopic water does not move through the soil but clings tightly to soil particles and is not available to plant roots. Gravitational water is the soil water in the process of percolating from the surface to the groundwater table. During heavy or prolonged rains, it fills the soil pores, except for occasional bubbles of trapped air. Once the rain stops, gravitational water may continue to move downward, but some of it adheres to the soil particles by surface tension, or capillary tension, and is called capillary water.

Evaporation and withdrawal by plants thins the film of water around the particles, increasing the capillary attraction. In the lower, wetter horizon, the capillary potential remains low. Since water moves from areas of low to areas of high capillary potential, water may move upward in the soil during periods of drought. Differences in capillary attraction can, of course, also cause downward or lateral movement through the soil. In humid regions, the supply of soil water is sufficient to ensure a net downward movement; consequently, the bases and salts in the soil are eventually leached and removed. In arid and semiarid regions, where the supply of soil water is not ordinarily sufficient to provide for deep gravitational movement, water may penetrate several feet into the soil. However, the water is usually brought back to the surface through capillary movement, carrying with it bases and salts, which then accumulate at the surface or in the upper layers of the soil.

Chemical Properties

Physical properties are important in describing soils, but their influence on soil productivity is indirect. The chemical properties of the soil — that is, its mineral content, organic content, and chemical reaction — have a more direct influence on soil productivity.

Mineral Components. The most abundant minerals in the soil are chemical combinations of silicon, oxygen, aluminum, and iron. They or their weathered products form the bulk of the inorganic matter in the soil. The elements most necessary for plant growth — calcium, potassium, phosphorus, nitrogen, and magnesium — are less abundant but are available from exchangeable ions on the surfaces of clay minerals and of humus particles or from the soil solution.

The exchange of ions between soil and plant roots occurs when some of the minerals in the parent material have been reduced to very fine particles of clay by weathering and by chemical change. Clay exhibits colloidal properties and has a layered crystalline form that resembles the pages of a book (Fig. 21-3). Clay colloids are extremely small and can combine to form a solid (gel) or be suspended in a liquid (sol) without changing their chemical composition. Because they

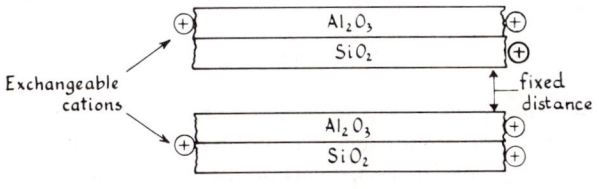

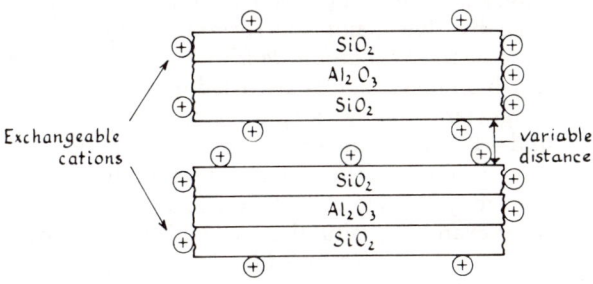

Fig. 21-3. Clay crystals. Note that Montmorillonite can have many more exchangeable cations in its variable spacing than can Kaolinite in its smaller fixed space. After R.L. Donahue, *Soils: An Introduction to Soils and Plant Growth*, 2nd Ed., © 1965. By permission of Prentice-Hall, Inc.

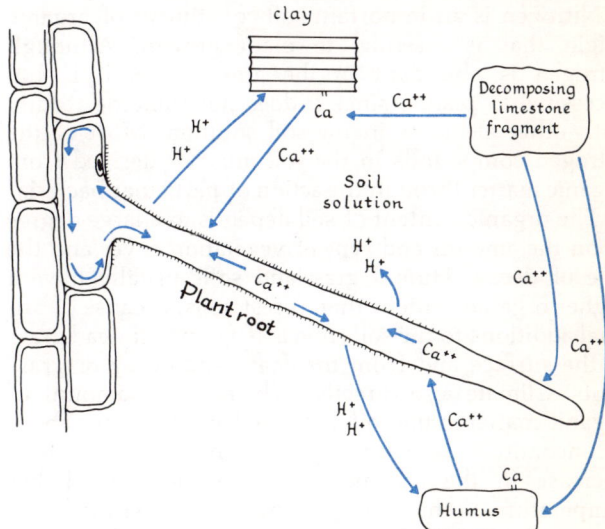

Fig. 21-4. Base exchange. After R.L. Donahue, *Soils: An Introduction to Soils and Plant Growth*, 2nd ed., © 1965. By permission of Prentice-Hall, Inc.

have a negative electric charge, they can hold positively charged basic ions (cations) of calcium, magnesium, and potassium. Through a process known as *cation exchange*, these bases can be given up by the clay colloids to plants, which require them for growth. The exchange takes place on the surfaces of the clay crystals and occurs between the clay and soil solution, between the clay and plant roots, and between mineral fragments and humus particles (Fig. 21-4).

Organic Matter. The chemical properties of soils depend on not only the characteristics of the weathered parent material but also the addition of minerals through the decomposition of organisms. Decay and action by live organisms eventually reduce plant remains to the colloidal state. At this stage, the organic matter is taken up by the soil solution or enters into a union with clay colloids to form a substance of complex chemical and physical properties.

The colloidal products of organic decay are called **humus.** The humus and the inorganic colloids increase the availability of water and nutrients to the plants. They indirectly restore the nitrogen, potassium, and calcium that are present in the decaying vegetation to living plants; they are essential to the survival of microorganisms that are necessary for plant growth; their decomposition yields organic acids that aid in the weathering of soil material; and they promote a soil structure that is favorable to plant growth. In general, the amount of organic matter in mid-latitude soils can be inferred from the darkness of the surface color.

Nitrogen is an important soil constituent of organic origin that is essential to plant growth. Although nitrogen is abundant in the atmosphere, it is not accessible to plants, since it does not combine readily with other elements in the soil solution. Most of the nitrogen compounds in the soil must be derived from organic matter through the action of particular bacteria.

The organic content of soil depends to a large degree upon the amount and type of vegetation cover and the rate of decay. Humid grassland soils usually have a higher organic content than forest soils, because of annual additions to the soil from a dense mat of dead grass at the surface and from the death and decay of grass roots. Climate also directly influences the amount of organic matter in the soil. Increased aridity brings about a concomitant decrease in plant growth and a general decrease in the organic matter in the soil. Higher temperatures, other things being equal, eventually increase the rate of decay to such an extent that much organic matter may be destroyed before it becomes incorporated in the soil so that the nitrogen content of the soil decreases as the temperatures rise. A similar relationship holds in forested areas, although at a lower level of organic content, since trees supply less organic debris to the soil than does grass. (Fig. 21-5).

Chemical Reaction. Productivity of a soil is also conditioned by its chemical reaction. Soil water, through the absorption of carbon dioxide from the air and the addition of acids from the products of plant decay, may become a weak acid that neutralizes or partly neutralizes the basic constituents in the soil. Depending upon the relative abundance of hydrogen ions and basic cations, soils are acid or alkaline. The degree of alkalinity or acidity is indicated by numbers prefaced by the symbol *p*H that indicate the hydrogen ion concentration in the soil solution. A *p*H of 7 denotes a neutral reaction — a balance between the hydrogen and the basic ions. Lower numbers denote a greater dominance of hydrogen ions in the soil solution and, therefore, greater soil acidity. A *p*H value above 7 indicates a dominance of basic ions and soil alkalinity. Most crops have a narrow tolerance for acidity or alkalinity; consequently, soils that have an abundance of the necessary minerals may support only a meager plant growth if the chemical environment is too extreme.

Factors Affecting Soil Formation

Among the weathering processes that lead to the formation of soil are those forming the regolith: **oxidation, hydration, hydrolysis, carbonation, solution,** and **leaching.** These preludes to soil origin can be called the geologic weathering processes to distinguish them from soil formation, which includes the additional process of *profile development*. These soil-forming processes are influenced by five major soil-forming factors: *parent material, climate, vegetation, landforms,* and *time*. The order in which these factors are listed in no way reflects their importance; any single factor may exert a profound effect upon the process of soil formation. The impact of a single factor on the development of soil properties may be determined by isolating its effect — that is, by examining situations in which all the other factors do not vary significantly. For example, if two soils are of the same age and have developed under similar climates and vegetation and on the same kind of terrain, any observable differences in their properties can be attributed to the influence of contrasting parent material on the soil-forming processes. If each factor is so isolated and studied, it is possible to describe with some confidence the influence of each upon the soil development.

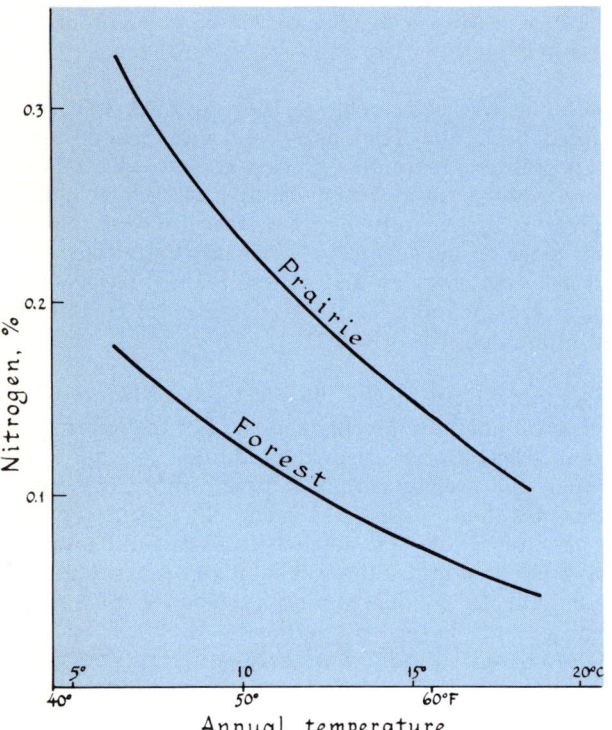

Fig. 21-5. The decrease in nitrogen content with increasing temperature. Prairie soils contain more nitrogen than forest soils, since grasses decompose into nitrogen more easily than forest litter does, but in each of the two soils the nitrogen content decreases as the temperature increases. This decrease is closely related to the fact that organic matter becomes so readily decomposed under higher temperatures that it may be washed away or otherwise decomposed before it can be incorporated into the soil. From Hans Jenny, *Factors of Soil Formation*, 1941. By permission of the author.

Parent Material. Parent material is the initial state of the soil system. It may be either unweathered rock or unconsolidated mineral matter from which the soil is derived. The parent rock may be recently cooled lava, newly exposed granite, or perhaps a mixture of debris that was newly deposited by streams, glaciers, or even wind.

Parent material influences soil formation mainly by contributing the original chemical and physical properties of the soil. Rocks rich in quartz, such as some granites or sandstones, produce acid soils with a sandy texture. Rocks containing minerals rich in iron and aluminum usually weather into red soils that contain oxides of iron and aluminum. The dark color of certain soils is determined by the mineral constitutents of their limestone parent rocks.

The degree to which parent material influences the soil properties varies with time. The older a soil, the less the influence of the original rock; in time, the soil properties derived from the parent material may be completely changed by weathering processes or by other factors, so that the parentage of soil from underlying rock is completely obscured.

Climate. The impact of temperature and rainfall on soil formation can be shown graphically. For example, the influence of temperature on soil formation can be shown by collecting soil samples along a line of increasing annual temperature and places where moisture availability, parent material, vegetation, age of the soil, and landforms are reasonably similar. Soil samples collected along such a transect, from Maine to Georgia, show decreasing nitrogen content with increasing temperature (Fig. 21-5). As the temperature increases, so does the rate of decay, so that more and more organic matter is destroyed before it can be incorporated in the soil.

Figure 21-5 should not be interpreted as indicating that the organic content of soils increases indefinitely as the tempreature drops. At some point, the temperature becomes too low for plant life, and there will be no organic matter in the soil. Even before this extreme is reached, the organic content of the soil drops off sharply because the cold temperature prevents complete decay of the organic debris. Consequently, much of the organic matter remains at the surface in an undecomposed state.

The influence of precipitation upon soil development can be shown in a similar way. Soil samples collected along an east-west transect across the United States show that decreasing rainfall results in a decrease in organic content and consequently a decrease in nitrogen content if other factors are held as constant as possible (Fig. 21-6). This decrease is clearly related to sparser plant cover in drier areas.

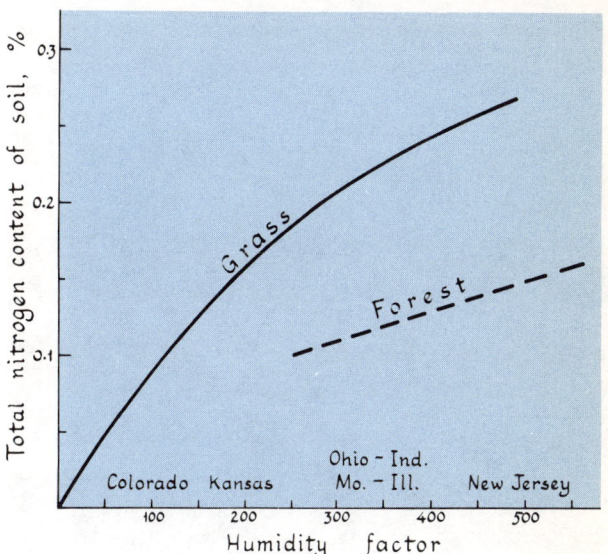

Fig. 21-6. The decrease in nitrogen content with decreasing precipitation. As this factor decreases (from right to left on the graph), so does nitrogen content, even though grasses provide consistently more nitrogen to their soils than forests do. From Hans Jenny, *Factors of Soil Formation*, 1941. By permission of the author.

Vegetation. Vegetation influences soil formation through the amount and character of the organic debris that it supplies to the soil and, to a lesser extent, through the biologic and physical effect of plant roots. An example from the American Middle West illustrates the strong role that vegetation may play. In the state of Illinois, the preagricultural vegetation consisted of prairie grass interfingered by broadleaf deciduous forest. This area was covered by **loess,** whose character varies mainly with the age of the deposit. Areas of similar age can be identified. Neither the form of the land, which is a gently rolling plain, nor the climate varies much over short distances.

In this setting, the effect of vegetation can be studied by comparing soil samples from grass and forest areas in which the other four soil-forming factors are reasonably similar. The difference in soils is striking. The forest soil has a profile with distinct horizons. The *A* horizon is gray-brown in color, the *B* horizon is generally orange in color, and the *C* horizon is weathered loess (Fig. 21-1). This forest-soil-profile is generally shallow (about 2 feet or 60 cm in depth), and the nitrogen content is relatively low. Grassland soils from the immediate vicinity have deeper profiles; there are no visible horizons, and the organic content is high. The surface soil is black when wet; with increasing depth, the color lightens and blends into the light buff of the weathered loess (Fig. 21-7). The influence of vegetation on soil formation is striking in this example.

Fig. 21-7. A prairie soil profile from Central Illinois. Note the dark surface becoming lighter in the lower layers. The depth shown is about 80 cm (about 30 inches). *Photograph by Lawrence C. Bliss.*

Landforms. Landforms affect soil evolution both directly and indirectly. Their indirect effect on microclimate has already been discussed with reference to the microhabitats of vegetation. Another indirect effect on soil development is through drainage. Flat land prevents or retards drainage and may result in permanently wet areas; therefore, the soil-forming processes, because of the exclusion of air, are largely those of reduction. Ferric compounds (Fe_2O_3) are reduced to ferrous compounds (FeO), which give a gray or blue-gray color to the waterlogged soil horizons. Steep slopes are normally better drained, to the point where very little water percolates directly into the soil, and the processes of soil formation are thereby greatly modified.

One of the direct effects of steep landforms is the retardation of the formation of mature soils. Mass wastage and slope wash on slopes that exceed a certain minimum inclination remove weathered material at a rate sufficient to prevent the development of a mature soil profile. As long as the steep slopes endure, the properties of their soil mantles do not change with age.

Time. As we have already seen, the effect of all the other soil-forming factors depends on the age of the soil. A new soil reflects the characteristics of its parent material. With increasing age, the impact of differences in climate, vegetation, or landforms gradually becomes evident. After some point is reached, no further changes take place, unless there is some change in the soil-forming factors; and the soil is then said to be mature. It is impossible to specify the time required to produce mature soils. First, since it is unlikely that all the soil-forming factors remain constant over extended periods, mature soils may be more theoretical than real. Second, available evidence indicates that the rate of development varies according to the nature of the parent material, the climate, and other factors.

In some cases, one can be reasonably specific. The soils that have developed on the limestone walls of a certain castle in the Ukraine can be fairly accurately dated as being no more than 300 years old, since the castle was in use until the middle of the 16th century. These soils, despite their relative youth, are remarkably similar to nearby soils that have developed on limestone bedrock, which is certainly several million years old. It would appear, therefore, that mature soils have developed on the castle walls within 300 years. The only factor that could conceivably affect this conclusion is a change in climate. Perhaps the climate has changed so much and so rapidly that the bedrock soil is not mature but only reflects the effect of the climatic environment of the past few hundred years, which, of course, might account for the similarity with the soils formed on the walls.

In other instances, soils have developed much more slowly. Soils have formed on marine clays of the North Sea floor after the Dutch built dikes and drained the area. Some of these soils are almost 300 years old; but although they are very different from the parent clay, they do not resemble much older soils of similar parent material. Marine clay soils apparently need much longer to reach maturity than do limestone soils.

Soil-Forming Processes

One major system of classifying soil-forming processes is based on soil properties that are mainly determined by large-scale differences in climate and vegetation. It assumes that on areas of nearly level to gently rolling terrain, the factors of climate and vegetation eventually produce mature soils whose properties no longer reflect those of the parent rock. Since the soils are mature, they are thought to be also independent of time in the sense that they no longer change with increasing age. Consequently, the effects of the factors of time and parent material can be minimized. Local contrasts in soil properties, resulting from differences in landforms, cannot be easily generalized; and these too are left out of consideration, except for the large-scale dichotomy between mountain soils and lowland soils. The emphasis on the factors of climate and vegetation leads to a zonal view of soils. The zonal concept implies that soils can be categorized into large groupings that have broad zonal distributions related to the zonal distributions of the major climatic types and vegetational associations. It also implies that the principal soil-forming processes are most logically described under climatic headings, since temperature and moisture determine the rate and the kinds of chemical weathering. Consequently, we can distinguish the soil-forming processes that occur in humid regions, such as **podzolization** (Russian *pod* soil + *zola* ashes), **ferralization** (Latin *ferrum* iron), and **laterization** (Latin *later* brick), from those that take place under arid climates, such as **calcification** (Latin *calcis* lime) and **salinization** (Latin *sal* salt).

Soil Formation in Humid Regions

Soil formation in humid regions is characterized by leaching (to extract by percolation) of the soil, but the nature and intensity of leaching varies with temperature and precipitation. In regions of coniferous forest cover where summers are short and cool, the surface organic matter is not completely reduced. Rainwater, filtering through the layer of forest litter, becomes acidic. As temperatures increase, the organic matter at the surface of the soil is increasingly reduced by chemical action; consequently, the water that filters into the soil becomes much less acidic. This gradual change in the character of soil water causes imperceptible changes in soil leaching, with resulting changes in soil properties. When the temperature contrasts are large, we can speak of a clear distinction between cold-humid and warm-humid processes of soil formation.

Soil Formation in Cold-Humid Climates. The soil-forming process in the cold-humid climates is called podzolization and consists essentially of leaching by acidic soil-water solutions. The soil solution contains organic acids, which are derived primarily from the products of incomplete organic decay. The following description of the development of a hypothetical soil illustrates the process of podzolization, as well as some processes common to many soils.

Assume that a portion of the bottom of Hudson Bay has been lifted above sea level, exposing parent material of mixed sand, silt, and clay, with considerable amounts of salt and bases. The parent material would first be changed as salt and bases were removed by percolating rainwater. This removal, or leaching, would be complete near the surface but would decrease downward; and an incipient soil profile would come into being with the development of horizons of contrasting salt and base content. Once the salt and bases were removed from the surface layer, the clay particles would become dispersed and could be moved downward, in suspension, by soil water. This process of eluviation would result in concentrating the clay particles at some point in the profile where the salt content would be high enough to cause them to flocculate, or group together, and precipitate out of the soil water. As a consequence, a surface horizon relatively free of clay would develop above a subsurface horizon rich in clay.

Podzolization. The removal of salts and bases and the downward movement of clay is not peculiar to podzolization but can occur in any soil-forming process. In podzolization, further changes take place. As the salt is being removed from the parent material, plants start to grow in the newly forming soil. In time, coniferous forests, the predominant vegetation of cold, moist climatic regions, are established on the slowly developing soil. The organic debris of such forests has a lower base content than that of most other vegetation, and it decays only partially because of the short, cool summers. In time, a mat of acidic organic matter collects at the surface. Organic acids are picked up by rainwater passing through the mat. In the presence of acidic soil water, the silicates break down and lose their iron and aluminum to the soil solution as sesquioxides of iron and aluminum (Fe_2O_3 and Al_2O_3). These combine with the organic solutions to form metal-organic colloids that are carried down by the percolating soil solution.

At the same time, the downward migration of clay increases because the acidic solution keeps the clay particles dispersed in a nonflocculated state. As the soil water percolates deeper into the soil, its acidity is counteracted by the increased base content in the lower parts of the profile. With decreasing acidity, the capacity of the soil solution to hold the metal-organic colloids and clay particles decreases and eventually causes their deposition in the *B* horizon. The mature soil resulting from this process is called a **podzol**.

Podzolization is occurring wherever acidic soils are found. It is usually associated with the short, cool summers and the long, cold winters of subarctic climates and with coniferous forests, but it can occur wherever the parent material is especially deficient in bases. Podzols or podzolic soils occur on quartz sands on the eastern coastal plain of the United States. This sandy parent material contains virtually no bases; consequently, it supports plants that require little basic material. The organic debris decays into highly acidic material which, in turn, leads to the development of soils similar to the podzols of the cold climates. Profiles similar to those of podzols have also been reported from the tropics, where they occur in soils rich in quartz sand. In the subarctic climate, however, where coniferous forests predominate, podzols will eventually develop on any parent material.

Soil Formation in Warm-Humid Climates. The organic debris supplied by the vegetation usually decays more rapidly and completely in the humid tropics and subtropics because of the higher annual temperature. It also has a higher base content, which is returned to the soil as the organic matter decomposes. As a result, the soil solution becomes less acidic, and podzolization gradually gives way to ferralization as one moves from subarctic to subtropic climates. There is an intermediate zone in which both processes are operating, one predominating over the other according to local circumstances of vegetation, terrain, or parent material. In the warm and humid tropics, however, the organic debris is nearly completely destroyed, and its bases are rapidly returned to the soil. The soil water is neutral or slightly basic in reaction, and ferralization is dominant.

Ferralization. Leaching by neutral soil water removes **silica** from the soil materials; the silica is carried downward and eventually out of the soil by way of the groundwater. Bases are also removed through leaching. This removal of silica creates a relatively greater concentration of iron and aluminum sesquioxides in the soil, which is the essence of ferralization. The sesquioxides are not mobile, as they are in the podzols, apparently because of the reduced acidity. Because of their stability, they accumulate throughout the soil profile.

High temperatures and abundant moisture lead to deep weathering, so that tropical soil profiles are very deep. The silica and the bases leached from the upper soil are not deposited in a B horizon but are carried out of the soil by the groundwater. The contrast between acidic surface layers and more basic lower layers, which is characteristic of podzolization, does not develop under ferralization, since the bases are constantly being removed from the entire profile. A distinctive result of complete ferralization is the absence of any horizon of deposition. Under these circumstances, tropical soils exhibit a gradual transition from the surface material to the parent rock, without any clear development of discontinuous horizons.

Ferralization, in its most advanced form, is characteristic of humid tropical climates and of broadleaf evergreen forests. The organic debris from the forest has a high base content, despite strong leaching in the soil. This condition is largely due to the

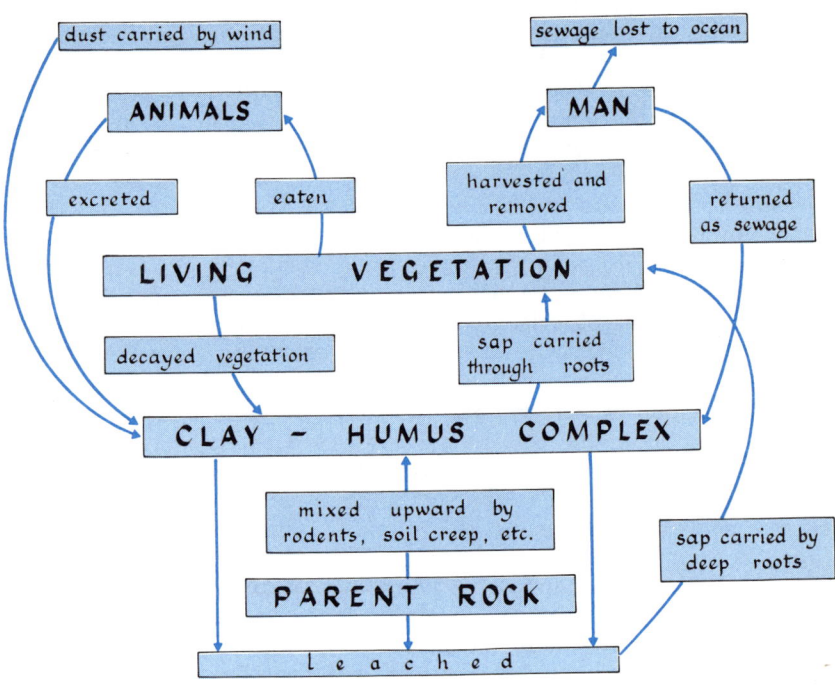

Fig. 21-8. The nutrient cycle. After S.R. Eyre, *Vegetation and Soils: A World Picture*, 1963. By permission of Aldine Publishing Company and Edward Arnold (Publishers) Ltd., London, England.

well-developed nutrient cycle that exists under a tropical forest (Fig. 21-8). The dead organic matter decays very rapidly under conditions of warmth and high humidity. In part, this rapid decay is the result of oxidation, but mostly, it is the product of the digestive processes of micro-organisms that are present in abundance on the humid tropical forest floor. The mineral substances produced by the rapid decay of organic matter include the bases necessary for the forest growth. These minerals are added to the soil, where they are taken up by roots, used by the trees, then returned to the soil. During this time the loss of minerals in each cycle is counter-balanced by minerals which are taken up by deep roots from the zone of decreased leaching.

Laterization. This process leads to the formation of an iron-rich horizon of unusual characteristics. The horizon is a porous, slaggy layer, mainly reddish in color with yellow or violet streaks. The pores may be filled with whitish aluminum hydroxide. Below the iron horizon is a blue-gray clay horizon, formed primarily by reduction. The iron horizon is soft when soaked, so that it can be easily cut with a spade. When dry, the material becomes very hard. Because of these characteristics, it has been used as a building material; hence the name "laterite."

Laterite develops just above the average position of the water table on parent material rich in iron, and the iron horizon may be the result of a fluctuating water table. Iron in the parent material is made soluble by reduction in the permanently saturated zone and is carried up by the rising water table and by capillary movement. Quantities of soluble iron are then deposited when the water subsequently recedes or evaporates. The iron, upon drying, changes to an insoluble form and becomes stabilized. In time, the persistent rise and fall of the water table forms a thick, reddish, iron-rich layer, or laterite. Below the layer of laterite, the dark-gray clay, or *glei*, zone is the product of the reduction taking place in the permanently saturated zone.

Laterites usually develop in the wet savanna and monsoon tropics just above the water table on smooth surfaces that have been exposed to weathering for long periods. Poor drainage, high temperature, and low organic soil content favor the concentration of iron and aluminum oxides, while the contrast between summer rain and winter drought causes fluctuation in the water table and capillary movement. Laterite also occurs on the arid margin of the savanna in Africa and in the semiarid and arid interior of Australia. Since these areas are presently too dry for laterization, the laterites are believed to be fossil horizons developed during past periods of greater rainfall.

Soil Formation in Semiarid Regions

The contrast between various soils in humid climates is clearly associated with changes in the soil-forming processes brought about by differing temperatures. Soils in semiarid regions differ mainly as a consequence of precipitation contrasts that cause variations in the amount of soil water and differences in the luxuriance of the grass cover and in the depth of grass roots. In areas of good drainage, the dominant soil-forming process is calcification; where drainage is poor, the soils may be subject to processes of salt accumulation known as salinization.

Calcification. This process results in the accumulation of bases, primarily calcium carbonate, in the *B* horizon. In dry regions, when water penetrates the soil, small amounts of calcium carbonate are dissolved and carried downward. As soon as the surface soil dries, capillary action brings water charged with calcium carbonate back to the surface, where lime is deposited as the water evaporates. During the next rain, the lime is again taken into solution and carried down, only to be brought back to the surface by the next capillary movement. The surface deposits of lime are always available to be carried downward by subsequent rains, but at depth a residue of lime is left behind each time the water returns to the surface. Eventually, there is enough accumulation of calcium carbonate to form a lime horizon. The depth of the lime horizon depends on the amount of rainfall and the depth of the grass roots. In the root zone, organic acids act to keep the lime in solution.

The grasses, characteristic of most semiarid regions, play a further role in the process of calcification. They draw up large quantities of calcium from the *B* horizon and deposit it in the *A* horizon. This process tends to stabilize the organic material and prevent it from entering the soil solution. Also, a large amount of organic matter is added directly to the soil below in the form of grass-root humus. This grass-root contribution and the slow rate of decay of organic matter in the dry environment account for the large accumulation of humus within semiarid grassland soils of the mid-latitudes.

Calcification develops best where there is good drainage and in steppe climates, where evapotranspiration exceeds precipitation. It is most effective near the boundary between steppe and humid climates, where grass is the predominant vegetation; where glacial till or lime-rich sediments are the parent material; and where the terrain is flat to gently rolling.

Salinization. Soils in dry regions with poor drainage are subject to the process of salt accumulation called salinization. Because of poor drainage and the excess of evaporation over precipitation, the occasional excess of water in the soil can be removed only through evaporation. Basins of interior drainage in desert regions usually have large salt flats that are the result of the evaporation of Pleistocene lakes and of periodic inflows of modern floodwater that subsequently evaporates. The minerals precipitated at the surface include carbonates, chlorides, and sulfates of sodium, calcium, and magnesium. Soils formed by salinization are called *solonchak* (Russian *solonchak* salt marsh) or *white alkali* soils (Arabic *alqili* soda ash).

Salinization can be a serious problem in the cultivation of semiarid and arid soils. Salt concentration can occur rapidly in dryland soils that are subject to irrigation without proper drainage. If there is inadequate drainage, the water table will rise when the soil is irrigated. In a relatively short time, capillary water may concentrate the salt in quantities toxic to cultivated plants. It is possible to reclaim the soil for farming, but the process of desalinization is costly and time-consuming.

Desalinization may occur under natural circumstances through a change in climate that brings increased rainfall or by the lowering of the water table through erosion of an outlet from a bolson. The calcium and magnesium salts are then removed by solution and leaching. As the salts are flushed from the soil, sodium and potassium carbonates may accumulate. These are alkaline salts; consequently, desalinization is frequently called **alkalinization.** Alkalinization can also be brought about by artificial flushing with irrigation water and by improved drainage. Sodium and potassium salts, the products of alkalinization, form strong hydroxides, which, if dissolved in water, disperse the organic matter in the soil and are toxic to plants. The diffusion of organic matter darkens the soil; hence it is frequently called *black alkali* or *solonetz* (Russian *solonetz* soil rich in sodium). The artificial addition of gypsum to the soil as it is being flushed prevents the development of the hydroxides and their harmful effects.

Desert Soil Formation

Clearly developed soil profiles are rare in the desert because rainwater is infrequent. The loose material at the surface is largely the product of mechanical weathering and is only slightly different from the parent material; therefore, desert soils may be considered permanently immature soils — some are called **lithosols** to indicate their incomplete weathering. Organic material is practically nonexistent. Hard crusts sometimes form as the result of the deposition of soluble salts at the surface.

The Soil System

We have learned something about soil properties, seen how the soil-forming factors influence soil development and studied some of the principal soil-forming processes. It is also worthwhile to view the soil as a system because such a view pulls the various topics discussed in the chapter into the unified whole.

The soil is a true system since it consists of factors which show clear relationships one with another and which operate together and with the soil-forming processes to produce distinct patterns of soil distribution. Like all physical systems the soil system is open since it exchanges mass and energy with its surroundings. The system has inflows of energy (solar and tectonic) and mass (precipitation, dissolved mineral elements and organic debris) and outflows of energy in the form of heat to the atmosphere and of mass (soil water) to streams, to ground water and to the atmosphere by evaporation.

Figure 21-9 shows that as energy and water as well as organic and inorganic compounds flow through the system, portions of these components are used by non-soil-forming processes as, for example, in the erosion of the land or the growth of vegetation. The kind of soil that may develop over time will depend upon how much energy and water are available for its formation. If the inflow of energy and precipitation is high and the remaining factors of landform, vegetation, and parent material do not absorb too much of the inflow, the soil-forming process will consist of ferralization, and a ferralite soil will develop.

Changes in energy or mass inflows, or their consumption by any of the factors will modify the resulting soil. If the land is very steep, the loss of water by runoff may be so great as to inhibit ferralization despite high temperatures. In cold climates, where the energy inflow is low, the principal soil-forming process is podzolization. Podzols may also form in warm, wet climates where the parent material is quartz sand. Here the energy required to weather the quartz is enormous and, at the same time, the nature of the parent material favors the development of vegetation rich in acids. The result is that despite great inflows of energy and mass, the soil-forming process is similar to that in a cold climate.

The soil system will tend towards a steady state if none of the factors change. But this is not likely to happen in nature, and even if soils are not subject to external change, such as changes in topography brought about by some tectonic event, modification will still occur in the system because of feedbacks from the soil to the soil-forming factors and processes. An example will demonstrate this occurrence.

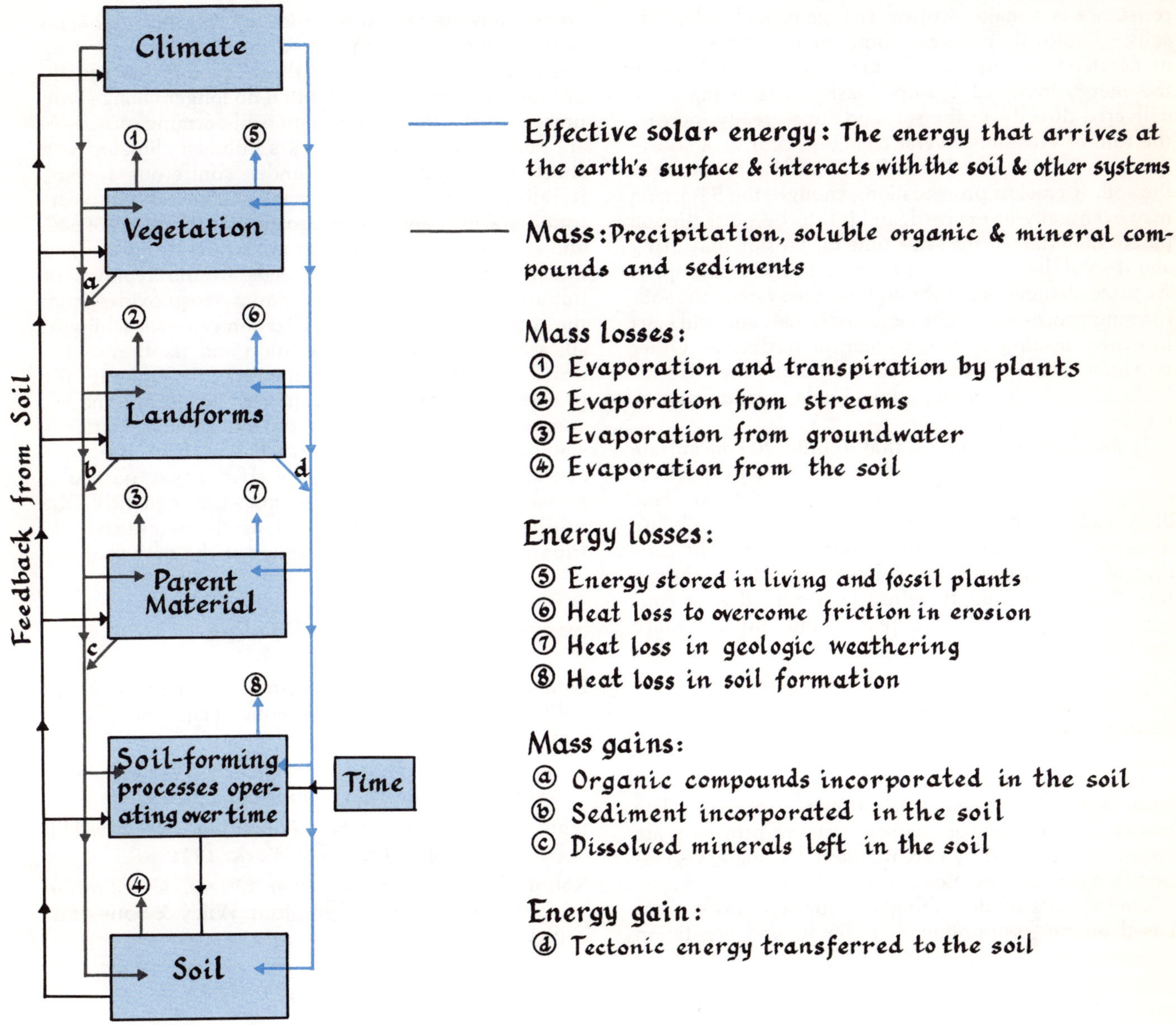
Climate
Vegetation
Landforms
Parent Material
Soil-forming processes operating over time
Time
Soil
Feedback from Soil
a
b
c
d
① ② ③ ④ ⑤ ⑥ ⑦ ⑧
Effective solar energy: The energy that arrives at the earth's surface & interacts with the soil & other systems
Mass: Precipitation, soluble organic & mineral compounds and sediments
Mass losses:
① Evaporation and transpiration by plants
② Evaporation from streams
③ Evaporation from groundwater
④ Evaporation from the soil
Energy losses:
⑤ Energy stored in living and fossil plants
⑥ Heat loss to overcome friction in erosion
⑦ Heat loss in geologic weathering
⑧ Heat loss in soil formation
Mass gains:
ⓐ Organic compounds incorporated in the soil
ⓑ Sediment incorporated in the soil
ⓒ Dissolved minerals left in the soil
Energy gain:
ⓓ Tectonic energy transferred to the soil

Fig. 21-9. The flow of energy and mass through the soil system. Note the feedbacks from the soil to the soil-forming factors and soil-forming processes.

Let us assume that the climate is warm and humid and that one of the soil-forming processes slowly removes clay from the evolving soil causing it to become gradually more sandy. As the soil becomes sandier, the micro-climate will change because the rate of water drainage through the soil will increase even though the general climate remains unchanged. This increased rate of drainage will lead in turn to a change in vegetation, consisting of the replacement of existing plants by species requiring less moisture. The increasing sandiness of the soil will also favor vegetation having low base requirements, since sand is a poor source of chemical bases. Eventually the changes may be so great that we can think of the original soil as the parent material in a new soil-forming process.

So far we have been discussing the soil system as it might operate without artificial inputs. But, as we have seen, human interference in natural systems tends to bring about changes in resistance to the flow of energy and mass through those systems. The reduction of resistance is a major artifical change brought about in soils. Removal of vegetation and cultivation of moderate slopes increase the rate of runoff and permit the energy involved in slope wash and gullying to be delivered directly to the soil and thus greatly increase the rate of erosion. Several changes occur as a consequence. The increased runoff means less moisture in the soil. If erosion proceeds long enough, the B horizon may eventually be exposed, and it thus becomes the soil parent material. At the same time the erosion may bring about small but significant changes in the topography. As these changes occur throughout the system, the soil-forming processes will be modified, a new soil will start to evolve leading to a new chain of feedbacks. Other potential changes can occur through the use of herbicides, fertilizers, or various kinds of land management.

It should be emphasized that in order to concentrate on obvious effects, only a few of the many possible changes throughout the system were described. This discussion of the soil system reaffirms that soils are dynamic systems subject to constant change. The concept of a mature soil, as we have suggested earlier in this chapter, should therefore be regarded as a convenient approximation to reality and not as reality itself.

Summary

A soil is an open system on the land surface resulting from weathering and soil-forming processes. The nature and the rate of the soil-forming processes are strongly influenced by parent material, climate, vegetation, landforms, and time.

One system of describing soils on a world scale is based on the assumption that climate and vegetation are the most important soil-forming factors and that on gently rolling, well-drained surfaces, the soil-forming processes influenced by these factors eventually produce mature soil profiles that no longer change with time. Among these important soil-forming processes are podzolization, which occurs in moist climates with short, cool summers and under coniferous forests; ferralization, which occurs in moist climates with year-round warmth and under broadleaf evergreen forests; and calcification, which occurs in semiarid climates and under grass. Podzolization leads to the removal of humus and of iron and aluminum sesquioxides from the A to the B horizon. Ferralization causes leaching of silica and a relative concentration and fixation of the iron and aluminum sesquioxides throughout the profile. Calcification is the process whereby lime accumulates in the B horizon and the organic matter is stabilized in the A horizon. These three processes operate with varying degrees of efficiency in particular zones of climate and vegetation. Consequently, the resulting soils are called zonal soils, the nature and distribution of which are the subject of the next chapter.

Suggestions for Further Reading

Donahue, R. L., *Soils: An Introduction to Soils and Plant Growth*, 2nd ed., Prentice-Hall, Inc., Englewood Cliffs, N.J., 1965.

Eyre, S. R., *Vegetation and Soils: A World Picture*, Aldine Publishing Company, London, 1963.

Jenny, H., *Factors of Soil Formation*, McGraw-Hill Book Company, Inc., New York, 1941.

Robinson, G. W. *Soils: Their Origin, Constitution and Classification*, 3rd ed., John Wiley & Sons, Inc., New York, 1951.

22 Soil Types and Their Distribution

Introduction

The properties of soils do not change abruptly but grade continuously and imperceptibly from one place to another. Nevertheless, the many varieties of soils can be classified into a small number of types that have distinctive characteristics. One widely used classification groups soils according to the soil-forming factors into three basic orders: **zonal soils, azonal soils,** and **intrazonal soils.** Zonal soils, developed on gently sloping land with good drainage, reflect primarily the influence of climate and vegetation. Azonal soils are characterized by incompletely developed profiles due to their youthfulness. Intrazonal soils reflect the influence of soil-forming factors that are not zonally distributed, such as parent material or landforms.

The zonal concept of soil formation was developed in Russia during the last half of the 19th century, when soil scientists noticed that the distribution of certain soils in European Russia coincided with climatic regions and that these soils had properties that cut across differences in the bedrock. Similar relationships between soils and climate were discovered in the United States. It was soon seen that these zonal soils, so called because they coincided with climatic zones, had properties that were strongly influenced by vegetation. The theory of zonal soils now includes the idea of coincident zones of climate and vegetation. The concept is one developed from study of mid-latitude soils and has been extended, but with some difficulties, to the tropics.

The zonal system of soil classification has some serious deficiencies mainly because it is based upon an inadequate knowledge of many of the world's soils. In order to overcome these deficiencies soil scientists of the U.S. Department of Agriculture have proposed a new system based primarily upon the descriptive properties of soils. The new system was published in 1960 and has been subject to intensive study and field checking since that date.[1] It is now being used to map soils in the United States, and it doubtless will become the major classification of the future. Unfortunately the new classification has such a formidable array of strange names that the novice may lose sight of the main objective of the system — an understanding of the nature of the world's soils. For this reason the discus-

---

[1] Soil Survey Staff, Soil Conservation Service, U.S. Department of Agriculture, *Soil Classification, A Comprehensive System, Seventh Approximation*, August, 1960.
———, Supplement to Soil Classification System (Seventh Approximation) March, 1967.

sion of some of the major aspects of the classification has been put at the end of this chapter. The simpler zonal concept will be used to describe the major soil types and their distribution.

Zonal Soils

The zonal soils are the most widespread, and their distribution lends itself most readily to the broad generalizations necessary for a description at a world scale. The zonal order has two suborders, the humid forest soils, or **pedalfers**, and the semiarid soils, or **pedocals**. "Pedalfer" and "pedocal" were formed by combining the Greek root *pedo* (ground) with the abbreviations of the Latin words *alumen* and *ferrum* (aluminum and iron) and with the abbreviation of the Latin word *calcis* (line). Pedalfers are soils of humid regions, distinguished by an absence of carbonates and by a concentration of iron and aluminum oxides in the soil profile. Pedocals are semiarid soils, which contain an accumulation of calcium carbonate or calcium-magnesium carbonate, mainly in the B horizon.

Pedalfers and pedocals are subdivided into great soil groups[2] — mature, relatively stable soils that exhibit fundamental differences in profile. These groups, which are discussed in detail below, are shown in tabular form in Table 22-1 and on the map (Fig. 22-1).

Pedalfers

Pedalfers include the following great soil groups: tundra soils, podzols, gray-brown podzolic soils, subtropical red and yellow soils, tropical red loams, ferralites, and laterites. These soil groups have developed their distinctive characteristics primarily as the result of different temperature regimes and vegetation cover. Moisture is available in all these soils, but particular temperature regimes have accelerated or retarded certain soil-forming processes and have created differences in the vegetation cover, which has also influenced soil formation.

Tundra Soils. Although the soils of the Arctic tundra are widespread, they are sometimes classed as intrazonal soils because they have properties that are frequently related to incomplete weathering and poor drainage. Their inclusion as a zonal soil under the heading of pedalfers is somewhat arbitrary.

---

[2]The great soil groups, in turn, can be further classified into soil series, the series into types, and the types into phases, which progressively reflect more local circumstances; but this further breakdown is beyond the scope of our discussion.

Table 22-1. Classification of Soils

Orders	Suborders	Great soil groups
Zonal soils	Pedalfers	Tundra soils
		Podzols
		Gray-brown podzolic soils
		Subtropical red and yellow soils
		Tropical red loams
		Ferralites
		Laterites
	Pedocals	Prairie soils
		Chernozems
		Brown steppe soils
		Sierozems
		Desert soils
Azonal soils		Dry sandy soils
		Alluvial soils
		Lithosols
Intrazonal soils		Hydromorphic soils
		Calcimorphic soils
		Halomorphic soils
		Planosols

Chemical alteration of the parent rock under the cold conditions of the tundra is slow, and consequently, much of the soil body consists of mechanically broken mineral particles. The development of distinct soil profiles is frequently inhibited by poor drainage due to the presence of permafrost. The effect of poor drainage, a dark-gray color usually associated with chemical reduction, is characteristic of many tundra soils. Vertical and lateral movement within the soil due to frequent freezing and thawing also interferes with the development of a soil profile.

The cold tundra climate leads to a slow rate of plant decomposition and to the formation of large amounts of humus that collect on the surface in the form of peat layers. Because of incomplete weathering and frost heaving, tundra soils usually lack distinct profiles and instead consist of layers of coarse mineral matter containing some silt mixed with layers of humus or peat.

Podzols. The podzol profile has pronounced and clearly definable horizons (Fig. 22-2). Typically, the undisturbed surface has a matted layer of organic litter that is highly acid and incompletely decomposed. This organic horizon is only slightly mixed into the soil, because burrowing insects and decay-promoting organisms cannot tolerate acidity and the coolness of the short summers. The A_2 horizon, below the layer of organic matter, is silvery gray and sandy. It is strongly leached; its color and texture result from the removal

Soil Types and Their Distribution 341

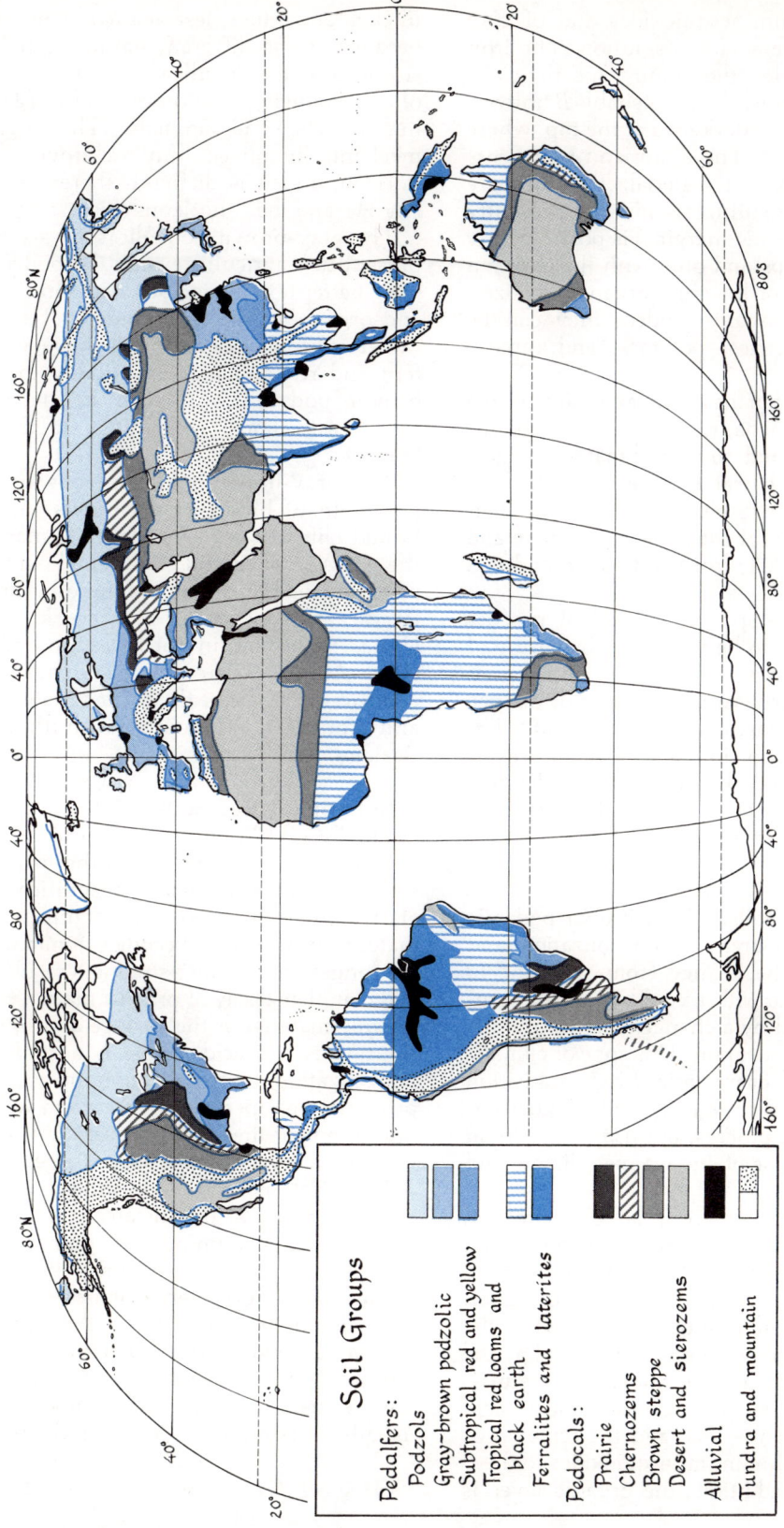

Fig. 22–1. The world distribution of soils. Note the arrangement of the Pedalfers from cold regions (Tundra soils and Podzols), through mesothermal regions (Podzolic and Subtropical soils), to warm regions (Tropical soils, Ferralites and Laterites). The Pedocals grade from more humid regions (Prairie soils and Chernozems) to drier ones (Steppe and Desert soils). Alluvial and mountain soils correspond to major river basins and mountain ranges respectively.

of the iron and aluminum sesquioxides and of clay through the action of the acid soil solution. The iron and aluminum sesquioxides, the humus, and the clays have accumulated in a readily distinguishable B horizon with a reddish-yellow color, darkened at the top, where the humus is concentrated. The C horizon represents the slightly weathered parent material. The soils are acid, and successful agriculture (which is presently possible only on their warmer margins) is possible only with quick-maturing crops and only with the addition of considerable amounts of lime and organic fertilizer. When these soils are properly tended, they can be highly productive of berries, potatoes, and rapidly growing grains.

Podzols occur in the northeastern extremes of the United States and, in Canada, northward to the boundary between subarctic and tundra climates. Podzols also occur throughout central European Russia and over much of Siberia, corresponding closely in their pattern to the distribution of the subarctic climate and coniferous forest (Figs. 22-1, front end paper, and Fig. 20-1(2)).

Variations in the development of podzol profiles within these areas appear to be related to the nature of the parent material. The most strongly developed podzol profiles occur in quartz sands; the least developed profiles are usually coincident with clay deposits. The clay belt of central Ontario, in Canada, is an area of such incompletely developed podzols; podzolization has been unable to remove all of the fine-textured material from the parent material, which is unusually rich in its clay content.

Gray-Brown Podzolic Soils. Gray-brown podzolic soils are found between areas of podzolization and areas of ferralization. The change from subarctic to humid continental climate with a cool but longer summer is often accompanied by a change from coniferous to broadleaf deciduous vegetation. This zone of climatic and vegetative change is characterized by a transition from podzols to gray-brown podzolic soils. Podzolization is still the dominant process, but it operates in an incomplete fashion because of the warmer climate and less acidic vegetation.

In general, the soils of this group are characterized by a gray-brown A horizon. Because of the longer and warmer summers, decay of vegetative debris is more complete, and the nitrogen content of the soil is higher. The acid content of the soil solution is weakened by the more complete breakdown of organic matter and by the fact that leaf fall from broadleaf deciduous trees returns basic constituents to the soil. Consequently, acid leaching of the A horizon is less, and the removal of humus and of iron and aluminum sesquioxides to the B horizon is reduced. In addition, the organic layer is thicker and much less matted than that on podzols, because of the different nature of the organic debris supplied by the broadleaf deciduous forest and because of the "plowing" action of micro-organisms, burrowing insects, and animals. The organic matter is mechanically mixed into the upper layers of the A horizon, and it is difficult, therefore, to separate the A_{00}, A_0, and the A horizons (Fig. 22-2).[3]

The gray-brown podzolic soils are more useful for present-day agriculture than the podzols because they have better texture and structure and a higher colloidal and organic or nitrogen content. The clay colloids of the gray-brown podzolic soils have a higher base content and respond more profitably to fertilizer. Gray-brown podzolic soils can be used for intensive agriculture almost indefinitely if they are carefully managed.

The gray-brown podzolic soils occur in areas of predominant broadleaf deciduous forest and of cool, humid climates — in mid-latitude marine climates, on the wet side of humid continental climates, and on the poleward side of the humid mesothermal climates. The largest areas of occurrence are in the northeastern United States and in northwestern and central Europe (Fig. 22-1). Other smaller areas include the northwestern part of the United States, northern China and southern Manchuria, southern Chile, New Zealand, and Tasmania.

Subtropical Red Soils. Gray-brown podzolic soils gradually give way to subtropical red soils where the climate becomes humid mesothermal. This change in soils is not accompanied by a significant change of vegetation; there is, however, a significant change in climate. Warm weather becomes dominant; in particular, the longer and warmer summers result in a more rapid and complete decay of organic matter. Since there is less organic material at the surface and in the soil, the soil solution is less acid. There is a gradual change from podzolization to ferralization in a zone where both processes are operating at once and where the domination of one or the other depends on the character of the landforms, parent material, or vegetation.

Ferralization in the subtropical red soils implies a partial removal of the silicates and a liberation of some of the iron and aluminum sesquioxides throughout the soil profile. At the same time, the soluble bases are removed. The prevalence of iron oxides produces the red color in the soils. Thus, the subtropical red soils have a dark reddish-brown A horizon, with a friable or crumbly structure and an intermediate texture but a low content of bases and nitrogen. The A horizon usually rests on a clay-rich, bright-red B horizon.

---

[3]See footnote 1, page 326.

Unfortunately, the high intensity of summer rainfall characteristic of the subtropical climate results in serious soil erosion, particularly where clean-field cultivation is practiced. The subtropical soils respond favorably to fertilization and, if cultivated conservatively, yield well for long periods. The red soils of southeastern China, for example, have been cultivated intensely for over 2,500 years. On the other hand, many areas of subtropical red soils in the United States have been damaged by careless management during the past two centuries.

Subtropical red soils are characteristic of most humid mesothermal climates with broadleaf deciduous or evergreen forests. In addition to the major areas in the southeastern part of the United States and in China, these soils are found in southern Brazil and in parts of South Africa, northeastern India, and eastern Australia (Fig. 22-1). Similar soils occur in the western foothills of the Sierra Nevada of California under a mixed forest of pine and live oak in a Mediterranean climate.

Noncalcic brown soils are related to subtropical red soils and usually occur downslope from them in certain areas of Mediterranean climates. These soils are associated with cool, moist winters and hot, dry summers and with a light forest vegetation mixed with brush or grass. They have a brown to reddish-brown A horizon and a red, somewhat compact B horizon. As the name suggests, the soil profile has been leached of carbonates, although it has a neutral or slightly alkaline reaction.

"Red earth," or "tierra roja," is the name given to a red soil that typically occurs on hard limestone in areas bordering the Mediterranean Sea (Fig. 22-2). Similar soils occur on the same parent material in the humid tropics and subtropics. In general, the soils have a simple profile consisting of red soil of variable depth resting directly upon the limestone, without any clear zone of transition. The soils are fairly rich in clay, have a relatively high iron-oxide content, and, considering the parent material, a surprisingly low base content. The origin of these soils is puzzling, because the pure limestones on which some of the tierra roja develops should yield abundant bases and very little iron-oxide clay. Despite this uncertainty about their origin, the red earths are included with the subtropical ferralitic soils because of their similar properties.

Subtropical Yellow Soils. Subtropical yellow soils are extensively developed only along the south Atlantic and Gulf coasts of the United States. They have a more scattered distribution throughout the rest of the humid subtropics. In the southeastern United States, the transition from subtropical red soils to subtropical yellow soils coincides with the change from the rolling and irregular plain of the Piedmont to the nearly flat surface of the Atlantic and Gulf coastal plains and with a change from a predominantly broadleaf deciduous cover to pine forests (Fig. 22-1).

This soil has a shallow, gray-white to yellow-white A horizon that is rich in sand but poor in nitrogen and bases (Fig. 22-2). The B horizon is deep, pale yellow to yellow-orange, and has a tendency to form a hardpan.[4] The soils are best developed where the terrain is flat and the water table is near the surface.

The difference between subtropical red and yellow soils may be due to differences in vegetation. The coniferous forest is an environment favorable to podzolic processes, which may produce the thin, gray-yellow A horizon and the thick, yellow B horizon. On the other hand, the deep and hardened B horizon may be due to the very flat terrain that results in a poorly drained subsoil. Because of this slow drainage, the iron oxide in the subsoil becomes strongly hydrated, forming limonite, a hydrated ferric oxide with a yellow color. If this is the case, the subtropical yellow soil should be considered intrazonal, since the dominant soil-forming factor is neither climate nor vegetation.

Tropical Red Loams. The tropical red loams associated with savanna climates occur under both grass and forest on well-drained areas with gently sloping terrain. These soils have deep profiles and no clearly distinguishable A and B horizons. They are intermediate in character between the ferralites of the tropical rainforest and the soils of the semiarid margins of the tropics.

The distinguishing characteristics of these soils are largely due to the strong seasonal contrast between wet summer and dry winter. Nodules of silica occur throughout the soil, because ferralization (or desilication) is seasonally interrupted. Bases are leached in the wet season, and the soil acquires an acidic character; but evaporation in the hot and dry winter stimulates an upward capillary movement, and some of the minerals are returned to the upper layers. Consequently, the surface is rich in colloidal clays, which can absorb and retain soil water and plant nutrients. The clay content causes the red loams to swell when wet and to crack when dry and makes them difficult to cultivate. Tropical red loams may contain a higher organic content than their red color suggests, since some of the humus components in the tropics are colorless.

The properties of tropical red loams vary greatly with climate and parent material. Red loams are best developed on rocks rich in iron and on some shales; they are poorly developed on granites and sandstones. Local terrain differences may also cause considerable

---

[4] A hardpan is an impervious layer created by an accumulation of clay or iron oxides.

variation in soil character. In regions of irregular topography, flat-topped ridges have deep profiles of red loam; downslope, there is a gradation of soil types, sometimes leading to thick, black soils that are rich in lime. Such a continuum of soils, which is derived from similar parent materials and under a single climate but which is differentiated on the basis of slope and drainage, is called a **catena** (Latin *catena* chain). Catenas may occur on any slope, but they were first recognized in the tropical red loams of Tanzania.

Tropical Black Earths. Several kinds of savanna soils with lime-rich horizons reminiscent of those developed by calcification, occur on the dry side of the savanna. They are sometimes called black cotton soils or tropical black earths. These soils frequently reflect the nature of their parent material. For example, one large area of black tropical soil has developed on the basalt lavas of India. Others occur on basic igneous rocks in southern Rhodesia and Tanzania or in shallow depressions in the East African Plateau (Fig. 22-1).

Ferralites. The ferralization process operates at full strength in the abundant rain and high temperature of the *Af* climates. Silicates are almost completely removed; silica nodules are replaced by concretions of iron and magnesium oxide. Bases and fine particles are leached, and the soil contains little nitrogen because of the rapid and complete decay of surface organic matter. Because of the loss of silica the ferralite clay has a simple structure and low water-holding capacity. Consequently water penetrates easily and quickly, which is an agricultural advantage in the tropical rainforest; the soils can therefore be tilled shortly after a rain. The texture and structure of the soil also make for easy tilling, but the meager base and nitrogen content is rapidly depleted if the forest cover is removed. Production frequently falls off rapidly after a year or two of cultivation and can be increased only through heavy applications of fertilizer.

The generalized distribution of ferralites (Fig. 22-1) corresponds closely to that of the *Af* climates, but these soils are not uniformly well developed in every part of the tropical rainforest. Poor drainage, rough terrain, and certain parent materials, such as soft limestone, inhibit ferralization and result in incompletely formed ferralites (Fig. 22-2). In any case, generalizations about their distribution and properties should be regarded with caution. Intensive soil surveys, similar to those of mid-latitude regions, have been carried out in only a few areas. Our knowledge of these and other tropical soils is far from adequate.

Laterites. Laterites occur mainly in areas of monsoon climate, such as Sierra Leone and Liberia in western Africa and peninsular India. Once believed to be very widespread in the tropics, they have been more closely identified with the tropical monsoon climate and with areas of poor drainage, where seasonal alternation of heavy rain and drought produce a rise and fall of the water table. The lateritic layer occupies the zone of fluctuating water table and apparently develops as a result of the alternation in moisture content. This notion is supported by the presence of a dark-gray clay layer formed in the deoxidized, saturated zone beneath the laterite (Fig. 22-2). If this theory of formation is correct, the laterites are intrazonal soils more closely related to poorly drained soils in other latitudes than they are to other zonal soils of the humid tropics.

Pedocals

Pedocals include the following great soil groups: **chernozems** (Russian *chernozem* black earth), **chestnut soils, brown steppe soils, sierozems** (Russian *sierozem* gray soil), and **desert soils. Prairie soils,** which occur in humid regions but under a grass vegetation, are usually also included, even though they lack a characteristic lime horizon, whose presence is implied by the name "pedocal." These great soil groups are distinguished principally on the basis of the impact of different amounts of precipitation on their development; temperature is of minor consequence in differentiating the various kinds of pedocals.

The importance of moisture to the soil-forming processes is critical in semiarid regions. All pedocals, except for prairie soils, have a lime horizon somewhere in their profiles, but the depth below the surface and the thickness of this concentration of calcium vary with precipitation. Furthermore, the density of the vegetation also depends on the amount of rain, and this, in turn, determines the amount of organic matter in the soil. A transect across the midwestern United States shows this relationship very clearly. The lime horizon appears at a depth of about 5 feet (1½ m) near the western boundary of Iowa. Eastward, in the prairie soils, there is no such horizon. Westward, the horizon gradually approaches the surface. At the same time, the organic content of the soil decreases with precipitation as one moves westward from chernozem to steppe soils.

Prairie Soils. Deep, black grassland soils occur along the boundary between semiarid and humid climates, especially in mid-latitudes (Fig. 22-1). On the humid side are the prairie soils, which do not have sharply defined horizons but are deep and dark. They are rich in organic matter and have a high nitrogen content (Fig. 22-2). Even though the soil has been leached of calcium, many other bases are present. The clay con-

Soil Types and Their Distribution 345

Fig. 22-2. Representative soil profiles. The individual profiles are shown at approximately the same scale. The markings on some of the profiles are depths in feet; they serve to indicate the scale for all the photographs. *Photographs by Guy D. Smith.*

Teirra roja

Laterite block

Podzol

Subtropical yellow soil

Prairie soil

Gray-brown podzolic soil

Ferralite

Chernozem

Brown steppe soil Hydromorphic soil Planosol

Sierozem Calcimorphic soil Alluvial soil

Red desert soil Halomorphic soil Lithosol

tent is high. Consequently, the water-holding capacity of the soil is large. The structure and the texture are favorable to current agricultural practices. For these reasons, prairie soils are now used for intensive grain production.

As the name suggests, prairie soils are found in regions of prairie grass under humid or subhumid climates. The major areas of occurrence are in the midwestern part of the United States, in the Ukraine, and in southwestern Siberia. Smaller areas occur in eastern Australia, northeastern Argentina and Uruguay, central Hungary and Romania, and south-central Manchuria. Soils with shallow profiles (18 to 24 inches or 45 to 60 cm deep) but otherwise similar to prairie soils occur in some areas of Mediterranean climates. There are several sizable areas of prairie soils in association with park-like grass and woodland vegetation in the Coast Ranges of California.

Some of the best examples of prairie soils are found on the most recent glacial and loess deposits. There the prairie vegetation can be interpreted as a relict from a drier period; the grass is thought to have persisted despite more humid conditions because of frequent fires, which prevent the growth of trees. The prairie soils can thus be viewed as young soils formed under conditions of fire that preserved a vegetation bearing no relation to the present climate. In that case, they should be considered azonal soils rather than pedocals.

Chernozems. The chernozems adjoin the dry margin of prairie soils and extend into semiarid or steppe climates (Fig. 22-1). In North America, they form a broad north-south belt in the eastern part of the Dakotas and Nebraska, with a westward prong in the prairie provinces of Canada. In Eurasia, they extend eastward from the northern Ukraine into central Siberia, with discontinuous patches in eastern Siberia and Manchuria. They are also found in northeastern Argentina and in eastern Australia. In the wetter part of their range, they are difficult to distinguish from the prairie soils; they are dark brown or black in the upper few feet of the profile but contain a lime horizon that decreases in depth with decreasing rainfall. The lime has a silty texture and occurs in an irregular layer of varying thickness (Fig. 22-2). In drier country, the lime horizon is about 2 feet (60 cm) below the surface, and the chernozems gradually merge into brown steppe soils. Chernozems are less leached than the prairie soils because of lower precipitation, and they are even more richly supplied with plant nutrients. But from the farmer's point of view, this advantage is offset by less rainfall and greater danger of drought.

Brown Steppe Soils. This great soil group includes chestnut and brown soils of the mid- and low-latitude semiarid climates. Because of sparser grass cover and less organic accumulation, the A horizon is lighter in color than that of the chernozems. Shades of gray occur in the brown color and become predominant as aridity increases. Enough rain falls to leach some soluble minerals from the surface layers. These minerals collect at a depth of a few feet and form compact B horizons consisting of a lime horizon over a layer rich in gypsum (hydrous calcium sulfate). The more arid brown steppe soils may have a lime-rich crust as well as a subsurface lime horizon.

The low nitrogen content of the chestnut and brown steppe soils is compensated for, to a substantial degree, by the incomplete leaching of nutrient bases. When irrigated, they yield very well, but rainfall is low and unpredictable. Consequently, dry farming is a risky enterprise.

The distribution of the brown steppe and chestnut soils is wider than that of the chernozems, since they include tropical variants (Fig. 22-1). In the United States they occur west of the chernozem soils and extend from the Canadian to the Mexican border. In the Soviet Union they are found south of the chernozems and extend from the Ukraine to Central Asia. They also occur in Manchuria, Argentina, and around the central core of desert soils in Australia. The sub-tropical and the tropical brown steppe soils have a redder color, possibly due to a lower humus content, which allows iron oxide a greater role as a coloring agent. In general, the lime horizon is somewhat deeper than in the mid-latitudes, and may contain some sodium salts (Fig. 22-2).

Sierozems. These semidesert and cool desert soils have only rudimentary profile development as a consequence of slight leaching. There is very little humus in the soil, which is gray in color. Lime materials and gypsum are fairly evenly distributed throughout the profile, but with a slight zone of concentration within about a foot of the surface. Occasionally, crusts may form at or near the surface. These crusts, which crack on drying, may be composed of a carbonate or gypsum. The profile generally has a uniformly fine texture (Fig. 22-2). Sierozems develop where chemical and biological weathering processes are slight, and they are usually restricted to alluvial or aeolian parent material. They are found in parts of the Great Basin, in the eastern part of Patagonia, and in the interior basins of central Asia (Fig. 22-1).

Sierozems have large reserves of plant nutrients, but they are deficient in nitrogen and low in colloidal content. To be agriculturally productive, these soils require large additions of organic material and nitrogen, as well as irrigation. Proper drainage is also needed to prevent the excessive accumulation of salts and bases in the A horizon through capillary rise.

Desert Soils. In desert soils, organic matter is virtually absent, and leaching is inconsequential. The parent material is little changed by weathering (Fig. 22-2). Soils rarely develop in areas of rough terrain and are only weakly developed on old alluvial deposits. But some of these incompletely developed soils may be quite productive under irrigation if nitrogen and organic materials are added. If the irrigated areas are not properly drained, the soils quickly become saline. Desert soils are more common in the subtropical deserts of the world; sierozems are more frequent in the mid-latitude drylands (Fig. 22-1).

Nonzonal Soils

The zonal soils just described are usually depicted as covering much of the earth's surface. Because their differing properties are mainly the result of climatic differences, the presence of a particular zonal soil may be inferred, even if soil surveys are lacking. In most cases, the map of zonal soils involves fairly gross conceptual and cartographic generalizations; local differences resulting from topography and parent material are too small to be shown on a world scale. However, some segments of the earth's surface that are covered by azonal and intrazonal soils are large enough to escape this generalization.

Azonal Soils

Azonal soils do not have soil profiles because they are incompletely developed for one reason or another. They are often found on steep slopes, on recent alluvial deposits, and on sand dunes. Azonal soils strongly resemble their parent material, and their distribution is unrelated to climate. Three main types can be recognized: **dry sandy soils, alluvial soils,** and **lithosols.**

Dry Sandy Soils. Dry sandy soils have immature profiles because of the chemical stability of the quartz sand grains characteristic of most sand deposits. They contain little or no clay or soluble bases and salts or humus. There is little material to be carried downward and concentrated in the B horizon. Soil development on dry sandy soils proceeds very slowly because of the resistant quality of the sand grains and the lack of colloid-forming minerals.

Alluvial Soils. These soils are associated with recent deposits of stream material that have not been significantly altered by soil-forming processes (Fig. 22-2). Alluvial soils are found in floodplains, in deltas, and in recent alluvial fans (Fig. 22-1). Such soils are occasionally of exceptional agricultural significance because they may represent a rich accumulation of mineral and plant material removed by erosion from many diverse soils and concentrated in one place through stream deposition.

Lithosols. Lithosols are very shallow, stony soils composed of partially weathered rock fragments (Fig. 22-2). They may be found on moderate to steep slopes in all climates but are most common in deserts, where soil-forming processes operate very slowly. They are least common in the humid tropics. As the name suggests, the soil is very similar to its parent rock material; consequently, there are few mineral elements available for plant growth and little organic content.

Intrazonal Soils

Intrazonal soils reflect the predominant influence of landforms or parent material as soil-forming factors. They generally occur in relatively small areas throughout the zonal soil regions. There are four principal categories: **hydromorphic soils, calcimorphic soils, halomorphic soils,** and **planosols.**

Hydromorphic Soils. Hydromorphic soils (Greek *hydror* water + *morphe* form) occur in swamps, marshes, or flat and poorly drained uplands. Depending upon the degree of drainage, there may be a bog layer of peat several feet thick over a sticky, impermeable layer of gray clay; half-bog soils, with peaty material over grayish and yellow-mottled soil in areas of permanently wet ground in the forest; or wiesenboden (German *wiesen* meadows + *boden* soils), with a thick, humus-rich layer overlying a heavy clay B horizon in areas of seasonal soaking and grassy vegetation (Fig. 22-2). The organic content is very high, and the soils are quite acid. Many areas of bog and marsh in the youngest glacial till of the midwestern United States have been drained because of their great potential value as farmland. These soils also occur along the coastal fringe of the recent marine plain along the Atlantic and Gulf coasts of the United States.

Calcimorphic Soils. The calcimorphic soils (calcium + *morphe*), which are derived from limestone parent material, fall into two groups. In one group, the soil color is predominantly red, and the soil profile has a very low base content and contains free sesquioxides. These tierra roja soils have been included with the pedalfers because of their high sesquioxide and low base content. In the second group, the soil color is brown to gray-black, and the soil profile is completely saturated with exchangeable bases and contains free calcium carbonate. A brown calcimorphic soil is shown in Fig. 22-2. Rendzina soils are important members of the second calcimorphic group. They show a strong affinity with their parent material; consequently, they are included with the intrazonal order.

Rendzina soils are dark gray or black and resemble prairie soils, except that they rest directly upon soft limestone parent materials. Soft limestone deposits break down readily, and the soil developing on them does not become acid even in humid climates. Grasses are the dominant vegetation, perhaps because of their strong tolerance for calcium. Rendzinas are naturally fertile; the black belt in Alabama, an excellent example of rendzina soils, was one of the richest cotton-producing areas in the Old South.

Halomorphic Soils. Halomorphic soils (Greek *halos* salt + *morphe*) result from the salinization and alkalinization associated with poor drainage in arid regions (Fig. 22-2). As we have mentioned, they may also develop under irrigation. Reclamation of this land requires the provision of an adequate drainage system and the application of large amounts of water and of gypsum.

Planosols. As the name suggests, planosols are developed in regions of extremely flat terrain that inhibits natural erosion (Fig. 22-3). The soil horizons are abnormally thick, since they are untouched by erosion. Leaching is very strong, however, and the leached material accumulates in the B horizon, which eventually becomes a hard, impermeable clay pan. Where the B horizon is well developed and water-tight, the planosols provide little water for plants, since rain cannot penetrate to the subsoil. The impoverished A horizon further detracts from their agricultural value (Fig. 22-2).

A New Soil Classification

The foregoing description of the world distribution of soils was based on a system which has been used in the United States since 1938; it presupposes that soils of identical or nearly identical characteristics will develop in different parts of the earth if they have similar physical environments. As we have learned, intensive studies of certain groups of soils lend support to such an impression. There are large areas, however, where soils have not been subject to careful study but have been described only in subjective terms, if at all. Equating soils from different areas on the basis of such preliminary studies must be regarded as a tentative undertaking, and the concept of zonal soils should be considered a useful but provisional way of viewing the soils of the world.

The zonal soil classification has also suffered from vague or contradictory definitions and from the idea that the system must be based on the properties of virgin soils. But virginity is as rare in soils as in vegetation; large areas of most soil regions have been used by man in a variety of ways so that certain soil properties have been lost or modified. Furthermore, many soils which seem to be in a natural state, have in fact been artificially modified in the past, and thus only appear to be virginal. To avoid these and other difficulties, soil scientists are developing a new taxonomic system of classification. The magnitude of this task is revealed

Fig. 22-3. The natural setting of a planosol in south-central Illinois. The exceedingly flat terrain is characteristic of these poorly drained soils. The picture was taken about 36 hours after a heavy summer shower.

by the fact that the current state of the classification is labeled the *Seventh Approximation*.

The new system differs from the zonal concept in several important ways. A fundamental notion of the scheme is that the soil is a three-dimensional continuum which can be divided in several ways and not just into vertically arranged soil horizons. The new system also emphasizes description rather than genesis. The definitions of all soil types in every category are expressed in terms of properties that can be observed and measured. The Seventh Approximation resembles the zonal system in being a classification that is based on descending levels of generalization as follows: orders, suborders, great groups, subgroups, families and series. In order to define these more precisely, small, basic entities of soil called *pedons* (Greek *pedon*, ground) have been established. A pedon is the smallest three-dimensional unit that contains all the characteristics of an individual soil. The area of any pedon is from one to ten square meters (about 10 to 100 square feet) and has a specific but undefined thickness which permits the study of all horizons and their interrelationships. A group of contiguous pedons having a range of characteristics which can be accommodated within a single soils series is called a *polypedon*.

Precision of description was one of the basic objectives of the Seventh Approximation; hence, nomenclature became an immediate problem in establishing the new scheme. At first it seemed desirable to adapt existing names by modifying and sharpening the definitions of the soils classes they represented and to coin new names only when new soils classes were recognized. However, this would place students of soils in the difficult position of using names with different derivations. A system of numbers rather than names was suggested as one solution but was rejected as being too abstract and difficult to remember. It was finally decided to use coined names, as short as possible and with mnemonic and other connotations to help users remember the names. Roots forming the words have been taken from several languages but most of them are Latin or Greek. For example, the names for the ten soil orders were created as shown in the Table 22-2. All orders have a common ending *sol* (L. *solum*, soil) with a connecting vowel *o* for Greek roots and *i* for other roots. Suborder names are the only ones that consist of exactly two syllables, the only name of that length. The first syllable suggests a property of the class; the second suggests the name of the order. Great group names are formed by prefixing one or more elements to the appropriate suborder name, while subgroup names consist of the name of the great group modified by one or more adjectives.

The orders can be distinguished in a very general nontechnical way by several simple descriptive criteria

Table 22-2. Soil Orders and the Formative Elements in their Names

Name of order	Formative element in name of order	Derivation of formative element	Mnemonicon and pronunciation of formative elements
Entisol	ent	Nonsense syllable	rec*ent*
Vertisol	ert	L. *verto*, turn.	inv*ert*
Inceptisol	ept	L. *inceptum* beginning	inc*ept*ion
Aridisol	id	L. *aridus*, dry	*arid*
Mollisol	oll	L. *mollis*, soft	*molli*fy
Spodosol	od	Gk. *spodos*, wood ash	p*od*zol; *od*d.
Alfisol	alf	Nonsense syllable	ped*alf*er
Utisol	ult	L. *ultimus*, last	*ulti*mate
Oxisol	ox	F. *oxide*, oxide	*oxi*de
Histosol	ist	Gk. *histos*, tissue	*histo*logy

that are frequently reflected in the order names. For example, the *entisols* (recent) have weakly developed profiles either because of resistant parent material, or because of deposition of fresh material, or because removal by erosion keeps pace with profile development. The *vertisols* (invert) are characterized by pronounced swelling when they get wet, and by contraction and deep cracking when they dry. Because material from the soil surface falls into the deep cracks formed by alternate wetting and drying, there is a tendency for a constant turnover of soil particles, hence the name *vertisol*. The *inceptisols* (incept) are soils with poorly developed profiles that show little sign of oxidation and contain very little clay or colloidal humus. The incomplete profile may be due to either poor drainage and waterlogging or to long, cold winters and short, cool summers that slow down chemical weathering.

Aridisols (arid) are also poorly weathered soils with a low colloidal humus content but differ from *inceptisols* in that they may have well-developed horizons of clay, lime or gypsum which result from soil-forming processes in arid climates. The *mollisols* (mollify) have dark, friable surface layers of considerable thickness that are rich in organic matter and bases. On the other hand *spodosols* (ashy) usually have well-defined horizons, one of which is a bleached A_2 horizon, whereas the B horizon is composed of iron and aluminum oxides and alluvial humus. The *alfisols* (pedalfer) usually have grayish brown A horizons and brown to orange-brown B horizons that have relatively high clay content and high base saturation. *Ultisols* (ultimate) usually have upper horizons of brown loams whereas the B horizons range from red to yellow, often with mottles of iron and clay concentrations. *Oxisols* (oxides) have very deep profiles rich in hydrated oxides of iron and aluminum along with varied amounts of kaolinite clay. *Histosols* (tissues) are organic soils, usually with twenty percent or more organic content. In some instances the soils may consist almost entirely of peat or muck.

The physical and chemical properties of soils as well as their relationships to the natural and cultural setting were studied carefully and reappraised so that the set of soils classes comprising a category were all included and defined at a consistent level of abstraction. Consequently soils types from the zonal classification can frequently only be approximately related to categories established by the new system. These approximate equivalents are shown in Table 22-3.

Some additional qualities of the Seventh Approximation can be illustrated by a few comparisons between soils classes of the old system and those of the new. The comparisons are imperfect since many of the new orders overlap a wide range of zonal soils. For example some but not all of the prairie, chernozem and chestnut soils are included in the order of Mollisols (Table 22-3). Young prairie soils, for instance are categorized as follows. The order is *Mollisol*, the suborder is *Udoll* (Latin *udus*, humid, connotation of a humid climate plus *oll*, syllable suggestive of order), the great group is *Hapludoll* (Greek *haplous*, simple, connotation of minimum horizon development, *hapl* plus *udoll*). On the other hand mature prairie soils with clay rich B horizons are more or less equivalent to the great group *Argiudoll* (Latin *argilla*, white clay, connotation of a clay rich horizon). In a similar manner young chernozems with minimum clay accumulation in the B horizon are included in the order *Mollisol* but the suborder is *Ustoll* (Latin *ustus*, burnt, connotation of a dry climate with a hot summer). The great group is *Haplustoll* (*haplous* and *ustoll*). Mature chernozems with clay rich B horizons are more or less equivalent to the great group *Argiustoll* (*argilla* and *ustoll*). Many chestnut soils are also classed as *Mollisols*, suborder *Ustolls*, great group *Argiustolls*. On the other hand the reddish chestnut soils of the Texas Panhandle are classed as *Alfisols*, suborder *Ustalf*, great group *Haplustalf*.

As a further example, some of the zonal order *podzols* are classified in the Seventh Approximation as follows: the order is *Spodosol*, the suborder is *Orthod* (Greek *orthos*, true, connotation of common; *orth* plus *od*, syllable suggestive of order), the great group is *Cryorthod* (Greek *kryos*, coldness, connotation of cold; *cry* plus *orthod*). The *Typic Cryorthods* are soils in arctic regions that include the podzols. The adjective *typic* is used for the subgroup believed to best typify the great group.

The gray-brown podzolic soils come under entirely different orders from the podzols. For example, some gray-brown podzolic soils are classed as *Alfisols*, suborder *Udalf* (*udus* — connotation of moist plus *alf*), great group *Fragiudalf* (*fragi* modified from Latin *Fragilis*, brittle, indicating the presence of a fragipan, a tight silt layer or pan in the B horizon). The ferralites are identified as *Oxisols*, suborder *Orthox* (Greek *orthos* — connotation of true or common), great soil group *Acrorthox* (Greek *akros* highest, most strongly weathered).

This brief introduction to a complex system makes evaluation difficult. It should, however, provide you with some insight to a complex but complete and well-structured classification that forms a comprehensive system of ordering ideas about soils, as well as pigeonholing individual soils into distinct categories.

Table 22-3. The Seventh Approximation Soil Orders and Their Approximate Equivalents in the Zonal Classification

Present order	Approximate equivalents
1. Entisols	Azonal soils, and some Low Humic Gley soils.
2. Vertisols	Tropical Black earths.
3. Inceptisols	Some Brown Forest, Low-Humic Gley, and Humic Gley soils.
4. Aridisols	Desert, Reddish Desert, Sierozem, Solonchak, some Brown and Reddish Brown soils, and associated Solonetz.
5. Mollisols	Chestnut, Chernozem, Prairie, Rendzinas, some Brown, Brown Forest, and associated Solonetz and Humic Gley soils.
6. Spodosols	Podzols, Brown Podzolic soils, and Ground-Water Podzols.
7. Alfisols	Gray-Brown Podzolic soils, Gray Wooded soils, Noncalcic Brown soils, Degraded Chernozem, and associated Planosols and some Half-Bog soils.
8. Ultisols	Red-Yellow Podzolic soils, Reddish-Brown Lateritic soils of the U.S., and associated Planosols and Half-Bog soils.
9. Oxisols	Ferralite soils, Latosols, Lateritic soils.
10. Histosols	Peat and Bog soils.

Summary

In this chapter, soils were classified and described on the basis of the character of their profiles. The classification involves three major orders: zonal, azonal, and intrazonal.

The zonal soils have properties that reflect the roles of climate and of vegetation as the principal soil-forming factors over much of the earth's surface. Zonal soils are of two kinds; pedalfers, which develop in humid regions; and pedocals, which evolve in more arid areas. The pedalfers are subdivided into great soil groups that differ mainly in terms of the effect of temperature on the soil-forming processes. Podzolization is of major importance in cold, wet climates; ferralization dominates in hot, wet climates.

The great soil groups of the pedocal suborder differ mainly on the basis of moisture availability and amount of vegetation. On the wetter edge of the semiarid climates, the characteristic lime horizon of pedocals is absent or far below the surface; and the vegetation is relatively luxuriant, so that the soils have a high content of organic matter. On the drier side, the lime horizon approaches the surface; the vegetation is sparse; and the organic content is correspondingly low.

Azonal and intrazonal soils have properties that are not determined by climate and vegetation. Young azonal soils reflect the character of their parent material to the extent that differentiation has not yet taken place. The properties of intrazonal soils reflect the predominance of soil-forming factors other than climate and vegetation, such as poor drainage or limestone parent material.

Problems arise whenever attempts are made to classify elements of the natural landscape. Soils are no exception. Difficulties associated with the zonal soil classification have prompted the U. S. Department of Agriculture to formulate a new classification, the Seventh Approximation. It is so called because it represents the seventh attempt to develop a precise and comprehensive soil classification. It has ten soil orders that are based on the internal characteristics of soil horizons rather than associations with geographic environments or soil genesis. The classification has a rather forbidding nomenclature and a highly structured organization. It does, however, have the advantage of not being tied too tightly to traditional concepts of soil origins. Consequently the classification should give greater flexibility to the study of soils and hence greatly increase our knowledge about them. Nevertheless to be usable the system requires considerable knowledge about soils. For this reason elements of the simpler zonal concept will probably continue to be used for introductory descriptions and explanations of soil distribution.

Suggestions for Further Reading

Bunting, Brian T., *The Geography of Soil*, Aldine Publishing Company, London, 1965.

Eyre, S. R., *Vegetation and Soils: A World Picture*, Aldine Publishing Company, London, 1963.

Robinson, G. W., *Soils: Their Origin, Constitution, and Classification*, 3rd ed., John Wiley & Sons, Inc., New York, 1951.

U.S. Department of Agriculture, *Soil Classification: A Comprehensive System, Seventh Approximation*, U.S. Government Printing Office, Washington, D.C., August 1960.

———, *Soils and Man*, Yearbook for 1938, U.S. Dept. of Agriculture, U.S. Government Printing Office, Washington, D.C., 1938.

23 Environmental Quality and Physical Geography

Introduction

On all sides today one hears cries to stop the plundering of resources and to man the ramparts against pollution. However overly dramatic these cries may appear they are based upon sound observation. The world may indeed be coming to an end — in fact it has already come to an end for the Barbary lion, Burchell's zebra and the Rufous gazelle in Africa; the Arizona wapiti, the Badlands bighorn and the Eastern cougar in the United States, to name a few victims of man's greed and carelessness. The final irony would be the extinction of mankind because it succeeded only too well in multiplying its numbers and in subduing the earth. The very notion of subduing reflects a failure to understand the nature of the physical environment and its interrelationships and hence an inability to cope with them. However, extinction need not, and may not be inevitable. The rapidly increasing concern over pollution and the concomitant increase in the appreciation and understanding of ecology (the study of how living and non-living environments function together) may bring mankind and the physical environment into some satisfactory stable state of equilibrium.

To control pollution we must appreciate and understand how the total environment works. At present it is deteriorating rapidly. The ailing parts must be identified and treated in such a fashion that the remaining parts, including man, are not harmed. This can be considered a rescue phase. The further and larger task is to maintain sound environmental quality indefinitely. Physical geography cannot provide the answers to all of the complexities of this task. However, a dominant theme of the text has been to develop an appreciation for the physical landscape as an open system, all of whose parts are affected by changes in any single one of them. The purpose of this final chapter, an outgrowth of that concern, is to examine some of the causes and consequences of a deteriorating environment.

The deterioration of the physical environment means either the concentration of matter above some unwanted level in the air, soil, or water; or else the depletion of matter below some desired level. This definition allows for both material and esthetic deterioration, but its meaning hinges on the words *unwanted* and *desired*. To put if differently, what is deterioration to one person may not be so to another. Belching smoke stacks can represent either payrolls or pollution, depending on the point of view; a dirty river may evoke either nostalgia for the old swimming hole or demands for the control of industrial and municipal effluents; a new suburban subdivision may produce feelings of progress and prosperity or visions of continuing encroachment of the city on its surroundings. Pollution is therefore easier to talk about than to define concretely. The one generalization that can be made is that deterioration of the physical environment often accompanies the conversion of energy by man for some specific purpose, such as burning of fuel, cultivation of crops, raising of livestock, or hunting and fishing. It may also involve depletion as exemplified by exhausted and eroded soils, impoverished or destroyed vegetation, or extinct species of birds, fishes or animals.

Water Pollution

Let us begin by examining how contaminating agents affect the landscape system. Air and water are contaminated directly, whereas soil, plants and animals are usually affected indirectly by means of tainted air or water. The pollution of these two fluids results from deleterious modification of surface conditions, such as overgrazing, or by direct addition of matter in excessive quantities, such as smoke from factories. Once material is introduced into the air or water it spreads by the random motion and collision of molecules, known as *Brownian movement*, when these fluids are still, or by turbulence and convection when they are in motion. In the instances of overgrazing and cultivation, pollution is a double-edged menace for these practices may not only lead to sediment clogged streams but they may also result in serious depletion of vegetation and soil.

The contamination of water occurs during the return portion of the hydrologic cycle (p. 175) when evaporating water leaves behind any contaminants it contained as a liquid. As water condenses, pollution begins, for the fresh water picks up some impurities as it falls through the atmosphere. However, the greatest addition of material occurs as the water flows across the land on its way to sea. The substances which commonly pollute water are sediments, nutrients, infectious agents, industrial wastes and heat. Because of its unique properties water is a powerful carrier and modifier of these substances.

Pollution by Sediments

This type of pollution results from the destruction or removal of the vegetation cover through careless cultivation, overgrazing, fire, and erosion by overland water flow. The rate at which overland flow removes sediment from the surface is controlled by the rate of runoff,[1] by the type of vegetation, by the structure and texture of the soil, and by the slope of the land. Tilling and overgrazing change the soil porosity by compacting and breaking up soil particles. At the same time the surface receives the immediate and full impact of the rain if there is no adequate shield of vegetation. Consequently, the rate of runoff increases sharply as does the rate of erosion, whenever the vegetal cover is reduced. When enough land surface becomes exposed, streams become polluted by an excessive content of sediment. Stream water then becomes opaque and absorbs the sun's heat more readily. The lack of light suppresses aquatic vegetation and the increased temperature brings about a deterioration of the fish population.

The increased rate of runoff also changes the frequency and magnitude of river flooding since more water is being dumped into the rivers at a faster pace. This increase in the amount of water which has already an unusually high sediment load produces clogged stream channels and more rapid aggradation on the floodplain. Swamps and marshes may enlarge or develop where none existed before. In other instances fertile flood plain soils may be buried by less useful subsoil sediment. Such events have plagued ancient as well as modern civilizations. The Romans, for example, had problems with expanding swamps and spreading malaria in many Italian valleys; and it is evident from descriptions of the landscape by Greek geographers and historians that by 500 B.C. or even earlier floods had greatly reduced the habitability of many areas in the eastern Mediterranean. Today many of the floodplains in the Piedmont and Coastal Plain of the southeastern part of the United States are being buried by sterile sand. Thus, continuing ignorance of and disregard for factors which control the rate of soil erosion have resulted in serious losses of landscape quality throughout history. Soil and vegetation depletion has brought ruin to the uplands, whereas excessive sedimentation has resulted in an expansion of swamps and a burial of soils in the lowlands.

Since the population of the world is increasing rapidly, it is not only difficult but very expensive to remedy the problems associated with accelerated erosion. A considerable part of the problem stems from a desire to maximize returns from crops and livestock. In farming this is done by crop specialization, enlarging fields (which makes for larger areas without plant cover), and increasing the use of fertilizers and machinery. The

---

[1] The rate of runoff is a function of the intensity and duration of rainfall, the porosity of the soil, and the length of the slope.

emphasis upon uniformity and bare soil over large areas is in direct contrast to the natural setting which emphasizes diversity and land cover. To restore the landscape to its pre-agricultural state would be impossible, but some land-use modifications will be necessary to establish a continuing, stable level of production. To obtain this equilibrium many strategically located parcels of land must be removed from cultivation and be allowed to return to a condition of forest, or small pond, or grass and swamp, thus providing a natural sponge to absorb the excess water whose surface flow erodes the land. Such a plan will reduce total crop yield and will increase food costs — an unthinkable solution in the short-term view, but a necessity in the long run.

Pollution by Nutrients

Nutrients for the aquatic vegetation which chokes streams and lakes originate from agricultural fertilizers, sewage and industrial wastes. Fertilizer use has increased about six times between 1935 and 1970, and the nitrogen, phosphorous and potassium in the fertilizers are washed into streams in ever larger quantities by surface runoff and groundwater flow. This influx favors the development of certain aquatic weeds and algae. The dense algae blossoms floating on the surface can cause the death of bottom plants by shutting off sunlight. Furthermore, when the algae die in late summer their decomposition not only releases hydrogen sulphide gas but also consumes great quantities of dissolved oxygen. The removal of oxygen from the water places severe restrictions on aqautic fauna, not only eliminating many desirable fish but other forms of water-associated wild life as well. Lake Erie is now in this unfortunate plight. Its rescue and restoration will require a great deal of study, research, money and time.

The state of productive abundance which emphasizes quantity and monotony of species rather than quality and diversity is called **eutrophication** (Greek, *eutrophos*, well nourished). Detergents, along with fertilizers, are regarded as one of the principal causes of eutrophication in streams and lakes. Household detergents contain phosphates to keep dirt in suspension, thus preventing it from returning to the clothes in the wash. As a result, sewer lines dump thousands of tons of phosphates into rivers and lakes, thus aggravating the state of eutrophication. It is easy to see the solution to this problem in terms of prohibiting the use of phosphates in detergents, but the link between the use of phosphates to wash clothes and the pollution of rivers and lakes is only dimly perceived by most people.

Pollution by Industrial Wastes

Industry uses water as a solvent, as a cleansing agent, as a mineral extractant, as a coolant and as a means of removing waste. This industrial use creates one of the greatest sources of pollution, including both organic and inorganic materials as well as heat. Any single industrial use of water will result in the transfer of concentrated industrial wastes to lakes or streams; multiple use of the water may produce extreme contamination.

The wastes are destructive of water quality and water life in many ways. Some have very high *biologic oxygen demands (BOD)*, that is, a great deal of dissolved oxygen is needed to decompose the waste. Others are toxic to aquatic life, some limit photosynthesis, and still others make the water smell or give it an ugly appearance. Only a few of these pollutants need be mentioned.

Organic wastes from food-processing plants, including distilleries, have very high BOD as do the residues from paper and pulp mills. Heavy metallic salts may be lethal to aquatic life, especially when they occur in combination. Small amounts of copper or zinc salts, for example, will not harm fish, but even very small amounts in combination may completely eradicate them.

The great danger posed by mercury in water is well known. It is an example of the dangerous assumptions that can be made about the environment without an understanding of the processes that operate in natural systems. The lethal qualities of two kinds of mercury have been known for a very long time. One kind of mercury poisoning, caused by metallic mercury or inorganic mercury compounds, does serious but curable damage to the alimentary tract and kidneys. The other, caused by organic mercury compounds, results in irreversible damage to the brain and central nervous system.

Many industrial processes require the use of metallic mercury compounds of which methylmercury is the most deadly. Until very recently it had been assumed that when these forms of mercury were deposited in water, they would remain relatively inert on the bottom of ponds, lakes or bays. Consequently, the first cases of mercury poisoning from eating fish tainted with methylmercury caused considerable astonishment in the 1960's. Several years passed before it was discovered that some micro-organisms living in the mud on the bottom of water basins could digest metallic and inorganic mercury and turn it into methylmercury. Algae ate the bacteria, little fish ate the algae, and big fish ate the little fish. Each transfer to a larger sized organism increased the concentration of the poison so that people who ate the big fish were getting sizable doses of the lethal material. Even though the danger has been recognized and ways are being considered to prevent tainted fish from being sold, problems remain. For instance, the highest allowable level of methylmercury in food is not yet known, nor have methods of neutralizing the cast-off mercury been discovered.

Thermal Pollution of Water

Perhaps the most dangerous thermal pollution involves the artificial warming of the atmosphere beyond its capacity to dispose of the excess heat without radical changes in its energy balance, but water is also being heated with similarly forseeable consequence to its heat budget. Nearly two-thirds of the water used in industry is earmarked for cooling purposes. The principal contributors to thermal pollution are electric power plants which after circulating great quantities of water through steam condensers, then return the heated water to its source. Because electric power production is expected to double in the next decade, the rate of thermal pollution will increase sharply unless suitable precautions are taken. It is estimated that, by the year 2000, the electric power industry will each day require about one third of the average daily stream runoff in the United States. During periods of low summer flow this amount could involve the total daily runoff.

Thermal pollution of such magnitude (and even of much smaller consequence) is a matter of urgent concern for a number of reasons. For example, warm water has a lower dissolved oxygen content than cold water. Therefore, the rate of decomposition of organic wastes discharged into streams decreases, thus intensifying the effects of these wastes upon water quality and increasing their duration. Fish and other cold-blooded animals are unable to regulate their body temperatures; hence their body temperature rises as the water warms up, and their need for oxygen increases. The metabolic rate of fish approximately doubles with each 10°C (18°F) rise in temperature but as the demand for oxygen increases its availability decreases. Game fish, which require relatively high amounts of oxygen, must migrate or die and be replaced by less demanding organisms until the entire aquatic life system, including plants, invertebrates and amphibians, is eventually degraded by the disappearance of one species after another.

Water in deep middle-latitude lakes is frequently thermally stratified (see p. 30). In such lakes the water usually mixes vertically in the spring so that much of the water mass is well supplied with oxygen. During the summer the surface water warms and the lake becomes stratified with a layer of light, warm water (epilimnion) floating on the colder, dense water below (hypolimnion) where most of the oxygen-using organisms are found. This stratification persists until late fall when mixing once again takes place, thus replenishing the oxygen in the hypolimnion. When water is needed as a coolant for industrial purposes it is pumped out of the hypolimnion and the warmed water is subsequently discharged at the surface of the lake. The added heat increases the duration of stable thermal stratification by delaying the fall cooling of the epilimnion and hastening the spring warming. At the same time the water from the hypolimnion is rich in plant nutrients so that warming it and discharging it in the upper, sunlighted zone favors the growth of algae. As the algal material dies it sinks into the hypolimnion and decomposes, using up the oxygen needed to maintain the abundant life in this zone. It is obvious that a very small change in the thermal structure can affect the life system in a lake, quickly and deleteriously.

Heat pollution is already modifying local climates. Recent studies indicate that river water may be warmed by 10°F (6°C) within thirty or forty miles (about 50 or 60 km) of some intensive industrial complexes. Not only are rivers affected but also lakes. The rapid expansion of power plants along the shores of a lake as large as Lake Michigan is already having sufficient effect on the temperature of the water to cause concern among those persons sensitive to future problems of environmental quality.

The modification of local climates by thermal pollution is an increasing threat to the environment. Great volumes of warm stream and lake water will raise the local air temperature and increase the rate of evaporation. Although these changes may be relatively small, the local effect may become significant. For example, one might expect the incidence of winter fogs over lakes and in river valleys to increase markedly. In colder climates the regimes of ice formation and ice melt would change. What impact these changes will have on aquatic life systems is not clearly understood.

Thermal pollution, like other kinds of contamination, threatens the quality and life of our water resources. Increasing discharges of sediments, nutrients, industrial wastes, plus the addition of great amounts of heat may be more than the aquatic systems can bear. Safe and efficient methods of dissipating heat from power plant cooling water must be developed to protect these systems. The use of industrially heated water for commercial purposes, such as heating buildings or using the water to aid evaporation in desalinization plants is not yet economically attractive. Thus, for some time in the future, heat will have to be dissipated to the atmosphere by means of cooling lakes and towers. Cooling lakes are constructed especially for transfering heat to the atmosphere. The recommended design of a lake for a large power plant calls for a man-made water body about a mile wide and three miles long that is shallow at one end and about 50 feet (15 m) deep at the other. The cooling water is drawn from the deep end at a depth of about 30 feet (9 m) and is discharged at the shallow end. Such lakes require a steady inflow of water to replace that lost by evaporation and to retard the accumulation of dissolved material.

The use of large lakes is not always possible in congested areas; here cooling towers can be used to treat industrial water. There are two basic types of towers; one, the wet tower, employs evaporation to remove heat, whereas the second, the dry tower, uses a radiator to transfer heat to the atmosphere without evaporation. In a wet tower water falls in a thin film over a series of baffles where it is exposed to a rising flow of air. Or else the water is sprayed into the tower where it evaporates quickly in the updraft. In many cases the additional water vapor in the air can be a local nuisance, especially in winter when thick fogs and ice films may develop. The dry tower avoids this problem by using a system similar to that of the automobile radiator. The heated water is pumped through a network of pipe in the tower around which air is circulated by a large fan. Cooling towers and lakes, while serving as efficient means of cooling water are, unfortunately, also effective means of heating the atmosphere.

Atmospheric Pollution

Most students of our planet's past agree that the earth's atmosphere was originally very different from present conditions. It is fairly certain that it was devoid of free oxygen and probably of free nitrogen as well. Today, as we have seen, the composition of the atmosphere by volume, at sea level, is 78 percent nitrogen, 21 percent oxygen, and one percent argon with traces of carbon dioxide and rare inert gasses. This composition is surprising since nitrogen is a rare element in the earth's crust, while oxygen although abundant, virtually never occurs in a free state outside the atmosphere. These two gasses are present in the atmosphere because living organisms, mainly plants, keep generating them. The amount of oxygen in the atmosphere, for example, is in equilibrium because it has been generated by photosynthesis as fast as it has been removed from the air by oxidation.

The first appearance of free oxygen in the air coincided with the start of photosynthesis, the process whereby green plants take in water and carbon dioxide and with the aid of solar energy combine them to form organic matter. An important by-product is oxygen which is transpired by plants to the air. Gradually the oxygen content of the air was built up through photosynthesis until it reached its present stable level nearly 400 million years ago. While this was going on, organic matter was being buried, mainly by marine sediments. As a result, the buried organic matter, prevented from recombining with oxygen, created a reserve of oxygen in the atmosphere, and stored energy in the form of fossil fuels. This energy, which accumulated slowly over hundreds of millions of years, is now being released to the atmosphere in a short period of time.

Free nitrogen is maintained in the atmosphere by the nitrogen cycle. Certain bacteria and algae in the soil convert the nitrogen present to ammonia and nitrate. In these forms it is used by higher plants to make plant proteins which are used by animals and man. When the plants and animals die, special bacteria break down their tissue and together with the action of other bacteria, release molecular nitrogen to the soil and air.

Calculations have been made in recent years of the annual oxygen balance in the United States. These estimates compare the amount of oxygen used in combustion of coal and petroleum to the amount of oxygen produced by photosynthesis. These calculations are admittedly rough, but they do indicate that more oxygen is consumed in the coterminous United States than is produced. It appears that we are dependent upon wind importing oxygen produced in other areas.

To a substantial degree the oxygen is consumed in the burning of fossil fuels in steel plants, in electric power plants, in automobiles and so forth. At the same time fuel burning is accompanied by the release of heat and particles, many of which contaminate the air more directly by way of smoke and fumes from industrial plants, auto exhausts, waste disposal by burning and jet air craft exhausts.

Thermal Pollution of the Air

It is useful to separate atmospheric heat pollution from that of water even though a good deal of the heat that is fed into the atmosphere comes from water heated by industrial processes. Aquatic life systems deteriorate promptly with increases in water temperature, but most air-breathing animals can regulate their body temperatures; hence they can tolerate relatively large, short-term increases in air temperature. However, long periods of high temperature will result in pronounced changes in climate which will quickly modify the other factors of the earth's surface system with accompanying changes in the process links between factors.

Virtually all the energy available at the earth's surface comes from the sun. Heat flow from the interior, although locally important, and that from tidal friction contribute less than 0.001 percent of the total. As we have already seen, all of this energy is returned to space in the form of radiant heat. A balance between solar income and terrestrial outgo of energy is maintained, for if the solar input increases the planetary temperature will also rise, and the higher temperature will bring about increased terrestrial radiation until a new balance is achieved.

The amount of energy being released by artificial burning at present has an insignificant effect on the annual average temperature of the atmosphere. Various estimates of the annual production of energy suggest that it comprises only about one fortieth of one percent of the amount radiated by the earth, but the release of heat to the atmosphere is increasing by about 8 percent per year. At that rate, the average temperature of the atmosphere would rise nearly 5°F (3°C) in less than 125 years. This warming is enough to melt the polar ice caps, thus causing sea level to rise about 100 feet (30 m). The 8 percent growth of energy transformation would exhaust the reserves of fossil fuel as well as the uranium and other ores used to generate atomic power long before such drastic results would occur. If controlled fusion reaction becomes available, however, then thermal stress may well increase to the degree indicated.

This discussion of temperature change does not take into account the effects of the particulate and gaseous matter released to the atmosphere as a result of the energy transformations. Some of these effects appear to cancel one another — for example, smoke tends to reduce the amount of solar energy reaching the earth but has little effect on outgoing earth radiation so the net effect is to lower temperature slightly. Carbon dioxide has an opposite effect and the great use of fossil fuels has caused a measurable increase in the concentration of that gas in the atmosphere, thus favoring a rise in atmospheric temperature. It is difficult to forecast the ultimate effect of these additions to the air, but they do have the potential for changing the earth's climate.

Non-Thermal Pollution of the Air

The major contaminants in the atmosphere include particulate matter, carbon monoxide, hydrocarbons, and oxides of nitrogen and sulfur. What are the effects of these additions to the air? What factors contribute to the non-thermal pollution of the air?

Particulate Pollution. This form of contamination results from the presence of solid or liquid particles in the atmosphere that settle out of the air fairly promptly in contrast to gaseous materials. Some of the most abundant of these particulates are smoke, lead, iron, asbestos, rubber and arsenic.

The major source of these particles is the automobile, although various industrial processes contribute their share. Lead is a source of major concern since it is a cumulative poison that may be inhaled or ingested in food and water.

Carbon Monoxide. This gas, coming mainly from auto exhausts, is perhaps the most abundant of the atmospheric pollutants. Since it combines more readily with hemoglobin than does oxygen it replaces oxygen in the blood stream. Exposure for about an hour to air with 120 parts per million of CO can lead to sleepiness and dizziness. Greater concentrations are often encountered in city traffic. Consequently, it is entirely possible that CO may be the cause of traffic accidents as well as contributing directly to the death of persons with respiratory ailments and heart disease.

Hydrocarbons. These are a complex lot of chemicals, mainly derived from auto exhausts, that are not harmful at the levels of concentration in which they usually occur. When abundant they may cause irritation to the eyes and respiratory system. However, some hydrocarbons react with nitrogen dioxide (NO_2) and sunlight to form photochemical smog. One of the principal elements of photochemical smog is ozone, which can be severely irritating to the eyes and respiratory system. Sunny weather and atmospheric conditions favorable to the concentration of hydrocarbons quickly produce serious discomfort to thousands of urban dwellers, and long continued exposure may cause serious respiratory diseases.

Oxides of Nitrogen and Sulfur. Nitric oxide (NO) forms when nitrogen in the atmosphere combines with oxygen. Again, this most often occurs during fuel combustion in automobile motors. The gas is relatively harmless, but it readily combines with oxygen to form NO_2 and, as noted earlier, NO_2 is a common catalyst for the formation of photochemical smog. At very high concentrations NO may become toxic because it combines with hemoglobin far more rapidly than oxygen does.

Oxides of sulfur form when fossil fuels are burned. Sulfur dioxide (SO_2) is a byproduct of coal burning and is an irritant to the eyes and throat and may induce tonsillitis and severe coughing. Sulfur dioxide will react with atmospheric oxygen to form sulfur trioxide. This will then combine with water droplets in the air to form sulfuric acid which causes damage to buildings and equipment as well as to lung tissue.

Pollution in the Atmosphere

This brief review of the major types of atmospheric contaminants indicates that motor vehicles are a primary source of emitted material. Also implicated are industrial processes which provide an abundance of particulate matter as well as gaseous forms of pollution. These sources alone cannot bear the complete responsibility for atmospheric contamination. Normally, gases and very fine particulate matter can be distributed through great volumes of air by wind and turbulence — a natural solution of pollution by dilution. Furthermore, great quantities of material are annually washed from the atmosphere by rain (although this may transfer the pollution to soils and streams). But over much of the world the situation is not normal — population is growing so fast that material is pumped into the air faster than it can be removed, particularly in areas of dense population where there is heavy motor vehicle traffic and major industry.

In addition to the pollution associated with areas of dense population, there is a regional and seasonal distribution which is related to the vertical temperature distribution in the atmosphere. Temperature inversions prevent upward movement of air (see Chapter 5). Consequently, places with high population and frequent inversions are optimum sites for smog, *smaze* (a combination of smog and haze) and other forms of air pollution. Mexico City, for example, is located in a deep, enclosed basin. Inversions act as a lid over the basin, preventing the polluted air from diluting upward. After several days of such conditions the air becomes dangerously polluted.

In the interior of the United States atmospheric pollution is generally more frequent and intense in winter because the cold land surface favors the development of stable air or ground inversions. But on the west coast of the United States, atmospheric contamination tends to be intensified in summer. The Pacific High becomes stronger in summer and moves poleward a few degrees. This change is sufficient to bring almost all of the west coast under the influence of the strong inversion on the eastern margin of the high. The inversion, at an elevation of 1000 to 5000 feet (about 300 to 1500 m) extends inland sufficiently far to cover the coastal cities for long periods in summer.

The Los Angeles lowland is especially noted for its intense air pollution. A nearly enclosed basin, frequent inversions, and cold air offshore on the one hand, and a thriving industry, a growing population and four million cars on the other, combine all the prerequisites for smog. The lowland is enclosed to the north, east and south by mountains. On the west, over the cold ocean, is a barrier of cool, southward flowing air. During the summer when inversions are prevalent over the lowland, warm air is trapped beneath the inversion and between the mountains surrounding the basin and the cold air barrier offshore. Under these conditions, air contamination soon becomes almost toxic. Since the trapped basin air cannot escape until the inversion weakens, the only solution is to reduce the amount of foreign matter released to the air.

Along the east coast air pollution is most pronounced from mid- to late summer. In summer the Atlantic High, with its inversion, extends westward, occasionally covering the eastern third of the United States. Megalopolis, between Boston and Baltimore, becomes covered by a vast blanket of smog. Farther inland, surface heat tends to weaken and lift the inversion and, since the wind is stronger and pollutants fewer, the air is relatively cleaner. The easiest solution, here as elsewhere, is to limit the quantity of material spewed into the atmosphere.

The most serious long-range problem is probably thermal pollution rather than contamination of water or air by unwanted matter. Dirty air can be cleaned by restraint in emission of contaminants and by devices to control those emissions. Unfortunately, thermal pollution cannot be remedied as easily since energy conversion is the foundation of the industrial society. A reduction in the use of energy does not seem likely in a world faced by increasing population and an ever greater demand for products requiring high energy use (such as preferences for aluminum goods over those made from iron or steel). Furthermore, the technical solutions to particulate and gaseous pollution require large amounts of energy, so that the expansion of industry demands more energy not only for the purposes of production but also to keep the air and water clean. This continuing expansion of the transformation of energy puts mankind on a collision course with thermal disaster. To be sure a safety margin now exists — a margin that may allow heat absorption from growing transformation of energy for the next thirty years. Beyond that, thermal pollution will begin to reach such a level that the climate will change drastically and affect all the other elements of the physical environment.

Pollution and the Environment

Throughout the book we have been concerned with various aspects of the earth's surface system; in this chapter our emphasis so far has been mainly on the types and nature of pollution. Let us now combine the two themes.

Man's modification of his environment has changed continually as a result of changing technology, differing attitudes and objectives, and increasing population. The world's population increased very slowly until about 200 years ago, but during the last 100 years the growth rate has been dramatic (Fig. 23-1). Estimates of the world's population a million years ago range from 150,000 to 500,000. These early hunters and gatherers exerted very little pressure on their natural environment (Fig. 23-2). By the time agriculture was invented, around 15,000 years ago, the world's population had increased to about ten million persons. Until then, the greatest pressure had been on the flora and fauna, especially on animals, although significant modifications of the vegetation cover had begun to occur in certain places. By the beginning of the Christian Era the population had risen to about 300,000,000. Farming had become more specialized and elementary forms of industry had developed. In specific places, as in the Mediterranean area, the Middle East and China, human influence upon soils and vegetation was now strong. This human impact produced feedback from all the other factors of the environment (Fig. 23-2).

During the Christian Era, the rate of population increase accelerated markedly, and by 1800 about one billion people inhabited the earth. At this time, the people of South and East Asia had perfected an intensive form of agriculture based on getting two high-yield crops of rice a year, or failing that, a crop of rice and a crop of some less heat demanding plant that would permit them to multiply during the next 170 years. Meanwhile, the Western world was entering the Industrial Revolution which not only permitted a vast increase of population, but also increased the transformation of energy at an unparalleled rate.

Since about 10,000 B.C., the earth's surface system has changed very little from natural causes. Man became better able to cope with what he viewed as a hostile environment, and so prospered and increased in number. His transition from a simple gatherer to a skillful hunter to a cultivator of plants and accomplished herdsman, and finally to the master of a complex technology, was accompanied by the development of complicated political systems, intricate communication networks, and sophisticated religions. However, this increasing ability to cope with the natural and supernatural was not accompanied by much awareness of the dangers of overpopulation or by attempts to understand the complex natural systems at the surface of the earth and man's impact on them.

Man has had reason to regret these oversights long before now. Carelessness with regard to the natural milieu brought troubles to the Greeks and Romans centuries ago. Probably other, and lesser known, cultures were dispersed or destroyed for reasons of ignorant or thoughtless abuse of the environment. The Maya are an example; they practised a conservative agriculture, slashing and burning a small plot in the tropical forest, and planting the same crops for a few years with the digging stick as their only implement until soil exhaustion led them to move to a new plot; but their ever-increasing numbers exerted considerable pressure on their fragile environment. Between 300 and 900 A.D., the Mayas prospered in the wet tropical areas of northern Guatemala and Tabasco. Toward the end of the tenth century they suddenly moved into the drier northern half of the Yucatan Peninsula where the bed rock is a porous limestone, and consequently there is virtually no surface drainage. The sole water supply for

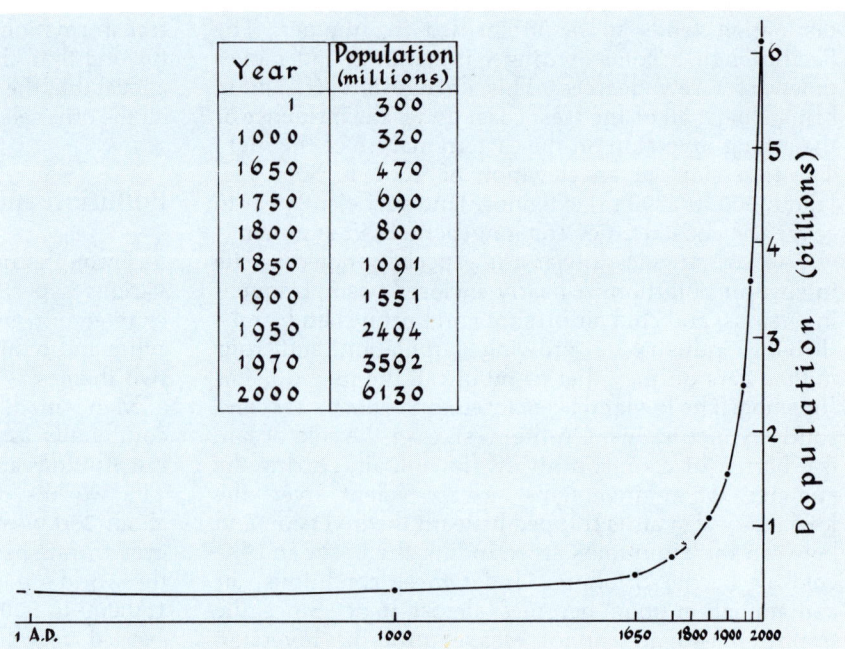

Fig. 23-1. Growth of world population during the Christian Era.

Year	Population (millions)
1	300
1000	320
1650	470
1750	690
1800	800
1850	1091
1900	1551
1950	2494
1970	3592
2000	6130

Environmental Quality and Physical Geography 361

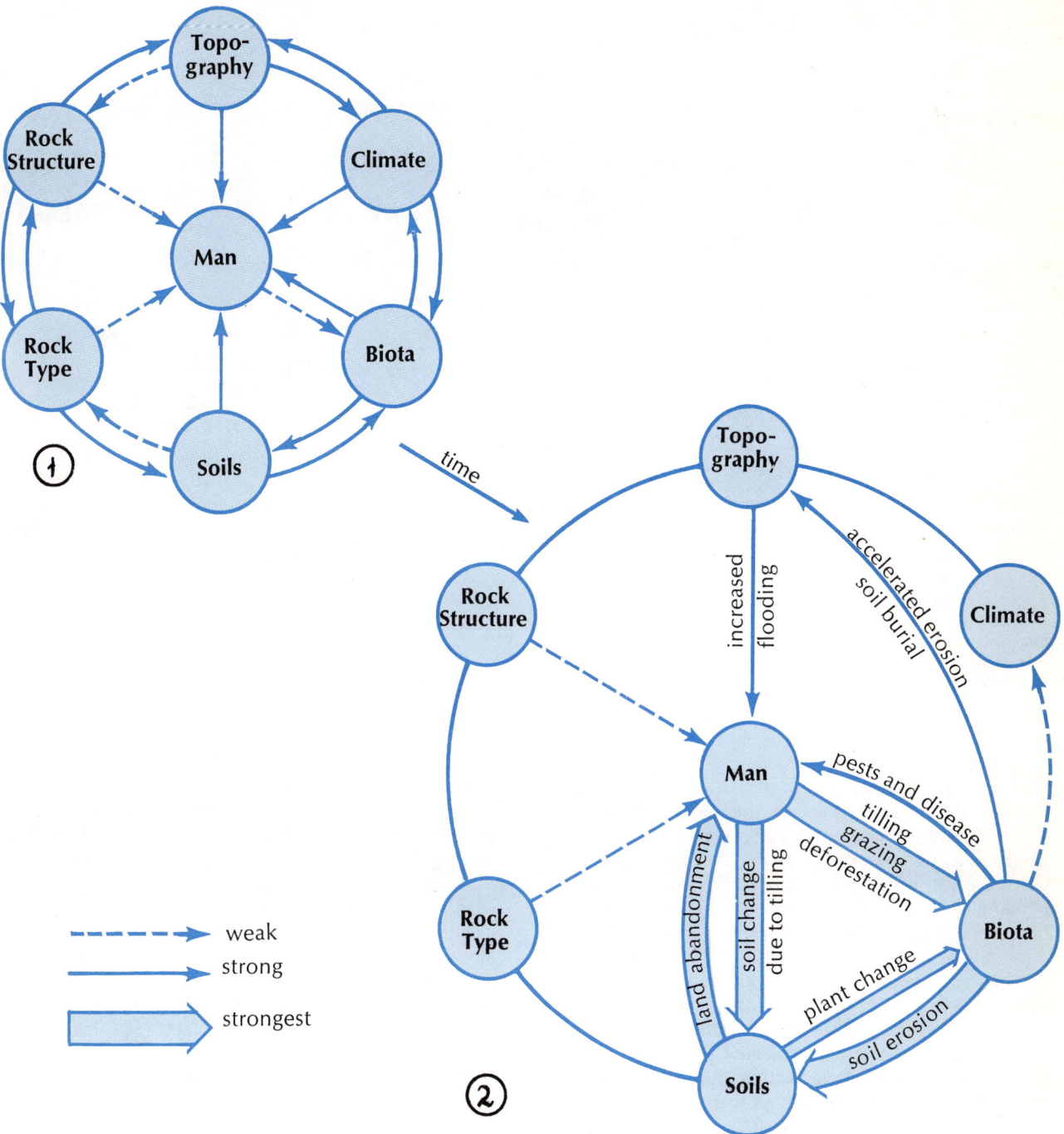

Fig. 23-2. Man and the landscape system a million years ago and at the beginning of the Christian Era. The influence of man upon the system was almost nil a million years ago. By the Christian Era, however, population and technology had reached levels sufficient to cause serious modifications of the landscape in some places. Tilling, grazing and deforestation caused serious soil erosion. The erosion was accentuated by changes in the texture and structure of the soils caused by tilling. Soil erosion resulted in land abandonment and deterioration of the biota. The impoverishment of the plant cover favored the growth of weeds, insect pests, and perhaps the spread of plant and animal disease. It also accelerated erosion and flooding.

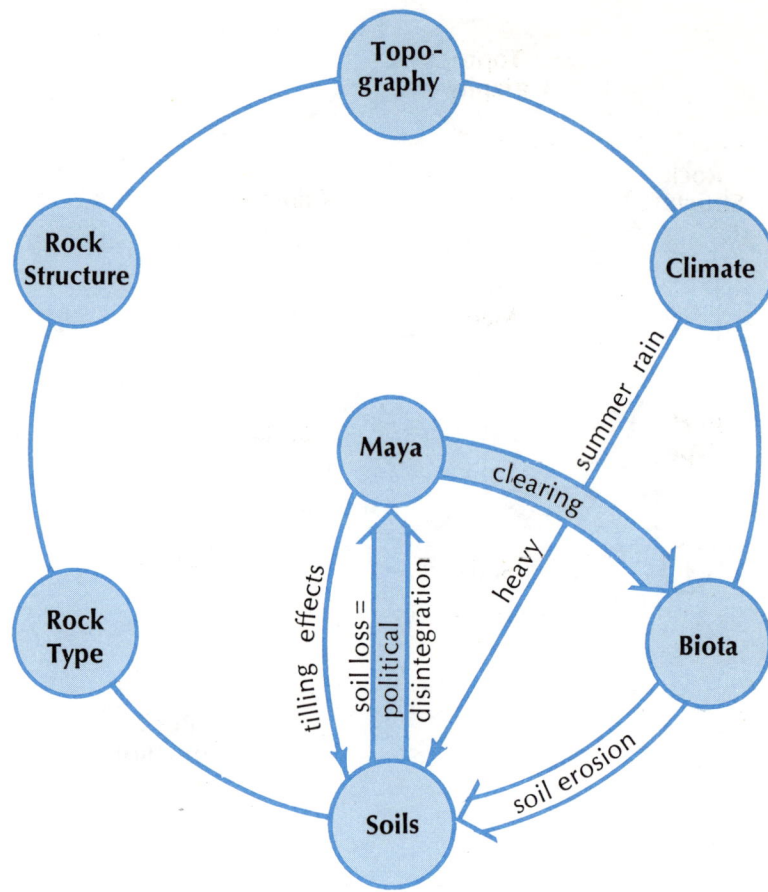

Fig. 23-3. The Maya and their environment. Extensive clearing in a region of heavy summer rains caused widespread soil erosion that resulted in scattered population and political disintegration.

the area came from partly water-filled limestone sinks called *cenotes* which became the focus of settlement and intensive agriculture. Here especially, extensive destruction of vegetation caused severe soil erosion. Once erosion became widespread, the Maya had to disperse even farther to seek areas of better soil. Population declined, and the great centers had to be abandoned. When the Spaniards arrived in the 1520's, most of the cities were in ruins, and the empire had disintegrated into 16 or so small states. Thus there is good reason to believe that a major factor prompting the move from their original homeland and the subsequent breakdown of the Maya Nation in Yucatan, was the erosive destruction of soil as a result of intensive cultivation (Fig. 23-3).

The story of the Maya is a small-scale example of what has happened all over the world. Like the Maya, people everywhere have been seeking new land when the old declined in quality. And like the Maya, who were trapped between ruined land behind them and the sea in front of them, mankind may eventually find itself trapped between a ruined planet and outer space.

What is modern man's place in the environmental system? Let us emphasize once again that *concentration* of population is more conducive to pollution than a mere general increase in population, although it is obvious that the two are related. Similarly, *changing emphasis* within the industrial complex results in greater emission of noxious wastes than does growth alone. For example, during the last thirty years the increasing industrial and domestic use of electricity has increased thermal and general pollution at a much greater rate than would have been expected from the rate of industrial growth alone.

The diversity of human activity brings about so many changes in the environment that the total impact of these activities and the multiple feedbacks between man and the elements of the environment are very difficult to describe in a simple way. Each activity and its effect on the surface system are most easily discussed individually. Since all possible activities everywhere on the surface of the earth cannot be investigated, we will concentrate on a few of the more obvious ones, such as agriculture, industry and urban growth, and we will limit ourselves entirely to the United States.

Environmental Quality and Physical Geography   363

Agriculture and the Environment

Agriculture in the United States is generally characterized by a high degree of mechanization (indeed, some critics claim that more fuel calories are devoted to raising the crops than the crops themselves produce), emphasis on single crops, large holdings, clean-field cultivation, heavy applications of fertilizers and extensive use of herbicides and insecticides. These practices combine to produce significant changes in the relations between the factors of the surface system (Fig. 23-4).

The extensive removal of vegetation leads to widespread erosion of soil and the loading of streams by sediment, both of which feed back to the biotic factor, modifying or eliminating some aquatic and terrestrial life (Fig. 23-4). At the same time the removal and change of the vegetation cover modify the micro-

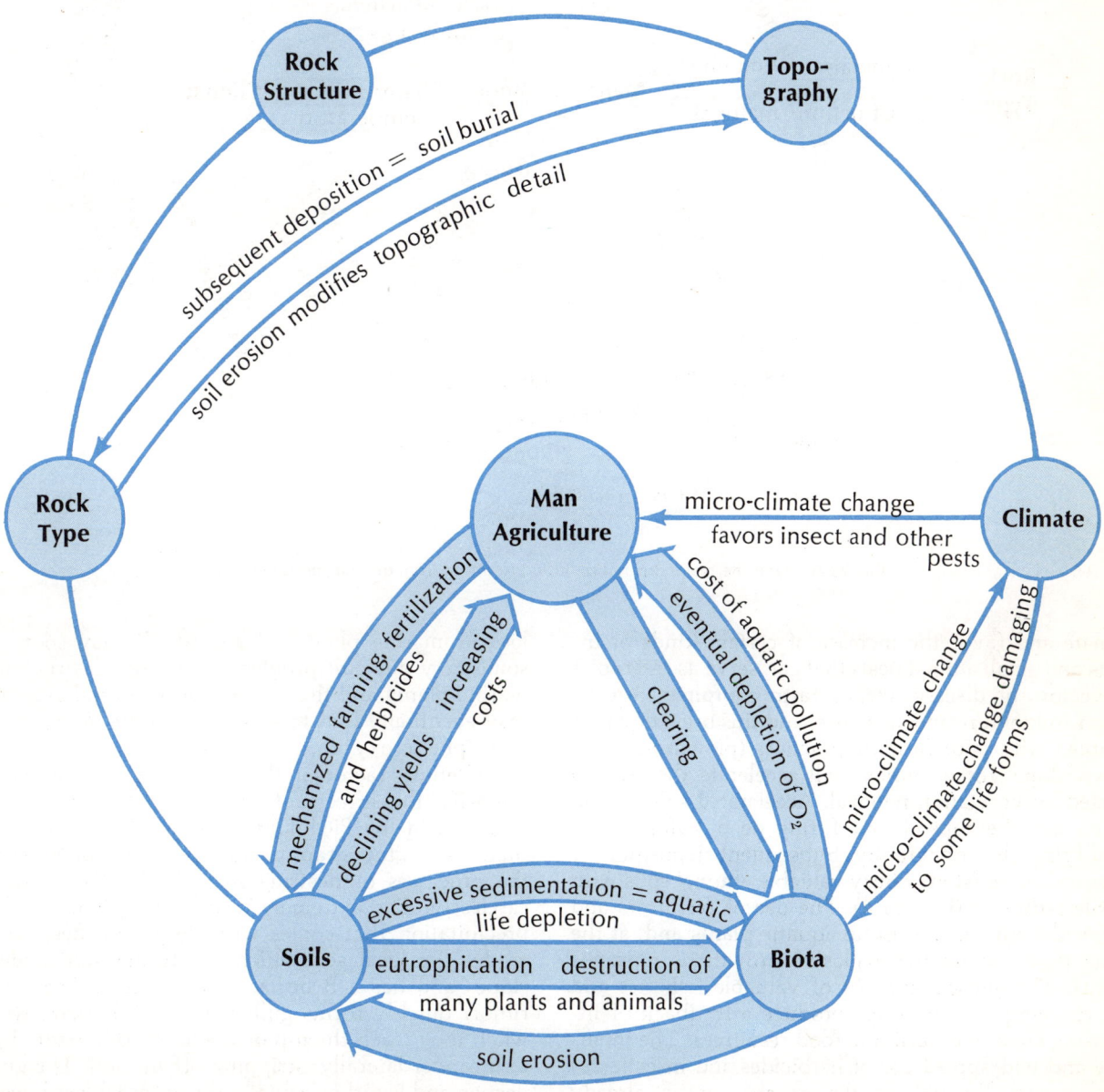

Fig. 23-4. Western agriculture and its effects on the landscape system.

364 Chapter 23

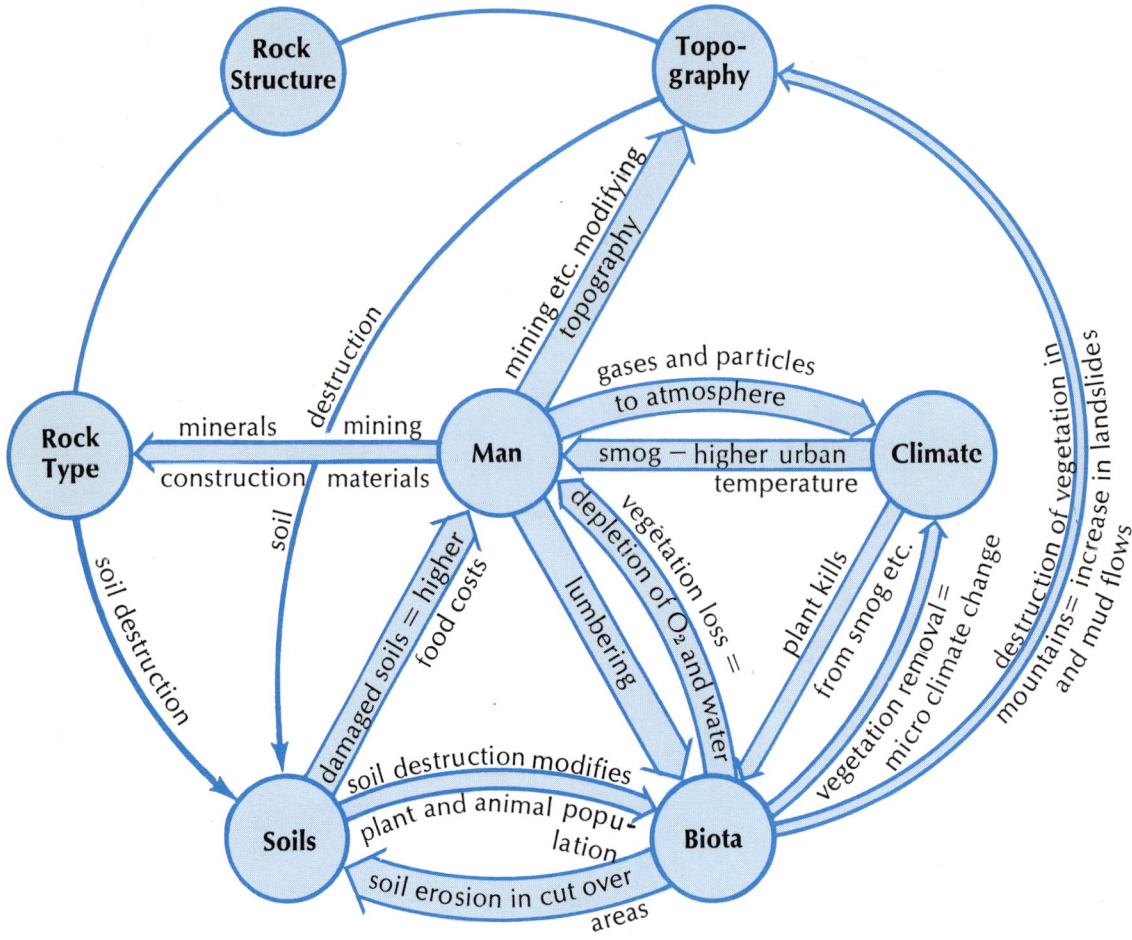

Fig. 23-5. Non-thermal industrial pollution resulting from mining and lumbering.

climate and favor the increase of certain kinds of insects and small animal pests that can serve as reservoirs or vectors of disease. Heavy farm equipment breaks down soil texture and structure, and this breakdown, if coupled with poor farming practices (plowing up and down slope, for example), will accelerate the erosion started by vegetation removal. Accelerated soil erosion strips off the topsoil and forms deep gullies, thus modifying the topography. Subsequent deposition of removed material may bury valuable alluvial soils with sterile sediment (Fig. 23-4). The use of fertilizers and herbicides kills many useful aquatic plants and, at the same time, favors the explosive growth of nuisance plants. The subsequent loss of valuable fisheries and the clogging of waterways produce a feedback of increasing costs and declining food resources. The intensive and widespread use of herbicides and insecticides may eventually pollute the oceans, as is already happening along the coast of southern California. The loss of marine plant and animal life may soon pose some very difficult problems, since the destruction of marine plants and the removal of terrestrial vegetation may eventually lead to a decrease in the production of atmospheric oxygen.

Sometimes factors in the natural system may serve to intensify the effects of man upon the system. For example, in the Piedmont of the southeastern United States, soils, topography, and climate combine to speed the processes of destruction when land is cultivated carelessly. The extremely heavy late summer and fall precipitation that comes with the hurricanes, rapidly erodes the local soil which has formed on moderate slopes and has a B horizon rich in clay. The intense runoff quickly forms gullies in the A horizon which, when they reach the top of the tough B horizon, begin to meander laterally, stripping off the soil. The gullies deepen and lengthen rapidly, the water table is lowered, and the potential for plant growth is greatly diminished.

Environmental Quality and Physical Geography 365

Land abandonment in the Piedmont began as far back as the beginning of the 19th century. And so the process of destruction continues, with damage spreading to all parts of the system, in many areas of the world, signalling an eventual catastrophic impact on man.

The Impact of Lumbering and Mining

Many industrial activities also contribute to declining soil resources and contaminate air and water with toxic substances. Figure 23-5 shows some of the interactions causing adverse changes in soils and vegetation. The diagram concentrates on some of the effects of mining and lumbering on the earth's surface system. To demonstrate the full gamut of change brought about by industrial activities alone would require a separate volume.

Mining removes certain types of rock, and accompanying spoil heaps of waste material modify the local topography. Strip and open-pit mining destroy many thousands of acres of soil. In addition to these direct effects additional land may be adversely affected by sediments, acids and other toxic materials derived from the spoil by sheet wash and spreading by streams. Aquatic life in the streams may be killed as well. Many streams, once known to be well populated with a variety of fish, are today virtually without a single species. Once mining is completed the pits eventually fill with water. Often they simply become a health menace as mosquito breeding areas, but, if carefully managed, they may become a local recreation resource.

Lumbering also often adds to the process of land degradation. Lumber is frequently clear-cut, meaning, all the trees in the area are removed. In addition, dry slash from the cutting operation may ignite and burn what little remains of the vegetation. The total removal of the plant cover not only increases runoff substantially, causing large-scale erosion and flooding, but also may bring about land slides and mud flows. The soil removed upstream may clog the lower reaches of the water courses. For example, the rivers in the northern coastal ranges of California used to have important salmon runs. Today, as a consequence of lumbering and the resulting heavy sediment load in the streams, few salmon return to spawn.

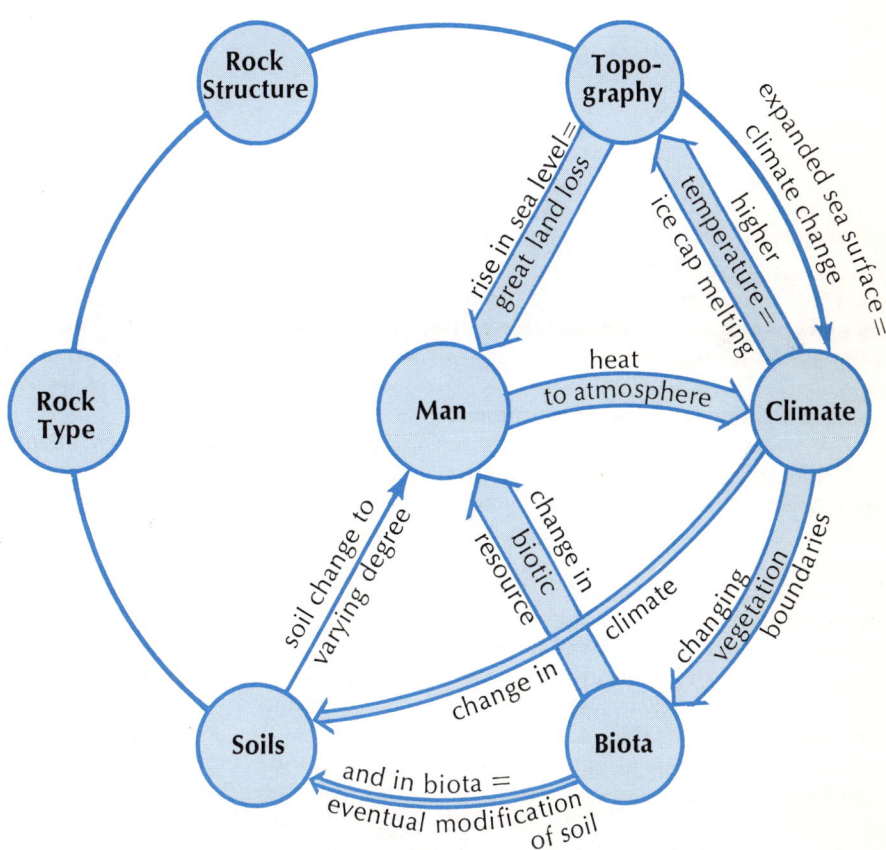

Fig. 23-6. Urbanization and the natural landscape.

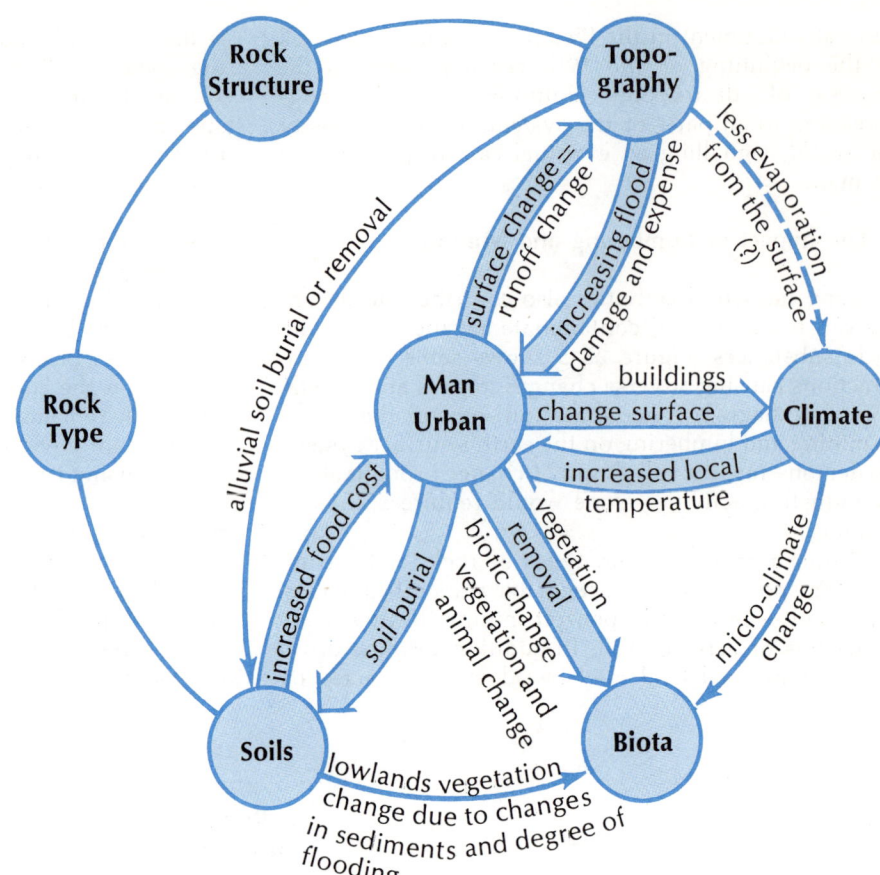

Fig. 23-7. Potential effects of thermal pollution of the atmosphere.

Soil destruction due to mining and lumbering modifies the local plant and animal life in such a way that the return of vegetation to heal the scars on the landscape is retarded, and the period of destructive erosion is thus prolonged. Soil damage also reduces available crop land, and inevitably causes an increase in the cost of living.

Urbanization and the Environment

Urban growth is related to industrial expansion, and its impact on the environment could well have been lumped with industrial pollution. However the threads of interaction become so complexly interwoven that Fig. 23-5 would have become impossible to read. Even so, simple expansion along with the accompanying growth of road and highway networks destroy vegetation and bury soil, thus greatly modifying the capacity of the earth's surface system to handle precipitation and solar radiation (Fig. 23-6).

Water-tight layers of pavement or roofing profoundly change the nature of topographic surfaces. Rainfall runoff is greatly increased because little water is lost by infiltration or transpiration. Runoff is more immediate and direct since the new surface presents far less friction than the natural surface. This increases the danger of flooding. The prompt removal of water from the surface also reduces the amount of water vapor returned to the atmosphere by evaporation and transpiration. Whether this will eventually exert some influence on the local climate is not clear; perhaps it will always be obscured by other effects of cities upon climate.

The low specific heat of material in buildings and pavements causes their surfaces in summer to become much hotter compared to the natural surface of the landscape. Multistory buildings also substantially increase the surface available for heating and form windbreaks, retarding the horizontal removal of heated air. Because of these effects the urban environment is substantially warmer in summer than the adjacent rural regions (see p. 169). This temperature contrast prevails, to a lesser degree, in winter as well. The increased input of heat from urban areas certainly brings about modifications of the local climate, but the nature and degree of these changes are warmly debated subjects.

The effect of thermal pollution upon local and world climate is as yet insignificant. As suggested earlier, however, the continued expansion of energy conversion at the present rate will soon lead to serious consequences. Figure 23-7 shows some of the potential effects of increasing release of heat to the atmosphere. Some authorities believe that only a small rise in temperature will be required to melt the polar ice. The consequent rise in sea level of several hundred feet would inundate great areas of productive soil and require the relocation of most of the world's cities. The expanded sea surface would bring about additional climate changes which would eventually alter present patterns of vegetation, including cropland. Changes in climate along with the change in vegetation would set in motion changes in the soil-forming processes, the nature and degree of which are difficult to predict.

Vegetation destruction, soil erosion, water and air pollution, and potential oxygen depletion are recurrent themes in the story of humanity's relationship to its environment. Thoughtless exploitation of the elements of the earth's surface brings about unwanted changes that are difficult and expensive to arrest. Remedial action is costly but there are no alternatives — no new lands to exploit, no unsullied waters to drink, no new atmospheres to breathe. We must do with what we have indefinitely.

Summary

The aim of this chapter has been to examine the processes that pollute the air, water and soil and to place them in the context of the interactions of the earth-atmosphere system. Most pollution comes from ignorance of the physical environment, and the first step in establishing an enduring equilibrium between people and their environment is to understand the functioning of that environment. It is only after that step has been taken that one can determine what actions are best suited to making or keeping the physical environment a fit place to live in.

Appendix: Location and Maps

Introduction

Geographers are concerned primarily with explaining the areal variations of phenomena on the surface of the earth. This concern implies an ability to locate phenomena and to map their distribution. The descriptive location of a place requires the existence of a known point of reference, a known system of directions, and a set of standard units for measuring lengths and angles. Even the simplest descriptions of location almost always involve a distance, a direction, and a point of reference: "It's 3 miles northeast of the university." In addition, it helps to know the size and shape of the surface on which the locations are to be found. This knowledge is particularly important in mapping, but it is also relevant to any standardized system of location.

The earth is a sphere, slightly flattened at the poles and bulging a bit at the equator (Fig. 1). Its diameter from a point on the equator to the **antipode** (Greek *anti* against + *pous*, *podos* foot) of that point on the opposite side of the earth is about 7,927 miles (12,756 km). The diameter from pole to pole is only about 27 miles (43 km) less, or about 1/295 smaller. Most commercial globes cannot possibly reproduce this flattening. Even on a globe with a 2-foot diameter the difference between polar and equatorial diameters is only a little more than 1/16 inch (about 2 mm on a globe with a radius of 60 cm); on Fig. 1, which is drawn true to scale, the flattening amounts to much less than 1/100 inch (less than a quarter of a millimeter) and is, of course, invisible. Technically, any north-south cross section of the earth is very nearly an ellipse, and the earth-globe is therefore called an *ellipsoid of rotation*. While satellite observations have uncovered irregularities in the shape of the earth, they are even slighter than the polar flattening. The measurements in Table 1, which summarize the size and shape of our planet, emphasize its slight departure from true sphericity.

Table 1. The Size of the Earth

Equatorial diameter	7,926.5 miles (12,756 km)
Polar diameter	7,899.6 miles (12,713 km)
Average radius*	3,958.7 miles (6,371 km)
Equatorial circumference	24,901.5 miles (40,075 km)
Area	about 200,000,000 square miles (510,000,000 km^2)

*This average is the radius of a true sphere of the same area as the earth.

Treating the earth as a sphere simplifies terminology. While angles and distances and the lines that mark them off can be described on an ellipsoid of rotation, the wording is cumbersome, unfamiliar, and complex. In contrast the corresponding properties of a sphere are much more familiar. One or two definitions will suit our purposes. A *great circle* is the circle produced by slicing a sphere into two equal halves. Any other slice will produce a smaller circle, appropriately called a

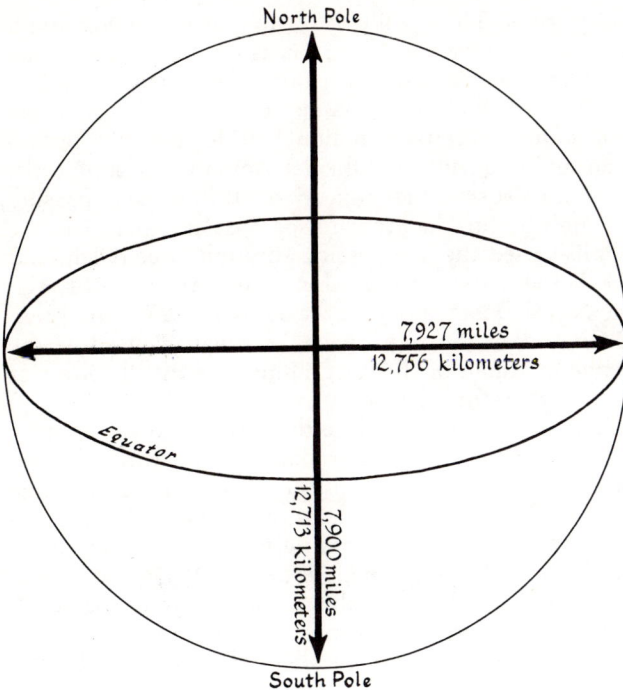

Fig. 1. The shape of the earth.

small circle. Through any two points on a sphere, you can draw as many small circles as you please, but only one great circle.[1] The arc of the great circle passing through any two points describes the shortest distance between them on the sphere. It is, therefore, the arc produced by stretching a string across the sphere between the two points. Angles on the sphere are measured between arcs of great circles or, more accurately, between the tangents to those great circles. The arcs of great circles are sometimes referred to as **orthodromes** (Greek *orthos* straight + *dromos* a course). In short, an arc of a great circle on the surface of a sphere is analogous to a straight line on a plane.

These arcs differ from straight lines in more than the obvious way, however. Since the shortest distance between two points on a sphere is determined by the arc of a great circle, and since we can indicate distance along the circumference of a circle in either linear units (for example, the circumference of a certain circle is 2.3 inches) or angular units[2] (a circle consists of 360°; a quarter circle consists of 90°, and so on), it follows that distances on a sphere can be indicated by either linear or angular measures (Fig. 2). We can say that two places are 69 miles apart or that they are 1° apart. (1° is equal to about 69 miles or 111 km on a sphere the size of the earth. The circumference of the earth, which is 24,901 miles or 40,075 km, represents 360°; 24,901 miles divided by 360 is about 69 miles.) This angular method of indicating distances is very useful in both location and mapping, as the angular distance between two points does not change when scales are changed.

Location

The Geographic Grid

The North and the South Poles are the only two points on the surface of the earth whose locations can be given without reference to any other places. Their location represents the intersection of the earth's axis of rotation with the surface. Since the Poles can also be used to determine directions, they form the basis of the simplest scheme of location, the *geographic grid*.

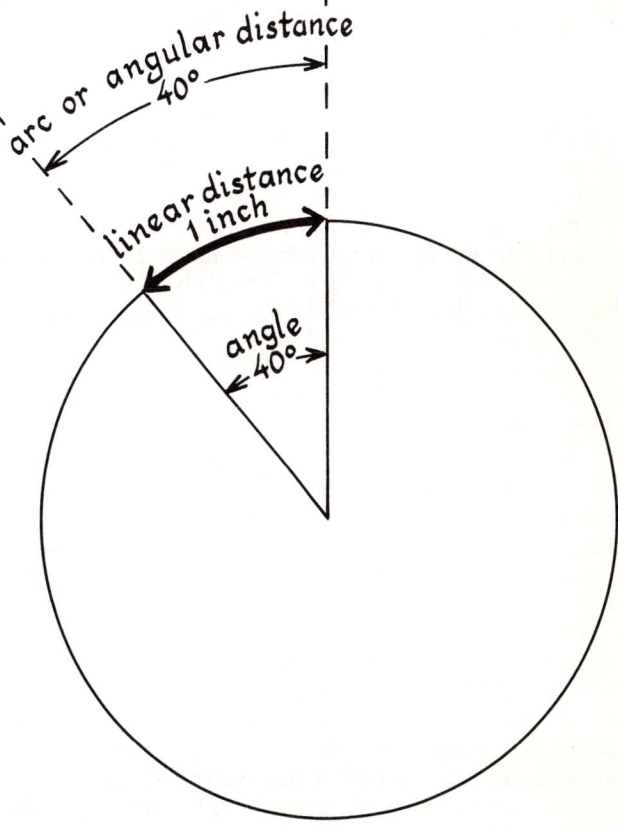

Fig. 2. Angular versus linear measure.

---

[1] Except that two antipodal points have an infinite number of great circles passing through them.

[2] Angular units are expressed in multiples of 60, an inheritance from the Babylonian numbering system. One degree of arc is subdivided into 60 minutes, and 1 minute consists of 60 seconds. The abbreviations for these units are ° for degrees, ' for minutes and " for seconds.

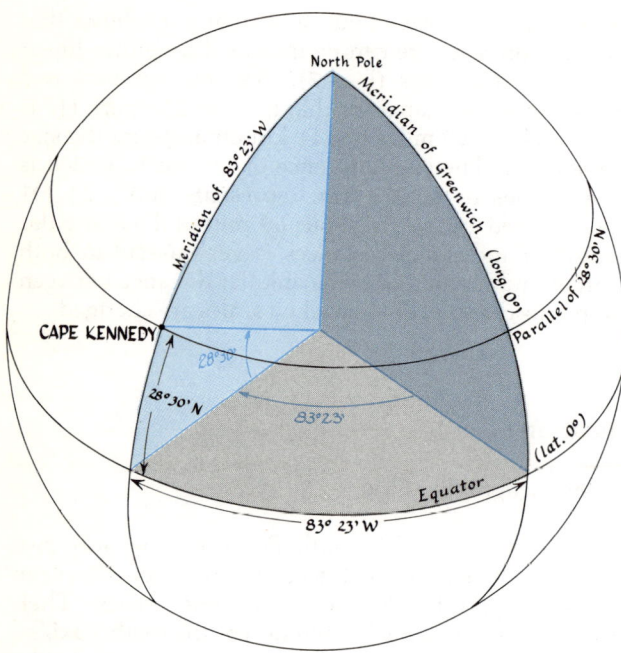

Fig. 3. The geographic grid of parallels and meridians.

The grid itself (Fig. 3) includes the great circle that is everywhere equidistant from the poles — the equator — and a series of small circles parallel to the equator. These are called *parallels of latitude*. On a globe or a map, they are shown only at some fixed interval of angular distance — say every 5°, or maybe every 10°. It is important to remember, however, that although only a few parallels are shown, an infinite number exist. Each parallel is named after its angular distance north or south of the equator. The most northerly parallel is the North Pole, which is a small circle by courtesy only, but is nevertheless referred to as the parallel of 90° N (90 degrees north) latitude. The South Pole has a similar role in its hemisphere. The latitude of the equator is, by definition, 0°. Every parallel thus has a designation of latitude ranging from 0° to 90°.

The position of a particular place is partly determined by the parallel of latitude on which it is located. When we say that the latitude of Cape Kennedy is 28° 30′ N, we simply mean that Cape Kennedy lies on the parallel of latitude whose angular distance north of the equator is 28 degrees, 30 minutes.

The other half of the geographic grid consists of an infinite series of great circles passing through the North and South Poles. These great circles are collectively called *meridians of longitude*, and they cross the parallels of latitude at right angles. Among the parallels of latitude, the equator stands out by its greater circumference and its position halfway between the poles. No meridian of longitude has any such claims to uniqueness. Thus, while it is obvious by analogy with latitude that we want to measure longitude as an angular distance east or west of a designated meridian, there is no logical candidate for the role of *prime meridian*. In the past, national pride caused a certain amount of confusion in the determination of longitude, since the Prussians reckoned from the meridian passing through Berlin, the French used the Paris meridian, the English used the one passing through Greenwich, and the United States used the meridian passing through the Naval Observatory on the outskirts of Washington, D.C. Today, the Greenwich Meridian is used as the prime meridian of the geographic grid by all countries and is, therefore, labeled 0°.

The longitude of a place then, refers to the angular distance between the meridian passing through that place and the Greenwich Meridian, and may range from 0° to 180° E in one direction and from 0° to 180° W in the other. Accordingly, when we say that the longitude of Cape Kennedy is 83° 23′ W, we mean that the meridian through the Cape is 83 degrees, 23 minutes west of the Greenwich Meridian.

Specifying both the latitude and the longitude of a place, as we have done for Cape Kennedy, locates that place on the surface of the earth. The accuracy of the location depends simply on the preciseness of the latitude and longitude we ascribe to the place.

Determination of Latitude and Longitude

There are a number of ways of determining latitude and longitude for a particular place. One method is described here because it emphasizes the relationship between location and other physical characteristics of places. As we have seen in Chapter 2, the angular distance between the noon sun and the horizon determines, in part, how much solar energy reaches the ground. This angle, called the *altitude of the noon sun*, can also be used to determine latitude.

At any given time of the year, the noon sun is directly overhead in one particular latitude. That latitude is called the *declination of the sun*. Its value for each day of the year is given in most nautical tables; it is also shown by a *nomograph*[3] called an *analemma* (Greek *analemma* sundial) (see Fig. 2-2). Therefore, if we know the day of the year, we can then find the latitude at which the noon sun is directly overhead. We can also measure the altitude of the noon sun by using a sextant or any other instrument that gives us the angular distance between the horizon and an object in the sky. Now if we know the noon sun's altitude at some place whose latitude is unknown, we can compute the

---

[3] A nomograph (Greek *nomos* law + *graphein* to draw) is a graphic representation of numerical relations.

latitude in a very simple way. Let us call the unknown place B and the place where the sun is directly overhead A. The latitude of B will differ from that of A by the same amount as the altitude of the noon sun at B differs from 90°. One other thing must be considered. The noon sun at B may be either in the north or in the south. If the sun is north of the zenith (the point directly overhead), B is south of A; if the sun is south of the zenith, B is north of A.

Two examples should make this procedure clear. Let us suppose that you measure the altitude of the noon sun at some place B, on March 21, and you find it to be 44° S. On that date the declination is 0°; the measured altitude differs from 90° by 46°; B is, therefore, 46° away from a place whose latitude is 0°; since the sun is to the south, B must be 46° north of 0°, or in latitude 46° N.

At some other place C, on June 22, the altitude of the noon sun is 88° N. On that date the declination is 23½° N; the measured altitude differs from 90° by 2°. C is therefore 2° away from 23½° N; since the sun is to the north, C must be 2° south of 23½°, or in latitude 21½° N.

This method depends, of course, on being able to see the sun at noon, but it has allowed very accurate measurements of latitude to be made for at least 500 years. The determination of longitude, on the other hand, while equally simple, could not be employed until the invention, late in the 18th century, of accurate, portable chronometers. The method is based on the fact that the earth rotates from west to east about its axis once every 24 hours. It follows that if it is noon in some particular longitude, it must be 6:00 A.M. 90° to the west, 6:00 P.M. 90° to the east, and midnight on the other side of the globe, which is both 180° to the east *and* 180° to the west.

Two conclusions can be drawn from these facts. First, a difference in longitude of 180° represents a time difference of 12 hours; therefore, a time difference of 1 hour corresponds to a difference in longitude of 180/12, or 15°; a time difference of 4 minutes represents a difference in longitude of 1°, and so on. Second, since time gets later as one goes eastward, it follows that if it is noon on a particular day, let us say Friday, it should be midnight between Friday and Saturday in a longitude 180° to the east. By the same token, it should be midnight between Thursday and Friday in a longitude 180° to the west. But a longitude 180° to the east is exactly the same as a longitude 180° to the west! Therefore, a place on that meridian would have two different times, 24 hours apart. To avoid this situation, we need to pick some meridian of longitude to separate one day of the week from the next, so that, for instance, if it is Tuesday on the western side of the meridian, it will be Wednesday on the eastern side. The line chosen to mark the beginning and ending of the day is called the *International Date Line*; it is approximately the meridian of 180° in the Pacific Ocean,

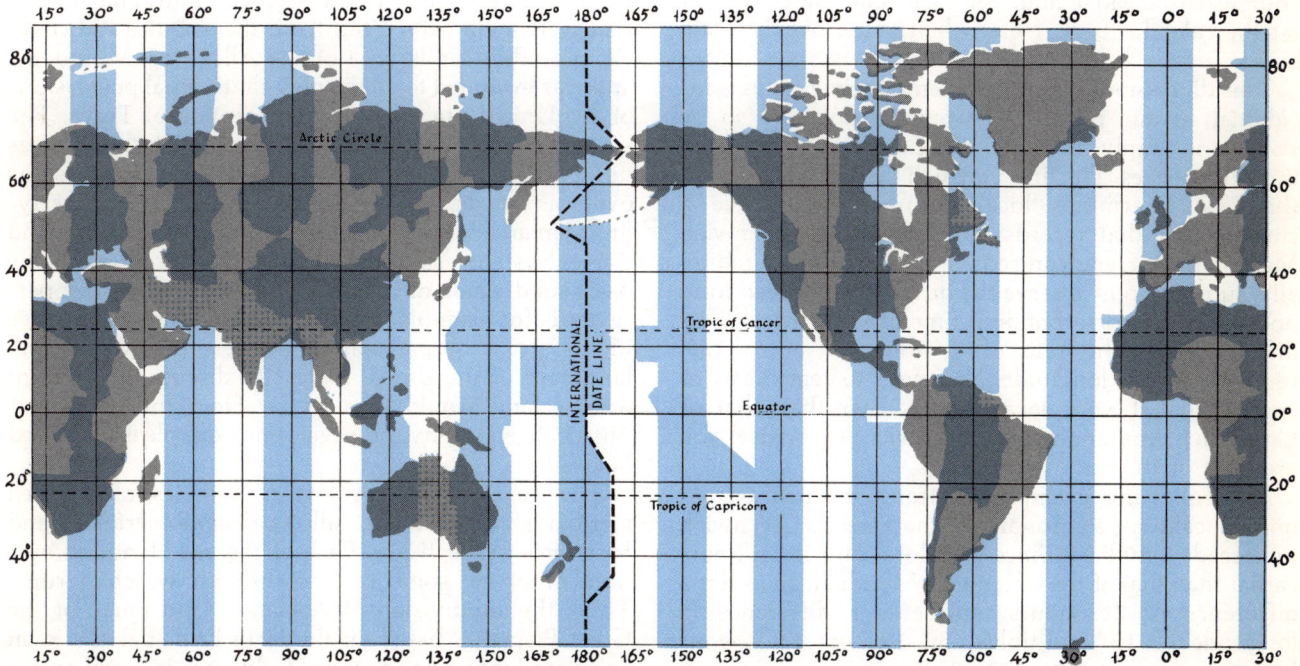

Fig. 4. World time zones.

except that certain departures from the meridian have been made so as not to bisect political units such as the Aleutian Island chain of Alaska or the island possessions of New Zealand (Fig. 4).

Another way of looking at the International Date Line is to consider the problem of standard time zones (see Fig. 4). Confusion would result from a situation in which every place measured time on the basis of the unique meridian of longitude on which it was located; one would have to change his watch with every step he took in an east or west direction. It is obviously much more convenient to advance or retard one's watch at 1-hour intervals only. Therefore, at the turn of the century, standardized time zones were established. Each zone is approximately 15° in longitudinal extent and straddles a central meridian from which the time of the whole zone is reckoned. A traveler moving from east to west has to set his watch back 1 hour each time he moves into a more westerly time zone. If this traveler were to return to his starting point after circumnavigating the earth, he would have crossed 24 time zones and would have set his watch back 24 hours. In order to compensate for this illusory gain of 24 hours, he has to set his calendar ahead by a day, somewhere along his route. The International Date Line is the place where these time discrepancies are straightened out. The practice is simple: When moving from east to west, you set your watch back 1 hour for each time zone, but you advance the calendar 1 day when you cross the Date Line; when moving to the east, you set your watch ahead 1 hour for each time zone, but you set the calendar back 1 day when you cross the Date Line.

This digression into the problems of time zones is incidental to our purpose, however. The point to be stressed is simply that longitude can be reckoned from time differences. The time at which the sun is highest in the sky determines local noon. If you have a chronometer that records the time at the Greenwich Meridian (or if you can receive time signals by radio), all you need do is observe the difference between local noon and the corresponding time at Greenwich. This difference, at the rate of 1 hour for every 15°, represents your longitude. You need to remember, of course, that if your local time is later than that of Greenwich, you are east of Greenwich; otherwise, you are west of it.

The following example illustrates the method. At noon, local time, a chronometer that records Greenwich time reads 2:30 P.M. Since your local time is 2½ hours earlier than that of the Greenwich Meridian, and since a difference of 2½ hours represents a difference in longitude of 15° multiplied by 2½, or 37½°, your longitude is, therefore, 37° 30′ W.

So much for the determination of latitude and longitude. Practical problems abound in this determination, but the method is theoretically very simple. Before leaving this subject, it is of some interest to note that these highly theoretical and completely imaginary meridians and parallels are represented on the surface of the earth by concrete, visible differences in the landscape. Many political boundaries are drawn along meridians or parallels. The border between Utah and Nevada, for example, is a meridian of longitude, specifically the 37th meridian of longitude west from Washington. The Washington Meridian, which passes through the old Naval Observatory, is itself 77° 03′ 02.3″ west of Greenwich. Hence, the boundary between Utah and Nevada lies just to the west of the meridian of 114° W longitude instead of being exactly on it. The town of Wendover lies on either side of this meridional boundary. The imprint of its presence is quite obvious as most of the stores are west of 114° 03′ 02.3″ W, since the absence of a sales tax and the presence of slot machines and relaxed liquor laws are all localized by the political boundary between the two states.

The U.S. Township and Range System

One other system of location is of great importance and interest, particularly to American geographers. This is the system of grid lines established by the Congress under the Articles of Confederation in 1785 for the purpose of subdividing the newly acquired public domain. It did not cover the original 13 states or the four (Kentucky, Tennessee, Vermont, and Maine) that were soon carved out from them, but it did cover all the public lands west of the Allegheny Mountains and was extended to all the new continental possessions of the United States, with the exception of Texas. The system was imposed on a small part of Alaska but was not carried over to Hawaii or to U.S. possessions overseas because of pre-existing systems of property-line demarcation. The interest of the Township and Range System, aside from the fact that it represents the basic subdivision of the land wherever it exists (property deeds for example, are identified by this system), lies in its great impact on the present-day cultural landscape of the United States. To discuss this imprint on the land would take us far from the themes of physical geography; however, one example is discussed below.

The basis of the system is a series of selected meridians and parallels, called *principal meridians* and *base lines*, respectively. Each of these is identified by a number or by some feature through which it runs. Thus, the subdivision of Arkansas (and much of the Great Plains) is based on the Fifth Principal Meridian and Base Line; that of central California is based on the Mt. Diablo Principal Meridian and Base Line, which intersect at the peak of that name. Figure 5 shows the

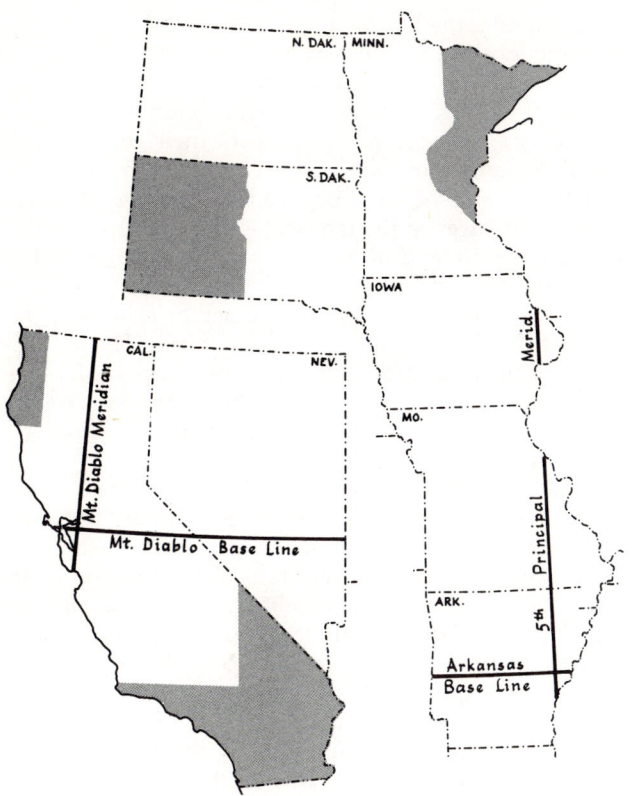

Fig. 5. Two examples of principal meridians and base lines of the U.S. Township and Range System.

position and areas governed by these two principal meridians.

The rest of the system is built up on a series of north-south and east-west lines at 6-mile intervals. Since north-south lines are, by definition, meridians and therefore converge toward the poles, they cannot in fact be drawn 6 miles apart through their length. In practice the spacing between lines is set at 6 miles along the base line; the north-south lines are allowed to converge for a certain distance to a *correction line* or *standard parallel* along which the 6-mile spacing is reestablished. In addition, other corrections such as those needed because of faulty surveying are also made along correction lines. Such correction lines are numerous on Fig. 6.

The system thus consists of north-south strips 6 miles wide, called *ranges*, which are divided by the east-west lines into blocks 6 miles on a side, called *townships*. The townships are numbered north or south of the base line, and the ranges, east or west of the principal meridian. This arrangement is illustrated in Fig. 6, which shows the location of Cape Kennedy with respect to the principal meridian and base line that govern the Township and Range System in the state of Florida. As you can see, the Cape is located in the 38th range east of the Tallahassee Meridian and in the 23rd township south of the Florida Base Line. This information is normally abbreviated to "T 23 S, R 38 E."

Townships have an area of approximately 36 square miles and are subdivided into 36 *sections*, each 1 mile

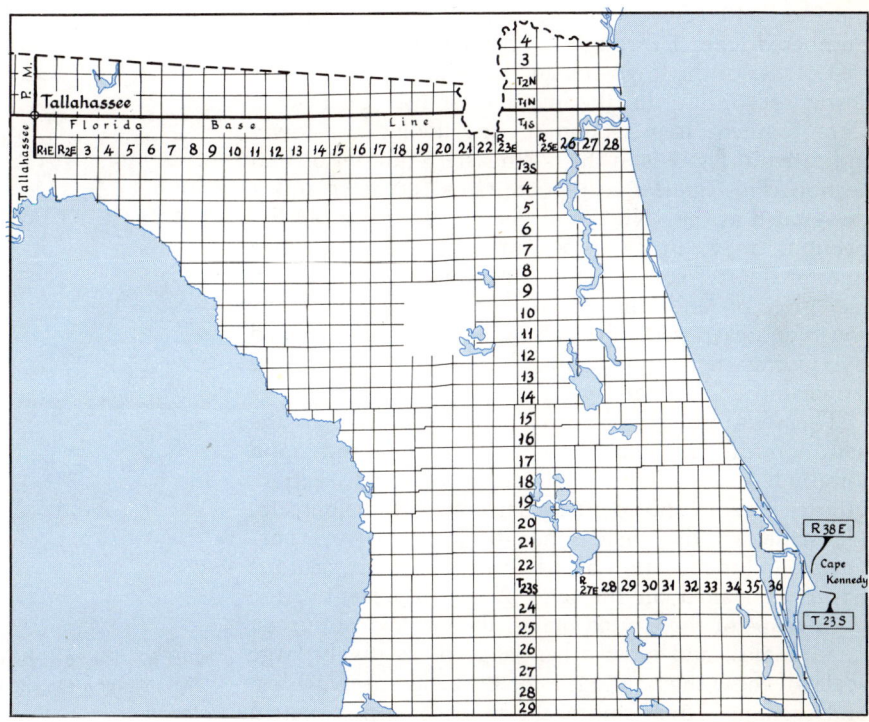

Fig. 6. The Township and Range System in Florida.

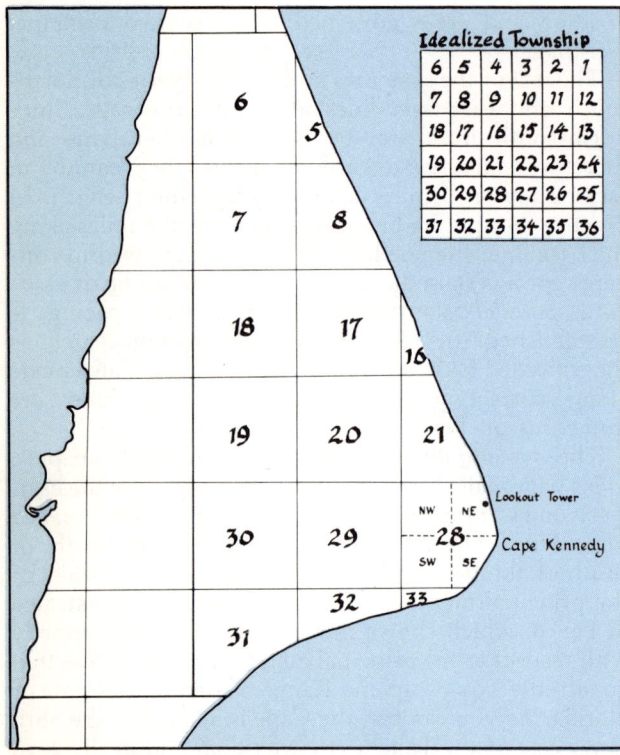

Fig. 7. Township 23 South (of the Florida Base Line), Range 38 East (of the Tallahassee Principal Meridian).

on a side. Each section thus has an area of 1 square mile, or 640 acres. The sections of a township are numbered from 1 through 36, beginning in the northeast corner of the township and ending in the southeast corner according to the method shown in Fig. 7. Cape Kennedy is in section 28, and the lookout tower indicated in Fig. 7 is in the northeastern quarter of that section. This quarter section contains 160 acres and is designated as the "NE 1/4 of Sec. 28, T 23 S, R 38 E." Sections may, in fact, be further subdivided into quarter quarter sections and even smaller units. Thus, any piece of land whose boundaries run along north-south and east-west lines can be identified and located by giving its appropriate designation under the Township and Range System.

There is no simple way corresponding to the astronomical methods of determining latitude and longitude to determine what range, township, section, quarter section, and so on a particular point belongs to. Nevertheless, the system has been extensively surveyed, and the location of the corners of many sections has been staked on the land. More importantly, the basic maps of the United States, the Geological Survey quadrangles, and most other maps of relatively large scale all show the boundaries of the ranges, townships, and sections. Furthermore, the imprint of this system

on the landscape has been so marked and so extensive that the location of the various lines of the system can easily be inferred, in many parts of the country, from property lines, particularly field boundaries; from boundaries of counties and of their minor civil divisions; and, most obviously, from the road network.

This close association between a location system and cultural features of the landscape is due to the fact that the Township and Range System was originated to assist the United States Land Office, under provisions of the various homestead acts, to sell or otherwise dispose of the public domain of the country. The location system thus preceded settlement, and the plans for new settlement very often placed property lines, political divisions, and a rectangular road network along the lines separating one section from another.

Figure 8 illustrates the results of the process in the state of Ohio. The southeastern part of the state was settled before the development of the Township and Range System, partly by veterans of the Revolutionary War. The road network exhibits a hexagonal pattern, which is a relatively efficient way of joining places more or less randomly distributed over the land but was not part of a grand design, consciously determined. On the other hand, the northwestern part of the state was settled after the Township and Range System was established. It is perfectly obvious that the rectangular

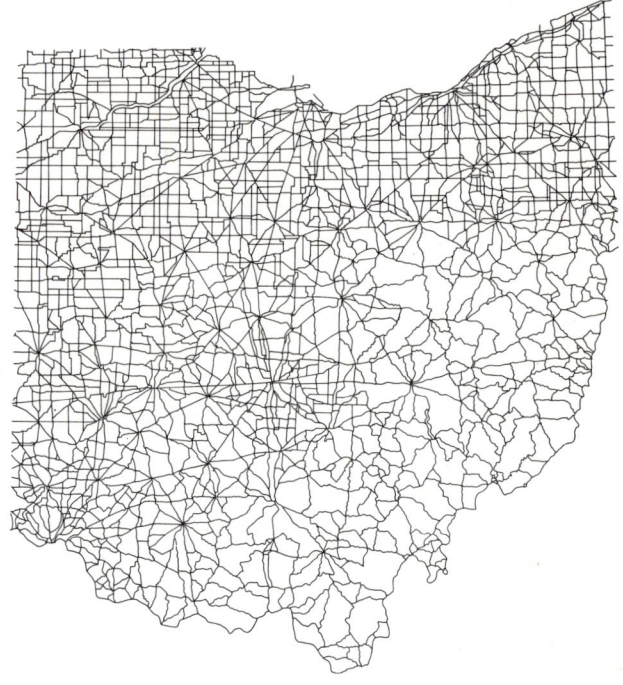

Fig. 8. The road pattern of Ohio. Note the contrast between the rectangular arrangement in the north and the spiderweb pattern in the south.

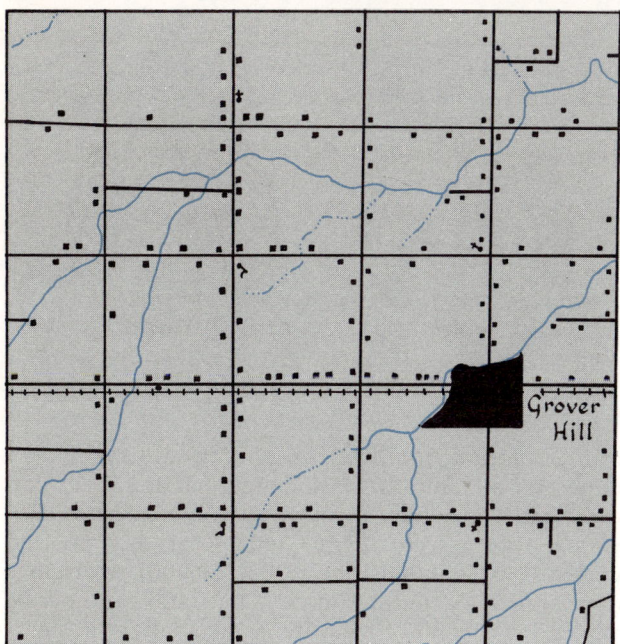

Fig. 9. The road pattern of T 1 N, R 3 E in northwestern Ohio. The road pattern in black faithfully reflects the section lines of the Township and Range System. The rivers, in blue, and the railroad and houses, in black, are added for reference.

network of roads owes its existence to the preestablished rectangular grid. The very close correspondence between roads and section lines is more clearly shown in Fig. 9 which is a larger-scale view of part of the state.

Maps

The Spherical Earth and Flat Maps

One of the first problems encountered in mapmaking is that of representing the spherical earth on a flat sheet of paper. As long as the area to be depicted is fairly small, the problem is largely academic. However, if one has to deal with an area as large as Wyoming, for example, there is no way to avoid the issue; and when one wants to show the entire globe on a flat map, he must resort to various expedient conventions, none of which is completely satisfactory. The conventions by which we transform a spherical surface into a flat one are called *map projections*; usually the transformation is described in terms of what it does to the parallels and meridians of the geographic grid. In general, therefore, we can define a map projection as any orderly, systematic and mathematically describable arrangement of the parallels and meridians on a flat surface.

Planes, cylinders, and cones are the only kinds of surfaces that are flat or can be made flat without distortion, and so all map projections can be classified into three corresponding categories: *planic*,[4] *cylindric*, and *conic*. Each of these surfaces can be made to coincide exactly with a sphere in a most limited way (Fig.10). A plane and the surface of a sphere coincide only at the point of tangency of the plane; a cylinder only along a single great circle; and a cone only along some small circle. These parts of the spherical surface that coincide exactly with the plane surface of the map are called the *elements* of the map projection.[5] In the planic, or azimuthal, projections, the element is a single point; the North or South Pole is very often chosen as the element, but any other point could be used as well. In the cylindric projections, the element is a great circle; most commonly the great circle chosen is the equator, but any other great circle (a meridian of longitude, for instance) could also serve as the element of the projection. In the conic projections, the element is a small circle; a parallel of latitude is normally picked as the element, although other small circles are sometimes employed.

Scale

Before proceeding further in the discussion of map projections, we need to investigate the concept of map scale. The scale of a globe is very easy to define if we consider the earth to be spherical. It is merely the ratio between the size of the globe and the size of the earth. To be more specific, the *linear scale* of the globe is the ratio between the length of any line on the globe and its corresponding length on the earth itself, and the *areal scale* refers to the ratio between areas. Normally the word "scale," unless otherwise qualified, is used to denote linear scales only, and we shall use it in that sense.

The scale of a globe, as just defined, is a ratio of two lengths. If we measure the equatorial circumference of a globe and find it to be 24.8 inches, how should we express the scale of that globe? We know that the corresponding circumference on the earth is about 24,800 miles. One way of indicating the ratio between the two

---

[4] Planic projections are more commonly called *azimuthal* projections for reasons discussed later in the chapter.

[5] Strictly speaking, a plane and a sphere can be made to coincide along a small or a great circle by passing the plane *through* the sphere. Similarly, secant cylinders or cones in cutting through a sphere coincide along two small circles. However, these loci of coincidence are very rarely used as elements of map projections. On the other hand, it is possible to derive a projection in which *two* small or great circles are *each* an element of the projection. The resultant map is somewhat different from, and more accurate than, the one obtained by using a secant cone or cylinder.

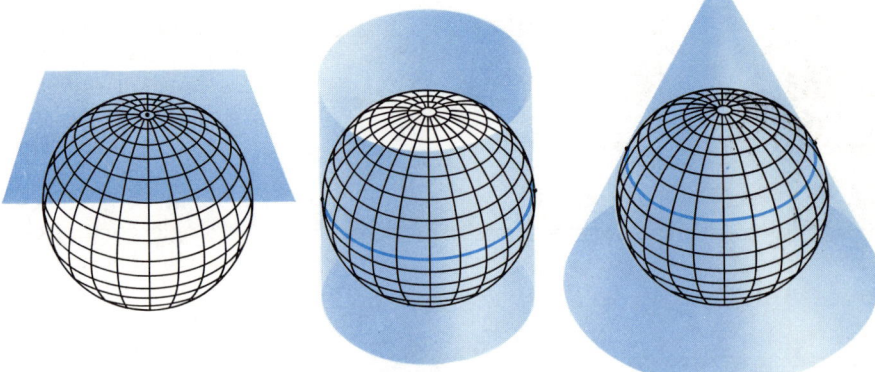

Fig. 10. Plane, cylinder, and cone tangent to a sphere. The elements shown in dark blue are respectively the North Pole, the equator, and a parallel of latitude.

lengths is simply to say that 24.8 inches on the globe represent 24,800 miles on the earth. It would be more elegant to say that one inch represents 1,000 miles. This kind of statement is called a *descriptive scale*.

A second and more useful way of representing scale is by a fraction. The equation

$$\text{Scale} = \frac{\text{globe distance}}{\text{earth distance}}$$

represents scale as a fraction and is also a convenient way of remembering the definition of scale as a ratio between distances. In our example the distances involved were 24.8 inches on the globe and 24,800 miles on the earth. We can therefore rewrite the equation as

$$\text{Scale} = \frac{24.8 \text{ inches}}{24,800 \text{ miles}} \quad \text{or Scale} = \frac{1 \text{ inch}}{1,000};$$

This looks very much like a descriptive scale; to make it into a *fractional scale*, we need the same kind of units in both numerator and denominator. A little arithmetic should convince you that there are 63,360 inches in 1 mile.[6] We can therefore rewrite

$$\text{Scale} = \frac{1 \text{ inch}}{1,000 \times 63,360 \text{ inches}}$$

or

$$\text{Scale} = \frac{1}{63,360,000}.$$

Notice that the fractional scale, also called the *representative fraction* (or *RF* for short), always has a numerator of 1 and that it no longer has any indication of the units in which it was originally computed. Put another way, a scale of 1/63,360,000 or, as it is often written, 1:63,360,000 (one to 63,360,000) refers to a numerical ratio, nothing more. That ratio can now be applied to any unit of measure, as we shall see in a moment.

A third way of indicating scale is by means of a linear diagram. It would be useful, for purposes of direct measurement, to have a line on the globe whose length represents some convenient unit of measurement, such as miles or kilometers. In the example we have been discussing, 1 inch on the globe represented 1,000 miles on the earth. Such a scale, called a *graphic scale*, is shown in Fig. 11.

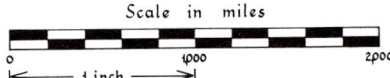

Fig. 11. A graphic scale.

So far, we have discussed only the scales of globes. What about map scales? The scale of a map is more difficult to define since a flat map must distort the spherical surface it represents. This means that the linear scale of a map changes from one part of the map to another, whereas on a globe the scale is constant and easy to define. We therefore imagine a globe that has the scale at which we wish to draw the map. For example, we want a map at a scale of 1:100,000. We imagine a globe of that scale — that is, a globe 1/100,000 the size of the earth. This is a very large, imaginary globe with a circumference of about 24,900 miles divided by 100,000, which is about 0.25 miles, or 440 yards (40,000 km divided by 100,000 or 400 m). The other dimensions of the globe will, of course, also be 1/100,000 of the corresponding dimensions on the

---

[6] It should also convince you of the enormous superiority of the metric system over the medieval conglomeration of units of measure with which we are needlessly saddled. In the metric system, a similar problem would be solved quite simply. The circumference of the earth is about 40,000 kilometers. If a certain globe had a circumference of 50 centimeters, its scale would be 50 centimeters/40,000 kilometers. But 1 kilometer is equal to 100,000 centimeters; therefore, the scale of the globe would be 50 centimeters/4,000,000,000 centimeters, or 1/80,000,000.

earth. This imaginary globe is called a *sphere of development*.

What do we mean, then, when we say that the scale of a map is 1:100,000? We have just said that the scale varies and that the plane on which we draw a map can coincide with a sphere only at the element. It follows that the scale of the map will be identical with the scale of the sphere *at the element*. Some kinds of projections are so contrived that the same scale is found along certain restricted lines as well as at the element, but generally the scale over most of the map will not be the same as that of the sphere of development. The scale of the sphere of development, which is applicable only at the element (and perhaps along a few other lines), is called the *nominal scale* of the map.

As you move away from the element, the true scale of the map differs more and more from the nominal scale. In other words, distortion is nil at the element and increases with distance away from it. If a map is of a fairly large scale,[7] it will not show much of the earth's surface. In other words, every part of the map will be close to the element if the map is centered on the element, and the distortion will be small even at the edges of the map. In such a case, true scale will differ only a little from the nominal scale. In contrast, a map of smaller scale such as a map of the whole world, obviously has to include parts of the earth that are very far from the element. The distortion may be enormous, and the difference between true and nominal scale will be correspondingly great.

In any case we can represent the true scale by an equation similar to the one we used for a globe:

$$\text{True scale} = \frac{\text{map distance}}{\text{earth distance}}.$$

But we need to bear in mind that the nominal scale indicated on a particular map will not be the same as the true scale except in the immediate neighborhood of the element. These statements, incidentally, apply to the descriptive and graphic scales as well as to the fractional scale (RF). The amount of distortion on a Mercator projection is shown in Fig. 12. The distortion is represented by a series of lines along which the ratio between the true scale and the nominal scale is constant. Thus, the line marked "4" is the line along which the true scale is four times as large as the nominal scale of the map.

Despite the fact that we always need to bear in mind the distinction between true and nominal scale, the nominal scale can be used in large-scale maps to solve

---

[7]Note that since scales are represented as fractions, a large scale is one with a low number in the denominator. A scale of 1/10,000 (1:10,000) is ten times *larger* than a scale of 1/100,000 (1:100,000).

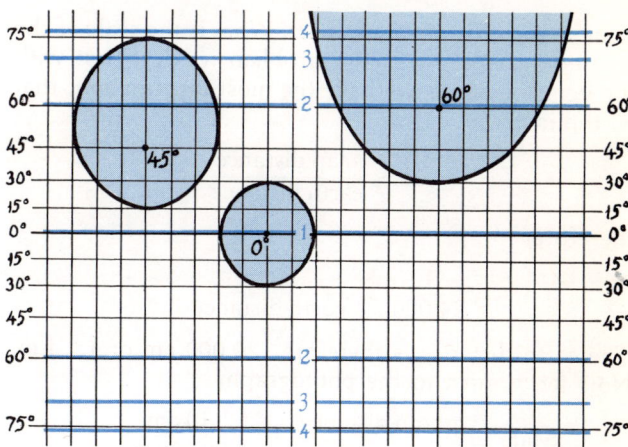

Fig. 12. Lines of equal linear distortion on a Mercator projection. Note that since the element is the equator, distortion is symetrical about that line. Hence places in the Northern Hemisphere that have a fourfold exaggeration of the linear scale each have their counterparts in the Southern Hemisphere, at the same distance from the equator. Note also that if the linear scale is four times too large, it follows that areas are exaggerated sixteen times (4 X 4 = 16). The enclosed figures show the distortion of shape since they each represent a circle on the globe.

certain recurring problems involving distances. The following examples illustrate some specific problems involved.

1. On a particular road map, the distance between Eugene, Oregon, and Portland, Oregon, is 9 inches. If you know that the two places are about 100 miles apart, what is the scale of the road map?

$$\text{Scale} = \frac{\text{map distance}}{\text{earth distance}} = \frac{9 \text{ inches}}{100 \text{ miles}} = \frac{9 \text{ inches}}{6,336,000 \text{ inches}}.$$

The map scale is therefore 1/704,000, or about 1:700,000.

2. On a map of scale 1:31,680, two places are 10 inches apart. How far are they on the ground?

$$\text{Scale} = \frac{\text{map distance}}{\text{earth distance}};$$

therefore,

$$\frac{1}{31,680} = \frac{10 \text{ inches}}{\text{earth distance}}.$$

By cross-multiplication we get

$$1 \times \text{earth distance} = 31,680 \times 10 \text{ inches}.$$

The earth distance is therefore 316,800 inches, or dividing by 63,360 (number of inches in 1 mile), 5 miles.

3. On an air photograph of unknown scale, two points are 10 cm apart. On a map of scale 1:24,000, the

same two points are 5 cm apart. What is the scale of the air photograph? In such a problem we deal first with the item for which we have the most information, that is, the map.

$$\text{Scale} = \frac{\text{map distance}}{\text{earth distance}};$$

therefore,

$$\frac{1}{24,000} = \frac{5 \text{ cm}}{\text{earth distance}}$$

The earth distance is therefore 120,000 cm or 1.2 km. Now let us turn to the photograph.

$$\text{Scale} = \frac{\text{photo distance}}{\text{earth distance}} = \frac{10 \text{ cm}}{120,000 \text{ cm}}$$

The scale of the air photograph is therefore 1:12,000.

Scale and Generalization. The implications of scale for geographers go considerably beyond the mere calculations of distance. Consider a map of vegetation for example. In a particular area there happen to be Douglas firs, big-leaf maples, and deciduous oaks. The firs and maples are pretty well intermixed; the oaks are in two solid patches, one about 500 feet (about 150 m) in diameter, the other with a diameter of about 1 mile (about 1.6 km). If the map you construct is of a very large scale, say about 1:1000 (1 inch represents about 80 feet or 1 cm represents 10 m), you can show every tree by using three small symbols (Fig. 13(1)). The map, although quite detailed, is nevertheless a generalizaton. There is no room to show anything but the trees, and each species is abstracted into a single symbol.

Suppose we reduce the scale of the map by a factor of 10. At 1:10,000 the two patches of oak are about 0.6 and 6 inches (1.5 and 15 mm) in diameter. You can still show the patches, but it is impossible to show individual trees. The usual scheme is to think of categories of trees and to show the categories rather than the individuals. One might, for instance, think of two categories in our example: (1) oaks and (2) mixed firs and maples. This is easily enough shown, as in Fig. 13(2). The kind of generalization involved in the second map is referred to as *conceptual generalization*, since we have created the concept of "a forest consisting predominantly of Douglas fir mixed with big-leaf maples."

If we decrease the scale of the map again, this time to 1:100,000, the two patches of oak have to be shown with diameters of 0.06 and 0.6 inches (0.15 and 1.5 mm), respectively. The bigger patch can be shown, but the smaller one is too small to map at its true scale. Two possibilities are open. You can show the small patch by a symbol that clearly has to be out of scale (Fig. 13(3)), or you can put that patch out of your mind and off the map on the supposition that it is too small to matter. If you use the out-of-scale symbol, you have made another conceptual generalization of a sort slightly different from the previous one involving firs and maples; if you simply drop the oaks out of the picture,

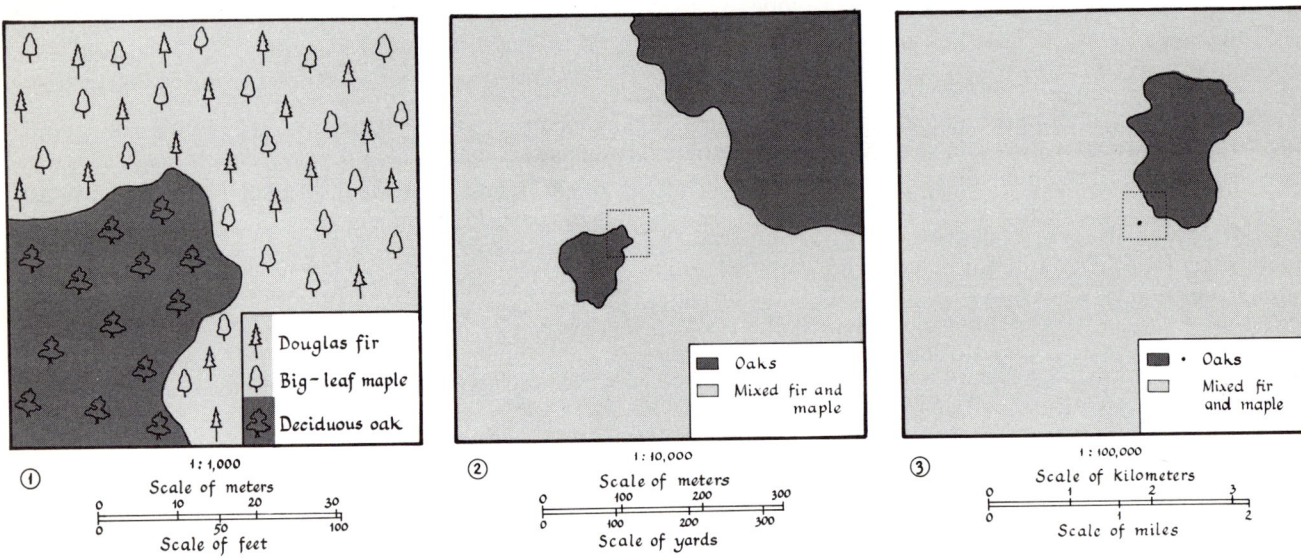

Fig. 13. Hypothetical maps of vegetation. (1) is at scale of 1:1,000, which allows each tree to be shown by a symbol. Only part of one of the oak patches can be shown on a map of this scale and size. (2) is at a scale of 1:10,000 and shows only two categories of vegetation. The dotted line represents the area shown on 2-13(1). (3) is at a scale of 1:100,000. At this scale the patches of oak must be shown by out-of-scale symbols. The dotted line represents the area shown on Fig. 13(2).

you have made a *cartographic generalization* — that is, you have simplified the lines on the map or omitted certain ones.

A further decrease of scale, to 1:1,000,000, would necessitate still more generalization. Neither of the patches of oak is big enough to show. We now have at least three choices open to us: (1) We can use out-of-scale symbols again, either for both patches of oak or only for the more important one; (2) we can leave the oaks entirely out of the picture; or (3) we can generalize our concepts even further by creating a new category to cover both fir-maple and oak. In the third possibility we can retain some specificity by naming the category "mixed Douglas firs and big-leaf maples with occasional patches of deciduous oaks," or we can be very general indeed with such a category as "woods."

This short digression into the relationship between scale and degree of generalization obviously represents a simple and somewhat pat example of the problems involved. Furthermore, we focused on the problems of the map maker. Similar but more difficult problems face the map reader as he tries to interpret maps of different scales and, therefore, of different degrees of generalization, both conceptual and cartographic.

Map Projections

Map Properties. We have already defined a map projection as any orderly, systematic, and describable arrangement of parallels and meridians on a flat surface. We have also seen that there are three classes of projections: planic, or azimuthal, cylindric; and conic. Before describing the projections, we need a slightly expanded definition of distortion on a map. While it is true that any map distorts the surface of the spherical earth, a map projection can have the property of maintaining certain features of the sphere more or less unchanged on the map. Two of the most important properties are called **equivalence** and **conformality**. An equivalent projection is one in which equality of area is maintained between the sphere and the map. Any bounded surface on the map will have exactly the same area as the corresponding bounded surface on the sphere of development. The two surfaces will no longer have the same shape — a circle on one may be an ellipse on the other; a square on one may look more like a parallelogram on the other — but the areas will be the same. In other words, the areal scale remains constant all over the map, though the linear scale must vary considerably. A conformal projection, on the other hand, distorts area but preserves angular relationships. Any angle on the sphere will be faithfully reproduced on the map; shapes and areas will be altered, but directions will be maintained.

These two important properties are, unfortunately, mutually exclusive. A map may be either equivalent or conformal; it cannot be both. Thus, in selecting a projection for a particular map, one has three choices: (1) maintaining area; (2) maintaining directions; or (3) keeping neither of these properties, either in some more-or-less happy compromise between equivalence and conformality or in favor of some altogether different property. These other properties include *equidistance*, in which distances are shown true to scale along certain selected lines (but never everywhere on the map), and a series of miscellaneous properties, among which is the property whereby certain important lines on the sphere (such as great circles) are shown as straight lines on the map. Various combinations of these properties are possible on a single map as long as they do not include both conformality and equivalence.

Planic, or Azimuthal, Projections. The principle of a planic projection illustrated in Fig. 14 is quite simple. The element of a planic projection is a point. Point A represents the element on the sphere of development: A' is the corresponding point on the map. The heavy line through A is a meridian, hence a north-south line. A similar line has been drawn through A'. Point B is any other point on the sphere. Where do we place the corresponding Point B' on the map? We first draw the great circle that connects A and B. We next measure the angle between the two lines. This angle, measured clockwise from north, is called the *azimuth* (Arabic *alsumut* the direction) of B with respect to A — hence the name azimuthal for planic projections. We also measure the distance from A to B, along the great circle, either in angular or linear measure. If we now transfer the azimuth and the distance to the map, we can locate B'. We can repeat this procedure for any other points we wish to show and thus build up a map on an azimuthal projection.

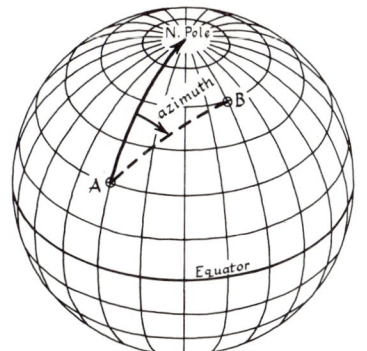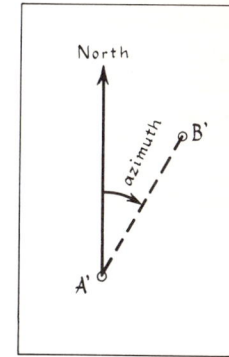

Fig. 14. The development of an azimuthal projection. On the sphere, at the left, the point A is the element, and B is some other point to be mapped. On the flat map, at the right, the same points are represented by A' and B' respectively.

In the process of transferring azimuths and distances from the sphere to the map, no advantage is gained by making the map azimuths different from those on the sphere. Changing azimuths would only add a needless distortion to the map. Distances, on the other hand, need not be shown as they are on the sphere; it is possible instead to use some function of the distance between A and other points, such as the square of the distance, or the square root, or any other mathematically definable relationship. The most commonly used functions are the trigonometric ones involving the sine and the tangent of the *angular* distance[8].

If, in fact, we reproduce distances on the map exactly as they are on the sphere, we get a particular kind of azimuthal projection, called the *azimuthal equidistant*, since distances are shown true to scale along all lines passing through the element (through A' in Fig. 14); along other lines the scale will differ from the nominal one. Although angles about the element are shown correctly, the projection is not conformal, because angles elsewhere on the map do not correspond to those on the sphere. Nevertheless, it has the properties of being equidistant and of showing true direction about a single point, the element. This makes it a good projection to use for a map centering on an airport, for instance, since it shows distances and directions from the airport as they actually are on the sphere of development.

If, instead of showing distances exactly as they are on the sphere, we use one of the trigonometric functions mentioned above, we get other kinds of azimuthal projections. For instance, if we plot distances from the element according to the formula

$$\text{Distance on the map} = 2R \sin\left(\frac{d}{2}\right),$$

where R is the radius of the sphere of development and d is the distance in angular measure, we get an azimuthal projection that is equivalent. This projection, first described by Johann Lambert in 1772, is called *Lambert's azimuthal equal-area* projection. This is an excellent projection for areas as large as a hemisphere because it shows areas true to scale, and while it therefore cannot be conformal, it does distort directions less over large areas than most other equivalent projections. It is used for the polar views in this book.

Using the formula

$$\text{Distance on the map} = 2R \tan\left(\frac{d}{2}\right)$$

---

[8] The sine of an angle, in a right triangle, is the ratio between the length of the side of the triangle opposite the angle and the length of the hypotenuse of the right triangle. The tangent is the ratio between the lengths of the opposite and adjacent sides.

yields an azimuthal projection that is conformal. It is usually called the *stereographic* projection for reasons having to do with its origin as a perspective projection described by the Greek geometer Hipparchus sometime around 150 B.C. Since its main property is conformality, it is most useful wherever directions need to be shown correctly. Wind direction, for example, is a very important aspect of weather maps. For this reason weather maps of the Northern Hemisphere are shown on a stereographic projection centered on the North Pole. The daily weather map of the U.S. Weather Bureau employs the same projection, since the Northern Hemisphere maps are composites of the various national ones.

By using another formula

$$\text{Distance on the map} = R \tan d,$$

we obtain a projection that is neither equivalent nor conformal and that does not preserve equidistance along any line drawn on it. It does, however, have the interesting and useful property of showing all great circles as straight lines on the map. A straight line between any two points on the map represents the great circle through those points on the sphere of development. This projection, called the *gnomonic*, is another perspective projection. It is probably the oldest known projection, dating back to about 500 B.C. Its uses lies particularly in the field of navigation, where knowledge of great circle routes is very important because these routes represent the shortest distances between places.

There are any number of other azimuthal projections of lesser usefulness. Among these is the *orthographic*, which shows parallels and meridians as they might be viewed from a great (in theory, an infinite) distance. It is of consequence to astronomers and astronauts because we see the moon from the earth as if it were drawn on an orthographic projection. Other azimuthal projections are of limited consequence or are as yet uninvented. Since there is no limit to the number of functions of distance that one can employ there is no end to the azimuthal projections.

Cylindric Projections. The procedure for building up cylindric projections is somewhat different from that for the azimuthals but the general principle remains the same. The element here is a great circle. Let us take as an example a cylindric projection whose element is the equator. We proceed by reproducing the equator of the sphere of development on the cylinder, which, when unrolled becomes our map. The parallels of latitude can then be shown either as straight lines parallel to the equator or as mathematically describable curves. The meridians can be shown as straight lines at right angles to the equator or again as curves of some sort. The spacing between the lines of the grid can be constant or variable, true to scale or otherwise. The choices between straight line and curve and between constant spacing and variable spacing determine the properties of the cylindric projection. In general, the cylindric projections lend themselves to showing the whole world, since the equator can be used as the element. We will mention several of the most important cylindric projections, together with their properties and major uses.

Only one important cylindric projection is conformal. This projection was constructed in 1569 by the Dutch cartographer Gerhard Kramer (better known by his latinized name, Mercator) as a base for navigational data. Navigators rely on compass directions, and the great value of the *Mercator* projection is that it not only shows directions truly, but it also shows compass

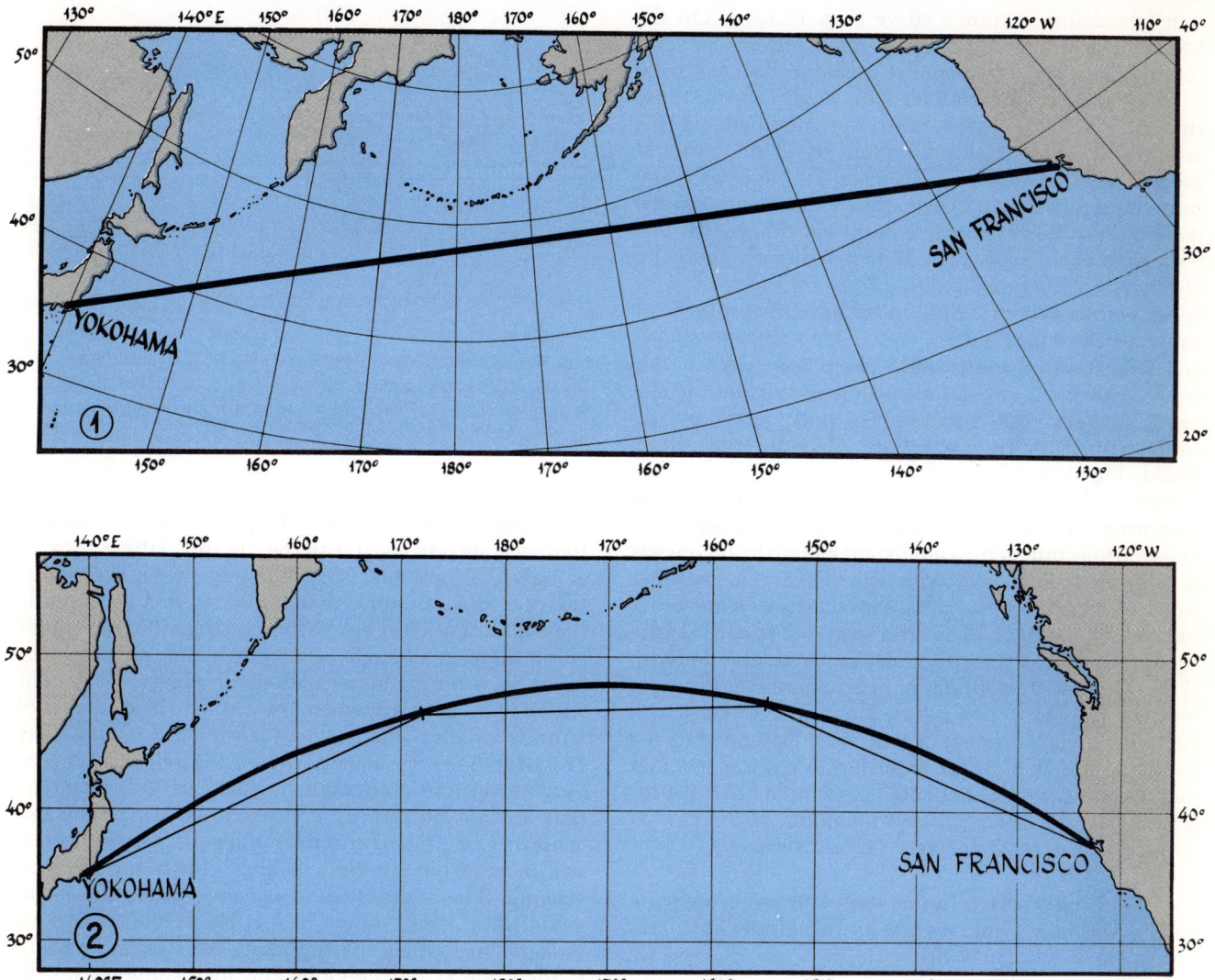

Fig. 15. The use of the gnomonic (1) and Mercator (2) projections in navigation. On the gnomonic projection, the straight line between San Francisco and Yokohama represents a great circle, hence the shortest route between them. When this line is transferred to a Mercator grid, it appears as a curved line. The straight line segments that approximate this curve are all at their proper angles with respect to north, because the Mercator is a conformal projection. Hence the navigator can easily measure the directions in which he must sail or fly to approximate the great circle route.

course lines on the sphere as straight lines on the map.[9] One use of the gnomonic (an azimuthal projection) and Mercator projections in navigation is shown in Fig. 15. The shortest route between San Francisco and Yokohama is the great circle passing through them. The easiest way to find that circle is to draw a straight line between the two places on a gnomonic projection (Fig. 15 (1)), since every straight line on that projection represents a great circle on the sphere. By noting where the line crosses parallels and meridians, we can transfer the line onto a Mercator projection. Notice that the transferred line is now a curve (Fig. 15 (2)). On the Mercator a straight line represents a line of constant compass bearing. We therefore break up the curve into a series of straight chords. The angle which those straight segments make with the meridians is the azimuth at which a ship's captain or jet pilot should proceed for the indicated distance if he wishes to go from San Francisco to Yokohama by the shortest route. It should not be necessary to point out that Mercator's projection is not equivalent. In fact, it distorts areas and distances greatly (see Fig. 12).

Numerous cylindric projections have the property of equivalence. Among them are *Goode's* the *sinusoidal*, and *Eckert's* (the basic world projection used in this book). However, angular distortion away from the element, usually the equator, is quite large on all equivalent cylindric projections showing the whole world (Fig. 16).

Some cylindric projections are neither equivalent nor conformal. The simplest, most naïve, and most easily drawn projection is the *equidistant cylindric*. It consists of a straight-line grid, with equidistance maintained along all meridians and along the equator or some other parallel of latitude. The grid will be square if the equator is shown true to scale, rectangular otherwise (Fig. 17). Note that although the equator is the element, we need not show a true scale along it. We can just as well, in this instance, show true scale along one of the parallels that is of some importance to the particular map we are concerned about. This simple kind of projection is a close enough approximation of the curved surface of the earth for very large-scale maps.

Conic Projections. In one sense, all projections are conic projections, since a plane and a cylinder are mere-

---

[9] A compass course line, or course line, or rhumb line, or loxodrome, to give it some of its names, is the line described on the earth by a person following a constant bearing on the compass, or by following a constant azimuth. The meridians and the equator are special cases of such lines. In general, however, course lines on the globe are spirals winding their way ever more tightly around one of the poles, since one cannot get to the poles without eventually going due north or due south; hence yet another alias — equiangular spiral.

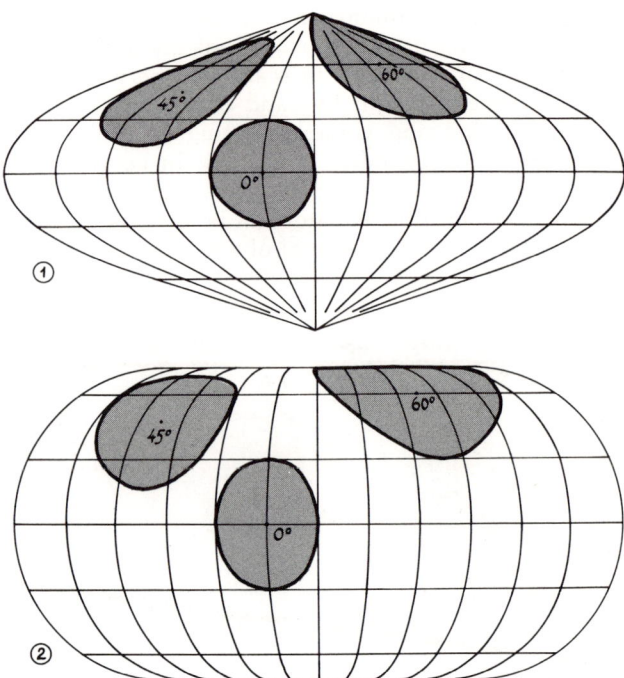

Fig. 16. The sinusoidal (1) and Eckert No. IV (2) projections. Distortion on these equivalent projections is indicated by the misshapen figures, all of which are true circles of identical size on the spherical earth.

ly limiting cases of a cone. In the former, the cone is flattened as far as it can be; in the latter, it is as elongated as possible. In building up a conic projection, the procedure is somewhat similar to that involved in the cylindrics. The cone, which is tangent to the sphere along some small circle, is unrolled, and that small circle appears on the map as an arc of a larger circle. The situation is simplest when the element is a parallel of latitude which is then called the *standard parallel* (Fig. 18). Meridians are usually shown as straight lines converging on one point; the parallels are arcs of circles that are not necessarily concentric (sharing the same center). Once again the properties of map projection will depend on the kind of lines drawn and on their spacing. The simplest kind of conic projection is the *equidistant conic*, shown in Fig. 19. The meridians are straight lines, their spacing along the standard parallel is the same as on the globe and they converge on the point that represents the apex of the cone on the map; the parallels are equally spaced, concentric circles with centers at the same point of convergence. This projection is seldom used, since it is neither conformal nor equivalent.

Commonly employed variations of conic projections include *Bonne's* which is used for continental areas because it is equivalent; and the *polyconic*, which has

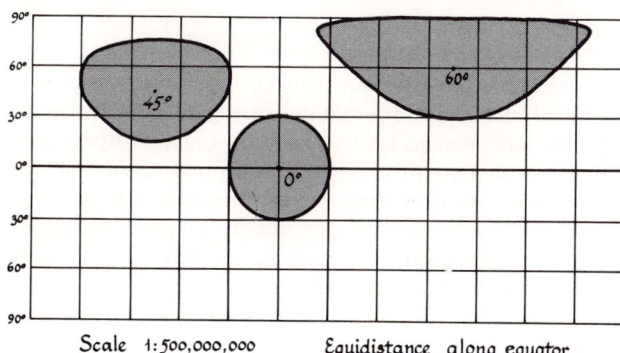

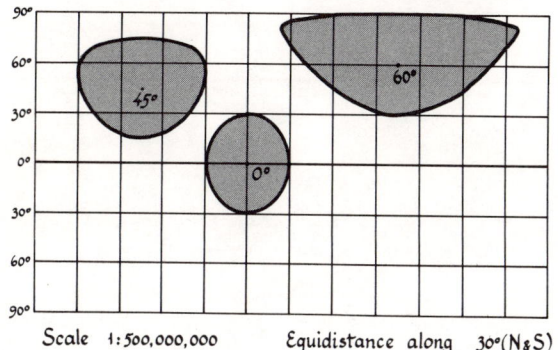

Fig. 17. Two variations of the equidistant cyclindric projection. Distortion is indicated in the same fashion as in Fig. 16. In (1), which is equidistant along the equator, there is distortion along all other parallels of latitude. In (2) equidistance is maintained along 30° N and 30° S, but there is distortion along the equator (and other parallels of latitude).

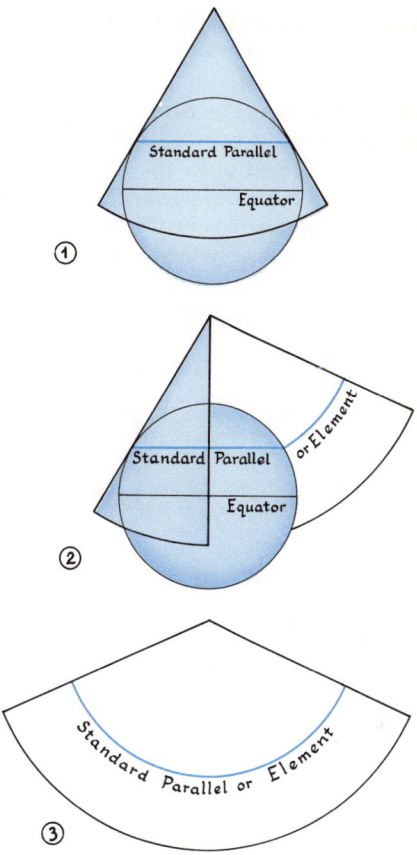

Fig. 18. The development of conic projections. The tangent cone (1) is partly unrolled in (2). Note the appearance of the element, in this case a parallel of latitude, when the cone is flattened (3).

equidistance along all parallels of latitude and is the basic projection for all quadrangle maps of the U.S. Geological Survey such as those reproduced in this text (Fig. 16-8, for example). Other commonly used conic projections are *Lambert's conformal conic* and *Albers's equivalent conic*. These projections are based on the fact that it is possible by certain mathematical manipulations to have not one standard parallel but two. The advantage of these projections is that they reduce distortion because there are two elements, which can be appropriately spaced with respect to the map, so that no part of the map is very far from one or the other of the two standard parallels. This projection is excellent for areas elongated in an east-west direction. It can also be used for areas drawn out in other directions, provided the standard parallels are chosen, not as parallels of latitude, but as arcs of small circles running through the central portion of the area to be mapped. If the major concern is area, Albers's is used; if directions are more important, Lambert's is preferable. Many maps of the United States and of Europe use one or the other of these conics.

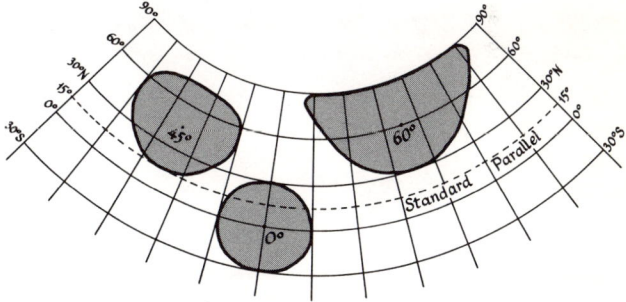

Fig. 19. The equidistant conic projection. Distortion is shown as in Figs. 16 and 17.

The cylindric and conic projections are much more recent than the azimuthals. Mercator's, to be sure, represents an earlier cylindric projection than Lambert's azimuthal equal-area, and the equidistant cylindrics are probably as old as the Christian Era, but most of the conics and cylindrics date from the end of the 18th century or later. The azimuthal projections are essentially based on very simple geometric considerations and, therefore, are appropriately associated with the ancient Greeks. The conics, on the other hand, had to await relatively advanced mathematical notions that were not developed until the eighteenth century. The cylindric projections are even later; they appear only during the nineteenth century, coinciding with the increasing interest in showing various distributions over the whole world, such as the areal extent of colonial empires or the distribution of population density.

Map Content

This discussion on map projections has touched only the highlights of mathematical cartography. Our main purpose has been to focus attention on the importance of scale and map properties in selecting the most appropriate projection and in interpreting a map. One reason for understanding as much as possible about maps and mapping techniques is that geographers heavily rely on maps to describe and analyze the surface of the earth. Air photographs have a similar func-

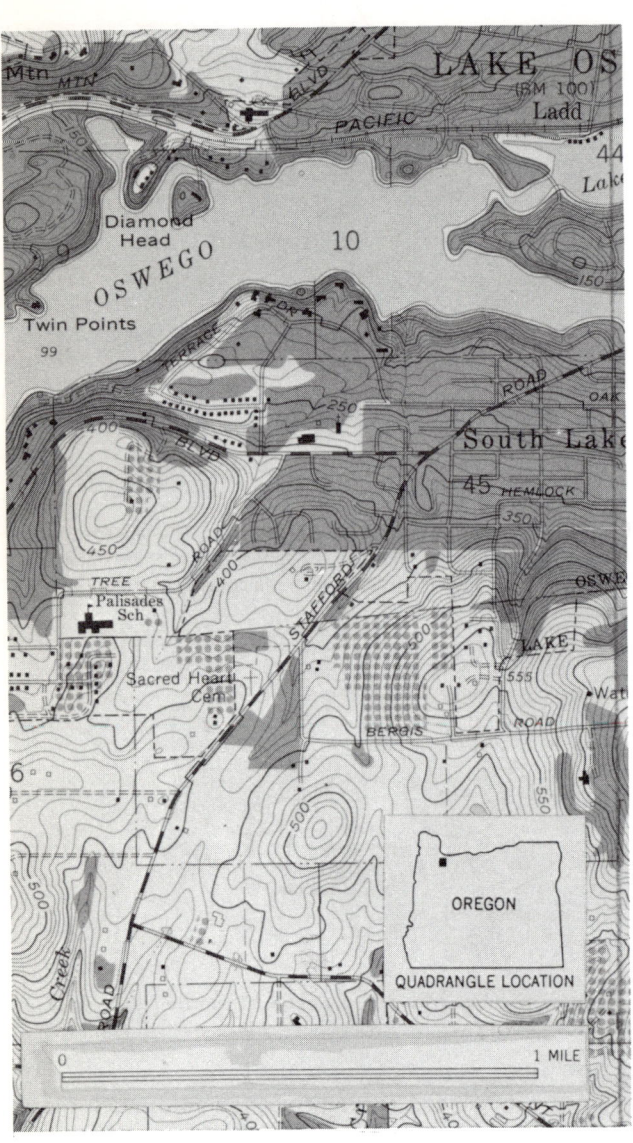

Fig. 20. Map and air photograph of Lake Oswego, Oregon. Note the striking difference in informational content between a selective and symbolic map and the nonselective air photograph.

tion, even though their characteristics differ somewhat (Fig. 20). A map can be defined as a graphic, selective, and symbolic representation of a portion of the earth's surface at a given scale and on a given projection. The vertical air photograph is neither selective — it shows everything that is "visible" to the photographic lens — nor symbolic — objects are shown as they appear and not by conventional symbols. Selectivity may, in a sense, be introduced into an air photograph by the choice of film and filter so that attention can be focused on some particular feature of the landscape. This kind of selectivity differs from the process where whole categories of phenomena are systematically left off a map while other characteristics of the land are shown in great detail. Air photographs also differ from maps in that scale and projection can be only indirectly manipulated. Despite these differences, however, air photographs and maps are closely associated in the minds of those who use them, if only because most of the world's maps are increasingly being constructed from the evidence of photography complemented by field work. Unfortunately, the special role of air photographs cannot be adequately treated in an introductory text. The following discussion, therefore, focuses entirely on maps.

One overriding problem that confronts the maker of a general-purpose, descriptive map is that of showing the uneven relief of the earth. How does one show elevation as well as latitude and longitude? Several ways have been developed, ranging from simple pictorial renderings of the terrain (Fig. 21), to representations of slope by means of lines of differing weight called hachures (Fig. 22), to abstract contour lines which give a more quantitative, if less visual, picture (Fig. 23). The contour line is usually the most useful analytically because it allows one to measure elevation as well as other characteristics dependent on elevation, such as slope and relative relief.

Elevation represents a feature of the landscape that can be expressed quantitatively. Such quantitative differences between places are commonly shown on maps by lines joining points of equal quantitative value. For example, the contour line, or **isohypse** (Greek *isos* equal + *hypsos* elevation), is a line joining points of equal elevation above some standard elevation, usually mean sea level (m.s.l.). One way of visualizing contour lines is to imagine an island. The normal coastal outline of the island shown on any map is the contour line, 0 feet above m.s.l. If the level of the water were to rise 10 feet, the island would have a new outline. If the water were then to recede back to its original level, the higher outline would represent the 10-foot contour line. All isohypses can be thought of in exactly this way. An example of this kind of represen-

Fig. 21. An early pictorial representation showing unequal relief of the terrain on a 16th-century map of the eastern Alps.

tation of the third dimension is shown in Fig. 23. Contour lines are always closed curves, even though the map area may be too small to show the whole extent of any particular isohypse. The spacing of the isohypses indicates the steepness of the slope of the land — close spacing means steep slopes, and wide spacing indicates gentle slopes.

Isohypses are not the only lines of this sort that can be shown on a map. Other items can also be treated as if they represented a third dimension. For example, the maps of pressure, temperature, precipitation, and others shown in this text use the same principle. We can think of the temperature at a particular point as a

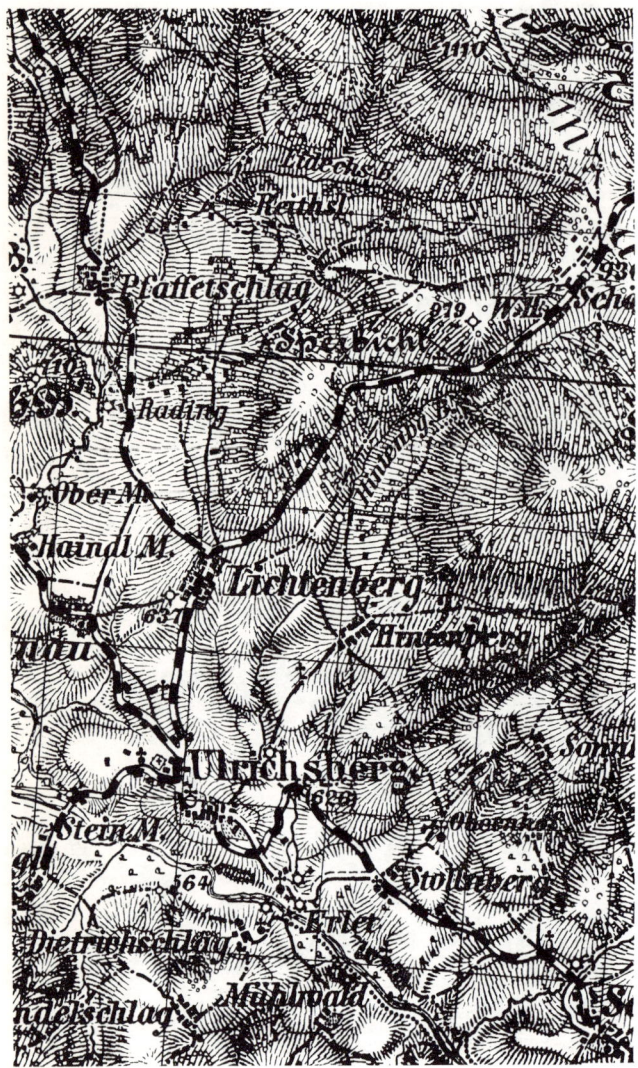

Fig. 22. Terrain representation by hachures on a late 19th-century Austrian topographic map of a portion of the Bohemian Forest.

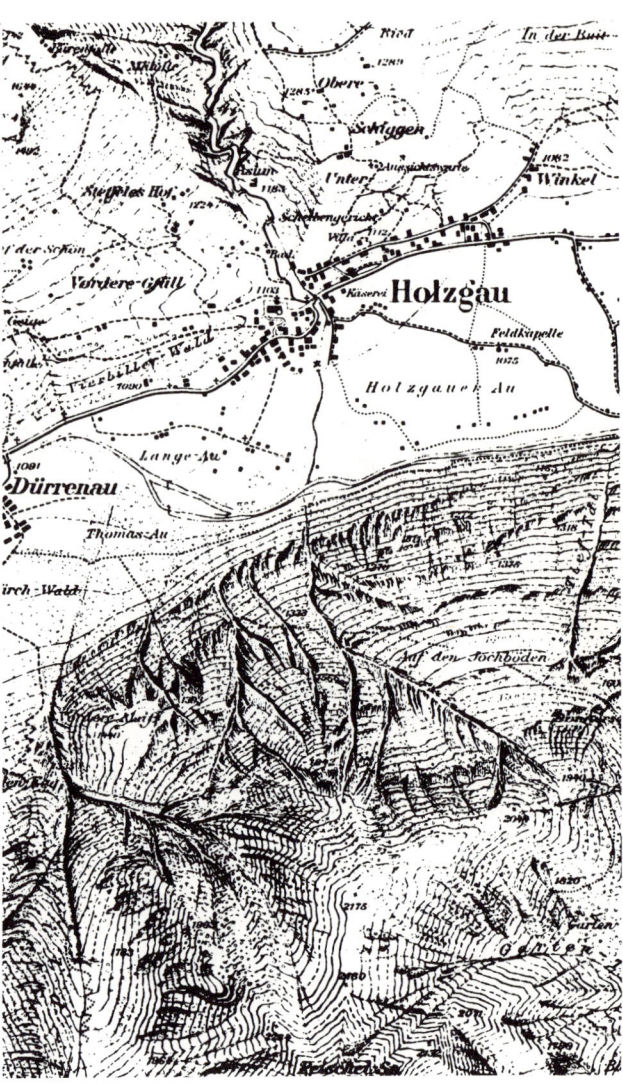

Fig. 23. Terrain representation by contour lines. Cliff hachures have been added to aid the reader.

quantity that can be graphed in a vertical direction. The third dimension is thus introduced into the problem, and the map shows a series of "contour lines" that are, in this case, not lines of equal elevation above sea level but rather lines of equal temperature.

Lines drawn on maps to represent a third variable (the other two are always latitude and longitude) are given the general name of **isolines** or sometimes *isograms* (Greek *isos* equal + *gramma* weight). There are two kinds of isolines, *isarithms* (Greek *isos* + *rithmos* number) and *isopleths* (Greek *isos* + *plethos* quantity). Isarithms are lines that join points of equal value where the value pertains to a single point on the surface of the earth. Thus, isolines of elevation, temperature, pressure, rainfall, and many others are all isarithms.[10] Isopleths, on the other hand, are lines joining equal values pertaining not to a single point but to an area. Lines of population density, for example, join areas where population density is the same (density, by definition, refers to a value over a given area, not at a single point). There are other kinds of isopleths as well, such as lines of equal relative relief, which depend on a difference between the highest and lowest elevation within a given area rather than a ratio, as in population density.

---

[10]Isolines of temperature, pressure, and rainfall are known as isotherms, isobars, and isohyets, respectively.

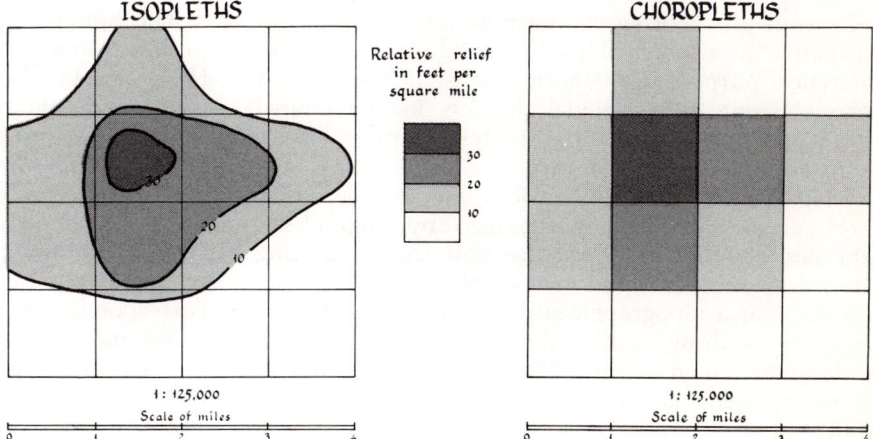

Fig. 24. Hypothetical isoplethic and choroplethic representations of relative relief. In the isoplethic map, relative relief is shown by lines of equal value; in the choroplethic map, by areas having the same value.

Quantitative characteristics of the land can also be shown by *choropleths* (Greek *choros* region + *plethos*) rather than isolines. A choroplethic map is one in which the lines are not drawn *through* points or areas of equal values, but are drawn as *boundaries* of areas. The areas bounded may have no direct relation to the values being mapped (counties or states in population density maps, for example), or they may be areas chosen because they are reasonably homogeneous with respect to the mapped quantity, or they may be perfectly arbitrary lines (a square-mile grid, for instance). In all three cases, categories of the value to be mapped are chosen, and the category to which each area belongs is indicated on the map by some symbol, usually by line shading of increasing darkness or perhaps by a color spectrum. Fig. 24 shows the same phenomenon, relative relief, by means of both an isoplethic and a choroplethic map.

Not all differences in the landscape can be quantified, of course. Soils, landforms, and vegetation are obvious features of the physical landscape that differ in *kind* rather than quantity. Such features are usually mapped by a modification of the choroplethic method. The areas are chosen so that their boundaries represent lines or zones where the phenomenon to be mapped changes rapidly in character over a short distance. In other respects the treatment is the same as with choroplethic maps, except that the categories are discontinuous (e.g., maples, firs, oaks) rather than by continuous quantities, as in a true choroplethic map. The various maps of Fig. 13 are illustrative of the numerous special problems posed by the mapping of differences in kind.

Some features of the earth's surface cannot be treated as exhibiting differences in either quantity or kind from one place to another. They are features discontinuous in distribution (as well as in character), such as cities, rivers, and highways. Such features are shown by discontinuous symbols.[11]

Maps can also be classified by the general purpose they serve. We may distinguish descriptive, analytical, recording, and special-purpose maps. The most common type, the descriptive map, shows a variety of features selected to give an overall general impression of an area. Such maps show settlements, parts of the transportation network, and some indication of relief, especially the river system. Such are the regional maps in atlases, road maps, and the U..S. Geological Survey quadrangles, to name only three important examples.

A second type of map is the analytical map. Its most characteristic feature is the selection of a very limited number of features, which it shows in great detail. The usual purpose of such maps is the analysis of the areal distribution of some phenomenon with a view to explaining some of the processes affecting its intensity or distribution. Fig. 4-11 is a relatively simple example of an analytical map. A more complicated kind would be the daily weather map (Fig. 4-2). There is practically no limit to the sophistication and usefulness (the two are not always related) of analytical maps.

A third kind of map is the field map, or recording map. This map is largely a record of as much pertinent information as possible in its proper areal relation. Its public appearance is relatively rare, since it is a preliminary step to the production of either a descriptive or analytical map.

The fourth category of maps consists of a mis-

---

[11] Figure 13 is an example of a feature that is treated as a continuous or discontinuous phenomenon depending on the scale and degree of generalization. In 13 (1) each tree is shown by a discontinuous symbol. In 13 (2) the entities mapped are no longer trees but rather types of forest assumed to be continuously distributed on the map.

cellaneous group of special-purpose maps, usually informative on a specific topic but not designed to serve analytical purpose in the normal sense of the term. A map showing congressional districts in the United States or the location of three-star restaurants in France is an example of this category.

Maps may also be classified according to scale. Three large categories are usually recognized by geographers. The largest-scale maps, and therefore those requiring the least conceptual and cartographic generalization, are maps at a topographic scale (Greek *topos* place + *graphein* to draw), arbitrarily defined as maps with scale greater than about 1:250,000. Next in order of necessary generalization are maps at chorographic scales (Greek *choros* region + *graphein*), between 1:250,000 and 1:5,000,000. The smallest-scale maps are essentially maps showing either the whole earth or so large a portion of it as to be classed as geographic (Greek *geos* the earth + *graphein*).

No matter how they are classified, maps are highly variable in content and purpose. The selection of phenomena to be shown, of symbols to depict them, of scale and projection, all depend on the particular purpose to be served. By the same token, the interpretation of any particular map requires close attention to the scale, projection, symbolization, and degree and kind of generalization employed by the cartographer.

Summary

This appendix has touched briefly on the many aspects of locating and mapping. The shape and size of the earth — a slightly bulging spheroid with a radius of about 400 miles or nearly 6400 km — is basic to all systems of location. The *geographic grid*, consisting of great circles passing through the poles called *meridians of longitude*, and of small circles "parallel" to the equator called *parallels of latitude*, is the most basic system. The *Township and Range* system is an important method of location in the United States. Its grid of north-south and east-west lines, running at intervals of one mile (about 1.6 km) has had a marked effect on the location of roads and political boundaries wherever it is found in North America.

The spherical shape of the earth creates certain problems in mapping on a flat surface. Only three surfaces can be flattened without distortion — planes, cones, and cylinders. They give rise to the three basic categories of projections — *azimuthal* or *planic*, *conic*, and *cylindric* — with *elements* (places of perfect accordance between the flat map and the *sphere of development*) respectively at a *point*, along one or more *small circles*, and along a *great circle*. The linear *scale* of a map, the ratio between distances on the map and their corresponding values on the earth, varies from one part of the map to another. The scale given for a map as a whole is called the *nominal scale* and is valid only at or along the element (and along a few other restricted lines). Away from the element there is increasing distortion, either of area or of directions or of both. One can construct maps that show areas correctly — they are known as *equivalent* — or that show direction correctly — they are said to be *conformal*. No map can be both equivalent and conformal; some are neither, but have other useful properties.

Map content can be as varied as map projections. *Descriptive, recording and analytical* maps make up the bulk of them, except for *special purpose* maps of limited and sharply focused interest. One over-riding concern in general descriptive maps is the portrayal of the relief of the earth. There are various methods of showing the third dimension (whether this be the concrete form of the land or an abstract notion such as a temperature surface) which range from the pictorial representation of relief to the use of contour lines to show slope and elevation.

Suggestions for Further Reading

Monkhouse, F. J., and H. R. Wilkinson, *Maps and Diagrams*, E. P. Dutton & Company, Inc., New York, 1963.

Raisz, E., *Principles of Cartography*, McGraw-Hill Book Company, Inc., New York, 1962.

Robinson, A. H., and R. D. Sale, *Elements of Cartography*, 3rd ed., John Wiley & Sons, Inc., New York, 1969.

Glossary

Ablation. The removal of ice or snow, usually from a glacier, by melting and evaporation.

Abrasion. The removal of bedrock material by the grinding action of particles transported by wind, waves, water, or ice.

Absolute humidity. The water vapor content of a mass of air, expressed as a given weight of water per unit volume of air, such as 4 grams per cubic meter. The term is also used in a more general sense to include all measures of water vapor content such as *specific humidity, mixing ratio* and *absolute humidity* to contrast them to *relative humidity.*

Adiabatic cooling. Reduction of temperature resulting from a pressure decrease in an air mass which takes place without exchange of heat between the air mass and its surroundings. In the same way, adiabatic heating occurs when pressure rises within an air mass.

Adiabatic heating. See **Adiabatic cooling.**

Adiabatic rate. The rate at which the temperature of a parcel of air will change as the pressure changes, under constant volume and without any exchange of heat between the parcel of air and its surroundings. Often expressed as approximately 5.5°F/1,000 feet, implying that the decrease in pressure associated with an increase in elevation of 1,000 feet will cause a drop in temperature of 5.5°F. This is the dry or unsaturated adiabatic rate; the wet or saturated adiabatic rate is about 3.0°F/1,000 feet. It differs from the dry rate by the amount of heat released by condensation within a rising and cooling air parcel. In metric units the two rates are expressed as 1°C/100 m and 0.55°C/100 m, respectively.

Adret. The sunnier, south-facing slope in mountain valleys of the Northern Hemisphere. The opposite north-facing slope which may be in the shade for most of the day in winter is called the *ubac.*

Aggradation. That part of the gradational processes that builds up landforms by depositing the products of degradation.

Aggregates. The arrangement and grouping of soil particles into units of characteristic and readily recognizable form.

Albedo. The reflective power of any body; when applied to the earth and atmosphere it is often restricted to the reflection of short-wave radiation. Expressed as a percentage, it indicates the proportion of the incoming light and other solar radiation that is reflected by clouds and dust in the atmosphere and by the surface.

Alkalinization. A soil-forming process in arid regions. Upward movement of capillary water brings alkaline salts to the surface where the alkali accumulates as the surface water evaporates.

Alluvial fans. Low cones of gravel, sand, and finer sediment usually deposited by intermittent streams whose emergence from a highland marks the apex of the cones, and whose action spreads the main body of the fans over the adjacent lowlands.

Alluvial soils. Azonal soils whose parent material is alluvium and whose recency precludes the development of soil horizons. They do not reflect the zonal character of climate and vegetation but are relatively similar wherever found.

Alluvium. Clay, silt, sand, or gravel transported by running water and deposited in the recent geologic past usually on alluvial fans or floodplains.

Alpine glaciers. Accumulations of ice in mountain valleys, also called *mountain glaciers.* The flowing tongue of ice may be anywhere from several meters to several hundred meters deep, and it may be very short or extend for dozens of miles in U-shaped canyons.

Alpine mountains. Mountain systems created by folding and uplift which occurred during the Alpine orogeny (from 60,000 to 2,000,000 years ago). They are the most recent, highest and most rugged of the world's mountains.

Altitude of the sun. Angular distance between the position of the sun and the earth's horizon. Sometimes also called *elevation of the sun.*

Anticlines. Upfolds of the earth's crust, strong enough to deform stratified rocks into arch-like structures, but gentle enough to leave the layers unbroken. Compare **Synclines.**

Anticyclones. High-pressure centers at or near the surface of the earth, characterized by clockwise and outward-moving winds in the Northern Hemisphere (counterclockwise in the Southern), subsidence, and usually stable lapse rates. Compare **Cyclones.**

Antipode. The point on the opposite side of the earth from any observer. It is obtained by imagining a straight line from the observer, passing through the center of the earth and emerging from the sphere at the antipode or antipodal point.

Arêtes. Sharp, narrow mountain ridges created by glacial erosion in the cirques on opposite sides of the arêtes.

Atmosphere. The gaseous envelope of any heavenly body. The earth's atmosphere consists of a mixture of 21 percent oxygen, 78 percent nitrogen, and other gases in fixed proportions; and, in addition, of variable amounts of water vapor. It extends upward indefinitely, but 97 percent of the mass of the earth's atmosphere is found in the lowest 18 miles or 29 km.

Atmospheric pressure. In climatology, refers to the force exerted in every direction by the weight of the overlying atmosphere.

Atoll. A low ring of coral reef or of small coral islands enclosing a central lagoon.

Autoconvectively unstable. Used to describe a lapse rate steeper than 18°F/1,000 feet (3.3°C/100 m). Such a steep lapse rate creates a vertical density gradient greater than the vertical pressure gradient and causes automatic overturn by exchange of the warm air at the ground with cold air aloft.

Azonal soils. One of the three basic orders of soils, which have weakly developed profiles or none at all, usually because of their youth; the most common azonal soil is formed in recent alluvium. Compare **Intrazonal soils, Zonal soils.**

Barchan dune. An isolated crescent-shaped sand dune whose horns project downwind.

Barrier reefs. Bands of coral lying parallel to, but at some distance from the shore, from which they are separated by a wide and deep lagoon.
Basement rock. The uppermost layer of igneous or metamorphic rock. Often covered by layers of sedimentary rock of variable depth.
Basins. (1) Depressions of structural origin enclosed by folded or faulted mountains. Mostly they have interior drainage so that the basins are being filled with alluvium. Also called *bolsons*. (2) A category of large-scale tectonic units, involving extensive surfaces almost entirely covered by nearly horizontal sedimentary deposits. Compare **Masses, Tables.**
Batholith. A very large, irregular, intrusive mass of igneous rock, extending downward to undetermined depth, which at one time rose through the earth's crust by melting and incorporating the surrounding rocks into its own body, but which stopped short of the surface and there became cooled. The core of the Sierra Nevada of California is a batholith.
Blowouts. Deflation hollows created by wind erosion in areas of sparse plant cover. The downwind depositional features are called *blowout dunes*.
Blowout dune. See **Blowouts.**
Bolsons. Basins of interior drainage, partly filled with alluvium from surrounding highlands. See **Basins.**
Broadleaf deciduous. Applied to trees that have leaves (in contrast to the needles of conifers) which they shed during a specific part of their annual growth cycles, usually in winter.
Broadleaf evergreen. Applied to trees with true leaves, a small proportion of which are being shed throughout the year. Such a forest and even its individual trees thus appear "evergreen" during the whole year.
Brown steppe soils. Soils formed under semiarid climates. The A horizon is brownish and slightly leached. The B horizon contains an accumulation of soluble bases, often in the form of a lime horizon.
Calcification. The accumulation of bases, primarily calcium carbonates in the B horizon of soils by upward, capillary movement of water charged with bases.
Calcimorphic soils. One of the intrazonal soils, distinguished by the predominant effect of the limestone parent material on its soil profile. *Tierra roja* and *rendzina soils* belong in this category.
California canyons. Submarine canyons on the continental shelf off the California coast. They have steep gradients and dendritic tributaries.
Calorie. The amount of heat required to raise the temperature of one gram of water by one degree, from 15° to 16°C and under standard atmospheric pressure.
Carbonates. A family of minerals composed mainly of calcite ($CaCO_3$) and dolomite ($CaMg(CaCO_3)_2$).
Carbonation. Combinations of calcium or potassium from feldspar minerals with carbon dioxide in the ground water, to form calcium or potassium carbonates.
Catena. A group of soils, usually formed from the same parent material, but having developed dissimilar profiles and other characteristics because of differences in slope and relief. Catenas are common in broad valleys, extending perpendicularly to the axis of the valley from the valley floor to the adjacent highlands.
Chaparral. An evergreen broadleaf forest of Mediterranean climes consisting of brushy thickets of mixed deciduous and evergreen bushes and low trees. Called *garique* or *maquis* in France, *Macchie* in Italy, and chaparral in Spain whence the name was transferred to California. The popular use of the term includes scrub forest in many areas without Mediterranean climate.
Chernozems. Soils formed at the transition between humid and dry climates. These pedocals are richly supplied with plant nutrients, have a dark brown or black A horizon and lime accumulation at depth.
Chestnut soils. Soils of semi-arid regions. They have a light brown A horizon overlying a compact B horizon consisting of a lime horizon on top of a layer rich in gypsum.
Chinook. A warm, dry wind that has been adiabatically heated by blowing down the slope of a major mountain range. The term is particularly common in the Great Plains and Pacific Northwest. The same phenomenon is called a *Föhn* in Switzerland, Austria and Germany.
Cinder cone. A conical hill built up by the ash and cinders ejected from a volcano.
Circle of illumination. The great circle that divides the sunlit half of the earth from the dark half.
Cirques. Deep, hollow amphitheaters wtih steep sides carved out of the sides of mountains by the headward part of glaciers. The broad floors often contain tarn lakes.
Clastic sedimentary rock. A rock, such as sandstone, developed from the breakup and transportation of preexisting rocks.
Closed system. A natural or artificial arrangement of processes and their consequences that does not exchange mass or energy outside the system.
Col. A high pass in a mountain range; the divide between two drainage systems.
Composite volcanoes. See **Stratovolcanoes.**
Condensation. The change in state of atmospheric water vapor from a gas to a liquid. Condensation requires a surface or nucleus on which the water vapor can condense. Dew represents the result of condensation directly on the surface of the earth; precipitation results from condensation on particles in the atmosphere called *nuclei of condensation*.
Conditional instability. The condition of an air mass whose lapse rate lies between 3.0° and 5.5°F/1,000 feet (0.55° and 1°C/100m). Such an air mass is unstable if saturated, but stable otherwise.
Conduction. The transfer of heat in a body by contact from one molecule to another.
Conformality. In a map projection, the property of making the angle between any two lines on the flat map equal to the angle between corresponding arcs of great circles on the sphere, thus showing directions correctly. Compare **Equivalence.**
Coniferous. Applied to cone-bearing trees carrying needle-shaped leaves. Also used to designate a forest of such trees. Most coniferous trees are evergreen, but certain larches are deciduous conifers.
Continental ice sheets. Large masses of ice, several thousand feet thick and extending over thousands of square miles, covering all of the surface except for the very highest peaks. Antarctica is the only presently existing continental ice sheet, but northeastern North America and northwestern Europe were several times overrun by huge ice sheets during the glacial periods of the Pleistocene.
Continental shelf. The portion of a continental block that is covered by the ocean. It extends to variable widths but usually terminates, in the upper edge of the continental slope, at a depth of about 500 feet below the level of the sea.
Convection. The transfer of heat by turbulent movement of a liquid or gas.

Convectional precipitation. Precipitation that is due to the convective ascendance of air in regions of strong surface heating and unstable lapse rates. See **Cyclonic** and **Orographic precipitation.**

Convergence. When applied to air movement, convergence refers to any differences in wind velocity that bring about a local accumulation of air at the surface and hence an upward movement of air. The Intertropical Convergence Zone and the Polar Front are two places where convergence and associated precipitation occur frequently.

Cooling by advection. The cooling from below which occurs whenever a relatively warm air mass passes over a colder surface. Advective cooling often produces fog but rarely results in significant precipitation.

Cordilleras. Systems of linear mountain ranges of considerable elevation and extent. The Rocky Mountains and the Andes are examples.

Coriolis force. The apparent horizontal deflection undergone by any freely moving body as seen from the earth's surface. In fact, it is the earth's surface that is rotating under the moving air, water, or other object, but we ascribe the deflection (clockwise in the Northern Hemisphere, counterclockwise in the Southern) to the object itself. The intensity of the deflection is at a maximum at the poles, and decreases to zero at the equator; it also declines with decreasing velocity of the moving particle.

Crenate bays. Bays in the form of rounded scallops. They usually occur along regular, cliffed shore lines.

Crevasses. Deep, vertical cracks in a glacier. They are formed where there is a sharp change in the slope of the underlying ground, as in the cirque area, or where the glacier makes a sharp turn.

Cuestas. Hills formed on gently tilted sedimentary structures capped by relatively resistant rock. The face of a cuesta is relatively steep because of the erosion of weaker layers below the caprock; the back of the cuesta consists of a gentle dip slope which corresponds to the top of the resistant layer.

Cyclogenesis. The formation of cyclones or local low-pressure systems. Cyclogenesis occurs wherever air converges and ascends, particularly along the Polar Front and in the Intertropical Convergence Zone.

Cyclones. Low-pressure centers at or near the surface of the earth, characterized by counterclockwise and inward moving winds in the Northern Hemisphere (clockwise in the Southern), ascending air, and often conditional or unstable lapse rates. Compare **Anticyclones.**

Cyclonic precipitation. Precipitation associated with cyclonic movement of air around a low-pressure center that occurs when polar and tropical air masses converge along a front. See **Convectional** and **Orographic precipitation.**

Deciduous. Applied to trees that shed all of their leaves (or needles) in a particular season of their annual growth cycle, usually in winter. Also applied to forests of such trees.

Declination of the sun. The latitude at which the noon sun is directly overhead on a particular day. On March 21 the sun is directly overhead at the equator and its declination is thus 0°.

Deflation. Wind erosion; the transport of weathered material through the action of the wind by surface creep, suspension, or saltation.

Degradation. That part of the gradational processes which tears down existing landforms by weathering and erosion.

Delta. Sediments deposited by a stream when its speed is checked by a larger river, a lake, or the ocean, gradually built up so that their surface lies above water in the form, roughly, of the Greek letter delta.

Deposition. The dropping in place of particles too large to be carried by air, water, or ice when the speed of the air or water begins to decrease or the ice is beginning to melt more rapidly than it can be supplied from its source of accumulation.

Desert soils. Poorly developed soils of arid climates; they lack organic matter, leaching is inconsequential, and the parent material is little changed by weathering.

Dew point. That temperature at which a parcel of air becomes saturated because of its reduced capacity to accommodate water vapor in gas form at lower temperatures.

Diastrophism. The process of deformation that produces landforms in the earth's crust by raising or lowering and by bending, breaking, or wrinkling the crust. Compare **Gradation, Vulcanism.**

Dikes. Wall-like bodies of igneous rock injected into vertical fissures. Compare **Sills.**

Divergence. When applied to the movement of air, divergence refers to any differences in wind velocity, which lead to dispersal of air at the surface and hence subsidence from aloft. The eastern parts of high-pressure centers are characterized by divergence and associated drought.

Doldrums. The zone of light, variable, but often squally winds of the Intertropical Convergence Zone.

Drainage basin. The area drained by any stream and all of its tributaries.

Drift. (1) The movement of the present continental blocks from their original consolidated position in one or two protocontinents, as in continental drift. (2) Unstratified or roughly stratified deposits of clay, sand, gravel, and boulders that have been transported by a glacier, as in glacial drift. (3) Ocean currents in the poleward parts of the major ocean basins, driven by the prevailing westerlies of these latitudes, as in west wind drift.

Drumlins. Elongated, spoon-shaped ridges less than a mile in length representing streamlined accumulations of till deposited by continental ice sheets; they often occur in clusters of several hundred, aligned in the direction of ice movement.

Dry adiabatic rate. See **Adiabatic rate.**

Dry sandy soils. Azonal soils with immature profiles resulting from the chemical stability of quartz sand grains.

Earthflows. Downslope flow of masses of water-soaked regolith. The earth slumps and slips at the top, thus creating a spoon-shaped scar upslope and an irregular bulge downslope.

Easterly waves. Waves in the pressure pattern moving from east to west, representing a poleward bulge of equatorial low pressure; divergence and clear weather precede the passage of the axes of the waves, convergence and rain showers occur to the west of the axes.

Ebb tides. Semidiurnal low tides. The corresponding high tides are called *flood tides.*

Ecosystem. An ecological community considered as a unit together with the nonliving factors of its environment.

Elongate bays. Deep, elongated bays that wind inland as narrow inlets.

Eluviation. Downward transport by water of suspended or dissolved material in the soil. Specifically, the movement out of the *A* horizon. Compare **Illuviation.**

Equatorial convergence precipitation. Precipitation produced by unstable lapse rates and upward movement caused by one of several factors creating convergence in the Intertropical Convergence Zone.

Equinoxes. Either of two positions of the earth with respect to the sun that results in the noon sun's being directly overhead at the equator, so that every place on the earth experiences twelve hours of daylight and twelve hours of darkness. Commonly, but inaccurately, used synonymously with the dates of these occurrences — *vernal equinox* on about March 21 and *autumnal equinox* on about September 23.

Equivalence. In a map projection, the property of making bounded regions on the flat map have the same areas as the corresponding regions on the sphere of development. Compare **Conformality.**

Erosion. The dislodgement or transportation of weathered material by wind, water, or ice.

Eskers. Long, narrow sinuous ridges of unsorted till deposited under a glacier or by the foundering of surface material when the glacier melts.

Eutrophication. The process whereby the upper portion of a water body becomes rich in dissolved plant nutrients but poor in oxygen.

Evaporation. The change of water from the liquid to the gaseous state. The rate of evaporation depends mainly on the temperature of the evaporating liquid.

Evergreen conifers. Needle-leaf trees whose seeds are carried on cones. Individual trees replace their needles continuously throughout the year, so that the trees and the forest they make up are evergreen.

Exfoliation. The breaking away and removal of the outer rock layers through removal of pressure or by thermal expansion and contraction.

Extratropical cyclone. A low-pressure center and convergent cyclonic winds formed by the outburst of a polar air mass along the polar front; also the associated warm and cold fronts and their respective condensation forms. See **Cyclonic precipitation.**

Extrusive rocks. Molten or only partly cooled igneous rocks that have been forced through the outer crust and flow out on the surface as lava or are blown out as volcanic ash or cinders. Compare **Intrusive rocks.**

Extrusive vulcanism. The process whereby extrusive rocks are formed at the surface. See **Extrusive rocks.**

Faulting. The breaking of the earth's crust and consequent vertical or horizontal dislocation of once contiguous materials on either side of the break. In a *normal fault*, tension in the earth's crust is relieved by the subsidence of a block of the crust, relative to another, along a fracture or fault. In a *reverse fault*, compression of the crust pushes one block above another along a zone of fracture.

Feldspars. A family of hard and light or medium-dark minerals. They contain silica, aluminum, and varying amounts of potassium, calcium and sodium.

Ferralites. Soil formed under hot and wet climates; it contains concretions of iron oxides and is low in clay and nitrogen because of strong leaching and chemical weathering.

Ferralization. The process whereby ferralites are formed.

Ferromagnesian minerals. A family of dark, easily weathered minerals. They are composed of iron, magnesium and silica in addition to calcium, potassium and sodium.

Firn. Glacial ice; the rounded granules of recrystalized snow that make up the surface of the upper part of a glacier. Also called *névé*.

Fjard coast. A noncliffed irregular coast characterized by an intricate shoreline with many long, branching bays, but without lagoons, usually associated with low hinterlands. Compare **Fjords.**

Fjords. Irregular cliffed coasts with elongated bays winding deeply inland in narrow inlets bordered by steep highlands. Compare **Fjard coast.**

Floccules. Small aggregates of clay particles formed in the presence of basic ions of calcium, magnesium or potassium.

Floodplains. The area covered by the alluvial debris deposited by a stream and its tributaries during flood stage.

Flood tides. Semidiurnal high tides. The corresponding low tides are called *ebb tides*.

Fringing reefs. A low ridge of coral built up parallel to the coast, often circling island atolls from which they are separated by a shallow lagoon.

Front. A line, or narrow zone, on the surface of the earth that represents the boundary between two distinct air masses. In a *cold front*, the cold air mass is underrunning the warm air in advance of it. In a *warm front*, the warm mass is impinging on the cold by overrunning it.

Frontal precipitation. Cyclonic precipitation along a front. See **Convectional, Cyclonic,** and **Orographic precipitation.**

Frost wedging. Cracking action of water which penetrates cracks in rocks and expands upon freezing.

Geographic factors. The descriptive elements of the landscape that also condition the rate at which basic processes operate to form the landscape. They include *climate, topography, rock type, rock structure, biota, soils,* and *time*.

Geostrophic wind. A wind whose speed and direction, parallel to the isobars, is determined by the balance of the pressure gradient and Coriolis forces; implies straight isobars (no centrifugal force) and no friction. Compare **Gradient wind.**

Glacial-lake plains. Flat plains that were once the bottoms of lakes fed by glacial melt at or near the ice front.

Glaciofluvial deposits. Material sorted by size and deposited by glacial melt waters.

Graben. A depressed block bounded by faults. A particularly long and continuous graben is called a *rift valley*. Compare **Horst.**

Gradient wind. A wind whose speed and direction, parallel to curved isobars, is determined by the balance of the pressure gradient, Coriolis, and centrifugal forces; implies curved isobars but no friction. Compare **Geostrophic wind.**

Gravity. The gravitational attraction by the earth's mass for bodies at or near the surface, as modified by the centrifugal force due to the earth's rotation.

Gray-brown podzolic soil. A soil developed in regions of transition between subarctic and humid continental climates and between northern coniferous and broadleaf deciduous vegetation. It has a gray-brown *A* horizon, moderately acid and leached, and some accumulation of iron and aluminum in the *B* horizon.

Greenhouse effect. The heating of the earth's surface brought about by the relative transparency of the atmosphere to incoming short-wave, solar radiation and its opacity to outgoing, long-wave, terrestrial radiation.

Halomorphic soils. One of the intrazonal soils; characterized by the predominant effect of the alkaline salt content of the parent material on the soil properties. *Solonchak* and *solonetz* soils belong to this group of soils.

Hanging valleys. Valleys formed by former tributaries to a main glacier. The smaller side glacier excavated a relatively shallow trough, so that present streams join the master stream by way of spectacular falls.

Heat domes. See **Urban heat islands.**

High-pressure center. See **Anticyclone**.
Hill. Landforms with moderate or steep slopes (slopes steeper than 6 percent), but with a local relief of less than 1,000 feet (300 m). Compare with **Mountain, Plain**.
Hogback ridges. Ridges formed by differential erosion in steeply dipping strata. Because of the steep dip the two sides of the hogback have nearly similar slopes in contrast to cuestas which have a gentle dip slope and a steep scarp.
Horns. Pyramidal peaks, glacially eroded into cusp-like faces. Many of the Alpine peaks of Switzerland bear the name, as in Matterhorn, Finsteraarhorn.
Horse latitudes. That part of the oceans, in approximately 30° latitude, covered by subtropical high-pressure centers. A zone of light, variable and weak winds.
Horst. A raised block bounded by faults on at least two sides; sometimes adjacent to a graben, and usually longest in the direction parallel to the faults. The Vosges and Black Forest are horsts on either side of the Rhine graben. Compare **Graben**.
Humus. The organic layer at the top of a soil composed of partially decomposed vegetable and animal matter; sometimes called the A_0 horizon.
Hurricanes. Tropical storms of intense winds (75 miles per hour or greater) centering on symmetrical centers of very low pressure, accompanied by heavy rain and thunderstorms. Also called *tropical cyclones*, *typhoons* in the Far East, and *willy-willies* in Australia.
Hydration. The adsorption of water by minerals through the adhesion of water molecules to the surfaces of mineral crystals.
Hydrologic cycle. The cycle whereby water leaves the surface of the earth, mainly from the oceans, by evaporation; condenses in the atmosphere; is returned to the surface by precipitation; and runs off the land back to the oceans.
Hydrolysis. Chemical weathering that involves the addition of the elements of water (hydrogen and oxygen) to the weathered material.
Hydrometeors. Any form of condensation in the atmosphere. They include clouds, fog, hail, rain and snow.
Hydromorphic soils. One of the intrazonal soils, characterized by the predominant effect of poor drainage on soil properties. Bog soils are an example.
Hydrophytes. Plants that thrive under conditions of plentiful soil moisture.
Hydrosphere. The watery part of the earth's crust. It includes all surface and subterranean water as well as the water vapor in the air.
Hygroscopic nuclei. Particles in the atmosphere with a strong affinity for water vapor so that they serve as a nuclei for condensation. Salt is mildly hygroscopic; silver iodide strongly so.
Igneous rock. Rock formed by the solidification, upon cooling, of molten magma. Compare **Metamorphic rock, Sedimentary rock**.
Illuviation. Accumulation of suspended or dissolved material in one soil horizon as a result of eluviation from another. Often accumulation in the *B* horizon. Compare **Eluviation**.
Insolation. Solar radiation or, more specifically, that portion of the solar radiation that is received at the surface of the earth.
Instability. The existence, in the troposphere, of a steep lapse rate (greater than 5.5°F/1,000 feet or 1°C/100m) that favors vertical movement of air and thus allows adiabatic cooling and saturation of the air mass and possibly condensation. See **Stability**.
Interfluves. The higher interstream areas that act as local divides and are being reduced by erosion.

Interfluvial ridge. A sharply defined interfluve.
Intermittent streams. Streams that are supplied with enough moisture to allow for stream flow only during a particular season of the year or even only once in a period of years.
Intertropical Convergence Zone (ITC). The zone of convergence between the northeast and southeast trades. Its average location is slightly north of the equator, but it moves northward in the Northern Hemisphere summer and southward in winter. It is the locus of moist and unstable air, thus a region of heavy year-round precipitation.
Intrazonal soils. One of the three basic soil orders, possessing profiles and other characteristics that can be ascribed less to climate and vegetation than to the predominance of one of the other soil-forming factors (often parent material). Compare **Azonal soils, Zonal soils**.
Intrusive rocks. Molten magmatic rock of coarse crystalline texture that has cooled below the surface and appears at the surface by subsequent upthrust or by the erosion of overlying materials. Compare **Extrusive rocks**.
Intrusive vulcanism. The process whereby intrusive rocks are formed. See **Intrusive rocks**.
Inversion. A temperature inversion refers to a localized but common inversion of the normal lapse rate, so that in a given layer, often at or near the ground, the temperature aloft is higher than it is near the ground. See **Lapse rate**.
Isobars. See **Isolines**.
Isobath. See **Isolines**.
Isohyets. See **Isolines**.
Isohypse. See **Isolines**.
Isolines. Lines joining points on the surface of the earth (or on some other datum plane, such as sea level) that have the same value for some quantifiable and continuous phenomenon. Thus, *isobars* are lines joining points of equal barometric pressure; an *isobath* joins points of equal depth below sea level; *isohyets* are lines of equal precipitation; *isohypses* are lines of equal elevation above sea level (often called contour lines); *isotachs* are lines of equal wind speeds; and *isotherms* are lines of constant temperature.
Isotachs. See **Isolines**.
Isotherms. See **Isolines**.
Jet Stream. Sinuous stream of rapidly moving west winds near the tropopause, often exceeding speeds of 250 miles (400 km) per hour, particularly in winter. There are often several jets in each hemisphere which have been given the names of polar jet, subtropical jet, etc.
Kames. Stratified drift deposited along the ice margin. Kames are mound-like hillocks of sand and gravel representing former deltas and alluvial fans built out from the flank of the ice. Kame terraces are accumulations of sand and gravel deposited by streams between the glacier and the adjacent valley wall.
Kame terraces. See **Kames**.
Karst. A limestone region or its landforms, characterized by the predominance of solution features such as sink holes and caverns and by the resultant subsurface drainage and associated relief forms.
Kettle holes. The holes in the smooth surface of an outwash plain that were created when blocks of ice, covered by drift, eventually melted.
Laccoliths. Masses of igneous material whose intrusions between originally horizontal sedimentary beds produce dome-like bulging of the overlying strata.
Landslide. Spectacular form of mass wastage in which large amounts of surface material slide rapidly downslope.

Langley. A gram calorie per square centimeter. See **Calorie.**
Lapse rate. The difference between the values of any two meteorological or climatic elements at different elevations above the surface, divided by the difference in elevation. Often employed for temperature and dew point. The normal lapse rate of temperature is about 3.5°F/1,000 feet (0.65°C/100m), implying that, on the average, there is a temperature drop of 3.5°F between the top and bottom of a layer of air 1,000 feet thick or a difference of 0.65°C between the top and bottom of a layer 100m thick. The lapse rate decreases aloft and becomes negligible at the tropopause.
Latent heat. The thermal energy in some heat exchanges that are not accompanied by a change in sensible temperature, as in the latent heat of evaporation or condensation, in which heat is transferred from the evaporating surface to the condensing nuclei.
Laterites. Soils of the monsoon climates formed in areas of poor drainage and seasonal alternation of low and high water tables; the *A* horizon is reddish, porous, slaggy, and extremely rich in iron.
Laterization. The process whereby laterites are formed.
Latitude. The angular distance of a point on the earth's surface north or south of the equator.
Lava. Fluid magma issuing from volcanic craters or other fissures in the earth's crust. An extrusive rock that cools rapidly after it is exposed.
Lava domes. See **Shield volcano.**
Leaching. The process whereby soil material is dissolved in percolating water and thus transported downward by gravitational pull, or occasionally upward by capillary action.
Levees. Low banks of coarse alluvium deposited on either side of a stream during flood. Also called *natural levees* in contrast to the artificial levees erected for flood protection.
Lithification. The conversion of unconsolidated sediments into solid rock.
Lithosols. One of the azonal soils having an imperfectly developed profile because of steep slopes or excessively resistant parent material.
Lithosphere. The outer crust of the earth, composed of granitic continental blocks (the sial) and basaltic ocean basins (the sima).
Local relief. The difference in elevation between ridge and adjacent valley or within a specified area, as in the local relief of a square mile. It is a measure of the ruggedness of the terrain and is often used as one of the criteria for distinguishing various kinds of landforms.
Loess. An unstratified deposit of calcareous loam transported by wind or water and forming deep layers of buff or yellowish soils.
Longitude. The angular distance of a point on the earth's surface east or west of some standard meridian, usually that of Greenwich.
Longitudinal dune. A long dune whose crest line is parallel to the predominant wind direction. The seif dune is a prominent example.
Low-pressure center. See **Cyclone.**
Magma. The molten or pasty rock in the lower part of the earth's crust, or immediately beneath it, from which igneous rock is derived by cooling and crystallization.
Mantle. The portion of the earth between the core and the crust; about 1,800 miles (2900 km) thick, it consists of dense olivine minerals in a solid state. Not to be confused with mantle rock, which refers to the weathered regolith of the earth's surface.

Mantle rock. See **Regolith.**
Masses. A category of tectonic structural units. Complex surfaces consisting of very old rock partly covered by a younger sedimentary cover. Compare **Basins, Tables.**
Mass movement. See **Mass wastage.**
Mass wastage. Downslope migration of weathered material by direct gravitational pull; includes slow migrating such as solifluction and soil creep, as well as the more rapid movement of rockfall, landslides, and earthflow. Sometimes called *mass movement.*
Meander. A tightly winding or smoothly looping turn in the channel of a stream.
Mesophytes. Plants adapted to moderate moisture conditions, in contrast to the drought-resistant xerophytes and the moisture-loving hydrophytes. Compare **Xerophytes, Hydrophytes.**
Metamorphic rock. Rock that has been hardened and usually rendered more crystalline by heat, pressure, or the addition of a new mineral. Compare **Igneous** and **Sedimentary rock.**
Millibar. Unit of pressure in the metric system. A millibar (mb) is equivalent to a pressure exerted by 1000 dynes over one square centimeter. (A dyne is the force needed to give a mass of one gram an acceleration of one centimeter per second per second.) The average sea-level pressure is 1013 mb.
Mixing ratio. The amount of moisture in the air expressed as the weight, in grams, of the water vapor per kilogram of dry air.
Monsoon. Either of the prevailing seasonal winds of roughly opposite flow in South Asia, as in "the Southwest monsoon" or "the winter monsoon," or the precipitation patterns associated particularly with summer flow, as in "the monsoon rains," or the whole circulation pattern and its climatic manifestation, as in "the Indian monsoon." Often also attributed to similar seasonal alternations of wind in other areas, as in "the monsoon climate of Sierra Leone."
Moraine. Unsorted material, ranging in size from clay to boulders, carried along by a glacier, and deposited in unstratified banks in the form of terminal, medial, lateral, recessional, or ground moraine.
Mountain. A landform with moderate or steep slopes (slopes greater than 6 percent), and with a local relief of more than 1,000 feet (300 m). Compare **Hill, Plain.**
Natural levees. See **Levees.**
Neap tides. Relatively low, high tides and correspondingly high, low tides that occur when the moon and the sun's gravitational pull is at right angles. Compare **Spring tides.**
Needle leaf. See **Evergreen conifers.**
Negative feedback. An arrangement of processes that are self-regulatory so that any change in one item results in other changes that bring the system back into equilibrium.
Noncalcic brown soils. Soils related to subtropical red soils and occurring downslope from them in Mediterranean climates. The *A* horizon is reddish-brown and overlies a red, compact *B* horizon. The soil profile has been leached of carbonates.
Normal fault. See **Fault.**
Occlusion. The complete displacement aloft of warm air that occurs in an extratropical cyclone when the cold front catches up to the warm front at the surface.
Open system. A natural or artificial system that is a sub-system of a larger arrangement of forces and processes and that exchanges mass and energy with the larger entity.
Orographic precipitation. Precipitation associated with the passage of air over a mountain chain or other highland. Com-

pare **Convectional** and **Cyclonic precipitation.**
Orthodrome. The arc of a great circle joining two points on the sphere representing the shortest distance between them; orthodromes appear as straight lines on the gnomonic projection.
Outwash plains. Plains of stratified deposits of glacial till laid down by glacial meltwater; also called *glaciofluvial drift*.
Oxidation. In weathering processes, the union of oxygen with other elements or minerals to form oxides which are often softer and bulkier than the original minerals.
Paleozoic mountains. Mountains formed during the Paleozoic orogenies. Caledonian mountains were formed about 425,000,000 years ago; Hercynian mountains about 230,000,000 years ago. The two systems, though dissimilar, are treated as one on the tectonic map of the world (back endpaper).
Parent material. The regolith, or weathered rock, that serves as the raw material for the formation of residual soils. The soil-forming processes change the top of the regolith into soil horizons which overlie an unchanged horizon that retains the characteristics of the parent material.
Pedalfers. One of the two suborders of zonal soils, pedalfers are formed in humid climates and are characterized by leaching of soluble bases and concentrations of Aluminum (Al) and iron (Fe) oxides in their profile.
Pediments. Gently sloping rock surfaces intermediate in slope and position between mountain front and valley, covered by a thin veneer of sand and gravel; common in semiarid and arid areas, particularly in basin and range country.
Pedocals. One of the two suborders of zonal soils, pedocals are formed in dry climates and are characterized by the presence of a calcium (Ca) horizon somewhere in the profile.
Peneplain. A nearly horizontal surface which will form by gradation across various types of rock and rock structure if erosion and deposition are allowed to proceed without disturbance by tectonic movement.
Perennial streams. Streams having a sufficient supply of water that they flow all year despite periods of larger and smaller than average flow.
Permafrost. Permanently frozen subsoil of arctic and subarctic climates.
Piedmont glaciers. Glaciers, intermediate between mountain valley glaciers and continental ice sheets, that form when mountain glaciers flow out of their valleys and coalesce to form broad foothill ice sheets.
Plain. A landform having more than half its surface area in gentle slopes (less than 6 percent) and with a local relief of less than 300 feet (100 m). Compare, **Hill, Mountain.**
Planosols. One of the intrazonal soils, characterized by a profile with a compacted accumulation of leached material in the B horizon that results from flat terrain and lack of erosion rather than from the climate and vegetation under which the soil is formed.
Playa. The dry lake floor of a desert basin.
Plucking. The action of glacial ice whereby rock is plucked from its location and incorporated into the flowing glacier.
Podzols. Soil formed under cold and wet climates, having clearly definable horizons consisting of an acidic layer of partly decomposed organic matter, a strongly leached A horizon, and an accumulation of iron and clay in the B horizon.
Podzolization. The process whereby podzols are formed.
Polar Front. The zone, at the surface (and aloft), along which polar and tropical air masses meet to create extratropical cyclones.

Positive feedback. An arrangement of processes that are unstable in that a change in one item will change other items in such a way as to intensify the original change. Nuclear explosions are examples of positive feedback.
Potential evapotranspiration. The amount of evaporation from the surface and transpiration from plants that would occur if the availability of moisture were not a limiting factor.
Prairie soils. Black soil, formed along the boundary between humid and semiarid climates under tall grass vegetation; usually deep and rich in organic matter, but without sharply defined horizons.
Precambrian shields. Extensive surfaces, low and nearly level, consisting of granites and metamorphosed sediments which represent the eroded remnants of the earliest orogenies.
Precipitation. Any form of water that is precipitated out of the atmosphere and reaches the surface of the earth. Snow and rain are the principal examples.
Pressure gradient force. The force exerted on any particle by the difference in pressure between one side of the particle and another. It can be thought of as a push from the direction of high pressure and a pull from the opposite direction.
Radiation. The emission of radiant energy in particle or wave form, requiring no intervening medium for transmission.
Radiational cooling. Cooling of a surface by greater outgoing radiation than incoming radiation. Particularly common in winter, at night, and when skies are clear.
Rain shadow. A position on the lee side of a mountain or highland that results in low precipitation particularly when compared to the abundant rain or snowfall on the exposed windward side of the barrier.
Reg. A natural mosaic of closely packed gravel and boulders resulting from the removal of all finer material by the wind in arid regions. Often called *desert pavement* or *desert armor*.
Regolith. The end product of the weathering process — a layer of decomposed rock and soil; and also called *mantle rock*.
Relative humidity. The amount of water vapor in the air; expressed as a percentage of the maximum amount of water that can be accommodated, in its gaseous state, at a given temperature.
Rendzinas. Intrazonal soils derived from limestone parent materials that are brown to gray-black and contain free calcium carbonate.
Resultant wind. The net flow of air at a particular place; arrived at by subtracting southerly flow from northerly, westerly from easterly, etc.
Reverse fault. See **Fault.**
Ria bays. Relatively shallow cone-shaped bays that penetrate deeply into the land with cliffed or noncliffed irregular shores.
Rill flow. Intermittent movement of water that is not clearly channeled and flows in a multitude of ill-defined and ephemeral channels or rills or as a sheet of water. Also called *sheet wash*.
Roches moutonnées. Rocks overrun by a continental ice sheet that have been rounded and smoothed by scour on the side exposed to the ice advance and roughened by plucking on the opposite side.
Rockfall. Downslope movement of loose material on steep slopes.
Salinity. The degree of salt concentration in a liquid. The average salinity of the oceans is between 35 and 37 parts of dissolved salts per 1000 parts of water.
Salinization. The process whereby salt accumulates near the surface of a soil in arid regions without good drainage.

Saltation. The downstream movement of large particles too heavy to be carried along in suspension but which bounce along the floor of the stream channel.

Saturated adiabatic rate. See **Adiabatic rate.**

Saturation mixing ratio. The particular mixing ratio, or water vapor content, of an air mass that is equal to the maximum possible content of water vapor by the air mass at its given temperature and pressure. Any further increase in water vapor will result in supersaturation and usually in condensation of the excess moisture.

Saturation vapor pressure. Same as saturation mixing ratio, except that the measure of moisture content is vapor pressure rather than the mixing ratio.

Savanna. A number of natural phenomena associated with essentially flat areas of the wet and dry tropics, including their predominantly grassy vegetation and climate marked by strong seasonal contrasts in precipitation.

Seamounts. Flat-topped, submarine mountain peaks resembling truncated volcanic cones rising from the ocean floor.

Sedimentary rock. Weathered fragments of other rocks deposited in originally horizontal strata by streams, winds, glaciers, and most commonly by shallow continental seas; sedimentary rock also includes the remains of organic material or products of chemical precipitation that are deposited in horizontal layers.

Seif. A large longitudinal dune, several hundred feet high and many miles long, whose crest is parallel to the wind direction but which has slip faces on alternate sides of the ridge as a result of periodic crosswinds.

Selva. The tropical rainforest or evergreen broadleaf tropical forest.

Semideciduous. Applied to tropical forests in which leaffall is limited to certain species of trees and does not mark the end of an active season or the beginning of a dormant one as in temperate forests.

Sheet wash. See **Rill flow.**

Shelf notches. Large steep-walled slots that have been eroded into the outer edge of the continental shelf for a few miles and extend down the continental slope to great depths.

Shield volcanoes. Huge accumulations of dark highly fluid basaltic lava that has spread gently from a central vent or fissure. The high but gently-sided volcanic peaks are also called *lava domes.*

Sial. The relatively light granitic masses rich in silicon (Si) and aluminum (A1) that make up the continental blocks. Compare **Sima.**

Sierozems. Semidesert soil with a rudimentary profile as a consequence of slight leaching; contains little humus, but lime is dispersed throughout the profile.

Silica. A combination of silicon and oxygen (SiO_2) which is the basis of one of the important families of durable minerals. Quartz is the major example.

Sills. A horizontal finger of igneous material intruded between sediments or other horizontal structures, usually as the offshoot of a vertical dike. See **Dikes.**

Sima. The relatively dense basaltic layer rich in silicon (Si) and magnesium (Ma) that makes up the oceanic basins and underlies the continental blocks. Compare **Sial.**

Sky radiation. That part of the solar radiation that does not reach the earth's surface directly, but is scattered downward by particles in the atmosphere. Also called *diffuse or scattered radiation.*

Slump. A form of landslide in which rock breaks along the edge of a cliff and slips downhill with a rotary movement (as seen from the side).

Soil creep. Slow, downslope movement of the entire regolith.

Solar constant. The amount of radiant solar energy that reaches the outer edge of the atmosphere at the average distance of the earth from the sun; it is equal to 2 langleys per minute.

Solar radiation. Radiation from the sun. See **Radiation.**

Solifluction. Earthflow in Arctic regions where the subsoil is permanently frozen.

Solstice. Either of the two positions of the earth with respect to the sun that result in the sun at noon being directly overhead at one of the tropics. Commonly, but inaccurately, used synonymously with the dates of these occurrences — *summer solstice* on about June 22 and *winter solstice* on about December 22.

Solution. The process whereby certain minerals are dissolved and homogeneously mixed in water.

Source region. The area over which an air mass originates and from which it acquires its characteristics.

Specific heat. The number of calories of heat needed to raise the temperature of 1 gram of a substance by 1°C; by the definition of a calorie, the specific heat of water is 1°. See **Calorie.**

Specific humidity. A measure of water vapor content; the weight of water vapor per weight of air (including the vapor). It differs little from the mixing ratio.

Spring tides. See **Neap tides.**

Stability. With reference to a layer of the troposphere, stability refers to the existence of a gentle lapse rate (less than 3°F/1,000 feet (0.55°C/100m) and including especially inversions) which inhibits the vertical movement of air and thus reduces the likelihood of precipitation. Compare **Instability.**

Stratified drift. Glacial debris that has been sorted and transported by running glacial meltwater.

Stratosphere. The part of the atmosphere lying above the troposphere, whose temperature is unaffected by daily or seasonal temperature changes emanating from the surface of the earth; it also differs from the troposphere in that temperature hardly changes with elevation, at least in its lower part. See **Troposphere.**

Stratovolcanoes. Volcanoes composed of alternate out-pourings of lava and ash or cinders, giving them slopes that are intermediate between the gentle sides of shield volcanoes and the steep flanks of a cinder cone; also called *composite volcanoes.*

Striations. Gouges or grooves cut into underlying rock by the debris incorporated in moving continental ice sheets.

Subsidence. In the atmosphere, refers to a massive down draft of air, often associated with a high-pressure center and outflowing of air at the surface, and resulting in stable lapse rates and drought.

Subtropical red and yellow soils. Soil formed in humid mesothermal climates where both podzolization and ferralization are operative, thus giving the soil properties intermediate between the high acidity and organic content of podzols and the lack of humus and iron concentrations of ferralites.

Summer solstice. See **Solstice.**

Surface creep. Slow downwind movement of sand under the constant impact of bouncing sand grains impelled by the wind.

Suspension. The state of particles when they are not dissolved but are mixed in air or water by its turbulent movement.

Swell. The smooth, relatively long, even-crested wave front that is produced as irregular individual waves approach the shore.

Synclines. Downfolds of the earth's crust, strong enough to deform stratified rocks into a trough-like structure, but gentle enough to leave the layers unbroken. Compare **Anticlines.**

Tables. A category of large-scale tectonic units, involving nearly horizontal surfaces partly stripped of their original sedimentary cover. Compare **Basins, Masses.**

Taiga. The northern coniferous forest of Eurasia, dominated by larch, spruce, fir, and pine in relatively sparse stands of low, thin trees.

Tectonic. Of or relating to the forces that create landforms by deforming the earth's crust (*disastrophism*) or by adding magmatic material to it (*vulcanism*).

Tectonic forces. See **Tectonic.**

Tectonic processes. See **Tectonic.**

Tertiary mountains. See **Alpine mountains.**

Thermocline. Any steep gradient of temperature in water bodies in summer between the warm surface isothermal layer and the colder isothermal layer at depth.

Tierra rojas. A red soil with a simple profile formed on resistant limestone in Mediterranean, and humid tropical and subtropical climates.

Till. The unstratified drift consisting of a heterogeneous mixture of rock fragments of all sizes, deposited by a glacier in the form of *moraines*, *till plains*, *drumlins*, and the likes.

Till plain. See **Till.**

Tornado. A violent destructive cyclonic wind of small size and narrow path, but which is accompanied by such a rapid drop in atmospheric pressure that buildings can be lifted and exploded by the intense pressure gradients created.

Traction. The state of particles being dragged, rolled or bounced along a stream bed by the current.

Trade winds. A system of planetary winds, particularly well marked over the oceans, that blow from the subtropical anticyclones of each hemisphere toward the equator — hence the name *northeast trades* and *southeast trades*.

Transverse dune. Steep-sided dunes in the shape of waves whose crests are at right angles to the direction of the wind which is generally moderate.

Tropical cyclones. Storms of weak or moderate intensity forming in the Intertropical Convergence Zone and producing substantial precipitation. This term is also applied to hurricanes in the Bay of Bengal and Arabian Sea.

Tropical red loams. Soils formed in savanna climates under conditions of good drainage. They have deep profiles with no clearly distinguishable *A* and *B* horizons.

Tropopause. The surface of demarcation between the troposphere and the atmosphere, where the normal decline of temperature with elevation found in the lower atmosphere gives way to the roughly isothermal conditions of the stratosphere; its average elevation is about eleven miles (18 km) over the equator and seven miles (11 km) over the poles, but its height fluctuates periodically and seasonally.

Troposphere. The lowest part of the atmosphere, below the tropopause, in which the temperature generally declines upward, convection is active, clouds form, and the effect of the surface can be observed in differences in moisture content and seasonal temperature changes; the part of the atmosphere in which weather occurs. Compare **Stratosphere.**

Turbidity. State of being muddy, impure, or unclear; the dirtiness of the atmosphere, its content of dust and other pollutants that decrease its transparency.

Turbidity currents. Submarine currents that consist of water which is denser than its surroundings by virtue of its greater load of sediment and which moves down the continental slope.

Ubac. See **Adret.**

Upwelling. The upward movement, near the shore, of cold subsurface water which replaces surface water displaced offshore by the Coriolis effect.

Urban dust dome. A dome-like zone of air in which dust and other pollutants produced in urban agglomerations are concentrated and sometimes trapped by the inversion which often develops as a result of the dust accumulation.

Urban heat island. The urban island over which temperatures are several degrees warmer than in the surrounding countryside because of the heating created by the physical character of the city and by its concentrated use of energy. Also called *heat domes*.

Valley Glaciers. Glaciers supplied by snow accumulation in mountain cirques and whose course extends down relatively narrow and deep U-shaped mountain valleys.

Vapor pressure. The partial pressure exerted by the water vapor present in the atmosphere.

Ventifacts. Cobbles whose sides have acquired an angular quality because of sand abrasion.

Volcanic ejecta. Cinder, ash, and other material forcibly ejected from a volcano.

Vulcanism. The landforming process whereby magmatic material is either intruded into the earth's crust or extruded at the surface. Compare **Diastrophism.**

Water gaps. Cuts, by narrow stream valleys, in linear ridges.

Water need. (1) The amount of precipitation necessary to create a humid climate, given the local circumstances of temperature and seasonality of rainfall. (2) Potential evapotranspiration.

Weathering. Physical disintegration and chemical decomposition of rock into regolith.

Wedging. See **Frost wedging.**

Westerlies. Prevailing west-to-east winds of the mid-latitudes, stronger in winter than summer and more consistent over the oceans than on land.

Wind gaps. Narrow gaps in linear ridges, abandoned by the streams that cut them during the course of uplift and erosion.

Wind rose. Graphic representation of strength, frequency and direction of winds at a particular place.

Winter solstice. See **Solstice.**

Xerophytes. Plants that are structurally adapted to dry climates, by having water-storing tissues, or waxy leaves that reduce evaporation, or other water-conserving mechanisms, including the ability to go through their reproductive life cycle in a very short period during which water is available.

Yardangs. Long, shallow grooves and sharp-crested ridges sculptured on bare surfaces by steady wind erosion.

Zonal soils. The most widespread of the three basic orders of soils; they have properties that reflect primarily the influence of climate and vegetation; where neither relief, poor drainage, nor peculiar parent material distorts the gentle gradients of climate and vegetation; these zonal soils cover large areas in which their properties change only very gradually and are closely related to the zonal distribution of the climate and associated plant cover. Compare **Azonal** and **Intrazonal soils.**

Index

ablation, 219
abrasion:
    by glaciers, 224-**225**
    by streams, 204
    by wind, 234
absolute humidity, 67
absolute zero, 23
absolute temperature scale, **23**
Adelaide, **159,**160
adiabatic process:
    cooling, 75, 76-79, 107
    dry adiabatic rate, 77, 79, **81,** 82-83
    heating, **78**
    Poisson's education, 76
    wet adiabatic rate, 77, 79, **81,** 83
Adirondack Mountains, 268, **270**
adret, 14, 310
Adria, Italy, **214**
Adriatic coast, 257
advection:
    advective cooling, 75, 76
    advective fog, 76, 85
Africa, **73,** 98 (see also northern, eastern, western, central, and southern Africa and individual countries)
Agadir, Morocco, **144, 145**
aggradation, 191
Agulhas Current (see Mozambique Current)
air masses:
    arctic, 91, **92**
    characteristics, 89-93
    classification, 89-91
    continental polar, 91, **92,** 93, 96, 103, 156, 160, 166
    continental tropical, **92,** 93, 103, 160
    definition, 89
    equatorial, 91, **92,** 93, 98
    importance to climate, 31-33
    maritime polar, 91, **92,** 93, 103, 158, 159, 160
    maritime tropical, **92,** 93, 96, 98, 99, 103, 156, 160, 166
    Northern Hemisphere, 91-93
    source regions, 89, 90, **92**
    summer, 93
    winter, 91-93
air pollution, 170-171, 357-359

Alabama, 259, 267, 288, 349
Alamosa, Colorado, **38,** 39
Alaska: 303, 372
    climate, 126, **141**
    coastline, 285
    glaciation, 218, **219,** 220, 221, **225, 228,** 231
    insolation, **28,** 29
    landforms, 273
    precipitation, 109
    temperature, **28,** 29, 39
albedo:
    effect on temperature, 29
    heat balance, **12**
    of snow and ice, 29, 138
    of vegetation, 168
Albers' equivalent conic projection, 383
Alberta, 145, **146**
Aletsch Glacier, Switzerland, **229**
Aleutian Islands, 51, 93, 114, **187,** 372
Aleutian low, 57, 62
Alexandria, Egypt, 128
Algiers, **118**
Alice Springs, Australia, 129
alkalinization of soils (see desalinization of soils)
alkali soils, 336
Allegheny: Front, 267
    Mountains, 265, **267,** 372
alluvial fans, **210-211**
alluvial soils, **346,** 348
alluvium, 210
alpine glaciers (see valley glaciers)
alpine mountains, 188, 189
Alps, **385,** 253
    climate, 78, 139, **140**
    glaciation, 224, **229,** 273, **274**
    orogeny, 188, 189, 224
    vegetation, 319
altimeter, 51
altitude of the sun, 9-10, 12, **13,** 370
altocumulus, 85, **86**
altostratus, 85, **86**
Amazon:
    Basin, 15, 72, 109, 148, 150, 250, 256, 314
    River, 250, 287
Amundsen-Scott Station, 127
analemma, 10, 370

Andes:
    climate, 31, 138
    glaciation, 221, 273, 274
    landforms, 198, 271
    precipitation, 109, 113, 114
Angara Shield, 189, 267, back endpaper
Angmagssalik, Greenland, **136,** 137
angular distance, **369,** 380
Ann Arbor, Michigan, 163, **164, 165**-166
Antarctica, 290, 299, 303
    climate, 57, 127, 134, 135, 138, 139, **140**
    glaciation, 217, 218, 219, 221
    landforms, 189
    precipitation, 113, 114, 115
Antarctic Bottom Water, 299, 300
Antarctic Intermediate Water, 299, 300
Antarctic Ocean:
    climate, 134, 135, 137
    currents, 300
    salinity, **293**
    temperature, 292, 296
    water masses, **302,** 303
Antarctic Oceanic Water, 301, **302,** 303
Antarctic Shield, 189
anticline, **185,** 186, **259**
    anticlinal axes, 263, **264**
    relationship to highlands, 263
anticyclone, **49,** 60, 62, 92
Antilles, **283**
antipode, 368
Appalachian Ranges, 189, 260, 264-65
Apulia, Italy, 258
Arabia, 113, 236, 254, **255**
Arabian Sea, 35
Arafura Sea, 304
Aral Sea, 254
Archangel, U.S.S.R., 132
Arctic Circle, 13, 14, 134
Arctic Continental Shelf Water, 303
arctic front, 134
Arctic Ocean, 6, 35
    air masses, 91, 93, 137
    climate, 134, 135, 137, 223
    condensation, 75

continental shelf, 287
    glaciation, **222,** 223, 224
    precipitation, 114, 115
    pressure, 57
    salinity, 292
    temperature, 292, 296
    water masses, 303
Arctic Oceanic Water, **302,** 303
arctic sea smoke, 75
arête, **227, 228,** 229
Argentina:
    climate, 147, 156, 158
    continental shelf, 303
    landforms, 237, 253, 259, 288
    soils, 347
    vegetation, 322, 323
Arica, Chile, 113, **118**
Arizona, 185, 239, 311, **323**
Arkansas, 265, 267, 372, **373**
Asia: 56, 57, 62, 108, (see also central, eastern, and southern Asia, Siberia, central asiatic Russia, and individual countries)
Asiatic low, 57
Astrakhan, U.S.S.R., **118**
Aswan, Egypt:
    climate, 142, **143, 145,** 160
    evaporation, 72
    heat balance, 19, **20,** 21
Atscama Desert, 113, 145, 254
Athens (Athinai), 158, **159,** 160
Atlantic Coastal Plain, 2, 247, 256, 286, 319, 334, 343, 348, 354
Atlantic Ocean:
    air masses, 91, 92, 93, 97, 137
    climate, 147
    continental shelf, **182,** 287
    currents, 6, 34, 35, 300, 301, 304
    glaciation, 223
    insolation, 15
    Mid-Atlantic Ridge, 182, 290
    precipitation, 108, 109, 113, 114, 115
    pressure, 57
    salinity, 292
    subsidence, 299
    temperature, 30-**31,** 36, 40, 100, 296
    tides, 298
    water masses, 301
    winds, 60, 62, **63,** 64

Index

Atlas Range, 265
atmosphere:
    composition, 10-11
    depletion of radiation by, 10ff, **12**
    general circulation, 64-65
    greenhouse effect, 15, 19
    heat budget, 11, **12**
    pressure, 41ff
    radiation by, 15, 18
    standard atmosphere, 37
    temperature, **37**
    vertical divisions, 37
atmospheric pollution:
    carbon monoxide, 358
    hydrocarbons, 358
    oxides of nitrogen and sulfur, 358
    particulate, 358
    thermal, 357-358, **366**
atolls, 280, **281**
Atrato Valley, Colombia, 109
Australia, 304
    air masses, 98
    climate, 97, 151, **152**, 154, 156, 158, 160
    continental shelf, 287
    hurricanes, 100
    landforms, 189, 236, 249, 254, 259, 260, 261
    precipitation, 113
    soils, 335, 343, 347
    vegetation, 315, 316, 318, 322, 323
Austria, 32, **33**
azimuth, 379, 380
azimuthal equal area projection, 380
azimuthal equidistant projection, 380
azimuthal projections, 375, **379-380**
azonal soils, 348
    alluvial soils, **346**, 348
    dry sandy soils, 348
    lithosols, 142, 146, 336, **346**, 348
Azores, 271
Azores high, 57, 62
backswamp, 211
badlands, 268, **269**
Baja California, 92, 113
bald cypress, 318, 319
Balkans, 274
Baltic Sea, 286, 296, 304
Baltic Shield, 189, back endpaper
Baltimore, 359
baobab, **317**
barchan dunes, 236-**237**
barometer, 41-**42**
barrier island, 283, **284**, 286
barrier reef, **280**
Barrow, Alaska, 126

bars:
    offshore, 283, 286
    point, 212
basalt, 177, **178**
base exchange in soils, 329
base line, 372-**373**
basin and range plains, 243, 244, 258-259
basins, 189, back endpaper
batholith, **187**, 271, **273**
Bay of Bengal, 35, 100
beach ridges, **277**, **286**
bedding planes, **179**, **269**
Benguela Current, **34**, 35, 113, 159, 300
Bergen, Norway, **118**
Bering Sea, 134, 287, 288
berm, **277**
Bernese Alps, 140, **299**, **273**, **274**
biologic oxygen demand (BOD), 355
Black Hills, South Dakota, 265
Black Sea, 304
blizzard, 140, 163
block mountain, 185
blowouts by wind, 235, 236, 237, **324**
Blue Ridge Mountains, 265, 268, 269
Bluff Springs, Texas, **185**
bog soils, (see hydromorphic soils)
bolson, **258**
Bombay, 118
Bonne's projection, 382
Boston, 169, 356
Boyle's Law, 43
Brazil:
    climate, 147, 148, 154
    evaporation, 72
    heat balance, 19, 20, 21
    insolation, 15
    landforms, 187, 193
    precipitation, 109, 112
    soils, 343
    vegetation, 316, 321, 322
Brazil Current, **34**, 300
Brazilian Planalto, 154
Brienz Range, **273**
British Columbia:
    climate, 109, 160, **162**
    coastline, **285**
    vegetation, 319
broadleaf forests:
    deciduous, 2, 156, 163, 166, 307-309, 318, **319**
    evergreen, 156, 160, 315-316, 318
    Mediterranean evergreen, 318
Bronson Deep, 290
brown steppe soils, 143, 156, **346**, 347
Brownian movement, 354
Buenos Aires, **118**

Burma, 62, 112, 151, 314
buttes, 260
Buys Ballot's law, 49
caatinga, 154, 155, 316
cactus, 310, 311, **323**, 324
Cairns, Australia, 151, **152**, 154
Cairo, Illinois, 214
calcification, 335
calcimorphic soils, **346**, 348-349
California, 365, 372, **373**
    alluvial fans, **211**
    climate, 84, 142, **143**, 158, **159**, 160
    insolation, **14**, 15, **18**
    landforms, 239, 251, 265, 269
    precipitation, 76, 109
    shore line, 281, **283**, 288, 290
    soils, 343, 347
    temperature, **32**, 33, 36, **39**, 40
    vegetation, **318**, 319, **322**
    vulcanism, 182, 189
California canyons, 289
California Current, **34**, 35, 36, 113, 158
California Plate, 182, back endpaper
calorie, 8
Cameroons, 109, 151
campos, 154, 155, 322
Canada (see also individual provinces and territories)
    climate, 146
    landforms, 189, 224, **230**, 255, 268
    soils, 342, 347
    vegetation, 309, 319
Canadian Rockies, 146, 161, 221, 273
Canary Current, **34**, 35, 113, 145, 159
Canton (Kuang-Chou), **157**
canyons, **261**
    submarine, 289
Cape Kennedy, 370, 373
capillary movement in soils, 328
carbonates, 177, 327
carbonation, 192
Caribbean Sea:
    climate, 109, 154
    coral reefs, 280
    islands, 282
    vegetation, 316
    water masses, 304
Carrollton, Missouri, 132-133
Cascades, 31, 32, 114, 271, 274
Caspian Sea, 145, 249, 254
Caspian-Turan Lowlands, 254
catena, 346
cation exchange, **329**
Caucasus, 189, 253, 274
Celsius temperature scale, **23**
cenotes, 362

centigrade temperature scale, **23**
central Africa, 15, 150, 260, 314, 315, 316, 321 (see also individual countries)
Central America:
    climate, 148
    landforms, 186, 271
    precipitation, 109, 112, 118
    shore line, 286
    vegetation, 314, 315, 317
central Asia:
    climate, 166
    landforms, 189, 237, 251, 274
    soils, 347
central asiatic Russia:
    heat balance, 19, **20**
    landforms, 249, 251, 254
    precipitation, **118**
    vegetation, 324
Central Oceanic Water, 301, **302**, 303
central United States (see also individual states):
    landforms, 237, 238, 253, 256, 259
    soils, 327, 344, 347, 348
    vegetation, 323
Cerro de Pasco, Peru, **136**, 138
Ceylon, 104, 314
chaparral, 168, **318**
Charles' law, 43
chemical weathering, 191-193
chernozems, 166, **345**, 347
Cherrapunji, 108, 152, **153**, 154
Chesapeake Bay, 286
chestnut brown soils, 347
Chile:
    mountains, 271, 273
    plains, 251, 258
    precipitation, 109, 113
    sand dunes, 235
    shore line, 285
    soils, 342
    vegetation, 163, 318
China:
    climate, 156, **157**, 158, 163, **164**, 166
    landforms, 189, 237, 250, 268, 274
    monsoon, 103, 104
    shore line, 285
    soils, 342, 343
    vegetation, 318, 319
China Sea, 104, 288, 304
chinook, **78-79**
choropleths, **387**
Churchill, Canada, 166, **167**
cinder cone, 186, **187**, 269, **271**
circle of illumination, 10, 11
circulation of the atmosphere, 64-65
cirques, **227**, **228**, **229**
cirrocumulus, 85

cirrostratus, 85
cirrus, 85, 86, 95
city climates, 169-171
Ciudad Juarez, 127, 128
classification:
    of air masses, 89-91
    of climates, 121ff, front endpaper
    of condensation forms, 84-85
    of landforms, 239ff
    of map projections, 379ff
    of soils, 339ff
clastic rocks, 178
climate:
    city, 169-171
    classification, 121ff
    definition, 7
    desert, 128, 129, 141-**147, 143, 144, 145, 146,** 154
    dry, 125, 127-129, 141-147, **143, 144, 145, 146,** 154
    effect on erosion, 180
    effect on landforms, 174
    effect on soils, 331
    effect on vegetation, 309, 310
    elements, 7
    forest, 168-169
    frost, 126, 138-141, **140**
    humid continental, 130, 131, 163-166, **164, 165**
    humid mesothermal, 130, 131, 152, **153,** 154, 156-158, **157**
    Mediterranean, 130, 131, 158-160, **159**
    mesothermal, 125, 130-131, 152, **153,** 154, 156-158, **157, 162**
    microthermal, 125, 126, 130-131, 163-**167, 164, 165**
    mid-latitude marine, 130-131, 160-163, **162**
    mid-latitude steppes and deserts, 145-146
    monsoon, 129-130, 150-154, **151, 152**
    non-zonal, 168-171
    polar, 125, 126-127, 134-138, **136, 137**
    rain shadow deserts, 146-**147**
    steppe, 128, 141-146, **143, 144, 145, 146,** 154
    subarctic, 130, 131, 166-**167**
    subtropical continental dry, 142-143
    subtropical oceanic dry, **144-145**
    tropical, 125, 129-130, 147-156, **148, 149, 151, 152, 155**
    tropical rainforest, 129, 130, 147-150, **148, 149, 151,** 154
    tropical savanna, 129, 130, 151, 154-156, **155**
    tundra, 126, 134-141, **136, 137, 140**
    world distribution, 134ff, front endpaper
climatic change, 221-224
climax vegetation, 307
closed system, 5
clouds:
    classification, 85, **86**
    convective, 85, **86,** 93, 95
    effect on annual heat balance, 11, 12
    effect on daily march of insolation, 14
    effect on temperature, 36-37
    limited convective, 85, **86**
    seeding, 74
    stratiform, 85, **86,** 91, 96
Coari, Brazil, 118
coast (see shore line)
coastal plains, 246-**247,** 279
Coast Ranges of California, 31, 32, 267, 271, **322,** 347, 365
cohesion in regolith, 196
Colombia, 109, 271, 314
Colorado:
    plateau, 260, **261**
    River, 235, 251, **261**
    state, **38, 39, 261**
col, **227,** 229
Columbia plateau, 260, 187
condensation:
    artificial, 74
    condensation forms, 84-85, **86**
    definition, 67, 72
    factors producing, 72ff
    hygroscopic nuclei, 72, 74
    latent heat of, 70, 71, 77
    measurement, 70
    nuclei, 72, 172
conformality, 379
conglomerate, 178, 179, 180
Congo, 147, **149,** 150
    Basin, 15, 109, 148
    lake plain, 247, 261
    River, 150, 247, 290
conic projections, 382-384, **383**
coniferous forest:
    mid-latitude, 163, 166, 319-320
    soils, 333
    taiga, 320
    tundra, 137
continental blocks, **180, 181**
continental drift, **181,** 183
continental glaciers, 218
continental shelf, 181, **182,** 276, 277, 287-289
    economic importance, 288-289
    fisheries, 288
    oil and gas resource 288-289
    water masses, **302,** 303-304
continental slope, **182,** 276, 289-290
contour lines, 385, **386**
convectional precipitation, 77, 79, 93, 98, 107, 150
convergence in atmosphere, 98, 107, 108-109
cooling lakes and towers, 356-357
Coppermine, Canada, 135, **136,** 137
coral reefs, 279-**280**
Coral Sea, 304
Coriolis force, 36, **46**ff, **48,** 59, 64-65, 100
Coriolis, G. G., 47
corrasion, 204
correction line, **373**
corrosion, 204
Corumbá, Brazil, 154, **155,** 156
Corvallis, Oregon, 170
crenate bays, 282
crevasses in glaciers, 219
crust of the earth, 180-184
    sial, 178, **180,** 181
    sima, 178, **180,** 181
crystal growth in weathering, 193, 194
Cuba, 257, 285
cuesta, **259, 260,** 264
cuestaform plains, 244, **256, 259, 260**
cumulonimbus, 85, **86,** 95, 97, 99
cumulus, 85, **86,** 150
cumulus convection, 91
Curaçao, Netherlands Antilles, **277, 283**
currents (see ocean currents)
cutoff meander, 212-213
cyclogenesis, 94, **95**
cyclone:
    extratropical, 94-96, **95,** 108, 109
    tropical, 99-102, **100, 101, 102,** 108, 112
    winds, **49,** 60, 62, 93
cyclonic precipitation, 77, 79, 97, 156
cyclonic shear, 99
cylindric projections, 375, 380-382
Darwin, Charles, classification of atolls, 280, **281**
date line, **371,** 372
Davis-Johnson classification of shore lines, 281
Death Valley, California, 142, **143**
De Candolle, Alphonse, 125-126
declination of the sun, 10, 368
deflation, 234
deflection due to earth's rotation (see Coriolis force)
degradation, 191
    rate of by streams, 204
deltas, 210, **214-215**
    delta plains, **250-251**
    deposits, **214-215**
    distributaries, **215,** 250, **251**
    submarine platform, 215
dendritic drainage, 243
density:
    of atmosphere, 51
    of ocean waters, 296
    of rocks, 178
deposition, 191
    by glaciers, 225, 229-232, **230**
    by running water, 205, **210-215, 211, 212, 213, 214**
    by wind, 235-**238, 236, 237**
    by waves and currents, 279
desalinization of soils, 336
desert climates, 128, 129, 141-147
    mid-latitude deserts, 145-146
    rain shadow deserts, 146-147
    subtropical continental deserts, 142-143
    subtropical oceanic deserts, **144-145**
desert pavement, **235**
desert shrub, **323-324**
desert soils, 336, **346,** 348
dew point, 69, 75
diastrophism, 184-186
    anticlines, **185,** 186
    block mountains, 185
    cuestas, **259, 260,** 264
    faulting, 184-186, **185**
    graben, **185**
    horst, **185**
    orogeny, 188
    synclines, **185,** 186
    warping, **186**
differential erosion, 191
differential weathering, 191
dikes, **187,** 188
dip slope, 259
disintegration of rock, 192
distributaries, **215,** 250, **251**
divergence in atmosphere, 99, 107
doldrums, 60, 62, 63, 64
dolomite, 177
dome, **259**
Donner Pass, California, 14
Douglas fir, 319
Doum palms, **321**
drainage basin, 202
drift, 33, 220, 229
drift plains, 253-254
drumlins, **230,** 231
dry climates, 125, 127-129, 141-147, 154 (see also desert and steppe climates)
dry sandy soils, 348
dry thermals, 91
dunes (see sand dunes)

dynamic equilibrium in slope formation, **209**, 241
dyne, 42
earth:
    axis inclination, 9
    internal structure, **180**-181
    movements of, 9
    size and shape, 368, **369**
earth flow, **198**
earthquakes, 271
easterly waves, **99**
eastern Africa (see also individual countries)
    climate, 109
    landforms, 185, 280, 284
    soils, 344
    vegetation, 316, 322
eastern Asia (see also individual countries)
    climate, 93, 102, **103**, 104, 112
    landforms, 189
    vegetation, 317, 318, 323
eastern Canada, 160, 166, 189, 224, 255, 268
eastern United States: 364
    climate, 93, 115, 156
    landforms, 265, 279
    soils, 327, 342
    vegetation, 318, 319
ebb tides, 297
Eckert's projection, **382**
Ecuador, 314
Egypt:
    climate, 128, 142, **143**, **145**, 160
    evaporation, 72
    heat balance, 19, **20**, 21
    landforms, 250, 254
Eire, 161, **162**
Eismitte, Greenland, 138, 139, **140**
elements of a projection, 375, 377, **379**
elevation:
    effect on temperature, 29, **37**
    orographic effect on precipitation, 107, 108, 109, 112
elongate bays, 283
eluviation, 326
enclosed seas, 304
England, 247, 269
entisol, 350, 351
epilimnion, 356
epiphytes, 314
equator, 13, 14, 15, 370, 375
equatorial convergence rainfall, 107, 108-109, 113
equatorial currents, 34, 301
equatorial low, 52-55, 56, 57, 65, 93
equidistance, 379
equidistant conic projection, 382, **383**

equidistant cylindric projection, 382, **383**
equinox, 10
equivalence, 379
erosion:
    by waves and currents, 276ff
    differential, 191
    effect of agriculture, 363-365
    glacial, 224-229
    mass wastage, 195-200
    stream, 203ff
    wind, 234-235
eskers, **230**, 231, **256**
estuaries, 214
Ethiopia, 185, 187
eucalyptus, 318
Eugene, Oregon, **118**, 132, 377
Eurasia, 57, 93, 163, 166, 250, 309, 320, 347
Europe, 32, 158, 161, 163, 220, 221, 237, 274, 318, 319 (see also northern and western Europe, European Russia)
European Russia:
    climate, 132, 163, **164**, 166
    continental shelf, 288
    glaciation, 221
    heat balance, 19, **20**
    landforms, 251, 255, 259
    orogeny, 189
    precipitation, **118**
    soils, 339, 342, 347
eutrophication, 355
evaporation:
    actual, 71-72, **73**
    definition, 67
    effect on temperature, 19-20, 29
    latent heat of, **12**, 19, 70, 71
    measurement, 70
    potential, 71, 72, **73**
    rate of, 71
    world distribution, 72, **73**
evaporimeter, 70
Everglades, 247
evergreen broadleaf forests, 156, 160, 315-316, 318
exfoliation, 193-194
extratropical cyclone:
    climatic importance, 96, 108
    cold front, 94, **95**, 96
    condensation forms, 94-96
    development, 94, **95**-96
    occluded front, **95**, 96
    warm front, 94, **95**, 96
    weather in, 94, **95**, 96
extrusive rocks, 177
Fahrenheit temperature scale, **23**
Fairbanks, Alaska, **28**, 29, 39
fans (see alluvial fans)
faulting, 184-186
    block mountain, 185
    compressional, 185-186

graben, **185**
horst, **185**
normal fault, **185**
reverse fault, **185**-186
tensional fault, **185**
fault scarps, 185
feedback, positive and negative, 5
feldspars, 177, 179, 192
Fergana Valley, 251
ferns, 318
ferralite, 150, 154, 156, 344, **345**
ferralization, 334-335, 342, 344
Ferrel cell, **65**
ferromagnesian minerals, 177, 192
fetch of water, 278
Finland, 255
firn, 218
fjard, 286
fjord, **285**
flood tides, 297
floodplains, 210, 211-215, **213**, **214**, 249-250
Florida, 371
    climate, 92, **102**
    diastrophism, 186
    landforms, 247, 257
    vegetation, **320**
fog:
    advective, 76
    city, 171
    garúa, 144
    ground, 75-76
    sea, 76, 144
Föhn (see chinook)
forests:
    broadleaf evergreen, 314-**315**, 318
    coniferous, 319-321
    deciduous, 315-317, 318-**319**
    high-latitude, 320-321
    mid-latitude, 307-308, 318-319
    semi-deciduous, 315-**316**
    taiga, 320
    tropical, 314-317
forest climates, 168-169
Fossil Butte, Wyoming, **260**
France, 42
    climate, 131
    landforms, **257**, 258, 265
    precipitation, 118
    temperature, 32, **33**
friction:
    in regolith, 196
    in winds, **48**, 49, 60
fringing reefs, **280**
frontal precipitation, 77, 79, 94-96, 107, 109-112
fronts:
    arctic, **92**, 96-97
    cold, 94, **95**, 96
    definition, 79
    occluded, **95**, 96

polar, **92**, 96-97
warm, 94, **95**, 96
weather along, 94ff
world distribution, **92**
frost climates, 126, 138-141, **140**
frost heaving, 199
frost wedging, 194
Fujiyama, 186, 271
gabbro, 177, 178
Ganges:
    River, 251, 253
    Valley, 103, 104
garigue (see chaparral)
garúa, 144
Geiger counter, 70
generalization in maps, 378-379
geographic grid, 369-**370**
geologic time table, 188
geomorphology, 3, 173
Georgia, 247, 268, 331
geostrophic wind, **48**
Germany, 221
Ghats, 112, 152
Gibraltar, **301**
glacial landforms:
    arête, **227**, **228**, 229
    cirques, **227**, **228**, 229
    col, **227**, 229
    drift, 220, 226, 229, **230**
    drift plains, 253-254
    drumlins, **230**, 231
    eskers, **230**, 231
    glacial lake plains, **230**, 231, 247, 254
    hanging valleys, **227**, 229
    kames, **230**, 231
    moraines, **220**, **221**, 229, **230**
    outwash plains, **230**, 231
    striations, **225**
    tarn lakes, 273
    till, 229, **230**, 231, 232
    till plains, **230**, 231
    U-shaped valleys, **228**, **229**
    varves, 231-232
glaciation:
    causes of, 221-224
    past, 219-221
    Pleistocene, **218**, 219-224
    present, 217, **218**
    Wisconsin, **218**, 232
Glacier National Park, **226**
glaciers:
    ablation, 219
    abrasion, 224, **225**
    continental, 218
    crevasses, 219
    deformation of ice in, 217, 218-219
    deposition, 225, 229-232
    erosion, 224-228
    firn, 218
    ice, 217, 218-219
    movement of, 217-218

piedmont, 218
plucking by, 224, **225**
transport by, 225
tributary, 217, **219**
valley or alpine, 218, **219**
glacio-fluvial deposition, 231
Glasgow, Montana, 32, **33**
glei, 335
gneiss, 180
gnomonic projections, 380, **382**
Goode's projection, 382
Goose Bay Labrador, 160, 166
graben, **185**
gradation:
 by glaciers, 217ff
 by sea, 276ff
 by streams, 202ff
 by wind, 233ff
gradational processes, 191
gradient wind, 49
Grand Banks, 288
granite, 177, 178, 187, 195
grasslands:
 cultural origin of, 322-323
 mixed-grass prairie, 322
 prairie, 322
 short-grass prairie, 322
 steppe, **322,** 323
 tall-grass prairie, 322
 tropical savanna grasslands, 321-322
gravel, 234
gravity, 2, 176, 191, 195, **203**
gray-brown podzolic soils, 163, 342, **345**
Great Basin, 114, 259, 324, 347
Great Britain, 171, 221
great circle, 368
Great Lakes, 76, 247, **259,** 319, 320
Great Plains, 3, **43,** 97, 319, 372
Great Sand Dunes, New Mexico, **236**
Great Salt Lake, 259
great soil groups, 340
 azonal, 339, 340, **346,** 348
 intrazonal, 339, **346,**348-349
 zonal, 339, 340-348, **345, 346**
Greece, 158, **159,** 160, 285
greenhouse effect, 15, 19
Greenland:
 climate, **136,** 138, 140
 glaciation, 217, 218, 219, 221
 sea fog, 76
 shore line, 285, 303
Greenland Current, **34,** 35
Greenland Ranch, California, 142, **143**
Greenwich:
 meridian, 370, 372
 time, 372
ground fog, 75-76
groundwater, 176
Guatemala, 360

Guernsey Island, 32, **33**
Guiana Uplands, 255
Gulf Coastal Plain:
 landforms, 247, 256
 soils, 345, 348
 vegetation, 2, 318, 319
Gulf of California, 189, 251
Gulf of Mexico, 114, 280, 288, 298, 304
Gulf of Siam, 288
Gulf Stream, 35, 76, 288, 300, 304
hachures, 385, **386**
Hadley cell, 64-**65**
hail, 163
Half Dome, 193
Half Moon Bay, California, 168
halomorphic soils, **346,** 349
hanging valleys, **227,** 229
Harbin, China, **164,** 166
hardpan, 343
Hawaii, 372
 insolation, **28**
 landforms, 187, 271
 precipitation, 108, 168
 temperature, **28,** 29, 39
 tides, **299**
heat:
 balance, 5-6, 7ff, **12**
 conduction, 8, 30
 convection, 8, 30
 latent, **12,** 19, 70
 radiation, 8
 specific, 22, 30
 turbulent transfer, 8, 18, 19, 21, 30
heat balance:
 annual march, 19, **20,** 21
 average annual, 11, 12, 19, 21
 daily march, 19, **20**
 long-wave component, **12,** 15, 18
 nonradiative component, 12, 18, 19, **20,** 21
 short-wave component, 11-**12,** 18
 summary of, 19
 world distribution, 19-21, **20**
heat dome, 221
hekistotherm, 126
High Plains of the United States, 253, **254**
high pressure centers (see subtropical high pressure)
hills and low mountains, 263ff
 badlands, 268, **269**
 conical, 244, 269
 distinguishing characteristics, 243-244
 linear, 244, 263-265
 non-linear, 244, 265-269
 rough, 244, 265-268
 rounded, 244, 268-269
 world distribution, **266**

Himalayas, 140
 climate, 62, 103, 104, 108, 112, 114, 154
 landforms, 189, 253, 270, 273
histosols, 350, 351
hogback ridges, 263-264
Honduras, 321
Honolulu, **28,** 29, 39, **299**
horn, **228, 229**
horse latitudes, 60, 62, 63
horst, **185**
Houston, Texas, 150
Hudson Bay, 86, 166
Hudson River, 288
Humbolt Current, 34, 109, 113, 159
humid continental climates, 130, 131, 163-166, **164, 165**
humidity:
 absolute, 67
 dew point, 69
 instruments, 69-**70**
 measurement, 69-70
 mixing ratio, 67-68, **69,** 74-75
 relative, 68-69, 71, 74-75
 saturation, 68-69
 specific, 67
 supersaturation, 68
 vapor pressure, 67, **68**
humid mesothermal climates, 130, 131, 152, **153,** 154, 156-158, **157**
humus, 329
Hungary, 347
hurricanes, **43, 101-102**
hydration, 192, 194, 204
hydrocarbons, 358
hydrologic cycle, **175**-176
hydrolysis, 192
hydrometeors (see condensation forms)
hydromorphic soils:
 bog soils, **346,** 348
 half-bog soils, 348
 wiesenboden, 348
hydrophytes, 310, 315
hydrosphere, 175, 176
hygrometer, 69
hygroscopic nuclei, 72, 74
hypolimnion, 356
Ibadan, Nigeria, 154, **155,** 156
ice:
 antarctic, 138
 icebergs, 135, 303
 land ice, 303
 movement in glaciers, 217-219
 pack ice, 303
 sea ice, 303
 shelf ice, 303
Iceland, 93, 112, 182
Icelandic low, 57
ice-scoured plains, 255, **256**
Idaho, 163, 237

igneous rock, 177-178, 180
Illinois, 6
 climate, **118,** 131, 165
 landforms, 214, 239
 soil, 331, **332,** 349
illuviation, 327
inceptisols, 350, 351
India:
 climate, 151, **152, 153,** 154
 jet stream, 62, 103, 104
 landforms, 187, 189, 250, 260
 monsoon, **45,** 102, 103, 104, 112
 orogeny, 187
 precipitation, 108, 112
 soils, 343, 344
 vegetation, 314, 315, 316
Indian Equatorial Water, 301, **302**
Indian Ocean:
 continental shelf, 287
 coral, 280
 currents, 35
 precipitation, 109, 113, 115
 salinity, **293,** 296
 subsidence, 299
 temperature, **295**
 tides, 298
 water masses, 301
 winds, 60, 62, 102
Indiana, 258, **319**
Indochina, 112, 156
Indonesia:
 climate, 109, 148
 monsoon, 104
 shore line, 280, 284, 304
 vegetation, 314, 315
Indus River, 250, 251, 253
industrial pollution, **364,** 365-366
inner core, **180**
insolation:
 annual cycle, 13, **14**
 daily march of, **14,** 18
 definition, 7-8
 effect of cloudiness on, 12, **14**
 factors effecting rate of, **9**ff
 gradients, summer and winter, 13
 latitudinal variation, **11,** 13-14, 15, 28-29
 measurement, 8
 solar constant, 9
 world distribution, 15, **16-17**
interfluves, 247, 253
interfluvial ridge, 202
interglacial periods, 220
intermittent stream flow, 202, 210-211
International Date Line, **371**-372
Intertropical Convergence Zone (ITC), 91, **92,** 98, **99,** 103, 104, 108, 109, 113, **114,** 115, 116, 117, 119, 147, 150
intrazonal soils, 339, 340, 348-349

calcimorphic, **346,** 348-349
halomorphic, **346,** 349
hydromorphic, **346,** 348-349
planosols, **346,** 349
intrusive rocks, 177, 178
Invercargill, New Zealand, 161, **162**
inversion of temperature, **39,** 83, 171, 359
Iowa, 259, 344
Iquique, Chile, **144, 145,** 160
Iran, 145, 259, 274
Iranian Plateau, 251
Ireland, 112
isarithm, 386
isobar, 42, **43, 52-55**
isobath, 181
isogram, 386
isohyet, 107
isohypse, 385, **386**
isoline, 15, 386
isopleth, 386, **387**
isotach, 43
isotherm, **24-27,** 28, 31
Israel, 185
Italy, 186, 258
Jamaica, 102, 108, 257
Japan: 171
landforms, 186, 250, 288, 290
monsoon, 62, 103, 104
vegetation, 318
Japan Current, **34,** 35, 36
Japan Sea, 104
Java, 186, 269
Java Sea, 288
jet stream, 59, 62
Jungfrau, **273**
Jungfraujoch, Switzerland, 139, **140**
Kalahari, 113
Kamchatka Current, **34,** 35
kame, **230,** 231
kame terrace, 231
kaolinite, 196, 199, 200, **329**
Karachi, **36,** 37
karst plain, 256-**258, 257**
Kauai, Hawaii, 108, 168
Kelvin temperature scale, **23**
Kentucky, 258, 265
Kenya, 240, 260
kettle holes, **230,** 231
Kharkov, U.S.S.R., **118**
Kilauea Point, Hawaii, 168
Kisangani (Stanleyville), Zaire, 130
Köppen, Wladimir, 125
Korea, 104, 285
Krasnovodsk, 145, **146**
Labrador, 134, 268, 285, 301
Labrador Current, **34,** 35, 288, 301
laccoliths, **187-**188
lagoons:
atoll lagoons, 280

lagoon coasts, 282-283, 284, **286-287**
Lake Erie, 355
Lake Eyre, 249, 259
Lake Michigan, **30,** 356
Lake Ontario, 259
Lake Oswego, Oregon, **384,** 385
lake plains, 247-**249**
Lake Superior, **256**
lakes: glacial, **226, 230,** 231, 247
oxbow, **212, 213**
playa, 258
Pleistocene, 247-248
Lambert, J., 380
Lambert's azimuthal equal-area projection, 380
Lambert's conformal conic projection, 383
land and sea breeze, 45
landforming factors, 174-175
landform properties, 242-243
constituent material, 240, 243
local relief, 242-243
slope, **240, 241,** 242
surface expression, 243
landforms: classification of, 239ff
constituent materials, 240, 243
effect on soil, 332
effect on vegetation, 310-311
geographic view of, 174
of glacial degradation, 226-229, **227**
of glacial deposition, 229-232, **230**
glacio-fluvial, 231-232
of stream degradation, 209-210
of stream deposition, 210-215
world distribution, **248, 266**
of wind gradation, 235-238
land ice, 303
landslides, 197
langley, 8
La Paz, Baja California, 92
lapse rate of temperature:
in air masses, 90
conditional, **80,** 83
effect on rising air, 77, 79ff
importance to precipitation, 84-85
normal, 38, **39**
stable, **80,** 83, 94, 96, 107
unstable, **80,** 82, 94, 95, 107
latent heat of condensation, 70, 77, 83, 100
latent heat of evaporation, **12,** 19, 70, 71, 75
laterite, 154, 156, 335, 344, **345**
laterization, 335
latitude: definition, 370
determination of, 370-371
parallel of, 370

variation of insolation with, **11,** 13-**14,** 15, 28-29
Laurentian Shield, 189, 224 back endpaper
Laurentian Uplands, 224
laurel, 318
lava, 177, 178, 186ff, 260
lava domes, 187
leaching, 193, 326
length of daylight, 10, 13, 137, 310
Leningrad, 19, **20**
levees, 211-**212,** 249, 250
lianas, 315
Liberia, 109, 151, 286, 344
Libyan Desert, 254
lichens, 137, 325
lime, 335, 340, 344, 347
limestone, 179, 180, 240
karst, 256-258, **258**
sinks, 257
Lisbon, 19, **20,** 72
lithification, 179
lithosols, 142, 146, 336, **346,** 348
lithosphere, 176, 180
Llano Estacado, 253
llanos, 155
Llanos of Venezuela, 154, 253, 321-322
local relief, 242-243
location, 368ff
loess, 236, **237, 238**
London, 161, **162,** 169, 170
longitude: definition, 370
determination of, 371-372
meridian of, 370, 372, 373
longitudinal sand dunes, 236
long-wave radiation, 8, **12,** 15, 18
Los Angeles, 159, 171
Los Angeles-San Bernardino Lowland, 251, 358
Louisiana, 288
Lourenco Marquez, Mozambique, **118**
low pressure centers, 50, 51, **52-55,** 57, 94-96
loxodrome, 382
lunitidal interval, 297
Maas (Meuse) River, 250, **251**
Madagascar, 112, 148, 150, 314, 315, 322
magma, 187
magnolia, 318, 320
Maine, 76, 286, 320, 331
Malabar coast, 112, 151, **152**
Malaysia:
climate, 148, **149,** 150, 151
monsoon, 62, 103, 104
precipitation, 112, **118**
vegetation, 314
Manaos, 19, **20,** 21, 72
Manchuria, 104, 163, **164,** 166, 342, 347
Mangalore, India, 151, **152**

mangrove forests, **317**
mantle, **180**
mantle rock, 195, 196
maple, 319
map projections:
azimuthal or planic, 375, **379-**380
azimuthal equidistant, 380
gnomonic, 380, **381**
Lambert's azimuthal equalarea, 380
orthographic, 380
stereographic, 380
conic, 375, 382-384
Albers' equivalent conic, 383
Bonne's, 382
equidistant conic, 382, **383**
Lambert's conformal conic, 383
polyconic, 382-383
cylindric, 375, 380-383
Eckert's, **382**
equidistant, 382
Goode's, 382
Mercator, **377, 381**-382
sinusoidal, **382**
definition, 375
distortion, 375, 377, 379
element, 375, 377
Map properties: conformality, 379
equidistance, 379
equivalence, 379
scale distortion, 377, 379
symbolism, **378-379, 385**
maquis, 318
marble, 179
Marianas Trench, 290
marine gradation, 276ff
marine plains, 244-245
marine terrace, 276-**277**
Marshall Islands, 280
Maryland, 269
masses, 189, back endpaper
mass wastage, 195-**200**
definition, 195
forms, 197-199
importance of, 200
processes affecting, 196
variations in rate, 199-200, 241
Mbandaka, Zaire, 147, **149,** 150
meander, 205, **212, 213**
Medicine Hat, Canada, 145, **146**
Mediterranean climates, 130, 131, 158-160, **159**
Mediterranean lands, 160, 186, 271
Mediterranean Sea, 119, 158, 160, 292, **301,** 304
megatherm, 125, 314
Mercator, 381
Mercator projection, **377, 381**-382

Index 403

mesa, **260, 261**
mesophyte, 310
mesothermal climates, 125, 130-131, 152, **153**, 154, 156-158, **157, 162**
mesotherms, 125
metamorphic rocks, 177, 179-180
Meuse (Maas) River, 250, **251**
Mexico: 360
    climate, 92, 113, 127, 128, 168
    mountains, 269, 274
    plains, 253, 259
    tectonic forces, 186
Mexico City, 359
Miami, 92
Michigan, 163, **164, 165**-166, **256**
Michoacan, 269
microthermal climates, 125, 126, 130-131, 163-**167, 164, 165**
microtherms, 125, 126
Mid-Atlantic Ridge, 182, 290
Mid-Latitude Continental Shelf Water, 303-304
mid-latitude low pressure, **52-55,** 56, 57, 60, 62
mid-latitude marine climates, 130, 131, 160-163, **162**
millibar, 42
mimosa, 316
minerals, 176-177
    carbonates, 177
    definition, 176
    feldspars, 177
    ferromagnesian minerals, 177
    mineral content of soils, 329
    silica, 176
Minneapolis, 128
Minnesota, 128, **226,** 320
Mississippi River, 149, 214, **215,** 237, 250, 251
Missouri:
    climate, 132-133, 156, **157**
    landforms, 267
    temperature, 30-**31, 32, 38,** 39, 40, 150
Missouri River, 221, 237
Mitchell, Montana, **179**
mixing ratio, 67-68, **69,** 74-75
Modesto, California, **32**
Moho, **180**
Mojave Desert, 114, 211, 254, 324
mollisols, 350, 351
Mongolia, 93, 114, 259
monsoon: currents, 35
    definition, **45,** 60, 62
    description, 102-104, **103**
    rainfall, 112, **118**
monsoon climates, 129-130, 150-154, **151, 152**
monsoon forest, 315-**316**
Montana, 32, **33, 252**
Montevideo, 156, **157,** 158
montmorillonite, 196, 199, **329**

moraine: lateral, **220**
    medial, **220, 229**
    recessional, **230**
    terminal, **221, 230**
Morocco, **144, 145**
Moscow (Moskva), 163, **164,** 166
Moscow, Idaho, 163
mosses, 137, 325
mountains: alpine, 188, 189
    distinguishing characteristics, **243,** 244, 269-270
    glaciated, 244, **273-274**
    nonglaciated, 244, 274
    orogeny, 188, 189
    volcanic, 244, **271-272, 273**
Mount Capulin, New Mexico, **186**
Mount Cass, Alaska, **220**
Mount Diablo Meridian, 372-373
Mount Hood, 186
Mount Mayon, 186
Mount Ranier, 186
Mount Shasta, 186, **272**
Mount Vesuvius, 186
Mount Waialeale, Hawaii, 108, 168
Mount Whitney, **195**
Mozambique, **118,** 322
Mozambique Current, **34,** 300
Muir Pass, California, **195**
natural levees, 211, **212,** 249, 250
natural systems, 3, **4,** 5-6
neap tides, 297, **298**
Near East, 1, 250
nearshore, **277**
Nebraska, 163, 253, 254, 322, 347
Netherlands, 250, 332
net radiation, 18
Nevada, 185, 249, 372
New England, 221, 268
Newfoundland, 287, 288
New Guinea, 213-214, 304, 314
New Jersey, 268, 286, 319
New Mexico, **186,** 235, **257**
New York, 15, 150, 169, 221, 268, **270**
New York City, 15, 150, 169
New Zealand, 97, 113, 161, **162,** 237, 285, 318, 322, 342, 372
Niagara Cuesta, 259
Niamey, Niger Republic, 142, **143,** 145, 152
Nicaragua, 247
Niger: Delta, 251, 286
    Republic, 142, **143,** 145, 152
    River, 250
Nigeria, 109, 154, **155,** 156
Nile: Delta, **215,** 250
    River, 247, 250
nimbostratus, 85
nitrogen content in soils, **330, 331**
North America:
    climate, 32, 56, 57, 92, 93, 114, 145, 149, 161, 163, 166
    continental shelf, 288

    glaciation, 220-221, 224
    landforms, 189, 247, 284
    upwelling, 36, 310
    vegetation, 163, 319, 320, 322
North Atlantic Deep Water, 299
North Carolina, **43,** 247, **287**
North Dakota, 347
northern Africa, 92, 113, 189, 236, 265
northern Canada, 90 (see also Northwest and Yukon Territories)
northern and western Europe:
    air masses, 92
    glaciation, 224, 254
    landforms, 189, 254
    orogeny, 189
    precipitation, 109, 112, 115
    soils, 342
    temperature, 32, **33**
    vegetation, 163, 323
North Pacific High, 51, 57, 62, 93
North Sea, 250, 287, 288, 298, 332
Northwest Territories, Canada, 135, **136,** 137
Norway, 112, 268, 273, 285
Oakland, California, 39, **322**
occluded front, **95,** 96
ocean currents:
    effects on precipitation, 113, 144
    effects on temperature, 33-36
    equatorial currents, 301
    lateral exchange at depth, 300
    perennial subsidence, 299
    relation to wind, 300
    schematic pattern, **35**
    speed, 300
    surface currents, 33-36, **34,** 300
    temperature, 33-36, 292, **295,** 296
    upwelling, 36, 76, 288, 298-299, **300**
    vertical exchange, 298-300
    West Wind Drift, 34, 35, 36, 92, 114, 137, 160, 299, 300, 303
    world distribution, **34**
    (see also individual currents)
oceanic basins:
    boundary, 289
    continental slope, **182,** 276, 289-290
    ocean floor, 276, 290
Oceans:
    currents, 33-36, **34,** 298-300
    effect on climate, 30ff, 113, 144
    physical characteristics, 292-296, **293, 294, 295**
    resources, 288-289
    subsidence, 35, 299-300
    swell, 279

    tides, 296-**298, 297, 299**
    upwelling, 36, 76, 288, 298-299, **300**
    water regions, 301-304, **302**
    waves, 278-279
offshore, **277**
Ohio, 267, **374, 375**
Ohio River, **212,** 221, **249**
Oimyakon, U.S.S.R., **137,** 138, 139, **140**
Oklahoma, 265
Olympic Peninsula, **182**
O'Neil, Nebraska, 163
Ontario, 342
Ontario Clay Belt, 247
open system, 5
Orcadas, South Orkney Islands, **136,** 137
Oregon, 289, 377, **384**
    climate, 36, 76, **118,** 132, 159, 171
    landforms, 186, **213, 258, 271**
    vegetation, **324**
organic activity in rocks, 193, 194
Orinoco River, 251
Orléans, France, 32, **33**
orogeny:
    alpine, 188, 189
    Caledonian, 188, 189
    Hercynian, 188, 189
    Precambrian, 188, 189
orographic precipitation, 77-79, **78,** 107, 108, 109, 112
orthodrome, 369
orthographic projection, 380
Ouachita Mountains, 265
outer core, **180**
outwash plains, **230,** 231
oxbow lakes, **212, 213**
oxidation, 192
oxisols, 350, 351
oxygen:
    in atmosphere, 11, 357, 364
    biologic oxygen demand (BOD), 355
    in lakes and streams, 355, 366
    in oceans, 304
ozone: layer, **37,** 38
    pollution, 358
Ozark Mountains, 267
Pacific Basin, 184
Pacific Equatorial Water, 301
Pacific Northwest, 187, 260, 271, 319, 342
Pacific Ocean:
    air masses, 32, 91, 92, 93, 97
    climate, 30-**31, 36,** 40, 150, 161, **162**
    condensation, 76
    continental shelf, **182,** 287
    coral, 280
    currents, **34,** 35
    deeps, 290
    insolation, 15

precipitation, 109, 113, 114, 115
pressure, 51, 57
salinity, 292, **293**
subsidence, 299
temperature, **294-295,** 296
tides, 288
vulcanism, 269, 271
water masses, 301, **302**
winds, 60, 62
pack ice, 303
Pakhta-Aral, U.S.S.R., 19, **20**
Pakistan, **36,** 37, 102, 145
Paleozoic mountains, 188, 189
Palmén, E., 65
Palmer Peninsula, 134
palmetto, **320**
Pamir Knot, 114
Pampa of Argentina, 156, 158, 237, 253, 322, 323
pantanal, 321-322
Papua Highlands, New Guinea, 213-214
Paraná River, 250
parent material, 327, 331
Paris, **118,** 131
Paris Basin, 259, **260**
particulate air pollution, 358
Pascal, Blaise, 41-42
Patagonia, 114, 147, 347
pedalfers, 340-344
pediments, 258-259
pedocals, 340, 344-348
pedon, 350
peneplain, 205
Pennsylvania, 221, **265,** 267
perennial stream flow, 202
permafrost, **137-138,** 166
Persian Gulf, 288, 296, 304
Perth, Australia, **118**
Peru, 144, 235
Peru Current, **34,** 35, 36, 300
pH of soils, 330
Philippine Islands: climate, 102, 104, 148, 149, 151
landforms, 186, 284, 290
photochemical smog, 358
photoperiodism, 310
photosynthesis, 308, 357
physical weathering, 192, 193-195
Picardy, **257**
piedmont alluvial plains, 251-253, **252**
piedmont glaciers, 218
pines, 2, 169, 319-**320**
plains:
    basin and range, 243, 244, 258-259
    cuestaform, 244, **256, 259**
    delta plains, **250, 251**
    drift plains, 253-254
    flood plains, 249-250
    glacial lake plains, **230,** 231, 247, 254

    with high relief features, **243,** 244, 258-261
    ice-scoured plains, 255, **256**
    irregular plains, 244, 253-258
    karst plains, 256-**258, 257**
    lake plains, 249-**251**
    of moderate relief, **243,** 244, 253-258
    outwash plains, **230,** 231
    piedmont alluvial, 251-253, **252**
    pitted, 244, 256-258
    recent marine, 246-**247**
    sand plains, 254-255
    of slight relief, **243,** 244, 246-253
    smooth plains, 244, 253
    stream eroded, 255-256, **257**
    tableland plains, 259-**261**
planic projections (see azimuthal projections)
planosols, **346,** 349
plant associations, 307
    distribution, **312-313**
    factors affecting the distribution of, 307-311, 314
plant communities, 307
plateaus, 242, 259-**261**
playa, 258
Pleistocene glaciation, 138, 219-224, 276
plucking by glaciers, 224, **225**
podzol, 163, 164, 333, 340-342, **345**
podzolization, 333-334
point bar deposits, 212
Point Delgada, California, **283**
Poisson's equation, 43, 76
Poland, 221
polar climates, 125, 126-127, 134-138, **136, 137**
polar front, 112, 115, 116, 118, 163
polar high pressure, **52-55,** 56, 57, 65
polder, 250
poles, 15, 18, 369, 370, 375
    North, 14, 223, 369, 370, 375
    South, 51, 138, 139, 369, 370, 375
pollution: agricultural, **363-365**
    of atmosphere: 357-359
    effect of urbanization on, **365,** 366-367
    industrial, 355-356, **364,** 365-366
    of water, 354-357
polyconic projection, 382-383
Popocatepetl, 168, 186
population explosion, **360**
Po River, 214, 251, 253
Port Étienne, Mauritania, **118**
Portland, Oregon, 377
Port Sudan, **118**

Portugal, 19, **20,** 72
potential evaporation, 71-72, **73**
potential evapotransporation, 71, 72, **73**
Potomac River, 265
prairie soils, 166, **332,** 344-347, **345**
Precambrian shields:
    Angara, 189, 267, back endpaper
    Antarctic, 189
    Baltic, 189, back endpaper
    Laurentian, 189, 224, back endpaper
precipitation:
    causes of high precipitation, 108-109
    causes of low precipitation, 112-114
    convectional, 77, 79, 93, 98, 107, 150
    cyclonic or frontal, 77, 79, 97, 156
    definition, 106
    effect on plant growth, 309-310
    effect on soil formation, **331**
    equatorial convergence, 107, 108-109, 113
    instruments, 106-107
    measurement, 107
    orographic, 77-79, **78,** 107, 108, 109, 112
    seasonal regimes, 114-120, **115, 117**
    world distribution, 108ff, **110-111**
pressure:
    Charles' law, 43
    definition, 41
    dynamically produced, 45-50
    horizontal distribution, **52-55,** 56
    horse latitudes, 60, 62, 63
    instruments, 41-**42**
    measurement, 42
    patterns aloft, 57, **58**
    Poisson's equation, 43, 76, **52-55,** 56
    reduction to sea-level, 42, 51
    relationship to wind movements, 45-50
    thermally induced, 43-45, **44**
    vertical distribution, 42, 76
    world distribution, 51-57, **52-55**
pressure gradient force, (PGF), 47ff
prime meridian, 370
Prince Rupert, British Columbia, 160, **162**
principal meridian, 372-**373**
projections (see map projections)
psychrometer, 69-70

Puerto Montt, Chile, **118**
Puerto Rico, 290
Pukapuka, 148, **149,** 150
Puy-de-Dôme, 42
Pyrenees, 253
pyrheliometer, 8
quartzite, 180, 195
quaternary, 188
Quebec, 319
radiation:
    atmospheric, 15, 18, 19
    effect of temperature on, 8
    importance to climate, 7-8
    measurement, 8
    net, 8
    radiation balance, **11-12,** 15, 18, 19
    short and long wave, 8
    solar, 7ff
    terrestrial, 15, 18, 19
radiational cooling, 75
radiosonde, 89
rain shadow, **78,** 79, 112, 113, 146-147
range (see township)
Rangoon, 130
raobs, 89
rawinsonde, 89
Red Sea, 185, 292, 296, 304
redwood, 319
Redwood City, California, 168
reef:
    barrier, **280,** 281, 284
    coral, 279-**280,** 284, 285
    formation, 279-**280**
    fringing, **280,** 281, 284
reg, 235
regolith, 195, 196
relative humidity, 68-69, 71, 74-75
relief: relative or local, 242-243
    representation of, **385, 386, 387**
rendzina soils, 240, 348-349
representative fraction, 376
Rhine Delta, 250, **251**
Rhodesia, **261,** 344
rhyolite, 177, 178
ria bays, 283, **284**
Richmond, Virginia, 160, 166
Rift Valley, Africa, 185, 189
rill flow, 202
Rio Grande, 251
Rio Negro, 150
Riverside, California, **18**
roches moutonnées, 226
rock: composition, 174
    igneous, 177-178, 180
    metamorphic, 177, 179-180
    resistance to weathering and erosion, 174, 180
    sedimentary, 177, 178-179, 180
    structure, 174

rockfall, 197
Rocky Mountains:
    climate, 31, **43**, 114, 138
    landforms, 187, 274
    vegetation, 322
Romania, 347
Rossby, C.G., 65
Ross Sea, 299
Royal Arches, Yosemite, 193
Royal Gorge of the Arkansas River, **210**
Rub'al Khali, 254
Ruby Range, Canada, **198**
runoff, 202
Sacramento River, 250
sagebrush, **318**, **324**
saguaro, **323**
Sahara:
    climate, 15, 91, 92, 93, 113
    landforms, 235, 254, 255
    vegetation, 324
Saigon, **155**, 156
St. Louis, 30-**31**, **32**, **38**, 39, 40, 150, 156, **157**
salinity of oceans, **292**, **293**, 296
salinization, 335-336
Salisbury Plain, **247**
saltation, **233**, 234
Salt Lake City, **28**
salt marsh, 250
Salton Sea, 251
Samarkand, 251
San Andreas fault, **189**
sandbars:
    offshore, 279, 286
    in river channels, 211
sand dunes:
    distribution, 235, 236
    slip face, 236
    types of, **236**, **255**
sand plains, 254-255
sandstone, 178, 179, 180
San Francisco, **381**
    climate, 84, 158, **159**, 160, 166
    insolation, 15
    temperature, 31, **32**, 40
San Joaquin River, 250
San Juan Mountains, 198
Santa Ana, **78-79**
Santa Cruz Mountains, 168
Santa Monica Mountains, **318**
Santiago, Chile, **118**
Sargasso Sea, 296
Sarmiento, Argentina, **147**
saturation: in atmosphere, 72
    dew point, 69
    relationship to temperature, **68**, **69**
    saturation mixing ratio, 68-**69**
    saturation vapor pressure, **68**
savanna climates (see tropical savanna climates)

savanna woodlands, **316**
scale: areal, 375
    chorographic, 388
    descriptive, 376
    determination of, 377-378
    distortion, 375, 377, 379
    geographic, 388
    graphic, **376**
    linear, 375
    linear diagram, **376**
    nominal, 377
    representative fraction, 376
    topographic, 388
Scandes, 268
Scandinavia, 224, 232, 255
schattenseite, 14
Scheldt River, 250, **251**
schist, 180
Scotland, 112
Scottish Highlands, 268
scree, 194
sea fogs, 76, 144
sea ice, 303
sea level, rise and fall, 6, **222**, 276, **278**
seamounts, 290
Seattle, **165-166**, **299**
sea water (see oceans)
section, 373-374, **375**
sedges, 137, 325
sedimentary rock, 177, 178-179, 180
sedimentary strata, 178, **179**, **259**, 263
seif, 236, **255**
selva, 150
Seventh Approximation soil orders, 350-351
shale, 178, 179, 180
Shantung Peninsula, 250
sheet wash, 174, 176, 200, 202, 241
shelf ice, 303
shelf notches, 289
Shenandoah Valley, **264**, 265
shield volcanoes, 187, 271
Shishaldin Volcano, Aleutian Islands, **187**
shore formation, **277**-280
shore line:
    advancing, 282
    classification, 280-283
    cliffed, 282, 283-285, **284**
    emergent, 281
    irregular, 282, 284-285, 286, **287**
    mobile crustal regions, 282
    neutral coasts, 282
    noncliffed, 282, 283-284, 285-**286**, 287
    regular, 282, 283, 286, **287**
    retreating, 282
    stable coasts, 282
    submergent, 281

shore zone:
    backshore, **277**
    foreshore, **277**
    formation, **277**-280
    rock types, 279-280
    waves, 278-279
short-wave radiation, 8, 11-**12**, 18
sial, 178, **180**, 181
Siberia:
    climate, 103, 126, 132, **137**, 138, 139, 140, 145, **146**, 166, **167**
    heat balance, 21
    landforms, 189, 267, 274
    precipitation, **118**
    soils, 342, 347
    vegetation, 319, 320
Siberian high, 102, 103, 104
Sicily, 186
sierozems, 146, **346**, 347
Sierra Leone, 109, 151, 344
Sierra Madre, Mexico, 31, 113, 274
Sierra Nevada, California, 31, 114, 187, 239, 274, 343
silica, 176, 177
sills, **187**, 188
silt, 178, 234
siltstone, 178, 179, 180
Silver Hill, Jamaica, 102, 108
sima, 178, **180**, 181
Singapore, **118**, 148, **149**, 150
Sinkiang, 114, 259
sinusoidal projection, **382**
Skagerrak, 304
sky radiation, 11
slip face, 236
slope:
    development, 205-209, **206**
    form, 242
    inclination, 242
    landform classification, 242
slumps, **197**
small circle, 368-369
snow, 70, 85, 89, 106, 107, 114, 140-**141**, 156, 163
snow line, 140-**141**
soil creep, 198-**199**
soil forming factors, 330-332
    climate, 331
    landforms, 332
    parent material, 331
    time, 332
    vegetation, 331
soil forming processes, 332-338
    alkalinization, 336
    in arid regions, 335-336
    calcification, 335
    ferralization, 334-335
    in humid regions, 333
    laterization, 335
    podzolization, 333-334
    salinization, 335-336
soil profile, 326-**327**, **332**, **345**, **346**

soils:
    azonal, 339, 340, 348
    base exchange, **329**
    chemical properties, 329-330
    chemical reaction, 330
    classification, 339ff
    color, 327
    effect on vegetation, 311
    eluviation, 326
    great soil groups, 340
    illuviation, 327
    intrazonal, 339, 340, 348-349
    mineral components, 329
    organic matter, 329-330
    parent material, 327, 331
    pedalfers, 340-344
    pedocals, 340, 344-348
    physical properties, 326-328
    porosity, 328
    soil-forming factors, 330-332
    soil-forming processes, 332-338
    soil profile, 326-**327**, **332**, **345**, **346**
    structure, 327-328
    texture, 327, **328**
    water, 328
    world distribution, 341
    zonal, 333, 339, 340-348
solar climate, 31
solar constant, 9
solar radiation, 7ff
solifluction, **198**, 199, 325
solonchak, 336
solonetz, 336
solstice, 10
solution, 192-193
    differential, 180, 193
    of rocks, 192-193
    in streams, 204
Somali coast, 109
sonnenseite, 14
Sonoran Desert, 114, 324
South America (see also individual countries):
    climate: 97, 109, 113, 134, 148, 151
    landforms, 198, 250
    shore line, 284, 288, 290
    upwelling, 301
    vegetation, 315, 317, 321, 322
South Carolina, 287
South Dakota, 265, 268, **269**, 347
southern Africa (see also individual countries):
    climate, 99, 113, 154, 156, 158
    landforms, 260-261
    soils, 343
    vegetation, 318, 322, 323, 324
southern Asia (see also individual countries):
    climate, 92, 100, 102, **103**, 112

Index 407

landforms, 236
vegetation, 315, 319
Southern Hemisphere Drift, 114
South Island, New Zealand, 237, 285, 322
Spain, 112, **284**
Spanish Sahara, 92
specific heat, 22, 30
specific humidity, 67
sphere development, 376-377
spodosols, 350, 351
spring tides, 297, **298**
spruce, 319, 320
stability in the atmosphere, **39**, 79, **80**, 83, 84, 92, 94, 106, 112, 113, 141, 144, 158
standard atmosphere, 37
standard parallel:
    projections, 382, **383**
    Township and Range System, 373
standard time, 372
steppe climates, 128, 141-146, **143, 144, 145, 146,** 154
stereographic projection, 380
Straits of Gibraltar, 304
stratocumulus, 85, **86**
stratosphere, **37**, 38
stratovolcano, 186
stratus, 85, **86**, 95
stream:
    channel, 202
    deltas, 210, **214-215**
    deposition, 210-215
    drainage basin, 202
    effect on alluvial fans, 210-211
    erosion, 202, 203ff
    floodplains, 210, 211-215, **213, 214, 249-250**
    gradation, 202ff
    gradient, 203
    intermittent, 202, 210-211
    meander, 205, **212, 213**
    peneplain, 205
    perennial, 202
    rate of incision, **207**
    systems, 202
    transportation, 204
    turbulence, 203
    velocity, 203, **204,** 205
striations, **225**
Subantarctic Water, **302,** 303
subarctic climates, 130, 131, 166-167
Subarctic Water, 303
submarine canyons, 289-290
    California canyons, 289
    dendritic, 289
    formation of, 289-290
    shelf notches, 289
    turbidity currents, 289
subsidence:
    of air, 44, 90, 92, 93, 98, 109, 113, 114
    of coral atolls, 280-281
    of water, 35, 299-300
subtropical high pressure, 56, **52-55,** 57, 59, 60, 62, 63, 65, 92, 98, 109, 113, 115, 156
subtropical red and yellow soils, 158, 342-343, **345**
Sudan, 154, 247, 261, 321
Sugar Loaf, Rio de Janeiro, 193
sun spots, 223
surface creep, wind induced, **233,** 234
surge of the trades, **98,** 99, 150
suspension:
    in streams, 204
    in wind, **233,** 234
Sweden, 286
swell, 279
Switzerland, 139, **140,** 229, 265, **273, 274**
syncline, **185,** 186
    relationship to lowlands, **259**
    synclinal axes, 263-264
Syr Darya, 250
Tabasco, Mexico, 360
tables, 189, back endpaper
taiga, 166, 320
Taipei, Formosa, 92
Tallahassee, **373**
talus cone, **195,** 197
talus slope, 197
Tamatave, Malagasy Republic, 148, **149,** 150
Tanzania:
    landforms, 185, 261, **284, 286**
    soils, 346
    vegetation, **316, 317, 321, 322**
Tarim Basin, 251, 259
tarn lake, 273
Tashkent, U.S.S.R., 251
Tasmania, 112, 342
Tatoosh Island, 32, **33**
Taxco, Mexico, 130
tectonic map, 188-189, back endpaper
tectonic plates, **183**
tectonic processes, 2-3, 184ff
tectonic uplift, 205-206
temperature:
    aloft, 37ff
    annual march, 30-**31, 32, 33,** 37
    contrasted to heat, 22
    daily march, **36,** 37
    definition, 22
    effect of air movement, 31-**33, 32**
    effect of cloudiness, **36-37**
    effect of elevation, **37, 38-39**
    effect of latitude, 28-29
    effect of mixing, 30
    effect of ocean currents, 33-**36**
    effect of specific heat, 30
    effect of surface, 29-31
    effect on plant growth, 309
    effect on soil formation, **330,** 331
    factors affecting, 28ff
    gradient, 7, 156, 160, 161
    instruments, 22-23
    lapse rate of, 38, **39**
    measurement, 23
    ocean waters, 292, **294, 295,** 296
    range, 30, **31, 32,** 137, 138, 139, 142, 145, 150, 157
    reduction to sea level, 29, **38-39**
    scales, 23
    world distribution, **24-27**
terraces, **277,** 282
tertiary mountains, 188, 189
Texas, 150, 253, 259, 288, 319, 372
texture:
    of landform elements, **242,** 243
    of soils, 327, **328**
Thailand, 104
thermal expansion and contraction, 193-194
thermocline, **30**
thermometer, 22-**23**
Thornthwaite, C.W., 72, 125, 164
thorn woodlands, 316-**317**
thunderstorms, 79
Tibet, 104, 114, 138
tidal wave, 296
tides, 296-**298, 297,** 299
Tien Shan, 114, 145
tierra rojas, 240
Tigris River, 250
till, 229, 230, 231, 232
till plain, **230,** 231
time zones, 371
Tobolsk, U.S.S.R., **118**
Tokyo, 171
Tongass National Forest, **228**
topography, 173
tornadoes, 93, **97-98**
Torricelli's experiment, 41, **42**
township, 373-**374**
township and range, 372ff
traction, 204
trade winds, 60, 62, 63, 64, 65, **98,** 148, 300
transverse dunes, 236, **237**
transverse waves (see easterly waves)
tropical atmospheric disturbances:
    easterly waves, **99**
    hurricanes, 99-**102, 100, 101**
    Intertropical Convergence Zone, 98, 99, 103, 104, 108, 109
    monsoons, 102ff
    surge of the trades, **98,** 99
    weak tropical cyclones, 99
tropical climates, 125, 129-130, 147-156, **148, 149, 151, 152, 155**
Tropical Continental Shelf Water, 304
tropical cyclones (see hurricanes)
tropical forests, 314-317
    deciduous, 315-317
    monsoon, 315-**316**
    savanna, **316**
    thorn, 316-**317**
    evergreen, 314-315
    mangrove, **317**
tropical rainforest (see broadleaf evergreen forest)
tropical rainforest climates, 129, 130, 147-150, **148, 149, 151,** 154
tropical red loams, 156, 343-344
    black cotton soils, 344
    tropical black earths, 344
tropical savanna climates, 129, 130, 151, 154-156, **155**
Tropic of Cancer, 13, **14,** 29, 98, 157, 314
Tropic of Capricorn, 14, 98
tropopause, **37,** 38
troposphere, **37,** 38-39
tsunami, 296
tundra:
    climates, 126, 134-141, **136, 137, 140**
    soils, 137-138, 340
    vegetation, 137, 324-325
turbidity currents, 289
turbidity of the air, 12
turbulence:
    in atmosphere, **12,** 162, 233
    in streams, 204
Turkestan, 19, 114, 251
Turkey, 271, 274, 285
Turukhansk, U.S.S.R., 19, **20,** 21
typhoons (see hurricanes)
Uaupés, Brazil, 147, **148,** 150
ubac, 14, 310
Uinta Mountains, 175, 187
Ukraine, 259, 342, 347
Ulan Bator, 145
ultisols, 350, 351
United States (see individual states and also Atlantic Coastal Plain, central United States, eastern United States, Great Basin, Great Lakes, Great Plains, Gulf Coastal Plain, Pacific Northwest, western United States)
unloading of rock, 193
upwelling, 36, 76, 288, **300,** 398-399
Urals, 189
Urbana, Illinois, **118,** 131
urban dust dome, 170
urban heat island, 169
Uruguay, 156, **157,** 158, 347

U-shaped valleys, **228, 229**
U.S.S.R., 256, 322, 323, 324 (see also central asiatic Russia, European Russia, Siberia)
Utah, 372
    insolation, **28**
    landforms, 175, 185, 197, **249,** 259
    temperature, **28**
Valentia, Eire, 161, **162**
valley cross sections, 207-210
    accelerating rate of stream incision, **208**
    decreasing rate of stream incision, 207-**208, 209**
    factors affecting 207-209
    steady rate of stream incision, **209,** 210
valley glaciers, 218, **219**
vapor pressure, 67
varves, 231-232
vegetation:
    classification, 314
    climax, 307
    description, 314-325
    effect of biotic factors, 311, 314
    effect of climate, 309-310
    effect of landforms, 310-311
    effect of soils, 311
    effect on landforms, 174-175, 196, 197
    effect on soils, 331
    hydrophytes, 310, 315
    man disturbed, 306, **363-365**
    mesophytes, 310
    natural, 306-307
    photoperiodism, 310
    photosynthesis, 308, 357
    plant associations, 307
    plant communities, 307
    world distribution, **312-313**
    xerophytes, 125, 310
veldt, 154, 155, 322, 323
Venezuela, 253, 274, 321
ventifacts, 235
Verkhoyansk, U.S.S.R., 126, **137,** 138, 139
vertisols, 350, 351
Vermont, 268, 372
Vestmannaeyjar, Iceland, 182
Vesuvius, 186
Victoria Falls, **261**
Vienna, 32, **33**
Vietnam, **155,** 156
Villa Cisneros, Spanish Sahara, 92
Virginia, 247, **264,** 265
Vladivostok, 132
volcanoes (see vulcanism)
Volga River, 251
Vostok, Antarctica, 139, **140**
vulcanism, 184, 186-188, 271-272
    extrusive, 186-**187**
        cinder cone, 186, **187,** 269, **271**
        composite volcano, 186, **187,** 271
        lava dome, 187
        shield volcano, 187, 271
        stratovolcano, 186
    intrusive, **187,** 271
        batholith, **187,** 271, **273**
        dikes, 187
        laccoliths, **187-188**
        sills, **187**
    landforms, 271-273
walnut, 319
warm front, **95,** 96
warping, **186**
Washington (state)
    landforms, 237, 289
    temperature, 32, **33**
    tides, **299**
    vegetation, **165,** 166
Washington, D.C., 31, 32, 40, 169
water:
    capillary, 328
    deficit and surplus, 164-166, **165**
    gravitational, 328
    hydrologic cycle, **175-176**
    hygroscopic, 328
    world distribution, 176
water balance, 72, 163-166, **165**
water gap, 264, **265**
water masses, 300-304, **302**
water pollution:
    by industrial wastes, 355-356
    by nutrients, 355
    by sediments, 354-355
    thermal pollution, 356-357
water table, 335, 336, 343, 344
water vapor, 11, 15, 67ff
waves:
    and coastal slope, 279
    and wind speed, 278
    oscillation, 279
    swell, 278-279
    translation, 279
weathering, 174, 191-195
Weddell Sea, 299
wedging, 194, **195**
Wenatchee, Washington, 32, **33**
Wendover, Nevada-Utah, 372
West Australia Current, 159, 300
westerlies, 60, 62, 64, 109, 112, 300
western Africa, 151, 255, 286, 290, 344, (see also individual countries)
western Canada, 160, **162**
western United States, 78, 273, 274, 279, 322, 342, 359 (see also individual states)
West Indies, 100
West Virginia, 265
West Wind Drift:
    North Atlantic, **34,** 35, 36, 92, 114, 137, 160, 299
    North Pacific, **34-35,** 92, 114, 160
    Southern Hemisphere, **34,** 114, 299, 300, 303
wet-bulb thermometer, 69-**70**
White Mountains, 268
Willamette Valley, 170-171, **213**
wind:
    aloft, **48,** 59
    convergence, 98, **99,** 107, 108, 109
    definition, 41
    divergence, **99,** 107
    doldrums, 60, 62, 63, 64
    effect on evaporation, 310
    equilibrium velocity, **48**
    fetch, 278
    geostrophic, **48**
    gradient, 49
    gravity, 140
    instruments, 42
    jet stream, 59, 62
    measurement, 42-43
    prevailing, 60
    resultant, **59,** 60
    rose, **59-**60
    trade winds, 60, 62, 63, 64, 65, **98,** 148, 300
    turbulence, 163, 233
    westerlies, 60, 62, 64, 109, 112, 300
    world distribution, 59ff, **64**
wind deposition:
    loess, 236, **237, 238**
    sand dunes, 236-237
wind erosion:
    abrasion, 234
    blowouts, 235, **324**
    deflation, 234
    optimum conditions, 234-235
    saltation, **233,** 234
    surface creep, **233,** 234
    suspension, **233,** 234
    transportation, 234
    turbulence, 233
    ventifacts, 235
    yardangs, 235
wind gap, 264
wind rose, **59**
Windward Passage, 304
Winnipeg, **118**
Wyoming, 187, **260**
xerophytes, 125, 310
Yakutsk, U.S.S.R., 166, **167**
Yangtze River, 250
yardangs, 235
Yellow River, 250
Yellow Sea, 288
Yenisey River, 21
Yentna Glacier, Alaska, **219**
Yokohama, **381**
Yosemite Valley, 193, **194**
Yucatan, 186, 257, 286, 360
Yugoslavia, 257
Yukon Territory, 57
Zaire, 147, **149**
Zambezi River, **261**
zonal soils, 333, 339, 340-348
    brown steppe soils, 343, **346**
    chernozems, 343, **345**
    chestnut soils, 347
    desert soils, **346,** 348
    ferralites, 334-335, 344, **345**
    gray-brown podzolic soils, 342, **345**
    laterites, 335, 344, **345**
    podzols, 333, 340-342, **345**
    prairie soils, **332,** 344, **345,** 347
    sierozems, **346,** 347
    subtropical red or yellow soils, 342-343, **345**
    tropical red loams, 343-344
    tundra soils, 340
    world distribution, **341**

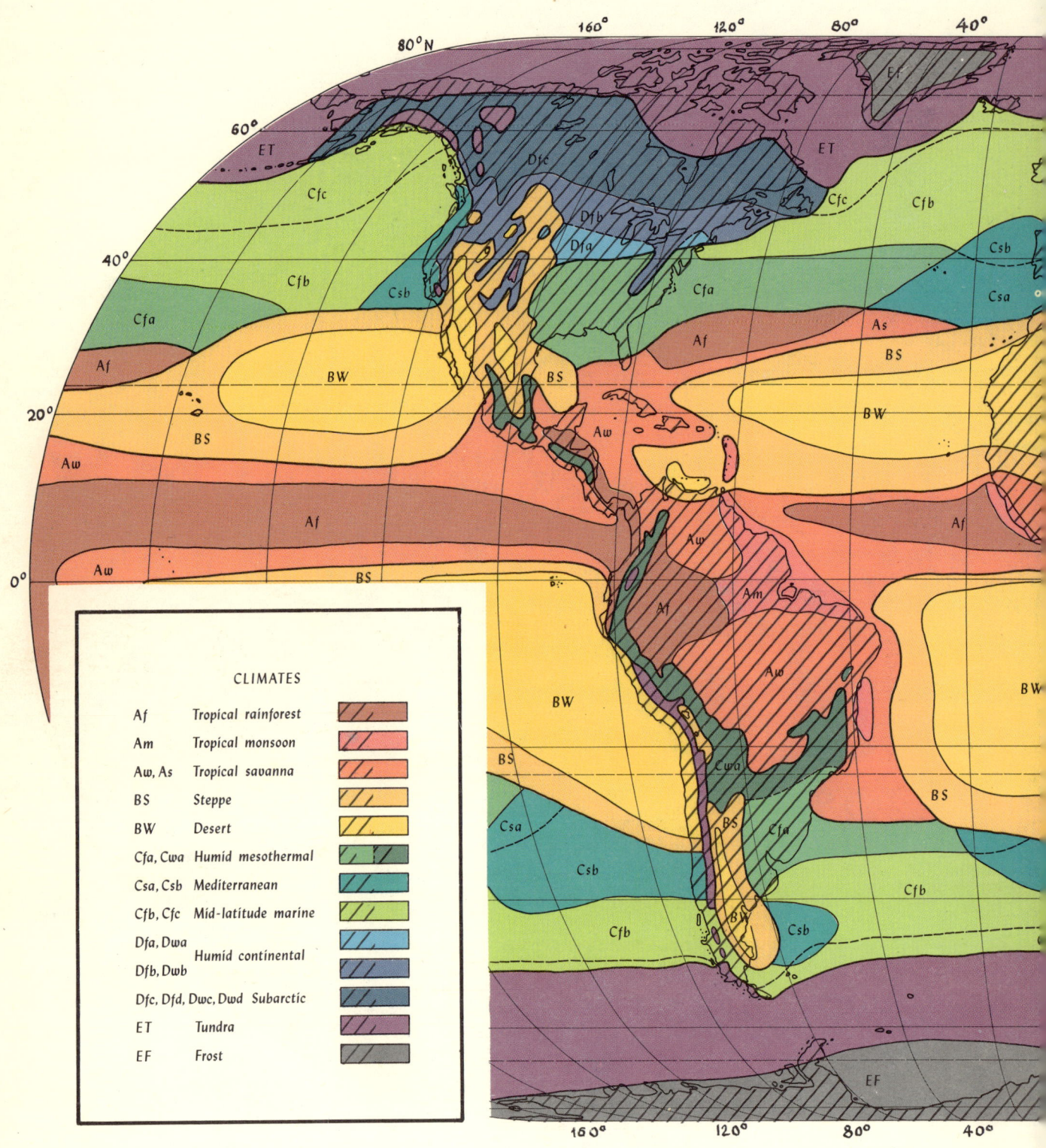